Fengel · Wegener
Wood

Dietrich Fengel
Gerd Wegener

Wood

Chemistry, Ultrastructure, Reactions

Walter de Gruyter
Berlin · New York 1989

Authors
Dietrich Fengel, Professor, Dr.rer.nat., Dipl.-Chem.
Gerd Wegener, Dr.rer.silv., Dipl.-Holzwirt
University of Munich
Institute for Wood Research
Division of Wood Chemistry and
Ultrastructure Research
Winzererstrasse 45
D-8000 München 40
F. R. of Germany

This paperback edition is identical to the hard-cover edition published in 1984
by Walter de Gruyter.

CIP-Titelaufnahme der Deutschen Bibliothek

Fengel, Dietrich:
Wood : chemistry, ultrastructure, reactions /
Dietrich Fengel ; Gerd Wegener. – Berlin ;
New York : de Gruyter, 1989
ISBN 3-11-012059-3
NE: Wegener, Gerd:

Library of Congress Cataloging in Publication Data

Fengel, Dietrich, 1931 –
Wood : chemistry, ultrastructure, reactions.

Bibliography: p.
Includes index.
1. Wood–Chemistry. I. Wegener, Gerd, 1945 –
II. Title.
TS932.F46 1983 674'.134 83-20497
ISBN 3-11-008481-3 (hard-cover edition)
ISBN 3-11-012059-3 (paperback)
ISBN 0-89925-593-0 (U.S.) (paperback)

Typesetting and Printing: Wagner GmbH, Nördlingen. Binding: Lüderitz & Bauer, Berlin.

Preface

One of the most important characteristics of wood is its renewability. It may even be inexhaustible providing that it is used with foresight and long-range planning. In an age of dwindling fossil-fuel resources, alternative resources such as wood which are continuously regenerated by nature become all-important.

To use wood wisely and judiciously we need to have a basic knowledge of its composition and structure as well as its behaviour under various external influences. Chemical studies of wood and its components may provide decisive factors not only for its applicability but also for the economic feasibility of many processes involving wood.

In the past several excellent books were published summarizing our knowledge of wood chemistry. Those by Hägglund, Wise and Jahn, Nikitin, Sandermann, Browning, and Kürschner deserve special mention. The latter two volumes appeared in 1963 and 1966, respectively. More recently a concise survey of wood chemistry was presented by Sjöström (1981). Progress in specific fields is described in various monographs such as 'Lignins' by Sarkanen and Ludwig (1971), 'Cellulose and Cellulose Derivatives' by Bikales and Segal (1971), and 'Pulp and Paper' by Casey (1980).

The aim of this book is to present a comprehensive account of progress and current knowledge in wood chemistry, drawing on the specialist literature from 1960 to 1982. For earlier publications the reader is generally referred to summarizing articles and books.

This volume falls implicitly, if not explicitly, into three large sections with a fluent transition between them. The first section describes the fundamentals of wood structure, analysis and components. Owing to the increasing interest being shown in bark, the structure and chemistry of this wood-joint product is summarized in a separate chapter. In the second section the reader will find reactions of wood and its components under degrading and changing conditions. The last section deals with the utilization of wood and wood components isolated by various chemical processes.

Since each component of wood has not only a chemical formula but also a molecular and supramolecular structure, and since most wood components are closely associated within the cell wall, the chemical compounds and reactions are also regarded from the standpoint of their ultrastructural aspects.

In order to visualize the structure of wood in diverse dimensions numerous scanning and transmission electron micrographs are presented in addition to light micrographs.

A great number of our colleagues followed the writing of this book with interest. Many of them supported our work by providing illustrations, granting permission for publication or making suggestions regarding the literature. This is sincerely appreciated. We are also greatly indebted to Drs H. von Aufsess, D. Grosser and M. Stoll for reading parts of the manuscript and offering valuable comments. We further wish to express our gratitude to all the coworkers in our laboratory who contributed with great enthusiasm and commitment to the preparation of the manuscript, above all Miss S. Hess. Thanks are also due to Mr J. B. Robinson for his conscientious revisions of the English. Last but not least we would like to thank the publishers for the attractive presentation of our book.

Munich, November 1983

Dietrich Fengel
Gerd Wegener

Contents

1. Introduction

Wood is a very old raw material. Thousands of years ago, when giant forest covered large areas of the earth's surface, primitive man used wood for fuel and tools. Because wood is natural in origin, serving to strengthen stems, branches and roots of trees and other plants, it returns to the natural cycle after having done its task, and is degraded into its basic elements. This explains why so little evidence of the early utilization of wood has survived, though some arrowheads, spears and tools up to 300 000 years old have been preserved under exceptional conditions in swamps and marshland.

During the prehistoric and historic periods wood was not only used as a building material but also increasingly gained importance as a chemical raw material for the production of charcoal (used in iron smelting), tar and pitch (useful for preserving and sealing ship hulls), and potash (used in glass production and as a bleaching agent for linen and cotton textiles).

But in another sense wood is a very modern raw material. Broad timber vaultings and precious furniture attest to its usefulness and beauty. Even in converted forms such as plywood, particleboard and fibreboard, wood has become a valuable building material. And, last but not least, wood is the basic substance for pulp and paper, fibres, films, additives, and many other products.

It is no exaggeration to say that wood is one of the most important products of nature. About one-third of the world's land surface is covered by forests containing a total growing stock of some 300 000 million m^3 of wood (Steinlin 1979). From this stock 2 600 million m^3 are harvested each year. This volume represents about 1 300 million tons of wood, an amount approximately equal to the world's production of corn ($1\,500 \cdot 10^6$ t), twice the production of steel ($700 \cdot 10^6$ t) or cement ($760 \cdot 10^6$ t), and 27 times the production of plastics ($48 \cdot 10^6$ t) (Statistisches Bundesamt 1981).

During the present century the world's consumption of wood increased considerably, and forecasts for the period up to the year 2000 predict a further rapid increase (FAO 1966, 1974, 1981; Hagemeyer 1976; Stone, Saeman 1977; Steinlin 1979). The estimates of the world's total requirements of roundwood in the year 2000 vary between 3 800 and 6 200 million m^3. Prognostications show that demand for industrial roundwood will nearly double, that for pulpwood nearly triple during the final 20 years of this century (Fig. 1–1). The relatively large amount of wood used for fuel (1 500 million m^3 in 1979) is expected to increase only slightly during this period.

These figures are offset by an estimated annual growth increment of 7 000 to 9 000 million m^3 (FAO 1966; Steinlin 1979). The annual growth of wood varies greatly

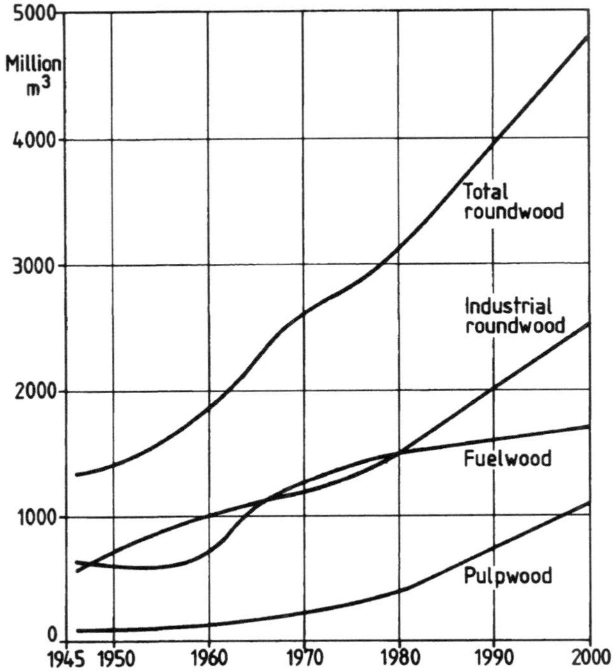

Fig. 1–1. World's consumption and requirements of wood 1946–2000 (according to FAO 1966, 1981; Hagemeyer 1976; Stone, Saeman 1977; Steinlin 1979).

according to climate and soil conditions. Whereas in temperate zones the increment is about 3 to 5 m^3 per hectare and year, tropical eucalyptus and pine plantations may produce 15 to 20 m^3 per hectare and year. Since world forest inventories also include forest-like vegetation, the mean increment in wood all over the world is no more than 1 to 2 m^3 per hectare and year (FAO 1966).

Sanderman (1973) has calculated that under optimal conditions a fast-growing pine may produce 13.7 g of cellulose per day. This quantity of cellulose corresponds to 8.2 g lignin, 6.5 g polyoses, and 0.3 g extractives, which results in a total of 27.7 g or about 56 cm^3 of wood substance being produced by one tree per day.

Nevertheless, a progressive reduction of the growing stock towards a value of about 23% of the earth's land surface in the year 2000 is predicted. This represents a decrease of 31%, with an above-average decrease of about 40% in the developing countries (Barney 1980). The situation is expected to stabilize in the year 2020, by which time all accessible forests in the developing countries will have been removed. This problem is closely connected with fuel consumption, which is predominantly based on wood and agricultural waste in these countries (Table 1–1). In some countries, non-commercial fuels constitute more than 90% of the total (Leach 1979). A restriction of the consumption of fuelwood is becoming more and more difficult with increasing prices for the commercial petroleum-based fuels.

Table 1–1: Fuel consumption in various world areas (in million tons coal equivalent) (Leach 1979)

Region	Commercial fuels	Fuel wood	Agricultural waste	Total	Fuel wood + agricultural waste as % of total
Africa*	66	116	22	204	68
Mideast	109	5	12	126	13
Far East**	247	143	100	490	50
Latin America	304	98	31	433	30
West Europe	1 568	18	6	1 592	2
USSR and Eastern Europe	1 771	43	24	1 838	4
North America	2 723	7	0	2 730	0
World***	7 885	498	167	8 550	8

* Excluding South Africa
** Excluding Japan
*** Including Japan, South Africa, Australia, etc.

The most important product of the chemical conversion of wood is pulp. All over the world 123 million tons of pulp were produced in 1980 (Fig. 1–2). During the same period the total consumption of paper and paperboard was 171 million tons, of which more than 25% were produced from wastepaper (VDP 1981). In some countries (e.g. Japan, Great Britain, F. R. Germany, G.D.R., Italy) wastepaper application exceeds 40–50%. This indicates that recycling is an important factor in the economic utilization of the raw material. Economic and environmental problems are reasons for a progressive change of pulping and bleaching processes.

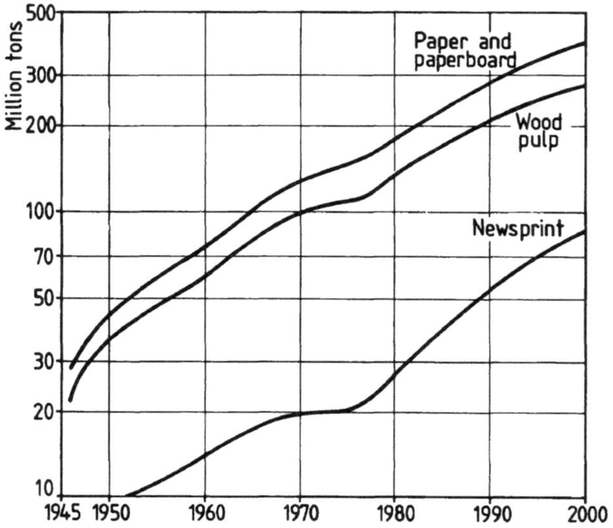

Fig. 1–2. World's production of wood pulp, paper and paperboard, and newsprint 1946–2000 (according to FAO 1966; Keays 1975; VDP 1981).

Table 1–2: Consumption of paper and paperboard in some countries and regions (kg/head) (VDP 1967, 1981; Hagemeyer 1976)

	1966	1979	1985	2000
U.S.A.	240	289	349 (U.S.A. and Canada)	566 (U.S.A. and Canada)
Canada	141	215		
Japan	50	151	284	558
U.S.S.R.	21	33	67	192
Western Europe				
E.G.	81	128	180 (E.G. and E.F.T.A.)	324 (E.G. and E.F.T.A.)
E.F.T.A.	105	135		
Other countries*		6	17	23
World	31	40	55	91

* Stands for the bulk of the developing countries

A survey of the development of paper and paperboard consumption shows that, despite high consumption in the highly industrialized countries, an increase of consumption is still possible (Keays 1975) (Table 1–2). On the other hand the consumption of paper in the developing countries is so low that a drastic increase seems to be unavoidable, certainly in the long run. We can expect the level of consumption in most developing countries to approach that of the industrial nations sometime well after the year 2000, particularly if the forecast is true that even in 2000 a one-dollar increase in the gross national product of the developing countries will be met by a 20-dollar increase in the industrial countries (Barney 1980).

In view of the future it is our task and duty to utilize wood economically so that it remains a significant resource. A knowledge of wood components and their chemical behaviour is more important now than ever. There is a need for effective protection against external influences (chemicals, enzymes, irradiation, temperature) on the one hand, and on the other for a careful isolation of wood components, and a search for new products based on them. For a better understanding of known technologies and further development of new processes, basic research into the isolation, characterization and reactions of wood components is still essential.

The chemistry of wood and its components cannot be regarded apart from its structure. Wood is not merely a chemical substance, or an anatomical tissue, or a material – it is a combination of all three. This entirely results from an intimate association of the chemical components which form ultrastructural elements, being combined into higher-order systems which in turn build up the walls of the cells that ultimately compose the wood tissue.

Literature

Barney, G. O. (Study Director) 1980, The Global 2000 Report to the President. Ed. by the Council on Environmental Quality and the US Foreign Office. US Government Printing Office, Washington

FAO 1966, World Forest Inventory 1963. FAO, Rome

FAO 1974, Yearbook of Forest Products, Review 1961–1972. FAO, Rome

FAO 1981, Yearbook of the Forest Products 1968–1979. FAO, Rome

Hagemeyer, R. W. 1976, Tappi *59,* No. 4, 46–49

Keays, J. L. 1975, Tappi *58,* No. 11, 90–95

Leach, G. 1979, Energy, Report Prepared for the Conference "Agricultural Production: Research and Development Strategies for the 1980's", Bonn, Oct. 8–12. German Foundation for International Development, German Agency for Technical Cooperation, Federal Ministry of Economic Cooperation, The Rockefeller Foundation

Sandermann, W. 1973, Holz, Roh- Werkst. *31,* 11

Statistisches Bundesamt 1981, Statistisches Jahrbuch 1981 für die Bundesrepublik Deutschland. Kohlhammer, Stuttgart, Mainz

Steinlin, H. 1979, Die Holzproduktion der Welt, ökologische, soziale und ökonomische Aspekte. In: Holz als Rohstoff in der Weltwirtschaft (Plochmann, R. and Löffler, H., Eds.). Landwirtschaftsverlag, Münster-Hiltrup, pp. 14–44

Stone, R. N. and Saeman, J. F. 1977, For. Prod. J. *27,* No. 10, 49–54

VDP 1967, Papier 1967, Ein Leistungsbericht. Verband Deutscher Papierfabriken e. V., Bonn

VDP 1981, Papier '81, Leistungsbericht der Deutschen Zellstoff- und Papierindustrie. Verband Deutscher Papierfabriken e. V., Bonn

2. Structure and Ultrastructure

2.1. Anatomical Aspects

Before reflecting on the microscopic and submicroscopic structure of wood a short survey of the anatomical features of softwoods and hardwoods is advisable. For more detailed information the reader is referred to the specialist literature: Wagenführ (1966), Jane (1970), Panshin and de Zeeuw (1970), Bosshard (1974), Wagenführ and Scheiber (1974), Grosser (1977), Schweingruber (1978), Sakai (1982). From the anatomical point of view wood is a perennial tissue resulting from the secondary growth in the stems, branches and roots of trees and shrubs.

The observation of wood without optical aids shows not only differences between softwoods and hardwoods as well as between various species, but also differences within one sample, such as sapwood and heartwood, growth rings, earlywood and latewood, the arrangement of pores etc. All these phenomena are the result of the development and growth of wood tissue. This tissue is constructed to meet the natural necessities of the tree, and consists therefore of strengthening, conducting and storing cells. Softwood obtained from coniferous trees and hardwood obtained from deciduous trees differ in cell type and cell function (Table 2–1).

The run and the arrangement of the cells can be recognized on the sections cut in the three main planes used for the anatomical characterization of wood: the cross or transverse section, the tangential section and the radial section (Fig. 2–1).

Softwood shows a relatively simple structure as it consists of 90–95% tracheids, which are long and slender cells with flattened or tapered closed edges (Fig. 2–2a). The tracheids are arranged in radial files, and their longitudinal extension is oriented in the direction of the stem axes (Fig. 2–3).

Table 2–1: Main functions of the various cell types in wood

	Mechanical function	Conducting function	Storing function	Secreting function
Softwoods	Latewood tracheids	Earlywood tracheids Ray tracheids	Ray parenchyma Longitudinal parenchyma (Resin canals)	Epithelial cells
Hardwoods	Libriform fibres Fibre tracheids	Vessels Vessel tracheids	Ray parenchyma Longitudinal parenchyma (Resin canals)	Epithelial cells

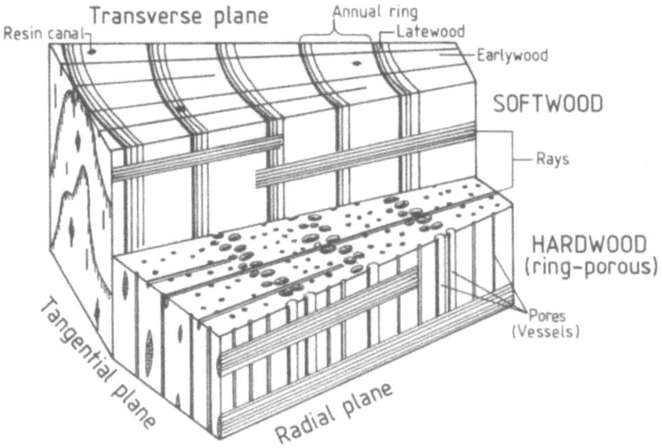

Fig. 2–1. Models of a softwood and a hardwood block, showing the main cutting planes for anatomical studies, and anatomical structures visible without optical aids.

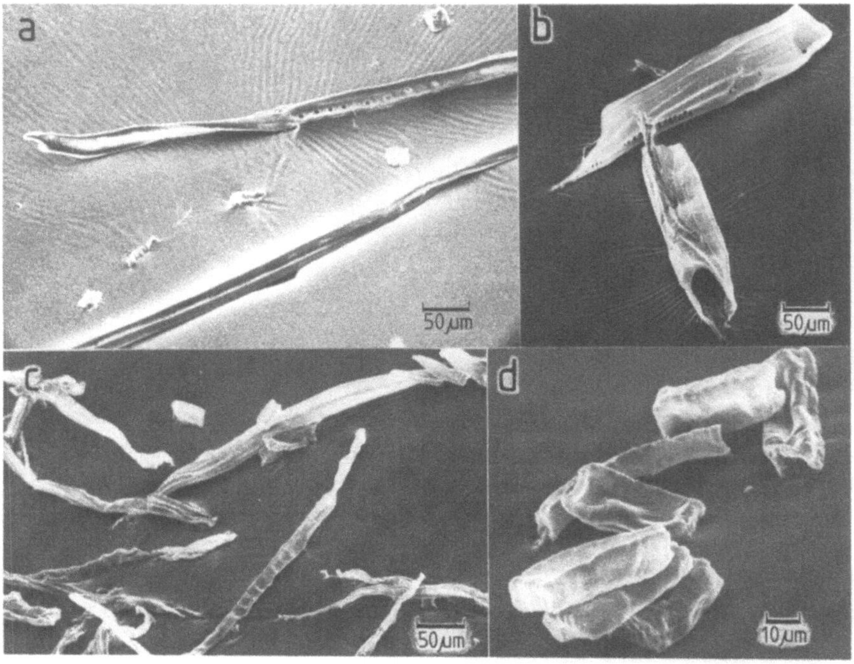

Fig. 2–2. Cells and cellular elements of softwoods and hardwoods. SEM micrographs, all at the same level of magnification, except for d).

a) Softwood tracheids *(Pinus sylvestris)*;

b) Hardwood vessel elements *(Fagus sylvatica)*;

c) Hardwood fibres *(Quercus robur)*;

d) Hardwood parenchyma cells *(Quercus robur)*.

In evolving from earlywood to latewood the cell diameters become smaller while the cell walls become thicker. At the end of the growth period tracheids with small lumina and small radial diameters are developed, while at the beginning of the subsequent growth period tracheids with large lumina and diameters are developed by the tree (Fig. 2–4). In spruce *(Picea abies)* minimal radial diameters of about 7 μm for the last latewood tracheids and maximal radial diameters of 32 μm for the first earlywood tracheids were determined (Fengel, Stoll 1973). This abrupt change is visible to the eye as an annual or growth ring.

The thick-walled latewood tracheids provide strength, while the spaceous earlywood tracheids predominantly conduct water and minerals within the tree. Various softwoods, such as larch (*Larix* spp.), spruce (*Picea* spp.), pine (*Pinus* spp.) and Douglas fir *(Pseudotsuga menziesii)*, also contain radially oriented tracheids accompanying the ray parenchyma cells.

The storage and the transport of assimilates take place within the parenchyma cells, which in softwoods are predominantly arranged in radially running rays

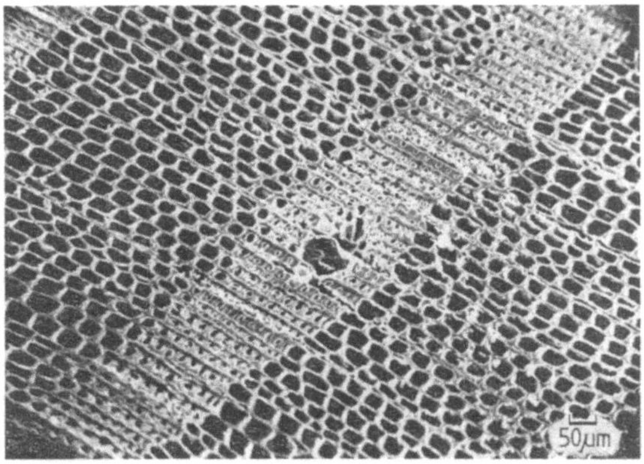

Fig. 2–3. Transverse plane of a softwood *(Pinus sylvestris)* with a whole annual ring, showing the transition between earlywood and latewood within the annual ring and at the ring border. In the latewood centre is a resin canal. SEM micrograph.

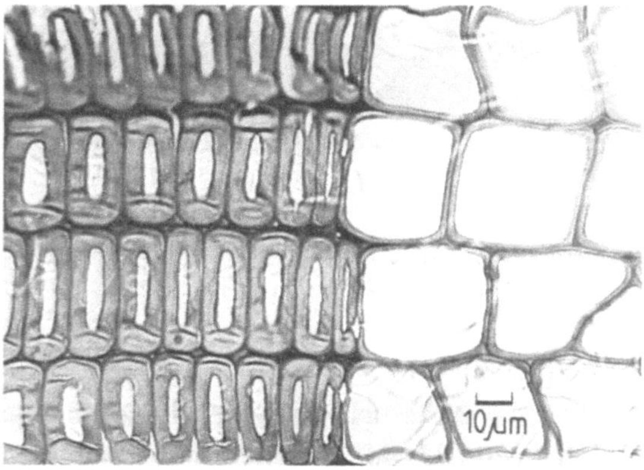

Fig. 2–4. Cross section of the annual ring border in a softwood *(Picea abies)*. Light micrograph.

(Fig. 2–5). The secreting elements are the epithelial cells, which surround the resin canals. These canals are the vertical and radial cavities within the tissue of most softwoods (Fig. 2–6). Softwoods without resin canals are, for example, firs (*Abies* spp.), yews (*Taxus* spp.), junipers (*Juniperus* spp.) and cedars (*Cedrus* spp.).

Hardwoods have a basic tissue for strength containing libriform fibres and fibre tracheids. Within this strengthening tissue are distributed conducting vessels, often with large lumina (Fig. 2–7). These vessels are long pipes ranging from a few centimeters up to some meters in length and consisting of single elements with open or perforated ends (Figs. 2–2b, 2–8).

Diffuse-porous and ring-porous hardwoods can be distinguished by the arrangement and the diameter of the vessels. Most of the hardwood species of temperate zones are diffuse-porous, e.g. maples (*Acer* spp.), birches (*Betula* spp.), beeches

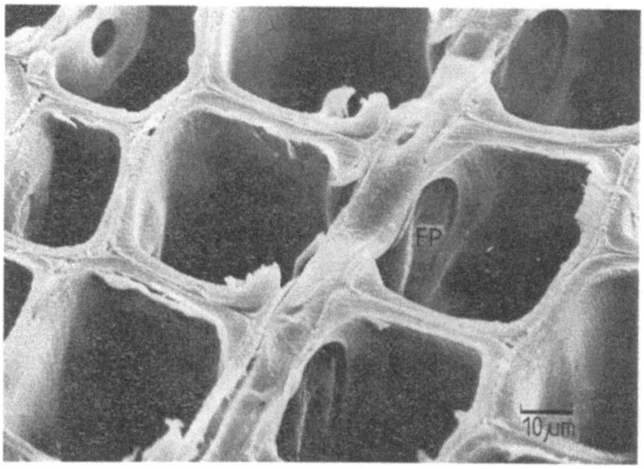

Fig. 2–5. Transverse plane of a softwood *(Pinus sylvestris)*. Spacious earlywood tracheids border on a ray. The adjacent radial cell walls are perforated by large fenestriform pits (FP). SEM micrograph.

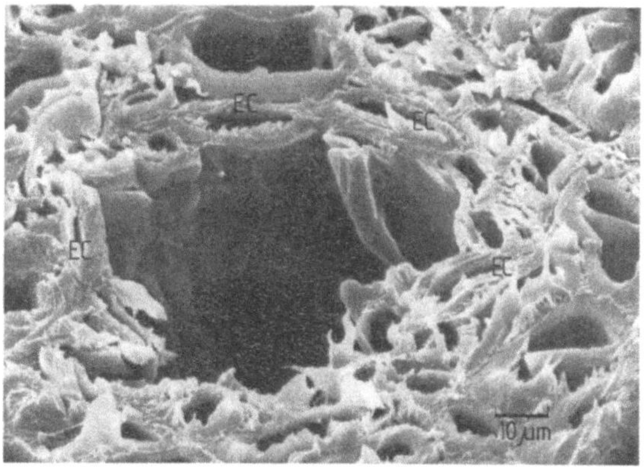

Fig. 2–6. Transverse plane of a softwood *(Pinus sylvestris)* with a resin canal which is surrounded by epithelial cells (EC). SEM micrograph.

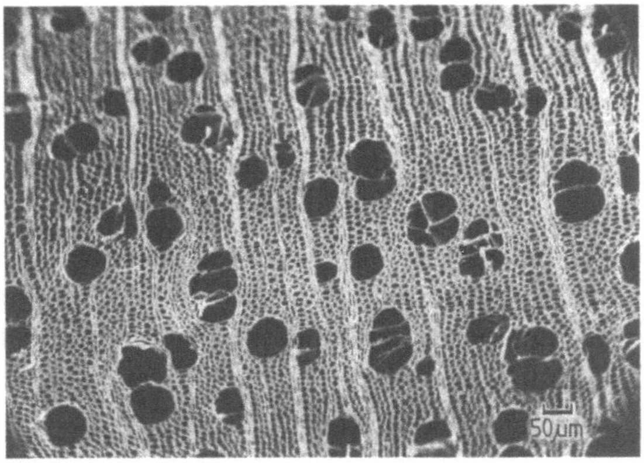

Fig. 2–7. Transverse plane of a diffuse-porous hardwood *(Betula verrucosa)*. Large-looking vessels are distributed within the basic tissue consisting of fibres. SEM micrograph.

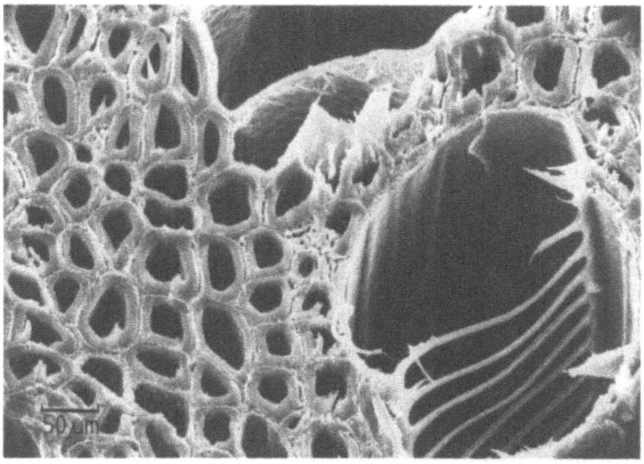

Fig. 2–8. Transverse plane of a hardwood *(Betula verrucosa)* showing the cross-cut fibres and a vessel with a ladder-like perforation. SEM micrograph.

(*Fagus* spp.), alders (*Alnus* spp.), hornbeam *(Carpinus betulus)*, and poplars (*Populus* spp.). These woods show no or only small differences in the diameter and the number of vessels over the whole growth ring (Figs. 2–7, 2–9). Ring-porous woods, such as oaks (*Quercus* spp.), elms (*Ulmus* spp.) and ashes (*Fraxinus* spp.), contain spacious vessels in the earlywood, and narrow vessels in the latewood after a most abrupt change (Fig. 2–10). There are also semi-ring-porous woods with a continuous range of diameters from earlywood vessels with large diameters to latewood vessels with small diameters, e.g. walnuts (*Juglans* spp.), or with an accumulation of vessels in the earlywood, e.g. cherry trees *(Prunus avium)*.

The dimensions of the hardwood fibres, which form the basic tissue, are smaller than those of the softwood tracheids. They have thicker cell walls and smaller lumina, and the differences in wall thickness and lumen diameters between earlywood and latewood are not as extreme as in softwoods (Fig. 2–2c).

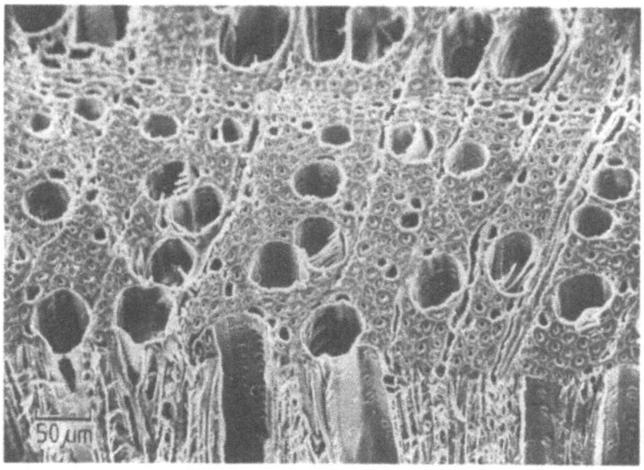

Fig. 2–9. Transverse and tangential plane of a diffuse-porous hardwood *(Fagus sylvatica)*. The tissue consists of thick-walled libriform fibres, thin-walled fibre tracheids, and relatively narrow vessels, some of them with ladder-like perforations, SEM micrograph.

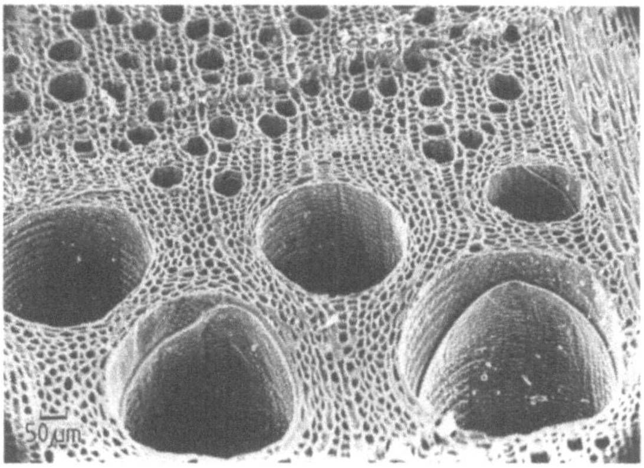

Fig. 2–10. Transverse plane of a ring-porous hardwood *(Quercus robur)*. The basic tissue contains spacious earlywood vessels (foreground) and narrow latewood vessels (background). SEM micrograph.

The parenchyma cells are short compact cells with stubby ends (Fig. 2–2d). The number of parenchyma cells in hardwoods is higher than in softwoods, which reveal large rays and longitudinal parenchyma. Particularly in tropical hardwoods there is a high percentage of longitudinal parenchyma (Fig. 2–11). Hardwoods from tropical and subtropical zones may also contain longitudinal and radial resin canals (e.g. *Shorea* spp.).

The wall thickness of the fibres or tracheids, the number and diameter of the vessels, as well as the number of parenchyma cells determine the density of woods. Table 2–2 summarizes the values of the density, cell dimensions and cell percentages of some softwoods and hardwoods.

Table 2–2: Density, cell dimensions and percentages of some wood species (Fengel, Grosser 1976)

		Softwoods (temperate zone)			Hardwoods (temperate zone)			Hardwoods (tropical zone)	
		Abies alba	*Picea abies*	*Pinus sylvestris*	*Fagus sylvatica*	*Quercus robur*	*Populus* spec.	*Ochroma lagopus*	*Tectona grandis*
Density	g/cm^3	0.32–0.41–0.71	0.30–0.43–0.64	0.30–0.49–0.86	0.49–0.68–0.88	0.39–0.65–0.93	0.40	0.05–0.13–0.41	0.44–0.63–0.82
Cell dimensions									
Tracheids/fibres									
Length	mm	3.4–4.3–4.6	1.7–2.9–3.7	1.4–3.1–4.4	0.6–1.3	0.6–1.6	0.7–1.6	1.1–2.2–3.6	0.7–1.4
Diameter	µm	25–50–65	20–30–40	10–30–50	15–20	10–30	20–40		
Vessels									
Length	mm				300–700	100–400	500		
Diameters	µm				5–100	10–400	20–150	130–200	50–370
Cell percentages (average values)									
Tracheids/fibres	%	90.4	95.3	93.1	37.4	$44.3^{+}/58.1^{++}$	61.8	4.0	66.3
Vessels	%				31.0	$39.5^{+}/.7^{++}$	26.9	3.0–4.5	11.6
Longitudinal parenchyma	%	scarce	1.4–5.8	1.4–5.8	4.6	4.9		74	11.6
Rays	%	9.6	4.7	5.5	27.0	$16.2^{+}/29.3^{++}$	11.3	17.0–19.0	15.5

$^{+}$ narrow growth rings
$^{++}$ wide growth rings

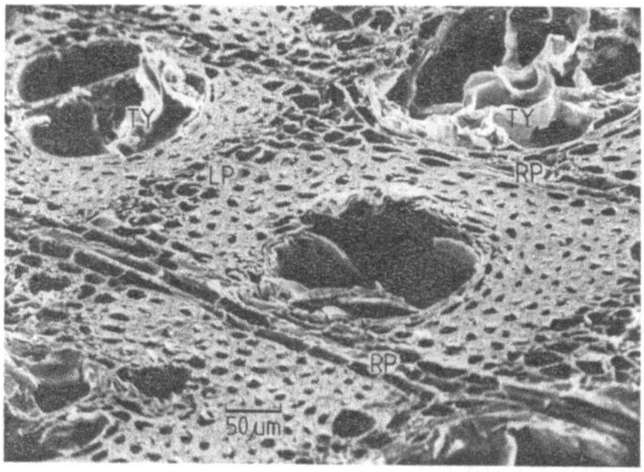

Fig. 2–11. Transverse plane of a tropical hardwood *(Tectona grandis)*. The basic tissue is interlaced with ribbons of longitudinal parenchyma (LP) and with ray parenchyma (RP). The vessels are closed by tyloses (TY). SEM micrograph.

2.2. Ultrastructure of the Cell Wall

Under strong magnification of visible light various layers can be recognized in the wood cell walls. A clear demarcation between the individual layers can be seen with the electron microscope (Fig. 2–12). With the aid of this instrument, the current knowledge of the structural composition of wood cell walls was obtained between 1950 and 1970. Details of the current image of the cell wall structure, particularly

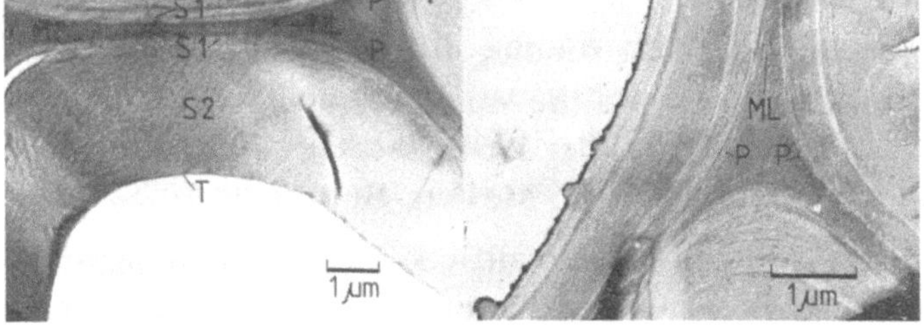

Fig. 2–12. Ultrathin sections of the walls of softwood tracheids (a) *(Picea abies)* and of a hardwood vessel (b) *(Fagus sylvatica)*, which show the various wall layers: ML = middle lamella, M = compound middle lamella, P = primary wall, S 1 = secondary wall 1, S 2 = secondary wall 2, T = tertiary wall, W = warts. TEM micrographs.

regarding its history, are described elsewhere (Wardrop 1964; Harada 1965; Côté 1968, 1977). Therefore we can confine ourselves to the results of this development.

The concentric arrangement of the cell wall layers is caused by the differences in the chemical composition and by different orientations of the structural elements. In this order of magnitude the components must be subdivided into a structural component, i.e. cellulose, and into sub-structural, blurring components, i.e. polyoses and lignin. When the polyoses and lignin are removed, the texture of the cellulosic elements, called fibrils, is visible (Fig. 2–13). Various electron microscopic observations gave rise to a model of the construction of wood cell walls (Fig. 2–14).

Between the individual cells there is a thin layer, the middle lamella, which glues the cells together to form the tissue. Though single fibrils may cross the middle lamella, this layer is in principle free of cellulose. The transition from the middle lamella to the adjacent cell wall layers is not very clear, so that for the middle lamella and both adjacent primary walls the term compound middle lamella is used.

In the primary wall (P) the cellulose fibrils are arranged in thin crossing layers. As the primary wall is the first layer deposited during the development of a cell, this system allows for an expansion of the young cell. Therefore the orientation of the fibrils in the outermost lamella is more oblique (Nečesaný 1966). Otherwise the amount of cellulose in the primary wall is very limited.

The subsequent wall layer is the secondary wall 1 (S 1) with a gentle helical slope of the fibrils. There are several lamellae with counter-running helices, and if these lamellae are very thin, the helices are visible as a crossed system (e.g. beechwood fibres).

The thickest wall layer is the secondary wall 2 (S 2). In this layer the fibrils run at a steep angle. Changes in the angle and differences in the packing of the fibrils result in a lamellar structure of the S 2 (Kishi et al. 1977, 1979). With the aid of the scanning transmission electron microscope (STEM), Ruel et al. (1978) detected an irregular lamellation in the S 2 of softwood tracheids. The interlamellar distance was 7.1 nm and 8.4 nm in spruce and fir, respectively.

In several studies, the angle and the winding direction of the fibrils in the S 2 were determined in order to demonstrate the variability and to find a correlation between cell size and development (Hiller 1964; Nečesaný 1966; Kantola, Seitsonen 1969; Ohtani, Ishida 1973; Chafe 1978; Meylan, Butterfield 1978).

A third secondary wall layer (secondary wall 3, S 3) is present in parenchyma cells, whereas fibres of monocotyledones, e.g. bamboo, may have four or more secondary wall layers.

The last fibrillar layer at the luminal border should be named tertiary wall (T), not secondary wall 3, as is sometimes done. It is different from the S 3 of the paren-

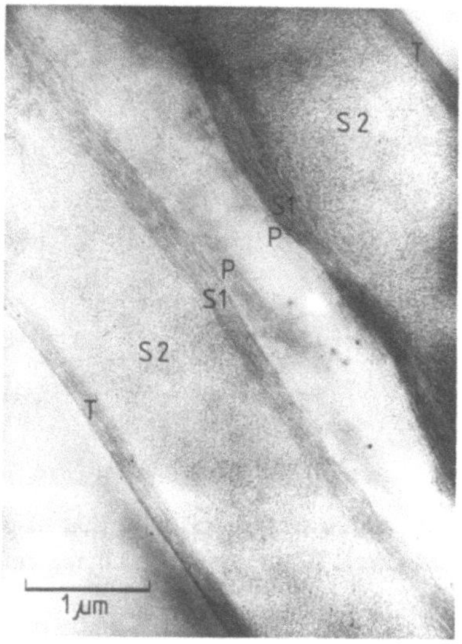

Fig. 2–13. Ultrathin section of the cellulose skeleton of tracheid walls in softwood *(Picea abies)*. The different orientations of the cellulose fibrils in the various wall layers are visible. TEM micrograph.

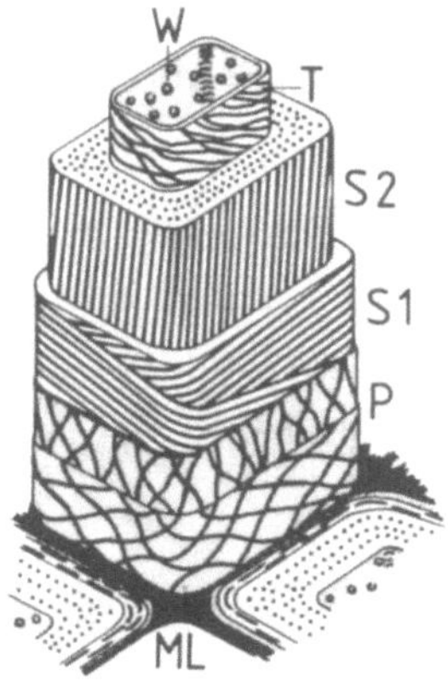

Fig. 2–14. Model of the cell wall structure of softwood tracheids and hardwood libriform fibres. ML = middle lamella, P = primary wall, S 1 = secondary wall 1, S 2 = secondary wall 2, T = tertiary wall, W = wart layer.

chyma cells (Fig. 2–15) and the other secondary wall layers. The fibrils are arranged under a gentle slope but not in a very strict parallel order. This layer has a higher concentration of non-structural substances, which gives the luminal surface a more or less smooth appearance.

In certain cases, e.g. the tracheids of fir and pine, libriform fibres and vessels of beech, the luminal surface is covered with warts (Fig. 2–12b). According to Parham and Baird (1974) there is a phylogenetic trend in the appearance of warts. Conifer tracheids and primitive hardwood cells are nearly always warted, but as the cell

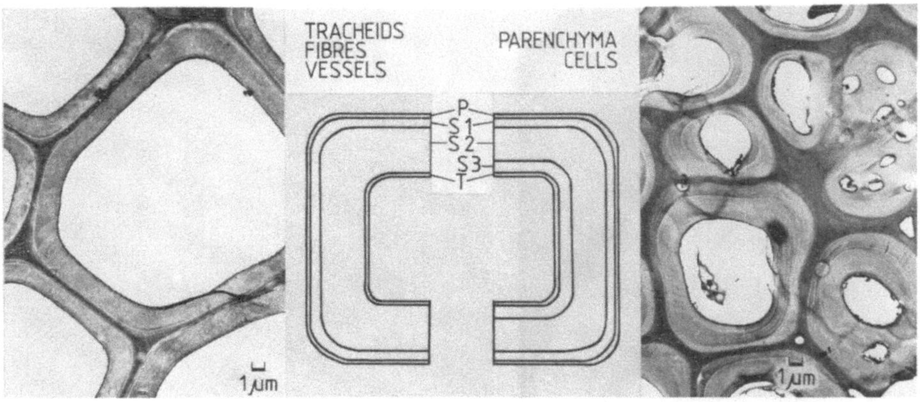

Fig. 2–15. Comparison of the wall structure of tracheids, fibres and vessels with that of parenchyma cells. Left: tracheids of umbrella tree *(Sciadopitys verticillata)*. Right: ray parenchyma cells of beech *(Fagus sylvatica)*. TEM micrographs.

types become more advanced or specialized, they become increasingly wart-free. The warts develop at about the time cell wall lignification nears completion, and consist largely of lignin-like material with some amounts of carbohydrates and pectic substance (Liese 1965a; Baird et al. 1974; Takiya et al. 1976). There may be a difference between earlywood and latewood in the size, number and distribution of warts, as shown in fir *(Abies alba)* (Verhoff, Knigge 1976).

The tracheids in some softwoods (*Pseudotsuga menziesii, Taxus* spp.) and the vessels of some hardwoods (*Acer* spp., *Tilia* spp., *Ulmus* spp.) exhibit helical thickenings as a part of their cell walls. These thickenings visible on the inner surface may or may not show a fibrillar structure, and are an integral part of the tertiary wall (Jutte, Levy 1973; Timell 1978).

The system of sloping fibrils in combination with the non-structural solidifying substances gives the wood cells a sturdy, but not inflexible construction which resists a wide range of forces acting on it. Due to the steep running fibrils the S 2 is provided with tensile resistance, whereas the S 1 with its gentle fibrillar slope is responsible for compression resistance, both along the cell axes.

In all types of cells the secondary wall 2 accounts for the main portion of the cell wall. Extremely high percentages are reached by the latewood tracheids of softwoods and by the libriform fibres of hardwoods, where the S 2 may take up more than 90% of the cell wall. The change in cell wall thickness from earlywood to latewood is determined by the S 2, while the S 1 and the tertiary wall contribute only minimally to this change. The data for the thickness of the wall layers of the tracheids in sprucewood *(Picea abies)* and of the cells in beechwood *(Fagus crenata)* are compiled in Tables 2–3 and 2–4.

Table 2–3: Average thickness and percentage of the wall layers in spruce tracheids *(Picea abies)* (Fengel, Stoll 1973)

Wall layer	Earlywood		Latewood	
	μm	%	μm	%
P M/2	0.09	4.3	0.09	2.1
S 1	0.26	12.4	0.38	8.8
S 2	1.66	79.0	3.69	85.8
T	0.09	4.3	0.14	3.3
Total wall	2.10		4.30	

Table 2–4: Average thickness and percentage of the wall layers in beech wood cells *(Fagus crenata)* (Harada 1962)

Wall layer	Vessels		Libriform fibres		Fibre tracheids		Longitudinal parenchyma		Ray parenchyma	
	μm	%	μm	%	μm	%	μm	%	μm	%
P	0.25	25	0.07	1	0.07	5	0.06	4	0.50	27
S 1			0.51	10	0.24	16	0.35	21		
S 2	0.50	50	4.32	87	0.99	67	0.78	48	0.92	50
S 3	–	–	–	–	–	–	0.37	22	0.37	20
T	0.25	25	0.10	2	0.17	12	0.09	5	0.07	3
Total wall	1.00		5.00		1.47		1.65		1.86	

2.3. Reaction Tissues

The shape of the cells, particularly of tracheids and fibres, is influenced not only by seasonal changes but also by mechanical forces. Trees react to strain forces acting on stems (e.g. by high wind or geotropic erection), boughs and branches (e.g. by their own weight) by forming reaction wood in the zones of compression or tension. Coniferous trees develop compression wood in the compressed ranges, and deciduos trees develop tension wood in the tensive ranges. Compression and tension tissues differ in anatomical, chemical and physical properties from each other as well as from the normal tissue of wood.

The main characteristics of compression wood are its dark colour, caused by a relatively high lignin content, rounded tracheids with intercellular spaces, absence of a tertiary wall, and a secondary wall 2 with helical cavities (Casperson 1965; Côté, Day 1965; Scurfield, Silva 1969; Timell 1973) (Fig. 2–16).

The angle of the fibrils and therefore also of the fissures in the S 2 of compression tracheids is about 45° in relation to the fibre axis. The round cross section of the cells and the formation of the cavities can be observed in the early stages of cell wall development (Wardrop, Davies 1964; Casperson, Zinsser 1965; Fujita et al. 1973, 1978; Timell 1973). In the compression wood of some conifers, such as yew *(Taxus baccata)* and ginkgo *(Ginkgo biloba)*, the helical cavities are absent (Timell 1978a, b).

Tension wood contains fewer and smaller vessels than normal wood, and the fibres are provided with a special wall layer, the so-called gelatinous layer or G-layer. Depending on the species involved, the G-layer may be present instead of the S 2, the tertiary wall or additionally to the normal wall layers (Casperson 1965; Côté, Day 1965; Jutte 1969) (Fig. 2–17).

The G-layer consists of concentric lamellae of cellulose fibrils aligned in the direction of the fibre axis (Casperson 1967; Mia 1968; Côté et al. 1969). The cellulose is highly crystalline, and the content of polyoses and lignin amounts only to a few

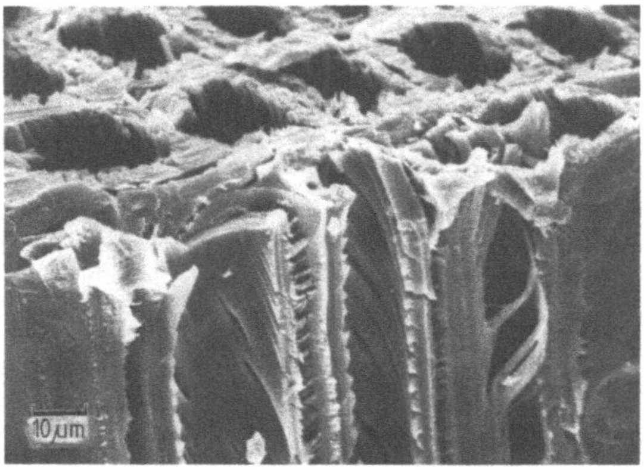

Fig. 2–16. Transverse and tangential plane of compression wood in spruce *(Picea abies)* showing the helical cavities in the secondary wall layer. SEM micrograph.

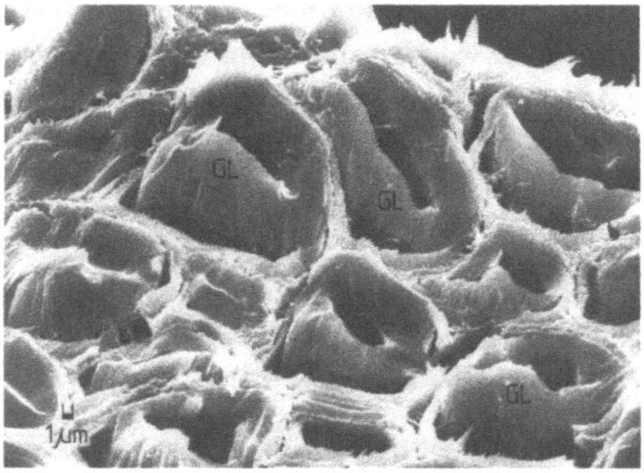

Fig. 2–17. Transverse plane of tension wood in poplar (*Populus* spec.). The thick G-layers (GL) of the tension wood fibres are partly separated from the residual cell walls. SEM micrograph.

percent (Kantola 1964; Norberg, Meier 1966; Harada et al. 1971). The thickness of the G-layer varies along a tension wood fibre, and is thickest in the central region (Okumura et al. 1977).

Reaction tissue was also detected in roots and barks (Höster, Liese 1966).

2.4. Functional Elements of the Conducting System

The conduction and distribution of aqueous solutions as well as the exchange of cell contents within the living part of wood are only made possible by the presence of openings in the cell walls. Apart from several variations there are only two principle types of openings: simple pits and bordered pits. The combination of both are called half-bordered pits.

These conspicuous details of the cell walls have been the object of many studies. For purposes of general description only a few publications, containing references to previous studies, shall be cited: Liese 1965b; Schmid 1965; Fengel 1966a; Harada and Côté 1967; Thomas 1968.

Simple pits are holes in adjacent cell walls, interrupted by a membrane in the region of the compound middle lamella. Simple pits appear only in parenchyma cells. For the exchange of plasmatic material the pit membranes are interlaced with plasmatic threads, the so-called plasmodesmata (Fengel 1966b; Imamura et al. 1974) (Fig. 2–18a).

The bordered pits belonging to the vascular cells (vessels, fibres, tracheids) have a somewhat different structure: the holes in both cell walls enlarge towards the pit membrane forming a cavity (Fig. 2–18c, d). The shape of the aperture of the bordered pits can be very different depending on the species, the cell type, earlywood and latewood etc. The pit apertures of hardwood vessels are often large and may have an oblong or oval shape; the borders are small. In general the membranes of hardwood pits are equal in thickness and consist of several lamellae of cellulose fibrils (Fengel 1966b; Schmid, Machado 1968; Czaninski 1972) (Fig. 2–18c). Few species of the dicotyledonous *Daphne* and *Osmanthus* were found to have tori in their intervascular pit membranes (Ohtani, Ishida 1978). In some hardwoods (e.g. tropical woods of the *Leguminosae* family, *Fagus crenata, Robinia pseudoacacia*) the apertures and borders are provided with vestures (Schmid, Machado 1964; Ishida, Ohtani 1970; Ohtani, Ishida 1973).

The bordered pits of softwood tracheids are characterized by a small aperture (porus) and large cavities. In earlywood the pit borders arch towards the cell lumen (Fig. 2–18d). The membranes of the bordered pits in softwood show a great variability of shape which is partly specific for the species and partly dependent on

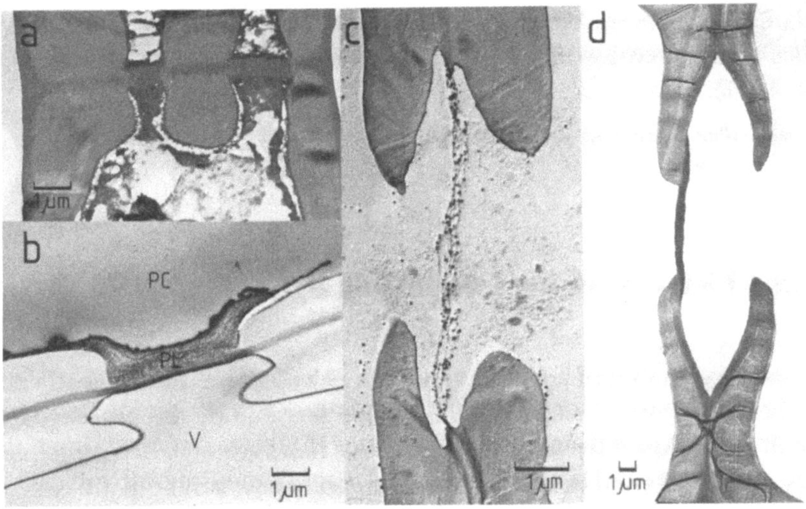

Fig. 2–18. The various types of pits in hardwoods and softwoods. Ultrathin sections, TEM micrographs.
a) Simple pits between ray parenchyma cells in beech *(Fagus sylvatica)*. The membranes are interlaced with plasmodesmata;
b) Half-bordered pit between a longitudinal ray parenchyma cell (PC) and a vessel (V) in oak *(Quercus robur)*. At the parenchyma side the pit membrane is covered with a protective layer (PL);
c) Bordered pit between two vessels in beech *(Fagus sylvatica)*. The several layers of the membrane are visible;
d) Bordered pit between two tracheids in spruce *(Picea abies)* in closed conditions. The disk-like torus lies tightly upon the porus.

earlywood and latewood, longitudinal and ray tracheids (Fengel 1968; Bauch et al. 1972). Some species (e.g. *Ginkgo biloba, Sciadopitys verticillata, Thuja* spp.) have membranes of uniform thickness (Fig. 2–19c). In other species the central area of the membranes is thickened and enforced with amorphous material. The central part is called torus and may be flat like a disk (e.g. earlywood of *Pinus* spp. and *Picea* spp.) (Figs. 2–18d, 2–19a) or like a convex lens (e.g. latewood of *Pinus* spp. and *Picea* spp., *Abies* spp., *Cupressus* spp.) (Fig. 2–19b).

Particularly the pits with a torus in their membranes can be closed by pressure differences within the adjacent cells. By pressing the torus against one porus the pit is closed irreversibly (Fig. 2–18d). During this process the outer part of the membrane (margo) is stretched by some re-organization of the fibrils which were aggregated to strands (Fengel 1972) (Fig. 2–20a). During the heartwood formation the margo can be incrusted with amorphous material so that a movement towards the porus is no longer possible (Fig. 2–20b).

Whenever a vascular cell borders on a parenchyma cell the conducting connection is made by half-bordered pits. These consist of one half of a simple pit on the parenchyma side and one half of a bordered pit on the vascular side (Fig. 2–18b). In softwoods half-bordered pits have a central thickening at the tracheid side (Harada

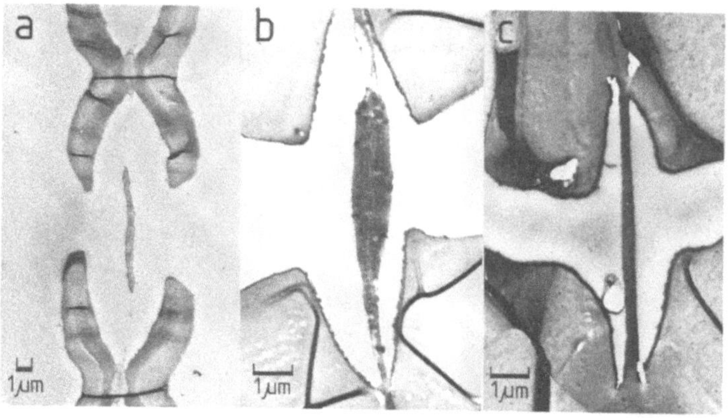

Fig. 2–19. Different membranes in bordered pits of softwoods. Ultrathin sections, TEM micrographs.
a) Membrane with a disk-like torus in earlywood of pine *(Pinus sylvestris)* in open condition;
b) Membrane with a convex lens-like torus in tamrit cypress *(Cupressus dupreziana)* in open condition;
c) Membrane without torus in Japanese umbrella tree *(Sciadopitys verticillata)*.

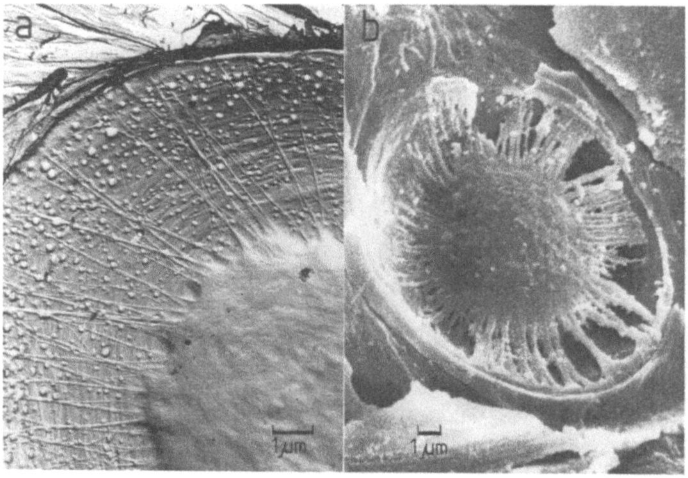

Fig. 2–20. Surface views of bordered pits in softwoods.
a) Closed pit in pine *(Pinus sylvestris)* showing a part of the torus and the fibrillar strands of the margo lying upon the pit border which is covered with warts. Surface replica, TEM micrograph;
b) Encrusted pit membrane in a heartwood pit of spruce *(Picea abies)*. SEM micrograph.

1964). A special form of half-bordered pits are the fenestriform and pinoid pits between ray parenchyma cells and longitudinal tracheids with large membranes, large apertures and small borders (Thomas, Nicholas 1968; Fengel 1970) (Fig. 2–21).

At the parenchyma side of half-bordered pits of hardwood vessels an additional layer is visible (Fig. 2–18b). This protective layer is deposited at the whole paren-

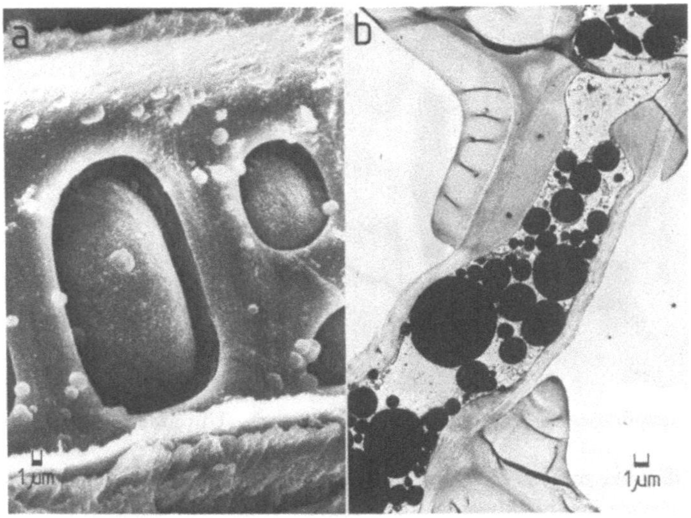

Fig. 2–21. Fenestriform pits in pine *(Pinus sylvestris)*.
a) View from the tracheid side onto the large apertures and the membranes. SEM micrograph;
b) Longitudinal section of a ray parenchyma cell with two fenestriform pits. The large membranes are relatively thick. The dark globules within the living content of the parenchyma cell are vacuoles containing phenolic substances. Ultrathin section, TEM micrograph.

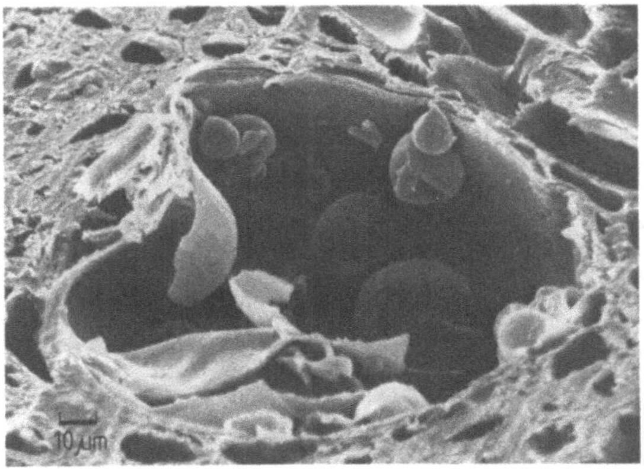

Fig. 2–22. Cross-cut vessel in teak *(Tectona grandis)* with balloon-like tyloses at the inner cell wall surface. SEM micrograph.

chyma wall adjacent to a vessel, and is the initial place of the tyloses formation (Foster 1967; Meyer 1967; Czaninski 1973).

The growth of tyloses is a natural physiological process combined with heartwood formation or with the death of sapwood (e.g. after felling). It may also be initiated by mechanical damages or fungus and virus infection (Koran, Côté 1965).

Tyloses are thin membranes which may interrupt the stream of water within the vessels. These membranes grow within the lumina beginning at the pits bordering on the associated parenchyma cells. After a partial dissolution of the pit membranes the tyloses extend like balloons into the vessel and may after a short time fill up the lumen (Koran, Côté 1964; Meyer 1967; Ishida, Ohtani 1968) (Figs. 2–22, 2–23). The tylosis walls consist of two or more layers, containing cellulose, polyoses and lignin, and in areas where two tylosis walls are in contact a middle lamella-like layer as well as simple pits are developed between them (Sachs et al. 1970; Murmanis 1975). In many wood species tyloses are also found in fibre tracheids (Gottwald 1972).

Fig. 2–23. Radial section of robinia *(Robinia pseudoacacia)*. The three vessels are filled out with tyloses. Light micrograph (by courtesy of D. Grosser).

Literature

Baird, W. M., Johnson, M. A. and Parham, R. A. 1974, Wood Fiber *6,* 114–125, 211–222

Bauch, J., Liese, W. and Schultze, R. 1972, Wood Sci. Technol. *6,* 165–184

Bosshard, H. H. 1974, Holzkunde. II. Zur Biologie, Physik and Chemie des Holzes. Birkhäuser, Basel, Stuttgart

Casperson, G. 1965, Svensk Papperstid. *68,* 534–544

Casperson, G. 1967, Holzforschung *21,* 1–6

Casperson, G. and Zinsser, A. 1965, Holz Roh-Werkst. *23,* 49–55

Chafe, S. C. 1978, Wood Sci. Technol. *12,* 203–217

Côté, W. A. 1968, The Structure of Wood and the Wood Cell Wall. In: Principles of Wood Science and Technology. Vol. I: Solid Wood (Kollmann, F. F. P. and Côté, W. A., Eds.). Springer, Berlin, Heidelberg, New York, pp. 1–54

Côté, W. A. 1977, Wood Ultrastructure in Relation to Chemical Composition. In: The Structure, Biosynthesis, and Degradation of Wood (Loewus, F. A. and Runeckles, V. C., Eds.). Plenum Press, New York, London, pp. 1–44

Côté, W. A. and Day, A. C. 1965, Anatomy and Ultrastructure of Reaction Wood. In: Cellular Ul-

trastructure of Woody Plants (Côté, W. A., Ed.). Syracuse University Press, Syracuse, N. Y., pp. 391–418
Côté, W. A., Day, A. C. and Timell, T. E., 1969, Wood Sci. Technol. *3,* 257–271
Czaninski, Y. 1972, C. R. Acad. Sci. Paris, Ser. D, *275,* 361–363
Czaninski, Y. 1973, Protoplasma *77,* 211–219
Fengel, D. 1966a, Svensk Papperstid. *69,* 232–241
Fengel, D. 1966b, Holz Roh-Werkst. *24,* 245–253
Fengel, D. 1968, Holz Roh-Werkst. *26,* 296–304
Fengel, D. 1970, Wood Sci. Technol. *4,* 176–188
Fengel, D. 1972, Holzforschung *26,* 1–9
Fengel, D. and Grosser, D. 1976, Holz, Morphologie und Eigenschaften. In: Ullmanns Encyklopädie der technischen Chemie, 4th Ed., Vol. 12. Verlag Chemie, Weinheim, pp. 669–679
Fengel, D. and Stoll, M. 1973, Holzforschung *27,* 1–7
Foster, R. C. 1967, Austral. J. Bot. *15,* 25–34
Fujita, M., Saiki, H. and Harada, H. 1973, Bull. Kyoto Univ. Forests No. 45, 192–203
Fujita, M., Saiki, H. and Harada, H. 1978, Mokuzai Gakkaishi *24,* 353–361
Gottwald, H. P. J. 1972, Wood Sci. Technol. *6,* 121–127
Grosser, D. 1977, Die Hölzer Mitteleuropas. Springer, Berlin, Heidelberg, New York
Harada, H. 1962, Mokuzai Gakkaishi *8,* 252–258
Harada, H. 1964, Mokuzai Gakkaishi *10,* 221–225
Harada, H. 1965a, Ultrastructure and Organization of Gymnosperm Cell Walls. In: Cellular Ultrastructure of Woody Plants (Côté, W. A., Ed.). Syracuse University Press, Syracuse, N. Y., pp. 215–233
Harada, H. 1965b, Ultrastructure of Angiosperm Vessels and Ray Parenchyma. In: Cellular Ultrastructure of Woody Plants (Côté, W. A., Ed.). Syracuse University Press, Syracuse, N. Y., pp. 235–249
Harada, H. and Côté, W. A. 1967, Holzforschung *21,* 81–85
Harada, H., Taniguchi, T. and Kishi, A. 1971, Bull. Kyoto Univ. Forests No. 42, 221–227
Hiller, C. H. 1964, Tappi *47,* 125–128
Höster, H. R. and Liese, W. 1966, Holzforschung *20,* 80–90
Imamura, Y., Harada, H. and Saiki, H. 1974, Memoirs Coll. Agricult. Kyoto Univ. No. 106, 11–26
Ishida, S. and Ohtani, J. 1968, Res. Bull. Coll. Exp. Forests *26,* 1–9
Ishida, S. and Ohtani, J. 1970, Res. Bull. Coll. Exp. Forests *27,* 347–354
Jane, F. W. 1970, The Structure of Wood, 2nd Ed.. A. and Ch. Black, London
Jutte, S. M. 1969, A Comparative Study on Normal and Tension Wood Fibres in Beech (*Fagus sylvatica* L.) and Ash (*Fraxinus excelsior* L.). Doctor Thesis, Rijksuniversiteit Leiden
Jutte, S. M. and Levy, J. F. 1973, Acta Bot. Neerl. *22,* 100–105
Kantola, M. 1964, Faserforsch. Textiltechn. *15,* 587–590
Kantola, M. and Seitsonen, S. 1969, Ann. Acad. Scient. Fenn., Ser. A, No. 300
Kishi, K., Harada, H. and Saiki, H. 1977, Bull. Kyoto Univ. Forests No. 49, 122–126
Kishi, K., Harada, H. and Saiki, H. 1979, Mokuzai Gakkaishi *25,* 521–527
Kórán, Z. and Côté, W. A. 1964, IAWA News Bull. 1964/2, 3–15
Kórán, Z. and Côté, W. A. 1965, The Ultrastructure of Tyloses. In: Cellular Ultrastructure of Woody Plants (Côté, W. A., Ed.). Syracuse University Press, Syracuse, N. Y., pp. 319–333
Liese, W. 1965a, The Warty Layer. In: Cellular Ultrastructure of Woody Plants (Côté, W. A., Ed.). Syracuse University Press, Syracuse, N. Y., pp. 251–269
Liese, W. 1965b, The Fine Structure of Bordered Pits in Softwoods. In: Cellular Ultrastructure of Woody Plants (Côté, W. A., Ed.). Syracuse University Press, Syracuse, N. Y., pp. 271–290
Meyer, R. W. 1967, For. Prod. J. *17,* No. 12, 50–56
Meylan, B. A. and Butterfield, B. G. 1978, Wood Sci. Technol. *12,* 219–222
Mia, A. J. 1968, Wood Sci. *1,* 105–115

Murmanis, L. 1975, Wood Sci. Technol. *9,* 3–14
Nečesaný, V. 1966, Drev. Výsk. *11,* 1–26
Norberg, P. H. and Meier, H. 1966, Holzforschung *20,* 174–178
Ohtani, J. and Ishida, S. 1973, Res. Bull. Coll. Exp. Forests Hokkaido Univ. *33,* 125–144
Ohtani, J. and Ishida, S. 1978, Mokuzai Gakkaishi *24,* 673–675
Okumura, S., Harada, H. and Saiki, H. 1977, Wood Sci. Technol. *11,* 23–32
Panshin, A. J. and de Zeeuw, C. 1970, Textbook of Wood Technology, 3rd Ed. Vol. 1. McGraw-Hill, New York, San Francisco, Toronto, London
Parham, R. A. and Baird, W. M. 1974, Wood Sci. Technol. *8,* 1–10
Ruel, K., Barnoud, F. and Goring, D. A. I. 1978, Wood Sci. Technol. *12,* 287–291
Sachs, I., Kuntz, J., Ward, J., Nair, G. and Schultz, N. 1970, Wood Fiber *2,* 259–268
Sakai, H. 1982, The Structure of Domestic and Imported Woods in Japan. Japan Forest Techn. Ass., Tokyo
Schmid, R. 1965, The Fine Structure of Pits in Hardwoods. In: Cellular Ultrastructure of Woody Plants (Côté, W. A., Ed.). Syracuse University Press, Syracuse, N. Y., pp. 291–304
Schmid, R. and Machado, R. D. 1964, Planta *60,* 612–636
Schmid, R. and Machado, R. D. 1968, Protoplasma *66,* 185–204
Schweingruber, F. H. 1978, Mikroskopische Holzanatomie. Zürcher AG., Zug
Scurfield, G. and Silva, S. 1969, Austral. J. Bot. *17,* 391–402
Takiya, K., Harada, H. and Saiki H. 1976, Bull. Kyoto Univ. For. *48,* 187–191
Thomas, R. J. 1968, Holzforschung *22,* 38–44
Thomas, R. J. and Nicholas, D. D. 1968, Tappi *51,* 84–88
Timell, T. E. 1973, Ultrastructure of the Dormant and Active Cambial Zones and the Dormant Phloem Associated with Formation of Normal and Compression Woods in *Picea abies* (L.) Karst.. Techn. Publ. No. 96, State Univ. New York, Coll. Environm. Sci. Forestry, Syracuse, N. Y.
Timell, T. E. 1978a, Wood Sci. Technol. *12,* 1–15
Timell, T. E. 1978b, Wood Sci. Technol. *12,* 89–103
Verhoff, S. and Knigge, W. 1976, Holz Roh-Werkst. *34,* 175–180
Wagenführ, R. 1966, Anatomie des Holzes. VEB Fachbuchverlag, Leipzig
Wagenführ, R. and Scheiber, C. 1974, Holzatlas. VEB Fachbuchverlag, Leipzig
Wardrop, A. B. and Davies, G. W. 1964, Austral. J. Bot. *12,* 24–38

3. Chemical Composition and Analysis of Wood

3.1. Survey of the Chemical Components

3.1.1. Macromolecular Substances

As far as the chemical components of wood are concerned, a distinction must be made between the main macromolecular cell wall components cellulose, polyoses (hemicelluloses) and lignin, which are present in all woods, and the minor low-molecular-weight components (extractives and mineral substances), which are generally more related to special wood species in kind and amount. The proportions and chemical composition of lignin and polyoses differ in softwoods and hardwoods, while cellulose is a uniform component of all woods.

A short introduction to the chemical wood components follows the general scheme in Fig. 3–1. A more detailed discussion is given in the analysis part of this chapter and especially in chapters 4 to 7.

In wood from temperate zones, the portions of the high-polymeric compounds building up the cell walls account for 97–99% of the wood material. For tropical woods this value may decrease to an average value of 90%. 65–75% of the wood consists of polysaccharides.

Cellulose is the major wood component, making up approximately one half of both softwoods and hardwoods. It can be briefly characterized as a linear high-molecular-weight polymer built up exclusively of β-D-glucose. Because of its chemical and physical properties as well as its supramolecular structure it can fulfil its function as the main structural component of the plant cell walls.

Polyoses (hemicelluloses) are in close association with cellulose in the cell wall. Five neutral sugars, the hexoses glucose, mannose, galactose and the pentoses xylose

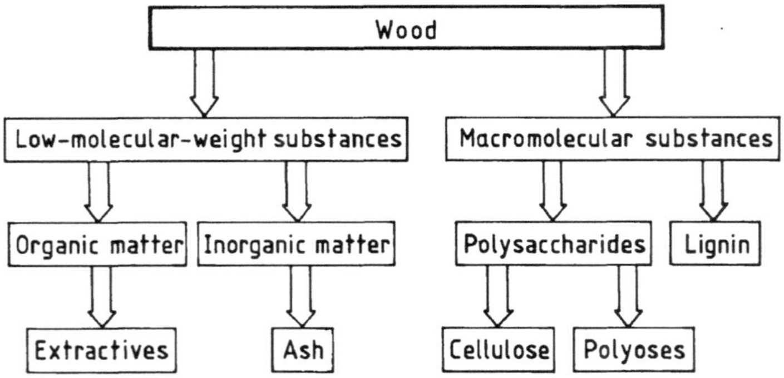

Fig. 3–1. General scheme of the chemical wood components

and arabinose are the main constituents of the polyoses. Some polyoses contain additionally uronic acids. The molecular chains are much shorter than in the case of cellulose, having side-groups and being branched in some cases. Hardwoods contain more polyoses than softwoods and the sugar composition is different.

Lignin is the third macromolecular wood component. The molecules of lignin are built up quite differently from those of the polysaccharides, as they consist of an aromatic system composed of phenylpropane units. There is more lignin in softwoods than in hardwoods and there are some structural differences between softwood and hardwood lignins. From a morphological point of view lignin is an amorphous substance located in the compound middle lamella as well as in the secondary walls. During the development of the cells lignin is incorporated as the last component into the cell walls, interpenetrating the fibrils and so strenghtening the cell walls.

Minor polymeric substances: These are found in wood in small amounts as starch and pectic substances. Proteins account for at most 1% of parenchyma cells of wood, but are mainly found in the non-wooden parts of the stem, i.e. cambium and inner bark.

3.1.2. Low-Molecular-Weight Substances

Besides the cell wall components there are numerous compounds which are called accessory or extractive material of wood. Though these components contribute only a few percent to the wood mass, they may have a great influence on the properties and processing qualities of woods. Some components, such as certain metal ions, are even essential for the living tree.

The low-molecular-weight substances belong to very different classes of chemical compounds and therefore it is difficult to find a clear but comprehensive classification system. A simple classification can be made by dividing them into organic and inorganic matter (Fig. 3–1). The organic matter is commonly called extractives. The inorganic part is summarily obtained as ash. With regard to analysis it is more useful to distinguish between substances on the basis of their solubility in water and organic solvents. These analytical aspects are discussed later in this chapter. Firstly, the main groups of chemical compounds comprising the low-molecular-weight wood components are briefly introduced.

Aromatic (phenolic) compounds: The most important substances of this group are the tanning compounds which can be divided in hydrolyzable tannins and condensed phlobaphenes. Other phenolic substances are for example the stilbenes, lignans and flavonoids, and their derivatives. Simple compounds deriving from the lignin metabolism also belong to this chemical group.

Terpenes are a wide-spread group of natural substances. Chemically, they can be derived from isoprene. Two or more isoprene units build up the mono-, sesqui-, di-, tri-, tetra- and polyterpenes.

Aliphatic acids: Saturated and unsaturated higher fatty acids are found in wood mostly in the form of their esters with glycerol (fats and oils) or with higher alcohols (waxes). Acetic acid is linked to the polyoses as an ester group. Di- and hydroxy-carboxylic acids mainly occur as calcium salts.

Alcohols: Most of the aliphatic alcohols in wood occur as ester components, while the aromatic sterols, belonging to the steroids, are mainly found as glycosides.

Inorganic substances: The mineral components of woods from temperate zones are predominantly the elements potassium, calcium and magnesium. In tropical woods other elements, for example silicon, may be the main inorganic component.

Other components: Mono- and disaccharides are found in wood only in minor amounts but they occur in higher percentages in the cambium and the inner bark. Small amounts of amines and ethene were also found in wood.

3.2. Analysis of Wood

3.2.1. Problems of Analysis

Wood analysis comprises the determination of wood composition as well as the isolation, purification and characterization of the wood constituents. As wood is a natural material, modified procedures and methods for wood and wood-related substances are applied besides the classical methods of analytical chemistry.

The methods of wood analysis are more or less standardized. A distinction can be made between methods which are mainly used in scientific research and those applied in the industrial production and control of wood-derived products such as pulp etc. They can differ with regard to the required precision and the special objective of analysis.

The main difficulty in general wood analysis is not the number of components, which are sometimes very different in their chemical composition and behaviour, but rather the fact that an intimate ultrastructural and chemical association exists between the macromolecules of the cell wall. The difficulties of a selective separation of the main components are demonstrated in Fig. 3–2. In the intermediate steps of wood analysis lignin portions remain with the isolated polysaccharides and even cellulose and polyoses can hardly be separated quantitatively without degradation and changes in molecular properties.

Wood analyses can be carried out very differently, e.g. by determining only the main cell wall components, namely the polysaccharides (holocellulose) and lignin in addition to the extractives and ash. On the other hand very detailed analyses including the determination of functional groups (e.g. acetyl groups) and of the individual sugar pattern of the polysaccharides are reported in the literature (e.g.

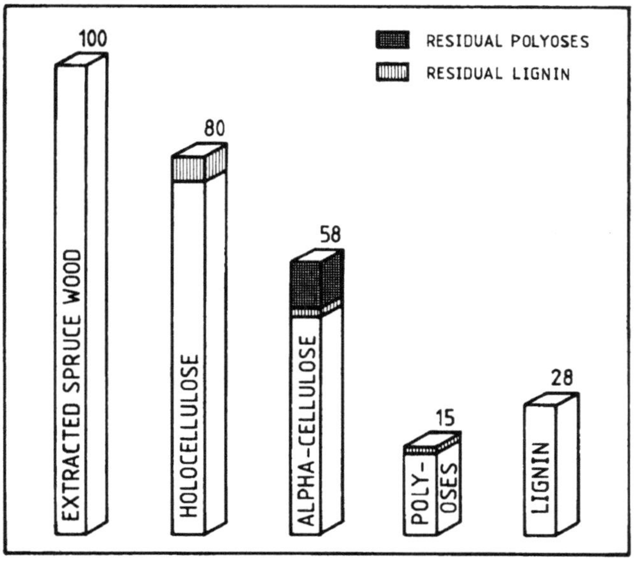

Fig. 3–2. The macromolecular cell wall components from the standpoint of wood analysis

Browning 1963; Timell 1967; Stewart et al. 1973; Smelstorius 1974; Fengel, Wegener 1979).

So-called summative analyses can be made to prove exactly how the individual components were separated and determined. In any case the aim of a satisfactory analysis is a sum of approximately 100% for all determined components. This aim is difficult to achieve, however, especially if the number of individual analyses increases causing lacking and/or overlapping results combined with the addition of individual errors (Browning 1967a). Values of 98–101% are generally acceptable. Four examples of different types of wood analyses are given in Fig. 3–3.

The following description of the analysis of wood components gives only a survey of several methods. Detailed descriptions are given in the specialist literature (e.g. Browning 1967a, b; Moore, Johnson 1967; Tappi Standards; ASTM Standards; SCAN Testing Standards; DIN-Normen, Zellcheming Merkblätter; ISO Standards; CPPA Standards; Appita Standards).

3.2.2. Sampling and Preparation of Samples

The kind of sampling and sample preparation depends on many factors and on the aim of analysis. Therefore our discussion here is limited to general remarks.

If the whole wood from a wood species is analyzed it is generally important to select a representative sample from this species. This requires the random selection of one or more representative trees as well as the selection of an average and normal part of the stem, i.e. without reaction wood, resin galls or accumulations of knots. A standardized sampling procedure is given in Tappi Standard T 257 os-76.

1	2	3	4
Holocellulose Residual lignin in holocellulose	*Alpha-cellulose* *Polyoses A, B* Residual lignin in cellulose and polyoses	*Cellulose* *Xylan, Mannan* *Pentosan* Residual polyoses in cellulose Residual lignin in cellulose and polyoses	Glucose Mannose Galactose Arabinose Xylose Rhamnose Uronic acids Acetyl
Lignin in wood Acid lignin	*Lignin in wood* Acid lignin Acid-soluble lignin	*Lignin in wood* UV spectrophotometric determination	*Lignin in wood* Calculation (by difference)
Extractives Solvent extract	Solvent-soluble substances Water-soluble substances	Ether extract Alcohol extract Steam extract Hot-water extract Cold-water extract	Fats, waxes Tannins Phenols Terpenes Proteins Monosaccharides Oligosaccharides Pectic substances
Ash in wood	*Ash in wood*	*Ash in wood*	Cations Anions

Fig. 3–3. Four examples of wood analyses

For the purposes of chemical analysis wood must be milled to achieve a complete penetration of the reagents and to ensure uniform reactions. The first step is chipping or comparable operations like sawing, shaving or the production of sticks or veneers. Further reduction of the particle size is obtained by grinding the wood in suitable mills like hammer mills, disk mills or wing beater mills. Heating must be avoided as well as overproduction of very fine and dusty material. For special purposes, e.g. the isolation of milled wood lignin (→ 3.2.9.), vibratory mills of different types and sizes with porcelain, glass or steel balls are used. Pulp and other fibrous material can be dispersed in pulp beaters or homogenizers. As the particle size after normal grinding is still non-homogeneous the sample has to be screened to eliminate the very fine material, which may cause trouble by clogging fine filters or passing through coarse filters. Additionally, atypical results are obtained with the finest fraction (Erickson 1962). The coarse material has to be ground once more. No general rule can be given for the best particle size to be used in wood analysis but a range of 40–80 mesh fractions or particle sizes between 0.05 and 0.4 mm are customary. The selected fraction should represent at least 90–95% of the origin sample to avoid discharging special parts of wood which are difficult to grind (e.g. heartwood or latewood).

3.2.3. Determination of the Water Content

As wood is a hygroscopic material the wood-water system is very important in many fields of wood technology, wood physics and wood chemistry. It is unusual to analyze water-free samples because of possible changes during drying and the difficulties of weighing dry samples without absorption of moisture. Therefore the samples are generally weighed in an air-dried condition, while the water content is determined in separate samples. The results of analyses are usually reported on a moisture-free, absolutely dry basis. Three important types of chemical water determination are:

- oven or vacuum drying
- titration with a selective reagent for water
- distillation with a water-immiscible solvent.

The simplest and most frequently applied method is oven drying at 105 ± 3 °C until constant weight is reached (Tappi Standard T 12 os-75; ASTM Standard D 2016–65). The determined value of water content can be falsified by volatile substances such as terpenes of coniferous woods. In this case the method of vacuum drying in a desiccator over phosphoric anhydride generally gives more exact results but has the disadvantage of very long drying times. A much shorter drying time can be obtained by using a vacuum oven and temperatures of about 60 °C, but also in this case highly volatile substances will escape.

The fastest and most reliable method is titration according to Karl Fischer (Fischer 1935). The Karl Fischer solution contains iodine, pyridine, sulfur dioxide and methanol, and reacts almost quantitatively with water. Titration can be performed best by potentiometric end-point determination. Attention must be paid to interfering substances which may additionally reduce iodine. The general method is standardized in ASTM Standards E 203 and D 1348. Many modifications were reported, mostly concerning the composition of the reagent (e.g. Verhoef 1977).

The determination of water in wood by distillation can be carried out with solvents such as xylene, toluene or trichlorethene by distilling the water in the sample in an azeotropic mixture. The resultant water is condensed and collected in a graduated trap. To obtain a sufficient amount of water large quantities of wood are required for this method.

A bibliography concerning determination of water in wood, pulp, paper and textiles was compiled by Weiner and Byrne (1961). Kollmann and Höckele (1962) made a critical comparison of 15 methods of water determination applied to several wood species. Special physical methods such as nuclear radiation techniques (Loos 1965), nuclear magnetic resonance (NMR) (Karras, Rahkamaa 1971; Magnusson et al. 1972; Nanassy 1978) and neutron moderation (Töppel 1969) are also applicable to moisture content determination of wood and pulp.

3.2.4. Extractives

As mentioned above the extractives cover a wide range of chemical compounds though they generally represent only a small part of wood. Amounts of extractives for different softwoods and hardwoods are listed in Tables 3–4, 3–5, and 3–6.

The extractives of a wood sample may be isolated for the purpose of a detailed examination of the structure and composition of one or more of their components. In the general wood analysis, only the quantity is determined after isolation. By this method extractive-free wood is obtained as starting material for the isolation and analysis of the macromolecular cell wall components.

The isolation of the extractives is carried out by extraction with neutral solvent mixtures and/or with single solvents in succession. According to the different solubilities of the extractive components many extraction schemes and sequences can be found in the literature (Browning 1967a). Fig. 3–4 gives an overview of the groups of extractives from an analytical standpoint, with examples of subgroups and individual compounds.

The volatile fraction containing for example terpenes in the case of softwoods is isolated mainly by steam distillation. The solvent extraction can be carried out with different solvents such as ether, acetone, benzene, ethanol, dichloromethane, or mixtures thereof. Fatty acids, resin acids, waxes, tannins and colouring matter are the most important substances extractable by solvents. The main components of the water-soluble part of wood consist of carbohydrates, proteins and inorganic salts.

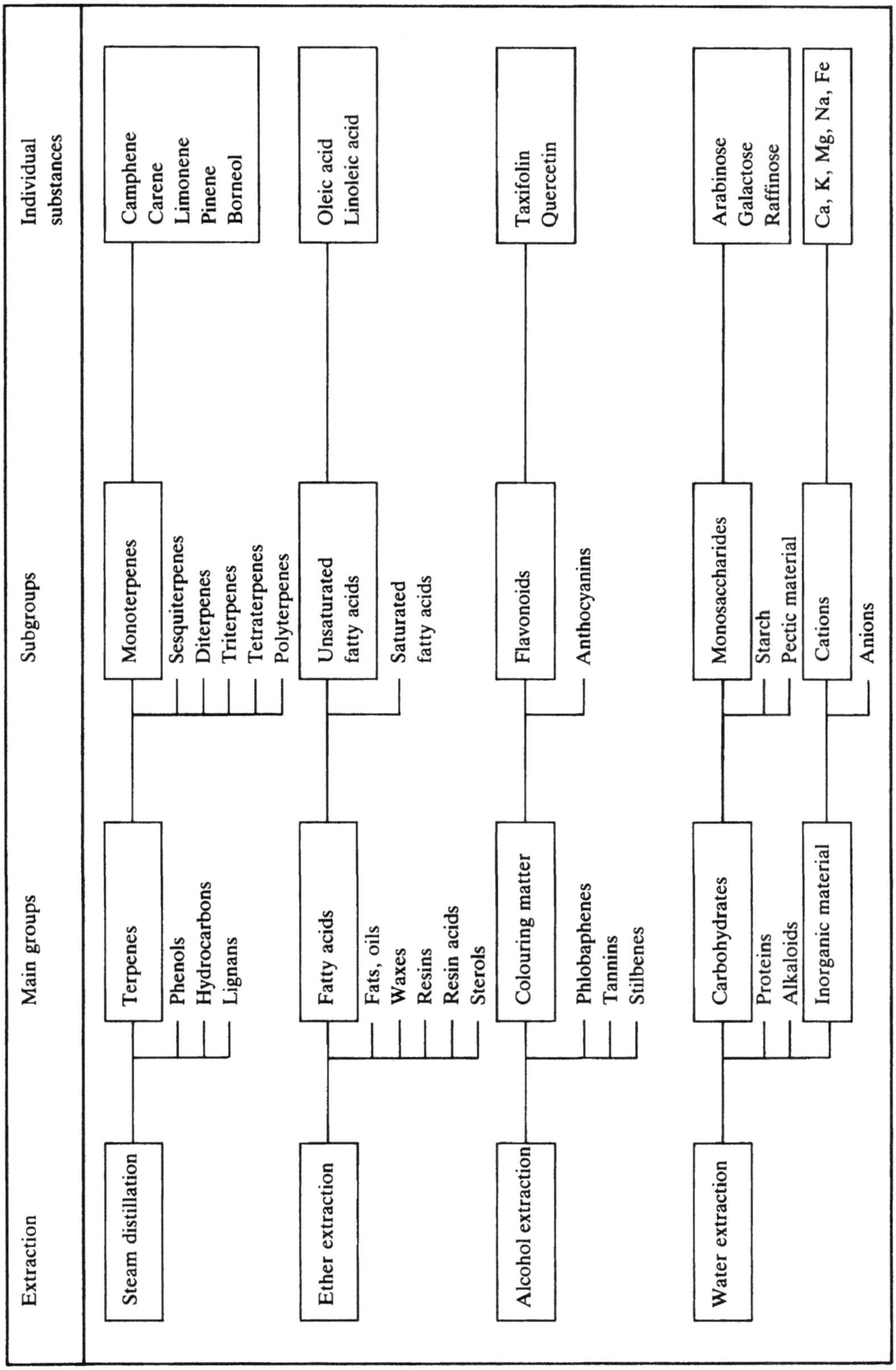

Fig. 3–4. Classification of the extractives with examples according to analysis groups

The distinction between the extractive components derived from individual extraction steps is not sharp in any case. For example tannins are principally soluble in hot water but are also found in alcohol extracts.

The preparation of wood for chemical analysis generally includes the removal of the extractives if the extraction procedure does not interfere with subsequent analysis. For example, the ash content is generally determined in unextracted wood because inorganic components may be removed during an extraction step applying water. A standard procedure which is often used for preparation of extractive-free wood is extraction with ethanol-benzene (1:2, 4 h) followed by extraction with 95% ethanol for 4 hours in a Soxhlet apparatus, and a final extraction with boiling water to remove residues of the solvents (e.g. Tappi Standard T 12 os-75; ASTM Standard D 1105–56). As benzene is known to be extremely injurious to health it was recently suggested to replace it with cyclohexane or toluene as a solvent component in combination with ethanol (Garves 1981; Goetzler 1982; Fengel, Przyklenk 1983).

Conventional techniques of isolation, determination and identification of extractive compounds are reviewed comprehensively by Browning (1967a), including crystallization, fractional distillation, saponification as well as spectroscopic (UV, IR) and chromatographic methods (paper, thin-layer, column, ion-exchange and gas chromatography). The application of different chromatographic methods on wood extractives is also described by Kholkin (1969).

Modern fractionation methods such as high-performance liquid chromatography (HPLC) or gel permeation chromatography (GPC) and the application of nuclear magnetic resonance (NMR) or gas chromatography followed by mass spectroscopy (GC-MS) are becoming increasingly important for the separation and identification of extractives (e.g. Weissmann 1973, 1974; Seligmann 1977; Hafizoglu 1979; Ekman 1980). Improved capillary gas chromatography techniques were successfully applied in the identification of hydrophilic and lipophilic extractives from spruce wood *(Picea abies)* as well as toxic compounds in pulp and paper mill waste waters (Holmbom 1977, 1980; Rogers et al. 1979; Theander 1981; Pensar et al. 1981).

3.2.5. Inorganic Material

The inorganic part of wood is analyzed as ash by incineration of the organic wood material at 600–850 °C. The amounts of ash are between 0.2 and 0.5% in the case of woods from temperate zones but often much higher in tropical woods (Tables 3–4, 3–5, 3–6).

The main components of wood ash are potassium, calcium, and magnesium as well as silicon in the case of some tropical woods. Errors in the determination of the ash content may derive from some losses of alkali metal chlorides and ammonia salts as well as from insufficient oxidation of carbonates of alkaline earth metals.

More reproducible and somewhat higher ash values result from the so-called sulfate ash determination. In this method the inorganic salts are converted to non-volatile sulfates by the addition of sulfuric acid (50%) during incineration. Details of methods for ash determination are described in Tappi Standard T 15 os-58 and ASTM Standard D 1102–56. To avoid loss of volatile ash components wet digestion or combustion with oxygen in a closed vessel is suggested.

A very rapid method (5 min) for ash determination in paper and paperboards for control analyses in the paper industry is described by Donetzhuber and Wallenius (1971).

Special identification of ash constituents can be accomplished by different methods such as flame spectroscopy, emission spectroscopy, X-ray analysis, atomic absorption spectroscopy and neutron activation analysis (Browning 1967a; Zicherman, Thomas 1972; Nagorny, Zietlow 1972; Clarke, Jayman 1975; Griebenow et al. 1977; Cutter, McGinnes 1980).

3.2.6. Delignification Methods (Preparation of Holocellulose)

Ritter and Kurth (1933) were the first to use the term holocellulose for the product obtained after the removal of lignin from wood.

An ideal delignification should result in a total removal of lignin without chemical attack on the polysaccharides, but there is no delignification procedure which can satisfy this requirement. Three important criteria can be defined for holocelluloses:

- low residual lignin content
- minimal loss of polysaccharides
- minimal oxidative and hydrolytic degradation of cellulose.

Two common methods are applied in the preparation of holocellulose on a laboratory scale:

- chlorination including alternating extraction with hot alcoholic solutions of organic bases (e.g. ASTM Standard D 1104–56)
- delignification with an acidified solution of sodium chlorite.

The first method using chlorine as a delignifying agent was originally described by Ritter and Kurth (1933) who applied an alcoholic solution of pyridine for the extraction of the solubilized lignin. Van Beckum and Ritter (1937) modified this method using monoethanolamine instead of pyridine. Other modifications are reported by Thomas (1945) and Timell and Jahn (1951) who used chlorination in carbontetrachloride and ice water, respectively. Kurth and Swelim (1963) replaced the chlorine with bromine.

Delignification with acidified sodium chlorite is based on the investigations of Jayme (1942) and Wise et al. (1946). The effective components of the delignifying solution are chlorine dioxide, chlorine and chlorate. The standard procedure comprises the treatment of pre-extracted wood with an acidified sodium chlorite solution (pH 4) at 70–80 °C for 3–5 hours. Generally softwoods require one hour more than hardwoods to reach a comparable low residual lignin content of the holocellulose.

Various modifications are described concerning reaction temperature, reaction time, amount of sodium chlorite and pH value (Timell, Jahn 1951; Klauditz 1957; Erickson 1962; Thompson, Kaustinen 1964; Wegener 1974; Uçar 1977; Plonka, Surewicz 1979). In particular, modified methods were applied to obtain largely unchanged polysaccharides for use in further investigations on cellulose and polyoses. These methods often combine low temperature with prolonged reaction times. A well-known procedure is delignification according to Klauditz (1957) at 35–40 °C for 40 hours. Thompson and Kaustinen (1964) delignified for 3–4 weeks at room temperature. Wegener (1974) varied the temperature between 70 and 20 °C, applying prolonged reaction times of up to 14 days. From the results of polysaccharide determinations in the chlorite liquors he concluded that the conditions at 50 °C and 25 h are most protective for the polysaccharides of spruce wood *(Picea abies)* (Wegener 1975). In the case of beech wood *(Fagus sylvatica)* the optimum conditions are a temperature of 30 °C combined with a reaction time of 4 days (Fengel et al. 1979). Plonka and Surewicz (1979) used different delignification methods and obtained best results with regard to summative analysis by a combination of the methods according to Wise et al. (1946) and Klauditz (1957). Fig. 3–5 shows the dependence of isolated polysaccharide and residual lignin amounts on the holocellulose yield for spruce and beech wood *(Picea abies, Fagus sylvatica)*.

There exist further investigations comparing especially chlorine and chlorite methods with regard to their efficiency in releasing lignin and preserving the remaining polysaccharides (Browning 1967b; Wegener 1974). But no general comment can be made because of some contradictory results and the fact that the criticism of delignification depends on many factors (e.g. wood species, residual lignin content of holocellulose). Towards the end of delignification losses of polysaccharides occur in any case.

A careful delignification can also be obtained by the use of dilute peracetic acid. The principle method for obtaining holocellulose was developed by Poljak (1948) and modified by Haas et al. (1955) and Leopold (1961). Investigations on the solubilized compounds during peracetic acid delignification were carried out by Albrecht (1971), Albrecht and Nicholls (1976), and Glinsky and Nicholls (1977). The delignification is generally claimed comparably protective for the polysaccharides as the chlorine and chlorite methods (Haas et al. 1955; Leopold 1961).

During delignification the lignin portion of wood is made soluble by altering and degrading substitution reactions as well as by oxidative reactions. Special investigations on the mechanisms of degradation of lignin and lignin model compounds

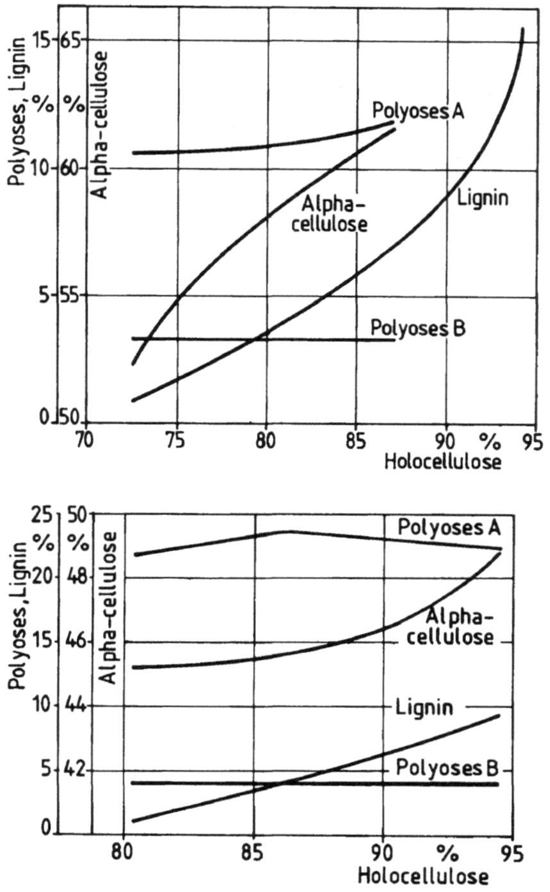

Fig. 3–5. Dependence of alpha-cellulose, polyoses and residual lignin in holocellulose on the degree of delignification of spruce wood (above) and beech wood (below) (Wegener 1974; Fengel et al. 1979)

during chlorine and chlorite delignifications were carried out by Sarkanen et al. (1962), Dence et al. (1962), Gierer and Sundholm (1971), Lindgren (1971, 1974), Vander Linden (1974), and Vander Linden and Nicholls (1976).

As already mentioned a small percentage of residual lignin generally remains within the holocellulose. Portions of this residual lignin are altered during the delignification, thus becoming soluble during the determination of acid-insoluble residual lignin by acid hydrolysis of the holocellulose (→ 3.2.9.). This acid-soluble lignin causes errors of up to 9% in the summative analysis of wood. Only if both acid-insoluble and acid-soluble lignin are determined will summative analyses with sums close to 100% be received.

The acid-soluble lignin part can be estimated spectrophotometrically (e.g. Schöning, Johansson 1965; Ahlgren, Goring 1971) as well as by determining the total residual lignin using the acetyl bromide method according to Johnson et al. (1961) (→ 3.2.9.). Wegener (1974) demonstrated that the acetyl bromide method is suitable for lignin determination in holocelluloses up to degrees of delignification of about 60%. For spruce chlorite holocellulose he agrees with Ahlgren and Goring (1971) and Maekawa and Koshijima (1980b) that the amount of acid-soluble lignin reaches a maximum at about 50% delignification (Fig. 3-6). At low lignin contents the determination becomes ambiguous.

The reactions of oxidative delignifying agents with the polysaccharides can be divided into oxidative conversion of functional groups and hydrolytic effects causing a cleavage of the polysaccharide chains. Generally aldehydic end-groups and hydroxyl groups are attacked in oxidative media, resulting in the formation of carboxyl and carbonyl groups. Uronic acid groups are formed during chlorite delignifications, as was proved by various authors (e.g. Timell 1961; Stewart, Smelstorius 1968; Miyazaki et al. 1971; Fengel et al. 1979). Gluconic acid groups were also identified (Alfredsson et al. 1961; Alfredsson, Samuelson 1974; Heikkila, Sjöström 1975).

Deacetylation up to 10% was reported by Lindberg and Rosell (1974) and Fengel et al. (1979). The loss of polysaccharides was studied under different aspects e.g. by Thompson and Kaustinen (1966), Thompson et al. (1968a, b), Brink and Pohlman (1972), and Maekawa and Koshijima (1980a, b). Ahlgren and Goring (1971) found losses of polysaccharides after 60% of the lignin had been removed during chlorite delignification of spruce wood. Mannan was first dissolved apart from small amounts of xylan. Polysaccharide losses already occur in the first stages of the delignification. This was proved by determination of a sugar content of 0.5% based on wood in chlorite liquors shortly after only 25% of the lignin had been removed (Wegener 1974, 1975). The highest values for the polysaccharide losses were 3.5%, based on total spruce polysaccharides. A considerable loss of hexosans was observ-

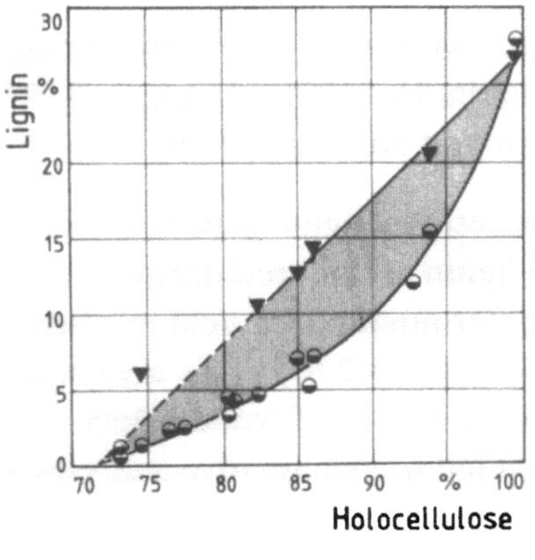

Fig. 3-6. Total lignin, acid-soluble and acid-insoluble lignin depending on the degree of delignification of spruce wood (Wegener 1974);
▼-▼ : total lignin (AcBr)
⊖-⊖ : acid-insoluble lignin (Runkel)
||||||||| : acid-soluble lignin

ed which are possibly dissolved as lignin-polysaccharide complexes (Wegener 1976) (→6.5.).

Depolymerization of cellulose occurs to a large extent during all types of delignification. Changes in the cell walls during chlorite delignifications were studied by several authors, including topochemistry as well as swelling and shrinking effects (Wood et al. 1972; Kibblewhite 1973; Lapinoja, Thompson 1973; Cinité et al. 1976; Stoll, Fengel 1977; Yang, Goring 1978).

The molecular size of the lignin dissolved during chlorite delignifications is relatively uniform compared with lignins from kraft and sulfite pulping (Ahlgren et al. 1971; Wegener 1976).

3.2.7. Isolation and Determination of Cellulose

There are three main types of methods for isolating and/or determining cellulose:
- separation of the main portions of polyoses and residual lignin from holocellulose
- direct isolation of cellulose from wood, including purification procedures
- determination of the cellulose content by total hydrolysis of wood, holocellulose or alpha-cellulose with subsequent determination of the resulting sugars.

In any isolation method cellulose cannot be obtained in a pure state, but only as a more or less crude preparation which is generally called alpha-cellulose. This term was coined by Cross and Bevan (1912) for wood cellulose which is insoluble in a strong sodium hydroxide solution. The portion which is soluble in the alkaline medium but precipitable from the neutralized solution was called beta-cellulose. Gamma-cellulose is the name for the portion which remains soluble even in the neutralized solution. In connection with the viscose process Jentgen (1911) suggested a solubility test for pulps in 17.5% sodium hydroxide. This method was modified in various manners and is now established as a standard method for the determination of alpha-, beta- and gamma-cellulose (Tappi Standard T 203 os-74; ASTM Standard D 1103-60). The corresponding German standards describe the alkali resistance of pulps to 18% and 10% NaOH (DIN 54355; Zellcheming Merkblatt IV/39/67).

The most common method of isolation and determination of cellulose on a laboratory scale was elaborated by Wise et al. (1946). Holocellulose is extracted under nitrogen in two steps with 5% and 24% potassium hydroxide. The procedure results in a cellulose still containing considerable amounts of residual polyoses and lignin. By repeated treatments, e.g. with different alkali solutions, the polyoses and the residual lignin content can be reduced. But simultaneously the degree of polymerization (→ 4.2.3.) and the yield of cellulose will decrease. Generally the yield of alpha-cellulose depends on the wood species and particularly on the isolation and determination procedure. Hence this value varies widely between 40–60% (Tables 3–4, 3–5, 3–6).

Apart from sodium and potassium hydroxide, lithium hydroxide was also used for the separation of polyoses and cellulose. Hamilton and Quimby (1957) found that sodium hydroxide and lithium hydroxide are more powerful than potassium hydroxide in removing polyoses, especially mannans. After extraction of spruce holocellulose with 5% and 17.5% sodium hydroxide Fengel (1980) obtained an alpha-cellulose still containing 10% mannan, 1.5% xylan and 1.5% residual lignin (→ 3.2.8.).

Another way of determining cellulose is the direct isolation of cellulose from wood. According to Kürschner and Hoffer (1929) wood is treated with nitric acid in ethanol. The application of 25% potassium hydroxide prior to the alcoholic nitration reduces the required time to one hour (Kürschner, Popik 1962). The resulting cellulose is relatively pure but degraded by hydrolytic effects.

Refluxing wood meal with a mixture of acetyl-acetone and dioxane, acidified with hydrochloric acid, also results in a very pure cellulose preparation (Seifert 1956, 1960). The cellulose values according to Seifert's method are some 10% lower than the yields of Kürschner-Hoffer celluloses but reproducibility is better (Kaszynska 1974).

Clermont and Bender (1961) obtained alpha-celluloses in high yields by treating wood with solutions of chlorine and nitrogen dioxide in dimethyl sulfoxide (DMSO) or of sulfur dioxide in DMSO.

Another possibility of a direct cellulose determination is the nitration of wood with acid mixtures as they are used for preparing cellulose nitrates for the viscosimetric determination of the degree of polymerization. The celluloses are comparable to those obtained by delignification and alkaline extraction (Timell 1957; Timell et al. 1959). From the yield of cellulose nitrate the cellulose content can be calculated or determined by sugar analysis after de-nitration.

Attempts to isolate cellulose from milled wood were reported by Hayashi and Kaji (1972). They extracted milled wood with 10% sodium hydroxide followed by a cuene treatment. A crude cellulose was obtained by neutralizing the cuene solution with acetic acid. The yields amounted to 20–30% of wood, corresponding to some 50% of the real cellulose content of wood.

General isolation procedures for polysaccharides and especially for cellulose are described elsewhere (e.g. Whistler, Wolfrom 1963, 1965).

The methods of cellulose determination without isolation by hydrolysis and subsequent sugar determination can be applied to wood as well as to holocellulose or alpha-cellulose. A general procedure is hydrolysis with concentrated acids and subsequent dilution steps to achieve secondary hydrolysis. Sulfuric acid is used frequently, starting with a concentration of 72% for the first hydrolysis step (Saeman et al. 1954). A method applying 100% trifluoroacetic acid in the first stage of hydrolysis with subsequent dilution steps was recently described (Fengel et al. 1978; Fengel, Wegener 1979).

Attention must be paid to a total hydrolysis of cellulose without considerable losses of sugars deriving from more easily hydrolyzable polyoses.

The identification and quantification of sugars after hydrolysis can be achieved by different chromatographic methods such as paper chromatography (PC) (e.g. Timell 1957; Stewart et al. 1973; Smelstorius, Stewart 1974; Tappi Standard T 250 pm-75), thin-layer chromatography (e.g. Wegener 1975) or gas chromatography (GC) partly combined with mass spectroscopy (GC-MS) (e.g. Borchard, Piper 1970; Whistler, BeMiller 1972; Stewart et al. 1973; Janson 1974; Tappi Standard T 249 pm-75). More recently automated sugar analysis by ion-exchange chromatography via sugar borate complexes was used to determine the polysaccharide components of wood and pulps (Sinner et al. 1975; Fengel et al. 1978; Fengel, Wegener 1979; Fengel et al. 1979). After correction of the glucose value by subtracting the glucose portion deriving from the glucomannan present in the sample, the calculated cellulose value will be close to the true cellulose content.
A fast spectroscopic method of sugar determination was described by Scott (1976; Scott, Green 1972, 1974), who measured the dehydrated products of sugars (furans) after total hydrolysis of wood and pulp at 322 and 380 nm.

An unconventional method for the quantitative determination of cellulose in wood was given by Jayme et al. (1974, 1976). The method includes a semichemical pulping step (NSSC vapour phase method), delignification with sodium chlorite, dissolution in EWNN (Iron-III-tartrate-sodium hydroxide complex) and calculation of the cellulose content from a differential distribution graph.

Apart from the yield and the amount of accompanying polysaccharides a cellulose sample can be characterized by determination of changes caused by the isolation procedure. One of the most important criteria is the molecular weight or the degree of polymerization, respectively. The molecular weight of macromolecules such as cellulose can be determined by absolute and relative methods. Absolute methods are:

- determination of the osmotic pressure (osmometer)
- determination of the sedimentation and diffusion constants (ultracentrifuge, diffusion cell)
- determination of the intensity of scattered light (light scattering photometer).

The calculation of the molecular weight (molecular mass) from the data obtained by absolute methods only needs universal constants (e.g. gas constant, Avogadro number) and material constants such as density, which can be determined easily (Vollmert 1973). Absolute determinations of molecular weights of celluloses by ultracentrifugation were carried out e.g. by Claesson et al. (1959) to obtain calibrating data for viscosity measurements of celluloses in EWNN solution.

The most frequently used relative method in macromolecular chemistry is measuring the viscosity, i.e. the tenacity of the solution of the respective polymer by the aid of a viscosimeter. Though by this method a property of the polymer is determined which depends on the molecular weight, the calculation is only possible with constants evaluated previously by absolute methods (e.g. Schurz 1972). Viscosity measurements of cellulose can be carried out in principle with solubilized celluloses in cellulose solvents (→ 4.2.2.) or after nitration and dissolution of the cellulose nitrate in a suitable solvent (e.g. acetone). Details of the preparation of cellulose solvents and solutions are described in detail elsewhere (e.g. Philipp, Linow 1965; Browning 1967b). Standardized procedures for capillary viscosimetry of cuene solutions of pulps are given in Tappi Standard T 230 os-76 and Zellcheming Merkblatt IV/36/61, and of EWNN solutions in Zellcheming Merkblatt IV/50/69, respectively.

The distribution of the DP values in a cellulose sample can be determined by fractional precipitation from solution (Kotera 1967; Philipp, Linow 1965) or by fractional dissolution (Elliott 1967). The resulting fractions contain portions of a certain DP range. From the weight and the DP value the distribution of different chain lengths can be calculated and plotted in distribution curves (Figs. 4–7, 4–8).

Other methods such as gel permeation chromatography (GPC) (Altgelt, Moore 1967; Minor 1979) and counter-current distribution (Cantow 1967) can also be used for the fractionation of cellulose (Krässig 1971; Valtasaari, Saarela 1975).

Oxidative changes in the cellulose molecules can be estimated by the determination of carbonyl (reducing) groups and carboxyl (acidic) groups. There are numerous methods for the determination of both types of groups (Browning 1967b; Pasteka 1972). For all methods the accuracy of determination is limited by the heterogeneity of the reactions and by interfering substances such as residual lignin.

The determination of carbonyl groups is generally based on oxidation (e.g. copper number method, Zellcheming Merkblatt IV/8/70), reduction (e.g. reaction with sodium borohydride) or derivation (e.g. hydroxylamine method). Samuelson (1970) reported an ion-exchange chromatographic method to characterize the carboxyl groups in bleached pulps.

The carboxyl groups in celluloses can also be determined by different methods (Browning 1967b; Dautzenberg, Philipp 1974a, b). Two well-known methods are the titration of the acidic groups in the presence of strong acid salts (e.g. sodium bicarbonate-sodium chloride method, Tappi Standard T 237 os-77) or the determination of the amount of cations bound to cellulose (e.g. methylene blue method; Wilson, Mandel 1961). Dautzenberg and Philipp (1974c) developed a method of direct titration in a non-aqueous homogeneous system (DMS/methylamine/ethanolamine).

Only relative amounts of carboxyl groups can be determined by infrared spectroscopy due to the absence of quantitative absorptivity values.

3.2.8. Isolation and Determination of Polyoses

This second group of cell wall polysaccharides (→ 5.) differs from cellulose from an analytical standpoint by its alkali solubility. Some polyoses such as the arabinogalactans from larch species *(Larix spec.)* and some hardwood polyoses are even water-soluble. Therefore the distinction between water-soluble polyoses, sugars and some water-soluble extractives (e.g. tannins, pectic material) is sometimes difficult if wood is pre-extracted with water. A part of the alkali-soluble polyoses may additionally become soluble during hot water extraction.

During the general wood analysis most of the polyoses are extracted from holocellulose with aqueous alkaline solutions of different concentrations. During extractions with strong alkali solutions, e.g. for the isolation of softwood glucomannan which is closely associated with the cellulose, some low-molecular-weight portions of cellulose may be dissolved together with the polyoses. On the other hand generally a few percent of residual polyoses remain in the alpha-cellulose (→ 3.2.7.). Especially for structural analyses the influences of the delignification procedure (loss of polyoses, oxidative and hydrolytic degrading reactions) can be avoided by extracting the polyoses directly from wood. However, the polyoses obtained do not represent the total polyoses of wood and must be purified from lignin impurities. Only arabinogalactans and hardwood xylans can be extracted in considerable yields from wood without a preceding delignification. In the case of hardwood xylans the yields mainly depend on the wood species. From trembling aspen *(Populus tremuloides)* nearly all the xylan could be extracted (Jones et al. 1961) while from beech wood *(Fagus sylvatica)* only about one half was obtained (Kürschner, Karácsony 1961; Fengel et al. 1979). Softwoods must be delignified in any case to obtain significant portions of the main polyoses.

Aqueous solutions of potassium and sodium hydroxide are mostly used as alkaline solvents for the extraction. The preferred hydroxide is potassium hydroxide mainly because the potassium acetate formed during the neutralization of the alkali extract is more highly soluble in the alcohol used for precipitation than the sodium acetate. Lithium and calcium hydroxide as well as quaternary ammonium hydroxides can also dissolve polyoses, but are generally not used. Liquid ammonia can be used for pre-swelling prior to an alkaline extraction or as an extractant as such. If deacetylation must be avoided dimethyl sulfoxide or hot water is used as a solvent prior to an alkali treatment or after a short swelling period in liquid ammonia. By adding boric acid or borates to sodium and potassium hydroxide the dissolving power for certain polyoses (especially glucomannans) is increased. In this case borate complexes are formed with hydroxyl groups in cis-position by linking borate ions ($B(OH)_4^-$) between the hydroxyl groups (Browning 1967b).

The dissolving power of potassium and sodium hydroxide at various concentrations is different and selective, respectively. This fact enables a first fractionating isolation of different groups of polyoses like mannans and xylans. Extraction with dilute alkali solutions (e.g. 5% KOH) removes the more soluble xylans and galactoglucomannans while most of the glucomannan can be removed only with higher alkali concentrations of 16 or 24% potassium hydroxide or 17.5% sodium hydroxide. In a comparison of the extracting power of sodium, potassium and lithium hydroxide an approximately equal effect of all three hydroxides was found for dissolving softwood xylans. More mannans were extracted with 10.7% LiOH or 10% NaOH in one stage than by a two-stage treatment with 14 and 25.2% KOH. Potassium hydroxide was most effective at a concentration of 45%. Most of the mannan part could be removed by a successive extraction with 10 and 18% NaOH (Hamilton, Quimby 1957). From similiar studies Fengel (1980) concluded that a

Table 3–1: Successive extractions of polyoses

1. step	2. step	3. step	4. step	5. step	further steps	References
5% KOH	24% KOH					Wise et al. (1946)
2% KOH	5% KOH	8% KOH	10% KOH	12% KOH	16, 24% KOH	Wise, Ratliff (1947)
Liq.NH_3	H_2O (hot)	1% Na_2CO_3	2% NaOH	5% NaOH		Björkqvist, Jörgensen (1951)
H_2O (hot)	5% KOH	16% KOH				Clermont (1955)
10% NaOH	18% NaOH					Hamilton, Quimby (1957)
14% KOH	25.2% KOH					Hamilton, Quimby (1957)
1% KOH	4% KOH	7% KOH	16% KOH			Most (1957)
1% NaOH	2% NaOH	3% NaOH	4% NaOH	6% NaOH	12% NaOH	Nelson, Schuerch (1957)
DMSO	H_2O (hot)	14% KOH	14% KOH + 3% H_3BO_3	24% KOH + 3% H_3BO_3		Lindberg, Meier (1957)
0.5% NaOH	10% KOH	10% NaOH				Hamilton, Thompson (1959)
H_2O (hot)	4% NaOH	10% NaOH	18% NaOH			Corbett, Ewart (1959)
DMSO	H_2O (hot)	14% KOH	24% KOH + 3% H_3BO_3			Croon et al. (1959)
5% NaOH	10% NaOH	18% NaOH				Painter, Purves (1960)
3% NaOH	10% NaOH	18% NaOH	18% NaOH + 5% H_3BO_3			Nelson (1960)
10% KOH	17% KOH	24% KOH + 4% $Na_2B_4O_7$				Bishop, Cooper (1960)
4% NaOH	17.5% NaOH					Laidlaw et al. (1961)
Liq. NH_3	H_2O (hot) or DMSO					Cafferty et al. (1964)
2.5% NH_4OH	4.5% NaOH	10% NaOH + 2% $Na_2B_4O_7$		1.8% HCl		Ebringerová et al. (1967)
$Ba(OH)_2$	10% KOH	1% NaOH	15% NaOH			Beélik et al. (1967)
5% NaOH	17.5% NaOH					Fengel (1980)

successive extraction of softwood holocellulose with 5% and 17.5% NaOH is most suitable for the separation of cellulose and polyoses. The alkali concentrations causing maximal swelling (LiOH: 10%, NaOH: 17.5%, KOH: 30%) were most effective, in each case releasing mannan.

For a more differentiated separation of polyoses several successive extraction sequences are reported in the literature. They generally include the use of increasing concentrations of alkali solutions. Some examples are listed in Table 3–1. More details about the fractional extraction of polyoses are given e.g. by Ward and Morak (1962) and Browning (1967b).

With most of the fractional extractions only a coarse separation of the main polyoses is achieved. The preparations are still mixtures of different types of polyoses and must be further fractionated and purified to obtain homogeneous fractions, particularly for structural analyses. This fractionation and purification generally requires several steps. One of the main methods is the fractional precipitation by acidification and addition of organic solvents like ethanol, methanol, or acetone to aqueous or dimethyl formamide solutions of polyoses (Browning 1967b). Acetylation of crude mixtures of polyoses can improve the separation as the solubility of the acetates, for example in chloroform, enables additional precipitation steps (Browning 1967b). A selective and almost quantitative precipitation of mannans is reached by using barium hydroxide, which forms insoluble complexes with the vicinal hydroxyl groups 2 and 3 of mannose units (Meier 1958). Fig. 3–7 shows a typical isolation and fractionation procedure for softwood polyoses using barium hydroxide.
Further precipitating agents are Fehling's solution or other alkaline copper salts as well as basic lead acetate and lead acetate (Lindberg 1962; Browning 1967b). Also the use of the iron-III-tartrate sodium complex (EWNN) was reported for fractionating polyoses (Jayme, Kringstad 1961). In particular, acidic polysaccharides form precipitable complexes with cetyltrimethylammonium bromide and hydroxide or cetylpyridinium chloride and bromide (Lindberg 1962; Browning 1967b). The effectivity of the precipitation depends on the length of the alkyl chain. A further description and discussion of principal fractionation methods is given by Browning (1967b).

Electrophoresis is mainly applied to prove the homogeneity of isolated polysaccharide fractions but was also used for experiments on fractionating the alkali extract (5%) from spruce holocellulose (Fengel 1976a). This method as well as gel permeation chromatography on polyacrylamide did not result in an acceptable fractionation. Gel permeation chromatography on dextran gels was used with good results for the separation of polyoses mixtures containing variable lignin portions (Kringstad, Ellefsen 1964; Kringstad, Cheng 1969). Ion-exchange chromatography on DEAE-columns (DEAE = diethylaminoethyl) is generally used for the purification of polyoses fractions (Applegarth, Dutton 1965; Hashi et al. 1970). But it was also successfully applied in separating the 5% KOH extracts from beech and spruce

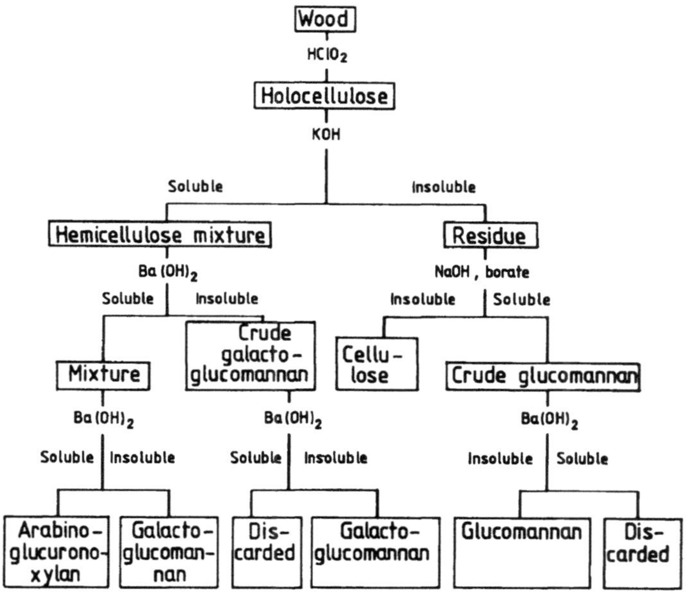

Fig. 3–7. Isolation and fractionation of softwood polyoses (Timell 1961)

holocellulose in a single-step procedure in relative pure mannan, xylan and arabinogalactan fractions as well as lignin-rich fractions (Fengel 1976b; Fengel, Przyklenk 1976).

In practice, investigations on polyoses usually combine the methods of successive extraction, fractionation and purification (e.g. Ebringerová et al. 1967; Béelik et al. 1967; Brink, Pohlman 1972; Dutton et al. 1973; Sears et al. 1978). An example of this type of investigation is demonstrated in Fig. 3–8.

The content of polyoses of a wood sample can be determined by different methods including the isolation of all or part of the polyoses as well as by indirect methods which only give a figure for the amount of polyoses without isolation.

In general wood analysis a standard procedure for the isolation and determination of polyoses is the extraction of chlorite holocellulose with 5% and 24% KOH successively (Wise et al. 1946). The alkaline solutions of polyoses are neutralized with acetic acid and treated with an excess of ethanol. The precipitating fractions are called polyoses A (from the 5% KOH extract) and polyoses B (from the 24% KOH extract). After determination of the ash content the sum of the two fractions represents most of the polyoses without yielding the exact amount of polyoses of the sample. This is caused by losses mainly of pentosans during delignification (→ 3.2.6.), by appreciable amounts of residual polyoses remaining in the alpha-cellulose (→ 3.2.7.) and by the fact that not the whole portion of the dissolved polyoses may precipitate from the alcoholic solutions (Smelstorius 1974). For the determination of softwood polyoses a modification of this method was recently suggested, calling for applications of 5 and 17.5% NaOH (Fengel 1980). A com-

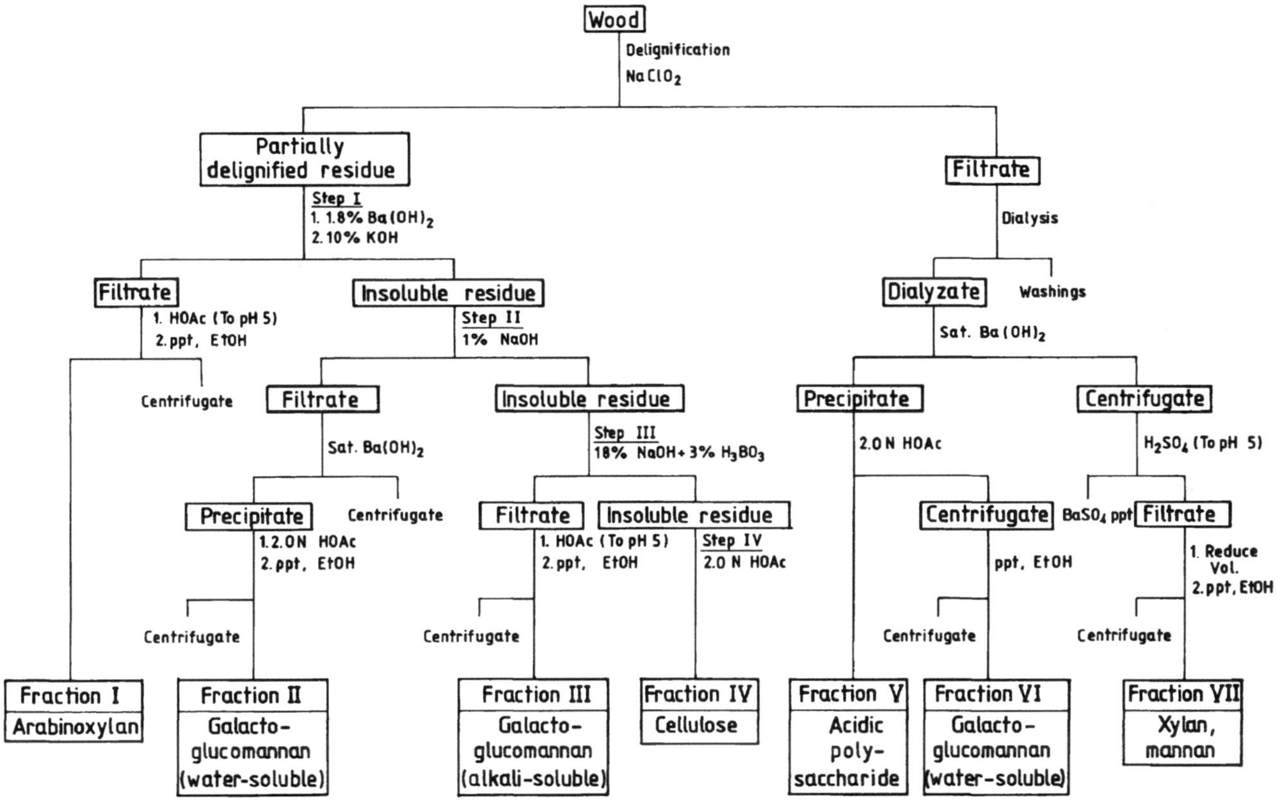

Fig. 3–8. Fractionation procedure for chlorite holocellulose and delignification liquor (Brink, Pohlman 1972)

bined method of polysaccharide isolation according to Wise et al. (1946) and additional direct determination of the content of pentosan, uronic acids and acetyl groups in water-extracted wood was described by Smelstorius and Stewart (1974). In principle any fractional isolation of polyoses can be used for their determination if the sum of the fractions represents the total amount of polyoses or dominant portions of them.

If no isolation but only the content of polyoses of a sample is to be determined the most exact and fastest method is the total hydrolysis of the polysaccharides and a subsequent sugar analysis (→ 3.2.7.). The hydrolysis can be carried out with sulfuric acid or with trifluoroacetic acid (TFA). The hydrolysis with TFA can be well adjusted to the respective starting material such as wood, holocellulose, alpha cellulose or pulp (Fengel, Wegener 1979). An example of automated sugar analysis of polyoses A from spruce and beech wood after hydrolysis with TFA (2 moles/l) is given in Fig. 3–9. Table 3–2 shows values for polyoses within an analysis of several woods and pulps.

A characteristic value, especially for pulps, is the so-called pentosan content, which estimates the total amount of pentosans without a determination of the individual sugar components (Tappi Standard T 223 os-78; ASTM Standard D 1787; Zellcheming Merkblatt IV/35/71). The principle of the method is the conversion of the pentosans to furfural by hydrochloric or hydrobromic acid. The furfural can be determined by gravimetric, volumetric, colorimetric or spectrophotometric methods (Browning 1967b).

An important value for bleached pulps is the alkali solubility in 10, 18 or 21.5% NaOH (S_{10}, S_{18}, $S_{21.5}$) under defined conditions (Tappi Standard T 235 os-76; DIN 54356; Zellcheming Merkblatt IV/44/67; ISO Standard 692–1974). The $S_{21.5}$-value can be related to the rayon yield of dissolving pulps.

By treating wood or pulp with boiling 1% NaOH some readily soluble polyoses are extracted together with degraded cellulose (Tappi Standard T 212 os-76; ASTM

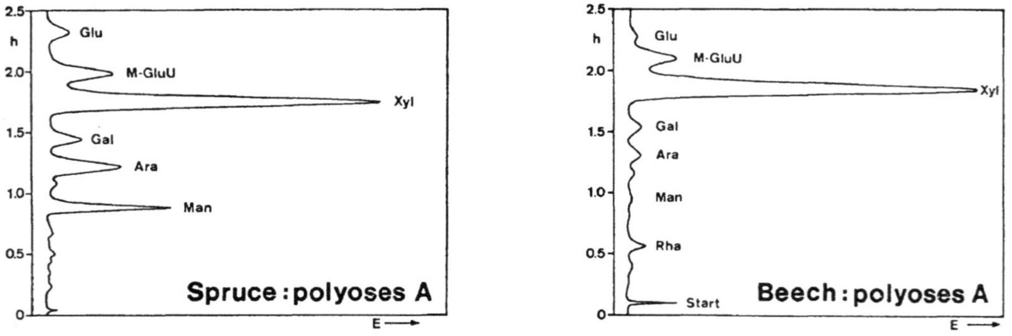

Fig. 3–9. Sugar analysis of polyoses A from spruce and beech wood (Fengel, Wegener 1979)

Table 3–2: Wood and pulp composition based on intensive hydrolysis with TFA (Fengel, Wegener 1979)

Sample	Mannan[a] %	Xylan[b] %	Galactan[c] %	Polyoses %	Cellulose %	Ash %	Lignin %
Picea abies	22.2	8.9	0.0	31.1	40.4	0.5[d]	28.0
Larix decidua	22.7	9.3	0.0	32.0	39.5	0.5[d]	28.0
Fagus sylvatica	1.4	27.8	2.6	31.8	43.3	0.5[d]	24.4
Betula verrucosa	4.8	32.5	1.7	39.0	34.2	0.5[d]	26.3
Fraxinus excelsior	5.7	28.3	2.0	36.0	37.9	0.5[d]	25.6
Spruce sulphite pulp (unbleached)	9.2	3.6	0.0	12.8	78.1	0.8[e]	8.3
Beech sulphite pulp (bleached, chem. grade)	1.8	2.6	0.0	4.4	97.3	0.0	0.0

[a] softwoods: galactoglucomannan with 1.3% acetyl; hardwoods: glucomannan
[b] softwoods: arabino-4-0-methylglucuronoxylan; hardwoods: 4-0-methylglucuronoxylan with 4% acetyl
[c] hardwoods: arabinogalactan
[d] assumed average content
[e] determined

Standard D 1109-56). This value indicates the degree of fungus decay or other degrading reactions, e.g. during pulping or bleaching.

An improved gravimetric determination of mannan by hydrolysis, neutralization with barium carbonate and precipitation with bromophenylhydrazine was described by Höpner (1966).

Simonson (1967) determined small amounts of polyoses in the presence of acid insoluble lignin in fractions of sulfite spent liquors from birch *(Betula spec.)*. He used a system involving hydrolysis with hydrochloric acid, followed by precipitation and removal of lignin by dialysis, and finally a colorimetric determination of the hydrolysis products.

3.2.9. Isolation and Determination of Lignin

Due to the properties of lignin resulting from its molecular structure (→ 6.3.) and its localization within the cell wall, the isolation of lignin in an unchanged form and its exact determination have not yet proved possible. All methods of isolation have the disadvantage of either fundamentally changing the native structure of lignin or releasing only parts of it relatively unchanged. Generally the lignin isolation methods can be divided into two large groups:

- methods yielding lignin as residue
- methods by which lignin is dissolved either without reacting with the solvent used for the extraction or by forming soluble derivatives.

A survey of important methods is given in Table 3–3. More detailed descriptions of methods and preparations are given elsewhere (e.g. Brauns 1952; Brauns, Brauns 1960; Pearl 1967; Browning 1967b; Lai, Sarkanen 1971).

Prior to lignin isolation the extractives must be removed to avoid the formation of condensation products with lignin during the isolation procedure. For the same reason, especially if strong mineral acids are involved in the isolation, solvents such as alcohol or acetone must be completely removed from the extracted wood. The first group of isolation methods yields so-called acid lignins by applying sulfuric or

Table 3–3: Lignin isolation methods and the resulting preparations

Remarks	Treatment	Preparation	References
	Lignin as residue		
Acid hydrolysis of polysaccharides	H_2SO_4	Sulfuric acid lignin (Klason lignin)	Klason (1906)*
	H_2SO_4/HBr	Sulfuric acid lignin (Runkel lignin)	Runkel, Wilke (1951)
	HCl	Hydrochloric acid lignin (Willstätter lignin)	Willstätter, Zechmeister (1913)*
	HCl/H_2SO_4	Hydrochloric acid lignin (Halse lignin)	Halse (1924)*
	HF	Hydrofluoric acid lignin	Fredenhagen, Cadenbach (1933)*
	CF_3COOH	Trifluoroacetic acid lignin	Fengel et al. (1978; 1981a, b)
Oxidation of polysaccharides	$Na_3H_2IO_6$	Periodate lignin (Purves lignin)	Purves et al. (1947)*
Hydrolysis/dissolution of polysaccharides	NaOH/H_2SO_4/ $Cu(NH_3)_4(OH)_2$	Cuoxam lignin, cuproxam lignin, cuprammonium lignin (Freudenberg lignin)	Freudenberg et al. (1929)*
	Lignin by dissolution		
No appreciable reactions between lignin and solvent	Alcohol extraction	Native lignin (Brauns lignin)	Brauns (1939)*
	Vibratory milling/dioxane-water-extraction	Milled wood lignin (MWL) (Björkman lignin)	Björkman (1956)
	Ball milling/H_2O-NaSCN-$C_6H_5CH_2OH$-DMF dissolution/extraction	Ball-milled wood lignin (BMWL)	Brownell (1965, 1968)
	Brown-rot fungi treatment	Enzymicly liberated lignin (ELL, EIL)	Schubert, Nord (1950)* Brown et al. (1968)
	Milling/enzymic treatment/solvent extraction	Cellulolytic enzyme lignin (CEL)	Pew (1957); Pew, Weyna (1962); Chang et al. (1975); Polčin, Bezúch (1978)
		Organosolv lignins	
Reactions between lignin and solvent	Alcohol/HCl	Alcohol lignin	Holmberg, Runius (1925)* Brauns, Hibbert (1935)*
	Dioxane/HCl	Dioxane acidolysis lignin	Storch (1936)* Freudenberg, Zechmeister (1954)*
	$CH_3COOH/MgCl_2$	Acetic acid lignin	Schutz, Knackstedt (1941)* Brauns, Buchanan (1945)*
	$HSCH_2COOH$/HCl	Thioglycolic acid lignin (TGA-L)	Holmberg (1930)*
	Phenol/HCl	Phenol lignin	Clark, Brauns (1944)*
	Mild hydrogenation	Hydrogenolysis lignin	Brewer et al. (1948)*
	Hydrotropic solvents	Hydrotropic lignin	Traynard, Eymery (1955)*

Remarks	Treatment	Preparation	References
		Derivatives by inorganic reagents	
Generally technical pulping processes	Sulfite/bisulfite	Lignin sulfonates (Lignosulfonates)	
	NaOH	Alkali lignin (Soda lignin)	
	Na_2S/NaHS	Thiolignin	
	NaOH/Na_2S	Kraft lignin (Sulfate lignin)	

* For original and further early literature see references in: Brauns (1952); Brauns, Brauns (1960); Browning (1967 b); Lai, Sarkanen (1971)

hydrochloric acid, mixtures thereof, or other mineral acids. In the case of sulfuric acid lignins, acid concentrations between 68 and 78% (mostly 72%) are used for the first hydrolysis stage followed by dilution steps to complete the polysaccharide hydrolysis with low-concentrated acids. The hydrochloric acid lignins obtained by treating wood with oversaturated hydrochloric acid are described as being less condensed than sulfuric acid lignins. All lignin preparations obtained by acid hydrolysis are changed in their structure and properties, predominantly by condensation reactions (Lai, Sarkanen 1971). The sulfuric acid and hydrochloric acid lignins additionally contain considerable amounts of sulfur and chlorine, respectively. Therefore these preparations are not usable for investigating structures but are mainly applied in estimating the lignin content.

Because of the disadvantages of the acid lignins several attempts have been made to obtain more carefully isolated lignins by removing the polysaccharides without acid hydrolysis. The oxidative degradation of the polysaccharide part of wood by the action of periodate ($Na_3H_2IO_6$) avoids condensation but causes some oxidative modifications of the lignin residue (periodate lignin). The dominant reaction is the oxidative conversion of sugar units to dialdehydes, thus making the polysaccharides hydrolyzable with boiling water. A proposed modification of this method calls for blocking the guaiacyl groups of lignin by acetylation, thus avoiding oxidation (Davydov et al. 1974).

Another means of avoiding considerable condensation reactions is the preparation of cuoxam lignin. The polysaccharides are removed by alternating steps of hydrolysis with boiling dilute sulfuric acid and extraction with cuprammonium hydroxide. The resulting lignin is more reactive than the acid lignins but due to the alternating acidic and alkaline conditions some structural changes take place in addition to incomplete lignin yields.

Within the group of dissolved lignins the most important method for obtaining relatively unchanged lignin is Björkman's procedure of vibratory milling and subsequent extraction of lignin with aqueous dioxane (milled wood lignin, MWL). On the basis of the original method several modifications were reported concerning the pretreatment of the starting wood material, the milling and extraction conditions (e.g. type and size of the mill, time of milling and extraction, amount of wood used for the milling procedure, extraction solvent) and the purification (e.g. Rezanowich et al. 1963; Bland, Menshun 1967; Lai, Sarkanen 1971; Chang et al. 1975; Salud, Faix 1980). By application of ultrasonics during the extraction step the isolation time could be essentially reduced (Wegener, Stoll 1976; Wegener, Fengel 1978; Wegener, Fengel 1979). The lignins obtained by this procedure proved to have characteristics comparable to milled wood lignins resulting from the standard method (e.g. methoxyl content, molecular weight distribution, polysaccharide residues) (Wegener, Fengel 1977). Though the amounts of crude milled wood lignin are up to 60% (based on lignin in wood), the yields of purified softwood milled wood lignins are 25% at most, and generally much lower. Higher yields were reported for hardwood lignins (Bland, Menshun 1971; Lai, Sarkanen 1971). Though milled wood lignins are probably not identical to the lignin in situ and these preparations may not even be representative of the total lignin of the cell wall they are regarded as especially suitable lignins for further investigation.

In the future improved results in the field of careful lignin isolation are to be expected by the combination of mechanical, chemical and enzymatical treatments as demonstrated e.g. by Chang et al. (1975) (cellulolytic enzyme lignin, CEL) and Polčin and Bezúch (1978) (enzymatically isolated lignin, EIL). The former authors obtained increased yields of lignin from sweetgum *(Liquidambar styraciflua)* and spruce *(Picea abies)* by treating ball milled wood with a commercial cellulase followed by a successive extraction with aqueous solutions of dioxane. A nearly complete isolation of the lignin of spruce *(Picea abies)*, birch *(Betula verrucosa)* and poplar *(Populus monilifera)* was achieved by Polčin and Bezúch (1978) by a multistep procedure of milling, extraction and enzymatic treatment.

A comparison of four different lignin isolation methods (MWL, CEL, thioglycolic acid lignin (TGA-L), hydroxymethylated TGA-L) applied to NSSC pulps (→ 16.3.1) from Douglas fir *(Pseudotsuga menziesii)* and red alder *(Alnus rubra)* was published by Glasser and Barnett (1979).

Organosolv lignins cover a wide group of analytical lignins isolated with various chemicals (Brauns 1952). They were used for numerous investigations without gaining general importance in wood and lignin analysis.

Organosolv lignins derived from ethanol-water delignification, which is expected to be a future pulping process (→ 16.7.) (Baumeister, Edel 1980), proved to be mostly unchanged, resembling to some extent analytical Björkman lignins (Schweers, Meier 1979; Fengel et al. 1981a, b).

A so-called autohydrolysis lignin was prepared by autohydrolysis of aspen wood meal *(Populus tremuloides)* at 195 °C and subsequent dioxane extraction (Chua, Wayman 1979).

Lignin sulfonates, alkali lignin, thiolignin and sulfate lignin are lignin derivatives predominantly obtained from the waste liquors of pulping processes (technical lignins) (→ 16.4., → 18.6.). Occasionally these lignins are also prepared in the laboratory.

The determination of the lignin content is important for the wood analysis as well as for the characterization of pulps. The methods of quantitative lignin determination may be subdivided as follows:

- direct methods, by which lignin is determined as residue
- indirect methods, by which the lignin content
  - is calculated after the determination of the polysaccharides
  - is determined by spectrophotometric methods
  - results from reactions of lignin with oxidizing chemicals.

Common to all lignin determination methods are problems arising from interfering substances (extractive compounds, polysaccharide degradation products) and/or the uncertainty as to whether the lignin content is completely recorded.

The direct methods are based in principle on the isolation and gravimetric determination of acid-insoluble lignins. The most firmly established method is the determination of lignin according to Klason. The hydrolysis is accomplished by treating pre-extracted wood or unbleached pulp with 72% sulfuric acid and a final hydrolysis step with 3% sulfuric acid under defined conditions. A detailed description of this method is given elsewhere (e.g. Browning 1967b, Tappi Standard T 22 os-74; ASTM Standard D 1106-56). Modified methods use for example hydrobromic acid in addition to 72% sulfuric acid to improve the precipitation of lignin during the dilution step (Runkel, Wilke 1951), or a mixture of 75% sulfuric acid and 89% phosphoric acid, which is especially recommended for pulps (Jayme et al. 1958). Slightly modified procedures for the determination of lignin in wood and pulp are used at the U.S. Forest Products Laboratory (Saeman et al. 1954; Moore, Johnson 1967; Effland 1977). Kürschner and Popik (1962) described a method applying 72% sulfuric acid on mercerized wood followed by cooking with 5% sulfuric acid. The method results in very low lignin values. This is caused by acid soluble lignin portions left out of consideration (Faix, Schweers 1972). Hydrochloric acid lignins and hydrofluoric acid lignin can also be used for quantification of the lignin content.

Errors within the acid lignin determinations may be caused by substances and reaction products remaining with the lignin in the non-hydrolyzable residue and causing deceptively high lignin values. On the other hand parts of lignin are soluble under acidic conditions, yielding too low lignin values. Amounts of about 1% of acid-soluble lignin are generally found for softwoods while up to 4% of acid-soluble lignin are reported for hardwoods. As already mentioned large portions of the residual lignin in holocelluloses are acid-soluble (→ 3.2.6., Fig. 3–6).

This acid-soluble lignin can be determined by ultraviolet spectrophotometric measurement at wavelengths of 280, 240 or 205 nm and comparison of the absorbancy with that of a reference lignin sample (e.g. Pearl, Busche 1960; Schöning, Johansson 1965; Ahlgren, Goring 1971; Bland, Menshun 1971; Merewether, Samsuzzaman 1972). Polysaccharide degradation products (e.g. furfural) are known to influence absorbancy at 280 nm and therefore measurements at 200 to 208 nm should be preferred (→ 6.4.2.).

With all indirect methods no lignin is isolated. A simple method is the calculation of the lignin amount from the holocellulose yield and the value of residual lignin in the holocellulose [100% – holocellulose yield (%) + residual lignin (%) = lignin in wood (%)]. But there are problems arising from the determination of exact residual lignin values (→ 3.2.6.).

A more exact lignin determination can be achieved by determining the polysaccharide portion of wood after total hydrolysis, and calculating the lignin content as difference from 100% (→ 3.2.7., Table 3–2).

To avoid the difficulties and uncertainties resulting from the acid-soluble lignin portions, the so-called acetyl bromide method can be applied. This spectrophotometric method allows the determination of the total lignin amount of a sample (Johnson et al. 1961; Marton 1967; Dietrichs, Zschirnt 1972; Wegener 1974; Ciarrocca, Bedetti 1976; van Zyl 1978). The determination is based on the fact that wood, pulp and holocellulose are each wholly soluble in a mixture of 25% acetyl bromide in glacial acetic acid. The lignin content is proportional to the absorbance at 280 nm. Problems arise if very low lignin amounts are to be determined in holocelluloses (Wegener 1974).

An UV spectroscopic method measuring the absorbance of the chlorite delignification liquor at 230 nm and using coniferyl alcohol as an internal standard is described by Schadenböck and Prey (1972).

Attempts to estimate lignin in wood quantitatively by the evaluation of infrared spectra are reported e.g. by Kolboe and Ellefsen (1962), and Karklins and Oklerina (1975). Similar studies were also carried out with pulps, bagasse and lignin-polysaccharide complexes without gaining a reliable and simple applicable method, however (Zhbankov et al. 1966; Marton, Spark 1967; Karklins et al. 1977; Saad et al. 1980; Feckl 1981).

All spectroscopic methods suffer from the problems of varying absorption coefficients of lignins from different samples, from the absence of suitable standard lignins for reference and from doubtful values caused by interfering substances.

The lignin content can also be quantified indirectly by determination of the characteristic methoxyl groups (Vieböck, Schwappach 1930; Hardell et al. 1980). The prerequisite for this determination is a knowledge of the methoxyl content of the respective lignin and a consideration of the methoxyl groups present in the polyoses of the respective sample (Besold, Fengel 1983).

The following indirect methods are used for the determination of lignin residues in pulp, the results of which are expressed in terms such as degree of delignification, cooking degree, hardness, bleachability etc. They are mainly used as control methods in pulping and bleaching. Generally they are restricted to pulps with yields lower than 70%, and thus are not applicable e.g. to thermomechanical pulps. On the other hand they become ambiguous in the case of bleached pulps with very low lignin contents.

There are two main reactions which are used for lignin determination under standardized conditions. The first one is the oxidation of lignin with potassium permanganate in acidic solution and is expressed as so-called permanganate numbers such as the kappa number or Johnson-Noll number (Tappi Standard T 236 os-76; Zellcheming Merkblatt IV/37/80; ISO Standard R 302–1963). The second one is the consumption of gaseous chlorine caused by substitution and oxidation reactions expressed as chlorine numbers such as the Roe number (Tappi Standard T 253 pm-75; Zellcheming Merkblatt IV/53/71; ISO Standard 3260–1975). The relationship between the results from these chemical reactions and the lignin content is described by empiric factors. For example, factors of 0.15–0.17 are used for converting the kappa number to the respective Klason lignin content. Detailed descriptions and discussions of these methods as well as conversion diagrams are given elsewhere (Rydholm 1965; Browning 1967b; Pearl 1967).

Dissolved lignins, lignin derivatives and lignin degradation products in pulping and bleaching spent liquors can be quantified by spectrophotometric measurements, precipitation or colour reactions (Browning 1967b). A fluorometric determination of lignin in pulping waste liquors by correlating the fluorescence intensity to the lignin concentration is described by Bublitz and Meng (1978).

3.3. Data of Wood Analyses

Apart from the isolation, determination and characterization of the compounds described above, wood analysis also covers the ultimate analysis which yields the amounts of the chemical elements forming wood material. The various wood species differ only slightly in this respect, with average weight percentages of 50% carbon, 43% oxygen and 6% hydrogen. The rest of about 1% is composed of nitrogen deriving from protein material (0.1–0.3%, Marutzky, Roffael 1977), and inorganic elements yielding ash (→ 3.2.5.). In atomic terms this composition corresponds to 47 hydrogen atoms, 28 carbon atoms and 24 oxygen atoms out of 100.

For many technological uses of wood (e.g. pulp and paper production, production of cellulose derivatives and regenerated cellulose products, fibreboard and particle-

Table 3–4: Chemical analysis of softwoods

Scientific name	Common name	Holocellulose %	Cellulose %	Polyoses %	Pentosan %	Lignin %	Ethanol-benzene extract %	Hot-water extract %	Ash %	References
Abies alba Mill.	Silver fir		42.3		11.5	28.9	2.3		0.8	Wagenführ, Scheiber 1974
Abies balsamea (L.) Mill.	Balsam fir	70.0	49.4	15.4	7.0	27.7	4.3	3.6	0.4	Clermont, Schwartz 1951
		72.8	44.8		5.3	29.4			0.2	Timell 1957 c
Abies sachalinensis Mast.	Japanese fir	70.4	40.9		12.5	28.7	2.9	4.5	0.5	Satonaka 1963
Araucaria angustifolia Ktze.	Parana pine		44.3		4.5	29.5	2.3		1.4	Wagenführ, Scheiber 1974
Cupressus dupreziana A. Camus	Tamrit cypress	67.7	46.7	17.6		36.8	17.8	6.1	0.4	Grosser et al. 1974
Ginkgo biloba L.	Maidenhair tree ♂		43.2		9.7	32.3		4.0	0.8	Timell 1960
	♀		42.1		10.0	33.8	3.7	1.1	0.5	
Larix laricina K. Koch	Tamarack larch	68.8	43.9		5.3	28.6			0.2	Timell 1957 c
Larix russica (Endl.) Sabine ex Trautv.	Siberian larch		41.5		8.9	26.4	2.8	13.8	0.2	Hutorsčikov et al. 1967
Libocedrus decurrens Torr.	Pencil cedar		47.9			39.4	13.1*	8.0	0.5	Wagenführ, Scheiber 1974
Metasequoia glyptostroboides Hu et Cherry	Metasequoia		48.3		8.7	31.5			1.0	Surminsky, Bojarczuk 1973
Picea abies Karst.	European spruce	80.9	46.0	15.3	8.3	27.3	2.0	2.0		Fengel 1966 a, b, 1967
		82.5	40.4	31.1		28.2	1.4		0.3	Wegener 1974, Fengel et al. 1978
Picea glauca (Moench) Voss	White spruce	75.7	44.8		9.8	27.1			0.3	Timell, Tyminsky 1975
Picea jezoensis (S. & Z.) Carr.	Yeso spruce	75.3	43.9		13.5	29.1	0.6	3.1	0.1	Satonaka 1963
Picea mariana (Mill.) B. S. P.	Black spruce	71.7	51.1	15.2	7.6	27.3	2.6	2.5	0.2	Clermont, Schwartz 1951
Picea omorika (Pančic) Purcyne	Serbian spruce		50.3		11.0	25.6			0.3	Markowić et al. 1966
Picea schrenkiana Fisch. et Mey.	Schrenk spruce		39.6	27.4	12.7	32.5	1.8*		0.3	Markowić et al. 1966
Pinus banksiana Lamb.	Jack pine	72.3	41.6		8.5	28.6			0.2	Timell 1957 c
		74.8	41.1		14.2	29.8	0.5	2.4	0.1	Satonaka 1963
Pinus nigra Arnold var. gotschensis	Austrian pine		49.5		11.0	27.2			0.2	Markowić et al. 1966
Pinus radiata D. Don	Radiata pine		45.5	16.3	9.3	26.8	1.5		0.2	Brasch, Wise 1956
Pinus strobus L.	Yellow pine		61.6		5.5	29.6	10.2	7.7	0.2	Wagenführ, Scheiber 1974
		70.6	41.4		13.2	27.6	6.6	4.1	0.5	Satonaka 1963
Pinus sylvestris L.	Scots pine	74.3	52.2	13.5	8.2	26.3				Kollmann, Fengel 1965
Pseudotsuga menziesii Mirb.	Douglas fir	67.0	50.4		6.8	27.2	4.4	5.6	0.2	Browning, Isenberg 1952
Sequoia sempervirens Endl.	Redwood	71.8	49.9	16.7		37.0	13.5	8.7	0.2	Fengel et al. 1973
Thuja plicata D. Don	Western red cedar		47.5	14.7	8.1	32.5			0.3	Wise, Ratliff 1947
						30.7	8.9*		0.2	Wagenführ, Scheiber 1974
Thujopsis dolabrata (L.f.) S.&Z	Japanese thuja	72.8	38.4		13.6	31.8	2.1	3.3	0.4	Satonaka 1963

*Ethanol extract

board) quantitative data on the chemical composition of wood species are often desirable or even necessary for many processes. As already mentioned (→ 3.2.1.), wood analyses can be carried out in very different manners. Apart from highly detailed analyses, which are reported only for few wood species, general analysis data from numerous wood species are known. Unfortunately knowledge of important technological wood species is still incomplete. The latest summary of analytical data from 153 softwood and hardwood species was given by Fengel and Grosser (1975). The following tables include the more recent data from wood analyses for softwoods (Table 3–4), hardwoods from temperate zones (Table 3–5) and some

Table 3–5: Chemical analysis of hardwoods from temperate zones

Scientific name	Common name	Holocellulose %	Cellulose %	Polyoses %	Pentosan %	Lignin %	Ethanol-benzene extract %	Hot-water extract %	Ash %	References
Acer japonicum Thunb.	Japanese maple	81.7	47.4		24.0	20.7	1.9	4.3	0.4	Satonaka 1963
Acer pseudoplatanus L.	Sycamore maple		38.3		20.3	25.3	2.5		0.4	Wagenführ, Scheiber 1974
Acer rubrum L.	Red maple	71.0	44.5		17.1	22.8	2.5	4.4	0.7	Browning, Isenberg 1952
		79.0	44.1		17.8	24.0			0.2	Timell 1957
Acer saccharum Marsh.	Sugar maple				15.6	22.7			0.3	Timell et al. 1958
			40.2			23.1			0.4	Timell 1969
Aesculus hippocastanum L.	European horse chestnut		47.5		18.1	26.2	2.3	2.3	1.0	Wagenführ, Scheiber 1974
Alnus glutinosa Gaertn.	Black alder		43.4		23.0	23.9	3.8		0.5	Wagenführ, Scheiber 1974
Betula papyrifera Marsh.	White birch		38.3			22.0			0.4	Timell 1969
Betula verrucosa Ehrh.	European birch		48.5		25.1	19.4		2.5	0.3	Varaksina et al. 1967
			45.3		25.3	23.9	2.1		0.4	Wagenführ, Scheiber 1974
Carpinus betulus L.	European hornbeam	85.7	46.4		23.9	17.8	4.4*	3.7	0.4	Kubaćková, Cetlová-Pranzková 1963
		79.4	42.8		20.1	21.2	2.0	3.9	0.5	Satonaka 1963
			43.0		27.0	20.9	2.0		0.4	Wagenführ, Scheiber 1974
Carpinus spec.	Hornbeam	79.8		35.5	19.3	20.0	2.0	3.9		Lal et al. 1977
Carya tomentosa Nutt.	Hickory		56.2		18.8	23.4	0.6		0.7	Wagenführ, Scheiber 1974
Castanea crenata Bl.	Japanese chestnut	72.8	40.3		22.7	25.9	2.7	9.5	0.3	Satonaka 1963
Castanea sativa Mill.	European chestnut		47.3		16.7	31.8	4.7		0.4	Wagenführ, Scheiber 1974
Fagus crenata Bl.	Japanese beech	81.0	44.7		20.7	20.6	1.3	3.6	0.7	Satonaka 1963
Fagus grandifolia Ehrh.	American beech		39.5			23.5			0.6	Timell 1969
Fagus sylvatica L.	European beech	85.6	49.1		22.0	23.8	0.8		0.3	Kürschner, Melćerová 1965
		85.8	44.5	30.2	20.6	22.2				Fengel et al. 1979
Fagus spec.	Beech	69.5		28.3	16.1	22.8	1.4	0.9		Lal et al. 1977
Fraxinus excelsior L.	European ash		37.9	36.0		25.6				Fengel et al. 1978
Juglans regia L.	European walnut		40.8		12.6	29.1	4.4		0.8	Wagenführ, Scheiber 1974
Platanus acerifolia Willd.	Sycamore		50.7		24.9	29.1		1.0	0.6	Wagenführ, Scheiber 1974
Populus alba L.	White poplar		49.0		25.6	23.1			0.2	Markowić et al. 1966
Populus tremuloides Michx.	Trembling aspen	80.3	49.4	21.2	17.2	18.1	3.8	2.8	0.4	Clermont, Schwartz 1951
			42.7			20.9			0.4	Timell 1969
			51.0		19.9	17.6				Clermont, Bender 1958
Populus spec.	Poplar	78.4		31.7	15.9	20.9	3.3	4.3		Lal et al. 1977
Quercus robur L.	English oak		41.1		22.2	29.6	0.4	12.2	0.3	Wagenführ, Scheiber 1974
Quercus rubra L.	American red oak	49.2			24.1	21.8		5.2	0.1	Wagenführ, Scheiber 1974
Quercus spec.	Oak	73.2	40.5	23.3	17.5	22.2				Kollmann, Fengel 1965
	sapwood	78.7	39.9	27.6		24.9	2.4		0.5	Bednar, Fengel 1974
	heartwood	77.0	37.6	28.6		24.5	4.4		0.3	
		76.0		34.7	13.9	25.1	3.2	0.9		Lal et al. 1977
Robina pseudoacacia L.	Black locust	81.7	50.1		23.7	20.6	2.8	4.6	0.3	Satonaka 1963
Salix alba L.	White willow		49.6		26.7	22.7			0.3	Markowić et al. 1966
	heartwood	89.2	48.7		16.3	25.7	3.2*	3.1	0.4	Kubaćkocá, Cetlová-Pranzková 1963
Salix spec.	Willow	77.8		24.5	17.0	25.4	2.0	5.3		Lal et al. 1977
Sorbus aucuparia L.	Europ. mountain ash		43.5		27.1	20.1	2.3	3.2	0.3	Wagenführ, Scheiber 1974
Ulmus americana L.	American elm		42.0		18.6	29.4				Clermont, Bender 1958
			48.5			21.7			0.6	Timell 1969
Ulmus carpinifolia Gled.	Dutch elm		43.0		21.8	27.3	1.6	0.6	0.8	Wagenführ, Scheiber 1974
Ulmus laevis Pall. (= *Ulmus effusa* Willd.)	Russian white elm		48.7		21.4	24.5			0.8	Marković et al. 1966

* Ethanol extract

Table 3–6. Chemical analysis of some important tropical woods

Commercial name	Scientific name	Holocellulose %	Cellulose %	Polyoses %	Pentosan %	Lignin %	Ethanol-benzene extract %	Hot-water extract %	Ash %	References
African mahogany, khaya	*Khaya anthotheca* C. DC.		43.9		16.0	28.2	3.5	3.3	1.1	Savard et al. 1960
Afrormosia	*Afrormosia elata* Harms		39.8		16.8	31.2	6.9	2.7	0.9	Savard et al. 1960
Afzelia	*Afzelia africana* Smith		35.8		18.1	28.0	9.1	5.3	1.9	Savard et al. 1960
Balsa	*Ochroma lagopus* Sw.		52.0		19.0	24.5	2.6	2.8	1.6	Wagenführ, Scheiber 1974
Bombay blackwood	*Dalbergia latifolia* Roxb.		35.1		19.5	34.2	4.4	5.9	1.0	Wagenführ, Scheiber 1974
Iroko	*Chlorophora excelsa* Benth. et Hook. f.		34.2		15.4	28.6	13.8	3.1	3.4	Savard et al. 1960
Kefe, awari	*Pterogota macrocarpa* K. Schum.	78.1	44.9	25.1	15.8	22.7	2.6	2.6	1.3	Lal et al. 1977
Limba, afara	*Terminalia superba* Engl. et Diels		44.2		16.4	29.2	3.4	3.4	3.1	Savard et al. 1960
Mansonia, bete	*Mansonia altissima* A. Chev.		37.7		19.9	26.3	8.2	1.9	1.8	Savard et al. 1960
Obeche	*Triplochiton sceroxylon* K. Schum.		44.4		15.5	31.7	2.0	2.2	1.7	Savard et al. 1960
		77.2	47.8	20.1	16.8	21.3	12.6	4.2	1.8	Lal et al. 1977
Okoume	*Aucoumea klaineana* Pierre		47.9		15.0	31.4	1.4	0.9	0.4	Savard et al. 1960
Red ironwood, ekki	*Lophira alata* Banks. ex Gaertn. f.		39.2		13.0	39.8	3.0	1.5	1.0	Savard et al. 1960
Teak (Congo)	*Tectona grandis* L.		39.1		13.0	29.3	13.0	1.8	0.7	Savard et al. 1960
(Thailand)			40.4		11.5	32.8	10.4	2.2	1.4	Savard et al. 1960
(India) sapwood			55.4		14.7	39.1	1.4*	6.2	0.7	Narayanamurti, Das 1955
(India) heartwood			57.2		7.7	34.2	1.0	3.7	0.9	Narayanamurti, Das 1955
Wenge, awong	*Millettia laurentii* De		38.8		16.9	31.5	6.7	3.0	0.2	Savard et al. 1960
	Wild.		47.2		11.7	26.4	6.8	3.8	0.4	Wagenführ, Scheiber 1974

* Ethanol extract

important tropical woods (Table 3–6). On the background to the discussion of wood analysis methods within this chapter, it is important to keep in mind that variations in reported data are caused in part by a variation in the respective wood composition but also result from different analysis methods.

The chemical behaviour of 97 commercial timbers was tabulated by Dietrichs (1972) according to important chemical properties such as extractive content, pH value, colouring reactions upon contact with metals or effects of different surface treatments.

Only a few recent investigations are concerned with the chemical differences between sapwood and heartwood, earlywood and latewood, and the chemical composition of reaction wood from different wood species. Generally speaking, sapwood from softwoods contains more lignin and cellulose and less extractives than heartwood, while in the case of hardwoods the respective values are comparable. Earlywood generally contains more lignin and less cellulose than latewood (Browning 1963). In a study of loblolly pine *(Pinus taeda)* – the most important pulp wood in the U.S.A. – the chemical analyses of earlywood and latewood revealed only small differences, apart from the higher extractive content in earlywood (Ifju, Labosky 1972).

The chemical composition of fast-grown juvenile wood and slow-grown mature wood of sycamore and cottonwood *(Platanus occidentalis, Populus deltoides)* was compared (Moore, Effland 1974). Some differences were found between the fast-grown and slow-grown wood of both species. For example, the xylose content was extremely high for the fast-grown woods (e.g. 29% higher in the case of cottonwood). The lignin content of both species was much higher in the slow-grown wood.

Chemical differences within a 35-year-old radiata pine wood *(Pinus radiata)* were reported by Uprichard (1971). The content of extractives, lignin and pentosan decreased with increasing distance from the pith, while the alpha-cellulose content increased. Sugar analyses showed that the xylan and mannan content of the young wood is higher, while later on the cellulose content increases (Harwood 1971).

It is a well-known fact that compression wood contains more lignin and less cellulose than normal wood, while the opposite holds true of tension wood. An interesting correlation between the tension wood amount and the galactose content of beech wood *(Fagus sylvatica)* was found by Ruel and Barnoud (1978).

The chemical analysis of different parts of a tree such as stump, roots or branches is becoming an important field of investigation with regard to full-tree, whole-tree and biomass utilization (Eskilsson 1972, 1974; Eskilsson, Hartler 1973; Keays 1974; Plummer 1978; Isebrands et al. 1979). A survey of this topic was given e.g. by Luger and Gampe (1978). Special investigations on parts of Scots pine trees *(Pinus silvestris)* and slash pine trees *(Pinus elliottii)* were carried out by Poller (1978; Poller et al. 1973) and Howard (1973).

Literature

Ahlgren, P. A. and Goring, D. A. I. 1971, Can. J. Chem. *49,* 1272–1275

Ahlgren, P. A., Yean, W. Q. and Goring, D. A. I. 1971, Tappi *54,* 737–740

Albrecht, J. S. 1971, Investigation of the Physical/Chemical Mechanism of Selective Delignification of Wood with Peracetic Acid. Doctor Thesis, Inst. of Paper Chem., Appleton, Wi.

Albrecht, J. S. and Nicholls, G. A. 1976, Pap. Puu *58,* 49–56

Alfredsson, B. and Samuelson, O. 1974, Svensk Papperstid. *77,* 449–452

Alfredsson, B., Czerwinsky, W. and Samuelson, O. 1961, Svensk Papperstid. *64,* 812–814

Altgelt, K. H. and Moore, J. C. 1967, Gel Permeation Chromatography. In: Polymer Fractionation (Cantow, M. J. R., Ed.). Academic Press, New York, London, pp. 123–179

Applegarth, D. A. and Dutton, G. G. S. 1965, Tappi *48,* 204–206

Baumeister, M. and Edel, E. 1980, Papier *34,* 10 A, V9–V18

Bednar, H. and Fengel, D. 1974, Holz Roh-Werkst. *32,* 99–107

Beélik, A., Conca, R. J. Hamilton, J. K. and Partlow, E. V. 1967, Tappi *50,* 78–81

Besold, G. and Fengel, D. 1983, Holz Roh-Werkst. *41,* 265–269

Bishop, C. T. and Cooper, F. P. 1960, Can. J. Chem. *38,* 793–804
Björkman, A. 1956, Svensk Papperstid. *59,* 477–485
Björkqvist, K. J. and Jörgensen, L. 1951, Acta Chem. Scand. *5,* 978–979
Bland, D. E. and Menshun, M. 1967, Appita *21,* 17–24
Bland, D. E. and Menshun, M. 1971, Appita *25,* 110–115
Borchardt, L. G. and Piper, C. V. 1970, Tappi *53,* 257–260
Brasch, D. J. and Wise, L. E. 1956, Tappi *39,* 581–588
Brauns, F. E. 1952, The Chemistry of Lignin. Academic Press Inc., Publ., New York
Brauns, F. E. and Brauns, D. A. 1960, The Chemistry of Lignin. Academic Press, New York, London
Brink, D. L. and Pohlman, A. A. 1972, Tappi *55,* 381–388
Brown, W., Cowling, E. B. and Falkehag, S. I. 1968, Svensk Papperstid. *71,* 811–821
Brownell, H. H. 1965, Tappi *48,* 513–519
Brownell, H. H. 1968, Tappi *51,* 298–300
Browning, B. L. 1963, The Chemistry of Wood. Intersci. Publ., New York, London
Browning, B. L. 1967a, Methods of Wood Chemistry. Vol. I. Intersci. Publ., New York, London
Browning, B. L. 1967b, Methods of Wood Chemistry. Vol. II. Intersci. Publ., New York, London
Browning, B. L. and Isenberg, I. H. 1952, Analytical Data and Their Significance. In: Wood Chemistry (Wise, L. E. and Jahn, E. C., Eds.), 2nd Ed., Vol. 2. Reinhold Publ. Corp., New York, pp. 1259–1277
Bublitz, W. J. and Meng, Y. T. 1978, Tappi *61,* No. 2, 27–30
Cafferty, P. D., Glaudemans, C. P. J., Coalson, R. and Marchessault, R. H. 1964, Svensk Papperstid. *67,* 845–849
Cantow, M. J. R. 1967, Additional Methods of Fractionation. In: Polymer Fractionation (Cantow, M. J. R., Ed.). Academic Press, New York, London, pp. 461–466
Chang, H., Cowling, E. B., Brown, W., Adler, E. and Miksche, G. 1975, Holzforschung *29,* 153–159
Chua, M. and Wayman, M. 1979, Can. J. Chem. *57,* 1141–1149
Ciarrocca, R. and Bedetti, R. 1976, Cell. Carta *27,* No. 9, 21–27
Cinite, V. A., Gravitis, J. A., Erin'š, P. P. and Rebjatnikova, A. F. 1976, Khim. Drev., No. 1, 12–20
Claesson, S., Bergmann, W. and Jayme, G. 1959, Svensk Papperstid. *62,* 141–155
Clarke, A. R. P. and Jayman, T. C. Z. 1975, Notes on Soil Techn.1975, 1–10, Tea Res. Inst. Talawakelle, Sri Lanka
Clermont, L. P. 1955, Pulp Pap. Mag. Can. *56,* No. 11, 107–114
Clermont, L. P. and Bender, F. 1958, Pulp Pap. Mag. Can. *59,* No. 7, 139–143
Clermont, L. P. and Bender, F. 1961, Pulp Pap. Mag. Can. *62,* No. 1, T 28
Clermont, L. P. and Schwartz, H. 1951, Pulp Pap. Mag. Can. *52,* No. 13, 103–105
Corbett, W. M. and Ewart, J. A. D. 1959, Svensk Papperstid. *62,* 277–283
Croon, I., Lindberg, B. and Meier, H. 1959, Acta Chem. Scand. *13,* 1299–1304
Cross, C. F. and Bevan, E. J. 1912, Researches on Cellulose. Vol. III. Longmans, Green, Co. London
Cutter, B. E. and McGinnes, Jr., E. A. 1980, Wood Fiber *12,* No. 2, 72–79
Dautzenberg, H. and Philipp, B. 1974a, Faserforsch. Textiltech. *25,* 469–475
Dautzenberg, H. and Philipp, B. 1974b, Faserforsch. Textiltech. *25,* 521–524
Dautzenberg, H. and Philipp, B. 1974c, Faserforsch. Textiltech. *25,* 422–425
Davydov, V. D., Tysjačnaja, G. J. and Uljašova, G. N. 1974, Khim. Drev. No. 2, 86–90
Dence, C. W., Gupta, M. K. and Sarkanen, K. V. 1962, Tappi *45,* 29–38
Dietrichs, H. H. 1972, Holz Zentralbl. *98,* No. 108, 1522–1525
Dietrichs, H. H. and Zschirnt, K. I. 1972, Holz Roh-Werkst. *30,* 66–74
Donetzhuber, A. and Wallenius, E. 1971, Svensk Papperstid. *74,* 238–240

Dutton, G. G. S., Joseleau, B. I. and Reid, P. E. 1973, Tappi *56,* 168–171

Ebringerová, A., Kramár, A., Rendoš, F. and Domanský, R. 1967, Holzforschung *21*, 74–77

Effland, M. J. 1977, Tappi *60,* No. 10, 143–144

Eiliott, J. H. 1967, Fractional Solution. In: Polymer Fractionation (Cantow, M. J. R., Ed.). Academic Press, New York, London, pp. 67–93

Ekman, R. 1980, Wood Extractives of Norway Spruce. Doctor Thesis, Abo Akademi, Finland

Erickson, H. D. 1962, Tappi *45,* 710–719

Eskilsson, S. 1972, Svensk Papperstid. *75,* 397–402

Eskilsson, S. 1974, Svensk Papperstid. *77,* 165–174

Eskilsson, S. and Hartler, N. 1973, Svensk Papperstid. *76,* 63–70

Faix, O. and Schweers, W. 1972, Holzforschung *26,* 9–10

Feckl, J. 1981, Gewinnung und Fraktionierung von Lignin-Polysaccharid-Komplexen. Ein Beitrag zur Aufklärung der Lignin-Polysaccharid-Bindung in der verholzten Zellwand. Doctor Thesis, Universität München

Fengel, D. 1966a, Holz Roh-Werkst. *24,* 9–14

Fengel, D. 1966b, Holz Roh-Werkst. *24,* 98–109

Fengel, D. 1967, Holz Roh-Werkst. *25,* 102–111

Fengel, D. 1976a, Holzforschung *30,* 73–78

Fengel, D. 1976b, Holzforschung *30,* 143–148

Fengel, D. 1980, Papier *34,* 428–433

Fengel, D. and Grosser, D. 1975, Holz Roh-Werkst. *33,* 32–34

Fengel, D. and Przyklenk, M. 1976, Wood Sci. Technol. *10,* 311–320

Fengel, D. and Przyklenk, M. 1983, Holz Roh-Werkst. *41*, 193–194

Fengel, D. and Wegener, G. 1979. Hydrolysis of Polysaccharides with Trifluoroacetic Acid and its Application to Rapid Wood and Pulp Analysis. In: Hydrolysis of Cellulose: Mechanisms of Enzymatic and Acid Catalysis (Brown, Jr., R. D. and Jurasek, L., Eds.). Adv. Chem. Ser. No. 181, ACS, pp. 145–158

Fengel, D., Grosser, D. and Wegener, U. 1973, Paleontogr. *144*, Abt. B, 31–43

Fengel, D., Uçar, H. and Wegener, G. 1979, Papier *33,* 233–239

Fengel, D., Wegener, G., Heizmann, A., and Przyklenk, M. 1978, Cell. Chem. Technol. *12,* 31–37

Fengel, D., Wegener, G. and Feckl, J. 1981a, Holzforschung *35,* 51–57

Fengel, D., Wegener, G. and Feckl, J. 1981b, Holzforschung *35,* 111–118

Fischer, K. 1935, Angew. Chem. *48,* 394–396

Garves, K. 1981, Holz Roh-Werkst. *39,* 253–254

Gierer, J. and Sundholm, L. 1971, Svensk Papperstid. *74,* 345 -351

Glasser, W. and Barnett, C. 1979, Holzforschung *33,* 78–86

Glinski, A. J. and Nicholls, G. A. 1977, Pap. Puu *59,* 745–760

Goetzler, M. 1982, Tappi *65,* No. 3, 149–150

Griebenow, W., Werthmann, B. and Töppel, O. 1977, Papier *31,* 503–508

Grosser, D., Fengel, D. and Schmidt, H. 1974, Forstwiss. Centralbl. *93,* 191–207

Haas, H., Schoch, W. and Ströle, U. 1955, Papier *9,* 469–475

Hafizoglu, H. 1979, Investigations on Turkish Tall Oil. Doctor Thesis, Abo Akademi, Finland

Hamilton, J. K. and Quimby, G. R. 1957, Tappi *40,* 781–786

Hamilton, J. K. and Thompson, N. S. 1959, Tappi *42,* 752–760

Hardell, H. L., Leary, G. J., Stoll, M. and Westermark, U. 1980, Svensk Papperstid. *83,* 44–49

Harwood, V. D. 1971, Holzforschung *25,* 73–78

Hashi, M., Teratani, F. and Miyazaki, K. 1970, Mokuzai Gakkaishi *1*6, 37–41

Hayashi, A. and Kaji, T. 1972, Mokuzai Gakkaishi *18,* 63–69

Heikkila, H. and Sjöström, E. 1975, Cell. Chem. Technol. *9,* 3–11

Holmbom, B. 1977, J. Am. Oil Chem. Soc. *54,* No. 7, 289–293

Holmbom, B. 1980, Pap. Puu *62,* 523–531

Höpner, T. 1966, Papier *20,* 122–125
Howard, E. T. 1973, Wood Sci. *5,* 312–317
Hutorščikov, I. S., Bujnickaja, M. I. and Zorina, G. A. 1967, Lesn. Zhur. *10,* 129–132
Ifju, G. and Labosky, Jr., P. 1972, Tappi *55,* 524–529
Isebrands, J. G., Sturos, J. A. and Christ, J. B. 1979, Tappi *62,* No. 7, 67–70
Janson, J. 1974, Faserforsch. Textiltechn. *25,* 375–382
Jayme, G. 1942, Cellulosechemie *20,* 43–49
Jayme, G. and Kringstad, K. 1961, Papier *15,* 127–130
Jayme, G., Knolle, H. and Rapp, G. 1958, Papier *12,* 464–467
Jayme, G., El-Kodsi, G., Stuber, M. and Böttcher, H. P. 1974, Papier *28,* 385–392
Jayme, G., El-Kodsi, G., Stuber, H. M., Böttcher, H. P. 1976, Appl. Polym. Symp. No. 28, 953
Jentgen, H. 1911, Kunststoffe *1,* 161–166
Johnson, D. B., Moores, W. E. and Zank, L. C. 1961, Tappi *44,* 793–798
Jones, J. K. N., Purves, C. B. and Timell, T. E. 1961, Can. J. Chem. *39,* 1059–1066
Karklins, V. B. and Okherina, E. E. 1975, Khim. Drev. No. 4, 3–10
Karklins, V. B., Belkova, L. P., Gromov, V. S. and Eidus, Y. A. 1977, Khim. Drev. No. 4, 91–96
Karras, M. and Rahkamaa, E. 1971, Pap. Puu *53,* 653–655
Kaszynska, J. 1974, Przegl. Papier. *30,* 87–89
Keays, Y. L. 1974, For. Prod. J. *24,* No. 11, 13–16
Kholkin, Y. I. 1969, Chromatography in Wood Chemistry (Sharkov, V. I., Ed.). Israel Program for Scientific Transl. Ltd.
Kibblewhite, R. P. 1973, Cell. Chem. Technol. *7,* 659–668
Klauditz, W. 1957, Holzforschung *11,* 110–116
Kolboe, S. and Eleffsen, Ø. 1962, Tappi *45,* 163–166
Kollmann, F. and Fengel, D. 1965, Holz Roh-Werkst. *23,* 461–468
Kollmann, F. and Höckele, G. 1962, Holz Roh-Werkst. *20,* 461–473
Kotera, A. 1967, Fractional Precipitation. In: Polymer Fractionation (Cantow, M. J. R., Ed.). Academic Press, New York, London 1967, pp. 43–66
Krässig, H. 1971, Papier *25,* 841–849
Kringstad, K. P. and Cheng, C. W. 1969, Tappi *52,* 2382–2385
Kringstad, K. and Ellefsen, Ø. 1964, Papier *18,* 583–591
Kubačková, M. and Cetlová-Pranzková, J. 1963, Drev. Vysk. *8,* 27–35
Kürschner, K. and Hoffer, A. 1929, Tech. Chem. Papier-Zellst.-Fabr. *26,* 125–139
Kürschner, K. and Karácsonyi, S. 1961, Holzforschung *15,* 107–114
Kürschner, K. and Melcerová, A. 1965, Holzforschung *19,* 161–171
Kürschner, K. and Popik, M. G. 1962, Holzforschung *16,* 1–11
Kurth, E. F. and Swelim, A. A. 1963, Tappi *46,* 591–593
Lai, Y. Z. and Sarkanen, K. V. 1971, Isolation and Structural Studies. In: Lignins. Occurrence, Formation, Structure and Reactions (Sarkanen, K. V. and Ludwig, C. H., Eds.). Wiley-Intersci., New York, pp. 165–240
Laidlaw, R. A., Farmer, R. H. and Smith, G. A. 1961, Holzforschung *15,* 15–18
Lal, N. K., Butnaru, R. and Simionescu, C. 1977, Cell. Chem. Technol. *11,* 59–66
Lapinoja, V. V. M. and Thompson, N. S. 1973, Pap. Puu *55,* 99–112
Leopold, B. 1961, Tappi *44,* 230–235
Lindberg, B. 1962, Methods of Isolating and Purifying Hemicelluloses. In: Proc. Wood Chem. Symp. Montreal. London Butterworth, pp. 67–75
Lindberg, B. and Meier, H. 1957, Svensk Papperstid. *60,* 785–790
Lindberg, B. and Rosell, K. G. 1974, Svensk Papperstid. *77,* 286–287
Lindgren, B. O. 1971, Svensk Papperstid. *74,* 57–63
Lindgren, B. O. 1974, Papier *28,* 10A, V29–V32
Loos, W. E. 1965, For. Prod. J. *15,* No. 3, 102–106

Luger, F. and Gampe, S. 1978, Potential forstlicher Reststoffe (Waldabfälle), Vol. 2. Bericht Forstl. Fak., Universität München, pp. 84–158
Maekawa, E. and Koshijima, T. 1980a, Mokuzai Gakkaishi *26,* 614–623
Maekawa, E. and Koshijima, T. 1980b, Mokuzai Gakkaishi *26,* 624–627
Magnusson, H., Eriksson, L. and Andersson, L. O. 1972, Svensk Papperstid. *75,* 619–622
Marković, N., Teržan, N. and Pjević, V. 1966, Sumarstvo *19,* 17–26
Marton, J. 1967, Tappi *50,* 335–337
Marton, J. and Spark, H. E. 1967, Tappi *50,* 363–368
Marutzky, R. and Roffael, E. 1977, Holz Zentralbl. *103,* No. 28, 424
Meier, H. 1958, Acta Chem. Scand. *12,* 144–146
Merewether, J. W. T. and Samsuzzaman, L. A. H. 1972, Holzforschung *26,* 11–18
Minor, J. L. 1979, J. Liq. Chrom. *2,* 309–318
Miyazaki, K., Smelstorius, J. A., Harwood, B. J. and Stewart, C. M. 1971, Appita *24,* No. 6, 452–454
Moore, W. E. and Effland, M. J. 1974, Tappi *57,* 96–98
Moore, W. E. and Johnson, D. B. 1967, Procedures for the Chemical Analysis of Wood and Wood Products. U.S. Forest Prod. Lab.
Most, D. S. 1957, Tappi *40,* 705–712
Nagorny, K. and Zietlow, K. 1972, Chem. Ing. Tech. *44,* 1201–1206
Nanassy, A. J. 1978, Wood Sci. *11,* 86–90
Narayanamurti, D. and Das, N. R. 1955, Holz Roh-Werkst. *13,* 52–56
Nelson, R. 1960, Tappi *43,* 313–317
Nelson, R. and Schuerch, C. 1957, Tappi *40,* 419–426
Painter, T. J. and Purves, C. B. 1960, Tappi *43,* 729–736
Pašteka, M. 1972, Determination of Carbonyl Groups in Oxidized Cellulose. In: Methods in Carbohydrate Chemistry (Whistler, R. L. and BeMiller, J. N., Eds.), Vol. V. Academic Press, New York, London, pp. 81–87
Pearl, I. A. 1967, The Chemistry of Lignin. Marcel Dekker Inc., New York
Pearl, I. A. and Busche, L. R. 1960, Tappi *43,* 961–970
Pensar, G., Ekman, R. and Peltonen, C. 1981, On the Distribution and Seasonal Variation of Lipophilic Non-Volatile Extractives in Stems of Norway Spruce (*Picea abies*). In: The Ekman Days 1981. Int. Symp. Wood Pulp. Chem., Stockholm, Vol. 1, pp. 52–54
Pew, J. C. 1957, Tappi *40,* 553–558
Pew, J. C. and Weyna, P. 1962, Tappi *45,* 247–256
Philipp, B. and Linow, K. J. 1965, Zellst. Pap. *14,* 321–326
Plonka, A. M. and Surewicz, W. 1979, Cell. Chem, Technol. *13,* 511–522
Plummer, G. M. 1978, Tappi *61,* No. 3, 45–47
Polčin, J. and Bezúch, B. 1978, Wood Sci. Technol. *12,* 149–158
Poljak, A. 1948, Angew. Chem. A *60,* 45–46
Poller, S. 1978, Holztechnologie *19,* 22–25
Poller, S., Scharf, P., Schultze-Dewitz, G. and Zenker, R. 1973, Holztechnologie *14,* 92–99
Rezanowich, A., Yean, W. G. and Goring, D. A. I. 1963, Svensk Papperstid. *66,* 141–149
Ritter, G. J. and Kurth, E. F. 1933, Ind. Engng Chem. *25,* 1250–1253
Rogers, I., Mahood, H., Servizi, J. and Gordon, R. 1979, Pulp Pap. Can. *80,* No. 9, T 286–T 290
Ruel, K. and Barnoud, F. 1978, Holzforschung *32,* 149–156
Runkel, R. O. H. and Wilke, K. D. 1951, Holz Roh-Werkst. *9,* 260–270
Rydholm, S. A. 1965, Characterization of Pulp Properties. In: Pulping Processes. Intersci. Publ. New York, London, Sydney, pp. 1107–1132
Saad, S. M., Issa, R. M. and Fahmy, M. S. 1980, Holzforschung *34,* 218–222
Saeman, J. F., Moore, W. E., Mitchell, R. L. and Millett, M. A. 1954, Tappi *37,* 336–343
Salud, E. C. and Faix, O. 1980, Holzforschung *34,* 113–121

Samuelson, O. 1970, Papier *24,* 671–678

Sarkanen, K. V., Kaheki, K., Murphy, R. A. and White, H. 1962, Tappi *45,* 24–29

Satonaka, S. 1963, Mokuzai Gakkaishi *9,* 26–34

Savard, J., Nicolle, J. and André, A. M. 1960, Analyse chimique des bois tropicaux. Centre Technique Forestier Tropical, Nogent-sur-Marne

Schadenböck, W. and Prey, V. 1972, Papier *26,* 116–118

Schöning, A. G. and Johansson, G. 1959, Svensk Papperstid. *62,* 646–648

Schöning, A. G. and Johansson, G. 1965, Svensk Papperstid. *68,* 607–613

Schurz, J. 1972, Viskositätsmessungen an Hochpolymeren. Verlag Berliner Union GmbH, Stuttgart

Schweers, W. and Meier, D. 1979, Holzforschung *33,* 15–18

Scott, R. W. 1976, Anal. Chem. *48,* 1919–1922

Scott, R. W. and Green, J. 1972, Tappi *55,* 1061–1063

Scott, R. W. and Green, J. 1974, Anal. Chem. *46,* 594–596

Sears, K. D., Alexander, W. J., Goldschmid, O. and Hamilton, J. K. 1978, Tappi *61,* No. 9, 105–110

Seifert, K. 1956, Papier *10,* 301–306

Seifert, K. 1960, Papier *14,* 104–106

Seligmann, O. 1977, Beiträge zur Strukturaufklärung polyphenolischer und 0-glykosidischer Pflanzeninhaltsstoffe durch Massenspektrometrie und GC-MS-Analytik. Doctor Thesis, Universität München

Simonson, R. 1967, Svensk Papperstid. *70,* 537–539

Sinner, M., Simatupang, M. H. and Dietrichs, H. H. 1975, Wood Sci. Technol. *9,* 307–322

Smelstorius, J. A. 1974, Holzforschung *28,* 67–73

Smelstorius, J. A. and Stewart, C. M. 1974, Holzforschung *28,* 204–211

Stewart, C. M. and Smelstorius, J. A. 1968, Chem. and Ind., 618–619

Stewart, C. M., Melvin, J. F., Tham, S. H. and Zerdoner, E. 1973, Cell. Chem. Technol. *7,* 371–386

Stoll, M. and Fengel, D. 1977, Wood Sci. Technol. *11,* 265–274

Surminski, J. and Bojaczuk, T. 1973, Rocznik Sek. Dendrol. Pol. *27,* 159–168

Theander, O. 1981, Hydrophilic Extractives from Scots Pine and Norway Spruce. In: The Ekman Days 1981. Int. Symp. Wood Pulp. Chem. Stockholm, Vol. 1, pp. 89–93

Thomas, B. B. 1945, Paper Ind. *26,* 1281–1284

Thompson, N. S. and Kaustinen, O. A. 1964, Tappi *47,* 157–161

Thompson, N. S. and Kaustinen, O. A. 1966, Tappi *49,* 83–90

Thompson, N. S., Kremers, R. E. and Kaustinen, O. A. 1968a, Tappi *51,* 123–126

Thompson, N. S., Kremers, R. E. and Kaustinen, O. A. 1968b, Tappi *51,* 127–131

Timell, T. E. 1957, Tappi *40,* 25–29, 30–33, 568–572

Timell, T. E. 1960, Svensk Papperstid. *63,* 652–657

Timell, T. E. 1961, Tappi *44,* 88–96

Timell, T. E. 1967, Wood Sci. Technol. *1,* 45–70

Timell, T. E. 1969, Svensk Papperstid. *72,* 173–181

Timell, T. E. and Jahn, E. C. 1951, Svensk Papperstid. *54,* 831–846

Timell, T. E. and Tyminski, A. 1957, Tappi *40,* 519–522

Timell, T. E., Glaudemans, C. P. J. and Gillham, J. K. 1958, Pulp Pap. Mag. Can. *59,* 242–252

Timell, T. E., Glaudemans, C. P. J. and Gillham, J. K. 1959, Tappi *42,* 623–634

Töppel, O. 1969, Papier *23,* 682–688

Uçar, H. 1977. Einfluß der Delignifizierung auf die Bestandteile des Buchenholzes. Doctor Thesis, Universität München

Uprichard, J. M. 1971, Holzforschung *25,* 97–104

Valtasaari, L. and Saarela, K. 1975, Pap. Puu *57,* 5–10

VanBeckum, W. G. and Ritter, G. J. 1937, Pap. Trade J. *105,* No. 18, 127–130

Vander Linden, N. G. 1974, Studies on Chlorine Dioxide Modification of Lignin in Wood. Doctor Thesis, Inst. of Paper Chem., Appleton, Wi.
Vander Linden, N. G. and Nicholls, G. A. 1976, Tappi *59,* No. 11, 110–113
Varaksina, T. N., Holkin, J. I. and Baženow, V. A. 1967, Lesn. Zhur. *10,* 137–139
Verhoef, J. C. 1977, J. Electroanal. Chem. *75,* 705–717
Vieböck, F. and Schwappach, A. 1930, Ber. *63,* 2818–2823
Vollmert, B. 1973, Polymer Chemistry. Springer-Verlag Berlin, Heidelberg, New York
Wagenführ, R. and Scheiber, C. 1974, Holzatlas. VEB Fachbuchverlag Leipzig
Ward, Jr. K. and Morak, A. J. 1962, Fractional Extraction and Properties of Hemicelluloses. In: Proc. Wood Chem. Symp. Montreal. London Butterworth, pp. 77–89
Wegener, G. 1974, Papier *28,* 478–486
Wegener, G. 1975, Papier *29,* 429–437
Wegener, G. 1976, Papier *30,* 177–185
Wegener, G. and Fengel, D. 1977, Wood Sci. Technol. *11,* 133–145
Wegener, G. and Fengel, D. 1978, Ultrasonics *16,* 186
Wegener, G. and Fengel, D. 1979, Tappi *62,* No. 3, 97–100
Wegener, G. and Stoll, M. 1976, Cell. Chem. Technol. *10,* 611–616
Weiner, J. and Byrne, J. 1961, Determination of Water (Moisture) in Wood, Pulp and Paper, and Textiles. Bibliographic Ser. No. 160, 2nd Ed., Inst. of Paper Chem., Appleton, Wi.
Weissmann, G. 1973, Holzforschung *27,* 193–197
Weissmann, G. 1974, Holzforschung *28,* 186–188
Whistler, R. L. and BeMiller, J. N. 1972, Methods in Carbohydrate Chemistry. Vol. VI. General Carbohydrate Methods; Fundamentals of Chromatographic Methods for Sugar Analysis. Academic Press, New York, London, pp. 3–75
Whistler, R. L. and Wolfrom, M. L. 1963, Methods in Carbohydrate Chemistry. Vol. II. Reactions of Carbohydrates; General Isolation Procedures. Academic Press. New York, London, pp. 3–62
Whistler, R. L. and Wolfrom, M. L. 1965, Methods in Carbohydrate Chemistry. Vol. V. General Polysaccharides; Cellulose, Preparation, Chemical and Physical Analysis. Academic Press, New York, London, pp. 3–189
Wilson, W. K. and Mandel, J. 1961, Tappi *44,* 131–137
Wise, L. E. and Ratliff, E. K. 1947, Anal. Chem. *19,* No. 7, 459–462
Wise, L. E., Murphy, M. and D'Addieco, A. A. 1946, Pap. Trade J. *122,* No. 2, 35–43
Wood, J. R., Ahlgren, P. A. and Goring, D. A. I. 1972, Svensk Papperstid. *75,* 15–19
Yang, J. M. and Goring, D. A. I. 1978, Holzforschung *32,* 185–188
Zhbankov, R. G., Garbuz, N. I., Derbentsev, F. F. and Krotkevich, V. P. 1966, Zh. Prikl. Spektrosk. No. 4, 442–445
Zicherman, J. B. and Thomas, R. J. 1972, Holzforschung *26,* 150–152
van Zyl, J. D. 1978, Wood Sci. Technol. *12,* 251–259

4. Cellulose

4.1. Occurrence

Cellulose is the structural basis of the plant cells and hence it follows that it is the most important natural substance produced by living organisms. This same statement applies to the quantitative occurrence of cellulose. In the biosphere $27 \cdot 10^{10}$ t of carbon are bound in living organisms, more than 99% of which are plants (Neumüller 1973). It can be assumed that about 40% of the plant-carbon is bound in cellulose (Fig. 4–1), which means that the total cellulose in the vegetable world amounts to about $26.5 \cdot 10^{10}$ t.

Cellulose is distributed in all plants from highly developed trees to primitive organisms such as sea-weeds, flagellates and bacteria. Cellulose can even be found in the animal kingdom: tunicin, the cuticular substance of tunicates, is identical to vegetable cellulose (Wardrop 1970). The percentage of cellulose in plant material varies depending on origin. There is high cellulose content in seed-hairs (cotton, kapok) and bast-fibres (ramie, flax, hemp); mosses, horse-tail, and bacteria contain less cellulose (Table 4–1). The isolation of cellulose is strongly influenced by the accompanying substances within the cell wall. Compounds such as fats, waxes, proteins and pectin can easily be removed by extraction with organic solvents and dilute alkali. This is done e.g. with cotton and ramie.

In wood, not only is cellulose accompanied by polyoses and lignin, it is also intimately associated with them, and the separation requires intensive chemical treatment. The isolated cellulose remains more or less impure. For many analytical purposes it is enough to determine the alpha-cellulose (→ 3.2.7.). To obtain 100% pure cellulose from wood alpha-cellulose must be subjected to further intensive treatments, such as partial hydrolysis, dissolution and precipitation, and the resulting product consists of very short molecular chains (Jayme, Knolle 1965a).

Cellulose is the basis of many technical products (paper, films, fibres, additives etc.), and is therefore predominantly isolated from wood by pulping processes on a

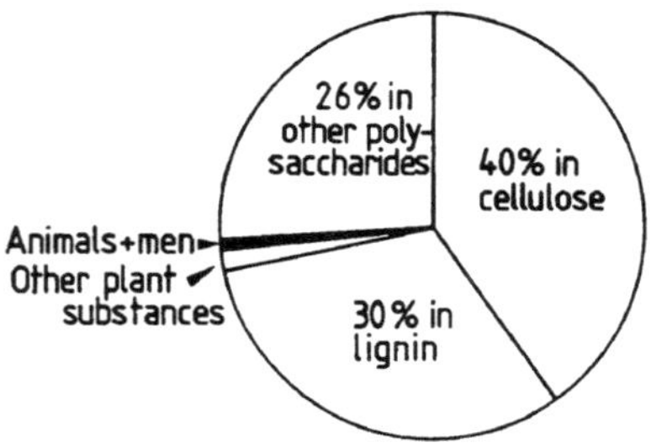

Fig. 4–1. Distribution of the carbon bound in organic matter (according to Gruber 1976).

Table 4-1. Cellulose content of various plant materials

Plant material	Cellulose [%]
Cotton	95–99
Ramie	80–90
Bamboo	40–50
Wood	40–50
Bark	20–30
Mosses	25–30
Horse-tail	20–25
Bacteria	20–30

large scale. By the use of various pulping chemicals, acidic, neutral, or alkaline conditions, and pressure, pulps of different properties are obtained (→ 16.). For several purposes the pulps must be purified by the additional process of bleaching (→ 16.7.). For the production of films, fibres and cellulose derivatives (→ 17.) a high degree of purity is necessary.

4.2. Molecular Properties

4.2.1. Constitution and Configuration

Cellulose consists of anhydroglucopyranose units which are joined to form a molecular chain. Therefore cellulose can be described as a linear-polymer glucan with a uniform chain structure. The units are bound by β-(1→4)-glycosidic linkages. Two adjacent glucose units are linked by elimination of one molecule of water between their hydroxylic groups at carbon 1 and carbon 4. The β-position of the OH-group at C1 needs a turning of the following glucose unit around the C1-C4 axis of the pyranose ring. Strictly speaking the repeating unit of the cellulose chain is a cellobiose unit with a length of 1.03 nm (Fig. 4-2a). Evidence for the uniformity of the units and linkages in cellulose was produced between 1920 and 1930 by methylation and other experiments (Freudenberg 1967).

Though there are OH-groups at both ends of the cellulose chain these OH-groups show a different behaviour. The C1-OH ist an aldehyde hydrate group deriving from the ring formation by an intramolecular hemiacetal linkage. That is why the OH-group at the C1-end has reducing properties, while the OH-group at the C4-end of the cellulose chain is an alcoholic hydroxyl and therefore non-reducing (Fig. 4-2b).

The cellulose chain is elongated, and the glucose units are arranged in one plane. There are three reasons for this arrangement. The first one is the β-glycosidic linkage. Only the β-position of the hydroxylic group at C1 allows an elongation of

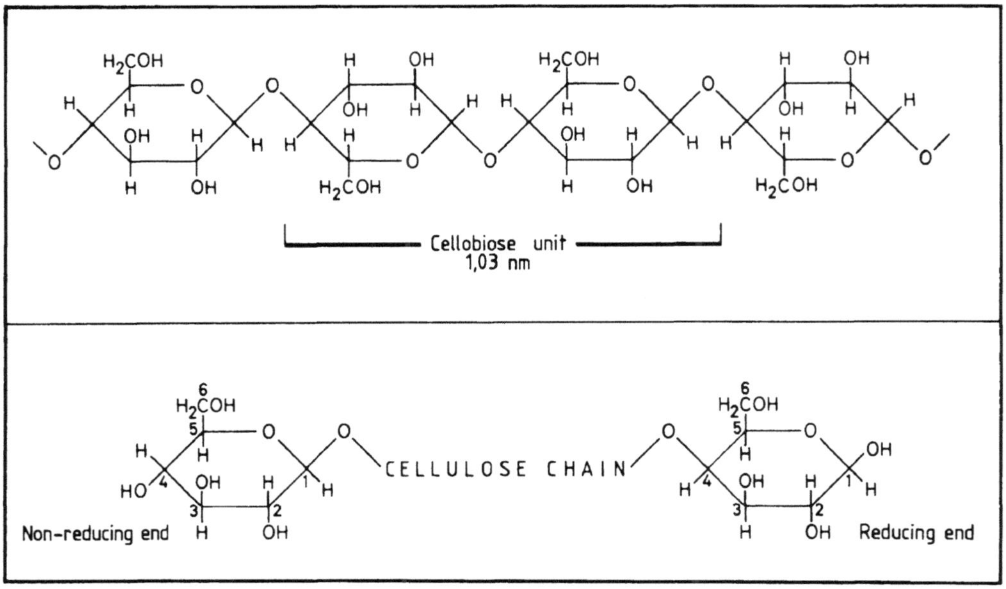

Fig. 4–2. Formula of cellulose.
a) Central part of the molecular chain.
b) Reducing and non-reducing end group of the molecule.

the molecular chain. An α-OH and an α-glycosidic linkage respectively lead to a helical molecular chain as it occurs with amylose in starch.

The second reason derives from the conformation of the pyranose ring. Bent hexagonal rings such as cyclohexane, pyran, and pyranose, may occur in various conformations, the limit forms of which are the chair and the boat form. The form lowest in energy, i.e. the most stable form, is the chair, while the forms highest in energy, i.e. labile forms, are the half-chair and the boat (Fig. 4–3). For cyclohexane the energy difference between chair and boat form amounts to 23.5 $kJ \cdot mol^{-1}$. At normal temperatures bent hexagonal rings take the most stable form so that the glucopyranose units of the cellulose have the chair conformation.

The third reason must be seen in connection with the ring conformation as there are two chair forms if OH-groups are considered. These hydroxylic groups may have a position above and below the ring (axial conformation) or in the plane of the ring (equatorial conformation) (Fig. 4–4). The latter has a lower energy content. Studies of methyl derivatives of glucose by infrared spectrometry confirm that the equatorial conformation of the pyranose ring is predominant (Michell, Higgins 1965). In this conformation the glucose units are arranged almost in a single plane. The relevant stereo-chemical formula is presented in Fig. 4–5. From the behaviour of cellulose in solution, i.e. a high intrinsic viscosity and a high negative temperature coefficient, it was concluded that a small amount of the glucose units (less than 2%) is in the more flexible boat and skew form (Goebel et al. 1976).

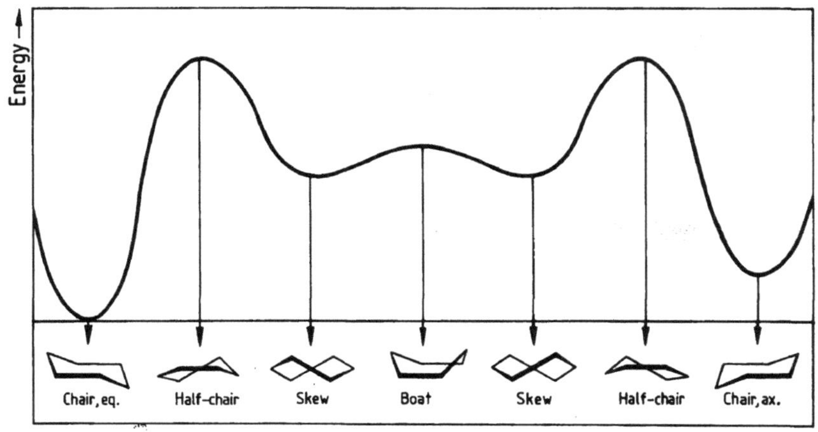

Fig. 4–3. Energy content of a hexagonal ring in various conformations.

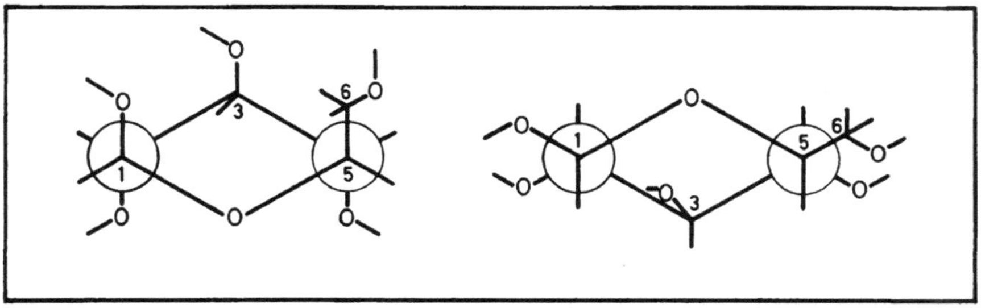

Fig. 4–4. Newman projection of a glucose ring with axial (a) and equatorial conformation (b).

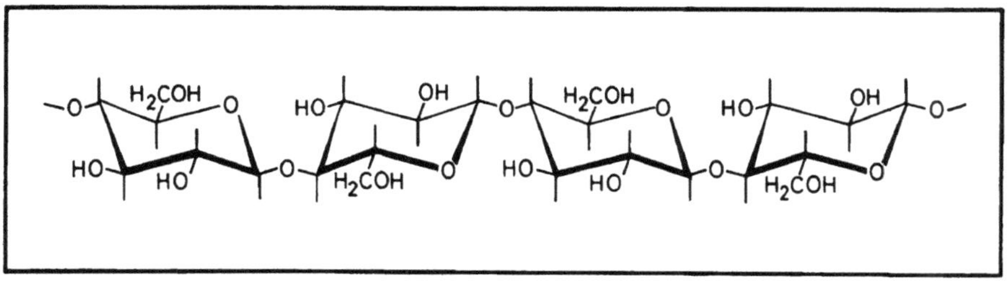

Fig. 4–5. Stereo-chemical formula of cellulose.

4.2.2. Cellulose in Solution

Cellulose is synthesized within the plant cell wall by joining glucose units to a macromolecular compound insoluble in all usual solvents. For the study of its molecular properties the dissolving of cellulose is necessary. A solution of cellulose is also needed for the realization of homogeneous reactions at the OH-groups and of structural transformations.

The dissolving of cellulose is possible by means of a heterogeneous conversion into esters (e.g. cellulose nitrate, → 17.3.2.; cellulose acetate, → 17.3.5.) or ethers (e.g. methylcellulose, → 17.4.3.; carboxymethylcellulose, → 17.4.4.). Cellulose esters are soluble in common solvents such as propanone (acetone) and ethyl acetate, whereas most of the cellulose ethers are soluble in water. A further ester sometimes used for viscosity measurements is cellulose xanthogenate (xanthate) (→ 17.3.4.), which is soluble in aqueous sodium hydroxide (Zellcheming Merkblatt IV/34/60; Elmgren 1965; Schurz, Schlor 1967). The most important purpose of cellulose xanthate is as a precursor for the production of rayon and cellophane (→ 17.3.4.).

Cellulose tricarbanilate was also used to study the behaviour of cellulose in solution. This compound is a very stable ester soluble in many esters, ethers and ketones (Burchard, Husemann 1961; Shanbhag 1968; Danhelka, Kössler 1976).

Cellulose can be directly dissolved in concentrated acids. Vink (1967) used phosphoric acid and Khripunov et al. (1975) trifluoroacetic acid (TFA) for the determination of molecular weight. Dissolving in acids however leads to a hydrolytic cleavage of the cellulose chains, so that with these solutions only the molecular weights of degradation products are obtained. The solution of cellulose in acids, e.g. in TFA, enables its total hydrolysis in a homogeneous reaction (Fengel et al. 1978). Furthermore, during the treatment with concentrated acids the cellulose molecules are transformed to derivatives, such as esters or addition compounds (Valtsaar, Dunlap 1952; Geddes 1956).

For a long time the copper complexes cuoxam (Schweizer 1857) and cuoxene or cuene (Traube 1911) were the only metal complexes known for dissolving cellulose. In a review Philipp et al. (1978) report on early experiments with mixtures of liquid ammonia and inorganic salts or of benzylpyridinium chloride and pyridine. Beginning with the discovery of the dissolving power of a cobalt ethylene diamine com-

Table 4–2. Metal complex solvents for cellulose

Name	Formula	Colour	Composition	Author	Year
Cuoxam	$Cu(NH_3)_4(OH)_2$	violet	$Cu+NH_4OH$	Schweizer	1857
Cuene	$Cu(en)_2(OH)_2$	violet	$Cu(OH)_2+H_2N-(CH_2)_2-NH_2$	Traube	1911
Cooxene	$Co(en)_3(OH)_2$	dark red	$Co(OH)_2+H_2N-(CH_2)_2-NH_2$	Jayme	1951
EWNN*	$Fe(C_4H_3O_6)_3Na_6$	green	$Fe(NO_3)_3$+Na−tartrate+NaOH	Jayme	1954
Nioxam	$Ni(NH_3)_6(OH)_2$	dark blue	$Ni(OH)_3+NH_4OH$	Jayme	1955
Nioxene	$Ni(en)_3(OH)_2$	violet	$Ni(OH)_3+H_2N-(CH_2)-NH_2$	Jayme	1955
Zincoxene	$Zn(en)_3(OH)_2$	transparent	$Zn(OH)_3+H_2N-(CH_2)-NH_2$	Jayme	1957
Cu-biuret-alkali	$Cu(H_2N-CO-NH-CO-NH_2)_2$	dark violet	$CuSO_4$+biuret+KOH	Jayme	1957
Cadoxene	$Cd(en)_3(OH)_2$	transparent	$CdO+H_2N-(CH_2)_2-NH_2$	Jayme	1957

* Modifications in preparing the complex: FeTNa (Valtasaari 1957)
$EWNN_{mod(NaCl)}$ (Jayme, El-Kodsi 1968)
modified $EWNN_{(NaCl)}$ (Achwal, Changule 1975)

plex for cellulose in 1951, Jayme developed several similar complexes containing Ni, Cd, Cu, Zn and Fe (Jayme 1961a, 1971a; Bergmann et al. 1964) (Table 4–2). From all these compounds the cadmium complex cadoxene and the iron tartratic sodium complex EWNN or FeTNa respectively, have proved best for the determination of the molecular weight or the degree of polymerization (DP) of cellulose (→ 4.2.3.), and the methods were improved during the following years. Modifications published by Valtasaari (1957), Jayme and El-Kodsi (1969), and Achwal and Changule (1975) simplified the preparation of the solvent EWNN (FeTNa) and improved its dissolving power. Apart from the green complex EWNN (FeTNa) the transparent cadoxene is the most important cellulose solvent (Brown 1967; Jayme 1971b, 1978).

There are only few results concerning the type of compound existing between metal complexes and cellulose. According to Vink (1964) in cuoxene solution strong copper-polysaccharide complexes are present, while in cadoxene solution only slight complexes are formed, and the binding of the Cd-cations is essentially due to an acid-base reaction between cellulose and cadoxene. Strong chelate linkages are assumed for complexes between cellulose and EWNN (Valtasaari 1971).

During the last few years some new solvent systems were developed containing a non-aqueous solvent and a component which modifies the cellulose molecule. Among these systems is dimethyl formamide (DMF) or dimethyl acetamide (DMAC) and N_2O_4 or NOCl (Schweiger 1969, 1974). In this case the cellulose derivative is cellulose nitrite (→ 17.3.3.). Another system is triethyl amine, urea, dimethyl sulfoxide (DMSO), and phthalic anhydride; the cellulose is esterified with phthalic acid (Garves 1972). The system DMSO-paraformaldehyde dissolves cellulose in the form of monomethylol cellulose, bearing a CH_3O-group at each C6-atom (Johnson et al. 1976; Swenson 1976; Guthrie, Hardcastle 1977). Solutions of methylol cellulose in DMSO however tend to aggregate, and the formation of gel particles increases with the age of the solution (Gruber, Gruber 1978).

Cellulose can also be dissolved in mixtures of hydrazine and water or DMSO under pressure and temperatures between 100 and 250 °C. The solution does not change its viscosity for several hours, which means that no degradation of the cellulose molecules takes place (Litt, Kumar 1977; Kolpak et al. 1977). A new solvent system also used for measuring the DP via viscosimetry consists of DMSO, chloral, and triethyl amine (Okajima 1978, 1979). The dissolution of cellulose occurs by means of the reaction of hydroxylic groups with chloral. The formation of hemiacetal and complex bonds leads to a high solvation of the cellulose molecules.

The dissolving process of cellulose starts with degradation of the fibrous and fibrillar structures and should result in complete disintegration into individual molecules unchanged in chain length. The degradation of the supramolecular structures occurs by swelling and insertion of chemical groups which break up the intermolecular linkages and solvate the single molecules. Philipp et al. (1975, 1978) observed a strong swelling in thickness while dissolving cellulose fibres in metal complexes, whereas in non-aqueous solvents containing amines and polar organic solvents, the dissolving process proceeds gradually beginning at the fibre surface.

In the first phase of dissolving the particles are still in intimate contact with each other because of the high concentration close to the dissolving place. Gel particles enclosing supramolecular structures of cellulose in various dimensions are also detached. These gel particles may survive a further dilution and subsequently disturb e.g. viscosimetric measurements (Dolmetsch, Dolmetsch 1967; Schurz 1969; Holt et al. 1976).

Leaving out the surviving gel particles, contact between the molecules relaxes as distance from the place of dissolving increases and as the dissolving process progresses. The behaviour of macromolecules depends very much on their concentration in solution. According to the scheme of Schurz (1977) the initial dense molecular network containing associated particles is dispersed during dilution to an entangled network and/or isolated associations (Fig. 4–6). An ideal separation into

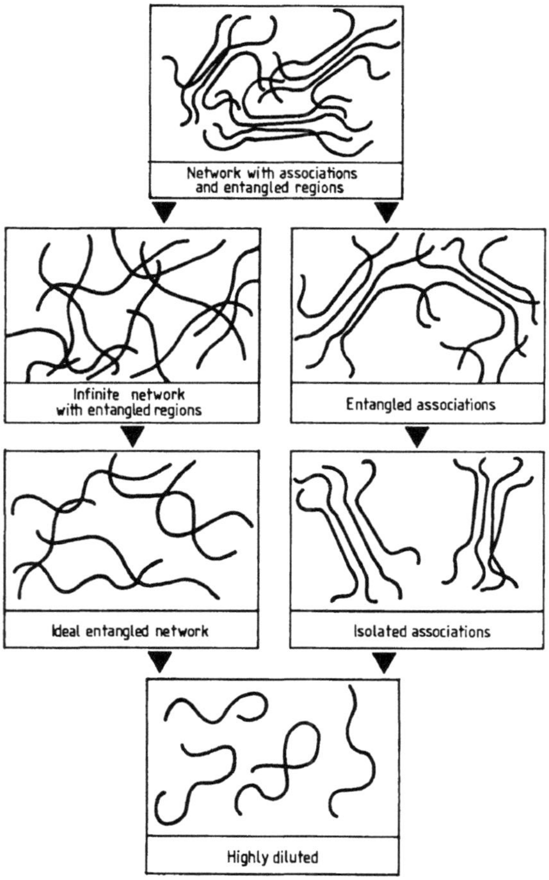

Fig. 4–6. Behaviour of a polymer in solution during dilution (according to Schurz 1977).

individual molecules occurs only in highly diluted solutions, i.e. for cellulose within 0.05%. Even in a solution of 0.1% cellulose nitrate (DP 6 000) the molecules still form a network (Schurz 1977).

Brown (1966) deduced that the cellulose molecules have the same flexibility in solution as other polymers. The greater extension of the molecular coils is caused by the greater bulk of the cellulose chain and the very favourable interaction with the solvent. In derivative cellulose the substituents have little steric influence on the extension of the macromolecules but determine the properties of the chains through the cellulose-solvent interaction.

These findings seem not to be true of metal complexes. From the large end-to-end length of cellulose chains in cadoxene Swenson (1973) concluded that the molecules are bent only in extended loops. The formation of polychelates in EWNN (FeTNa) is the reason for an increase of the stiffness of cellulose chains (Valtasaari 1971).

4.2.3. Molecular Weight and Chain Length

The molecular weight of cellulose varies widely (50 000–2.5 million) depending on the origin of the sample. As cellulose is a linear polymer with uniform units and bonds the size of the chain molecule is usually spezified as degree of polymerization (DP):

$$DP = \frac{\text{molecular weight of cellulose}}{\text{molecular weight of one glucose unit}}$$

The various degrees of polymerization of plant-derived cellulose as well as of technical cellulose products are summarized in Table 4–3. The data vary from cotton cellulose in still closed capsules with DP 15 300 to rayon fibres with DP 305. Except for the lower plants horse-tail and bacteria the DP-range is between 7 000 and 15 000 for plant cellulose. An intensive chemical treatment such as pulping, bleaching and transformation decreases the DP-value strongly. But also a careful delignification or extraction, even the influence of atmospheric oxygen as with cotton after the opening of the capsules, reduces the degree of polymerization. There is also evidence that the DP of cellulose in wood is reduced during aging of a living tree, i.e. the DP is highest in cells adjacent to the cambium and decreases towards the pith (Shimizu et al. 1970).

The reduction in DP is not uniform. The resulting molecular chains vary in length. Such cellulose is polydisperse and the symbol for the degree of polymerization should correctly be written $\overline{DP}$ or $\overline{P}_w$. The latter symbol characterizes also the mode of determination, namely using data connected with the weight of the molecules. These data are received from light scattering or ultracentrifugation, and are also used for the calculation of DP with viscosimetry.

Apart from the weight average degree of polymerization $\bar{P}_w$ there is the number average degree of polymerization $\bar{P}_n$, which can be determined by osmometry. The determining factor for $\bar{P}_n$ is the number of molecules with the respective degree of polymerization. The polydispersity of a cellulose sample can be indicated by the relation between $\bar{P}_w$ and $\bar{P}_n$. In a polydisperse system is $\bar{P}_n < \bar{P}_w$, and the polydispersity U is defined as

$$U = \frac{\bar{P}_w}{\bar{P}_n} - 1$$

The higher the value for U the higher is the degree of polydispersity.

It is assumed that cellulose in native state is very uniform in its degree of polymerization. This is indicated by the high DP of 15 300 for cotton (Table 4–3). The fractionation of mature cotton cellulose from still closed capsules resulted in one narrow and high peak containing 90% of the weight of the fractions between DP 13 500 and 14 500 and a low peak between DP 1 000 and 2 000 (Marx-Figini 1964; Marx-Figini, Schulz 1966b) (Fig. 4–7a). In commercial cotton the main portion of the cellulose also has a DP-value of more than 10 000, but two further peaks are visible in the differential distribution curve (Fig. 4–7b).

What was shown for cotton cellulose is certainly also valid for native cellulose in wood. The evidence however is difficult or even impossible to obtain because of the intimate association of the cell wall components. The intensive chemical treatment necessary for the separation of cellulose particularly from lignin reduces the chain length. An index for a high and relatively uniform degree of polymerization in wood

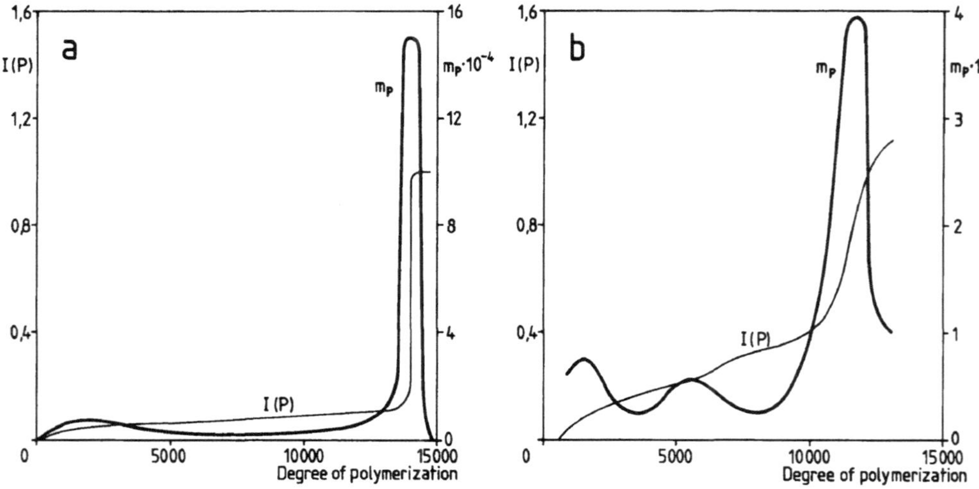

Fig. 4–7. Integral (I(P)) and differential (m_P) distribution of the degree of polymerization in cellulose of cotton from still closed seed capsules (a), and of commercial cotton (b) (according to Marx-Figini and Schulz 1966b).

Table 4–3: Average degrees of polymerization of celluloses from various materials as determined by measuring the viscosity

Material	Degree of polymerization	Mode of solution	References
California cotton, unopened *(Gossypium)*	15.300	Nitrate	Goring, Timell 1962
California cotton, opened	8.100	Nitrate	Goring, Timell 1962
Kapok *(Ceiba pentandra)*	9.500	Nitrate	Goring, Timell 1962
Textile flax *(Linum usitatissimum)*	8.800	Nitrate	Goring, Timell 1962
Ramie *(Boehmeria nivea)*	10.800	Nitrate	Goring, Timell 1962
Trembling aspen wood *(Populus tremuloides)*	10.300	Nitrate	Goring, Timell 1962
White birch wood *(Betula papyrifera)*	9.400	Nitrate	Goring, Timell 1962
White birch bark	7.500	Nitrate	Goring, Timell 1962
Jack pine wood *(Pinus banksiana)*	7.900	Nitrate	Goring, Timell 1962
Engelmann spruce wood *(Picea engelmannii)*	8.000	Nitrate	Goring, Timell 1962
Engelmann spruce bark	7.100	Nitrate	Goring, Timell 1962
Cinnamon fern *(Osmunda cinnamomea)*	8.300	Nitrate	Goring, Timell 1962
Horse-tail *(Equisetum arvense)*	2.400	Nitrate	Goring, Timell 1962
Bacteria *(Acetobacter xylinum)*	4.000–6.000	Nitrate	Husemann, Werner 1963
Cotton linters, bleached	1.000–5.000	Nitrate	Temming et al. 1972
Sulphite pulp (spruce), bleached	1.255	Cadoxene	Jayme et al. 1969
Sulphate pulp (pine, spruce), unbleached	975	Cadoxene	Jayme et al. 1969
Sulphate pulp (pine, spruce), bleached	965	Cadoxene	Jayme et al. 1969
Chemical pulp (beech)	715	Cadoxene	Jayme et al. 1969
Rayon fibres	305	Cadoxene	Jayme, Hasvold 1968

cellulose can be seen in the DP of 10 300 for cellulose from aspen wood (Goring, Timell 1962) (Table 4–3), a wood known to be easily delignified. Considering the degradation rate during nitration of wood Patscheke and Poller (1980) calculated a DP of about 12 000 for native cellulose in spruce wood.

The delignification of wood by analytical or technical methods results in celluloses with several peaks in the differential distribution curves of DP. An example for spruce cellulose isolated by chlorite delignification is shown in Fig. 4–8. Other studies of pulp celluloses resulted in two or three peaks in the low DP-range (Schurz et al. 1970; Schempp 1975). Treatments at high temperatures (→ 12.4.3.) or with γ-rays (→ 13.2.4.) not only reduce the average value of DP but also change the number and range of the peaks in the DP-distribution curve (Fengel 1967a, 1970; Polčin 1966; Kusama et al. 1976).

The uniform composition of the cellulose molecule simplifies the description of its size and shape. The diameter of the composing unit, the anhydroglucose is given in

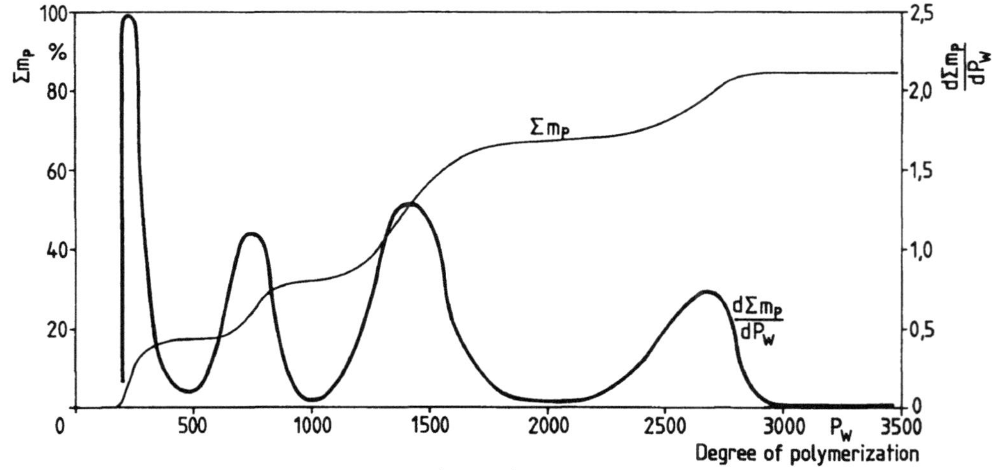

Fig. 4–8. Integral (mP) and differential $\left(\frac{d\ m_P}{d\bar{P}_w}\right)$ distribution of the degree of polymerization in alpha-cellulose from spruce wood.

the literature as 515 pm, which is true of the distance between the center of C1-oxygen and C4-oxygen and approximately true of the diameter perpendicular to the plane of the pyranose ring (Fig. 4–9). The width in plane and perpendicular to the chain direction is about 1 nm, i.e. about twice the diameter of the perpendicular direction.

The length of a single cellulose chain with a DP of 14 000 is 7.2 μm, i.e. 14 000 times or 7 000 times respectively the diameter of the chain. Thus the cellulose consists of extremely long molecules with a ribbon-like diameter.

4.2.4. Hydrogen Bonds

The stabilization of long molecular chains in order systems, i.e. the formation of supramolecular structures, originates in the presence of functional groups which are able to interact with each other. The functional groups of the cellulose chains are the hydroxyl groups, three of them being linked to each glucose unit. The surfaces of cellulose chains are – so to speak – studded with OH-groups. These OH-groups are not only responsible for the supramolecular structure but also for the chemical and physical behaviour of the cellulose.

OH-groups as well as NH-groups are able to interact with each other or with O-, N- and S-groups forming a particular linkage, the hydrogen bond (H-bond). Most of the supramolecular structures of natural and synthetic polymers are based on the H-bonds. Furthermore, the supramolecular arrangements in liquid water, the clusters, and the crystalline structure in ice are the results of hydrogen bonds between H_2O-molecules.

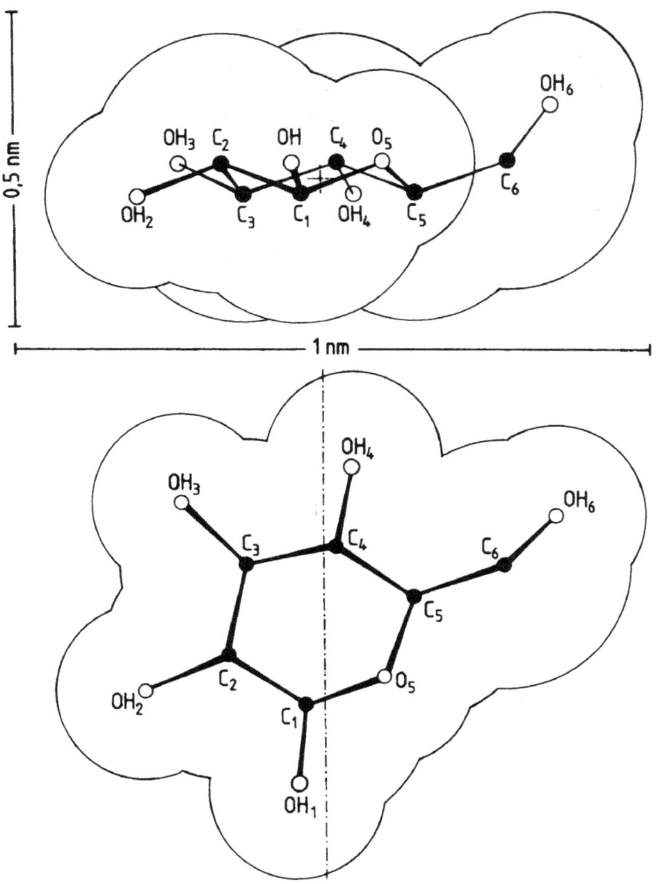

Fig. 4–9. Spatial dimension of a glucose molecule according to the van-der-Waals radii of its atoms.

The H-bonds take place by the approach of the H-atom of an OH- or NH-group to the lone electron pair of another O-atom. In the linked state the H-atom is coordinatively divalent. The distance between the two oxygen atoms linked by an H-bond is 275 pm instead of 350 pm, assuming theoretically the presence of a van-der-Waals bond.

A hydrogen bond is characterized by:
- the strength of the binding energy, which depends on the density of charge and the angle between the atoms linked with each other;
- sterical factors which cause an asymmetrical distribution of the electrons;
- the kinetics of the H-bridges, i.e. the frequency with which the OH- or NH-groups oscillate and the protons change position (Luck 1965, 1967).

A comparison of the binding energies between various atoms (Table 4–4) shows that the hydrogen bonds are about one power of ten weaker than coordinative bonds, but about two powers of ten stronger than van-der-Waals forces. It can be assumed that the binding energy between cellulosic OH-groups is about the same or

Table 4–4: Energy of various linkages (Stuart 1967; Cotton, Wilkinson 1974)

Linkage	Compound	Energy $kJ \cdot mol^{-1}$
v. d. Waals	$H_2O_{liq.}$	0.155
O–H···O	$H_2O_{liq.}$	15
O–H···O	$C_nH_{2n+1}OH$ (Cellulose?)	28
N–H···N	Melamin	25
O–H		460
C–O		356
C–H		414
C–C		347

somewhat higher than the binding energy of OH-groups in alcohols. The energy of H-bonds between water and cellulose was calculated to be 25 kJ · mol^{-1}. (Morrison, Dzieciuch 1959).

The OH-groups of cellulose molecules are able to form two types of hydrogen bonds depending on their site at the glucose units (Fig. 4–10). There are hydrogen bonds between OH-groups of adjacent glucose units in the same cellulose molecule (intramolecular linkages). These linkages give a certain stiffness to the single chains. There are also hydrogen bonds between OH-groups of adjacent cellulose molecules (intermolecular linkages). These linkages are responsible for the formation of supramolecular structures.

The primary structures formed by hydrogen bonds are the fibrils, which make up the wall layers and finally the whole cell wall. Moreover the surfaces of isolated wood cells or fibres in the non-dried state are able to form hydrogen bonds with

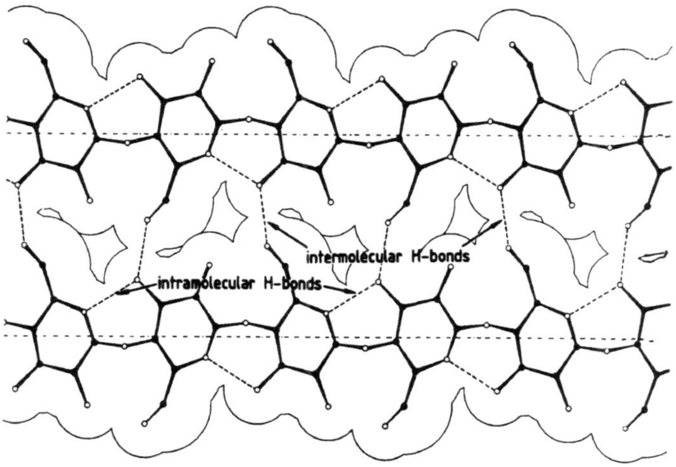

Fig. 4–10. Intramolecular and intermolecular hydrogen bonds in two adjacent cellulose molecules of the 002 plane.

each other. The mechanical properties of a pulp or paper sheet are determined by fibre-fibre bonds which are the result of H-bonds between fibre surfaces (Jayme 1961b; Nissan, Sternstein 1964). The surface properties of fibres, above all the number of OH-groups able to form fibre-fibre bonds, determine the strength of a sheet and depend on the isolation process (Davison 1972; Mohlin 1974). Studies of the interaction of various liquids with cellulose fibres gave evidence that there are further intermolecular forces apart from H-bonds which may influence the strength properties of a paper sheet (Robertson 1970).

Several studies by light and electron microscopy showed the very fine fibrillar bridges which join the cellulosic surfaces of contact (Page 1960; Jayme, Hunger 1961).

Hydrogen bonds do not only exist between cellulose OH-groups but also between cellulose-OH and water-OH. Depending on the water content single water molecules or clusters can be linked to the cellulosic surfaces (Dobbin 1970). The absorption of water by a cellulose sample depends on the number of free OH-groups or rather on the cellulose OH-groups not linked with each other. This can be demonstrated by the absorption and desorption isotherms. Fig. 4–11 shows that isolated wood cellulose absorbs more water than cotton cellulose at the same relative humidity. Additionally the difference of the hysteresis indicates the presence of fewer free OH-groups in cotton than in wood cellulose.

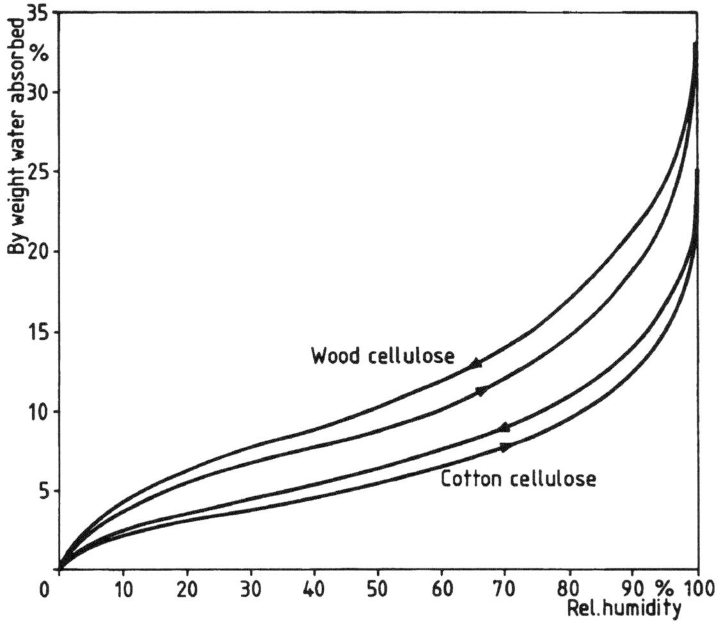

Fig. 4–11. Water adsorption and desorption isotherms (20 °C) of wood and cotton cellulose (according to Christensen and Kelsey 1959, and Jeffries 1960).

Morrison and Dzieciuch (1959) calculated the differential enthalpy and entropy of the adsorption of water by cotton cellulose (Fig. 4–12). At low water content, i.e. up to the formation of a monomolecular layer, there are high negative values for enthalpy and entropy which mark a linkage of water molecules by two or more H-bonds to the cellulose surface. After the complete occupancy of the cellulose surface by water molecules (about 0.2 moles per 100 g cellulose) there is a change in the linkage between water-OH and cellulose-OH, indicated by kinks in the curves. With the formation of further layers of water molecules the relative number of hydrogen bonds between water and cellulose decreases and this process is expressed by a decrease in the differential enthalpy and entropy of adsorption.

The entrance of water into the cellulosic structure means a swelling of the structure. But other solvents apart from water can also be adsorbed and linked by H-bonds, e.g. dimethyl sulfoxide (SO_2-groups) and pyridine (NH-groups). Solvents of that type result in a swelling dependent on temperature. Other solvents, such as dioxane or benzene, cannot be linked. They are only included in the structure, and result in a swelling independently of temperature (Richter et al. 1957; Chitumbo, Brown 1974) (Fig. 4–13). The presence of nonpolar solvents in cellulose impedes the formation of intermolecular H-bonds during drying. Solvents such as cyclohexane or benzene cannot be completely removed by drying even in a high vacuum, so it was

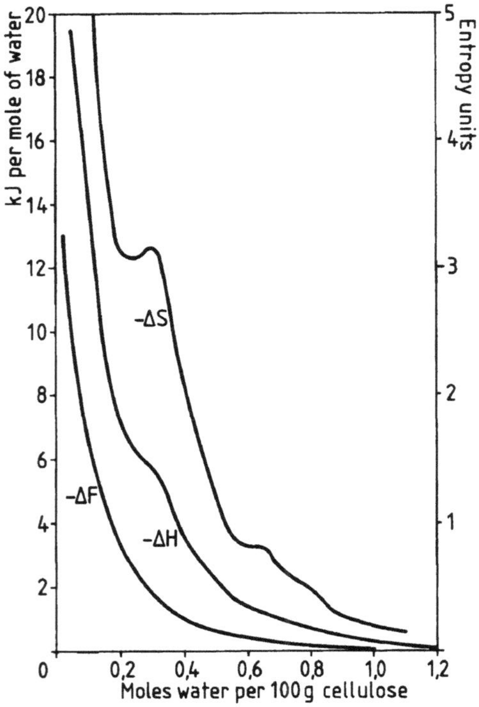

Fig. 4–12. Free energy ($-\Delta F$), differential enthalpy ($-\Delta H$), and entropy ($-\Delta S$) of the adsorption of water by cotton cellulose (according to Morrison and Dziecuich 1959).

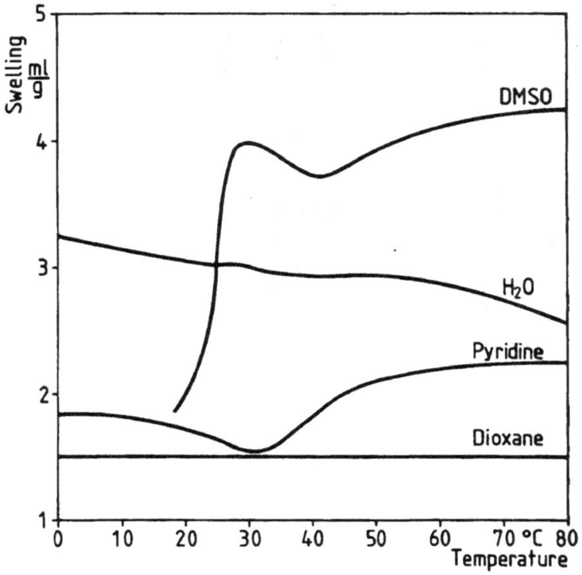

Fig. 4–13. Swelling-temperature curves for a cellulose gel in various solvents.

assumed that solvent molecules are wedged between cellulose surfaces (Staudinger et al. 1953; Richter et al. 1957). Celluloses containing residues of a nonpolar solvent are highly reactive, e.g. they can be easily acetylated.

The reverse process of adsorption of water and swelling is the removal of water and shrinkage of the cellulose. This process of drying, though continuous, can be resolved into individual stages (Fig. 4–14). The first of them is the cleavage of H-bonds between water molecules, which are the bonds of lowest energy in the system of cellulose-water. Parts of the water are removed and the cellulose surfaces approach each other. This process goes on until only a monomolecular layer of water remains between two cellulose surfaces. Then H-bonds between water-OH and cellulose-OH are cleft and hydrogen bonds between the cellulose surfaces are formed.

In the 1950's infrared spectroscopy was introduced to cellulose chemistry. Several studies resulted in the attribution of the various absorption bands to the respective atomic groups (Tsuboi 1957; Liang, Marchessault 1959; Higgins et al. 1961; Hummel 1965). Not only can atomic groups such as CH_2, CH, C=O and C-O-C be

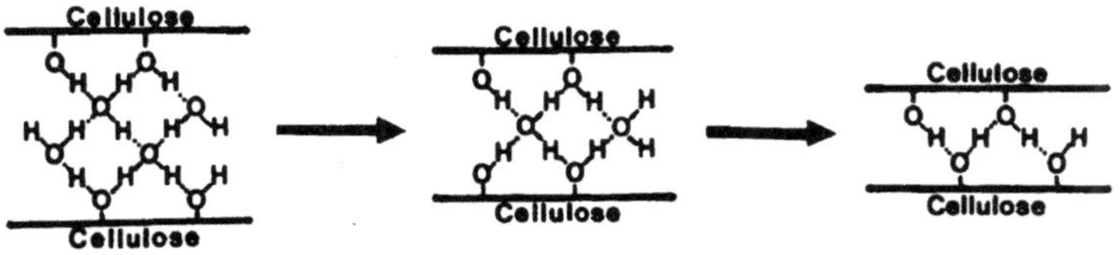

Fig. 4–14. Change of the hydrogen bonds during the removal of water from two adjacent cellulose surfaces.

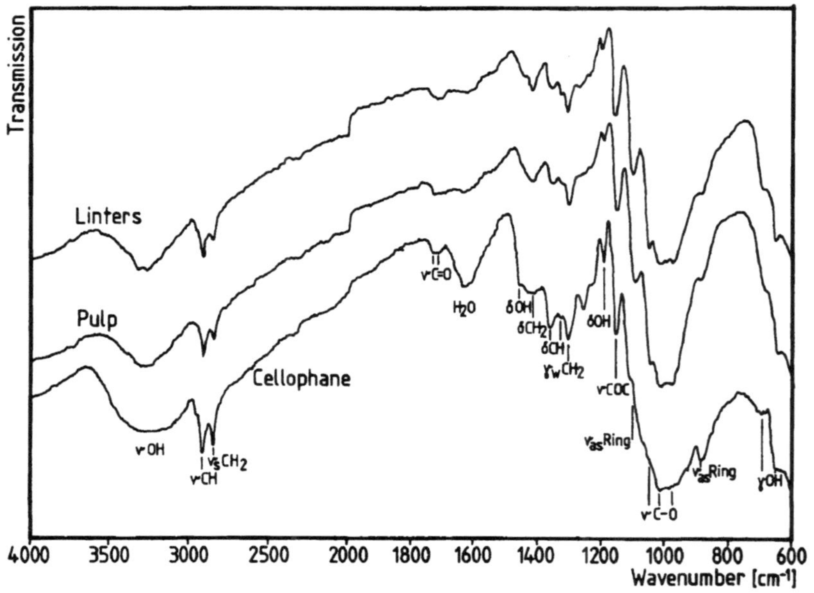

Fig. 14–15. Infrared spectra of three celluloses. The important oscillation bands are marked.

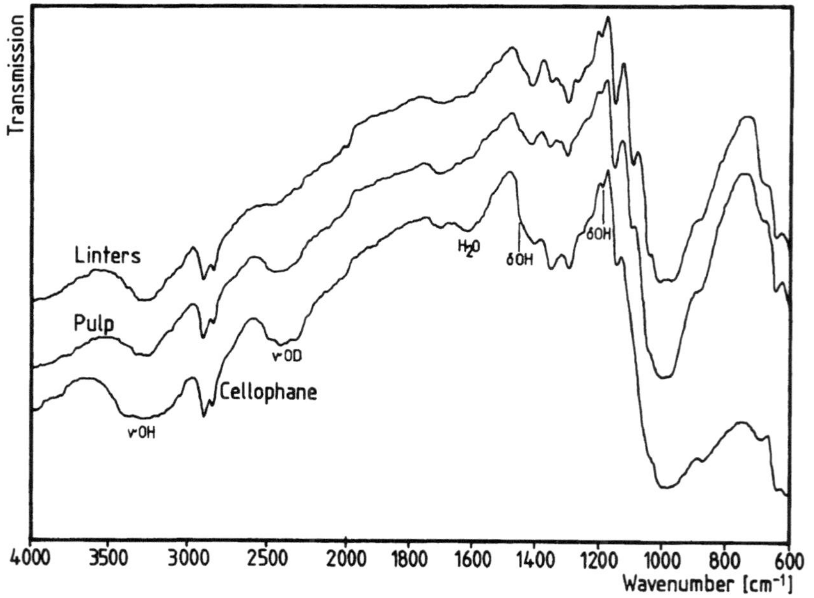

Fig. 4–16. Infrafred spectra of three celluloses after an exchange with deuterium oxide.

identified but also OH-groups and linked water. In Fig. 4–15 the infrared spectra of three celluloses are compared. Though dried under the same conditions a different portion of linked water (1 595 cm^{-1}) can be observed. The broadened band of the OH-stretching oscillation (3 200–3 300 cm^{-1}) is a further indication of a higher portion of water linked to cellulose-OH. From the three samples in Fig. 4–15 cotton linters has the lowest accessibility for water and cellophane has the highest one. This shows that more internal H-bonds exist between the OH-groups of cellulose in cotton linters than in pulp or in cellophane, and this implies a superior supramolecular organization.

The accessibility of the cellulose-OH for water can also be demonstrated by the adsorption of deuterium oxide (D_2O) or tritiated water (THO). The linkage of D_2O to cellulose-OH is indicated by the absorption band at 2 500 cm^{-1} (Fig. 4–16). The absorption in the OD-range is higher for cellophane than for pulp and for cotton linters. The accessibility can be changed by a chemical treatment of the cellulose (Jeffries 1963, 1964; Rånby 1964).

4.3. Supramolecular Structures

4.3.1. The Crystalline Lattice of Cellulose I

In the solid state the hydrogen bonds between the cellulose molecules are not arranged irregularly or at random, but a regular system of the H-bonds results in an order system with crystal-like properties. These properties were first detected in 1913 by Nishikawa and Ono with the aid of X-ray diffraction. Further studies using this method were carried out by several scientists during the following years. The results led to several models of the crystalline unit of cellulose, of which the final model deriving from Meyer and Misch (1937) is valid even today in its principal features. A survey of these early studies can be read in the book by Meyer and Mark (1953).

Because of the existence of various polymorphous forms (→ 4.3.2.) native cellulose has to be termed cellulose I in respect to its crystalline lattice. This lattice as evaluated by X-ray diffraction is monoclinic, i.e. it has three axes of different lengths and one non-90° angle (symbol C_2^2); additionally it is of a primitive type, i.e. only the corners of the crystalline unit are occupied (symbol P), and its translation is 1/2, i.e. the chains are screwed around the longitudinal axis by 180° (symbol 2_1). Thus the complete description of the space unit of cellulose is $C_2^2P2_1$. In Fig. 4–17 it can be recognized that the central chain of the space unit runs counter to the chains arranged at the corners. As each chain forms a corner of a crystal unit every second cellulose chain is reversed. Thus it can be said that the cellulose lattice consists of counter-running chain pairs. Additionally the chains are staggered by a distance of 0.25 along the b-axis.

The terms of the crystal axes a, b and c are used in the same manner as in most current literature. In recent years the longitudinal axis has sometimes been called c, but there is no agreement in the labels of the a- and b-axes (Gardner, Blackwell

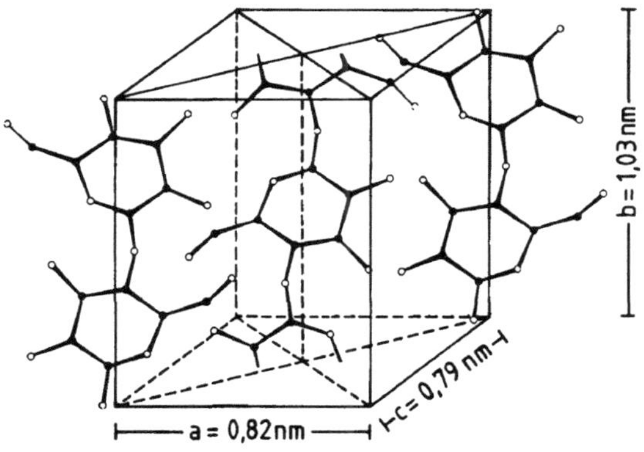

Fig. 4–17. Monoclinic space unit of cellulose I.

1974; Sarko 1976). It has to be noticed that a change in the terms of the crystal axes has consequences for the indices of the crystal planes. The main crystal planes for the cellulose space unit are shown in Fig. 4–18. These planes are represented by peaks of different intensities in the diagrams of X-ray or electron diffraction (Fig. 4–19).

The results of the diffraction measurements however are not very definite as some reflections overlap and different interpretations are possible. Hence space units with four cellulose chains and therefore with a double extension in the direction of the a- and c-axes have also been described (Honja, Watanabe 1958; Fisher, Mann 1960; Gardner, Blackwell 1974). Fisher and Mann (1960) restrain the larger unit to *Valonia* and bacterial cellulose. Blackwell et al. (1978) returned to the two-chain unit. A cell arranged diagonally to the Meyer-Misch model was proposed by Ellis and Warwicker (1962).

The diffuseness of the X-ray pattern of cellulose in wood complicates the interpretation, and thus the crystal structure of wood cellulose has been interpreted in terms of the lattice parameters of cellulose I determined from *Valonia* or ramie (Nishimura et al. 1981). The peak position of the 002 reflection in wood is found to shift

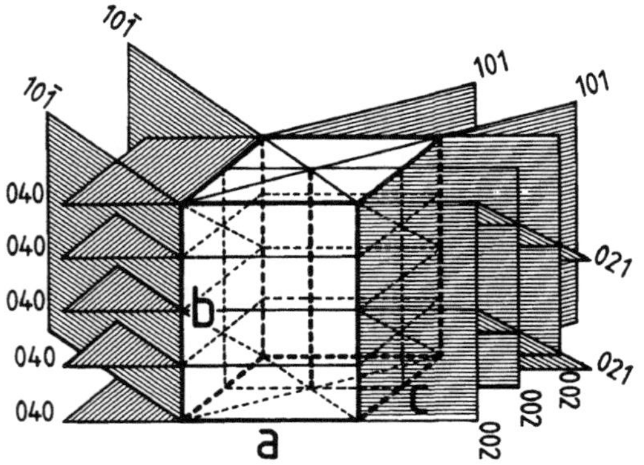

Fig. 4–18. The important lattice planes of the space unit of cellulose I.

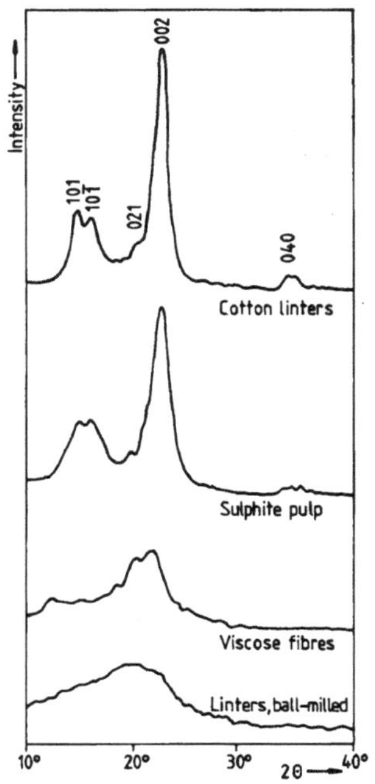

Fig. 4–19. Diffractograms of various celluloses (according to Ant-Wuorinen and Visapää 1965).

towards lower 2 Θ values than those of *Valonia* or ramie. This shift may be caused by lattice deformation or by the crystallite size.

The H-bonds are indeed an important factor for the arrangement and coherence of the cellulose chains in the crystal lattice. The arrangement of the two chains (002 plane) in Fig. 4–10 showing hydrogen bonds between O3 and O5′ (intramolecular) and between O6 and O3″ (intramolecular) was constructed with the atomic van-der-Waals radii in mind. The picture agrees with the findings of Liang and Marchessault (1959). In this case there is a possibility for the formation of an H-bridge to the next 002 plane by a second valence at the OH6-group (O6 . . . O4′). The CH_2OH-group is the only group which can rotate and connect two 002 planes. Fig. 4–20 demonstrates the dense package of the cellulose molecules in the crystal lattice.

In the model the OH2-group is not linked. This is contrary to the findings of Mann and Marrinan (1958) that all OH-groups should be linked by H-bonds. Other authors assume a second intramolecular linkage (O6 . . . O3′) with the consequence that no H-bonds to the chains of the adjacent 002 planes are possible (Gardner, Blackwell 1974; Sarko 1976, 1978). In this case the connection of the planes should occur by van-der-Waals forces. Another variant is given by Quivoron et al. (1967) who propose an O1 . . . O2 intramolecular linkage; thus the OH6 would be free for H-bonds in two dimensions. But in such a system the pyranose rings would be subjected to an extreme deformation which would adversely affect the stability of the system.

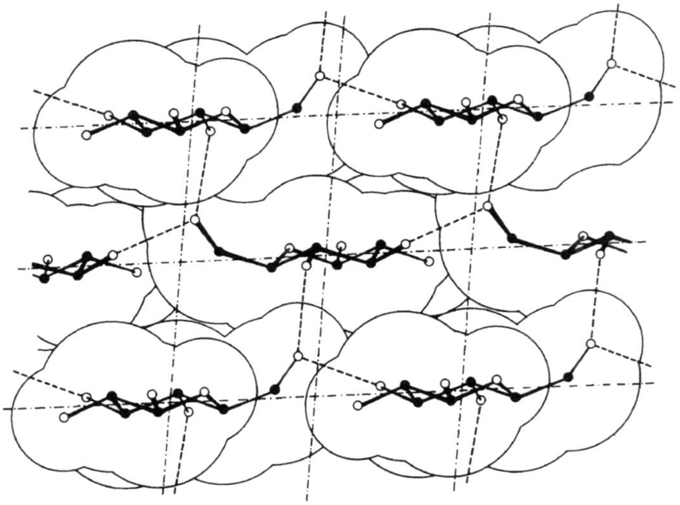

Fig. 4–20. Package of the cellulose molecules in the crystalline lattice; view onto the 040 plane.

In recent years some doubts have arisen as to whether cellulose chains are in fact counter-running. Models have been proposed in which the cellulose chains of the adjacent 002 planes are arranged in the same direction. Calculating the packing energy Sarko (1976) found a lower energy value for a parallel than an antiparallel arrangement of the chains. According to Gardner and Blackwell (1974) there is closer agreement with the diffraction intensity data if parallel cellulose chains are assumed. Viswanathan and Shenouda (1971) came to the opposite conclusion and interpreted their results from X-ray diffraction as being coincident with antiparallel cellulose chains; additionally these authors assume a helical form of the chains alternating in direction with seven cellobiose residues per turn.

From the standpoint of the sterical extension and the space requirement of the cellulose molecules an antiparallel arrangement in the unit cell seems to be possible. This arrangement allows H-bonds within the 002 planes as well as between adjacent 002 planes. The package of the lattice is very dense, so that within the crystal no hydroxyl group is accessible for water even if this group (OH2) is not linked by a hydrogen bond (Fig. 4–20).

4.3.2. The Polymorphous Lattices of Cellulose

The diffraction diagrams of various cellulosic materials (Fig. 4–19) show differences in the intensity of the peaks which are particularly striking between cotton linters and sulphite pulp on the one hand and viscose fibres and ball-milled linters on the other. Ball-milling destroys the crystalline lattice completely, but dissolving and precipitation changes the crystal lattice of cellulose. Many chemical and thermal treatments cause changes in the lattice, some of which are presented in the scheme

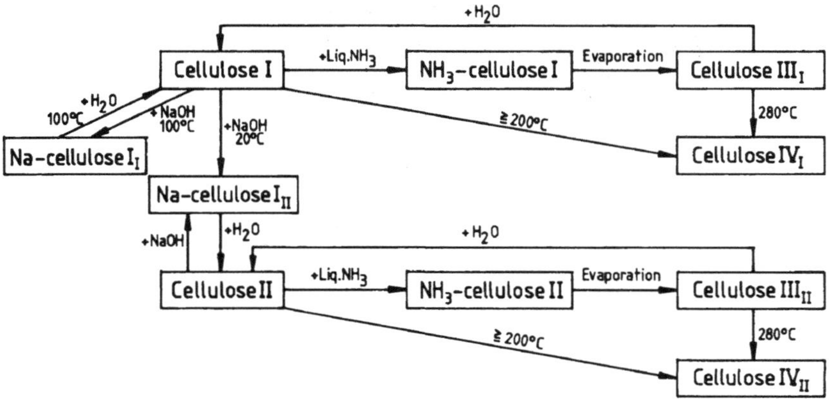

Fig. 4–21. Transformation of cellulose into its various lattice modifications.

of Fig. 4–21. As far as technical utilization is concerned the most important of these polymorphous forms are Na-cellulose I and cellulose II apart from cellulose I.

The path from cellulose I to cellulose II goes by way of Na-cellulose I. If a cellulose sample is treated with alkali solution the cellulose swells to various extents depending on the type and the concentration of alkali, but also on temperature. The progress of swelling with increasing concentration of alkali (NaOH, KOH, LiOH) is demonstrated in Fig. 4–22. The concentration is given as volume of the hydrated cations, which is a measure of their space requirement. At low concentration only the large pores in the cellulose structure are occupied, so that all three alkalis have the same swelling effect. With increasing concentration the smaller cations (K^+ = 0.232 nm, Na^+ = 0.276 nm) can more easily advance into smaller pores than Li^+ (= 0.340 nm). Na^+ seems to have a favourable diameter which is able to widen

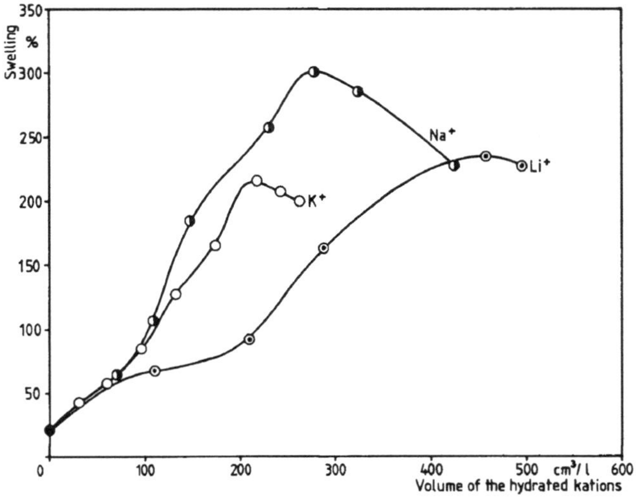

Fig. 4–22. Swelling of spruce cellulose (bleached sulphite pulp) in NaOH, KOH, and LiOH of increasing concentrations, expressed as volume of the hydrated cations (Fengel 1980).

the smallest pores down to the space between the lattice planes and advance into them. Consequently it results in the highest swelling (Fengel 1980). With increasing alkali concentration the OH-groups become more and more accessible for water (Rånby 1964).

This process means a complete de-crystallization of the cellulose and the degree of swelling determines the degree of lattice transformation. The most complete transformation of cellulose I to cellulose II is reached with NaOH, while other alkalis produce only a partial transformation or none at all (Vigo et al. 1969). Alkali treatment and particularly treatment with sodium hydroxide causes not only a widening up of the lattice but also a change in the conformation and a shift of the lattice planes (Petitpas et al. 1963; Takal, Colvin 1978). After the removal of surplus NaOH a new lattice, that of sodium cellulose I, can be determined (Fig. 4–23). In this lattice relatively large distances occur between the cellulose molecules; the interspaces contain water molecules (Bartunek 1956). It should be noticed that in this state the OH-groups of the cellulose molecules are transformed to ONa-groups and therefore the dimensions of the cellulose molecules are enlarged (Table 4–5).

According to Hayashi et al. (1976) two forms of Na-cellulose I exist which are obtained by different temperatures from cellulose I (Fig. 4–21). Both differ only slightly in the dimensions of the space unit. Na-cellulose I_I can be re-transformed to cellulose I, whereas Na-cellulose I_{II} can only be transformed to cellulose II, from which it also can derive.

During an intensive washing the linked Na-ions are removed and another lattice is formed, the lattice of cellulose II (Fig. 4–23). Apart from a change in the dimensions of the space unit (Table 4–5) the planes of the cellulose molecules are aligned approximately in the $10\bar{1}$ direction. These changes are associated with a change in the hydrogen bonds. From IR spectra it has been deduced that the OH6-groups are not shared in intramolecular linkages (Marchessault, Liang 1960). But there is no agreement in the opinions of several authors about the linkage of the cellulose chains in the cellulose II lattice. While Sarko (1978) describes intraplane and in-

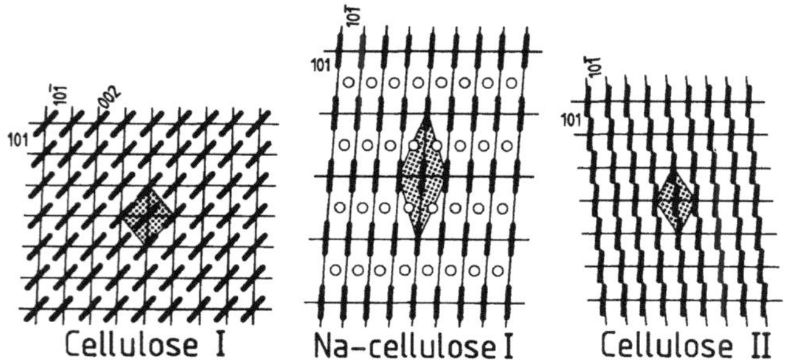

Fig. 4–23. Crystalline lattices of cellulose I, Na-cellulose I, and cellulose II, showing the change of direction of the molecular chains. The dotted areas indicate the position of the space units.

Table 4–5: Dimensions of the space units of cellulose polymorphs

Typ	Source	Dimensions a [nm]	b [nm]	c [nm]	β [Degree]	References
Cellulose I	Ramie	0.817	1.031	0.784	84.1	Kiessig 1950
	Ramie	0.817	1.034	0.785	83.6	Wellard 1954*
	Cotton, pulp	0.821	1.03	0.790	83.3	Ellefsen 1960
	Valonia	0.817	1.038	0.786	83.0	Blackwell et al. 1978
Na-cellulose I	Pulp	1.28	2.05	1.32	40.0	Bartunek 1956
Na-cellulose I_I	Ramie, cotton	2.460	2.049	0.892		Hayashi et al. 1976
Na-cellulose I_{II}	Rayon	2.479	2.045	0.896		Hayashi et al. 1976
Cellulose II	Merc. ramie	0.801	1.030	0.913	62.6	Kiessig 1950
	Merc. ramie, rayon	0.792	1.034	0.908	62.7	Wellard 1954*
	Merc. cotton, merc. pulp	0.802	1.03	0.903	62.8	Ellefsen 1960
	Rayon	0.801	1.036	0.904	62.9	Kolpak, Blackwell 1976
	Merc. cotton	0.802	1.036	0.899	63.4	Kolpak et al. 1978
Cellulose III		0.774	1.03	0.99	58	Wellard 1954*
Cellulose IV		0.811	1.03	0.79	90	Wellard 1954*
		0.812	1.03	0.799	90	Ellefsen 1960

* according to Marchessault and Sarko (1967)

terplane H-bonds, the model of Blackwell et al. (1978) has a sheet-like structure with H-bonds only in the $10\bar{1}$ plane. Both however assume an antiparallel orientation of the cellulose chains.

From the thermodynamic point of view the lattice of cellulose II is more stable than that of cellulose I, and all kinds of regenerated cellulose show the cellulose II lattice. The packing energy is lowest for an antiparallel orientation of the chains in the cellulose II lattice (Sarko 1976). Coming from dissolved cellulose a change in orientation of the chains may be conceivable. But also in the solid state the transformation from cellulose I to cellulose II is possible, e.g. during mercerization of cotton. In this case a shifting and a slight turning around the longitudinal axis but no inversion of the long molecules is imaginable (Takal, Colvin 1978). Roche (1979) observed no change in the fibrillar structure of ramie cellulose after transformation to cellulose II as well as to cellulose III and IV. Another problem originates from the volume of the cellulose molecules which do not allow placement within the cellulose II lattice as evaluated by X-ray diffraction.

Further polymorph forms are cellulose III and cellulose IV, both obtainable from cellulose I as well as from cellulose II. The treatment of cellulose I and II with liquid ammonia results in ammonia cellulose from which the ammonia can be removed by evaporation and/or heating. The two modifications remaining after this procedure differ only slightly in their space unit dimensions, but the treatment with water transforms cellulose III_I only to cellulose I, and cellulose III_{II} only to cellulose II

(Fig. 4–21). Cellulose IV is obtained from cellulose III as well as from cellulose I and II by high temperature treatment (more than 200 °C, in glycerol) (Ellefsen et al. 1964; Sarko 1978). There are also two slightly differing modifications, cellulose IV_I and IV_{II}.

A so-called cellulose x was detected after the treatment of cotton and pulp cellulose with strong acids. The space unit of cellulose x differs from the space unit of cellulose IV only by a shift of the molecular chains (Ellefsen 1960; Ellefsen et al. 1964).

4.3.3. Crystallinity and the Size of Crystallites

By looking at an X-ray diffractogram it can be seen that the peaks deriving from the various lattice planes are based not on the zero line but on a certain background which has to be attributed to a non-crystalline or amorphous portion in cellulose. The amorphous background varies in its level depending on the origin of the cellulose (Fig. 4–19).

The degree of crystallinity (cristallinity index, state of order) which stands for the crystalline portion in a cellulose sample can be calculated by subtraction of the background from the whole curve using either the height (002 peak) or the whole area (Visapää 1964; Knolle, Jayme 1965). Another method for determining the degree of crystallinity is by IR spectrometry using ratios of certain absorption bands (Nelson, O'Connor 1964; Basch et al. 1974; Ferrús, Pagés 1977). Jayme and Roffael (1970) developed a method of determination combining swelling in NaOH with the water retention value (WRV).

The degree of crystallinity varies from 80 to 70% for seed hairs (cotton) and bast fibres (ramie), and from 70 to 60% for wood pulp; regenerated cellulose (rayon) has a crystallinity of about 45% (Table 4–6). A peculiar position is occupied by Fortisan with 74% crystallinity, a fibre produced by deacetylation of cellulose acetate. In a study of the specific surface of the crystalline portions Scallan (1971)

Table 4–6: Crystallinity of various celluloses (Hindeleh, Johnson 1972, 1974; Jayme 1975)

Cellulose	Crystallinity [%]
Ramie	72
Cotton linters	71.3
Chemical grade pulp (spruce)	68.8
Bleached sulphate pulp (spruce, pine)	68.0
Bleached sulphite pulp (birch)	67.9
Bleached sulphite pulp (spruce)	67.8
Bleached sulphate pulp (birch)	65.1
Bleached sulphate pulp (bamboo)	59.9
Rayon	45
Fortisan	74

obtained much higher values for the degree of crystallinity (native celluloses 96–89%, regenerated celluloses 85–65%).

Intensity and width of the peaks in a diffractogram make a determination of the dimensions of the ordered regions (crystallites) possible. Overlapping peaks can be corrected and separated by mathematical methods (Gjønnes et al. 1958; Hindeleh, Johnson 1972). According to determinations of that kind the crystallites are about 50 nm (ramie), and 10–20 nm (viscose fibres) in length (Kiessig 1958; Haase et al. 1973; Schurz, John 1975). For *Valonia* cellulose longitudinal periods of 100 nm were obtained (Bourret et al. 1972).

A dependency of the length of crystallites on the angle of the fibrils and the distance from the core in wood was deduced by El-Osta et al. (1974). According to these findings the average crystallite length becomes smaller with the increase of the angle and the decrease of the radial distance.

The crystalline regions resist the attack of dilute acids, so that even a long-time treatment with acid can reduce the DP only to a certain value, called the leveling-off degree of polymerization (LDP or DP_L) (Battista 1950). The DP_L, too, depends on the origin and on the pretreatment of the cellulose. From Fig. 4–24 the DP_L can be seen to be about 300 for cotton linters, 140 for spruce alpha-cellulose, and 50 for viscose fibres. The values found in the literature range between 200–350 for cotton and ramie, 150–300 for wood cellulose and 15–50 for viscose fibres (Battista 1965; Scallan 1971). A treatment with NaOH (mercerization) reduces the DP_L of linters and pulp to about 60 (Ruck 1962). The DP values at which the fibre strength

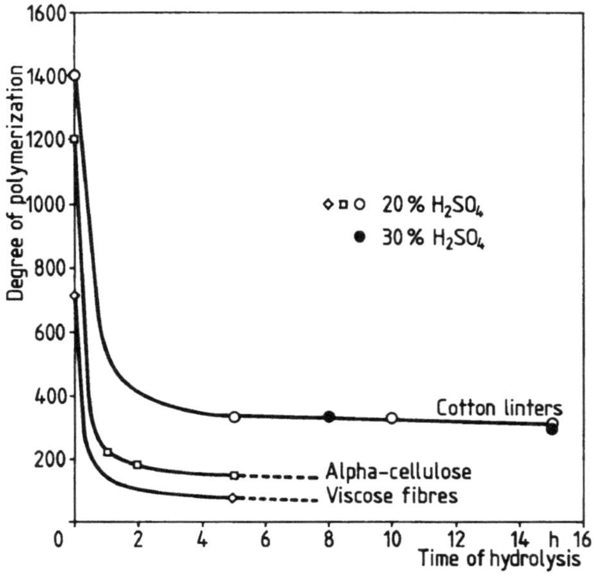

Fig. 4–24. Influence of the time of hydrolysis with sulfuric acid on the degree of polymerization in three celluloses.

approaches zero is in the same range as the leveling-off degree of polymerization, i.e. 200 for cotton, 55 for regenerated fibres (Krässig 1976).

In the electron microscope cellulose partially hydrolized with dilute acids shows short, spindle-like particles which tend to form larger aggregates (Fig. 4–25) (Rånby 1951; Krässig, Käppner 1961; Jayme, Knolle 1965b). The measurement of the length of particles from cotton and regenerated cellulose (Fortisan) resulted in broad distribution ranges with a maximum at about 40 nm for cotton and about 25 nm for Fortisan (Krässig 1976).

An intensive mechanical treatment of cellulose reduces its crystallinity, and samples which were ball-milled for several hours are almost completely disordered (Ant-Wuorinen, Visapää 1965; Stewart, Forster 1976). When dry ball-milled cellulose is subsequently exposed to moisture its crystallinity increases. If the decrystallization was incomplete the crystalline structure of cellulose I is restored. Completely decrystallized native cellulose recrystallizes in the lattice of cellulose II when exposed to moisture (Caulfield, Steffes 1969). The increase of crystallinity during wetting of cellulose is explained by a progress of water from less ordered regions to the ordered regions, causing a reduction of the dimensions of the crystalline portions but an increase of the state of order (Ray, Bandyopadhyay 1975).

Acidic hydrolysis also causes an increase of crystallinity. Jayme and Roffael (1969) treated various celluloses with 3 n H_2SO_4 and obtained an increase in the degree of crystallinity e.g. from 68 to 93% for cotton linters after 11 hours of treatment while during this time the recovery was reduced only to 90%.

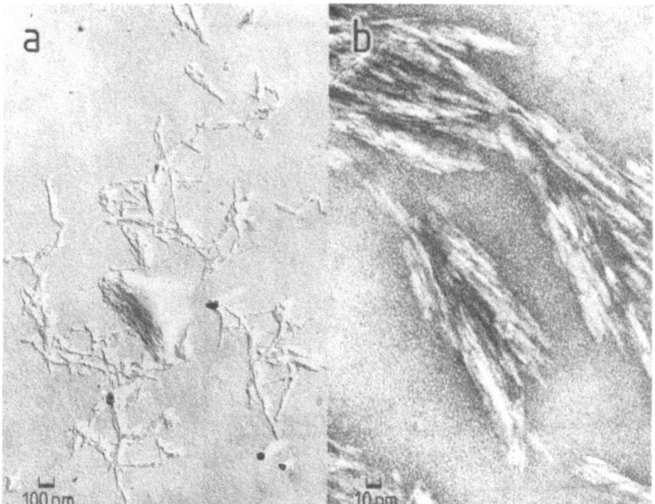

Fig. 4–25. Electron micrographs of partially hydrolized celluloses.
a) Alpha-cellulose from spruce wood, boiled with 20% H_2SO_4 for 5 hr.
b) Alpha-cellulose from beech wood, boiled with 23% trifluoroacetic acid (TFA) for 2 hr, and subsequently with 57% TFA for 5 hr.

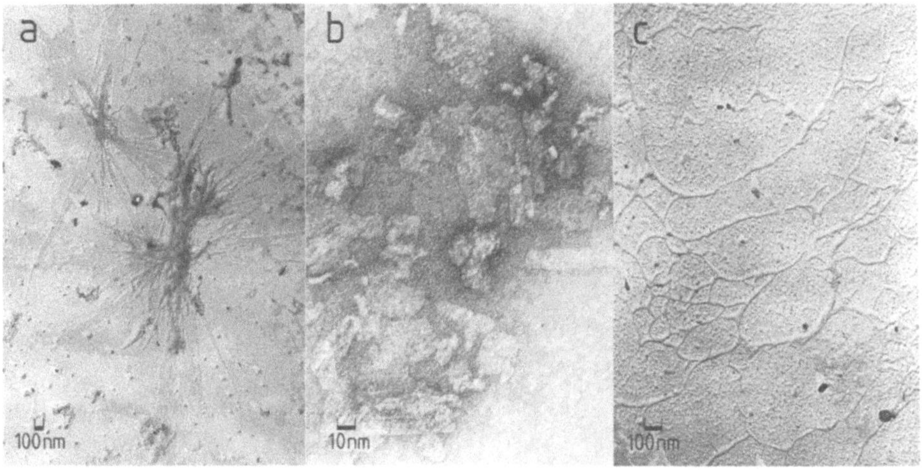

Fig. 4–26. Electron micrographs of regenerated cellulose structures.
a) Whiskers from cellulose, boiled with 20% H_2SO_4 for 5 hr and subsequently dispersed in 80% ethanol.
b) Crystalline platelets from cellulose, dissolved in 100% trifluoroacetic acid (TFA), precipitated, and subsequently boiled with 23% TFA for 6 hr.
c) Fibrils from cellulose, dissolved in EWNN, and precipitated from a highly diluted solution.

The recrystallization of cellulose can be made visible in the electron microscope. After being degraded by an intensive acid treatment cellulose is able to rebuild crystalline structures in the form of whiskers or platelets (Fengel 1967b). Furthermore, highly diluted solutions of cellulose yield crystalline structures of submicroscopic dimensions after precipitation (Fengel 1967b; Macchi et al. 1968; Maeda et al. 1970) (Fig. 4–26). Starting from cellulose acetate the precipitate has the lattice of cellulose II or even cellulose IV, the supramolecular structures of which depend on the solvent used and on the temperature of precipitation (Bittiger, Husemann 1972; Ramesh et al. 1977). Fibrils with the lattice of cellulose I could be precipitated from cuoxam solution (Macchi, Palma 1969).

Buleon and Chanzy (1978) succeeded in producing cellulose II monocrystals from heavily degraded cellulose acetate (DP 15). The growing plane of these crystals was the 101 plane. Previously Chanzy and Roche (1976) had shown a lateral growth of cellulose II crystals onto fibrils of *Valonia* cellulose forming a shish-kebab structure.

4.3.4. The Fibrillar Structure

As described above (→ 2.2.) the structured skeleton of the wood cell wall is made up of cellulose fibrils. These thread-like units are present in all cell walls containing cellulose, i.e. in bacteria and algae as well as in bast fibres and seed hairs. The animal tunicin is also organized in fibrils (Wardrop 1970; Manley 1971). The fibrils represent the association of cellulose molecules and contain ordered and less or-

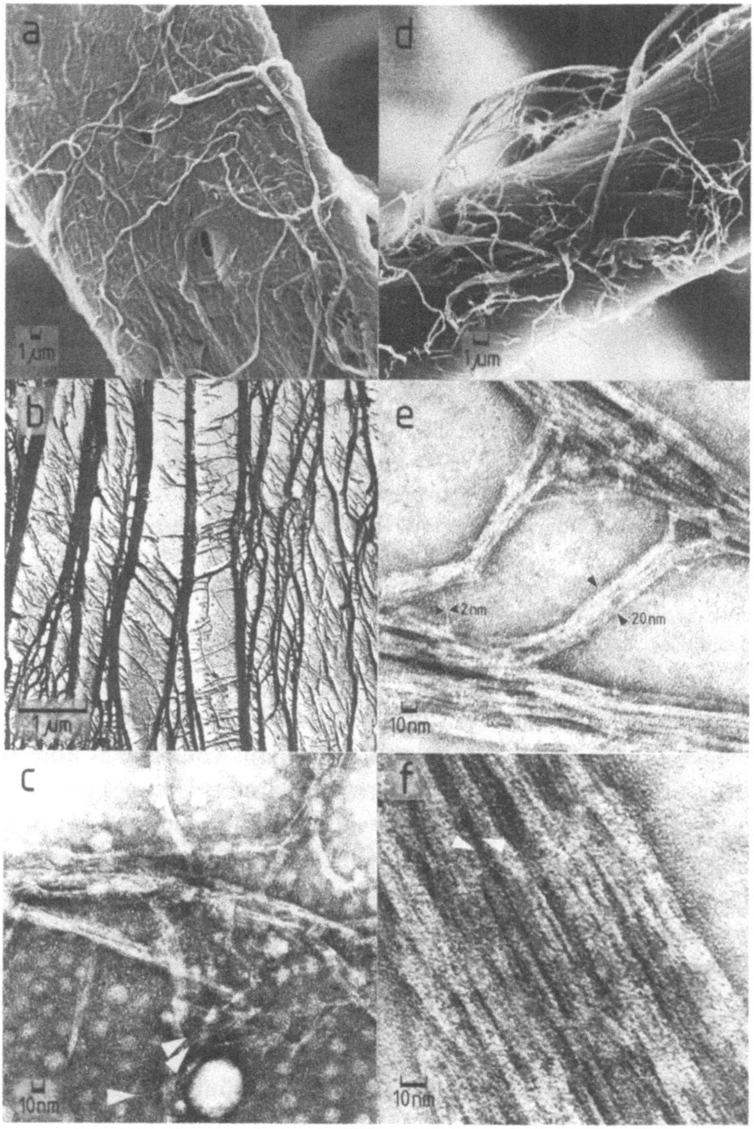

Fig. 4–27. Various electron microscopical shapes of cellulose fibrils from wood (a–c) and cotton linters (d–f).

a) Fibrillar bundles on the surface of a tracheid in spruce sulphite pulp.

b) Highly swollen secondary wall layer of a tracheid in spruce sulphite pulp; the thinnest visible fibrils are 20–30 nm in diameter.

c) Mechanically disintegrated alpha-cellulose from beech wood; the thinnest visible fibrils are 1–1.5 nm in diameter (arrows).

d) Fibrillar bundles and lamellae on the surface of a cotton hair after disintegration.

e) Mechanically disintegrated cotton cellulose showing fibrils of various diameters.

f) Fibrillar bundle, consisting of fibrils of various diameters; single molecules are obviously running diagonally from one fibril to another one (arrows).

dered regions. Due to their small diameter a detailed study of their external structure was not possible until electron microscopy was introduced. The progress in resolution power of the microscopes and in preparation methods resulted in the detection of smaller and smaller fibrillar units (Fig. 4–27).

While in early studies the smallest units, called microfibrils, were assumed to have a diameter ranging from 10 to 25 nm (Vogel 1953; Frey-Wyssling 1954; Rånby 1954), later even smaller units, called elementary fibrils, were detected (Mühlethaler 1965; Heyn 1966, 1969). For the elementary fibrils an average diameter of 3.5 nm was assumed, though some measurements of fibrils from bacteria, algae and coleoptiles of maize ranged from 1.5 to 3.0 nm (Ohad et al. 1962; Ohad, Danon 1964; Schnepf 1965).

There are some indications that the fibrils are not very uniform in diameter, probably depending on the origin and certainly on the treatment of the sample. Jayme and Koburg (1959) found different fibrillar diameters (10–30 nm) in pulps of various woods, prepared by different methods. Fibrils of 1.8 to 3.8 nm in thickness were determined in pulps and holocelluloses of several woods by Sullivan (1968), and he suggested that there is no constancy of fibrillar diameters within a wood species and the mode of treatment. Relatively uniform fibrils with the same width as in wood (3.5 nm) were measured in softwood sulphate pulp by Heyn (1977). By treating cellulose of the same origin with various chemicals it could be shown that fibrillar units can be split more or less completely into sub-units (Fengel 1962, 1967b; Erinš, Odincov 1965).

Intensive disintegration in an homogenizer can split the fibrils down to molecular diameters (Fengel 1974) (Fig. 4–27c). The comparison of fibrillar diameters in holocellulose and alpha-cellulose from spruce wood gives evidence that the accompanying compounds (polyoses, residual lignin) may conserve the fibrillar dimensions of the cellulose. There is a relatively narrow range of diameters with a maximum at 2.5 nm in holocellulose and a broad distribution of diameters (1.2–4.8 nm) in alpha-cellulose (Fengel 1978) (Fig. 4–28). In the cambial cell wall of wood Hanna and Côté (1974) detected fibrillar units of 1.0 to 1.5 nm in diameter, which they called sub-elementary fibrils.

Using dark field electron microscopy Goto et al. (1975) found a rectangular cross-sectional shape of fibrils with an edge length of 2–6 nm in the gelatinous layer of poplar tension wood.

The width of the crystallites as obtained from diffractograms approximately corresponds to the values obtained from electron micrographs. In several publications the diameters of crystallites in natural and regenerated celluloses are given as 4–6 nm (Hindeleh, Johnson 1972, 1974; Haase et al. 1973; Jayme et al. 1973; Schurz, John 1975). Diameters in the range of 2.7–3.3 nm (wood, cotton, ramie cellulose) and 1.6–1.8 nm (regenerated celluloses) have been calculated by Scallan (1971).

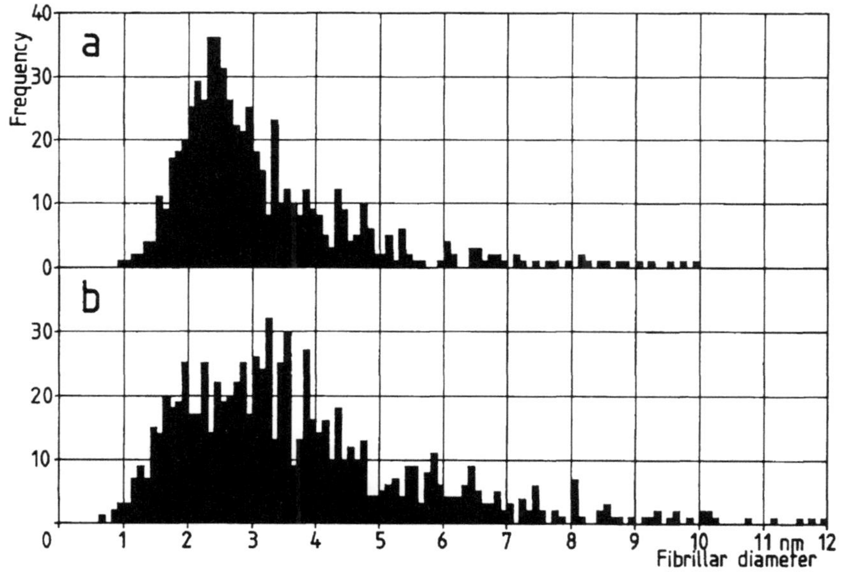

Fig. 4–28. Distribution of fibrillar diameters in holocellulose (a) and alpha-cellulose (b) from spruce wood.

From the different data on the fibrillar width in *Valonia* it can be concluded that a dense package of the smallest units exists in this cellulose. In some studies diameters of 10–14 nm and cross-sectional areas of 15 to 20 × 7 to 10 nm are described (Caulfield 1971; Goto et al. 1973; Lazaro, Chiaverina 1973); in others smaller units between 6.3 × 6.8 nm and 2 × 3.5 nm have been observed (Manley 1971; Blackwell, Kolpak 1975; Macchi 1976). Bourret et al. (1972) found a splitting of fibrils into units with diameters of 3–4 nm at some places.

As we progress to molecular sizes the observations in the electron microscope become more and more difficult. The borders of the fibrils become blurred with the increase of magnification because of the noise of the background. A bright field observation needs a negative staining, i.e. a deposit of staining compounds between the fibrillar units or at their borders. Therefore the resolution of single fibrils is influenced by the space between the fibrils and the grain size of the staining compound. The interfibrillar spaces within the cell wall range between 1.2–5 nm in the wet state and about 1 nm in the dry state (Stone, Scallan 1965; Nelson, Oliver 1971). The grains of the finest staining compounds (uranyl acetate, thorium nitrate) have diameters of 0.6–1 nm (Fengel 1978). Grains of this size are able to enter the interfibrillar voids, but they limit the resolution of the fibrils to about 1 nm.

The results of electron microscopic observations can be summarized as follows: The cellulose within the cell wall is organized in fibrils, the basic units of them having diameters in the range of 2–4 nm. It is probable that these basic units (elementary fibrils, protofibrils) are associated in higher systems with diameters of 10–30 nm (microfibrils). The fibrillar units can be split longitudinally into sub-units and single molecular chains by chemical and mechanical treatment. This fact is important for the critical examination of the internal organization of the fibrils.

4.3.5. The Internal Structure of Fibrils

The results of diffraction studies, degradation experiments, electron microscopic observations and several other investigations on cellulose entailed various concepts of the arrangement of the molecules within the fibrillar units. Common to all models described in the literature are ordered regions formed by longitudinally arranged chains assuming a parallel as well as an antiparallel orientation. Thus the models differ mainly in the presentation of the less ordered regions. All models can be reduced to three basic principles (Fig. 4–29):

- the longitudinally arranged molecules change from one ordered region to the subsequent one, the transition areas being the less ordered regions (fringed micelle system);
- the fibrillar units are individual cords consisting of longitudinally arranged molecules and sequences of ordered and disordered regions;
- the ordered regions are packages of chains folded in a longitudinal direction, the areas containing the turns between adjacent chain packages being the less ordered regions.

The fringed micelle system is the oldest one to date. It is based on the first model of Gerngross et al. (1932) originally proposed for gelatine and caoutchouc. When cellulose research was still in its early stages the real length of the molecular chains was underestimated, so that each micelle with its fringes on both ends was seen to

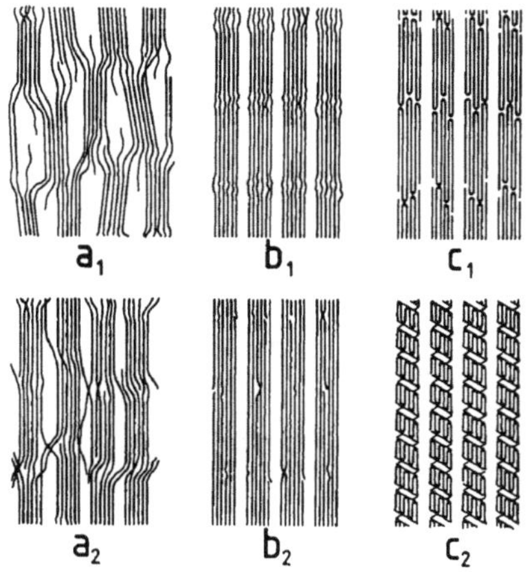

Fig. 4–29. Three basic models for the arrangement of cellulose molecules within the fibrillar units (a_1–c_1), and respective variations (a_2–c_2).

a) Ordered regions with transitional molecular chains changing from one ordered region to another.

b) Individual fibrillar units with a sequence of ordered and less ordered regions.

c) Fibrillar units consisting of folded cellulose chains.

be an individual unit. Chains running through several micelles were later described by Kratky and Mark (1937) (Fig. 4–29a$_1$). Subsequently Hess and Kiessig (1943) proposed a similar structure, the micelles and transition regions being in a dense package, for synthetic fibres such as polyamide, polyester and polyethylene. A new proposal came from H. and H. Dolmetsch (1968) consisting of individual fibrillar cords which are connected by transitional molecules (Fig. 4–29a$_2$). Scallan's (1971) calculation of the crystallinity and crystallite dimensions is based on a similar model. Fringed models of larger dimensions, i.e. where the fringes are not single molecules but micellar cords, derive from Krässig and Kitchen (1961) and Hearle (1963).

In a structural concept of fringed micelles the cellulose chains can run in a parallel or antiparallel direction. Considering electron microscopic observations the model can be accepted only as defined by H. and H. Dolmetsch (1968) and Scallan (1971). It has the advantage of a simple explanation of the variation in fibrillar diameters.

A system of individual fibrillar units consisting of long periods of ordered regions interrupted by completely disordered regions was developed by Hess et al. (1957) (Fig. 4–29b$_1$). By using this model differences in DP_L and length of the crystallites can easily be explained. According to Stöckmann (1971) the cellulose fibrils can be deformed in an entropy-elastic manner, a behaviour for which a sequence of well ordered and disordered regions is necessary. Mühlethaler (1969) varied this model by reducing the disordered regions to lattice defects and ends of molecular chains (Fig. 4–29b$_2$). Haase et al. (1973, 1975), too, suggest from X-ray studies that the less ordered regions consist of paracrystalline lattice defects with crystalline bridges running through. Fibrils of this kind would be of a uniform diameter and highly crystalline. The less ordered regions could probably be formed by a layer of disor-

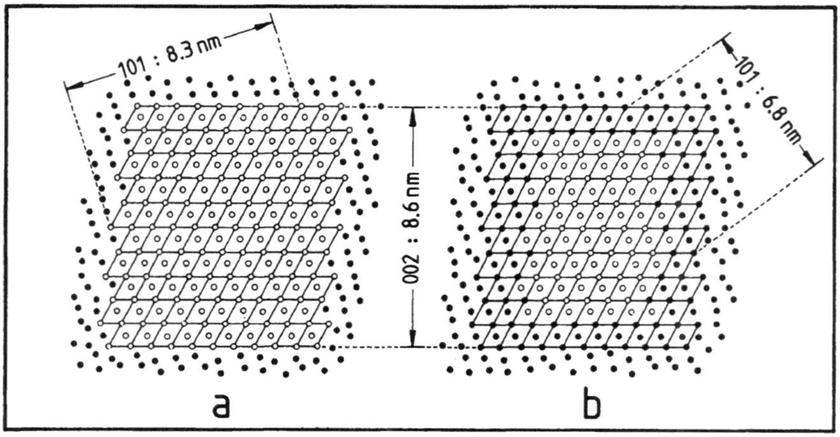

Fig. 4–30. Cross-sectional model proposed for regenerated cellulose fibres, provided with a sheath of less ordered cellulose molecules (a), and an outer layer of the ordered core accessible to water or deuterium oxide respectively (Krässig 1976).

dered cellulose molecules coating the crystalline cords in a manner similar to the earlier model of Frey-Wyssling (1954). A corresponding presentation for regenerated cellulose fibres is given by Krässig (1978; Siesler et al. 1975). In this model a distinction is made between a disordered layer and an outer layer of the ordered cord which is still accessible to water (Fig. 4–30). Leppard and Colvin (1978) purportedly made visible a less ordered layer during the development of fibrils. But these electron microscopical observations did not remain unchallenged (Willison et al. 1980). Ohad and Mejzler (1965) discussed a fibrillar model which was completely ordered with randomly distributed chain ends. The variation of the fibrillar diameters and changes in the crystalline amount is explained by Beall and Murphey (1970) by a possibly hexagonal shape of the cellulose fibrils.

The detection of a folding process during the crystallization of synthetic polymers (Fischer 1959; Keller 1959) was a stimulus to the development of various models with folded cellulose chains. Such an arrangement automatically resulted in a reversed run of adjacent chains and simplified the image of the biosynthesis of antiparallel cellulose molecules (Marx-Figini, Schulz 1966a). To obtain a general overview of all these models can be subdivided according to three principles:

- the chains are folded in one lattice plane only (101) so that the whole crystallite consists of layers of folded chain molecules (Tønnessen, Ellefsen 1960; Marx-Figini, Schulz 1966a, b; Chang 1971; Watanabe et al. 1974) (Fig. 4–29c_1);
- the chains are folded throughout the whole crystallite so that the crystalline planes are connected not only by hydrogen bonds but also by the turns of chains (Dolmetsch 1961, 1962; Battista 1965; Nissan 1971);
- the chains are folded in the form of a ribbon, and the ribbon is wound helically forming the fibril (Manley 1964, 1971) (Fig. 4–29c_2).

Though several phenomena, e.g. during degradation of cellulose, are explained by chain folding (Butnaru, Simionescu 1973; Chang 1974; Jovanovic et al. 1974), these models are strongly disputed. The determination of the DP of cellulose cut transversally to sections of 2 μm in thickness gave no indication for a chain folding (Muggli 1968). Rees and Skerrett (1968) consider a chain folding, particularly of the type proposed by Manley (1964), to be unlikely, as it would involve such an enormous expense of energy that it is difficult to imagine any source of sufficient intermolecular compensation. From mechanical tests of single wood fibres Mark (1967, 1971) concluded that physical and mechanical properties contradict a folding of cellulose molecules, and thermodynamic reasons are advanced by Stöckmann (1972) against the folding model.

The formation of the supramolecular structure of cellulose occurs during biosynthesis within the cell wall. The most plausible mechanism for this process is a simultaneous polymerization and crystallization (Stöckmann 1972). But supramolecular structures are also built up from dissolved and/or degraded cellulose independently of genetic influences. This phenomenon indicates a mechanism which

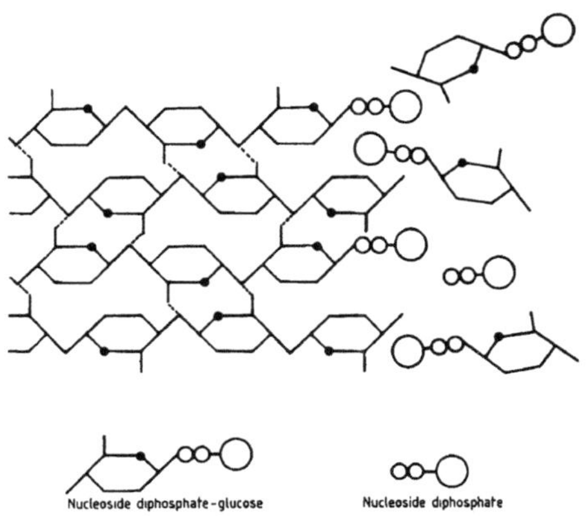

Fig. 4–31. Possible mechanism of biosynthetic growth of cellulose chains in antiparallel arrangement.

compels the molecules into an order system. The mechanism may be caused by the sterical conditions of the molecules which restrict the possibility of the formation of intermolecular hydrogen bonds. The cellulose molecules fit into one another only in a definite position so that an internal matrix mechanism for the formation of the supramolecular structure of cellulose is probable.

This mechanism should also be effective during biosynthesis of cellulose, and if the mechanism prescribes an antiparallel arrangement of the molecular chains this arrangement seems to be possible without folding. The biosynthetic growth of a polysaccharide chain occurs by linking a nucleoside diphosphate-glucose to the C4-end of the chain, and by cleaving nucleoside diphosphate (Lehninger 1979; Franz 1978). But the glucan chain also has a C1-end (with a nucleoside diphosphate-group) and at this end the linkage of a nucleoside diphosphate-glucose with its C4-end is imaginable. In this way a glucan chain can also grow in the reverse direction (Fig. 4–31).

Literature

Achwal, W. B. and Chaugule, D. A. 1975, Angew. Makromol. Chemie *42,* 77–89
Ant-Wuorinen, O. and Visapää, A. 1965, Pap. Puu *47,* 311–322
Bartunek, R. 1956, Kolloid-Z. *146,* 35–43
Basch, A., Wasserman, T. and Lewin, M. 1974, J. Polym. Sci., Polym. Chem. Ed. *12,* 1143–1150
Battista, O. A. 1950, Ind. Eng. Chem. *42,* 502–507
Battista, O. A. 1965, J. Polym. Sci., Pt. C, No. 9, 135–155
Beall, F. C. and Murphey, W. K. 1970, Wood Fiber *2,* 282–284
Bergmann, W., Hasvold, K. K. and Tröften, J. 1964, Papier *18,* 150–158
Bittiger, H. and Husemann, E. 1972, J. Polym. Sci., Pt. B, *10,* 549–553
Blackwell, J. and Kolpak, F. J. 1975, Macromolecules *8,* 322–326
Blackwell, J., Kolpak, F. J. and Gardner, K. 1978, Tappi *61,* No. 1, 71–72
Bourret, A., Chanzy, H. and Lazaro, R. 1972, Biopolymers *11,* 893–898
Brown, W. J. 1966, Tappi *49,* 367–373
Brown, W. 1967, Svensk Papperstid. *70,* 458–461

Buleon, A. and Chanzy, H. 1978, J. Polym. Sci., Polym. Phys. Ed. *16,* 833–839
Burchard, W. and Husemann, E. 1961, Makromol. Chemie *44/46,* 358–387
Butnaru, R. and Simionescu, C. 1973, Cell. Chem. Technol. *7,* 641–651
Caulfield, D. F. 1971, Textile Res. J. *41,* 267–269
Caulfield, D. F. and Steffes, R. A. 1969, Tappi *52,* 1361–1366
Chang, M. 1971, J. Polym. Sci. Pt. C, No. 36, 343–362
Chang, M. M. Y. 1974, J. Polym. Sci., Polym. Chem. Ed., *12,* 1349–1374
Chanzy, H. D. and Roche, E. J. 1976, Appl. Polym. Symp. No. 28, 701–711
Chitumbo, K. and Brown, W. 1974, J. Polym. Sci., Symposia No. 47, 261–268
Christensen, G. N. and Kelsey, K. E. 1959, Holz Roh-Werkst. *17,* 189–203
Cotton, F. A. and Wilkinson, G. 1974, Anorganische Chemie. Verlag Chemie, Weinheim, Intersci. Publ., New York, London, p. 153
Danhelka, J. and Kössler, I. 1976, J. Polym. Sci., Polym. Chem. Ed., *14,* 287–298
Davison, R. W. 1972, Tappi *55,* 567–573
Dobbins, R. J. 1970, Tappi *53,* 2284–2290
Dolmetsch, H. 1961, Kolloid-Z. *176,* 63–64
Dolmetsch, H. and Dolmetsch, H. 1962, Kolloid-Z. *185,* 106–119
Dolmetsch, H. and Dolmetsch, H. 1967, Papier *21,* 53–62
Dolmetsch, H. and Dolmetsch, H. 1968, Papier *22,* 1–11
Ellefsen, Ø., 1960, Norelco Rep. *8,* 104–107
Ellefsen, Ø, Kringstad, K. and Tønnesen, B. A. 1964, Norsk Skogind. *18,* 419–430
Ellis, K C. and Warwicker, J. O. 1962, J. Polym. Sci. *56,* 339–357
Elmgren, H. 1965, Arkiv Kemi *24,* 237–282
El-Osta, M. L. M., Kellogg, R. M., Foschi, R. O. and Butters, R. G. 1974, Wood Fiber *6,* 36–45
Erinš, P. P. and Odincov, P. N. 1965, Sow. Beitr. Faserforsch. Textiltechn. *2,* 120
Fengel, D. 1962, Elektronenoptische Untersuchungen zur Kenntnis des Feinbaus von Nadelholztracheiden unter Anwendung der Ultradünnschnitt-Technik. Doctor Thesis, T. H. Darmstadt
Fengel, D. 1967a, Holz Roh-Werkst. *25,* 102–111
Fengel, D. 1967b, Papier *21,* 635–645
Fengel, D. 1970, Tappi *53,* 497–503
Fengel, D. 1974, Naturwissenschaften *61,* 31–32
Fengel, D. 1978, Holzforschung *32,* 37–44
Fengel, D. 1980, Papier *34,* 428–433
Fengel, D., Wegener, G., Heizmann, A. and Przyklenk, M. 1978, Cell. Chem. Technol. *12,* 31–37
Ferrús, L. and Pagés, P. 1977, Cell. Chem. Technol. *11,* 633–637
Fischer, E. W. 1959, Z. Naturforsch. *14a,* 584–587
Fisher, D. G. and Mann, J. 1960, J. Polym. Sci. *42,* 189–194
Franz, G. 1978, Papier *32,* 521–525
Freudenberg, K. 1967, Chem. Ber. *100,* 172–188
Frey-Wyssling, A. 1954, Science *119,* 80–82
Gardner, K. H. and Blackwell, J. 1974, Biopolymers *13,* 1975–2001
Garves, K. 1972, Tappi *55,* 263
Geddes, A. L. 1956, J. Polym. Sci. *22,* 31–39
Gerngroß, O., Herrmann, K. and Lindemann, R. 1932, Kolloid-Z. *60,* 276–288
Gjønnes, J., Norman, N. and Viervoll, H. 1958, Acta Chem. Scand. *12,* 489–494
Goebel, K. D., Harvie, C. E. and Brant, D. A. 1976, Appl. Polym. Symp. No. 28, 671–691
Goring, D. A. I. and Timell, T. E. 1962, Tappi *45,* 454–460
Goto, T., Harada, H. and Saiki, H. 1973, Mokuzai Gakkaishi *19,* 463–468
Goto, T., Harada, H. and Saiki, H. 1975, Mokuzai Gakkaishi *21,* 537–542
Gruber, E. 1976, Papier 30, 533–537
Gruber, E. and Gruber, R. 1978, Cell. Chem. Technol. *12,* 345–353

Guthrie, J. T. and Harcastle, S. V. 1977, Polymer *18,* 203–204
Haase, J., Hosemann, R. and Renwanz, B. 1973, Kolloid-Z., Z. Polymere *251,* 871–875
Haase, J., Hosemann, R. and Renwanz, B. 1975, Cell. Chem. Technol. *9,* 513–527
Hanna, R. B. and Côté, W. A. 1974, Cytobiol. *10,* 102–116
Hayashi, J., Yamada, T. and Kimura, K. 1976, Appl. Polym. Symp. No. 28, 713–727
Hearle, J. W. S. 1963, J. Appl. Polym. Sci. *7,* 1175–1192
Hess, K. and Kiessig, H. 1943, Z. Phys. Chemie *193,* 196–217
Hess, K., Mahl, H. and Gütter, E. 1957, Koloid-Z. *155,* 1–19
Heyn, A. N. J. 1966, J. Cell Biol. *29,* 181–197
Heyn, A. N. J. 1969, J. Ultrastruct. Res. *26,* 52–68
Heyn, A. N. J. 1977, Tappi *60,* No. 11, 159–161
Higgins, H. G., Stewart, C. M. and Harrington, K. J. 1961, J. Polym. Sci. *51,* 59–84
Hindeleh, A. M. and Johnson, D. J. 1972, Polymer *13,* 423–430
Hindeleh, A. M. and Johnson, D. J. 1974, Polymer *15,* 697–705
Holt, C., Mackie, W. and Sellen, D. B. 1976, Polymer *17,* 1027–1034
Honja, G. and Watanabe, M. 1958, Nature *181,* 326–328
Hummel, D. O. 1965, Melliand Textilchem. *1,* 108–121, 139–145
Husemann, E. and Werner, R. 1963, Makromol. Chemie *59,* 43–60
Jayme, G. 1961a, Tappi *44,* 299–304
Jayme, G. 1961b, Papier *15,* 581–600
Jayme, G. 1971a, New Solvents. In: Cellulose and Cellulose Derivatives (Bikales, N. M. and Segal, L., Ed.). Wiley- Intersci., New York, London, Sydney, Toronto, pp. 381–410
Jayme, G. 1971b, Mitt. Fachaussch. Zellcheming *19,* No. 1/2, 17–19
Jayme, G. 1975, Cell. Chem. Technol. *9,* 477–492
Jayme, G. 1978, Papier *32,* 145–149
Jayme, G. and El-Kodsi, G. 1968, Papier *22,* 120–124
Jayme, G. and El-Kodsi, G. 1969, Papier *23,* 487–495
Jayme, G. and Hasvold, K. K. 1968, Papier *22,* 875–878
Jayme, G. and Hunger, G. 1961, Electron Microscope 2- and 3-Dimensional Classification of Fibre Bonding. In: Formation and Structure of Paper, Transactions Oxford Symposium. Techn. Sect. Brit. Paper and Board Makers' Association, London, pp. 135–170
Jayme, G. and Knolle, H. 1965a, Holz, Roh-Werkst. *23,* 438–440
Jayme, G. and Knolle, H. 1965b, Makromol. Chemie *82,* 190–204
Jayme, G. and Koburg, E. 1959, Holzforschung *13,* 37–43
Jayme, G. and Roffael, E. 1969, Papier *23,* 1–7
Jayme, G. and Roffael, E. 1970, Papier *24,* 614–618
Jayme, G., Roffael, E. and Hasvold, K. K. 1969, Papier *23,* 597–600
Jayme, G., Roffael, E. and Islam, A. 1973, Papier *27,* 589–591
Jeffries, R. 1960, J. Textile Inst., Transactions *51,* T339–T374
Jeffries, R. 1963, Polymer *4,* 375–389
Jeffries, R. 1964, J. Appl. Polym. Sci. *8,* 1213–1220
Johnson, D. C., Nicholson, M. D. and Haigh, F. C. 1976, Appl. Polym. Symp. No. 28, 931–943
Jovanovic, S., Palma, A. and Schulz, G. V., 1974, Makromol. Chemie *175,* 3503–3522
Keller, A. 1959, Makromol. Chemie *34,* 1–28
Khripunov, A. K., Plisko, E. A., Laius, L. A., Baklagina, Y. G., Petrova, V. V. and Gerasimova, V. A. 1975, Vysomol. Soed. *17b,* No. 8, 600–603
Kiessig, H. 1950, Z. Elektrochemie *54,* 320–325
Kiessig, H. 1958, Papier *12,* 117–122
Knolle, H. and Jayme, G. 1965, Papier *19,* 106–110
Kolpak, F. J. and Blackwell J. 1976, Makromolecules *9,* 273–278
Kolpak, F. J., Blackwell, J. and Litt, M. H. 1977, J. Polym. Sci. *15,* 655–658

Kolpak, F. J., Weih, M. and Blackwell, J. 1978, Polymer *19,* 123–131
Krässig, H. 1976, Appl. Polym. Symp. No. 28, 777–790
Krässig, H. 1978, Tappi *61,* No. 3, 93–96
Krässig, H. and Käppner, W. 1961, Makromol. Chemie *44/46,* 1–7
Krässig, H. and Kitchen, W. 1961, J. Polym. Sci. *51,* 123–127
Kratky, O. and Mark, H. 1937, Z. Phys. Chemie *B36,* 129–139
Kusama, Y., Kageyama, E., Shimida, M. and Nakamura, Y. 1976, J. Appl. Polym. Sci. *20,* 1679–1688
Lazaro, R. and Chiaverina, J. 1973, Cell. Chem. Technol. *7,* 269–282
Lehninger, A. L. 1979, Biochemie. Verlag Chemie, Weinheim, p. 451
Leppard, G. G. and Colvin, J. R. 1978, J. Microscopy *113,* Pt. 2, 181–184
Liang, C. Y. and Marchessault, R. H. 1959, J. Polym. Sci. *37,* 385–395, *39,* 269–278
Litt, M. H. and Kumar, N. G. 1977, US-Pat. 4 028 132
Luck, W. 1965, Naturwissenschaften *52,* 25–31, 49–52
Luck, W. A. P. 1967, Naturwissenschaften *54,* 601–607
Macchi, E. M. 1976, Appl. Polym. Symp. No. 28, 763–776
Macchi, E. and Palma, A. 1969, Makromol. Chemie *123,* 286–288
Macchi, E., Marx-Figini, M. and Fischer, E. W. 1968, Makromol. Chemie *120,* 235–237
Maeda, H., Kawada, H. and Kawai, T. 1970, J. Polym. Sci., Pt. C., No. 30, 543–549
Manley, R. S. J. 1964, Nature *204,* 1155–1157
Manley, R. S. J. 1971, J. Polym. Sci., Pt. A-2, *9,* 1025–1059
Mann, J. and Marrinan, H. J. 1958, J. Polym. Sci. *32,* 357–370
Marchessault, R. H. and Liang, C. Y. 1960, J. Polym. Sci. *43,* 71–84
Marchessault, R. H. and Sarko, A. 1967, Adv. Carbohydr. Chem. *22,* 421–482
Mark, R. E. 1967, Cell Wall Mechanics of Tracheids. Yale Univ. Press, New Haven, London, pp. 265–272
Mark, R. E. 1971, J. Polym. Sci., Pt. C, No. 36, 393–406
Marx-Figini, M. 1964, Papier *18,* 546–549
Marx-Figini, M. and Schulz, G. V. 1966a, Biochim. Biophys. Acta *112,* 81–101
Marx-Figini, M. and Schulz, G. V. 1966b, Naturwissenschaften *53,* 466–474
Meyer, K. H. and Mark, H. 1953, Makromolekulare Chemie. Akadem. Verlagsges., Leipzig, pp. 365–377
Meyer, K. H. and Misch, L. 1937, Helv. Chim. Acta *20,* 232–244
Michell, A. J. and Higgins, H. G. 1965, Tetrahedron *21,* 1109–1120
Mohlin, U. B. 1974, Svensk Papperstid. *77,* 131–137
Morrison, J. L. and Dzieciuch, M. A. 1969, Can. J. Chem. *37,* 1379–1390
Muggli, R. 1968, Cell. Chem. Technol. *2,* 549–567
Mühlethaler, K. 1965, The Fine Structure of the Cellulose Microfibril. In: Cellular Ultrastructure of Woody Plants (Côté, W. A., Ed.). Syracuse University Press, Syracuse, N. Y., pp. 191–198
Mühlethaler, K. 1969, J. Polym. Sci., Pt. C, No. 28, 305–316
Nelson, M. L. and O'Connor, R. T. 1964, J. Appl. Polym. Sci. *8,* 1311–1324
Nelson, R. N. and Oliver, D. W. 1971, J. Polym. Sci., Pt. C, No. 35, 305–320
Neumüller, O. A. 1973, Römpps Chemie-Lexikon. 7th Ed., Vol. 3. Franckhsche Verlagshandlung, Stuttgart, p. 1812
Nishimura, H., Okano, T. and Asano, I. 1981, Mokuzai Gakkaishi *27,* 709–715
Nissan, A. H. 1971, according to Jones, D. W. in: Cellulose and Cellulose Derivatives (Bikales, N. M. and Segal, L., Eds.). Wiley-Intersci., New York, London, Sydney, Toronto, pp. 118–121
Nissan, A. H. and Sternstein, S. S. 1964, Tappi *47,* 1–6
Ohad, I. and Danon, D. 1964, J. Cell. Biol. *22,* 302–305
Ohad, I. and Mejzler, G. 1965, J. Polym. Sci., Pt. A, *3,* 399–406
Ohad, I., Danon, D. and Hestrin, S. 1962, J. Cell. Biol. *12,* 31–46

Okajima, K. 1978, 1979, Chemica Scripta *13,* 102–107, 113–119, 120–122

Page, D. C. 1960, Paper Techn. *1,* T165–T169

Patscheke, G. and Poller, S. 1980, Cell. Chem. Technol. *14,* 3–12

Petitpas, T., Oberlin, M. and Mering, J. 1963, J. Polym. Sci., Pt. C, No. 2, 423–427

Philipp, B., Schleicher, H. and Wagenknecht, W. 1975, Cell. Chem. Technol. *9,* 265–282

Philipp, B., Schleicher, H. and Wagenknecht, W. 1978, Cell. Chem. Technol. *12,* 529–552

Polčin, J. 1966, Faserforsch. Textiltechn. *17,* 128–132

Quivoron, C., Néel, J. and Champetier, G. 1967, Cell. Chem. Technol. *1,* 3–10

Ramesh, R., Patel, C. K. and Patel, R. D. 1977, Makromol. Chemie *178,* 227–233

Rånby, B. G. 1951, Disc. Faraday Soc. No. 11, 158–164

Rånby, B. G. 1954, Makromol. Chemie *13,* 40–52

Rånby, B. G. 1964, Papier *18,* 593–600

Ray, P. K. and Bandyopadhyay, S. B. 1975, J. Appl. Polym. Sci. *19,* 729–733

Rees, D. A. and Skerrett, R. J. 1968, Carbohyd. Res. *7,* 334–348

Richter, G. A., Herdle, L. E. and Wahtera, W. E. 1957, Ind. Eng. Chem. *49,* 907–912

Robertson, A. A. 1970, Tappi *53,* 1331–1339

Roche, E. 1970, Textile Res. J. *49,* 309–312

Ruck, H. 1962, Papier *16,* 703–712

Sarko, A. 1976, Appl. Polym. Symp. No. 28, 729–742

Sarko, A. 1978, Tappi *61,* No. 2, 59–61

Scallan, A. M. 1971, Textile Res. J. *41,* 647–653

Schempp, W. 1975, Zellcheming-Mitt. *23,* No. 1/2, 17–21

Schnepf, E. 1965, Planta *67,* 213–224

Schurz, J. 1969, Faserforsch. Textiltechn. *20,* 481–486

Schurz, J. 1977, Cell. Chem. Technol. *11,* 3–28

Schurz, J. and John, K. 1975, Cell. Chem. Technol. *9,* 493–501

Schurz, J. and Schlor, M. 1967, Cell. Chem. Technol. *1,* 529–544

Schurz, J., Haas, J. and Müller, K. 1970, Papier *24,* 691–698

Schweiger, R. G. 1969, Chem. Ind. *1969,* 296

Schweiger, R. G. 1974, Tappi *57,* No. 1, 86–90

Schweizer, E. 1857, J. Prakt. Chemie *72,* 109, 344

Shanbhag, V. P. 1968, Arkiv Kemi *29,* 1–22, 33–45, 163–177

Shimizu, M., Inoue, M. and Yamashita, T. 1970, Bull. Fac. Agricult., Shizuoka Univ. No. 20, 37–40

Siesler, H., Krässig, H., Grass, F., Kratzl, K. and Derkosch, J. 1975, Angew. Makromol. Chemie *42,* 139–165

Staudinger, H., In den Birken, K. H. and Staudinger, M. 1953, Makromol. Chemie *9,* 148–187

Stewart, C. M. and Foster, R. C. 1976, Appita *29,* 440–448

Stöckmann, V. E. 1971, J. Polym. Sci., Pt. C, No. 36, 363–381

Stöckmann, V. E. 1972, Biopolymers *11,* 251–270

Stone, J. E. and Scallan, A. M. 1965, J. Polym. Sci., Pt. C, No. 11, 13–25

Stuart, H. A. 1967, Molekülstruktur. Springer, Berlin, Heidelberg, New York, p. 52

Sullivan, J. D. 1968, Tappi *51,* 501–507

Swenson, H. A. 1973, Tappi *56,* No. 2, 106–110

Swenson, H. A. 1976, Appl. Polym. Symp. No. 28, 945–952

Takal, M. and Colvin, J. R. 1978, J. Polym. Sci., Polym. Chem. Ed. *16,* 1335–1342

Temming, H., Grunert, H. and Huckfeldt, H. 1972, Temming-Linters. P. Temming AG, Glückstadt, pp. 116–117

Tønnesen, R. S. and Ellefsen, Ø. 1960, Norsk Skogind. *14,* 266–269

Traube, W. 1911, Chem. Ber. *49,* 3319

Tsuboi, M. 1957, J. Polym. Sci. *25,* 159–171

Valdsaar, H., Dunlap, R. 1952, 122nd ACS-Meeting, Atlantic City, N. J., Abstracts of Papers, p. 3D

Valtasaari, L. 1957, Pap. Puu *39,* 243–252
Valtasaari, L. 1971, Makromol. Chemie *150,* 117–126
Vigo, T. L., Wade, R. H., Mitscham, D. and Welch, C. M. 1969, Textile Res. J. *89,* 305–316
Vink, H. 1964, Makromol. Chemie *76,* 66–81
Vink, H. 1967, Svensk Papperstid. *70,* 734–736
Visapää, A. 1964, Faserforsch. Textiltechn. *15,* 579–582
Viswanathan, A. and Shenouda, S. G. 1971, J. Appl. Polym. Sci. *15,* 519–535
Vogel, A. 1953, Makromol. Chemie *11,* 111–130
Wardrop, A. B. 1970, Protoplasma *70,* 73–86
Watanabe, S., Hayashi, J. and Akahori, T. 1974, J. Polym. Sci. *12,* 1065–1087
Willison, J. H. M., Brown, R. M. and Mueller, S. C. 1980, J. Microscopy *118*, 177–186

5. Polyoses (Hemicelluloses)

5.1. Nature and Classification

In addition to cellulose a number of various polysaccharides called polyoses or hemicelluloses are present in wood as well as in other plant tissues. The name hemicelluloses can be traced back to E. Schulze (1891), and it is based on the assumption that these polysaccharides are precursors of cellulose. Though in the scientific sector the term hemicelluloses is well defined, in the technical sector misunderstandings still exist. Thus the extractives in the alkalizing liquor from chemical grade pulp consisting of polyoses and short-chain cellulose are called hemicelluloses, or the term "hemicelluloses" is explained as "low molecular cellulose" (Bauer 1970). To avoid ambiguity in this book the non-cellulosic polysaccharides are called polyoses. This term ("Holzpolyosen") was first used by Staudinger and Reinecke (1939).

The polyoses differ from cellulose by a composition of various sugar units, by much shorter molecular chains, and by a branching of the chain molecules. The sugar units (anhydro-sugars) making up the polyoses can be subdivided into groups such as pentoses, hexoses, hexuronic acids and deoxy-hexoses (Fig. 5–1). The main chain of a polyose can consist of only one unit (homopolymer), e.g. xylans, or of two or more units (heteropolymer), e.g. glucomannans. Some of the units are

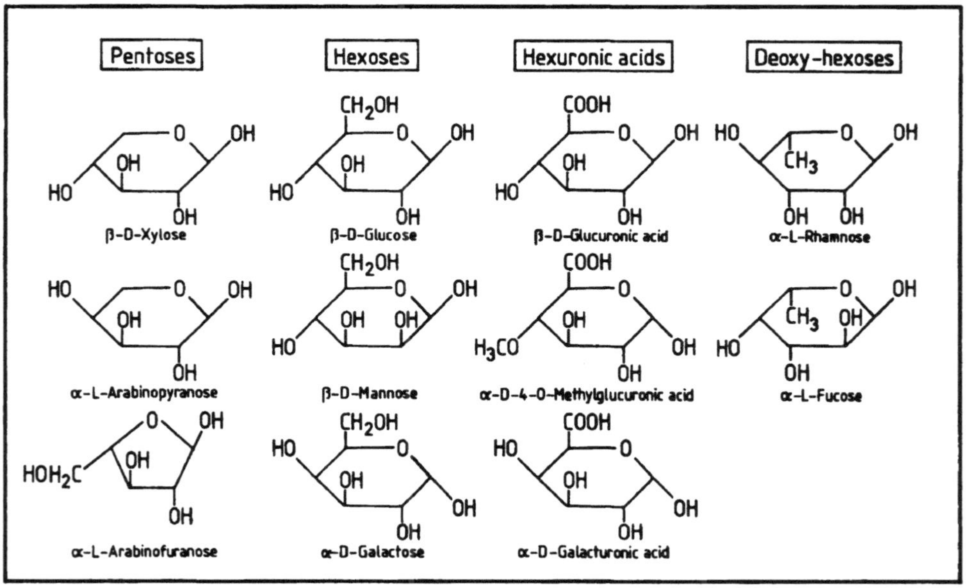

Fig. 5–1. Formulas of the sugar components of polyoses.

always or sometimes side groups of a main chain (backbone), e.g. 4-O-methylglucuronic acid, galactose.

As with amino acids in protein chemistry, abbreviations have also been introduced for sugars and sugar units. These generally consist of the first three letters of the name: Glu for glucose, Xyl for xylose, Rha for rhamnose etc. Other types of abbreviations are used for the uronic acids, e.g. GalA or GalU for galacturonic acid, Me-GluA or Me-GluU for 4-O-methylglucuronic acid. We will use the abbreviations with U instead of A as these clearly show that the compound is a uronic (alduronic) acid and not an aldonic or aldaric acid. The latter are also derivatives of sugars (aldoses) which may occur during an oxidative degradation of polysaccharides (Fig. 5–2). A fourth letter is often added to the abbreviated names of the sugars to indicate whether the sugar is in the pyranose or furanose form, e.g. Arap, Araf.

Monosaccharides as well as the polyoses they compose contain asymmetrical carbon atoms, and therefore show an optical rotation in solution. The optical rotation is an important property of all carbohydrates and is mostly used for their characterization. The specific rotation of the polyoses results from the basic framework and from the nature and frequency of the side chains and groups. All natural xylans and most of the mannans have a negative specific rotation, whereas the polygalacturonans have positive specific rotations, as determined commonly in 6% sodium hydroxide solution. Marchessault et al. (1963) and Gutmann and Timell (1965) developed equations for the estimation of the specific rotation particularly of xylans from their chemical structure.

The classical classification of polyoses is into hexosans, pentosans and polyuronides. This is a rough classification which does not take into consideration that the sugar units from different groups are mixed in most polyoses. A classification scheme based on the behaviour in relation to the separation from cellulose was proposed by Stewart (1954). The polyoses which can be extracted from holocellulose are called non-cellulosic glycosanes; the residue is called cellulosic glycosanes, and is subdivided into cellulose and non-glucosic cellulosic glycosanes. The classification according to the main components of the respective polyose has proved

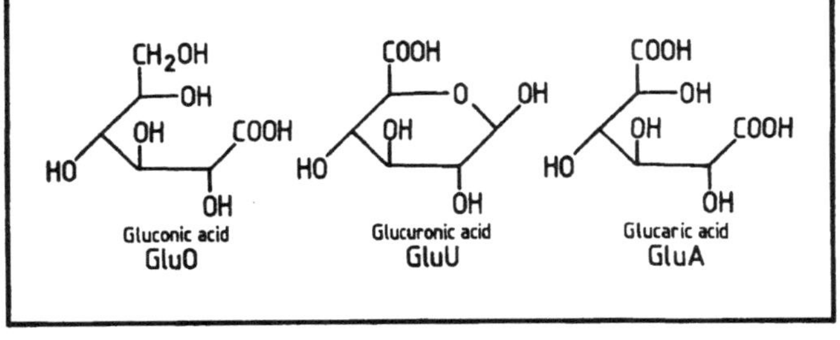

Fig. 5–2. Formulas of oxidation products of glucose.

useful for many years. In this system the polyoses are classified as xylans, mannans, galactans etc.

A general classification comprising all plant carbohydrates was presented by Aspinall (1973). His system consists of the following groups:
- cellulose;
- hemicelluloses: xylans, glucomannans;
- pectic substances: galacturonans, arabinans, galactans and/or arabinogalactans I (essentially linear chains)
- other polysaccharides: arabinogalactans II (highly branched chains), fuco-(or galacto-)xyloglucans;
- glycoproteins.

Softwoods and hardwoods differ not only in the percentages of total polyoses but also in the percentages of individual polyoses and composition of these polyoses. Regarding the non-glucosic sugar units present in wood it can be noticed that softwoods have a high proportion of mannose units and more galactose units than hardwoods, and hardwoods have a high proportion of xylose units and more acetyl groups than softwoods (Table 5–1).

Previous results in the field of polyoses chemistry have been summarized by Schuerch (1963), Timell (1964a, 1965, 1967) and Aspinall (1964a, 1973).

Table 5–1. Non-glucosic units of the polyoses in various woods

Species	Man %	Xyl %	Gal %	Ara %	Uron.A. %	Rha %	Acetyl %	References
Abies balsamea	10.0	5.2	1.0	1.1	4.8		1.4	Côté et al. 1966
Larix decidua	11.5	5.1	6.1	2.0	2.2*	0.0		Fengel et al. 1978
Larix laricina	12.3	6.0	2.4	1.3	2.8		1.6	Côté et al. 1966
Picea abies	13.6	5.6	2.8	1.2	1.8*	0.3		Fengel et al. 1978
Picea glauca	12.0	7.0	1.9	1.1	4.4		1.2	Côté et al. 1966
Picea mariana	9.4	6.0	2.0	1.5	5.1		1.3	Côté et al. 1966
Pinus strobus	8.1	7.0	3.8	1.7	5.2		1.2	Côté et al. 1966
Pinus sylvestris	12.4	7.6	1.9	1.5	5.0		1.6	Côté et al. 1966
Tsuga canadensis	10.6	3.3	1.8	1.0	4.7		1.4	Côté et al. 1966
Thuja occidentalis	7.4	3.8	1.5	1.7	5.8		0.9	Côté et al. 1966
Acer rubrum	3.3	18.1	1.0	1.0	4.9		3.6	Timell 1969
Betula alleghaniensis	1.8	18.5	0.9	0.3	6.3		3.7	Timell 1969
Betula papyrifera	2.0	23.9	1.3	0.5	5.7		3.9	Timell 1969
Betula verrucosa	3.2	24.9	0.7	0.4	3.6*	0.6		Fengel et al. 1978
Fagus grandifolia	1.8	21.7	0.8	0.9	5.9		4.3	Timell 1969
Fagus sylvatica	0.9	19.0	1.4	0.7	4.8*	0.5		Fengel et al. 1978
Fraxinus excelsior	3.8	18.3	0.9	0.6	6.0*	0.5		Fengel et al. 1978
Populus tremuloides	3.5	21.2	1.1	0.9	3.7		3.9	Timell 1969
Robinia pseudoacacia	2.2	16.7	0.8	0.4	4.7		2.7	Timell 1969
Ulmus americana	3.4	15.1	0.9	0.4	4.7		3.0	Timell 1969

* 4-O-Methylglucuronic acid

5.2. Xylans

5.2.1. Hardwood Xylans

Xylans are polyoses generally with a homopolymer backbone of xylose units which are linked by β-(1→4)-glycosidic bonds. In the case of hardwoods the xylan chains are laced, at irregular intervals, with groups of 4-O-methylglucuronic acid with an α-(1→2)-glycosidic linkage at the xylose units (Timell 1964a). Many of the OH-groups at C2 and C3 of the xylose units are substituted by O-acetyl groups. A segment of O-acetyl-4-O-methylglucuronoxylan from hardwood is represented by the formula in Fig. 5–3.

Most of the xylans isolated from various hardwood species have a ratio of about 10:1 (Xyl:Me-GluU), i.e. on the average every tenth xylose unit is linked to a side group of 4-O-methyglucuronic acid (Table 5–2). The analysis of xylans from tropical hardwoods resulted in a Xyl:Me-GluU ratio of about 6:1 for *Shorea* and *Dipterocarpus* species whereas mangrove (*Rhizophora* spec.) had a ratio of 11:1 (Lee et al. 1981). Arabinose in amounts of about 2% (based on xylan) was present forming non-reducing end groups.

The proportion of acetyl groups varies slightly in hardwoods of temperate zones. From the data in Table 5–1 a molar ratio of 1:0.5 to 1:0.6 (Xyl:acetyl) can be calculated. The acetyl groups are obviously distributed equally at C2 and C3. Lindberg et al. (1973a) determined the ratio of xylose units bearing no acetyl groups, acetyl groups at C2, C3, and both C2 and C3 to be 44:24:22:10 in birch xylan.

From methylation and degradation experiments the distribution of the side groups as well as the degree of branching have been determined. Thus it was found that hardwood xylan has two or three branching points with obviously very short side chains which are linked at C3 of the backbone (Koshijima et al. 1965; Zinbo, Timell 1965). A xylan without branches and a ration of 11:1 (Xyl: Me-GluU) was isolated from red cotton wood *(Bombax malebaricum)* by Garegg and Han (1968).

The average degree of polymerization of the xylan backbone varies between about 100 and 200 depending on wood species and mode of isolation (Table 5–2). Fractionations of birch xylan resulted in DP distributions between 20 and 200 (LeBel, Goring 1963; Klevinskaja et al. 1974). Fractions of alkali extracts from birch and beech wood received by precipitation, extraction or ion exchange chromatography had various ratios of xylose and 4-O-methylglucuronic acid (Wikström 1967; Kirilova, Beinart 1974; Fengel, Przyklenk 1976). Fractions with ratios up to 3:1 (Xyl:Me-GluU) have been isolated from beech wood.

Besides the main units hardwood xylans contain small amounts of rhamnose and galacturonic acid (Samuelson, Wictorin 1966; Shimizu, Samuelson 1973). Further studies showed that the reducing end of xylans consists of a combination of xylose,

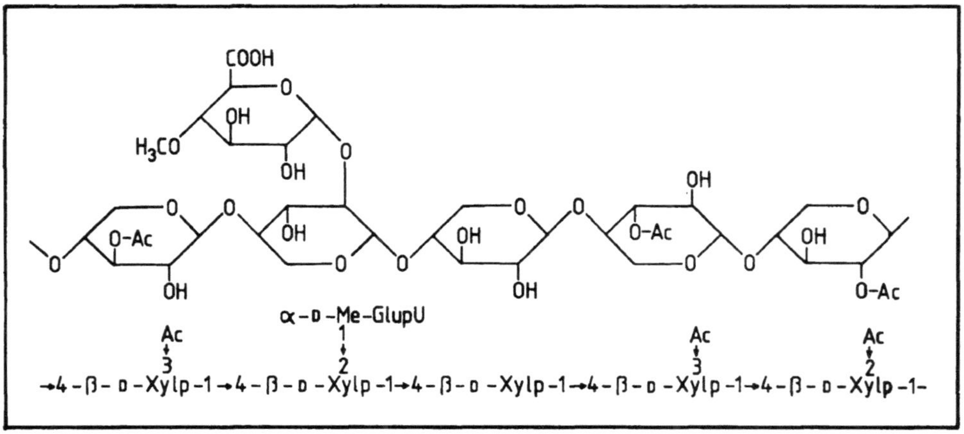

Fig. 5–3. Partial chemical structure of O-acetyl-4-O-methylglucuronoxylan from hardwood.

Table 5–2. Characterization of 4-O-methylglucuronoxylans isolated from various hardwood species

Species	Yield %	Xyl:Me-GluU	Acetyl %	$\bar{P}_n$	$[\alpha]_D$ Degree	References
Acer saccharum	12.4	10:1		205	−80	Timell (1959a)
Betula lutea	17.5	10:1		192	−81	Timell (1959b)
Betula papyrifera	16.5	10–11:1	5.65	180	−86	Timell (1960c)
Betula verrucosa	15.0	10–11:1	4.5	169	−80.4	Han, Swan (1968)
Carpinus betulus	17.3			86	−68	Ebríngerová et al.
		9.8:1				(1967, 1969)
Eucalyptus globulus	4.4	10–11:1	3.2	157	−74.2	Han, Swan (1968)
Eucalyptus regnans		11:1		209	−90	Dudkin et al. (1971)
Fraxinus excelsior	17.1	8:1		154	−68	Dudkin et al. 1972)
Populus nigra		9.7:1		218	−73	Dudkin et al. (1971)
Populus tremuloides	21.4	9:1		212	−75	Jones et al. (1961)
Ulmus americana	5–6	7:1		185	−70	Gillham, Timell (1958)

rhamnose and galacturonic acid units with the following sequence (Johansson, Samuelson 1977):

β-D-Xylp-1⟶4-β-D-Xylp-1→3-α-L-Rhap-1→2-α-D-GalpU-1→4-β-D-Xyl

This structure is seen to be responsible for the alkali resistance of the xylan molecule, as the galacturonic acid makes it stable after the removal of the reducing xylose unit (Ericsson et al. 1977; Andersson, Samuelson 1978).

The xylans present in roots are obviously different from those in stem wood. In the roots of sugar maple Bardalaye and Hay (1974) detected 4-O-methylglucuronoglucoxylans with a heteropolymer backbone consisting of glucose and xylose units.

5.2.2. Softwood Xylans

In general softwood xylans differ from hardwood xylans by the lack of acetyl groups and by the presence of arabinofuranose units linked by α-(1→3)-glycosidic bonds to the xylan backbone. Thus the softwood xylans are arabino-4-O-methylglucuronoxylans. A segment of the molecule is shown in Fig. 5–4.

Softwood xylans have a higher portion of 4-O-methylglucuronic acid than hardwood xylans. In most softwood xylans investigated the ratio of Xyl:Me-GluU is 5 to 6:1 (Timell 1967; Harwood 1972), sometimes even 3 to 4:1 (Gvozdeva et al. 1971; Sharkov et al. 1974). The ratio of Xyl:Ara is in the range of 6 to 10:1. The mean ratio for the three components of softwood xylan can be given as 8:1.6:1 (Xyl:Me-GluU:Ara). The xylan molecules from softwoods seem to be shorter than hardwood xylans as DP determinations resulted in values of 70 to 130 (Timell, Zinbo 1967; Zinbo, Timell 1967; Harwood 1972). The chains are slightly branched with one to two side chains per molecule (Table 5–3).

The fractionation of the alkali-soluble polyoses of spruce showed that the softwood xylans are also mixtures of compounds with various ratios. Fractions with ratios between 10:3:1 and 2.5:0.8:1 (Xyl:Me-GluU:Ara) were isolated (Fengel 1976). Zinbo and Timell (1967) found an increase of arabinose units and more branching points in fractions of higher molecular weight.

Recently the same sugar sequence as described for the alkali-resistant end group of hardwood xylan (→5.2.1.) was detected in spruce xylan *(Picea abies)* by Andersson et al. (1983).

From maritime pine *(Pinus pinaster)* an arabino-4-O-methylglucuronoxylan with additional side-groups of glucuronic acid was isolated (Roudier 1964).

The xylan of compression wood is similar to that of normal wood except for the proportion of arabinose, which is less in compression wood (Côté et al. 1966; Hoffmann, Timell 1972b).

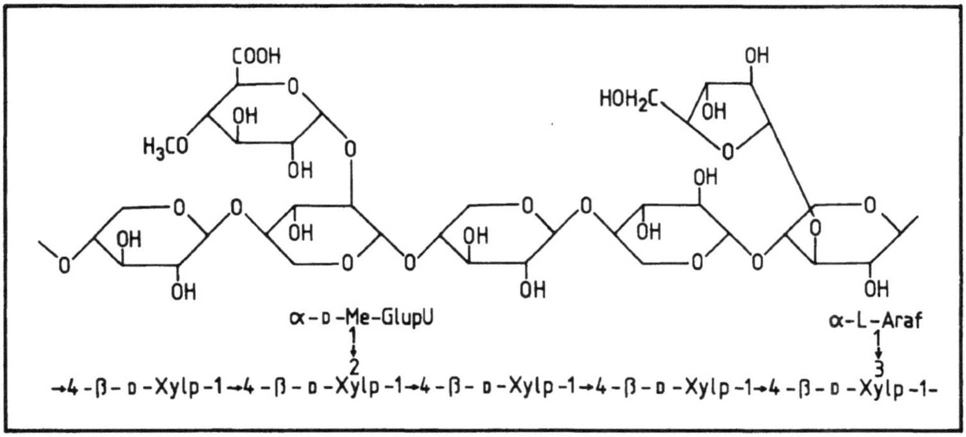

Fig. 5–4. Partial chemical structure of arabino-4-O-methylglucuronoxylan from softwood.

Table 5–3. Characterization of arabino-4-O-methylglucuronoxylans isolated from various softwood species

Species	Yield %	Xyl:Ara:Me-GluU	$\bar{P}_n$	$[\alpha]_D$ Degree	Branches per molecule	References
Ginkgo biloba	4	9:1:1.5	185	−35	1.3	Mian, Timell (1960b)
Larix gmelinii		10.5:1:2.7	152	−38.2		Gvozdeva et al. (1971)
Larix sibirica		10:1:3.9	100	−54.3		Gvozdeva, Levanova (1979)
Picea abies	2.7	7.4:1:1.25	128	−60.3	1.8	Zinbo, Timell (1967)
Picea engelmanii	8	7:1:1.2	92	−35	0	Mills, Timell (1963)
Pinus cembra var. *sibirica*	3.1	10:1		−34.2		Sharkov et al. (1971)
Pinus radiata		5.3:1:0.9	78	−61.2	0.7	Harwood (1972)
Pinus strobus	5	9:1:1.3	≥100	−65	0.7	Banerjee, Timell (1960)
Pinus sylvestris		11.6:1:3.7	108	−36.5		Gvozdeva et al. (1971)
Tsuga canadensis	4.6	6:1:1.2	73	−50	1.7	Timell, Zinbo (1967)

5.2.3. Xylans from Other Plants

The composition of xylans from various plants and particularly of grasses has been investigated, too. The simplest structure was found in xylan from esparto grass *(Stipa tenacissima)* which consists of β-(1→4)-glycosidic linked xylan chains with a DP of 65 and one branching point per molecule (Croon, Timell 1960). Other grasses like bamboo, barley and Guinea grass contain arabino-4-O-methylglucurono-xylans (Karnik et al. 1963; Buchala 1973, 1974). In addition, galactose units were also detected in xylans from barley *(Hordeum vulgare)* and Guinea grass *(Panicum maximum)* having degrees of polymerization of about 90. Maekawa (1976) found acetyl groups in addition to arabinose and 4-O-methylglucuronic acid units in bamboo *(Phyllostachys reticulata);* therefore he places bamboo xylan between hardwood and softwood xylan.

An arabino-4-O-methylglucurono-xylan with a ratio of 25:5:1 (Xyl:Me-GluU:Ara) and a DP of 120–130 was isolated from cinnamon fern *(Osmunda cinnamomea)* by Timell (1962a). 4-O-methylglucuronoxylans are also present in seed hairs of milkweed *(Asclepias syriaca)* and kapok *(Ceiba pentandra)* (Barth, Timell 1958; Currie, Timell 1959).

Other types of xylan are present in seaweed. Xylans from *Rhodymenia palmata* are linear polymers with β-(1→3) and β-(1→4) linked xylose units (Björndal et al. 1965).

5.2.4. Supramolecular Structures

The studies of Marchessault (1964; Marchessault et al. 1961; Marchessault, Settineri 1964) showed that xylan, particularly if it is deacetylated and uronic acid groups are absent, is able to crystallize in the form of hexagonal platelets with various layers of about 5 nm in thickness (Fig. 5–5). A chain folding of the xylan molecules seems to be possible.

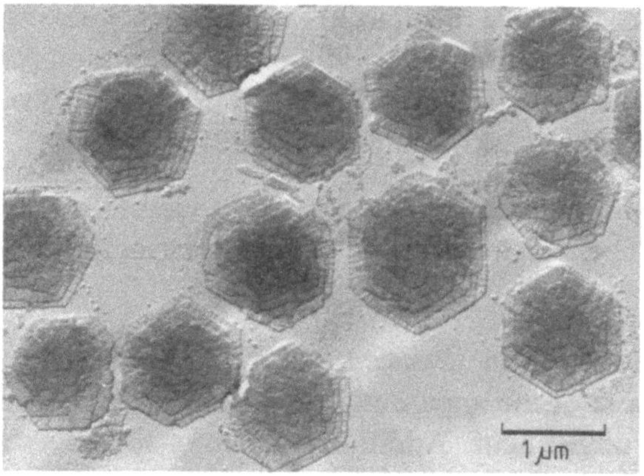

Fig. 5–5. Hexagonal crystals of xylan from esparto (by courtesy of H. Chanzy).

Water plays an important role in the formation of xylan crystals. From X-ray diffraction a trigonal unit cell for xylan monohydrate has been determined with the following dimensions:

a = c = 0.916 nm
b = 1.485 nm
β = 60°

With increasing humidity an increase of cell size was observed, the dimensions at 100% humidity being:

a = c = 0.964 nm
b = 1.500 nm
β = 60°

There are six monomer units in one space unit, and the xylan chains are arranged in a threefold screw axis. The molecules are stabilized by intramolecular H-bonds

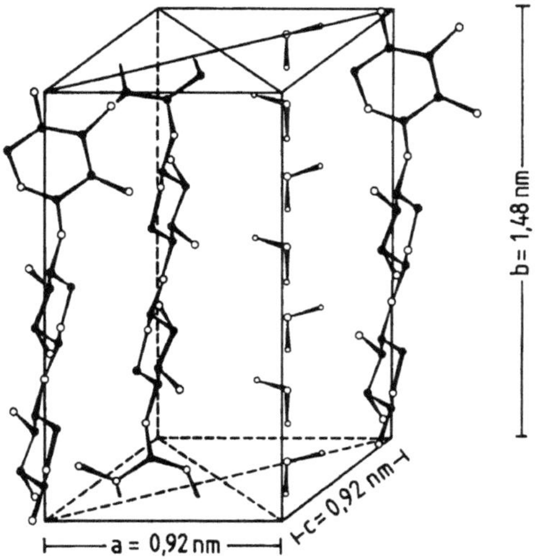

Fig. 5–6. Hexagonal unit cell of xylan hydrate (according to Nieduszynski and Marchessault 1971).

(O5 . . . O3); a direct connection between adjacent xylan chains by intermolecular H-bonds seems to be impossible, and it is assumed that the water molecules incorporated in the lattice stabilize the structure (Fig. 5–6) (Settineri, Marchessault 1965; Rees, Skerrett 1968; Nieduszynski, Marchessault 1971).

It could be shown that acetyl-4-O-methylglucuronoxylan added to holocellulose is able to crystallize within the fibres in the direction of the fibre axes (Marchessault et al. 1967).

From the molecular structure it can be deduced that a strict order over a longer distance is not possible because of the apparently irregularly arranged side groups of acetyl or arabinose and uronic acid. Additionally the xylan chains are more flexible than hexosan chains because of the lack of a CH_2OH-group (C6) (Swenson et al. 1965; Wikström 1968).

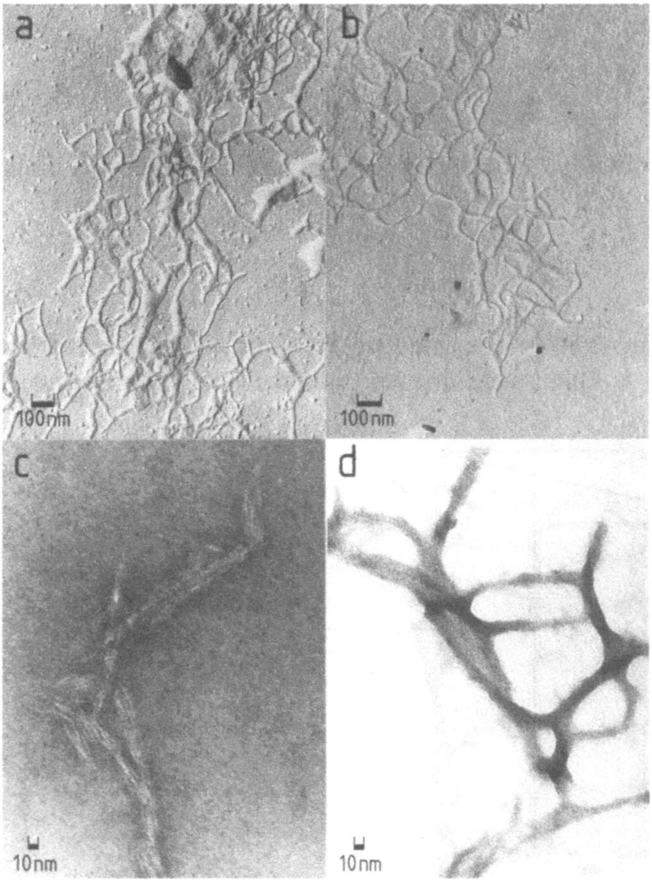

Fig. 5–7. Electron micrographs of xylans from hardwood and softwood at different magnifications:
a, c) Xylan from beech wood;
b, d) Xylan from spruce wood.

The flexible xylan molecules can be arranged to form fibrils with a flexible structure which is different from the structure of cellulose fibrils. Precipitates of xylan fractions both from hardwood and softwood show a loose arrangement of flexible fibrils in the electron microscope (Fig. 5–7) (Fengel 1967, 1976; Fengel, Przyklenk 1976). At high magnification it can be seen that the fibrils consist of fibrillar subunits with diameters of 2–3 nm. An association with lignin influences the appearance of the fibrils (→ 6.5.).

5.3. Mannans

5.3.1. Hardwood Mannans

Mannans from wood are characterized by a heteropolymer backbone consisting of mannose and glucose units. Thus the wood mannans are correctly referred to as glucomannans.

The simplest structure is shown by the hardwood glucomannans as they consist only of glucose and mannose units forming chains which are slightly branched. The mannose and glucose units are linked by β-(1→4)-glycosidic bonds. The ratio of mannose and glucose units is about 1.5–2:1 in most hardwoods investigated *(Populus tremuloides, Fagus grandifolia, Ulmus americana, Acer rubrum, Carpinus betulus)* (Timell 1960a, 1967; Ebringerová et al. 1972). The highest portion of glucose was found in the glucomannan of birchwood *(Betula papyrifera, B. lutea)* with a ratio of about 1:1 (Timell 1960a, b), and the lowest in the glucomannan of sugar maple *(Acer saccharum)* with a ratio of 2.3:1 (Man:Glu) (Adams 1961).

The degree of polymerization ($\overline{P}_n$) of hardwood glucomannans is about 60–70 (Mian, Timell 1960a; Ebringerová et al. 1972). Compared to the portion of xylan the glucomannans (3–5%) are of minor importance. This may explain why this component is so seldom studied.

5.3.2. Softwood Mannans

Softwoods contain about 20–25% mannans consisting of a glucomannan backbone to which acetyl groups and galactose residues are attached. Hence these polyoses are O-acetyl-galactoglucomannans (Fig. 5–8).

The ratio of mannose and glucose units is about 3:1, but these units are obviously distributed randomly within the molecule since mannobiose, mannotriose, mannotetraose, mannosylglucose, glucosylmannose and cellobiose were obtained by means of partial hydrolysis (Mian, Timell 1960b). The portion of galactose units joined by α-(1→6)-linkages is different in mannans isolated by water extraction and

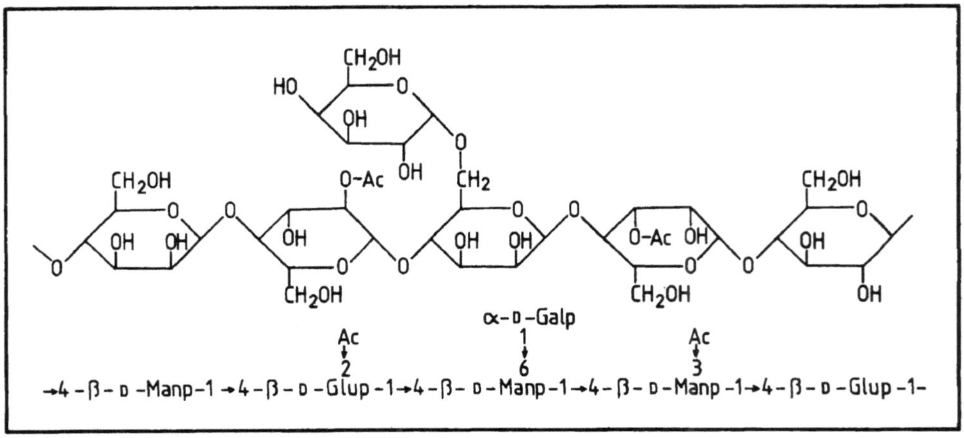

Fig. 5–8. Partial chemical structure of O-acetyl-galactoglucomannan from softwood.

those isolated by alkali extraction. Water-soluble galactoglucomannan has a ratio of 3:1:1 (Man:Glu:Gal), alkali-soluble galactoglucomannan a ratio of 3:1:0.2 (Timell 1967).

The ratio of mannose, glucose and galactose must be regarded as average values as there are variations probably depending on the mode of separation. The ratios found in galactoglucomannans of various softwoods are summarized in Table 5–4.

Fractionation studies of alkali extracts from spruce and larch resulted in several fractions containing galactoglucomannans of varying composition (Rogers, Thompson 1969; Hashi et al. 1970; Fengel 1979). Fengel (1979) obtained fractions from the 5% alkali extract of spruce holocellulose containing galactoglucomannans with ratios between 7.1:1:0.5 and 1.8:1:1.1.

The alpha-glycosidic linkage of the galactose units is a very sensitive linkage which may be cleaved during the alkaline extraction. This may be the reason why a minor portion of glucomannans with few or no galactose residues is obtained during certain fractionations (Timell 1961). Rogers and Thompson (1969) assume that a cleaving enzyme system within the wood is responsible for the loss of galactose units.

The acetyl groups seem to be distributed equally on the C2 and C3 of the mannose units (Hashi et al. 1971; Lindberg et al. 1973b). In Parana pine *(Araucaria angustifolia)* one third of the acetyl groups were found to be linked to glucose units (Katz 1965).

While for galactoglucomannans of some softwoods, e.g. black spruce and red pine, a slight branching of the backbone has been found (Linnell, Swenson 1966; Hoffmann, Timell 1970a), for other softwoods, e.g. Scots pine and radiata pine, linear main chains are described (Lindberg et al. 1973b; Harwood 1973).

Table 5–4. Characterization of galactoglucomannans isolated from various softwood species

Species	Yield %	Man:Glu:Gal	Acetyl %	$\bar{P}_n$	$[\alpha]_D$ Degree	References
Abies amabilis	4.0	3:1:1		76	−40	
	8.1	3:1:0.1		95	−38	Schwarz, Timell (1963)
Araucaria angustifolia	1.5	3.8:1:0.34	5.9	78	−28.7	Katz (1965)
Ginkgo biloba	5	3.6:1:0.2		96	−36	Mian, Timell (1960b)
Larix gmelinii		2.8:1:0.25		142	−29.2	Gvozdeva et al. (1971)
Larix laricina		2.8:1:0.12		35	−30	Kooiman, Adams (1961)
Larix leptolepsis		3.9:1:0.4	4.3		−24.1	Hashi et al. (1970)
Larix sibirica		3:1:1		70	−29.7	
		3:1:0.1		112	−30.5	Gvozdeva, Levanova (1979)
Picea engelmannii	1.0	3:1:1		100	− 7	
	8.1	3:1:0.2		107	−40	Mills, Timell (1963)
Picea engelmannii	0.1	4.3:1:1.3		64	+2.8	Rogers, Thompson (1969)
Picea mariana		3:1:1				Thompson, Kaustinen
		4.3:1:1				(1964)
Pinus banksiana		2.9:1:0.1		18–21	−26	Bishop, Cooper (1960)
Pinus densiflora		3.1:1:0.1	6.26	81	−28.2	Koshijima, Tanaka (1970)
Pinus radiata	6.8	3.7:1:0.13		45	−20.4	Harwood (1973)
Pinus resinosa	4.0	2.35:1:0.26		73	−10.3	Hoffmann, Timell (1970a)
Pinus sylvestris		4.3:1:0.26	5.95			Meier (1961)
Pinus sylvestris		3.3:1:0.17		75	−33.8	Gvozdeva et al. (1971)
Pinus sylvestris		2.8:1:0.2	8.8		−28	Lindberg et al. (1973b)
Tsuga canadensis	4.8	3:1:1		44	− 8.2	
	6.2	3:1:0.1		100	−40	Timell (1962c)

Though the proportion of galactoglucomannan is quite different in normal (22%) and compression wood (8–9%) of tamarack *(Larix laricina)*, no difference could be detected in composition and structure of galactoglucomannans from normal and compression wood (Hoffmann, Timell 1972c).

There is evidence that parts of the galactoglucomannans are linked to lignin (Linnell et al. 1966; Koshijima, Tanaka 1970) (→ 6.5.).

An alkali-soluble galactoglucomannan similar to that of xylem was isolated in a yield of 2.6% from the secondary phloem of Scots pine *(Pinus sylvestris)* (Fu, Timell 1972a) (→ 9.2.2.).

5.3.3. Other Mannans

Like xylans, mannans are widespread in the vegetable kingdom. Galactoglucomannans with structures similar to those in softwood have been detected in stems and leaves of several plants. Thus red clover *(Trifolium pratense)* contains a mannan with a ratio of 1.1:1:0.25 (Man:Glu:Gal) (Buchala, Meier 1973a). The mannan of cinnamon fern *(Osmunda cinnamomea)* has a ratio of 2:1:0.1 and a $\bar{P}_n$ of about 100 (Timell 1962b). Similar galactoglucomannans have been isolated from primitive plants such as horsetails *(Equisetum spp.)*, club-mosses *(Lycopodium spp.)*, and

Psilotum nudum (Timell 1964b). Another type of mannan consisting of β-(1→4)-linked mannose units with only about 5% glucose units was extracted from green seaweed *(Codium fragile)* by Love and Percival (1964).

Mannans are particularly important as the main components of seeds and tubers. Earlier studies showed that the mannans isolated from seeds of ivory nut *(Phytelephas macrocarpa)* and date palms *(Phoenix dactylifera)* consist exclusively of mannose units; some mannan fractions contain a few galactose residues (Aspinall et al. 1953, 1958; Meier 1958). The units of the backbone are β-(1→4)-linked; some form side groups by means of a β-(1→6)-linkage. The degree of polymerization of the mannan fractions was relatively low ($\overline{P}_n$ = 15–80). The mannans in the seeds of the guar plant *(Cyamopsis tetragonoloba)* and the locust bean *(Ceratonia siliqua)* were studied by Hui and Neukom (1964). These are mixtures of galactomannans with various Man:Gal-ratios between about 7:1 and 1:1. The $\overline{P}_w$ is in the range of 900–1500. The locust mannan consists of β-(1→4)-linked mannan chains with α-(1→6)-linked galactose residues (Seiler et al. 1973). Mannans of the same type were also isolated from seeds of other leguminose species, e.g. *Gleditsia triacanthos, Trigonella foenum-graecum* (Leschziner, Cerezo 1970; Reid 1971; Reid, Meier 1973).

The seed mannans desposited in the endosperm tissue are reserve polysaccharides and suffer a mobilization by enzymatic depolymerization during the germination of the seeds (Keusch 1968).

The reserve polysaccharides in orchid tubers are also mannans consisting of mannose and glucose units (salep mannan). In the water extract of *Orchis morio* a glucomannan with a Man:Glu-ratio of 3.3:1 was isolated (Buchala et al. 1974). The units are linked by β-(1→4)-bonds and there are about 7 branching points per molecule which has a $\overline{P}_n$ of 665. Acetyl groups are also present (5.3%), being linked at the C2 and C3 of the mannose units.

5.3.4. Supramolecular Structures

Native mannans from wood seem to be unable to crystallize in a lattice structure. X-ray diagrams with more or less distinct peaks have been obtained from mannans of guar seeds, of ivory nuts and of seaweeds (Palmer, Balantyne 1950; Meier 1958; Frei, Preston 1968). These mannans have backbones composed entirely of mannose residues. Ivory nut and seaweed mannans contain no or only a few galactose units, whereas mannan from guar seeds has a high portion of galactose units, so that probably each second mannose unit bears a galactose residue. From the X-ray data it is assumed that the galactose units are arranged in the 002 plane, which results in a distance between two mannose chains in the a-axis of 1.35 nm (Table 5–5). The dimensions of the unit cell grow with increasing humidity, i.e. water molecules are incorporated in the crystal lattice (Marchessault, Sarko 1967).

Table 5–5. Dimensions of the cell unit of various mannans

Source	Dimensions a nm	b nm	c nm	References
Guar seeds	1.35	1.03	0.87	Palmer, Balantyne (1950)
Ivory nuts	0.76	1.09	0.88	Meier (1958)
	0.722		0.892	Chanzy et al. (1979)
Seaweed	0.721	1.027	0.882	Frei, Preston (1968)

Mannan of seaweed has an orthorhombic unit cell. The pyranose rings of the corner chains are positioned in the 102 plane and the central chain approximately in the 002 plane so that hydrogen bonds to the corner chains are possible (Frei, Preston 1968; Niduszynski, Marchessault 1972) (Table 5–5; Fig. 5–9). The mannan lattice can be changed in a similar manner as is known from cellulose by treatment with alkali. After treating *Codium* mannan with 12–14% KOH and washing with water Frei and Preston (1968) determined the cell unit to be monoclinic with the dimensions a = 1.88 nm, b = 1.02 nm, c = 1.87 nm, and β = 57.5° (mannan II).

Meier (1958) showed that only one part of the ivory nut mannan is crystalline (mannan A) in the native state while the other part (mannan B) is amorphous or paracrystalline. Various methods were used for the crystallization of low molecular mannans from ivory nuts (Bittiger, Husemann 1972; Chanzy et al. 1979). The results were rhombic crystals made up of several layers showing sharp diffraction patterns (Fig. 5–10). Ivory nut mannan can also be recrystallized on cellulose, where it grows perpendicularly to the fibrillar axes of the substrate (Chanzy et al. 1978). By this process a shish-kebab structure is obtained (Fig. 5–11).

After selective hydrolysis of pine polyoses Yundt (1951) succeeded in the crystallization of glucomannan. In the electron microscope he saw rod- and whisker-like crystals. A mannan fraction from spruce wood was redissolved in KOH and preci-

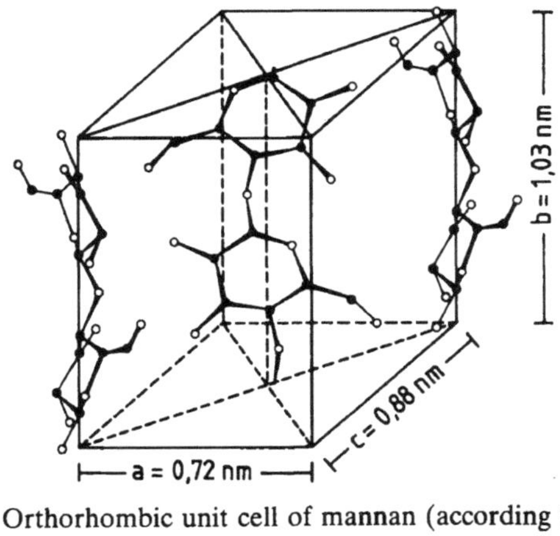

Fig. 5–9. Orthorhombic unit cell of mannan (according to Frei and Preston 1968).

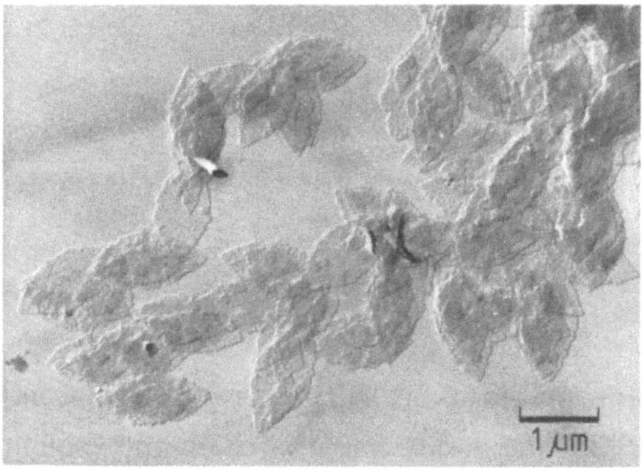

Fig. 5–10. Rhombic crystals of mannan from ivory nuts. Electron micrograph (by courtesy of H. Chanzy).

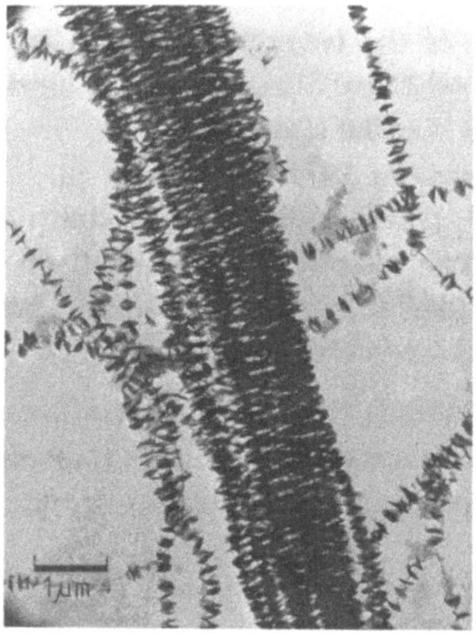

Fig. 5–11. Ivory nut mannan recrystallized on cellulose fibrils forming a shish-kebab structure. Electron micrograph (by courtesy of H. Chanzy).

pitated with ethanol by Fengel (1965). The electron microscopic picture showed lateral aggregates of relatively stiff fibrils (Fig. 5–12). The O-acetyl groups of native softwood glucomannans seem to prevent a molecular orientation as birefringence cannot be observed shortly until deacetylation (Katz 1965). Deacetylated glucomannans from pine *(Pinus sylvestris)* and redwood *(Sequoia sempervirens)* crystallize in the same shape as ivory nut mannan and can also form a shish-kebab structure with cellulose (Grosrenaud 1980).

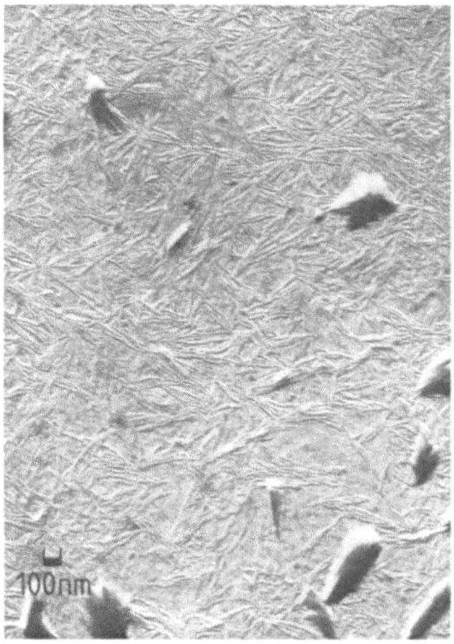

Fig. 5–12. Glucomannan from spruce wood recrystallized after dissolution in KOH. Electron micrograph.

Galactoglucomannans as separated by fractionation of alkali extracts from holocellulose show no defined structure in the electron microscope at lower magnification (Fig. 5–13a). But at high resolution conditions very fine fibrils with diameters of 1–2 nm can be detected (Fengel 1966a, 1979) (Fig. 5–13b). In some fractions an intimate association between galactoglucomannan and 4-O-methylglucuronoxylan seems to exist. Furthermore, an association with lignin in the supramolecular range cannot be excluded (→ 6.5.).

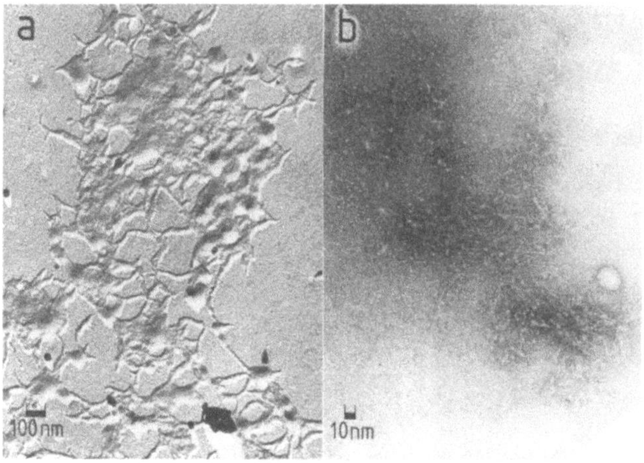

Fig. 5–13. Electron micrographs of galactoglucomannan from spruce wood at different magnifications:

a) Mixture of xylan and mannan. The xylan fibrils are partially covered by amorphous mannan.

b) Very thin fibrils of galactoglucomannan at high magnification.

5.4. Glucans

Apart from cellulose there are some other polysaccharides consisting of glucose units in wood as well as in other plant tissue. Among them starch is the most important reserve polysaccharide present in fruits, seeds and other storing tissues. Thus starch is also present in the parenchyma cells of wood tissue.

Starch consists of various components differing in molecular weight and molecular structure. There are the linear amyloses A, B, V, and the branched amylopectin. The glucose units of amyloses are linked by α-(1→4) glycosidic bonds; in amylopectin α-(1→6)-linkages are additionally present. The α-glycosidic linkages can easily be split, a fact which is important for mobilization during metabolic processes. With this linkage the pyranose rings are arranged beneath an angle of about 120° which results in a helical structure of the starch molecules with 6 glucose units per turn. Starch therefore exists only in the form of granules and not as fibrils. Nevertheless the various amyloses are able to crystallize. The probable crystal lattices are discussed by Marchessault and Sarko (1967).

Another glucan in wood is callose. Callose is known above all as a substance present in sieve cells of phloem. But it is also a component of parenchyma cells in xylem. Here it forms protecting layers on the membranes of half-bordered pits, which probably seals the pits in order to protect the plasmatic cell content against the water-conducting vascular cells. Callose consists of β-(1→3)-linked glucose units (Fu et al. 1972). The callose molecules are able to join together forming fibrillar structures (Fengel 1966b) (Fig. 5–14).

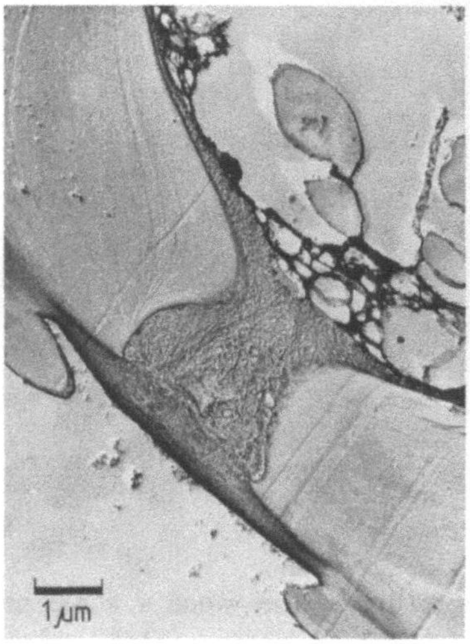

Fig. 5–14. Fibrillar callose seals a half-bordered pit at the parenchymous side. Ultrathin section of beech wood, electron micrograph.

From the compression wood of tamarack *(Larix laricina)*, red spruce *(Picea rubens)*, balsam fir *(Abies balsamea)*, and red pine *(Pinus resinosa)* and the normal wood of the latter an acidic glucan in a percentage of 2–4% was isolated by Hoffmann and Timell (1972a, d, e). This glucan is called laricinan and consists of about 200 glucose units linked by β-(1→3)-glycosidic bonds, of which few (6–7%) are β-(1→4)-linked. The glucose backbone, which has about 8 branching points, bears several glucuronic and a few galacturonic acid groups.

β-(1→3)-linked glucans are widely distributed in the plant kingdom. Hoffmann and Timell (1970b) give a survey of a number of publications concerning glucans of this type in bacteria, algae, and fungi. According to Meier et al. (1981) β-(1–3)-glucan is probably an intermediate in the biosynthesis of cellulose at least in cotton.

Glucans consisting of β-(1→3)- and β-(1→4)-linked glucose units in ratios of about 1:1 to 1:2 have been found in mango fruits *(Mangifera indica)*, oats *(Avena sativa)*, wheat *(Triticum aestivum)*, maize plants *(Zea mays)* and mung beans *(Phaseolus aureus)* (Das, Rao 1964; Buchala, Wilkie 1971, 1973; Buchala, Meier 1973b; Buchala, Franz 1974).

Xyloglucans with a β-(1→4)-linked backbone of glucose units to which single α-xylose units are attached, some of them bearing residues of galactose and fucose, are present in many plants. Furthermore a fucoxyloglucan has been found in cultured cells of sycamore *(Acer pseudoplatanus)* (Albersheim et al. 1973; Bauer et al. 1973 (Fig. 5–15).

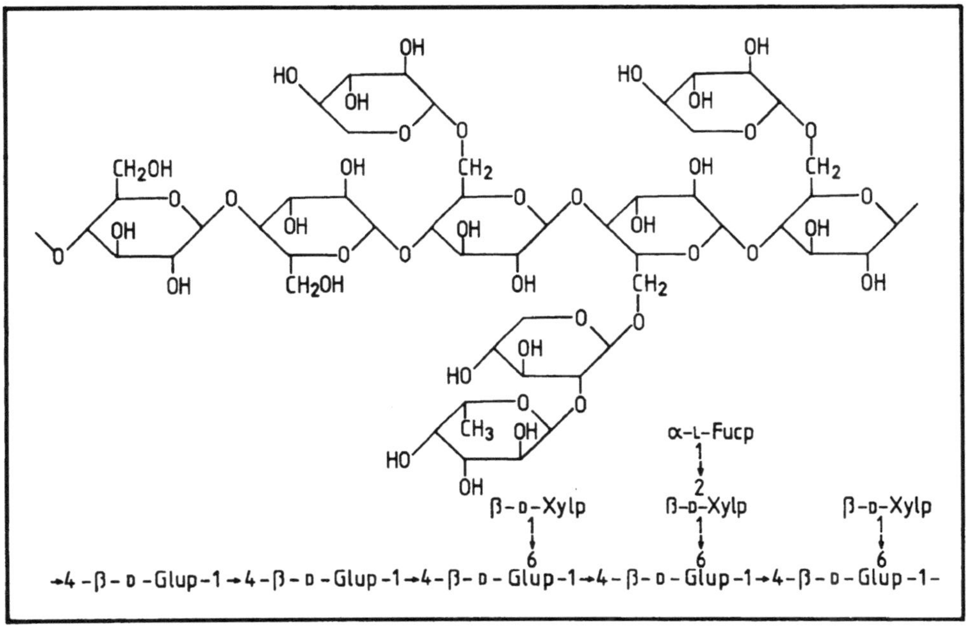

Fig. 5–15. Partial chemical structure of fucoxyloglucan from a cell culture of sycamore.

5.5. Galactans

The polyose group of galactans has been well known for a long time, particularly the arabinogalactans from larchwood. These polyoses are water-soluble and can be isolated in amounts of 10–25% (Adams, Douglas 1963; Aspinall 1964b; Timell 1967). In percentages of 0.5–3% galactans are also present in the wood of other species such as *Pinus, Araucaria, Acer, Fagus* and *Betula* (Brasch, Jones 1959; Aspinall 1964b; Adams 1964; Kuo, Timell 1969; Shimizu 1975). An increased percentage of galactans is found in compression wood as well as in tension wood (Meier 1964; Ruel, Barnoud 1978).

In general the galactans are highly branched. The arabinogalactan from larchwood has a backbone of β-(1→3)-linked galactose units, and side chains of β-(1→6)-linked galactose units, galactose and arabinose units, single arabinose units, and single glucuronic acid units (Jones, Reid 1963; Lynch et al. 1968; Fu, Timell 1972b) (Fig. 5–16). The ratio of galactose and arabinose units is about 6:1, and about one third of the arabinose units are in the pyranose, two thirds in the furanose form. The molecular weight ($\bar{M}_n$) of the arabinogalactans of various larch species varies from 29 600 to 58 500. But even within a single larch species arabinogalactan may vary in molecular weight and probably also in the Gal:Ara-ratio (Simson et al. 1968; Nikolaeva et al. 1971).

Larch galactan is mainly a heartwood component, and its content increases from the lower part of the stem to the top of a tree as well as from the core to the sapwood border (Côté, Timell 1967).

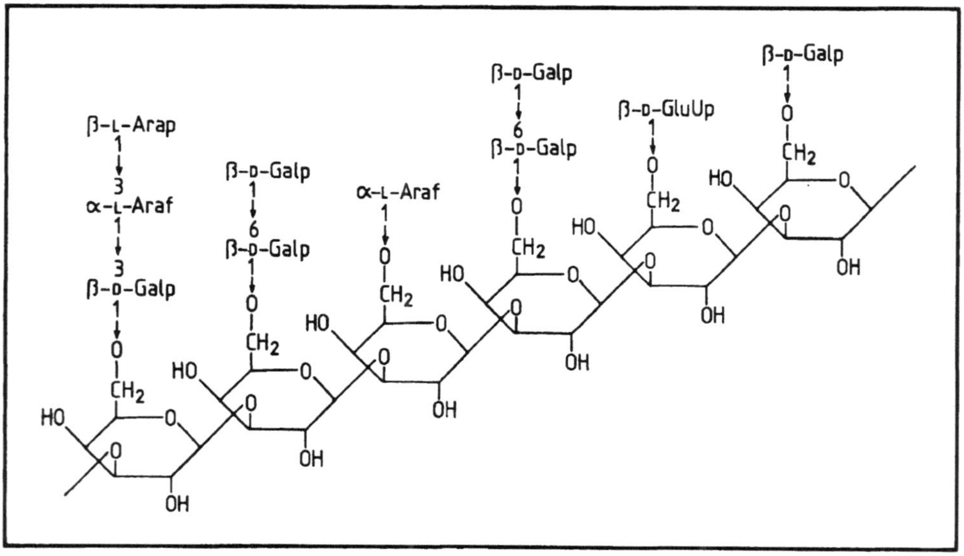

Fig. 5–16. Partial chemical structure of arabinogalactan from larch wood.

Apart from normal wood arabinogalactan Jiang and Timell (1972) isolated a slightly branched galactan from the compression wood of tamarack *(Larix laricina)*. The molecules consist of 200–300 β-(1→4)-linked galactopyranose units. One out of 20 galactose units carries a single, terminal residue of galacturonic acid, attached to C6 of a backbone unit.

A similar galactan with a β-(1→4)-linked backbone and 6 to 8 branches was found in the compression wood of red spruce *(Picea rubens)*. The molecule consists of at least 300 galactopyranose units and bears some residues of galacturonic and glucuronic acid (Schreuder et al. 1966). The same type of galactan is also present in the compression wood of balsam fir *(Abies balsamea)* (Côté et al. 1967).

The galactans of hardwoods are characterized by the content of rhamnose units. Adams (1964) found a water-soluble rhamnoarabinogalactan with a molar ratio of 1.7:1:0.2 (Gal:Ara:Rha) in sugar maple *(Acer saccharum)*. The molecule is slightly branched, and the galactose units are joined by β-(1→3)-linkages.

The galactan from tension wood of beech *(Fagus sylvatica, F. grandifolia)* is highly branched with a β-(1→4)-linked backbone (Fig. 5–17). Apart from galactopyranose and rhamnopyranose units there are units of arabinofuranose, 4-O-methylglucuronic, and galacturonic acid in the molecule which has a $\overline{P}_n$ of 350 to 400 (Meier 1962; Kuo, Timell 1969).

Arabinogalactans with a β-(1→3)-linked backbone of galactose units and various side chains of arabinose, rhamnose, galacturonic and 4-O-methylglucuronic acid were isolated from maple sap *(Acer saccharum)* and from gums of bunya pine *(Araucaria bidwillii)*, acacia *(Acacia mearnsii)*, cotton *(Gossypium arboreum)*, cholla *(Opuntia fulgida)* and other plants (Adams, Bishop 1960; Aspinall 1964b; Aspinall, McKenna 1968; Aspinall et al. 1968a; Parikh, Jones 1965; Buchala, Meier 1981).

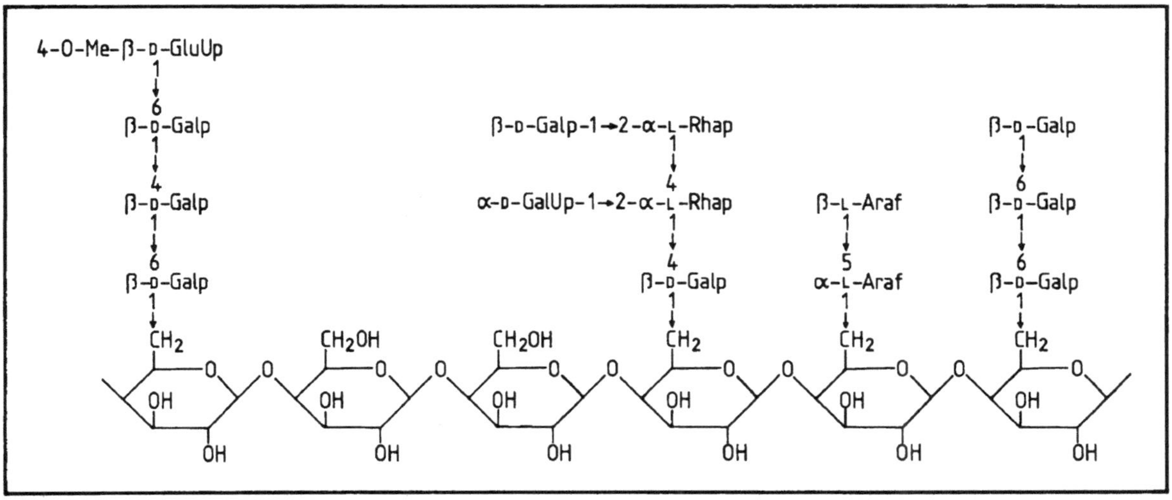

Fig. 5–17. Partial chemical structure of galactan from tension wood of beech.

5.6. Pectins

The group of pectin compounds comprises the galacturonans, the galactans and the arabinans. The galactans have been described in the preceding chapter.

Galacturonans of various composition are components of many plants, and occur particularly in fruit peels and gums. The content of galacturonans in soft- and hardwoods is less than 1% (Jayme, Hahn 1960). They are predominantly deposited in the middle lamellae and the tori of bordered pit-membranes (Bauch et al. 1968).

The pectin isolated from a cell culture of sycamore *(Acer pseudoplatanus)* consisted of an arabinan and a rhamnogalacturonan (Talmadge et al. 1973). The latter compound has a backbone of α-(1→4)-linked galacturonic-acid units which contains a rhamnose unit at regular intervals of about 8 units; about half of them carry a galactan side chain. The rhamnose units are linked by α-(1→2)- and α-(1→4)-bonds to the adjacent galacturonic-acid units (Fig. 5–18).

An arabinan was also isolated from the wood of maritime pine *(Pinus pinaster)* in a yield of 0.31% (Roudier 1964). The polyose contains more than 90% arabinose and small percentages of galactose, xylose and glucose units. The arabinose units are joined by α-(1→5)-linkages. The backbone bears arabinose side chains which are α-(1→3)-linked at the branching points (Fig. 5–19).

In other plant tissues and gums the galacturonans consist of homopolymer or of heteropolymer backbones (rhamnogalacturonans) with side chains of galactose, arabinose, xylose and fucose units (Aspinall 1964a, 1973; Aspinall et al. 1968b). The backbone of rhamnogalacturonan from karaya gum *(Sterculia urens)* additionally contains galactose units. Apiose, an unusual pentose (Fig. 5–20) was detected in the side chains of galacturonan from duckweed *(Lemna minor)* (Aspinall 1973).

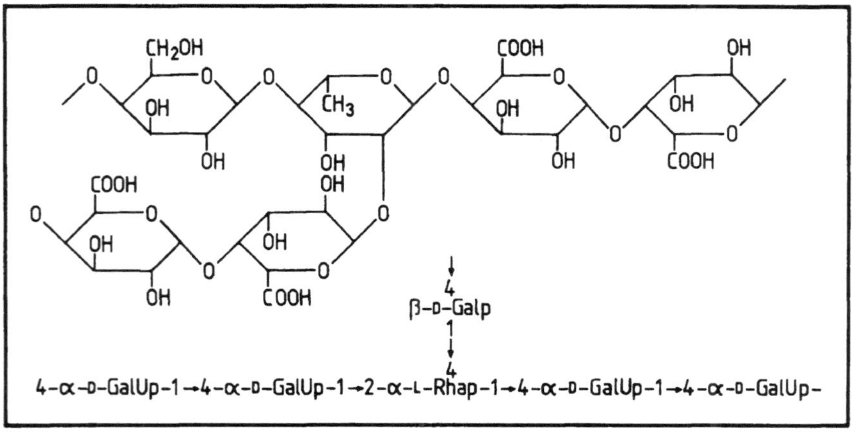

Fig. 5–18. Partial chemical structure of rhamnogalacturonan from a cell culture of sycamore.

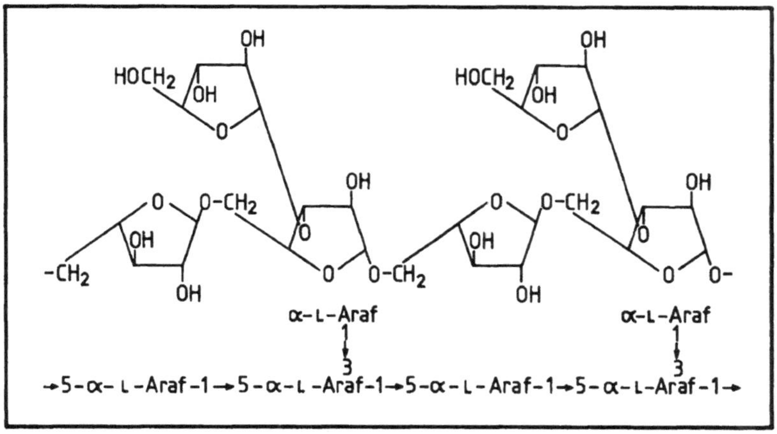

Fig. 5–19. Partial chemical structure of arabinan from maritime pine.

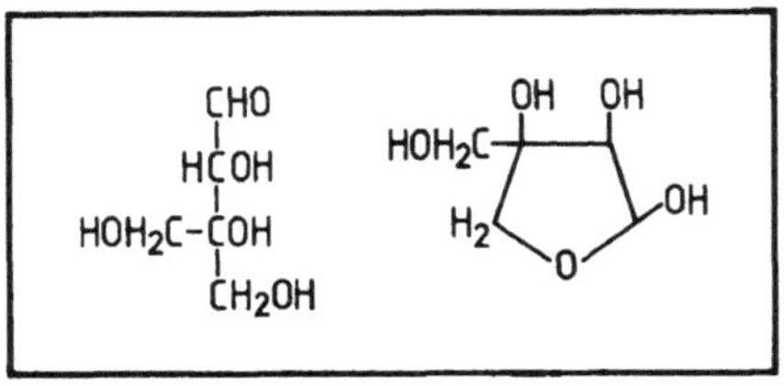

Fig. 5–20. Formula of apiose in chain and ring form.

Literature

Adams, G. A. 1961, Can. J. Chem. *39,* 2423–2429

Adams, G. A. 1964, Svensk Papperstid. *67,* 82–88

Adams, G. A. and Bishop, C. T. 1960, Can. J. Chem. *38,* 2380–2386

Adams, M. F. and Douglas, C. 1963, Tappi *46,* 544–548

Albersheim, P., Bauer, W. D., Keegstra, K. and Talmadge, K. W. 1973, The Structure of the Walls of Suspension-Cultured Sycamore Cells. In: Biogenesis of Plant Cell Wall Polysaccharides (Loewus, F., Ed.). Academic Press, New York, London, pp. 117–147

Andersson, S. I. and Samuelson, O. 1978, Svensk Papperstid. *81,* 79–84

Andersson, S. I., Samuelson, O., Ishihara, M. and Shimizu, K. 1983, Carbohyd. Res. *111*, 283–288

Aspinall, G. O. 1964a, Progress and Prospects in the Chemistry of Cell Wall Polysaccharides. In: Chimie et Biochimie de la Lignine, de la Cellulose et des Hémicelluloses. Actes du Symposium International de Grenoble. Les Imprimeries réunies, Chambéry, pp. 421–433

Aspinall, G. O. 1964b, Some Recent Developments in the Chemistry of Arabinogalactans. In: Chimie et Biochimie de la Lignine, de la Cellulose et des Hémicelluloses. Actes du Symposium International de Grenoble. Les Imprimeries réunies, Chambéry, pp. 89–97

Aspinall, G. O. 1973, Carbohydrate Polymers of Plant Cell Walls. In: Biogenesis of Plant Cell Wall Polysaccharides (Loewus, F., Ed.). Academic Press, New York, London, pp. 95–115

Aspinall, G. O. and McKenna, J. P. 1968, Carbohyd. Res. *7,* 244–254

Aspinall, G. O., Hirst, E. L., Percival, E. G. V. and Williamson, I. R. 1953, J. Chem. Soc. *1953,* 3184–3188

Aspinall, G. O., Rashbrock, R. B. and Kessler, G. 1958, J. Chem. Soc. *1958,* 215–221

Aspinall, G. O., Carlyle, J. J. and Young, R. 1968a, Carbohyd. Res. *7,* 421–430
Aspinall, G. O., Graig, J. W. T. and Whyte, J. L. 1968b, Carbohyd. Res, *7,* 442–452
Banerjee, S. K. and Timell, T. E. 1960, Tappi *43,* 849–857
Bardalaye, P. C. and Hay, G. W. 1974, Carbohyd. Res. *37,* 339–350
Barth, F. W. and Timell, T. E. 1958, J. Am. Chem. Soc. *80,* 6320–6325
Bauch, J., Liese, W. and Scholz, F. 1968, Holzforschung *22,* 145–153
Bauer, R. 1970, Chemiefaser-Lexikon. Deutscher Fachverlag, Frankfurt/M.
Bauer, W. D., Talmadge, K. W., Keegstra, K. and Albersheim, P. 1973, Plant Physiol. *51,* 174–187
Bishop, C. T. and Cooper, F. P. 1960, Can. J. Chem. *38,* 793–804
Bittiger, H. and Husemann, E. 1972, Polym. Lett. *10,* 549–553
Björndal, H., Eriksson, K. E., Garegg, P. J., Lindberg, B. and Swan, B. 1965, Acta Chem. Scand. *19,* 2309–2315
Brasch, D. J. and Jones, J. K. N. 1959, Can. J. Chem. *37,* 1538–1545
Buchala, A. J. 1973, Phytochemistry *12,* 1373–1376
Buchala, A. J. 1974, Phytochemistry *13,* 2185–2188
Buchala, A. J. and Franz, G. 1974, Phytochemistry *13,* 1887–1889
Buchala, A. J. and Meier, H. 1973a, Carbohyd. Res. *31,* 87–92
Buchala, A. J. and Meier, H. 1973b, Carbohyd. Res. *26,* 421–425
Buchala, A. J. and Meier, H. 1981, Carbohyd. Res. *89,* 137–143
Buchala, A. J. and Wilkie, K. C. B. 1971, Phytochemistry *10,* 2287–2291
Buchala, A. J. and Wilkie, K. C. B. 1973, Phytochemistry *12,* 499–505
Buchala, A. J., Franz, G. and Meier, H. 1974, Phytochemistry *13,* 163–166
Chanzy, H., Dube, M. and Marchessault, R. H. 1978, Tappi *61,* No. 7, 81–82
Chanzy, H., Dube, M., Marchessault, R. H. and Revol, J. F. 1979, Biopolymers *18,* 887–898
Côté, W. A. and Timell, T. E. 1967, Tappi *50,* 285–289
Côté, W. A., Simson, B. W. and Timell, T. E. 1966, Svensk Papperstid. *69,* 547–558
Côté, W. A., Pickard, P. A. and Timell, T. E. 1967, Tappi *50,* 350–356
Croon, I. and Timell, T. E. 1960, J. Am. Chem. Soc. *82,* 3416–3418
Currie, A. L. and Timell, T. E. 1959, Can. J. Chem. *37,* 922–929
Das, A. and Rao, C. V. N. 1964, Tappi *47,* 339–343
Dudkin, M. S., Tatarkina, G. V. and Kazanskaya, I. S. 1971, Khim. Drev. No. 8, 23–26
Dudkin, M. S., Shkantova, N. G. and Parfenteva, M. A. 1972, Khim. Drev. No. 12, 55–58
Ebringerová, A., Kramár, A., Rendoš, F. and Domanský, R. 1967, Holzforschung *21,* 74–77
Ebringerová, A., Kramár, A. and Domanský, R. 1969, Holzforschung *23,* 89–92
Ebringerová, A., Kramár, A. and Domanský, R. 1972, Holzforschung *26,* 89–92
Ericsson, T., Petersson, G. and Samuelson, O. 1977, Wood Sci. Technol. *11,* 219–223
Fengel, D. 1965, Naturwissenschaften *52,* 641–642
Fengel, D. 1966a, Naturwissenschaften *53,* 18–19
Fengel, D. 1966b, Holz Roh-Werkst. *24,* 245–253
Fengel, D. 1967, Svensk Papperstid. *70,* 70–77
Fengel, D. 1976, Holzforschung *30,* 143–148
Fengel, D. 1979, Cell. Chem. Technol. *13,* 279–297
Fengel, D. and Przyklenk, M. 1976, Wood Sci. Technol. *10,* 311–320
Fengel, D., Wegener, G., Heizmann, A. and Przyklenk, M. 1978, Cell. Chem. Technol. *12,* 31–37
Frei, E. and Preston, R. D. 1968, Proc. Roy. Soc. *B 169,* 127–145
Fu, Y. L. and Timell, T. E. 1972a, Cell. Chem. Technol. 6, 517–519
Fu, Y. L. and Timell, T. E. 1972b, Svensk Papperstid. *75,* 680–682
Fu, Y. L., Gutmann, P. J. and Timell, T. E. 1972, Cell. Chem. Technol. *6,* 507–512
Garegg, P. J. and Han, M. 1968, Svensk Papperstid. *71,* 331–334
Gillham, J. K. and Timell, T. E. 1958, Can. J. Chem. *36,* 1467–1474

Grosrenaud, A. 1980, Mannanes et Glucomannanes: Cristallisation et Polymorphisme. Doctor Thesis, l'Université Scientifique et Medicale de Grenoble
Gutmann, P. J. and Timell, T. E. 1965, J. Polym. Sci., Pt. C, No. 11, 281–292
Gvozdeva, E. N. and Levanova, V. P. 1979, Khim. Drev. No. 1, 3–5
Gvozdeva, E. I., Sharkov, V. I. and Kujbina, N. I. 1971, Khim. Drev. No. 8, 35–41
Han, M. and Swan, B. 1968, Svensk Papperstid. *71,* 552–557
Harwood, V. D. 1972, Svensk Papperstid. *75,* 207–212
Harwood, V. D. 1973, Svensk Papperstid. *76,* 377–379
Hashi, M., Teratani, F. and Miyazaki, K. 1970, Mokuzai Gakkaishi *16,* 37–41
Hashi, M., Teratani, F. and Miyazaki, K. 1971, Mokuzai Gakkaishi *17,* 405–410
Hoffmann, G. C. and Timell, T. E. 1970a, Tappi *53,* 1896–1899
Hoffmann, G. C. and Timell, T. E. 1970b, Wood Sci. Technol. *4,* 159–162
Hoffmann, G. C. and Timell, T. E. 1972a, Svensk Papperstid. *75,* 135–142
Hoffmann, G. C. and Timell, T. E. 1972b, Svensk Papperstid. *75,* 241–242
Hoffmann, G. C. and Timell, T. E. 1972c, Svensk Papperstid. *75,* 297–298
Hoffmann, G. C. and Timell, T. E. 1972d, Tappi *55,* 733–736
Hoffmann, G. C. and Timell, T. E. 1972e, Tappi *55,* 871–873
Hui, P. A. and Neukom, H. 1964, Tappi *47,* 39–42
Jayme, G. and Hahn, G. 1960, Holzforschung *14,* 52–56
Jiang, K. S. and Timell, T. E. 1972, Svensk Papperstid. *75,* 592–594
Johansson, M. H. and Samuelson, O. 1977, Wood Sci. Technol. *11,* 251–263
Jones, J. K. N., Purves, C. B. and Timell, T. E. 1961, Can. J. Chem. *39,* 1059–1066
Jones, J. K. N. and Reid, P. E. 1963, J. Polym. Sci., Pt. C, No. 2, 63–71
Karnik, M. G., Morak, A. J. and Ward, K. 1963, Tappi *46,* 130–134
Katz, G. 1965, Tappi *48,* 34–41
Keusch, L. 1968, Planta *78,* 321–350
Kirillova, V. V. and Beinart, I. I. 1974, Khim. Drev. No. 15, 27–31
Klevinskaja, V. J., Katkevič, R. G. and Gromov, V. S. 1974, Khim. Drev. No. 1, 18–25
Kooiman, P. and Adams, G. A. 1961, Can. J. Chem. *39,* 889–896
Koshijima, T. and Tanaka, R. 1970, Mokuzai Gakkaishi *16,* 399–400
Koshijima, T., Timell, T. E. and Zinbo, M. 1965, J. Polym. Sci., Pt. C. No. 11, 265–279
Kuo, C. M. and Timell, T. E. 1969, Svensk Papperstid. *72,* 703–716
LeBel, R. G. and Goring, D. A. I. 1963, J. Polym. Sci., Pt. C, No. 2, 29–48
Lee, J. Y., Cho, N. S., Ishizu, A. and Nakano, J. 1981, Mokuzai Gakkaishi *27,* 745–749
Leschziner, C. and Cerezo, A. S. 1970, Carbohyd. Res. *15,* 291–299
Lindberg, B., Rosell, K. G. and Svensson, S. 1973a, Svensk Papperstid. *76,* 30–32
Lindberg, B., Rosell, K. G. and Svensson, S. 1973b, Svensk Papperstid. *76,* 383–384
Linnell, W. S. and Swenson, H. A. 1966, Tappi *49,* 444–446
Linnell, W. S., Thompson, N. S. and Swenson, H. A. 1966, Tappi *49,* 491–493
Love, J. and Percival, E. 1964, J. Chem. Soc. *1964,* 3345–3350
Lynch, R. S., Stilmann, J. E. and Timell, T. E. 1968, Svensk Papperstid. *71,* 890–891
Maekawa, E. 1976, Wood Res. No. 59/60, 153–179
Marchessault, R. H. 1964, Texture and Morphology of Xylans. In: Chimie et Biochimie de la Lignine, de la Cellulose et des Hémicelluloses. Actes du Symposium International de Grenoble. Les Imprimeries réunies, Chambéry, pp. 287–301
Marchessault, R. H. and Sarko, A. 1967, Adv. Carbohydr. Chem. *22,* 421–482
Marchessault, R. H. and Settineri, W. J. 1964, Polym. Lett. *2,* 1047–1051
Marchessault, R. H., Morehead, F. F., Walter, N. M., Glaudemans, C. P. J. and Timell, T. E. 1961, J. Polym. Sci. *51,* S66–S68
Marchessault, R. H., Holova, H. and Timell, T. E. 1963, Can. J. Chem. *41,* 1612–1618
Marchessault, R. H., Settineri, W. and Winter, W. 1967, Tappi *50,* 55–59

Meier, H. 1958, Biochim. Biophys. Acta *28,* 229–240

Meier, H. 1961, Acta Chem. Scand. *15,* 1381–1385

Meier, H. 1962, Acta Chem. Scand. *16,* 2275–2283

Meier, H. 1964, On the Chemistry of Reaction Wood. In: Chimie et Biochimie de la Lignin, de la Cellulose et des Hémicelluloses. Actes du Symposium International de Grenoble. Les Imprimeries réunies, Chambéry, pp. 405–412

Meier, H., Buchs, L., Buchala, A. J. and Homewood, T. 1981, Nature *289,* 821–822

Mian, A. J. and Timell, T. E. 1960a, Can. J. Chem. *38,* 1511–1517

Mian, A. J. and Timell, T. E. 1960b, Svensk Papperstid. *63,* 884–888

Mills, A. R. and Timell, T. E. 1963, Can J. Chem. *41,* 1389–1395

Nieduszynski, I. and Marchessault, R. H. 1971, Nature *232,* 46–47

Nieduszynski, I. and Marchessault, R. H. 1972, Can. J. Chem. *50,* 2130–2138

Nikolaeva, G. V., Levin, E. D. and Ioffe, G. M. 1971, Khim. Drev. No. 8, 155–158

Palmer, K. J. and Ballantyne, M. 1950, J. Am. Chem. Soc. *72,* 736–741

Parikh, V. M. and Jones, J. K. N. 1965, J. Polym. Sci., Pt. C, No. 11, 139–148

Rees, D. A. and Skerrett, R. J. 1968, Carbohyd. Res. *7,* 334–348

Reid, J. S. G. 1971, Planta *100,* 131–142

Reid, J. S. G. and Meier, H. 1973, Caryologia *25,* 219–222

Rogers, J. K. and Thompson, N. S. 1969, Svensk Papperstid. *72,* 61–67

Roudier, A. J. 1964, Les hémicelluloses du bois de Pin maritime des Landes: Revue des résultats antérieurs, exposé des traveaux récents. In: Chimie et Biochimie de la Lignine, de la Cellulose et des Hémicelluloses. Actes du Symposium International de Grenoble. Les Imprimeries réunies, Chambéry, pp. 113–125

Ruel, K. and Barnoud, F. 1978, Holzforschung *32,* 149–156

Samuelson, O. and Wictorin, L. 1966, Svensk Papperstid. *69,* 777–782

Schreuder, H. R., Côté W. A. and Timell, T. E. 1966, Svensk Papperstid. *69,* 641–657

Schuerch, C. 1963, The Hemicelluloses. In: The Chemistry of Wood. Intersci. Publ., New York, London, pp. 191–243

Schulze, E. 1891, Ber. *24,* 2285

Schwarz, E. C. A. and Timell, T. E. 1963, Can. J. Chem. *41,* 1381–1388

Seiler, A., Reid, J. S. G. and Meier, H. 1973, Verh. Schweiz, Naturforsch. Ges. *1973,* 73–75

Settineri, W. J. and Marchessault, R. H. 1965, J. Polym. Sci., Pt. C, No. 11, 253–264

Sharkov, V. I., Kujbina, N. I., Soloveva, J. P. and Borzakovskaja, B. S. 1971, Khim. Drev. No. 8, 27–34

Sharkov, V. I., Kujbina, N. I. and Gvozdeva, E. N. 1974, Khim. Drev. No. 1, 5–7

Shimizu, K. 1975, Bull. Gov. Forest Exp. Station, Meguro, No. 272, 1–78

Shimizu, K. and Samuelson, O. 1973, Svensk Papperstid. *76,* 150–155

Simson, B. W., Côté, W. A. and Timell, T. E. 1968, Svensk Papperstid. *71,* 699–710

Staudinger, H. and Reinecke, F. 1939, Holz Roh-Werkst. *2,* 321–323

Stewart, C. W. 1954, Holzforschung, *8,* 46–48

Swenson, H. A., Schmitt, C. A. and Thompson, N. S. 1965, J. Polym. Sci., Pt. C, No. 11, 243–252

Talmadge, K. W., Keegstra, K., Bauer, W. D. and Albersheim, P. 1973, Plant Physiol. *51,* 158–173

Thompson, N. S. and Kaustinen, O. A. 1964, Pap. Puu *46,* 637–650

Timell, T. E. 1959a, Can. J. Chem. *37,* 893–898

Timell, T. E. 1959b, J. Am. Chem. Soc. *81,* 4989–4994

Timell, T. E. 1960a, Svensk Papperstid. *63,* 472–476

Timell, T. E. 1960b, Tappi *43,* 844–848

Timell, T. E. 1960c, J. Am. Chem. Soc. *82,* 5211–5215

Timell, T. E. 1961, Tappi *44,* 88–96

Timell, T. E. 1962a, Svensk Papperstid. *65,* 122–125

Timell, T. E. 1962b, Svensk Papperstid. *65,* 173–177

Timell, T. E. 1962c, Tappi *45,* 734–738, 799–802
Timell, T. E. 1964a, Adv. Carbohydr. Chem. *19,* 247–302
Timell, T. E. 1964b, Svensk Papperstid. *67,* 356–363
Timell, T. E. 1965, Adv. Carbohydr. Chem. *20,* 409–483
Timell, T. E. 1967, Wood Sci. Technol. *1,* 45–70
Timell, T. E. 1969, Svensk Papperstid. *72,* 173–181
Timell, T. E. and Zinbo, M. 1967, Tappi *50,* 195–198
Wickström, R. 1967, Svensk Papperstid. *70,* 819–821
Wickström, R. 1968, Svensk Papperstid. *71,* 399–404
Yundt, A. P. 1951, Tappi *34,* 94–95
Zinbo, M. and Timell, T. E. 1965, Svensk Papperstid. *68,* 647–662
Zinbo, M. and Timell, T. E. 1967, Svensk Papperstid. *70,* 597–606, 695–701

6. Lignin

6.1. Significance and Occurrence

Next to cellulose lignin is the most abundant and important polymeric organic substance in the plant world. The incorporation of lignin into the cell walls of plants gave them the chance to conquer the Earth's land surface. Lignin increased the mechanical strength properties to such an extent that huge plants such as trees with heights of even more than 100 m can remain upright.

Lignin is a characteristic chemical and morphological component of the tissues of higher plants such as pterodophytes and spermatophytes (gymnosperms and angiosperms), where it typically occurs in the vascular tissues, specialized for liquid transport and mechanical strength (e.g. xylem, → 2.1.). Primitive plants such as fungi, lichens and algae are not lignified, while it is still questionable whether mosses contain true lignin or only phenolic compounds which also yield a non-hydrolyzable residue during acid treatment (Freudenberg 1968; Sarkanen, Hergert 1971). More recent investigations on several mosses (e.g. *Sphagnum magellanicum*) indicate that they contain no lignin and that lignin occurrence is indeed restricted to vascular plants (Erickson, Miksche 1974; Miksche, Yasuda 1978; Nimz, Tutschek 1977).

The amounts of lignin present in different plants is quite variable. While in wood species the lignin content ranges from 20 to 40% (→ 3.3.) aquatic and herbaceous angiosperms as well as many monocotyledons (e.g. horse-tail species) are less lignified (Neish 1968; Sarkanen, Hergert 1971; Krüger 1976).

Additionally the distribution of lignin within the cell wall (→ 8.2.) and the lignin contents of different parts of a tree are not uniform. For example high lignin values are characteristic for the lowest, highest and inner parts of the stem, for softwood branches, bark (→ 9.2.) and compression wood (→ 2.3.). The lignin contents of needles and leaves are described inconsistently as high or low, possibly depending on their state of development (Wardrop 1971; Uprichard 1971; Howard 1973; Miksche, Yasuda 1977; Luger, Gampe 1978).

In most cases of wood utilization lignin is used as an integrated part of wood. Only in the case of pulping and bleaching (→ 16.) is lignin more ore less released from wood in degraded and altered forms, representing a large potential carbon source of more than 35 million tons of carbon per year world-wide for chemical and energy purposes (→ 18.6.).

6.2. Lignification of the Wood Cell Wall

6.2.1. Synthesis of the Monomeric Lignin Units

The building-up of the lignin macromolecules by the plant comprises complicated biological, biochemical and chemical systems which have been extensively studied and repeatedly reviewed (Brauns, Brauns 1960; Neish 1964; Freudenberg 1964a; Schubert 1965; Pearl 1967; Freudenberg, Neish 1968; Sarkanen 1971; Weissenböck 1976; Wardrop 1976, 1981; Grisebach 1977; Adler 1977; Gross 1977, 1978; Glasser 1980).

Numerous studies with radioactive carbon (^{14}C) have confirmed that the p-hydroxy-cinnamyl alcohols p-coumaryl alcohol (I), coniferyl alcohol (II) and sinapyl alcohol (III) are the primary precursors and building units of all lignins (Fig. 6–1).

Fig. 6–2 gives a survey of the main formation steps of the lignin precursors (Neish 1968; Sarkanen 1971; Grisebach 1977; Gross 1977, 1978). The biosynthesis of lignin starts with glucose (I) deriving from photosynthesis. It is converted to shikimic acid (II), the most important intermediate substance of the so-called shikimic acid pathway. The two aromatic amino acids L-phenylalanine (IV) and L-tyrosine (V) are formed by reductive amination via prephenic acid (III) as the final compounds of that pathway. On the other hand they are the starting substances ("amino acid pools") for the enzymatic phenylpropanoid metabolism (cinnamic acid pathway) leading not only to the three cinnamyl alcohols via activated cinnamic acid derivatives, but also to extractive components like flavonoids or stilbenes (→ 7.2.3.). The amino acids are deaminated by deaminases (phenylalanine ammonia lyase and tyrosine ammonia lyase) to their corresponding cinnamic acids (VI, VII). The dominant further steps are successive hydroxylation (by phenolases (hydroxylases)) and methylation (by O-methyltransferases) leading to p-coumaric acid (VI), caffeic acid (VIII), ferulic acid (IX), 5-hydroxy-ferulic acid (X), and sinapic acid (XI).

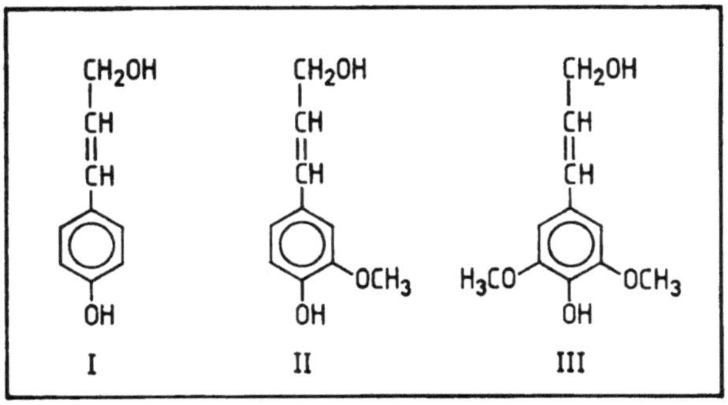

Fig. 6–1. The building units of lignin: p-coumaryl alcohol (I), coniferyl alcohol (II), sinapyl alcohol (III).

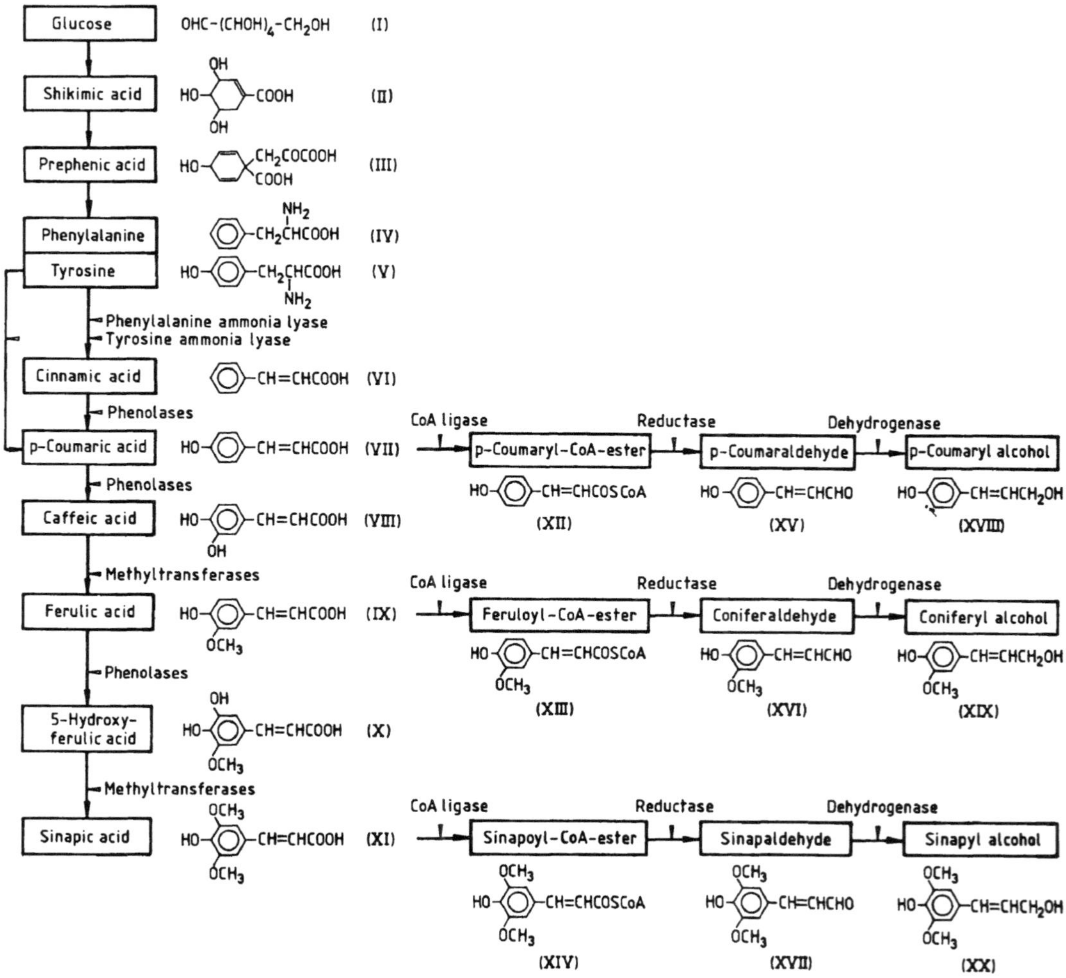

Fig. 6–2. Metabolic pathway from glucose to phenylpropane lignin precursors.

One reason for the higher proportions of syringyl units in hardwood lignins than in softwood lignins (→ 6.4.1.) is expected to be the higher affinity of the angiosperm 4-O-methyltransferase for 5-hydroxy-ferulic acid as compared to the respective transferase of gymnosperms. The regulation of lignin production is evidently also controlled very early by the different activities of the deaminases under varying conditions with regard to factors such as light and hormone supply (Grisebach 1977; Higuchi et al. 1977).

The cinnamyl alcohols (XVIII–XX) are finally formed by enzymatic activation (CoA ligase) and reduction (NADP reductase, NADP hydrogenase) of the corresponding acids via coenzyme-A thioesters (XII–XIV) and aldehydes (XV–XVII) (Gross et al. 1973, 1975; Gross 1977, 1978; Hahlbrock, Grisebach 1979; Alibert, Boudet 1979; Grisebach et al. 1981).

From a biochemical standpoint it is important that all enzymes involved in the formation of the cinnamyl alcohols are tissue-specific and are located predominantly or even exclusively in the lignifying xylem cells. Therefore the translocation of water-soluble glycosides of the cinnamyl alcohols, e.g. coniferin, the β-D-glucoside of coniferyl alcohol, (Fig. 6–3), from sites of precursors synthesis (cambial zone) to those of lignin deposition (lignifying xylem) (Freudenberg, Neish 1968) is no longer expected today as a necessary step in lignin incorporation. A more probable view is to regard the cinnamyl alcohol glucosides, whose existence was proved in the cambial sap of several conifers but not in angiosperms (Freudenberg, Neish 1968), as a reservoir of precursors for the lignifying cells (Grisebach 1977; Gross 1978). The glucosides must be reconverted to alcohols by β-glucosidase before the polymerization reactions (Fig. 6–3).

The presence of the hydrolase β-glucosidase in lignifying cells and its absence in cambium cells not involved in lignification was proved histochemically by the well-known indican colour reaction (Freudenberg 1968). Though this reaction had proved the presence of glucosidase, only recently the isolation and characterization of a cell wall-bound glucosidase from spruce seedlings was achieved (Marcinowski, Grisebach 1978; Marcinowski et al. 1979). The mechanism of transporting coniferin through the plasma membrane to reach the glucosidase in the cell wall is still uncertain, though some results hint at a participation of the Golgi vesicles (Pickett-Heaps 1968) (→ 6.2.3.).

6.2.2. Formation of Lignin Macromolecules

The biosynthesis of lignin from the monomeric phenylpropane units can be generally described as a dehydrogenative polymerization. The principal ideas about that pathway were elaborated by Freudenberg and coworkers. They were the first to produce an in-vitro lignin, called dehydrogenation polymer (DHP), by treating

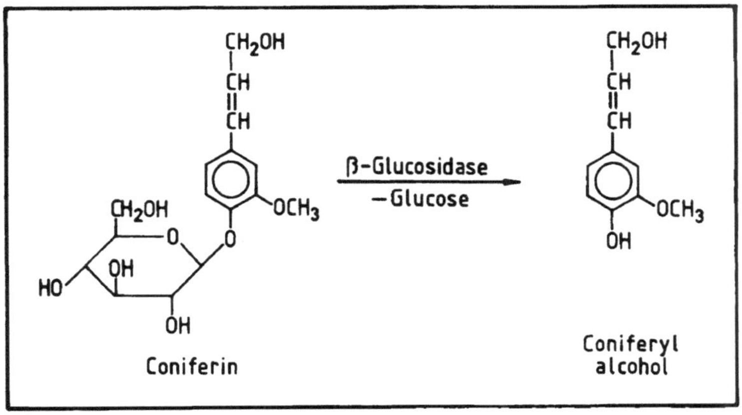

Fig. 6–3. Hydrolysis of coniferin by β-D-glucosidase.

coniferyl alcohol with a fungal laccase from the mushroom *Psalliota campestris* or with a horse-radish peroxidase and hydrogen peroxide (Freudenberg, Neish 1968; Sarkanen 1971; Adler 1977).

The first step of the biochemical pathway for building up lignin macromolecules is the enzymatic dehydrogenation of the p-hydroxycinnamyl alcohols, yielding mesomeric ring systems with a loosened proton. Fig. 6–4 shows the formation of the resonance-stabilized phenoxy radicals from coniferyl alcohol by a one-electron transfer.

Though the catalytic action of laccase in the presence of air (Freudenberg, Neish 1968) or several chemical oxidation substances (Sarkanen 1971; Stoldere et al. 1978) are able to form phenoxy radicals from cinnamyl alcohols, the most frequently suggested catalysts for initiating the polymerization reactions are cell wall peroxidases in combination with hydrogen peroxide as an oxidant (peroxidase/H_2O_2-system) (Lai, Sarkanen 1975; Mader 1976; Yamasaki et al. 1976; Grisebach 1977; Gross 1978). The participation of these enzymes was proved additionally to Freudenberg's experiments by a histochemical method (Harkin, Obst 1973).

The origin of the hydrogen peroxide was cleared up by discovering cell wall-bound enzyme systems able to deliver H_2O_2 (Elstner, Heupel 1976; Halliwell 1978). In contrast a calcium peroxide-generating enzyme acting outside the cell wall was recently proposed to produce phenoxy radicals. These findings would avoid the problems of explaining how enzymes such as peroxidases can enter and act within the lignifying xylem cell wall (Westermark 1982).

Only 4 phenoxy radicals (I–IV, Fig. 6–4) are actually involved in lignin biosynthesis, V sterically hindered or thermodynamically disfavoured (Glaser 1980). The principal coupling modes of the radicals are shown in Table 6–1. The relative electron densities determine the frequency of different sites involved in coup-

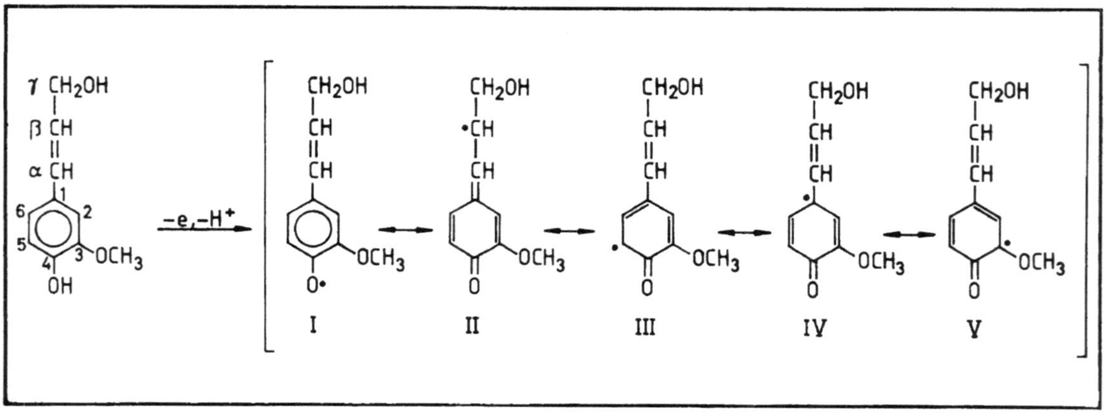

Fig. 6–4. Enzymatic dehydrogenation of coniferyl alcohol yielding phenoxy radicals.

Table 6–1. Coupling modes of phenoxy radicals from p-hydroxycinnamyl alcohols (Glasser 1980)

	I	II	III	IV
I	unstable peroxide	β-O-4	4-O-5	1-O-4*
II	β-O-4	β-β	β-5	β-1*
III	4-O-5	β-5	5-5	1-5*
IV	1-O-4*	β-1*	1-5*	1-1*

* Other options possible

ling reactions. From quantum mechanical calculations it was deduced that e.g. all phenoxy radicals have the highest π-electron densities at the phenolic oxygen atom, thus favouring the formation of aryl ether linkages such as the β-O-4 linkage, the most frequent type of bond in softwood and hardwood lignins (Glasser 1980) (Table 6–2).

The polymerization of the monomeric precursors by random coupling reactions cannot be studied in vivo but it is known from numerous in-vitro experiments to run without enzymatic control as a spontaneous process. The first step in polymerization is the formation of dimeric structures. Some prominent so-called dilignols are shown in Fig. 6–5. The further polymerization is called ›end-wise‹ polymerization, involving a coupling of monolignols with the phenolic end-groups of di- or oligolignol or a coupling of two end-group radicals, yielding a branched polymer via tri-, tetra-, penta- and oligolignols. The limited number of coniferyl alcohol and coniferaldehyde end-groups present in isolated Björkman lignin from spruce indicated that this type of polymerization is favoured over the coupling of monolignols with

Table 6–2. Types and frequencies of interunit linkages in lignin models (Number of linkages per 100 C_9-units)

	Glasser, Glasser (1981)	Erickson et al. (1973c)	Nimz (1974)
β-O-4	} 55	49–51	} 65
α-O-4		6–8	
β-5	16	9–15	6
β-1	9	2	15
5-5	9	9.5	2.3
4-O-5	3	3.5	1.5
β-β	2	2	5.5
β-β (THF)*	–	–	2
α/γ-O-γ	10	–	–
α-β	11	–	2.5
β-6, 6-5,	2	} 4.5-5	–
1-O-4, 1-5	(only 1–5)		
Total	117	85.5–96	99.8

* THF: Tetrahydrofuran structure

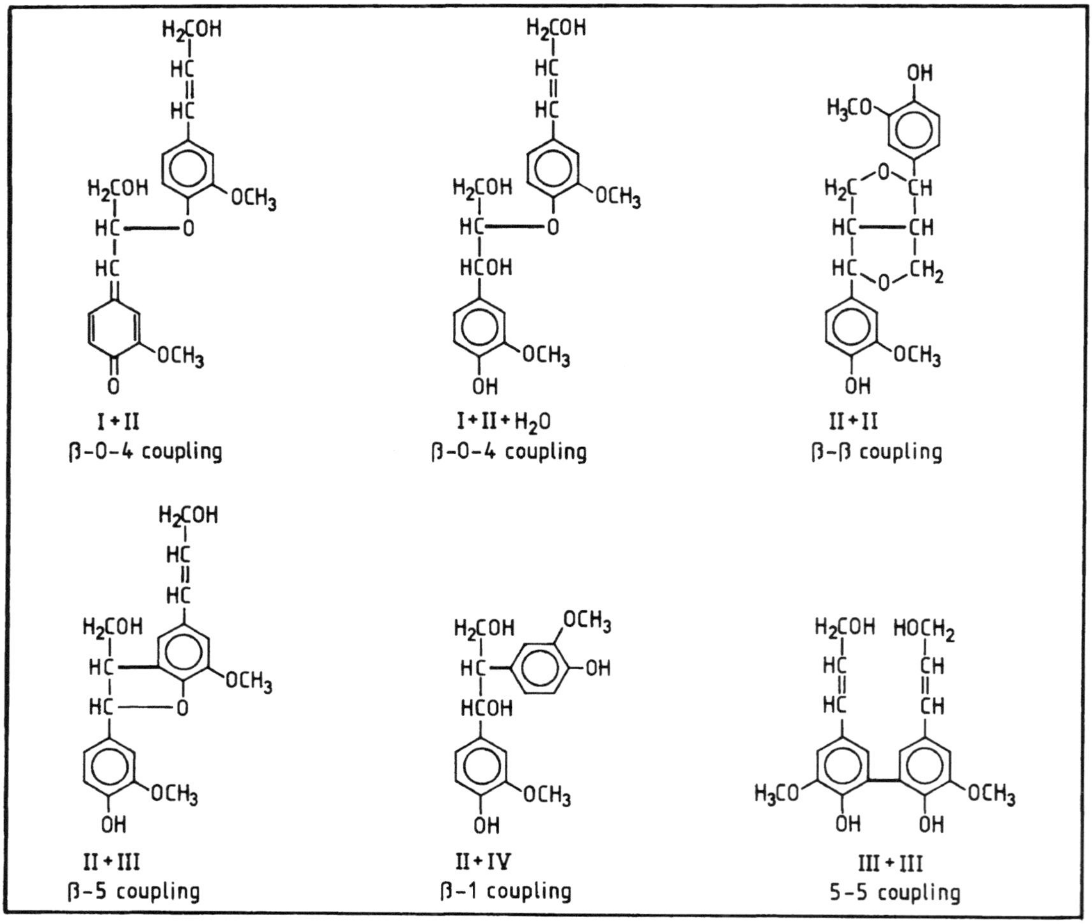

Fig. 6–5. Typical dilignol structures. Roman numerals from Fig. 6–4.
I+II: Quinone methide
I+II+H_2O: Guaiacylglycerol-β-coniferyl ether
II+II: D,L-Pinoresinol
II+III: Dehydroconiferyl alcohol
II+IV: 1,2-Diguaiacylpropane-1,3-diol
III+III: Dehydrobisconiferyl alcohol

each other to form dimeric structures. This is probably caused by a restricted supply of the monomers in the lignifying cell (Adler 1977). Fig. 6–6 gives two examples of higher structural units within the polymerization process. The tetralignol (guaiacylglycerol-β-coniferyl-γ-dehydrodiconiferyl ether) is formed by coupling dehydrodiconiferyl alcohol with a dimeric quinone methide (Fig. 6–5). The pentalignol represents the guaiacylglycerol ether of the tetralignol.

A non-radical, ionic coupling principle in lignin formation is the addition of transient quinone methides to water (Fig. 6–5) or phenolic groups (e.g. tetralignol, Fig. 6–6). Other quinone methide-derived intermediates may build up yet unknown

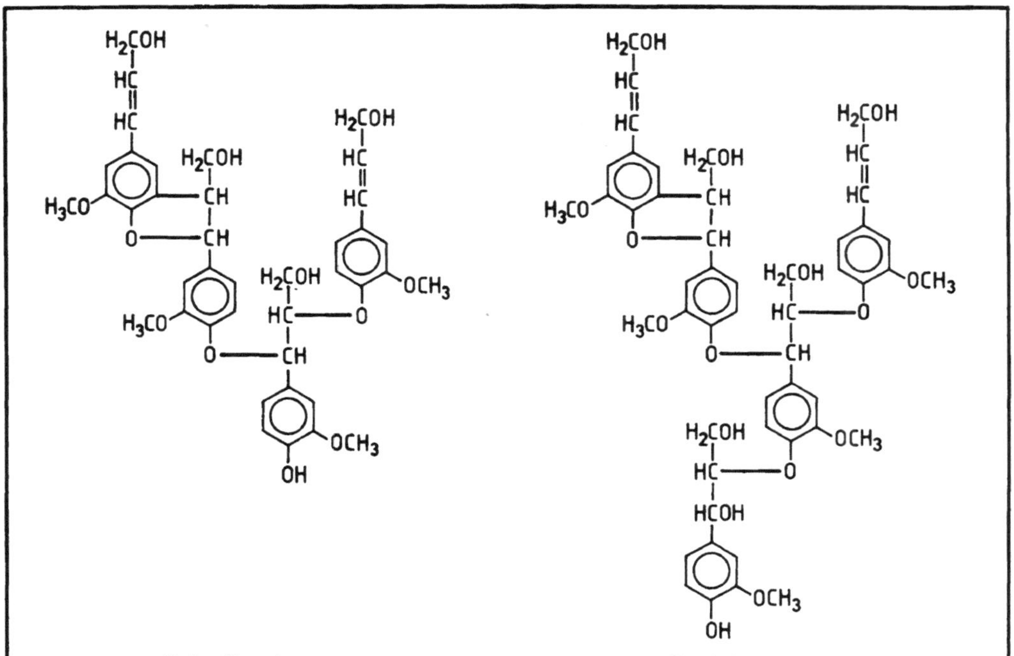

Fig. 6–6. Examples of higher lignol structures:
Tetralignol: guaiacylglycerol-β-coniferyl-γ-dehydrodiconiferyl alcohol
Pentalignol: guaiacylglycerol ether of the tetralignol.

dialkyl- and alkyl-aryl ether structures (Leary, Thomas 1975; Nakatsubo et al. 1976; Leary 1980; Glasser 1980). Additionally quinone methides are regarded as conceivable reaction partners in the formation of lignin-polysaccharide linkages (Freudenberg 1968; Leary et al. 1981) (→ 6.5.). Another non-dehydrogenative, hypothetical lignification mechanism based on cationic vinyl polymerization was described by Glasser et al. (1976), leading to compounds comprising α-β and α-O-γ linkages.

Summarizing the formation of lignin it is evident that these macromolecules are not formed by a genetically prescribed, regular mechanism, but by a random coupling of lignols to a non-linear polymer. The final constitution of lignin is therefore determined mostly by the reactivity and frequency of the building units involved in the polymerization. From a morphological standpoint the growing lignin molecules are forced to fill up the spaces between the preformed polysaccharidic fibrillar elements of the cell wall. The incorporation of the hydrophobic lignin causes a de-swelling of the cell walls.

6.2.3. Aspects of Lignin Deposition in the Differentiating Cell Wall

The incorporation of lignin within the polysaccharide cell wall frame work is generally seen as the final phase of the differentiating process of the typical secondary xylem cell. It gives the wood cell walls their characteristics with regard e.g. to strength, density or swelling properties.

Though there are still many open questions concerning details about the exact beginning of lignin deposition numerous results have cleared up the course of lignification. Results from ultraviolet, fluorescence and light autoradiographic microscopic studies as well as from electron microscopic investigations have confirmed that lignin is most probably deposited initially in the cell corners when the surface enlargement of the cell is finished and just before the secondary wall 1 (S 1) starts thickening. The lignification proceeds in the intercellular layer (middle lamella, ML) and the primary wall (P), starting at the tangential walls and spreading centripetally. The lignification of the compound middle lamella (ML + neighboured primary walls) continues during the differentiation of the S 1 and S 2 layers, and even until the formation of the tertiary wall (T). The lignification of the secondary wall layers proceeds slowly in a first stage but becomes more rapid after the thickening of the tertiary wall has been completed (Wardrop 1957, 1971, 1976, 1981; Saleh et al. 1967; Kutscha, Schwarzmann 1975; Fujita et al. 1978a; Fujita, Harada 1979; Fujita 1981; Tabake et al. 1981). These findings indicate a permanent lignification process throughout the whole time of cell wall differentiation, with a considerable delay to the synthesis of cellulose and polyoses.

As already mentioned, in contrast to Freudenberg's results proving an intercellular lignification process with participation of the cambium cells (Freudenberg, Neish 1968), today the lignification is considered as a process controlled by the individual cell, i.e. as an intracellular process. By means of this assumption it is possible to explain the partial lignification of phloem fibres and sclereids even a considerable time after those cells were produced by the cambium (Srivastava 1966; Wardrop 1976). In tissues in which lignifying tracheids are bordered by other tracheids the intercellular layer is lignified, while if a non-lignifying parenchyma cell is adjacent to a tracheid the lignification stops at the primary wall (Wardrop 1976; 1981).

Other hints toward an intracellular formation of lignin precursors were obtained by discovering different types of lignin in different cells of the same tissue (Fergus, Goring 1970), and by proving the existence of phenylalanine ammonia lyase (Fig. 6–2) in the cell wall of a lignifying xylem cell but not in the cambium (Rubery, Northcote 1968). The hypothesis still lacks an explanation as to how lignin or its precursors reach the middle lamella and the outer regions of the cell wall.

Concerning cytological aspects, earlier findings that certain vesicles fuse with the plasma membrane of lignifying cells were recently confirmed and specified (Pickett-Heaps 1968; Hepler et al. 1970; Wardrop 1976; Fujita et al. 1978b; Fujita 1981). Evidently the lignin precursors are synthesized and stored in vesicles deriving mainly from Golgi vesicles and partly from the endoplasmatic reticulum (ER), and finally supplied to the cell wall. The further transport within the cell wall is unknown.

As factors influencing lignification the contents of mineral elements (mainly calcium), physiologically active compounds (e.g. auxin) and genetic factors have been assumed without definite results (Wardrop 1981).

The final distribution of lignin within the different cell wall layers is discussed in chapter 8.

6.3. Structure and Constitution of Lignin

6.3.1. Studies Elucidating Lignin Structure

As outlined in the previous sections lignin is one of the most difficult natural polymers with regard to its structure and heterogeneity. The unique biogenesis results in polyphenolic, branched polymers which lack regular and ordered repeating units such as in the case of cellulose or proteins.

In addition to the outstanding contributions of Freudenberg and coworkers to our knowledge of dehydrogenation reactions (Freudenberg, Neish 1968) numerous other analytical studies of model compounds, synthetic lignins (DHP) and isolated lignins were necessary to complete our picture of the principal structure of lignin. These investigations may be grouped as follows:

- degradation experiments, partly combined with tracer studies, including: ethanolysis, acidolysis, hydrogenolysis, mild hydrolysis, thioacetolysis, oxidation (→ 10.)
- elemental analysis
- determination of functional groups.

Various lignin preparations have been studied with regard to their suitability for structural analyses, all of them revealing both advantages and drawbacks (Glasser 1980). Investigations of monomeric and dimeric model compounds are generally carried out in homogeneous phase, avoiding the problems involved in heterogeneous reactions, surface properties etc. Dehydrogenation polymers (DHP) can be regarded as lignin polymers with an exactly determinable composition (Schweers, Faix 1973; Faix, Besold 1978). But it is evident that they are not uniform with regard to molecular weight, ignore the role of carbohydrates in the lignification process and show structural differences compared with mildly isolated lignins (Evliya, Olcay 1974; Lai, Sarkanen 1975; Faix, Schweers 1975; Brunow, Lundquist 1980; Brunow, Wallin 1981; Feckl 1981). Isolated lignins suffer from degradation and changing effects, impurities and the difficulty of isolating reproducible and identical lignin samples from wood. Finally it is still questionable whether carefully isolated lignins are representative of the total lignin in wood (→ 3.2.9.) (Chang et al. 1975; Wegener, Stoll 1976; Adler 1977).

Ethanolysis, the hydrolytic treatment of wood or lignin with dilute alcoholic hydrochloric acid under pressure, was the original method for obtaining defined phenylpropanoid ketones (Hibbert's ketones) by splitting β-aryl ether linkages. Together with mild catalytic hydrogenolysis experiments, yielding mainly propylcyclohexane units, the phenylpropane nature of lignin was proved. This matter is extensively described and reviewed elsewhere (Neish 1968; Wallis 1971; Hrutfiord 1971; Kachi, Terashima 1974; Hoffmann, Schweers 1975a, b, 1976; Sakakibara 1977; Sudo et al. 1978, 1979; Sano, Sakakibara 1980; Hwang, Sakakibara 1981; Hwang et al. 1981).

In addition to classical ethanolysis, so-called acidolysis was used for degradation, applying acidified dioxane-water mixtures (9:1) to model compounds and isolated lignins as well. Numerous monomeric and dimeric degradation products (mainly ω-hydroxyguaiacylacetone) were isolated and characterized in addition to Hibbert's ketones (Adler et al. 1957; Adler 1961, 1977; Lundquist, Miksche 1965; Lundquist 1973, 1976; Lundquist et al. 1977).

Mild hydrolysis of wood for structural studies of lignin was either performed with dioxane-water mixtures (1:1) at 180 °C for 20 min or by percolation with water at 100 °C for several weeks (Nimz 1966, 1974; Sano, Sakakibara 1974; Sakakibara 1977, 1980).

A special technique of degrading lignin was worked out by Nimz (1969; 1974; Nimz, Das 1971). The treatment of spruce and beech wood with thioacetic acid in the presence of borontrifluoride and subsequent alkaline hydrolysis with sodium hydroxide (thioacetolysis) yielded mixtures of mono- to tetrameric degradation products in high yields of 91% and 77% based on beech and spruce lignin, respectively. The most important advantage of this degradation is its selectivity for cleaving α- and β-aryl ether bonds.

The oxidative degradation of lignin for structural studies must preserve the aromatic rings. Suitable methods are oxidation with potassium permanganate following hydrolysis and methylation, nitrobenzene oxidation, and oxidation with metal oxides (mostly CuO), in each case in combination with alkali. Numerous modifications, e.g. the application of sodium hydroxide and cupric oxide in the hydrolysis step, were elaborated mainly with the aim of increasing the yield of monomeric and dimeric carboxylic acids. Both aromatic degradation acids and aldehydes give valuable information about the structure of lignin, e.g. about etherified and free phenolic hydroxyl groups (Freudenberg, Neish 1968; Chang, Allen 1971; Erickson et al. 1973a; Glasser, Glasser 1976; Adler 1977). Separation and identification of the degradation products are predominantly performed by means of gas chromatography (GC) after derivatization (Larsson, Miksche 1971; Brink et al. 1972; Naveau et al. 1972; Erickson et al. 1973b; Morohoshi, Glasser 1979).

Elemental analysis together with the determination of the methoxyl content give information about the average composition of the C_9-units in lignin. Using analytical data obtained from Björkman lignin of spruce Freudenberg (1968) described the average lignin unit with the formula $C_9H_{7.12}O_2(H_2O)_{0.40}(OCH_3)_{0.92}$, assuming the loss of approximately two hydrogen atoms and addition of 0.4 molecules of water if compared to the average elemental composition of coniferyl alcohol in softwood lignin of $C_9H_{9.1}O_2(OCH_3)_{0.92}$. The composition of the average C_9-unit in lignin models (→ 6.3.2.) can be judged in comparison with corresponding data from isolated lignins.

The determination of functional groups such as free aliphatic and aromatic hydroxyl groups, benzyl alcohol or ether groups, carbonyl and methoxyl groups for the structural elucidation of lignin can be performed by means of numerous chemical and physical methods or combinations of both. The non-destructive physical methods include UV and IR spectroscopy as well as nuclear magnetic resonance spectroscopy (^{1}H-NMR, ^{13}C-NMR), electron spin resonance spectroscopy (ESR), and mass spectroscopy (MS), partly in combination with gas chromatography (GC-MS). Analytical techniques and results were described and reviewed extensively (Lai, Sarkanen 1971; Goldschmid 1971; Hergert 1971; Ludwig 1971; Nimz, Das 1971; Nimz 1974; Faix, Schweers 1974a, b; Robert, Gagnaire 1981).

Apart from structure elucidation most of the methods mentioned in the previous sections are also used for general characterization and comparison of different isolated lignins, and to determine alterations of lignins by chemical and physical treatments, e.g. during pulping processes (→ 6.4.1., 6.4.2., 10., 11.).

6.3.2. Structural Models of Lignin

As the lignin macromolecule cannot be described by a simple combination of one or a few monomeric units by one or a few types of linkages as in the case of cellulose (→ 4.) or polyoses (→ 5.) lignin structure is still a matter of models.

The first lignin model was designed by Freudenberg (1964b, 1968), based on the dehydrogenative polymerization concept and fulfilling all analytical data available at that time. This scheme for spruce lignin represents 18 phenylpropane units as a section of the total molecule which was assumed to consist of more than 100 units in the native state.

Later on, Adler (1977) gave a structural scheme for spruce lignin comprising 16 prominent C_9-units, mainly derived from the results of oxidative degradation experiments (Larsson, Miksche 1971; Erickson et al. 1973c) (Fig. 6–7). In such a fragmentary scheme it is unavoidable that certain structural units and linkages cannot be accounted for exactly. Thus, for example, the appearance of one syringyl unit (13) is not quantitative, and the pinoresinol unit (10,11) probably overemphasizes this structural element (Adler 1977).

A softwood lignin model based on the evaluation of degradation products deriving from mild hydrolysis with dioxane-water and from hydrogenolysis was recently

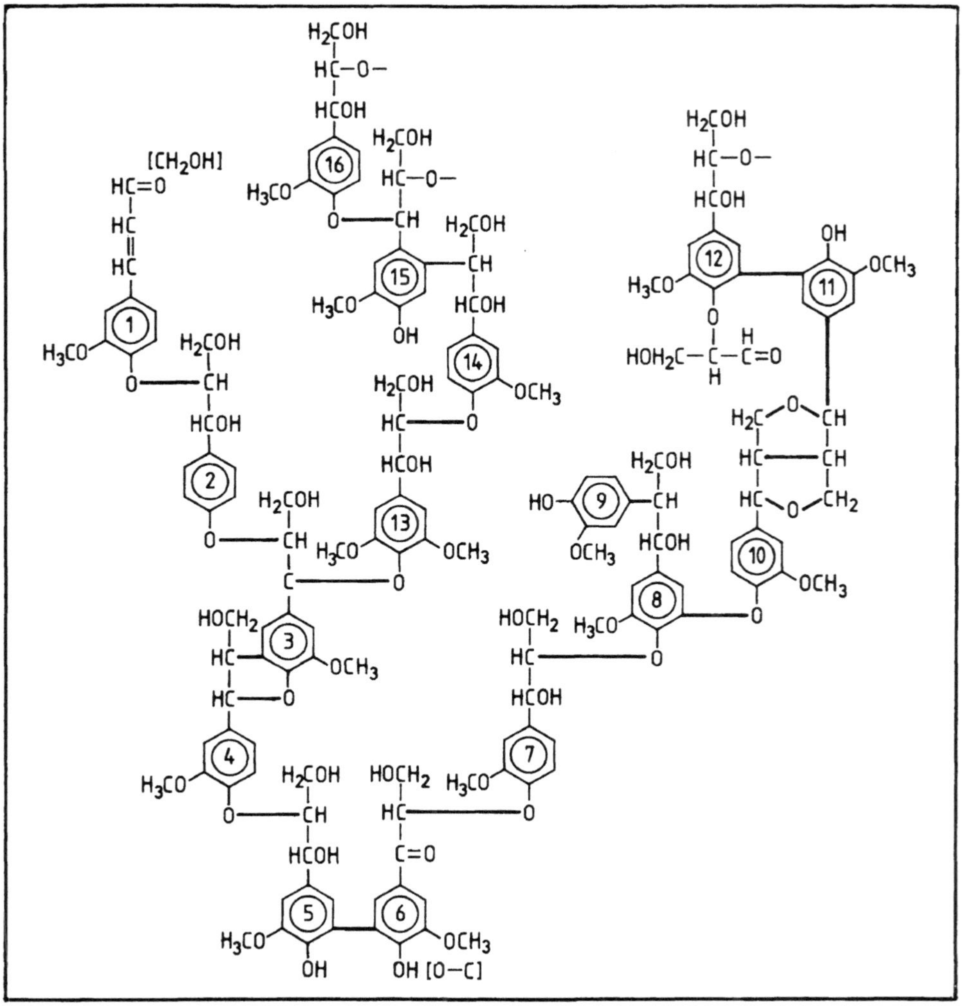

Fig. 6–7. Structural scheme of spruce lignin (according to Adler 1977).

described by Sakakibara (1980). It demonstrates a section of 28 C_9-units with several alternative structural elements, and agrees with most of the available analytical data. As in the case of the other above-mentioned models its evidence suffers from the relatively small number of units.

Apart from the described lignin models composing a part of the lignin molecule by random manual coupling of substructures, functional groups and identified linkages, the largest softwood lignin structural model was evaluated by simulating the lignin formation by computer. The original model reported by Glasser and Glasser (1974a, b) was mainly based on the simulation of radical coupling reactions of the p-hydroxycinnamyl alcohols, and was composed of 80 phenylpropane units. Later on this model was extended and improved in details (Glasser, Glasser 1976; Glasser et al. 1976). The most recent concept (Glasser, Glasser 1981) (Fig. 6–8) comprises

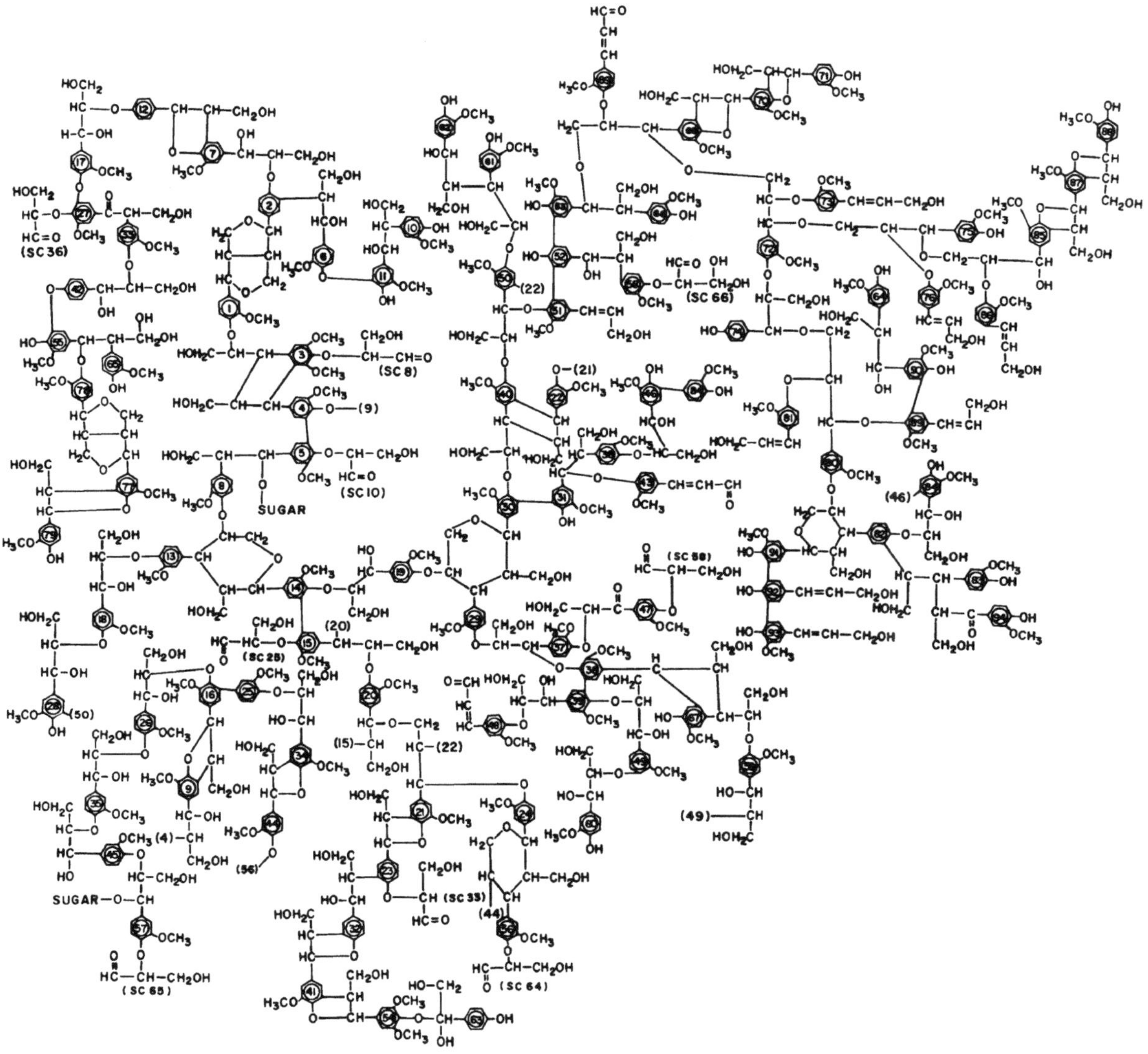

Fig. 6–8. Softwood lignin model designed by computerized evaluation (by courtesy of W. G. Glasser).

94 units, corresponding to a total molecular weight of more than 17 000, and is based on a wide array of analytical information obtained from investigations of milled wood lignin from loblolly pine *(Pinus taeda)*. Among the methods involved were elemental analysis, sugar and ash determination, ^{1}H-NMR spectroscopy for determination of functional groups, permanganate oxidation followed by GC/MS, and gel permeation chromatography (GPC). A reversed analysis program allows an adjustment between a distinct lignin structure and additional analytical results, thus obtaining a structural scheme which is in high accordance with the analytical reality of isolated lignin samples (Glasser, Glasser 1981; Glasser et al.

1981). The concept of the system for evaluating lignin structure is outlined in Fig. 6–9.

A valuable approach for simplifying lignin schemes was elaborated by Faix (1974, 1976; Faix et al. 1977), using graphs which consider important analytical lignin data such as the ratios of the different phenylpropane units, the types and frequencies of interunit linkages, functional groups etc. Fig. 6–10 shows the evaluation of the ratio of the three phenylpropane units in graphic form. Easily determinable analytical data such as methoxyl content and UV and IR data are involved as characteristic lines in a triangular coordinate system. The percentages of guaiacyl (G), syringyl (S) and p-hydroxyphenyl units (H) can be estimated by projecting the corresponding intersection of the lines onto the coordinates.

Studies of hardwood lignins were carried out by Larsson and Miksche (1971) on the basis of oxidation experiments with milled wood lignin from birch *(Betula verrucosa)* and by Nimz (1974), applying thioacetolysis on beech wood *(Fagus sylvatica)*. The latter obtained monomeric (49.7%), dimeric (25.0%), trimeric (10.9%) and tetrameric (5.3%) degradation products besides 9.1% oligomeric compounds. As the thioacetic acid treatment is known to cleave C-α and C-β linkages selectively, Nimz (1974) succeeded in calculating the proportions of ten types of linkages between the C_9-units in beech lignin from the relative yields of monomeric and dimeric

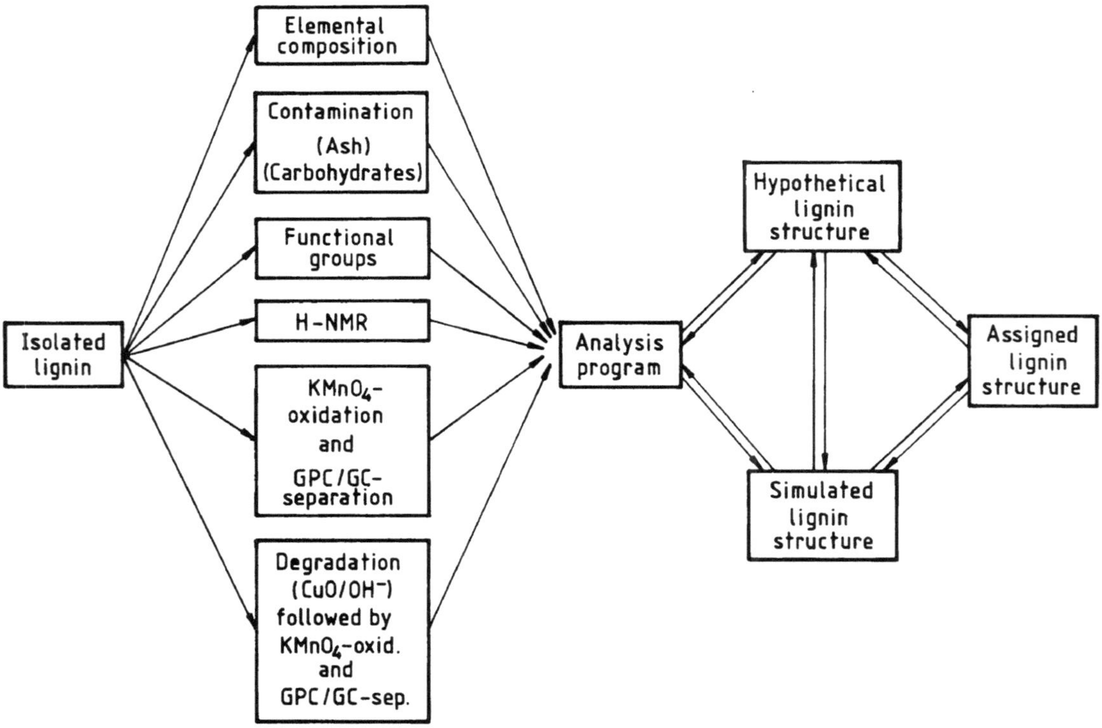

Fig. 6–9. Concept for evaluation of lignin structure (according to Glasser, Glasser 1981).

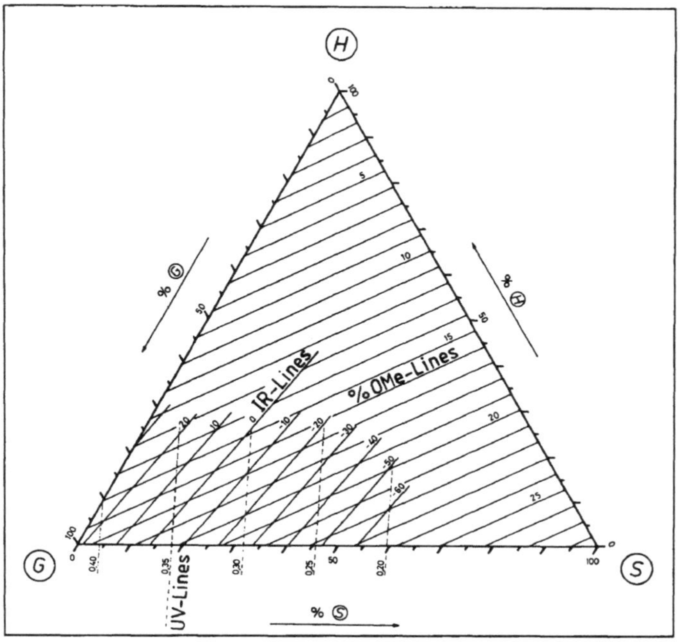

Fig. 6–10. Graphical evaluation of the ratio of the three phenylpropanoid units of lignins (by courtesy of O. Faix).

compounds (Table 6–2). Based on these figures and UV, IR, H-NMR and ^{13}C-NMR data a structural scheme was designed for beech lignin (Fig. 6–11). It consists of 25 phenylpropane units including six units which may be replaced by modified structures. The composition can be described by the formula $C_9H_{7.16}O_{2.44}(OCH_3)_{1.36}$ which is rather close to the corresponding formula of isolated beech milled wood lignin (Freudenberg, Sidhu 1961). In addition to structural elements known from softwood lignin a dibenzyltetrahydofuran unit (9′, 10′) was found, comprising a γ-O-γ and a β-β linkage.

Due to improved analytical techniques information about the frequency of the different types of linkages in the lignin molecule was substantially increased during the last ten years. The most prominent type of linkage both in softwood and hardwood lignins is the β-O-4 linkage with more than one half of all interunit linkages. Other important types of linkages (e.g. 5-5, β-5, α-O-4) and their proportions in softwood and hardwood lignin schemes are given in Table 6–2.

Though there are differences between the reported percentages of linkages in different lignin models, and though the occurrence of the long-familiar non-cyclic α-O-4 benzyl ether bonds is controversial even today (Leary 1980; Nimz 1981; Leary 1982) the types and frequencies of the most important linkages can be considered quite firmly established. Additionally numerous miscellaneous linkages and minor structures are known, which help to interpret discrepancies between lignin

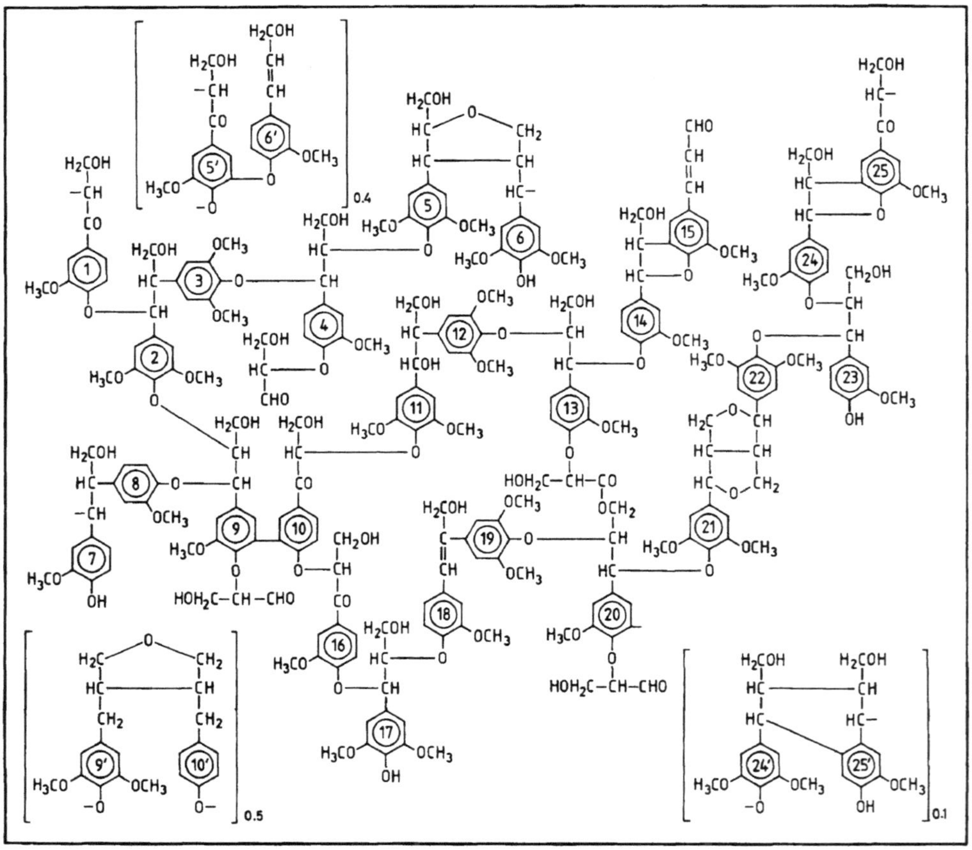

Fig. 6–11. The structural concept of beech lignin (according to Nimz 1974).

models and analytical results, but do not change the idea of the principal construction of the lignin molecules.

In contrast to the generally accepted concept of lignin structure based on random coupling of the precursors, there are some other concepts which assume ordered structures with repeating units, allowing the formulation of a degree of polymerization (DP) for lignin. A lignin model of this type was worked out by Forss (Forss, Fremer 1965, 1981; Forss et al. 1966), based on fractionation of defined lignin sulfonates. It consists of an ordered system of repeating units with a DP of 18.

The proved heterogeneity of lignins, the morphological aspects of lignin incorporation within the polysaccharidic structures in the cell wall, and the non-homogeneity of isolated lignin-polysaccharide complexes (→ 6.5.) favour the idea of non-ordered lignin macromolecules.

6.3.3. Lignin Heterogeneity

The general term lignin does not consider differences in the composition of this cell wall component. In the following a survey is given of lignin heterogeneity which was proved for numerous plants of different botanical sections, classes, orders and genera as well as for different wood tissues, cells and even cell wall layers of one species (Sarkanen, Hergert 1971).

It has long been known that the lignins of softwoods (gymnosperms), hardwoods (dicotyledons, angiosperms) and grasses (monocotyledons, angiosperms) differ with regard to their content of guaiacyl (G), syringyl (S) and p-hydroxyphenyl (H) units, as is provable e.g. by nitrobenzene oxidation yielding different amounts of the corresponding aldehydes (vanillin, syringaldehyde, p-hydroxybenzaldehyde). Other chemical methods used for the determination of lignin composition are acidolysis, permanganate oxidation and methoxyl determination (e.g. Kratzl, Claus 1962; Higuchi et al. 1972; Erickson et al. 1973a, b, c). All chemical degradation methods suffer more or less from the fact that the degradation products are received only from non-condensed C_9-units, thus depressing the percentages of mainly condensed p-hydroxyphenyl units (Nimz et al. 1981).

Physical methods for the determination of G-, S- and H-units in lignins include UV and IR spectroscopy (Kawamura, Higuchi 1964; Sarkanen et al. 1967a, b, c; Fergus, Goring 1970; Goldschmid 1971; Hergert 1971; Faix, Schweers 1974a, b), and more recently ^{13}C-NMR spectroscopy (Lüdemann, Nimz 1974; Nimz et al. 1974; Nimz, Lüdemann 1976; Nimz, Tutschek 1977; Nimz et al. 1981; Robert, Gagnaire 1981).

Histochemical methods (Mäule reaction; Wiesner reaction, → 10.3.1.) were also used for the differentiation of lignins (Srivastava 1966; Sarkanen, Hergert 1971; Meshitsuko, Nakano 1978).

As the classification into softwood, hardwood and grass lignins does not satisfy all results obtained from numerous lignins, a more reliable classification system was established in which all lignins are divided into two major groups: guaiacyl lignins (G-lignins) and guaiacyl-syringyl lignins (GS-lignins) (Gibbs 1958). Guaiacyl lignins are predominantly polymerizates of coniferyl alcohol while guaiacyl-syringyl-lignins are composed of varying parts of the aromatic nuclei guaiacyl and syringyl in addition to small amounts of p-hydroxyphenyl units. This simple classification scheme was refined by Kawamura and Higuchi (1964) by differentiating both main classes into several subgroups related to botanical plant groups. Based on extensive ^{13}C-NMR spectroscopic investigations on softwood, hardwood and grass lignins Nimz et al. (1981) also favour a more refined classification scheme, distinguishing e.g. all monocotyledon lignins or at least grass lignins as GSH-lignins from the lignins of dicotyledons (GS-lignins).

Most of the gymnosperm lignins are typical guaiacyl lignins with minor amounts of syringyl- and p-hydroxyphenylpropane units. Though softwood (coniferous) lignins are generally described as rather uniform (Sarkanen et al. 1967a; Erickson, Miksche 1974) no general G:S:H-ratio can be given for different softwood species. Distinct ratios were only reported for spruce *(Picea abies)* lignin (G:S:H = 94:1:5;

Erickson et al. 1973c) and for pine *(Pinus taeda)* lignin (G:S:H = 86:2:13; Glasser, Glasser 1981).

Exceptions to these typical guaiacyl lignins within the gymnosperms were found in species from the genus *Podocarpus* with slightly higher syringyl percentages of 2–6% and in the species *Tetraclinis articulata (Cupressaceae)* showing typical characteristics of GS-lignins. The same is true of certain species of *Ephedra* and *Stangeria* (Erickson, Miksche 1974).

The reaction wood of softwoods (compression wood, → 2.3.) has not only a higher lignin content than normal wood but also a considerably higher percentage of H-units (up to 70%) than normal softwood G-lignins, and may therefore be classified as a GH-lignin (Sarkanen, Hergert 1971; Erickson et al. 1973d; Yasuda, Sakakibara 1975; Nimz et al. 1981).

The variability of lignin composition is much greater in hardwoods than in softwoods. The syringyl content of the typical hardwood SG-lignins varies between 20 and 60%, and the range is even wider if herbaceous plants are also included (10–65%) (Sarkanen et al. 1967b; Erickson et al. 1973a). A distinct ratio for beech *(Fagus sylvatica)* lignin was given by Nimz (1974) (G:S:H = 56:40:4).

An interesting finding is the fact that the lignin of the European mistletoe *(Viscum album)* is a GS-lignin regardless of whether the host plant is a coniferous or deciduous tree (Nimz et al. 1974; Becker, Nimz 1974).

Other examples of lignin heterogeneity are the higher syringyl amount found in heartwood of hardwoods as compared to the corresponding sapwood (Parameswaran et al. 1975) and the higher guaiacyl content of beech-root lignin in comparison with the corresponding xylem lignin (Erickson et al. 1973a). The composition of lignin is influenced by its age, as shown in the case of poplar wood *(Populus nigra)* (Venverloo 1971; Sarkanen, Hergert 1971). The ratio of syringaldehyde to vanillin obtained by nitrobenzene oxidation was high in the mature xylem and decreased in the younger xylem and in the phloem. The primary xylem yielded only vanillin.

Investigations of bark lignins from several softwoods *(Picea abies, Taxus baccata)* and hardwoods *(Betula verrucosa, Fraxinus excelsior, Vitis vinifera)* showed that the softwood bark lignins are typical G-lignins with increased amounts of H-units as compared to the corresponding wood lignins. The hardwood bark lignins are GS-lignins with higher proportions of G-units than the wood lignins (Anderson et al. 1973) (→ 9.2.4.).

A very important aspect of lignin heterogeneity was elaborated by Fergus and Goring (1970) who were the first to show that the composition of birch *(Betula papyrifera)* lignin differs depending on its location. UV microscopy applied to ultrathin sections revealed that the lignin in the secondary walls of the vessels and in the middle lamellae is predominantly a G-lignin. The secondary wall lignin of fibres and ray parenchyma however is mainly composed of syringyl units. Lignin from the middle lamellae and cell corners of the fibre and ray cell tissue is a typical mixed GS-lignin. Later on Musha and Goring (1975) extended these results by correlating the distribution of S- and G-residues in hardwood lignins from different morphological regions with the ratio of methoxyl groups per C_9-unit (OCH_3/C_9) of the overall wood lignin.

The principal findings of Fergus and Goring (1970) were confirmed by results obtained by isolation and characterization of lignin fractions assigned to certain cell wall regions (Kolar et al. 1979; Cho et al. 1980; Lapierre, Monties 1981). A highly sophisticated method was used by Hardell et al. (1980) who were successful in isolating samples of different cells (ray cells and vessels) and cell wall layers (S 1, S 2) from birch wood *(Betula papyrifera)* for lignin analysis. Based on results from comparable investigations of different morphological elements of white oak *(Quercus alba)* doubt was recently cast on the heterogeneity of hardwood lignin composition in different structural elements (Obst 1982).

The proved differences in the lignin composition of certain cells and cell wall layers substantially support the idea of an intracellular biosynthesis of lignin (→ 6.2.3.).

6.4. Characterization and Properties of Lignins and Lignin Derivatives

6.4.1. Chemical Composition and Molecular Weight

In connection with pulping and bleaching (→ 16.) and new technologies of lignin utilization (→ 18.6.) the analytical characterization of lignins has become increasingly important.

A first chemical characterization of lignins can be reached by elemental analysis and determination of methoxyl groups. Additionally the nonlignin components are considered by determination of ash and polysaccharide moieties. Further analytical characteristics are the content of other functional groups (e.g. phenolic and aliphatic hydroxyl groups, carbonyl, carboxyl groups) indicating changes of lignin structure due to the isolation procedure or chemical treatments (Lai, Sarkanen 1971; Glasser, Barnett 1979b; Meier et al. 1981). Lignin degradation or condensation reactions (→ 10., 11.) can also be proved e.g. by determination of the average molecular weight or preferably of the molecular weight or size distribution (Goring 1971).

Table 6–3 gives a compilation of the elemental composition and the methoxyl contents of several lignins from different softwood and hardwood species, and from three nonwood sources. The variations of results within one species may be caused

Table 6–3. Elemental composition and methoxyl content of analytical and technical lignins

Lignin	Species	C %	H %	O %	OCH_3 %	others %	References
Softwood analytical lignins							
MWL	*Picea abies*	63.8	6.0	29.7	15.8		Björkman, Person 1957
MWL	*Picea abies*	63.8	6.0	29.7	15.4		Marton, Marton 1964
MWL	*Picea abies*	62.7	5.7	31.6	15.2		Chang et al. 1975
MWL	*Picea abies*	62.8	5.9	31.3	15.8		Faix et al. 1980
MWL	*Picea abies*	62.3	5.7	32.0	15.6		Fengel et al. 1981a
BMWL	*Picea abies*	62.8	5.9	31.2	11.5		Bezúch, Polčin 1978
EIL*	*Picea abies*	60.6	6.1	33.3	14.6		Bezúch, Polčin 1978
CEL*	*Picea abies*	61.2	5.5	33.3	15.2		Chang et al. 1975
Ethanol lignin	*Picea abies*	62.7	5.8	31.5	15.6	5.1 OC_2H_5	Fengel et al. 1981a
H_2SO_4 lignin	*Picea abies*	62.3	5.5	32.2	14.1		Fengel et al. 1981a
TFA lignin	*Picea abies*	63.2	5.7	26.3	13.0		Fengel et al. 1981a
MWL	*Picea mariana*	63.7	6.3	29.4	15.4		Björkman, Person 1957
MWL	*Picea jezoensis*	61.5	5.8	32.7	15.5		Sakakibara et al. 1981
MWL	*Pinus sylvestris*	64.0	6.1	29.8	15.7		Björkman, Person 1957
MWL	*Pinus ponderosa*	62.5	6.0	29.9	15.0		Sarkanen et al. 1967a
MWL	*Pinus taeda*	61.6	5.9	32.5	14.0		Glasser, Glasser 1981
MWL	*Thuja plicata*	63.8	6.1	30.1	16.1		Björkman, Person 1957
MWL	*Thuja plicata*	64.1	6.0	29.9	15.0		Sarkanen et al. 1967a
MWL	*Larix occidentalis*	63.7	6.1	30.2	12.9		Sarkanen et al. 1967a
MWL	*Tsuga heterophylla*	63.4	6.3	29.8	15.7		Björkman, Person 1957
MWL	*Tsuga heterophylla*	63.0	5.6	31.4	14.3		Sarkanen et al. 1967a
MWL	*Pseudotsuga menziesii*	64.6	5.8	29.6	12.5		Sarkanen et al. 1967a
MWL	*Pseudotsuga menziesii*	60.9	5.4	34.7	13.9		Glasser, Barnett 1979a
MWL	*Araucaria angustifolia*	59.1	5.6	35.3	17.8		Fengel et al. 1983
Hardwood analytical lignins							
MWL	*Fagus sylvatica*	60.3	6.3	33.4	21.4		Björkman, Person 1957
MWL	*Fagus sylvatica*	58.4	6.3	35.3	19.5		Fišerová, Šutý 1980
MWL	*Fagus sylvatica*	60.7	6.0	33.3	21.5		Faix et al. 1980
MWL	*Fagus sylvatica*	60.2	5.9	33.9	21.7		Fengel et al. 1981a
MWL	*Betula verrucosa*	58.8	6.5	34.0	21.5		Björkman, Person 1957
MWL	*Betula* spec.	59.7	6.1	34.2	21.4		Faix et al. 1980
MWL	*Populus tremuloides*	60.0	6.1	33.9	21.5		Faix et al. 1980
MWL	*Alnus rubra*	58.5	5.8	35.7	19.8		Glasser, Barnett 1979a
MWL	*Acer macrophyllum*	60.4	5.7	33.9	20.0		Sarkanen et al. 1967b
MWL	*Liquidambar styraciflua*	57.6	5.6	35.1	21.4		Chang et al. 1975
CEL*	*Liquidambar styraciflua*	56.6	5.6	36.9	21.5		Chang et al. 1975
MWL	*Shorea polysperma*	60.6	5.9	33.5	19.3		Salud, Faix 1980
MWL	*Gmelina arborea*	58.7	5.8	35.5	20.8		Sosanwo, Lindberg 1976
MWL	*Eucalyptus regnans*	59.2	6.3	33.6	22.9		Bland, Menshun 1970
MWL	*Ochroma lagopus*	57.3	5.9	36.8	14.6		Fengel et al. 1983
MWL	*Lophira alata*	58.8	5.4	35.8	16.6		Fengel et al. 1983
Nonwood analytical lignins							
'MWL'	*Bambusa* spec.	62.0	5.7	32.3	17.9		Faix et al. 1980
'MWL'	*Calamus rotang*	60.8	5.1	34.1	17.2		Faix et al. 1980
MBL	*Saccharum officinarum* (bagasse)	59.5	5.4	35.0	17.6		Fernandez, Šutý 1981
Technical lignins							
Eth./H_2O lignin	*Picea abies*	68.0	5.7	26.3	13.3		Fengel et al. 1981a
Eth./H_2O lignin	*Picea abies*	67.8	6.2	26.1	15.2	2.1 OC_2H_5	Meier et al. 1981
Eth./H_2O lignin	*Fagus sylvatica*	61.0	6.0	33.0	20.1	1.4 OC_2H_5	Meier et al. 1981
Eth./H_2O lignin	*Betula verrucosa*	64.2	6.1	29.7	20.7	1.3 OC_2H_5	Meier et al. 1981
Kraft lignin	*Pinus* spec.	65.9	5.8	25.9	14.0	1.6 S	Marton 1964
Kraft lignin	*Pinus* spec.	63.4	5.7	30.9	13.2	0.5 S	Fengel et al. 1981a
Kraft lignin	*Pinus taeda*	58.3	4.6	21.6	13.4	2.1 S	Yamasaki et al. 1981
Lignin sulfonate	*Picea abies*	51.5	5.9	37.7	11.4	7.0 S	Fengel et al. 1981a
Lignin sulfonate	*Fagus sylvatica*	46.8	5.4	43.3	14.3	6.0 S	Fengel et al. 1981a

* Preparations isolated by enzymatic attack

both by non-uniform lignin samples and methodical differences and errors. The values for the analytical lignins demonstrate that the carbon content of softwood lignins (60–65%) is generally higher than that of hardwood lignins (56–60%). This is due to the higher oxygen content of hardwood lignins, which is caused by their higher methoxyl content (18–22%) as compared to softwood lignins (12–16%).

The nonwood lignin samples have methoxyl contents ranging between those of softwood and hardwood lignins. Both acid lignins show essentially lower methoxyl contents, probably due to the severe chemical action during isolation. Accordingly kraft and sulfite lignins show low methoxyl contents. The data for kraft lignin and lignin sulfonates are only intended as examples because there is a large variability of lignin samples, depending on the processing conditions. Generally the sulfur content of lignin sulfonates is much higher (6–7%) than in kraft lignins (0.2–2%). The sulfur in lignin sulfonates is bound in the sulfonate groups (→ 10.3.2.) occurring as salts, e.g. as calcium salts.

Polysaccharides are a common contaminating component of isolated lignins. The amounts of residual polysaccharides are highly dependent on the type of lignin isolation and purification, but also to some extent on the wood species. The values for analytical spruce lignins, representative of softwood lignins, range on an average between 0.6 and 2% polysaccharides (Björkman 1956; Brown et al. 1968; Wegener, Fengel 1977, 1979; Fengel et al. 1981a). Higher values were reported for spruce MWL (4.1%) by Chang et al. (1975), and for Douglas fir MWL (4.8%) by Glasser and Barnett (1979b). By means of special purification techniques the residual polysaccharide content can be reduced to less than 1% (Lundquist, Simonson 1975; Lundquist et al. 1977). By lowering the polysaccharide content of analytical lignins the composition of the residual polysaccharides is changed, e.g. the glucose content is decreased and the arabinose portion increases, possibly due to a participation of this sugar in a lignin-polysaccharide linkage (→ 6.5.). A typical composition of the polysaccharides in an average spruce MWL is shown in Table 6–4 in

Table 6–4. Sugar composition of MWL polysaccharides and polyoses from spruce *(Picea abies)* (according to Wegener, Fengel 1979)

Sugar	MWL polysaccharides Mole %	Polyoses Mole %
Rhamnose	4.1	1.0
Mannose	21.4	44.6
Arabinose	13.5	4.8
Galactose	13.7	7.2
Xylose	15.9	22.3
4-O-MGluU	1.6	5.1
Glucose	30.1	14.9
Xyl:Ara	1.2:1	4.6:1
Glu:Man	3:2	1:3

comparison with the corresponding sugar composition of spruce polyoses. There is no conformity between the two polysaccharide compositions. The differences are especially evident with regard to the higher amounts of arabinose and glucose in MWL residual polysaccharides, revealing that these polyoses parts remain with the lignin with higher priority during isolation and purification procedures (Wegener, Fengel 1979).

The polysaccharide amounts in analytical hardwood lignins are higher on an average, ranging between 3 and 9% (Björkman, Person 1957; Brown et al. 1968; Lundquist et al. 1979; Glasser, Barnett 1979b; Fengel et al. 1983).

The content of residual polysaccharides in technical lignins is generally low, about 1%. But in particular the polysaccharide moieties in lignin sulfonates can be influenced by purification (e.g. dialysis), fractionation and modification procedures to yield special marketable products (→ 18.6.1.; Anonymous 1981, 1982).

The claim of a typical or average molecular weight of lignin or lignin derivatives is still a questionable matter. This is caused by several factors:
- the multiplicity of lignin isolation procedures
- degradation of the macromolecules during isolation
- condensation effects, especially under acidic conditions (→ 10.)
- the pronounced polydispersity of all solubilized lignins
- insufficient determination methods with regard to the polydisperse character of isolated lignins
- uncertainties about the behaviour of lignins in solutions, thus complicating calibration systems.

Due to the above-mentioned factors a comparison of literature data is rather difficult and questionable.

Polydispersity is a property of all isolated lignins whether obtained by analytical or technical procedures. The most obvious explanation for this phenomenon, which is not necessarily valid for all natural polymers, is the random degradation of the native cell wall lignin by chemical attack during isolation, yielding soluble fragments of different sizes but rather uniform chemical composition (Goring 1971).

From an experimental standpoint the polydispersity makes it necessary to consider both number and weight average values of the molecular weight ($\bar{M}_n$ and $\bar{M}_w$), determinable by different absolute and relative methods, with the ratio $\bar{M}_w/\bar{M}_n$ expressing the degree of polydispersity (Vollmert 1973). Frequently applied methods include osmometry, light scattering and ultracentrifuge techniques (Goring 1971; → 3.2.7.), and more recently gel permeation chromatography (GPC) and high pressure (performance) liquid chromatography (HPLC) in combination with a calibration of the columns with suitable standards or by ultracentrifuge measurements of separated fractions. A substantial characteristic of lignins in solutions is their low viscosity, causing quite different behaviour during determination of molecular weights as compared to polysaccharides or synthetic polymers (Goring 71).

Björkman (1956; Björkman, Person 1957) was the first to determine an average molecular weight ($\bar{M}_w$) of 11 000 for spruce milled wood lignin *(Picea abies)*. A comparable $\bar{M}_w$-value of 11 000 ($\bar{M}_w/\bar{M}_w = 3.1$) was recently reported for pine MWL *(Pinus taeda)* (Yamasaki et al. 1981). Rezanowich et al. (1963) obtained a weight average value of 7 100 from viscosity and ultracentrifuge measurements, which is in accordance with the $\bar{M}_n$-value of 2 100 for a pine MWL, considering a $\bar{M}_w/\bar{M}_n$-ratio of about 3.4 (Marton, Marton 1964). Higher softwood MWL molecular weights ($\bar{M}_w$) of 15 000 and in the magnitude of 21 000 were also reported (Soundararajan, Wayman 1970; Wayman, Obiaga 1974; Chang et al. 1975). The highest $\bar{M}_w$-values reported for analytical lignins are 77 000 for an enzymatically isolated MWL of Eastern hemlock *(Tsuga canadensis)*, and 85 000 for a dioxane spruce lignin fraction *(Picea glauca)* (Wayman, Obiaga 1974; Rezanowich et al. 1963).

One fraction from the high-molecular-size region of a GPC separation of spruce MWL *(Picea abies)* had a molecular weight ($\bar{M}_w$) of about 40 000, determined by ultracentrifuge measurements (Wegener, Fengel 1977). A broad distribution of $\bar{M}_n$-values between 2 000 and 19 000 for fractions from ball-milled wood lignin (BMWL) was obtained by Brownell (1970).

Very low molecular weights for numerous milled wood lignins were recently determined by Faix et al. (1980, 1981) with $\bar{M}_w$-values ranging between 2 800 for spruce ($\bar{M}_w/\bar{M}_n = 3.65$) and average values for the hardwoods beech, aspen and birch between 3 700 and 5 500, with degrees of dispersity between 2.3 and 2.6. (Table 6–3).

As already pointed out there are also no uniform values for molecular weights of technical lignins due to their heterogeneity depending on the pulping process and the influence of different purification procedures such as dialysis or ultrafiltration. The values from the literature for lignin sulfonates vary between about 1 000 and more than 100 000 or even 10^6 (Goring 1971; Lin, Detroit 1981; Lindberg, Törmälä 1981). For kraft lignins generally lower average values are reported. Marton and Marton (1964) measured e.g. $\bar{M}_w$-values of 3 500 and 2 900 for pine and hardwood kraft lignin, respectively. Weight average values for fractions obtained from different stages of kraft delignification revealed an increase from 1 800 for the first to 51 000 for the final fraction taken at the end of the cook, all fractions being polydisperse systems (McNaughton et al. 1967). $\bar{M}_w$-values of dialyzed kraft pine lignin fractions obtained by chromatographic separation, and measured by ultracentrifuge sedimentation ranged between 370 and 44 300 (Connors et al. 1980). Very low $\bar{M}_w$-values of 800 and 2 500 for spruce and beech alkali lignins, respectively, were determined by Meier et al. (1981).

The estimation of molecular weights of insoluble lignins was recently performed by thermal softening measurements, based on a linear relation between log $\bar{M}_w$ and the thermal softening temperature T_s (→ 12.4.4.) (Tanahashi et al. 1982).

By means of improved column chromatographic systems and calibration techniques more detailed information was obtained in recent years regarding lignin molecular weight or size distributions, demonstrating clearly the typical polydispersity of all types of lignins (Böttger et al. 1976; Forss et al. 1976; Froment, Robert 1977; Wegener, Fengel 1977; Hüttermann 1978; Connors 1978; Connors et al. 1978, 1980; Chua, Wayman 1979; Baumeister, Edel 1980; Faix et al. 1980, 1981; Yamasaki et al. 1981; Lin, Detroit 1981; Månsson 1981; Concin et al. 1981; Budin, Susa 1982). Bimodal patterns of the molecular weight distributions covering large ranges, as shown for example in Fig. 6–12, are typical of most of the investigated lignins. Standardized chromatographic column systems are especially suitable for observing changes in the molecular weight distribution of lignins by chemical treatments (Kringstad et al. 1981; Dix, Roffael 1981).

The question as to whether association effects between lignin fragments, elution solvent and column bed material influence the elution profiles is still unclarified, as is the assumption that polystyrene molecules, which are often used for calibration, have a comparable elution behaviour as lignin fragments (Faix et al. 1981; Connors et al. 1980). However, the multitude of results reveal that the broad molecular weight distributions of lignins are not an effect of separation methods, but essentially a demonstration of the polydisperse character of isolated lignins.

With the exception of the above-mentioned method using thermal softening, it is essential for molecular weight determinations and UV spectral studies as well (→ 6.4.2.) to obtain isolated lignins totally in solution. Due to the multitude of

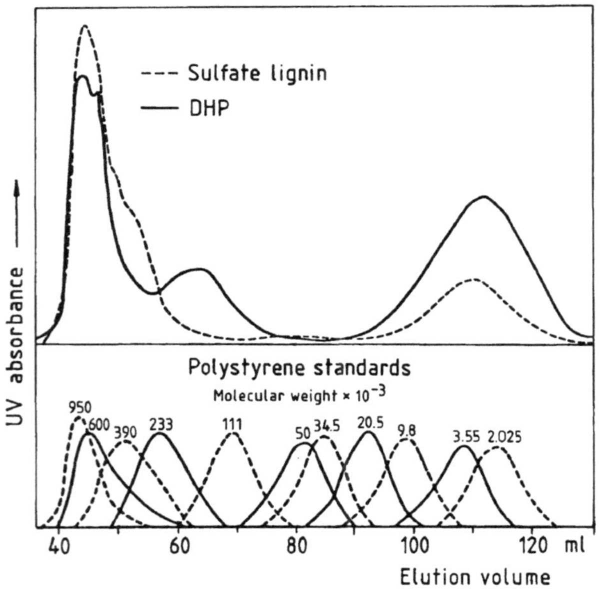

Fig. 6–12. Molecular weight distributions of lignins, calibrated with polystyrene standards (according to Connors et al. 1980).

analytical and technical lignins, as well as fractions thereof with different properties and molecular weights, only some general remarks can be given about the solubility of lignins.

The two principle solvent parameters for lignin-dissolving properties are the hydrogen bonding capacity and the cohesive energy density (Hildebrand's solubility parameter) (Marton 1971; Goring 1971). Suitable solvents for analytical lignins isolated with organic solvents are e.g. dioxane, dimethylsulfoxide (DMSO), formamide, dimethylformamide (DMF), tetrahydrofuran (THF), pyridine, dichloroethane and ethyleneglycol-monomethylether (methyl cellosolve).

Further good solvents for lignins are acetyl bromide in acetic acid (Johnson et al. 1961) and hexafluoropropanol (Fengel et al. 1983; Wegener et al. 1983).

With the exception of acid lignins (→ 3.2.9.), which are practically insoluble in all solvents, analytical lignins or fractions thereof may also be partially soluble in methanol, ethanol, acetone or solvent mixtures (Rezanowich et al. 1963; Goring 1971). Technical alkali lignins and lignin sulfonates are generally soluble in water, dilute alkali, salt and buffer solutions as well as in some basic and neutral polar solvents (Marton 1971). Especially for UV spectroscopic studies the choice of a suitable solvent is restricted by the limited transmission of most of the solvents in the lower UV wavelength range (→ 6.4.2.).

6.4.2. Ultraviolet and Infrared Spectroscopic Behaviour

Ultraviolet absorption is a well-known and widely used tool for lignin identification, qualitative and quantitative determination, as well as characterization of changes in lignin structure and properties (→ 3.2.9., 6.4.1.) (Brauns, Brauns 1960; Goldschmid 1971).

The distinct absorption of lignin in the ultraviolet range is based on its aromatic character, i.e. the sum of phenylpropane units, and on several chromophoric structural elements (→ 13.1.3.).

The typical lignin spectrum comprises a maximum at about 280 nm followed by a slope to lower wavelengths, with a more or less pronounced shoulder in the region of 230 nm. A second typical extinction maximum with a high absorptivity value appears in the range between 200 and 208 nm. Though there are some differences in the UV spectra of different lignins, as demonstrated for various analytical and technical lignins in Figs. 6–13 and 6–14, the course of the absorption curves emphasizes the above-mentioned general characteristics.

Nevertheless, the fine resolution of UV spectra reveals changes in the spectral behaviour caused by structural differences in lignins (Björkman, Person 1957; Sarkanen et al. 1967a, b, c; Norrström 1972; Norrström, Teder 1971, 1973; Nilsson-Idner, Norrström 1974; Nilsson-Idner et al. 1972, 1974; Faix, Schweers 1974a; Lin, Detroit 1981).

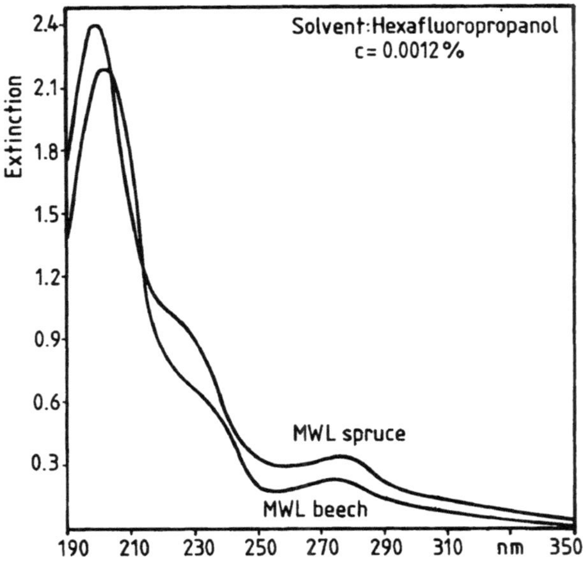

Fig. 6–13. UV spectra of milled wood lignins.

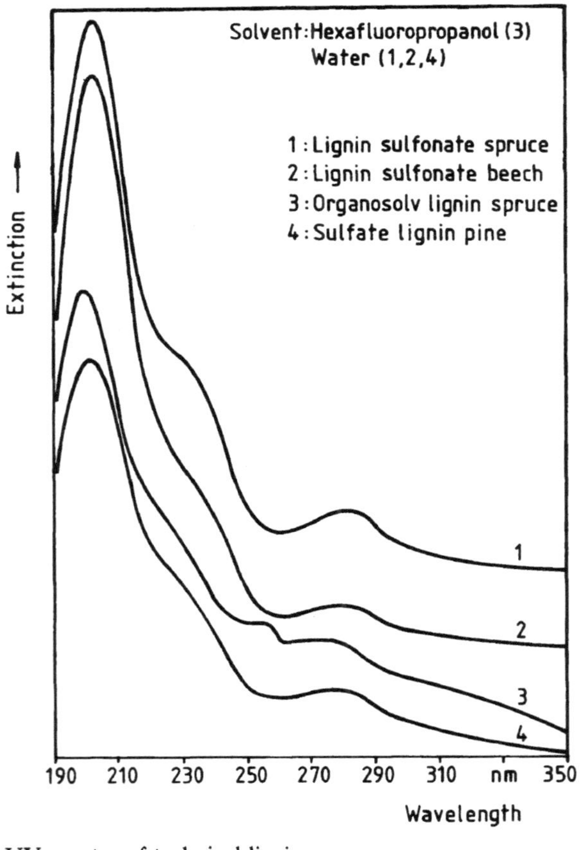

Fig. 6–14. UV spectra of technical lignins.

In contrast to softwood lignins with a maximum at about 280 nm or only slightly lower, hardwood lignins show a shift of this maximum to somewhat shorter wavelengths in the range of 275–277 nm. This fact contributes to the higher symmetry of the phenylpropane units in hardwood lignins, caused by the higher amounts of syringyl units. Additionally the absorptivities of hardwood lignins are generally somewhat lower than those of softwood lignins, with decreasing values at increasing OCH_3/C_9-ratios (Sarkanen et al. 1967a; Goldschmid 1971; Fengel et al. 1981b). Table 6–5 gives a survey of extinction coefficients of various lignins measured in different solvents at 280 nm. The term extinction coefficient is synonymous with the terms absorptivity, absorbancy index and specific extinction (Browning 1967a). The values demonstrate the generally higher extinction coefficients of softwood lignins (19.7–20.7 $l \cdot g^{-1} \cdot cm^{-1}$) as compared to hardwood lignins (12.6–14.2 $l \cdot g^{-1} \cdot cm^{-1}$). The only exception is the MWL from the tropical hardwood lauan *(Shorea polysperma)*, whose high absorptivity may be caused by an extremely low methoxyl content (Salud, Faix 1980). The lignin sulfonate values both of softwoods and hardwoods are low (in the range of 10–12 $l \cdot g^{-1} \cdot cm^{-1}$), while kraft lignin and an organosolv lignin from a semi-technical ethanol/H_2O-cook have high values at about 24–26 $l \cdot g^{-1} \cdot cm^{-1}$.

Table 6–5. Extinction coefficients of lignins

Lignin	a_{280} $l \cdot g^{-1} \cdot cm^{-1}$	Solvent	References
MWL spruce	16.7	Methyl cellosolve	Fengel et al. 1981b
MWL spruce	19.0	Alkali/H_20	Chang et al. 1975
MWL spruce	20.7	Formamide	Brown et al. 1968
EIL spruce	12.5	Dioxane/H_2O	Bezúch, Polčin 1978
MWL spruce	19.5	Dioxane	Rezanowich et al. 1963
MWL pine	18.8	Methyl cell./ethanol	Sarkanen et al. 1967a
MWL hemlock	17.7	Methyl cell./ethanol	Sarkanen et al. 1967a
MWL Douglas fir	19.7	Methyl cell./ethanol	Sarkanen et al. 1967a
MWL larch	20.2	Methyl cell./ethanol	Sarkanen et al. 1967a
MWL spruce	19.2	AcBr/acetic acid	Wegener 1974
MWL beech	13.0	Methyl cellosolve	Fengel et al. 1981b
MWL beech	13.3	Formamide	Brown et al. 1968
MWL poplar	14.2	Methyl cell./ethanol	Sarkanen et al. 1967b
MWL maple	12.9	Methyl cell./ethanol	Sarkanen et al. 1967b
MWL lauan	17.0	Methyl cell./ethanol/H_2O	Faix et al. 1980
Dioxane lignin poplar	12.6	Dioxane	Rezanowich et al. 1963
Ethanol/H_2O lignin spruce	25.8	Methyl cellosolve	Fengel et al. 1981b
Lignin sulfonate spruce	11.9	H_2O	Fengel et al. 1981b
Lignin sulfonate beech	10.4	H_2O	Fengel et al. 1981b
Lignin sulfonate softwood	11.9	H_2O	Lin, Detroit 1981
Kraft lignin pine	24.6	H_2O	Fengel et al. 1981b
Kraft lignin pine	26.4	Methyl cell./H_2O	Lin, Detroit 1981

The absorptivity of lignin samples at 280 nm may be strongly influenced by contaminating degradation products of polysaccharides, e.g. furfural or hydroxymethylfurfural, which also have their absorption maxima at about 280 nm (Kleinert 1971). On the other hand the evaluation of the second maximum between 200 and 208 nm requires a suitable solvent, and is susceptible to errors due to the high extinction coefficient at this wavelength.

Slight but measurable changes in the UV spectroscopic behaviour are caused by varying amounts of chromophoric groups, e.g. phenolic hydroxyl groups, carbonyl groups etc. (Goldschmid 1971). These changes can be used in the so-called difference spectroscopy, e.g. for estimation of free and etherified phenolic hydroxyl groups ($\Delta\varepsilon$-ionization curves). Structural changes can also be visualized after a reductive treatment or catalytic hydrogenation (reduction $\Delta\varepsilon$-curves, hydrogenation $\Delta\varepsilon$-curves) (Browning 1967b; Goldschmid 1971). UV spectroscopy was also used for characterizing the chemical heterogeneity of technical lignins with regard to their potential utilization (Lin, Detroit 1981).

One of the most significant changes in the UV spectrum of lignin is caused by the loss of the aromatic character of lignin, e.g. by chlorination or chlorite treatments (→ 10.3.3.). With increasing lignin degradation the extinction values decrease, the maxima flatten out, and finally at least the 280 nm maximum disappears (Ivancic, Rydholm 1959; Wegener 1975).

Besides the usual excitation spectra lignins also show emission spectra (fluorescence spectroscopy) suitable for characterization and quantification (Lundquist et al. 1978; Lundquist 1981).

In sum, the predicative ability of UV spectroscopic data for lignin characterization is mostly limited to the comparison of different lignins, the evaluation of some special functional groups, and the determination of severe structural changes by chemical treatments.

Infrared spectroscopy in the near IR region (wavelength range: 2.5–15 μm, wavenumbers: 4 000–600 cm^{-1}) is another useful physical method for characterizing lignin and lignin derivatives.

While the infrared spectrum is a characteristic property of compounds with exactly known structures, there are several uncertainties with the interpretation of lignin IR spectra. This is mainly caused by two factors. Firstly, there are large variations in lignin structures and compositions, depending on the origin of the sample and the special isolation procedure; secondly variations are caused by different techniques of measuring lignins in suitable solvents, in the form of films, or in the most frequently applied form of potassium bromide pellets (Zbinden 1964; Hergert 1971).

Additionally infrared spectroscopy of lignins is nearly exclusively a qualitative characterization, and only few attempts with ambiguous results have been made to evaluate special bands quantitatively (Wada 1961; Kolboe, Ellefsen 1962; Michell 1966; Sarkanen et al. 1967a, b; Faix, Schweers 1974b; Karklins, Okherina 1975). A quantitative lignin determination by evaluation of the aromatic ring vibration bands at 1 510 and 1 600 cm^{-1} was recently applied to lignin model compounds and milled wood lignin (Feckl 1981).

IR spectra of lignins show several major absorption bands which can be assigned empirically to structural groups, based on a multitude of results obtained both from model compounds and lignins (Ekman, Lindberg 1960; Wada 1961; Michell 1966; Higuchi, Kawamura 1966; Sarkanen et al. 1967a, b, c; Hergert 1971; Faix, Schweers 1974b). Typical IR bands with their most probable assignments are listed in Table 6–6. The assignment of a band cannot be deduced from a single spectrum, but must be proved by measuring derivatives of lignin model compounds and lignin samples, thus shifting the band position of structural elements or eliminating their bands. Suitable derivation methods are methylation, acetylation, reduction, sulfonation or conversion to salts, enabling the determination of functional groups, e.g. hydroxyl or carbonyl groups (Hergert 1971).

IR spectra of different lignins are shown in Figs. 6–15 and 6–16. The acid lignins were measured in KBr-pellets while from all other lignins films were prepared on germanium disks (Hofmann, Zundel 1971; Fengel et al. 1981b).

The most characteristic infrared bands of lignin are found at about 1 510 and 1 600 cm^{-1} (aromatic ring vibrations) and between 1 470 and 1 460 cm^{-1} (C-H deformations and aromatic ring vibrations). The first mentioned wavenumber region is poor in additional bands and can therefore be used to prove the existence of lignin in unknown preparations. The intensity of these bands, however, is strongly influenced by structures bordering on the aromatic nuclei (Katrizki, Topsom 1977; Feckl 1981). The different relations between the intensities of the bands at 1 510 and 1 600 cm^{-1} can be used for a differentiation of softwood and hardwood lignins.

Table 6–6. Important infrared absorption bands of lignin (according to Hergert 1971)

Position cm^{-1}	Band origin
3 450–3 400	OH stretching
2 940–2 820	OH stretching in methyl and methylene groups
1 715–1 710	C=O stretching nonconjugated to the aromatic ring
1 675–1 660	C=O stretching in conjugation to the aromatic ring
1 605–1 600	Aromatic ring vibrations
1 515–1 505	Aromatic ring vibrations
1 470–1 460	C-H deformations (asymmetric)
1 430–1 425	Aromatic ring vibrations
1 370–1 365	C-H deformations (symmetric)
1 330–1 325	Syringyl ring breathing
1 270–1 275	Guaiacyl ring breathing
1 085–1 030	C-H, C-O deformations

In unconjugated syringyl model compounds and hardwood lignins the intensities of these bands are nearly the same (e.g. MWL beech, Fig. 6–15), while in unconjugated guaiacyl compounds and softwood lignins the intensity of the 1 510 cm^{-1} band is considerably higher (e.g. MWL spruce, Fig. 6–15). The typical guaiacyl and syringyl bands (ring breathing) are located at about 1 270 and 1 330 cm^{-1}, respectively.

The intensive bands of the carbonyl groups appear in the range between 1 660 and 1 725 cm^{-1}, allowing significant conclusions about this functional element in lignin structure. The exact position of the band is dependent on whether the C=O groups are in conjugation with the aromatic ring (position below 1 700 cm^{-1}) or not (position above 1 700 cm^{-1}). Intensive bands in this region are evident in the ethanol lignin (Fig. 6–15) and in the organosolv and sulfate lignins (Fig. 6–16). Distinct carbonyl bands were also found earlier in other thiolignin samples (Field et al. 1958; Marton 1964). Carbonyl bands originating from carboxyl groups can be identified after an alkaline treatment, as the C=O band disappears and is replaced by the carboxylate ion band at 1 600 cm^{-1}. Acetyl and uronic ester groups of polysaccharide residues absorb at 1 735–1 740 cm^{-1}, and may interfere with carbonyl bands (Hergert 1971).

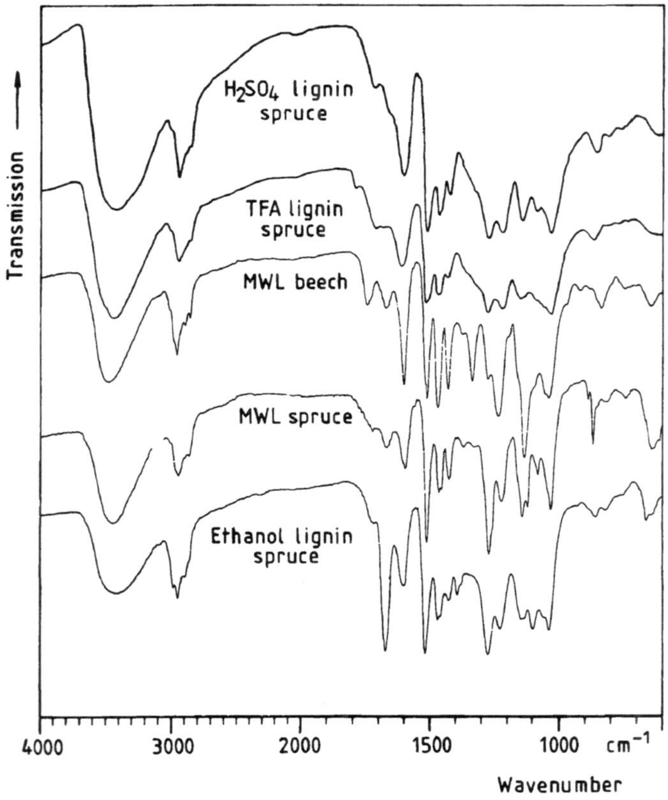

Fig. 6–15. IR spectra of analytical lignins.

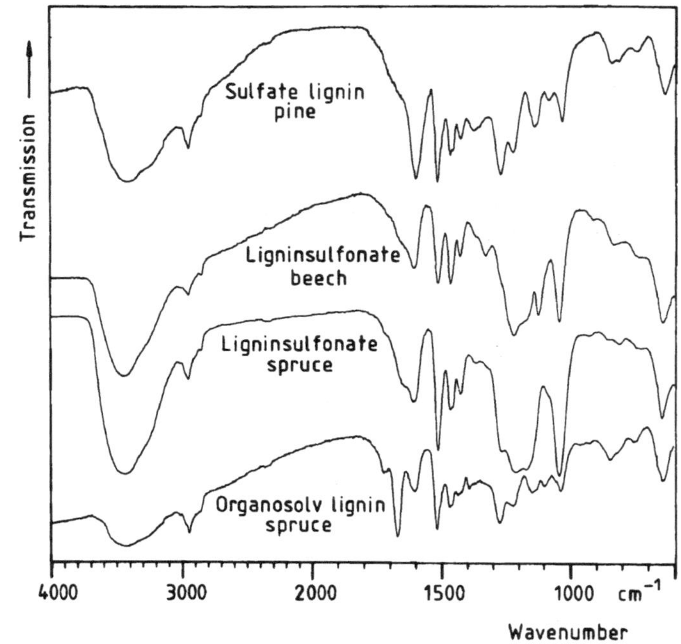

Fig. 6–16. IR spectra of technical lignins.

All spectra in Figs. 6–15 and 6–16 show the broad band of the hydroxyl stretching in the region of 3 400 cm^{-1}, which cannot be used for structural elucidation of complex molecules such as lignin. The same is true of the C-H stretching bands at about 2 800–3 000 cm^{-1} (partially also caused by OH stretching) and the bands in the region of 1 000–1 400 cm^{-1}, caused by combination and overlapping of C-O stretching bands, and by several deformations. The width and intensity of the bands between 1 000 and 1 100 cm^{-1} give hints concerning sugar or polysaccharide moieties, while the absence of the hydroxyl bands above 3 000 cm^{-1} indicates the absence of polysaccharidic components (Feckl 1981).

The sulfonic acid groups in technical lignin sulfonates appear at about 1 200 cm^{-1} (Fig. 6–16). The acid lignins show generally broader bands than the milled wood lignins, probably due to condensation reactions during lignin preparation (Fig. 6–15).

IR spectra were reported for numerous lignins from a multitude of wood species, showing additional differences caused by the origin of the lignins and the kind of isolation as well (Kawamura, Higuchi 1964; Michell et al. 1965; Higuchi, Kawamura 1966; Kawamura, Bland 1967; Hergert 1971; Sarkanen et al. 1967a, b; Bezúch, Polčin 1978; Fišerová, Suty 1980; Meier et al. 1981). The results of Kawamura and Higuchi (1964) and Sarkanen et al. (1967a, b) allowed a classification of lignin spectra according to the ratio of guaiacyl-, syringyl- and p-hydroxyphenyl units. Generally, however, the interpretation of lignin IR spectra aiming at a distinct differentiation of lignin samples or for proving their identity must be undertaken very carefully.

6.4.3. Ultrastructural Appearance

Only a few investigations are concerned with lignin structures visible in the electron microscope. Electron micrographs of fractionated lignin sulfonates show spherical shaped particles of different sizes (Rezanowich et al. 1964). Though even the smallest particles observed do not represent single lignin molecules, the hydrodynamic properties of soluble lignins, elucidated by viscosity, sedimentation, diffusion and relaxation experiments, support the idea that lignin appears in solution as a compact-shaped microgel (Rezanowich, Goring 1960; Goring 1971; Lindberg, Törmälä 1981).

Fig. 6–17 gives an impression of beech lignin sulfonate particles at high magnification with diameters between 20 and 50 nm. Besides the typical globular structures larger irregular-shaped bodies are also visible. Possibly the pronounced polyelectrolytic character of lignin sulfonates supports the formation of larger associations.

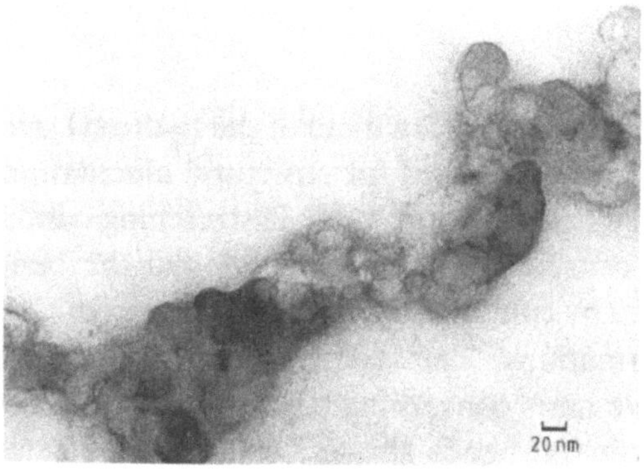

Fig. 6–17. Spruce lignin sulfonate. TEM micrograph.

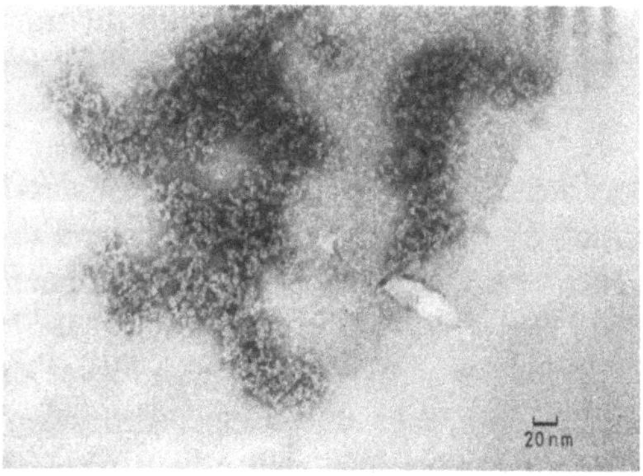

Fig. 6–18. Pine kraft lignin. TEM micrograph.

In contrast, a sample of pine kraft lignin shows much smaller structural elements which are not sharply defined, and lack the tendency of forming globular aggregates (Fig. 6–18). The sizes of the loose particles are in the range of 5–10 nm.

Electron microscopical observations on analytical lignins (spruce MWL, spruce ethanol lignin, → 3.2.9.) also revealed mostly spherical particles in addition to deformed elements of different sizes (Figs. 6–19, 6–20). The globular character was proved to be independent from the preparation procedure preceding the microscopic observation. Therefore it is probable that these lignins also occur as statistical coils, in which the molecules may form associations, possibly by means of hydrogen bonds (Fengel 1976a). In both lignins the particle sizes vary between 10 and 100 nm, and some of the particles show structured surfaces. Observations on spruce MWL and Brauns native lignin as well as methanol lignin from beech showed comparable particle sizes and shapes (Košiková et al. 1978).

A significant loosening and dispersing effect was reached by a Cl/ClO_2-treatment of spruce MWL (Fig. 6–21) and by a treatment with boiling water (Fig. 6–22) (Fengel 1976a; Wegener, Fengel 1979). Evidently chemical treatments of isolated lignins

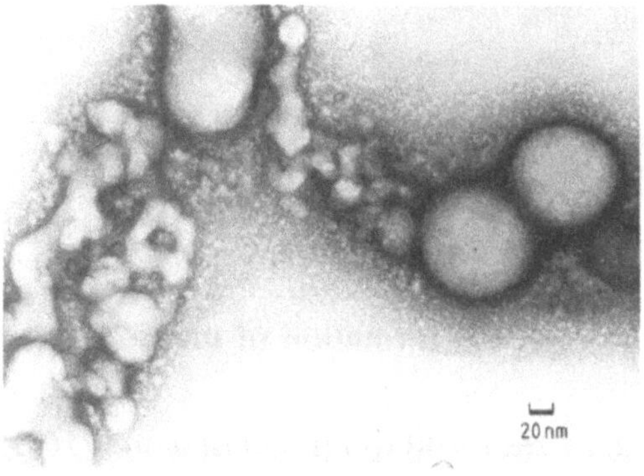

Fig. 6–19. Spruce MWL. TEM micrograph.

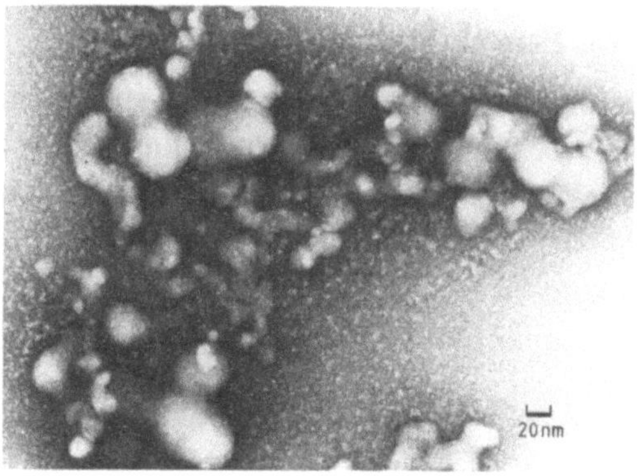

Fig. 6–20. Spruce ethanol lignin. TEM micrograph.

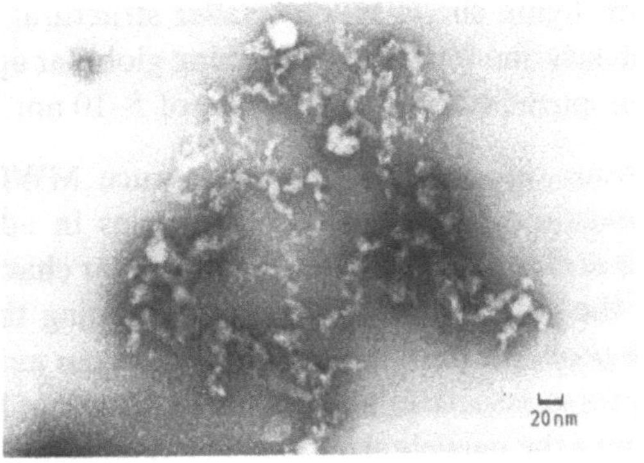

Fig. 6–21. Spruce MWL after Cl/ClO_2-treatment. TEM micrograph.

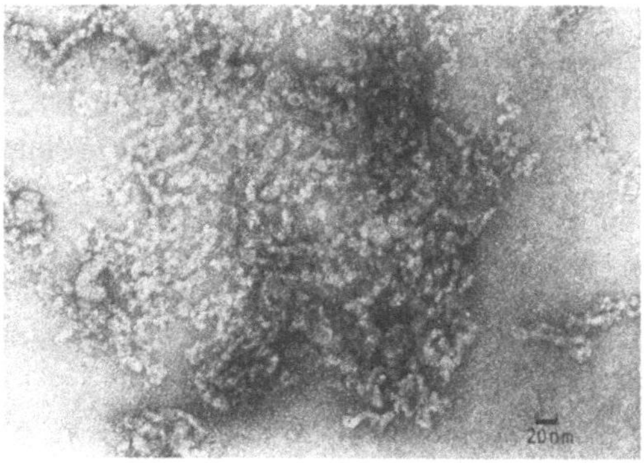

Fig. 6–22. Spruce MWL after hot-water treatment. TEM micrograph.

cause a loss of the spherical shape, and the formation of more or less loosened structures of smaller sizes.

Analytical lignins obtained as residues after acid treatment of wood (H_2SO_4 lignin, TFA lignin; → 3.2.9.) show more or less irregular particles (Fig. 6–23). Results from acid treatments of ultrathin cell wall sections, yielding so-called lignin skeletons, indicate that the particle sizes of acid lignins may depend on the isolation conditions (Fig. 8–1) (Fengel 1976a; Fengel, Stoll 1975).

Electron microscopic observations on ethanol/H_2O lignins from spruce and poplar reveal regular and globular lignin associations with average diameters in the magnitude of 100 nm, and for the most part with structured surfaces (Fig. 6–24) (Fengel et al. 1981b).

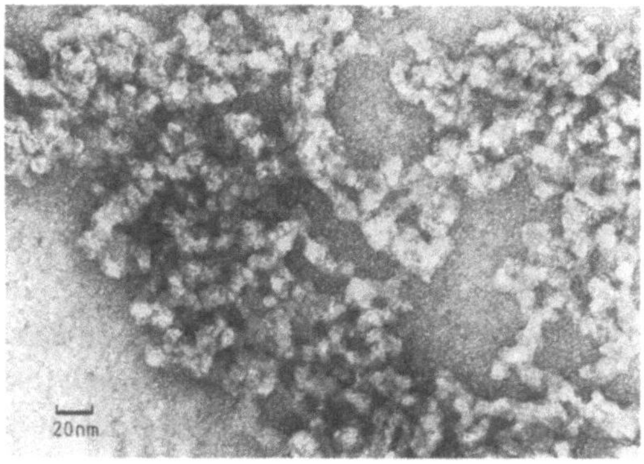

Fig. 6–23. Spruce H_2SO_4 lignin. TEM micrograph.

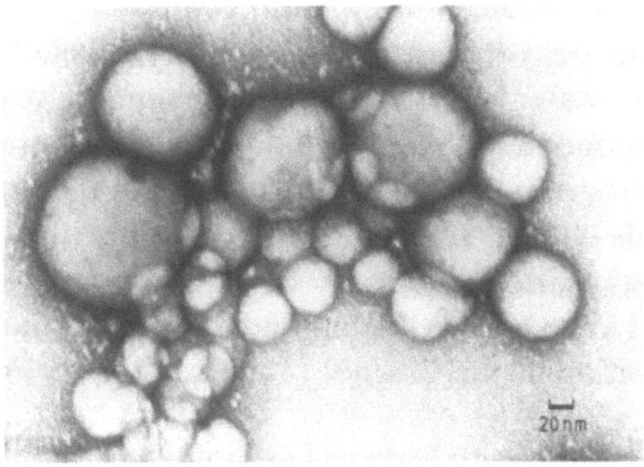

Fig. 6–24. Poplar organosolv lignin. TEM micrograph.

6.5. Lignin-Polysaccharide Complexes

It is generally accepted today that lignin is not simply deposited between the cell wall polysaccharides, but is linked and associated with at least part of them. According to Freudenberg (1968) the presence of carbohydrates is even a prerequisite for the formation of lignin macromolecules within in the plant cell walls.

Nevertheless our knowledge of the molecular and supramolecular connections between the wood cell wall components cellulose, polyoses and lignin is still far from settled. The phenomenon of the intimate association between the polysaccharide and lignin part of the cell wall is described by the terms lignin-polysaccharide complex (LPC) or lignin-carbohydrate complex (LCC). In a more practical sense the terms describe the fact that in very different fractions isolated from wood, containing variable lignin and polysaccharide portions, the components cannot be totally separated by selective chemical treatments or special separation and purification techniques. Even in highly purified cellulose there will remain some residual polyoses and residual lignin (→ 3.2.1.). On the other hand attempts to purify a carefully isolated lignin sample failed to obtain a preparation free of polysaccharide residues.

The early results and ideas about the existence and nature of lignin-polysaccharide complexes in wood were reviewed by Merewether (1957). Later on, Lai and Sarkanen (1971) and Wegener (1975) summarized the knowledge of lignin-polysaccharide complexes. From numerous experimental facts it is evident that apart from other possible types of associations (hydrogen bonds, van-der-Waals forces, chemosorption) chemical linkages are very probably involved in the interaction of lignin and polysaccharides.

Freudenberg (Freudenberg, Grion 1959; Freudenberg, Harkin 1960) was the first to demonstrate the formation of an addition compound composed of a quinone methide and sucrose by enzymatic dehydrogenation of coniferyl alcohol in a concentrated sucrose solution in water and dimethylformamide, respectively (Fig. 6–25). The corresponding monoglucose compound was recently prepared by Katayama et al. (1980) by dehydrogenation of coniferyl alcohol in the presence of horse-radish peroxidase or a crude enzyme from the fruit body of the fungus *Agaricus bisporus* in saturated aqueous glucose solutions. The linkage of D-glucuronic acid to a quinone methide proved to be a benzyl ester bond, combining the carboxyl group of the acid to the α-C-position of the dilignol (Tanaka et al. 1976; 1979).

The high reactivity of quinone methides with hydroxyl groups was demonstrated by Leary (1980; Leary et al. 1981) who showed by ^{13}C-NMR spectroscopy that etherification of vanillyl alcohol with sugars occurs at any of the free sugar hydroxyl groups (preferably at the C6-OH), yielding different p-hydroxylbenzyl ethers as model compounds for lignin-polysaccharide complexes.

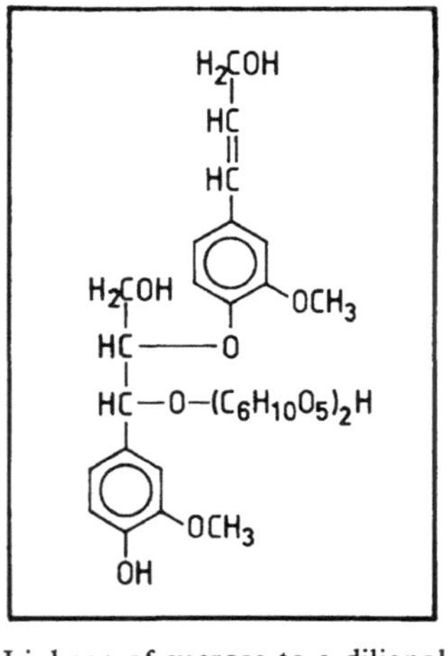

Fig. 6–25. Linkage of sucrose to a dilignol (according to Freudenberg, Harkin 1960).

The problems of model experiments simulating the synthesis of linkages between coniferyl alcohol and sugars were recently described by Feckl (1981; Feckl, Fengel 1982a). He obtained very different products from the catalytic dehydrogenation of coniferyl alcohol in the presence of arabinose, depending on the reaction conditions. Only in a non-aqueous medium (dimethylsulfoxide, DMSO) low-molecular-weight compounds were formed, from which arabinose could be split off by hydrolysis. While a non aqueous hydrophobic medium may also be present within the cell wall membrane during the enzyme-catalyzed in-vivo reaction, the concentrations of the reaction compounds used in all model experiments do not represent the probable situation within the cell walls. Feckl and Fengel (1982a) used an excess of twenty times of arabinose in relation to coniferyl alcohol, while Katayama et al. (1980) applied an excess of even five hundred times of glucose in an aqueous system. The few synthesis experiments to clear up the formation and the character of the lignin-polysaccharide linkages reveal some possible mechanisms between the lignin precursors and polysaccharide components. The results, however, are mainly dependent on the catalysts used and the reaction conditions, thus giving no uniform picture of the in-vivo connections of lignin and polysaccharides in the cell wall.

The same is true of the more numerous experiments carried out with the aim of isolating and characterizing degradation products representing or containing lignin-polysaccharide complexes in a more or less native state. Those complexes were isolated from very different materials:

- finely divided wood (milled wood) after extraction of milled wood lignin (Björkman 1957; Brownell 1965, 1970; Tanaka, Koshijima 1972; Koshijima et al. 1972, 1976; Yaku et al. 1976)
- milled wood lignin (Lundquist 1976; Lundquist et al. 1979, 1980; Wegener, Fengel 1979)
- enzymatically and chemically degraded wood (Krejcberg et al. 1974; Sergeeva et al. 1979)
- methylated and acetylated wood and holocellulose (Košiková et al. 1969, 1972; Košiková 1973)
- chlorite delignification liquors (Wegener 1976; Glinski, Nicholls 1977)
- alkali extracts from holocellulose (Kringstad, Ellefsen 1964; Kringstad 1965; Linell, Swenson 1966; Linell et al. 1966; Kringstad, Cheng 1969; Fengel, Przyklenk 1975, 1976; Fengel 1976b, c, 1979; Feckl 1981; Feckl, Fengel 1982b, c)
- technical pulps and pulping liquors (Pearl, Beyer 1964; Simonson 1971; Meshitsuka et al. 1975; Polčin, Bezúch 1977; Ohara et al. 1980).

The methods of isolation, fractionation and purification are numerous, depending on the type and composition of the starting material. An effective fractionation of alkali extracts from spruce and beech holocelluloses was achieved e.g. by anion-exchange chromatography (Fengel 1976b; Fengel, Przyklenk 1976). The fractions obtained from the 5% KOH extract from spruce holocellulose and their composition are shown in the Figs. 6–26 and 6–27.

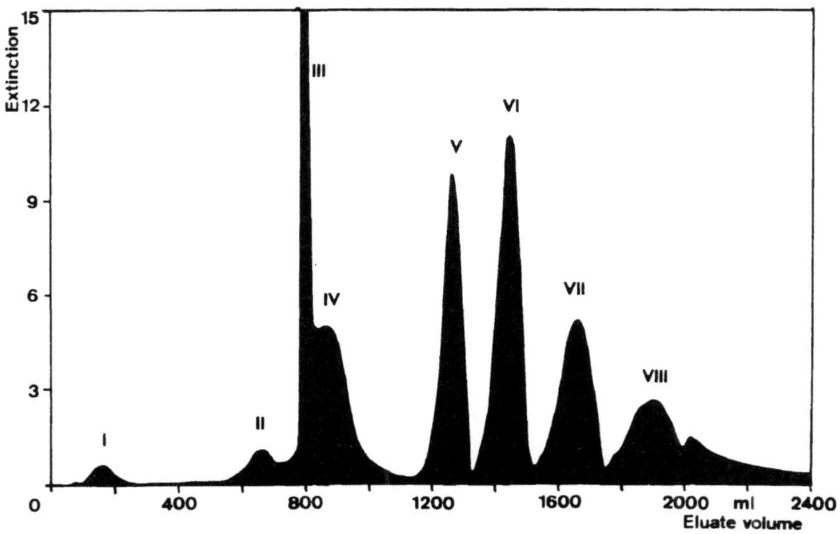

Fig. 6–26. Fractionation of a spruce alkali extract by ion-exchange chromatography.

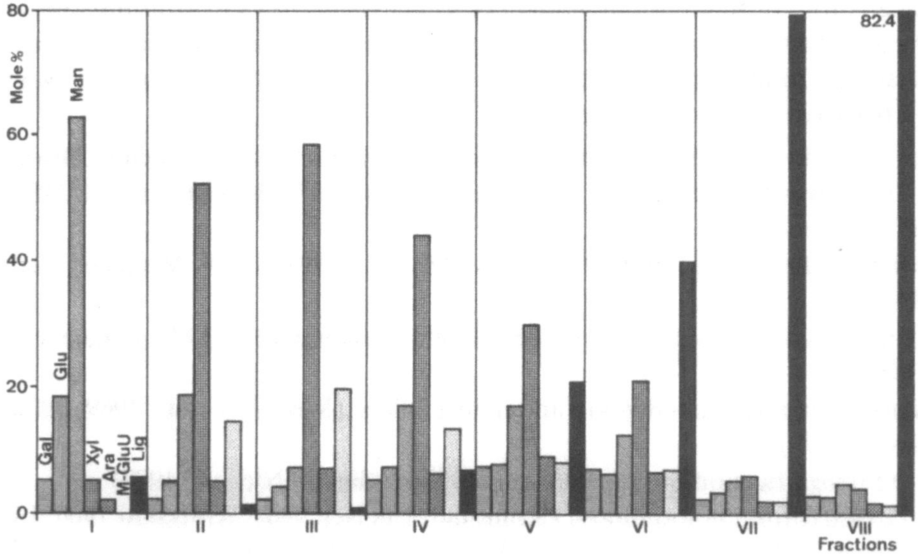

Fig. 6–27. Composition of the fractions from ion-exchange chromatography of a spruce alkali extract.

Gel filtration chromatography is a frequently applied fractionation method. Fig. 6–28 gives an example of separating a fraction from spruce milled wood lignin into 7 fractions, containing different lignin and polysaccharide portions (Wegener, Fengel 1979).

Other methods used are electrophoresis, cation-exchange chromatography, and chromatography on hydroxyl apatite or polyamide (Lindgren 1958; Fengel 1976b;

Pivoverova, Bogomolow 1977; Feckl 1981; Feckl, Fengel 1982b; Wegener 1983). In particular the fractionation of alkali holocellulose extracts by cation-exchange and hydroxyl apatite chromatography revealed that at least part of the isolated lignin-polysaccharide complexes may only be associations of lignin and polysaccharide fragments.

The molecular weights of isolated complexes range from about 600 up to 15 000, determined by gel permeation chromatography as well as by ultracentrifuge and ebulliometric measurements (Björkman 1957; Koshijima et al. 1972; Wegener 1976; Feckl 1981; Feckl, Fengel 1982a, b).

Generally the polyoses are expected to be linked to lignin though bonds between cellulose and lignin cannot be excluded (Pew, Weyna 1962; Wegener, Fengel 1979; Erickson et al. 1980). The polyoses fragments in lignin-polysaccharide complexes may be residues from xylan as well as from mannan. Lignin-xylan complexes were isolated from hardwoods (Bolker, Wang 1969; Merewether et al. 1972; Stewart 1973, Fengel, Przyklenk 1976), while from softwoods lignin-mannan and lignin-xylan complexes were obtained (Meier 1958; Linell, Swenson 1966; Linell et al. 1966; Kringstad, Cheng 1969; Smelstorius 1974; Fengel 1979).

Fig. 6–29 gives an impression of the ultrastructural appearance of a spruce lignin-polysaccharide complex, containing both mannan and xylan portions besides about 70% lignin. The polyoses are embedded in lignin and are intensively coiled and wound. Some look like tied-up cords.

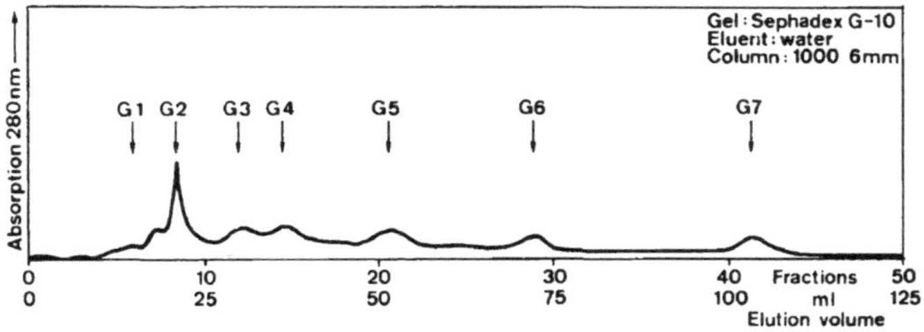

Fig. 6–28. Fractionation of a spruce MWL fraction by gel filtration chromatography.

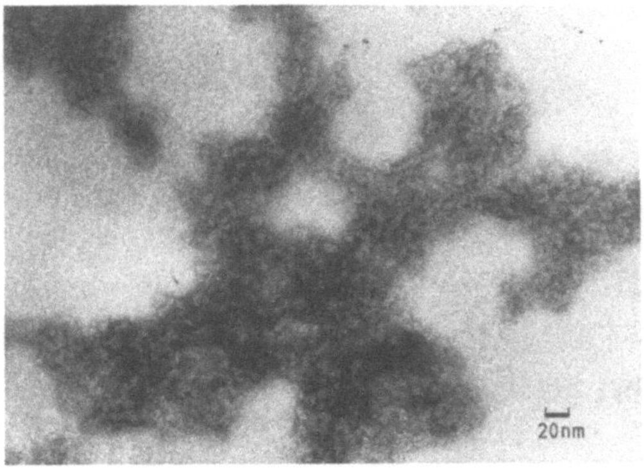

Fig. 6–29. Spruce lignin-polysaccharide complex. TEM micrograph.

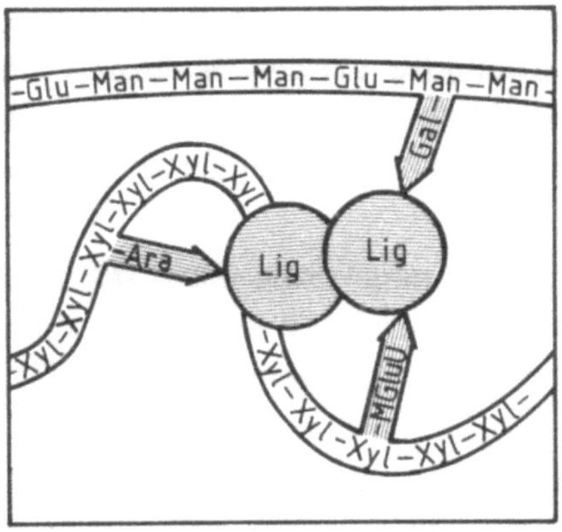

Fig. 6–30. Schematic illustration of connections between softwood polyoses and lignin.

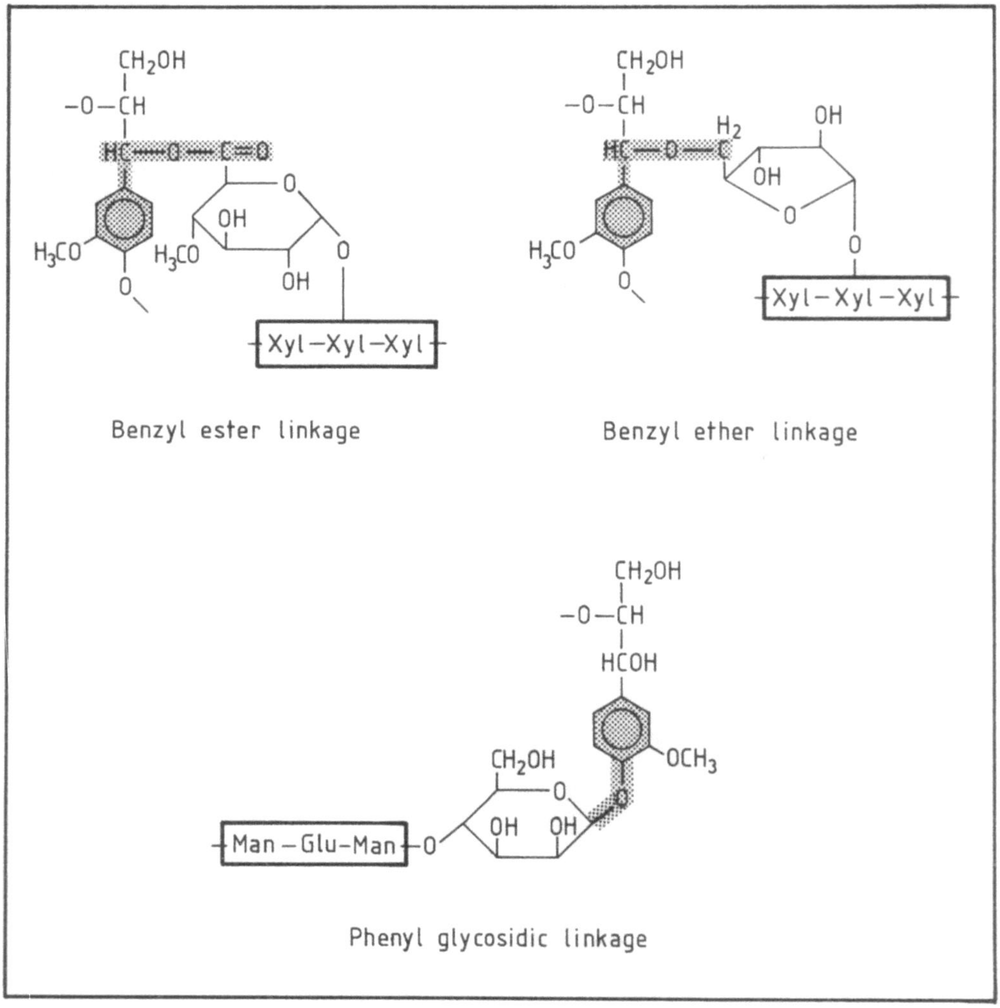

Fig. 6–31. Most frequently suggested types of lignin-polysaccharide linkages.

As connecting links to lignin the polyoses side-groups arabinose, galactose and 4-O-methylglucuronic acid (→ 5.2., 5.3.) are most frequently suggested, due to their sterically favoured positions and to the fact that these sugars were enriched in many different lignin-rich complexes (Kringstad, Cheng 1969; Ezzat 1973; Fengel, Przyklenk 1975; Wegener 1976; Eriksson, Lindgren 1977; Wegener, Fengel 1979; Lundquist et al. 1980; Feckl 1981; Feckl, Fengel 1982a, b). Furthermore, the recently shown stability of these sugars against an oxidative glycol cleavage supports this suggestion (Eriksson et al. 1980). Fig. 6–30 gives a simplified illustration of the probable connections between polyoses and lignin.

Our knowledge of the probable types of covalent linkages between lignin and polyoses is mainly derived from degradation experiments, mostly performed as mild alkaline, acidic or enzymatic hydrolysis. The most frequently suggested types are ether linkages (alkali stable), ester linkages (alkali lable) as well as glycosidic bonds (Bolker, Wang 1969; Košiková et al. 1972, 1975, 1977, 1979; Brownell 1971; Merewether, Samsuzzaman 1972; Košiková, Joniak 1976; Eriksson, Lindgren 1977; Lundquist et al. 1978; Bezúch, Polčin 1978; Joseleau, Gancet 1981) (Fig. 6–31).

Principal structural possibilities of lignin-polyoses complexes are shown in Fig. 6–32. The models include covalent and physical (hydrogen) bonds between one lignin element and one polysaccharide element (a), lignin elements bound to a polysaccharide element and polysaccharide elements bound to a lignin element (b), and finally a network of polysaccharide and lignin elements linked by chemical and physical linkages (c).

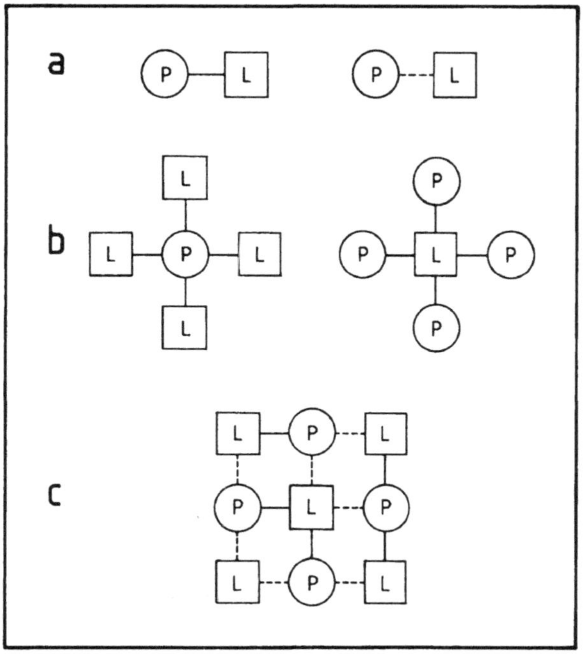

Fig. 6–32. Models of the possible structure of LPCs. P: polysaccharide element; L: lignin element.

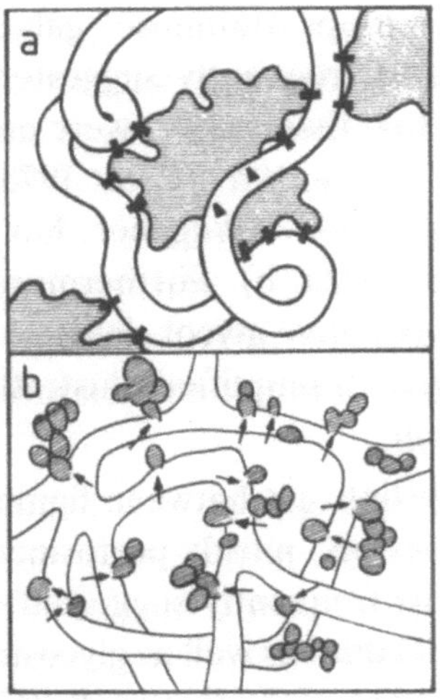

Fig. 6–33. Ultrastructural models for lignin-polysaccharide complexes. a: Fengel 1976c; b: Košiková et al. 1978.

Electron microscopic observations combined with chemical results led to more predictive model ideas about lignin-polysaccharide complexes on the supramolecular level (Fengel 1976c; Košiková et al. 1978) (Fig. 6–33). The two models differ with regard to the diameter of the polysaccharide fibrils (a: 50–80 nm; b: 100–200 nm), and indicate different standpoints about the frequency of bonds between lignin and polysaccharides. In the first model (Fengel 1976c) coiled polyoses fibrils are repeatedly connected with large lignin particles, while in the second model (Košiková et al. 1978) single lignin particles or aggregates of particles are linked once to the surface of polysaccharide fibrils. In the latter case the number of chemical bonds related to the mass of polysaccharides and lignin would be very low.

Though our knowledge of the associations between lignin and polysaccharides in the wood cell wall has been significantly improved, the questions about the chemical homogeneity of isolated lignin-polysaccharide complexes and the frequency of chemical and other bonds involved are still unsolved. Finally, special types of native covalent linkages have not yet been definitively proved.

Literature

Adler, E. 1961, Pap. Puu *43,* 634–643

Adler, E. 1977, Wood Sci. Technol. *11,* 169–218

Adler, E., Pepper, J. M. and Eriksoo, E. 1957, Ind. Eng. Chem. *49,* 1391–1392

Alibert, G. and Boudet, A. 1979, Physiol. Veget. *17,* 67–74, 75–79

Anderson, A., Erickson, M., Fridh, H. and Miksche, G. E. 1973, Holzforschung *27,* 189–193

Anon. 1981, Westvaco Lignochemicals and Their Applications. Westvaco Chem. Div.; Charleston Heights, S.C., U.S.A.

Anon. 1982, Lignin Chemicals. Borregaard A. S., Spec. Chem. Sect., Sarpsborg, Norway

Baumeister, M. and Edel, E. 1980, Papier *34,* No. 10A, V9–V18
Becker, H. and Nimz, H. 1974, Z. Pflanzenphysiol. *72,* 52–63
Bezúch, B. and Polčin, J. 1978, Cell. Chem. Technol. *12,* 473–482
Björkman, A. 1956, Svensk Papperstid. *59,* 477–485
Björkman, A. 1957, Svensk Papperstid. *60,* 243–251
Björkman, A. and Person, B. 1957, Svensk Papperstid. *60,* 158–169
Bland, D. E. and Menshun, M. 1970, Appita *23,* 427–433
Bolker, H. I. and Wang, P. Y. 1969, Tappi *52,* 920–923
Böttger, J., Krause, T. and Schurz, J. 1976, Holzforschung *30,* 41–44
Brauns, F. E. and Brauns, D. A. 1960, The Chemistry of Lignin. Supplement Volume. Academic Press, New York
Brink, D. L., Wu, Y. T., Naveau, H. P., Bicho, J. G. and Merriman, M. M. 1972, Tappi *55,* 719–721
Brown, W., Cowling, E. B. and Falkehag, S. I. 1968, Svensk Papperstid. *71,* 811–821
Brownell, H. H. 1965, Tappi *48,* 513–519
Brownell, H. H. 1970, Tappi *53,* 1278–1281
Brownell, H. H. 1971, Tappi *54,* 66–71
Browning, B. L. 1967a, Methods of Wood Chemistry, Vol. I. Intersci. Publ., New York
Browning, B. L. 1967b, Methods of Wood Chemistry, Vol. II. Intersci. Publ. New York
Brunow, G. and Lundquist, L. 1980, Pap. Puu *62,* 669–672
Brunow, G. and Wallin, H. 1981, Studies Concerning the Preparation of Synthetic Lignin. In: The Ekman Days 1981. Int. Symp. Wood Pulp. Chem., Stockholm, Vol. 4, pp. 125–127
Budin, D. and Susa, L. 1982, Holzforschung *36,* 17–22
Chang, H. M. and Allan, G. G. 1971, Oxidation. In: Lignins. Occurrence, Formation, Structure and Reactions (Sarkanen, K. V. and Ludwig, C. H., Eds.). Wiley-Intersci., New York, pp. 433–485
Chang, H. M., Cowling, E. B., Brown, W., Adler, E. and Miksche, G. 1975, Holzforschung *29,* 153–159
Cho, N. S., Lee, J. Y., Meshitsuka, G. and Nakano, J. 1980, Mokuzai Gakkaishi *26,* 527–533
Chua, M. G. S. and Wayman, M. 1979, Can. J. Chem. *57,* 1141–1149
Concin, R., Burtscher, E. and Bobleter, O. 1981, Holzforschung *35,* 279–282
Connors, W. J. 1978, Holzforschung *32,* 145–147
Connors, W. J., Lorenz, L. F. and Kirk, T. K. 1978, Holzforschung *32,* 106–108
Connors, W. J., Sarkanen, K. V. and McCarthy, J. L. 1980, Holzforschung *34,* 80–85
Dix, B. and Roffael, E. 1981, Holzforschung *35,* 189–193
Ekman, K. H. and Lindberg, J. J. 1960, Pap. Puu *42,* 21–22
Elstner, E. F. and Heupel, A. 1976, Planta *130,* 175–180
Erickson, M. and Miksche, G. E. 1974, Phytochemistry *13,* 2295–2299
Erickson, M., Miksche, G. E. and Somfai, I. 1973a, Holzforschung *27,* 113–117
Erickson, M., Larsson, S. and Miksche, G. E. 1973b, Acta Chem. Scand. *27,* 127–140
Erickson, M., Larsson, S. and Miksche, G. E. 1973c, Acta Chem. Scand. *27,* 903–914
Erickson, M., Larsson, S. and Miksche, G. E. 1973d, Acta Chem. Scand. *27,* 1673–1678
Eriksson, Ö. and Lindgren, B. O. 1977, Svensk Papperstid. *80,* 59–63
Eriksson, Ö., Goring, D. A. I. and Lindgren, B. O. 1980, Wood Sci. Technol. *14,* 267–279
Evliya, H. and Olcay, A. 1974, Holzforschung *28,* 130–135
Ezzat, S. 1973, Holzforschung *27,* 16–20
Faix, O. 1974, Holzforschung *28,* 222–225
Faix, O. 1976, Papier *30,* No. 10A, V1–V10
Faix, O. and Besold, G. 1978, Holzforschung *32,* 1–7
Faix, O. and Schweers, W. 1974a, Holzforschung *28,* 94–98
Faix, O. and Schweers, W. 1974b, Holzforschung *28,* 50–54
Faix, O. and Schweers, W. 1975, Holzforschung *29,* 48–55

Faix, O., Gyzas, E. and Schweers, W. 1977, Holzforschung *31,* 137–144

Faix, O., Lange, W. and Beinhoff, O. 1980, Holzforschung *34,* 174–176

Faix, O., Lange, W. and Salud, E. C. 1981, Holzforschung *35,* 3–9

Feckl, J. 1981, Gewinnung und Fraktionierung von Lignin-Polysaccharid-Komplexen. Doctor Thesis, Universität München

Feckl, J. and Fengel, D. 1982a, Unpublished results

Feckl, J. and Fengel, D. 1982b, Holzforschung *36,* 233–237

Feckl, J. and Fengel, D. 1982c, Holzforschung *36,* 269–273

Fengel, D. 1976a, Holzforschung *30,* 1–6

Fengel, D. 1976b, Holzforschung *30,* 73–78, 143–148

Fengel, D. 1976c, Svensk Papperstid. *79,* 24–28

Fengel, D. 1979, Cell. Chem. Technol. *13,* 279–297

Fengel, D. and Przyklenk, M. 1975, Svensk Papperstid. *78,* 17–21, 617–620

Fengel, D. and Przyklenk, M. 1976, Wood Sci. Technol. *10,* 311–320

Fengel, D. and Stoll, M. 1975, Instrument u. Forschung *4,* 1–14

Fengel, D., Greune, A. and Wegener, G. 1983, Holzforschung *37,* 119–122

Fengel, D., Wegener, G. and Feckl, J. 1981a, Holzforschung *35,* 51–57

Fengel, D., Wegener, G. and Feckl, J. 1981b, Holzforschung *35,* 111–118

Fergus, B. J. and Goring, D. A. I. 1970, Holzforschung *24,* 113–117

Fernandez, N. and Šutý, L. 1981, Studies in Sugar Cane Bagasse Lignin. In: The Ekman Days 1981. Int. Symp. Wood Pulp. Chem., Stockholm, Vol. 5, pp. 52–60

Field, L., Drummond, P. E., Riggins, P. H. and Jones, E. A. 1958, Tappi *41,* 721–727

Fišerová, M. and Šutý, L. 1980, Cell. Chem. Technol. *14,* 243–252

Forss, K. and Fremer, K. E. 1965, Pap. Puu *47,* 443–454

Forss, K. and Fremer, K. E. 1981, Some Properties of Spruce Lignosulphonic Acid. In: The Ekman Days 1981. Int. Symp. Wood Pulp. Chem., Stockholm, Vol. 4, pp. 29–38

Forss, K., Fremer, K. E. and Stenlund, B. 1966, Pap. Puu *48,* 565–574, 669–676

Forss, K. G., Stenlund, B. G. and Sågfors, P. E. 1976, Appl. Polym. Symp. No. 28, 1185–1194

Freudenberg, K. 1964a, The Formation of Lignin in the Tissue and in vitro. In: The Formation of Wood in Forest Trees (Zimmerman, M. H., Ed.). Academic Press, New York, pp. 203–218

Freudenberg, K. 1964b, Holzforschung *18,* 3–9

Freudenberg, K. 1968, The Constitution and Biosynthesis of Lignin. In: Constitution and Biosynthesis of Lignin (Freudenberg, K. and Neish, A. C., Eds.). Springer-Verlag Berlin, Heidelberg, New York, pp. 47–122

Freudenberg, K. and Grion, G. 1959, Chem. Ber. *92,* 1355–1363

Freudenberg, K. and Harkin, J. M. 1960, Chem. Ber. *93,* 2814–2819

Freudenberg, K. and Neish, A. C. 1968, Constitution and Biosynthesis of Lignin. Springer-Verlag Berlin, Heidelberg, New York

Freudenberg, K. and Sidhu, G. S. 1961, Holzforschung *15,* 33–39

Froment, P. and Robert, A. 1977, Cell. Chem. Technol. *11,* 691–696

Fujita, M. 1981, Deposition of Major Cell Wall Components in the Differentiating Tree Xylem Cells. Doctor Thesis, University of Kyoto

Fujita, M. and Harada, H. 1979, Mokuzai Gakkaishi *25,* 89–94

Fujita, M., Saiki, H. and Harada, H. 1978a, Mokuzai Gakkaishi *24,* 158–163

Fujita, M., Saiki, H. and Harada, H. 1978b, Mokuzai Gakkaishi *24,* 353–361

Gibbs, R. D. 1958, The Mäule Reaction, Lignins, and the Relationships between Woody Plants. In: The Physiology of Forest Trees (Thimann, K. V., Ed.). Ronald Press Comp., New York, pp. 269–312

Glasser, W. G. 1980, Lignin. In: Pulp and Paper. Chemistry and Chemical Technology (Casey, J. P., Ed.), Vol. I., 3rd Ed. Wiley-Intersci., New York, pp. 39–111

Glasser, W. G. and Barnett, C. A. 1979a, Holzforschung *33,* 78–86

Glasser, W. G. and Barnett, C. A. 1979b, Tappi *62,* 101–105

Glasser, W. G. and Glasser, H. R. 1974a, Macromolecules *7,* 17–27

Glasser, W. G. and Glasser, H. R. 1974b, Holzforschung *28,* 5–11

Glasser, W. G. and Glasser, H. R. 1976, Cell. Chem. Technol. *10,* 39–52

Glasser, W. G. and Glasser, H. R. 1981, Pap. Puu *63,* 71–83

Glasser, W. G., Glasser, H. R. and Nimz, H. H. 1976, Macromolecules *9,* 866–867

Glasser, W. G., Glasser, H. R. and Morohoshi, N. 1981, Macromolecules *14,* 253–262

Glinski, A. J., Nicholls, G. A. 1977, Pap. Puu *59,* 745–760

Goldschmid, O. 1971, Ultraviolet Spectra. In: Lignins. Occurrence, Formation, Structure and Reactions (Sarkanen, K. V. and Ludwig, C. H., Eds.). Wiley-Intersci., New York, pp. 241–266

Goring, D. A. I. 1971, Polymer Properties of Lignin and Lignin Derivatives. In: Lignins. Occurrence, Formation, Structure and Reactions (Sarkanen, K. V. and Ludwig, C. H., Eds.). Wiley-Intersci., New York, pp. 695–768

Grisebach, H. 1977, Naturwissenschaften *64,* 619–625

Grisebach, H., Lüderitz, T. and Schmid, G. 1981, Enzymes Involved in Biosynthesis of Lignin Precursors. In: The Ekman Days 1981. Int. Symp. Wood Pulp. Chem., Stockholm, Vol. 3, pp. 51–54

Gross, G. G. 1977, Biosynthesis of Lignin and Related Monomers. In: The Structure, Biosynthesis and Degradation of Wood. Recent Adv. in Phytochem. *11* (Loewus, F. A. and Runeckles, V. C., Eds.). Plenum Press, New York, pp. 141–184

Gross, G. G. 1978, Recent Advances in the Chemistry and Biochemistry of Lignin. In: Biochemistry of Plant Phenolics. Recent Adv. in Phytochem. *12* (Swain, T., Harborne, J. B. and Van Sumere, C. F., Eds.). Plenum Press, New York, pp. 177–220

Gross, G. G., Stöckigt, J., Mansell, R. L. and Zenk, M. H. 1973, FEBS Lett. *31,* 283–286

Gross, G. G., Mansell, R. L. and Zenk, M. H. 1975, Biochem. Physiol. Pflanz. *168,* 41–51

Hahlbrock, K. and Grisebach, H. 1979, Ann. Rev. Plant. Physiol. *30,* 105–130

Halliwell, B. 1978, Planta *140,* 81–88

Hardell, H. L., Leary, G. J., Stoll, M. and Westermark, U. 1980, Svensk Papperstid. *83,* 44–49, 71–74

Harkin, J. M., Obst, J. R. 1973, Science *180,* 296–298

Hepler, P. K., Fosket, D. E. and Newcomb, E. H. 1970, Amer. J. Bot. *57,* 85–96

Hergert, H. L. 1971, Infrared Spectra. In: Lignins. Occurrence, Formation, Structure and Reactions (Sarkanen, K. V. and Ludwig, C. H., Eds.). Wiley-Intersci., New York, pp. 267–297

Higuchi, T. and Kawamura, I. 1966, Holzforschung *20,* 16–21

Higuchi, T., Tanahashi, M. and Sato, A. 1972, Mokuzai Gakkaishi *18,* 183–189

Higuchi, T., Shimada, M., Nakatsubo, F. and Tanahashi, M. 1977, Wood Sci. Technol. *11,* 153–167

Hoffmann, P. and Schweers, W. 1975a, Holzforschung *29,* 18–24

Hoffmann, P. and Schweers, W. 1975b, Pap. Puu *57,* 581–592

Hoffmann, P. and Schweers, W. 1976, Pap. Puu *58,* 227–244

Hofmann, K. P. and Zundel, G. 1971, Rev. Sci. Instrum. *42,* 1727

Howard, E. T. 1973, Wood Sci. *5,* 312–317

Hrutfiord, B. F. 1971, Reduction and Hydrogenolysis. In: Lignins. Occurrence, Formation, Structure and Reactions (Sarkanen, K. V. and Ludwig, C. H., Eds.). Wiley-Intersci., New York, pp. 487–509

Hüttermann, A. 1978, Holzforschung *32,* 108–111

Hwang, B. H. and Sakakibara, A. 1981, Holzforschung *35,* 297–300

Hwang, B. H., Sakakibara, A. and Miki, K. 1981, Holzforschung *35,* 229–232

Ivancic, A. and Rydholm, S. A. 1959, Svensk Papperstid. *62,* 554–566

Johnson, D. B., Moore, W. E. and Zank, L. C. 1961, Tappi *44,* 793–798

Joseleau, J. P. and Gancet, C. 1981, Svensk Papperstid. *84,* R123–R127

Kachi, S. and Terashima, N. 1974, Mokuzai Gakkaishi *20,* 453–459

Karklins, V. and Okherina, E. E. 1975, Khim. Drev., No. 4, 3–10

Katayama, Y., Morohoshi, N. and Haraguchi, T. 1980, Mokuzai Gakkaishi *26,* 358–362, 414–420

Katrizky, A. R. and Topsom, R. D. 1977, Chem. Rev. 1977, 639–658
Kawamura, I. and Bland, D. E. 1967, Holzforschung *21,* 65–74
Kawamura, L. and Higuchi, T. 1964, Comparative Studies of Milled Wood Lignins from Different Taxonomical Origin by Infrared Spectroscopy. In: Chimie et Biochimie de la Lignine, Cellulose et des Hémicelluloses. Les Imprimeries, Réunies de Chambéry, pp. 439–456
Kleinert, T. N. 1971, Papier *25,* 65–67
Kolar, J. J., Lindgren, B. O. and Roy, T. 1979, Cell. Chem. Technol. *13,* 491–499
Kolboe, S. and Ellefsen, Ø. 1962, Tappi *45,* 163–166
Koshijima, T., Taniguchi, T. and Tanaka, R. 1972, Holzforschung *26,* 211–217
Koshijima, T., Yaku, F. and Tanaka, R. 1976, Appl. Polym. Symp. No. 28, 1025–1039
Košiková, B. 1973, Drev. Vysk. *18,* 25–31, 83–90, 139–143
Košiková, B. and Joniak, D. 1976, Cell. Chem. Technol. *10,* 695–701
Košiková, B., Polčin, J. and Dandárová-Vašátková, H. 1969, Holzforschung *23,* 37–43, 43–47
Košiková, B., Joniak, D. and Skamla, J. 1972, Cell. Chem. Technol. *6,* 579–588
Košiková, B., Joniak, D. and Košik, M. 1975, Cell. Chem. Technol. *9,* 61–70
Košiková, B., Joniak, D. and Kosáková, L. 1979, Holzforschung *33,* 11–14
Košiková, B., Polčin, J. and Joniak, D. 1977, Holzforschung *31,* 191–193
Košiková, B., Zakutna, L. and Joniak, D. 1978, Holzforschung *32,* 15–18
Kratzl, K. and Claus, P. 1962, Monatshefte Chem. *93,* 219–229
Krejcberg, Z. N., Aroncik, B. M. and Sergeeva, V. N. 1974, Khim. Drev. No. 1, 74–77
Kringstad, K. 1965, Acta Chem. Scand. *19,* 1493–1494
Kringstad, K. P. and Cheng, C. W. 1969, Tappi *52,* 2382–2385
Kringstad, K. and Ellefsen, Ø. 1964, Papier *18,* 583–591
Kringstad, K. P., Månsson, P. and Mörck, R. 1981, Changes in the Molecular Weight Distribution of Kraft Lignins Resulting from Various Chemical Treatments. In: The Ekman Days 1981. Int. Symp. Wood Pulp. Chem., Stockholm, Vol. 5, pp. 91–93
Krüger, G. 1976, Chem. Unserer Zeit *10,* 21–29
Kutscha, N. P. and Schwarzmann, J. M. 1975, Holzforschung *29,* 79–84
Lai, Y. Z. and Sarkanen, K. V. 1971, Isolation and Structural Studies. In: Lignins. Occurrence, Formation, Structure and Reactions (Sarkanen, K. V. and Ludwig, C. H., Eds.). Wiley-Intersci., New York, pp. 165–240
Lai, Y. Z. and Sarkanen, K. V. 1975, Cell. Chem. Technol. *9,* 239–245
Lapierre, C. and Monties, B. 1981, Evidence of Lignin Heterogeneity by Progressive Extraction of Poplars. In: The Ekman Days 1981. Int. Symp. Wood Pulp. Chem., Stockholm, Vol. 5, pp. 35–39
Larsson, S. and Miksche, G. E. 1971, Acta Chem. Scand. *25,* 647–662
Leary, G. J. 1980, Wood Sci. Technol. *14,* 21–34
Leary, G. J. 1982, Wood Sci. Technol. *16,* 67–70
Leary, G. and Thomas, W. 1975, Tetrahedron Lett. *42,* 3631–3634
Leary, G. J., Sawtell, D. A. and Wong, H. 1981, The Formation of Model Lignin-Carbohydrate-Compounds in Aqueous Solution. In: The Ekman Days 1981. Int. Symp. Wood Pulp. Chem., Stockholm, Vol. 1, pp. 63–66
Lin, S. Y. and Detroit, W. J. 1981, Chemical Heterogeneity of Technical Lignins – Its Significance in Lignin Utilization. In: The Ekman Days 1981. Int. Symp. Wood Pulp. Chem., Stockholm, Vol. 4, pp. 44–52
Lindberg, J. J. and Törmälä, P. 1981, Dynamic Properties of Lignins. In: The Ekman Days 1981. Int. Symp. Wood Pulp. Chem., Stockholm, Vol. 4, pp. 59–65
Lindgren, B. O. 1958, Acta Chem. Scand. *12,* 447–452
Linell, W. S. and Swenson, H. A. 1966, Tappi *49,* 444–446, 494–497
Linell, W. S., Thompson, N. S. and Swenson, H. A. 1966, Tappi *49,* 491–493
Ludwig, C. H. 1971, Magnetic Resonance Spectra. In: Lignins. Occurrence, Formation, Structure and Reactions (Sarkanen, K. V. and Ludwig, C. H., Eds.). Wiley-Intersci. New York, pp. 299–344

Lüdemann, H. D. and Nimz, H. 1974, Makromol. Chem. *175*, 2393–2407
Luger, F. and Gampe, S. 1978, Potential forstlicher Reststoffe (Waldabfälle), Vol. 2. Bericht Forstl. Fak., Universität München, pp. 84–158
Lundquist, K. 1973, Acta Chem. Scand. *27*, 2597–2606
Lundquist, K. 1976, Appl. Polym. Symp. No. 28, 1393–1407
Lundquist, K. 1981, Analysis of Lignins by Spectral Methods. In: The Ekman Days 1981. Int. Symp. Wood Pulp. Chem., Stockholm, Vol. 1, pp. 81–85
Lundquist, K. and Miksche, G. E. 1965, Tetrahedron Lett. *25*, 2131–2136
Lundquist, K. and Simonson, R. 1975, Svensk Papperstid. *78*, 390
Lundquist, K., Josefsson, B. and Nyquist, G. 1978, Holzforschung *32*, 27–32
Lundquist, K., Ohlsson, B. and Simonson, R. 1977, Svensk Papperstid. *80*, 143–144
Lundquist, K., Simonson, R. and Tingsvik, K. 1979, Svensk Papperstid. *82*, 272–275
Lundquist, K., Simonson, R. and Tingsvik, K. 1980, Svensk Papperstid. *83*, 452–454
Mader, M. 1976, Planta *131*, 11–15
Månsson, P. 1981, GPC of Kraft Lignins. In: The Ekman Days 1981. Int. Symp. Wood Pulp. Chem., Stockholm, Vol. 5, pp. 94–95
Marcinowski, S. and Grisebach, H. 1978, Eur. J. Biochem. *87*, 37–44
Marcinowski, S., Falk, H., Hammer, D. K., Hoyer, B. and Grisebach, H. 1979, Planta *144*, 161–165
Marton, J. 1964, Tappi *47*, 713–719
Marton, J. 1971, Reactions in Alkaline Pulping. In: Lignins. Occurrence, Formation, Structure and Reactions (Sarkanen, K. V. and Ludwig, C. H., Eds.). Wiley-Intersci., New York, pp. 639–694
Marton, J. and Marton, T. 1964, Tappi *47*, 471–476
McNaughton, J. G., Yean, W. Q. and Goring, D. A. I. 1967, Tappi *50*, 548–553
Meier, H. 1958, Acta Chem. Scand. *12*, 1911–1918
Meier, D., Faix, O. and Lange, W. 1981, Holzforschung *35*, 247–252
Merewether, J. W. T. 1957, Holzforschung *11*, 65–80
Merewether, J. W. T. and Samsuzzaman, L. A. M. 1972, Holzforschung *26*, 11–18
Merewether, J. W. T., Samsuzzaman, L. A. M. and Calder, I. C. 1972a, Holzforschung *26*, 180–185
Merewether, J. W. T., Samsuzzaman, L. A. M. and Cooke, R. G. 1972b, Holzforschung *26*, 193–197
Meshitsuka, G. and Nakano, J. 1978, Mokuzai Gakkaishi *24*, 563–568
Meshitsuka, G., Shirogane, H. and Nakano, J. 1975, Mokuzai Gakkaishi *21*, 33–38
Michell, A. J. 1966, Aust. J. Chem. *19*, 2285–2298
Michell, A. J., Watson, A. J. and Higgins, H. G. 1965, Tappi *48*, 520–531
Miksche, G. E. and Yasuda, S. 1977, Holzforschung *31*, 57–59
Miksche, G. E. and Yasuda, S. 1978, Phytochemistry *17*, 503–504
Morohoshi, N. and Glasser, W. G. 1979, Wood Sci. Technol. *13*, 165–178, 249–264
Musha, Y. and Goring, D. A. I. 1975, Wood Sci. Technol. *9*, 45–58
Nakatsubo, F., Sato, K. and Higuchi, T. 1976, Mokuzai Gakkaishi *22*, 29–33
Naveau, H. P., Wu, Y. T., Brink, D. L., Merriman, M. M. and Bicho, J. G. 1972, Tappi *55*, 1356–1361
Neish, A. C. 1964, Cinnamic Acid Derivatives as Intermediates in the Biosynthesis of Lignin and Related Compounds. In: The Formation of Wood in Forest Trees (Zimmerman, M. H., Ed.). Academic Press, New York, pp. 219–239
Neish, A. C. 1968, Monomeric Intermediates in the Biosynthesis of Lignin. In: Constitution and Biosynthesis of Lignin (Freudenberg, K. and Neish, A. C., Eds.). Springer-Verlag, Berlin, pp. 5–43
Nilsson-Idner, K. and Norrström, H. 1974, Svensk Papperstid. *77*, 99–101
Nilsson, K., Norrström, H. and Teder, A. 1972, Svensk Papperstid. *75*, 733–738
Nilsson-Idner, K., Norrström, H. and Teder, A. 1974, Svensk Papperstid. *77*, 60–62
Nimz, H. 1966, Holzforschung *20*, 105–109
Nimz, H. 1969, Chem. Ber. *102*, 3803–3817
Nimz, H. 1974, Angew. Chem. Int. Ed. Engl. *13*, 313–321

Nimz, H. H. 1981, Wood Sci. Technol. *15,* 311–316
Nimz, H. and Das, K. 1971, Chem. Ber. *104,* 2359–2380
Nimz, H. H. and Lüdeman, H. D. 1976, Holzforschung *30,* 33–40
Nimz, H. H. and Tutschek, R. 1977, Holzforschung *31,* 101–106
Nimz, H. H., Lüdemann, H. D. and Becker, H. 1974, Z. Pflanzenphysiol. *73,* 226–233
Nimz, H. H., Robert, D., Faix, O. and Nemr, M. 1981, Holzforschung *35,* 16–26
Norrström, H. 1972, Svensk Papperstid. *75,* 611–618, 891–899
Norrström, H. and Teder, A. 1971, Svensk Papperstid. *74,* 85–93, 337–344
Norrström, H. and Teder, A. 1973, Svensk Papperstid. *76,* 458–462
Obst, J. R. 1982, Holzforschung *36,* 143–152
Ohara, S., Hosoya, S. and Nakano, J. 1980, Mokuzai Gakkaishi *26,* 408–413
Parameswaran, N., Faix, O. and Schweers, W. 1975, Holzforschung *29,* 1–4
Pearl, I. A. 1967, The Chemistry of Lignin. Marcel Dekker Inc., New York
Pearl, I. A. and Beyer, D. L. 1964, Tappi *47,* 458–462, 779–782
Pew, J. C. and Weyna, P. 1962, Tappi *45,* L47–L56
Pickett-Heaps, J. D. 1968, Protoplasma *65,* 181–205
Pivovarova, V. A. and Bogomolov, B. D. 1977, Khim. Drev., No. 2, 48–51
Polčin, J. and Bezúch, B. 1977, Wood Sci. Technol. *11,* 275–290
Rezanowich, A. and Goring, D. A. I. 1960, J. Colloid Sci. *15,* 452–471
Rezanowich, A., Yean, W. Q. and Goring, D. A. I. 1963, Svensk Papperstid. *66,* 141–149
Rezanowich, A., Yean, W. Q. and Goring, D. A. I. 1964, J. Appl. Polym. Sci. *81,* 1801–1812
Robert, D. and Gagnaire, D. 1981, Quantitative Analysis of Lignins by ^{13}C-NMR. In: The Ekman Days 1981. Int. Symp. Wood Pulp. Chem., Stockholm, Vol. 1, pp. 86–88
Rubery, P. H. and Northcote, D. H. 1968, Nature *219,* 1230–1234
Sakakibara, A. 1977, Degradation Products of Protolignin and the Structure of Lignin. In: The Structure, Biosynthesis and Degradation of Wood. Rec. Adv. in Phytochemistry *11* (Loewus, F. A. and Runeckles, V. C., Eds.). Plenum Press, New York, pp. 117–139
Sakakibara, A. 1980, Wood Sci. Technol. *14,* 89–100
Sakakibara, A., Miki, K. and Takahashi, H. 1981, Lignans, Brauns' Lignin and Cell Wall Lignin. In: The Ekman Days 1981. Int. Symp. Wood Pulp. Chem., Stockholm, Vol. 1, pp. 73–80
Saleh, T. M, Leney, I. and Sarkanen, K. V. 1967, Holzforschung *21,* 116–120
Salud, E. D. and Faix, O. 1980, Holzforschung *34,* 113–121
Sano, Y. and Sakakibara, A. 1974, Mokuzai Gakkaishi *20,* 584–591
Sano, Y. and Sakakibara, A. 1980, Res. Bull. Coll. Exp. For. Hokkaido Univ. XXXVII, 241–260
Sarkanen, K. V 1971, Precursors and Their Polymerization. In: Lignins. Occurrence, Formation, Structure and Reactions (Sarkanen, K. V. and Ludwig, C. H., Eds.). Wiley-Intersci., New York, pp. 95–163
Sarkanen, K. V. and Hergert, H. L. 1971, Classification and Distribution. In: Lignins. Occurrence, Formation, Structure and Reactions (Sarkanen, K. V. and Ludwig, C. H., Eds.). Wiley-Intersci., New York, pp. 43–94
Sarkanen, K. V., Chang, H. M. and Allan, G. G. 1967a, Tappi *50,* 583–587
Sarkanen, K. V., Chang, H. M. and Allan, G. G. 1967b, Tappi *50,* 587–590
Sarkanen, K. V., Chang, H. M. and Ericsson, B. 1967c, Tappi *50,* 572–575
Schubert, W. J. 1965, Lignin Biochemistry. Academic Press, New York
Schweers, W. and Faix, O. 1973, Holzforschung *27,* 208–213
Sergeeva, L. L., Semenovskii, A. V., Maltseva, A. L. and Shorygina, N. N. 1979, Khim. Drev., No. 3, 35–40, 41–44
Simonson, R. 1971, Svensk Papperstid. *74,* 153–165, 268–270
Smelstorius, J. A. 1974, Holzforschung *28,* 97–101
Sosanwo, O. and Lindberg, J. J. 1976, Pap. Puu *58,* 201–206
Soundararajan, T. N. and Wayman, M. 1970, J. Polym. Sci., Part C, No. 30, 521–531

Srivastava, L. M. 1966, Tappi *49,* 173–183
Stewart, C. M. 1973, Cell. Chem. Technol. *7,* 691–701
Stoldere, I. A., Gravitis, J. A. and Erins, P. 1978, Khim. Drev., No. 2, 20–26
Sudo, K., Hwang, B. H. and Sakakibara, A. 1978, Mokuzai Gakkaishi *24,* 424–425
Sudo, K., Hwang, B. H. and Sakakibara, A. 1979, Mokuzai Gakkaishi *25,* 61–66
Takabe, K., Fujita, M., Harada, H. and Saiki, H. 1981, Mokuzai Gakkaishi *27,* 813–820
Tanahashi, M., Aoki, T. and Higuchi, T. 1982, Holzforschung *36,* 117–122
Tanaka, R. and Koshijima, T. 1972, Mokuzai Gakkaishi *18,* 403–408
Tanaka, K., Nakatsubo, F. and Higuchi, T. 1976, Mokuzai Gakkaishi *22,* 589–590
Tanaka, K., Nakatsubo, F. and Higuchi, T. 1979, Mokuzai Gakkaishi *25,* 653–659
Uprichard, J. m. 1971, Holzforschung *25,* 97–104
Venverloo, C. L. 1971, Holzforschung *25,* 18–24
Vollmert, B. 1973, Polymer Chemistry. Springer-Verlag, Berlin
Wada, S. 1961, Chem. High Polym. Japan *18,* 617–628
Wallis, A. F. A. 1971, Solvolysis by Acids and Bases. In: Lignins. Occurrence, Formation, Structure and Reactions (Sarkanen, K. V. and Ludwig, C. H., Eds.). Wiley-Intersci., New York, pp. 345–372
Wardrop, A. B. 1957, Tappi *40,* 225–243
Wardrop, A. B. 1971, Occurrence and Formation in Plants. In: Lignins. Occurrence, Formation, Structure and Reactions (Sarkanen, K. V. and Ludwig, C. H., Eds.). Wiley-Intersci., New York, pp. 19–41
Wardrop, A. B. 1976, Appl. Polym. Symp. No. 28, 1041–1063
Wardrop, A. B. 1981, Anatomical Aspects of Lignin Formation in Plants. In: The Ekman Days 1981. Int. Symp. Wood Pulp. Chem., Stockholm, Vol. 1, pp. 44–51
Wayman, M. and Obiaga, T. I. 1974, Tappi *57,* 123–126
Wegener, G. 1974, Papier *28,* 478–486
Wegener, G. 1975, Beitrag zur Charakterisierung der Natriumchlorit-Delignifizierung von Fichtenholz. Doctor Thesis, Universität München
Wegener, G. 1976, Papier *30,* 177–185
Wegener, G. 1983, Pap. South. Africa *3,* No. 5, 11–26
Wegener, G. and Fengel, D. 1977, Wood Sci. Technol. *11,* 133–145
Wegener, G. and Fengel, D. 1979, Tappi *62,* 97–100
Wegener, G. and Stoll, M. 1976, Cell Chem. Technol. *10,* 611–616
Wegener, G., Przyklenk, M. and Fengel, D. 1983, Holzforschung *37,* 303–306
Weissenböck, G. 1976, Biol. Unserer Zeit *6,* 140–147
Westermark, U. 1982, Wood Sci. Technol. *16,* 71–78
Yaku, F., Yamada, Y. and Koshijima, T. 1976, Holzforschung *30,* 148–156
Yamasaki, T., Hata, K. and Higuchi, T. 1976, Mokuzai Gakkaishi *22,* 582–588
Yamasaki, T., Hosoya, S., Chen, C. L., Gratzl, J. S. and Chang, H. M. 1981, Characterization of Residual Lignin in Kraft Pulp. In: The Ekman Days 1981. Int. Symp. Wood Pulp. Chem., Stockholm, Vol. 2, pp. 34–42
Yasuda, S. and Sakakibara, A. 1975, Mokuzai Gakkaishi *21,* 363–369
Zbinden, R. 1964, Infrared Spectroscopy of High Polymers. Academic Press, New York

7. Extractives

7.1. Importance

The term wood extractives covers a large number of different compounds which can be extracted from wood by means of polar and non-polar solvents. In a narrow sense extractives are those compounds which are soluble in organic solvents, and in this sense the term extractives is used in wood analysis (→ 3.2.4.). But water-soluble carbohydrates and inorganic compounds also belong to the extractable substances.

The portion soluble in organic solvents amounts to only few percent in the wood of trees from temperate zones, but the concentration can be much higher in certain parts of a tree, e.g. in branch bases, heartwood, roots, areas of sore irritation. Relatively high amounts of extractives are found in certain tropical and subtropical woods (Table 3–6).

Various studies are concerned with the artificial induction of a high extractive production (up to 15%) in pines. In the case of treatment with the herbicide paraquat or similar compounds before harvest, areas soaked with resin (lightwood) are formed within the stem (Conner et al. 1977; Zinkel, McKibben 1978) (→ 18.7.).

Table 7–1: Percentage of the acetone extract and its fractions (based on dry wood), and the composition of the petroleum ether solubles (based on the fraction) of soft- and hardwoods (Assarsson, Åkerlund 1966)

	Picea abies %	*Pinus sylvestris* %	*Betula verrucosa* %	*Populus tremula* %
Acetone extract	2.22	3.10	3.46	4.53
Ethyl ether insolubles	0.95	0.68	1.43	2.01
Ethyl ether solubles	1.24	2.42	2.03	2.30
Petroleum ether insolubles	0.15	0.08	–	0.04
Petroleum ether solubles	1.04	2.29	2.03	2.27
– Free fatty acids	7.52	7.55	–	4.67
– Resin acids	27.37	29.03	0.30	0.75
– Neutral compounds	62.13	60.59	99.70	94.58
– Hydrocarbons	1.84	3.21	1.99	5.01
– Waxes	8.41	3.64	9.67	13.53
– Triglycerides	18.67	38.17	39.88	45.68
– Higher alcohols	9.55	6.06	7.68	10.59
– Diglycerides	5.26	1.39	8.48	2.55
– Monogylcerides	5.26	0.79	11.07	5.20
– Oxidized compounds	13.05	1.89	14.46	6.72

The content and composition of extractives vary among wood species (Table 7–1). But there are also variations depending on the geographical site and the season (Swan 1968; Dahm 1970; Snajberk, Zavarin 1976; Su et al. 1981). On the other hand, the composition of the extractives of certain woods which are difficult to distinguish anatomically can be used for the determination (Seikel et al. 1965). The extractives are concentrated in the resin canals and the ray parenchyma cells; lower amounts are also found in the middle lamellae, intercellulars and cell walls of tracheids and libriform fibres (Back 1960; Paasonen 1967b; Grosser et al. 1974). Variations in the composition of extractives from resin canals and from ray cells were found by Back (1960) and Kimland and Norin (1972). Tsoumis (1965) observed a correlation between extractives and warts at the inner face of tracheids.

The composition of the extractives is changed during seasoning of wood; particularly unsaturated compounds, fats and fatty acids are degraded (Donetzhuber, Swan 1965; Assarsson 1966; Assarsson, Åkerlund 1966, 1967). This fact is important for the production of pulp as certain extractives in fresh wood may cause yellow specks (pitch troubles) or a yellowing of the pulp (Bergman 1965; Croon 1965; Tachibana, Sumimoto 1980). Extractives may also influence the strength of refiner pulp, the gluing and finishing of wood as well as the drying behaviour (Sandermann, Puth 1965; Gardner 1965; McMillin 1969a; Meyer, Barton 1971; Roffael, Rauch 1974; Popper 1975).

Several woods contain extractable substances which are toxic or deterrent for bacteria, fungi and termites (Becker 1966; Weissmann, Dietrichs 1975a; Bauch et al. 1977). Other extractives give colour and odour to the wood. Despite these characteristics Sandermann (1966) holds that most of the extractives are non-participating substances or cellular slip products which are not of essential importance. The number of publications concerned with the utilization of wood extractives, predominantly for the substitution of petrochemicals, is large (e.g. Zinkel 1975; Lange 1976a; Collier 1977; Traitler, Kratzl 1980a) (→ 18.7.).

One part of the extractives is termed resin, a term which does not characterize certain chemical compounds but rather a physical condition. Resin is to be seen as a mixture of different compounds which mutually inhibit crystallization (Sandermann 1960). Nevertheless the following compounds can be attributed to the resin components:

– terpenes, lignans, stilbenes, flavonoids and other aromatics.

Apart from these substances other organic compounds are present in the extractives:

– fats, waxes, fatty acids and alcohols, steroids, higher hydrocarbons.

Earlier publications summarizing the knowledge of the various kinds and behaviours of the extractive components of wood include Sandermann (1960, 1966), Hillis (1962) and Buchanan (1963).

7.2. Extractives of Softwoods

7.2.1. Terpenes and Terpenoids

The terpenes and their derivatives are a large group of compounds which are widespread in the plant and animal kingdoms. More than 4000 different terpenes have been isolated and identified (Cordell 1974). Many fragrant, odoriferous and flavoured substances of flowers and spices belong to these compounds. Moreover, oleoresins and other plant secretions contain terpenes. Thus the oleoresin of wood also contains compounds belonging to the various classes of terpenes. The common chemical characteristic of terpenes is their composition of isoprene (2-methyl butadiene) units (Newman 1972).

According to the number of isoprene units linked in a terpene, the terpenes are subdivided into several classes: monoterpenes (2 units), sesquiterpenes (3 units), diterpenes (4 units), sesterterpenes (5 units), triterpenes (6 units). The isoprene units are linked according to the isoprene rule which means that the tail of one unit is linked to the head of the next one. This rule is strictly followed for up to 5 units,

Name	Number of 5C-units	Structure
Isoprene (basic unit)	1 × 5C	
Monoterpenes	2 × 5C	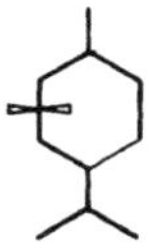
Sesquiterpenes	3 × 5C	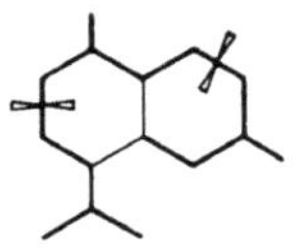
Diterpenes	4 × 5C	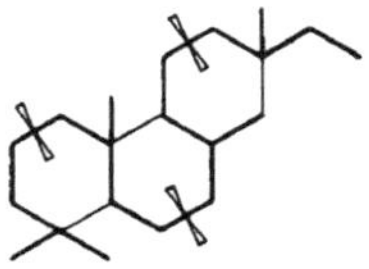
Triterpenes	6 × 5C	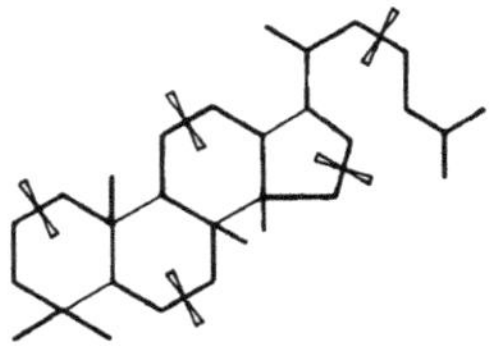

Fig. 7–1. Basic structures of the various terpenes.

whereas the structure of many triterpenes must be explained by a composition of two sesquiterpenes with tail-tail linkage (Fig. 7–1).

In general the terpenes and their derivatives are written according to a certain scheme as most of them are linked intramolecularly, forming one or more rings. The ring formation results in a removal and/or change of double bonds. The terpenes are pure hydrocarbons whereas the terpenoids bear functional groups such as OH, C=O, COOH etc.

The extractives of softwoods contain all classes of terpenes from monoterpenes to tri- and tetraterpenes, except for sesterterpenes which are a very rare class. In hardwoods mainly higher terpenes are present, monoterpenes are found only in some tropical hardwoods (Sandermann 1966).

The monoterpenes can be subdivided into acyclic, monocyclic and bicyclic compounds (Figs. 7–2, 7–3). All types can be found in the volatile fraction of softwoods isolated by steam distillation. The volatile wood oil (turpentine) consists mainly of monoterpenes. The most important ones are α- and β-pinene and limonene, which are obviously present in all softwoods; but Δ^3-carene, camphene, myrcene and β-phellandrene are also wide-spread (Table 7–2). Though the composition of turpentine is seen to be characteristic for the species, variations between populations

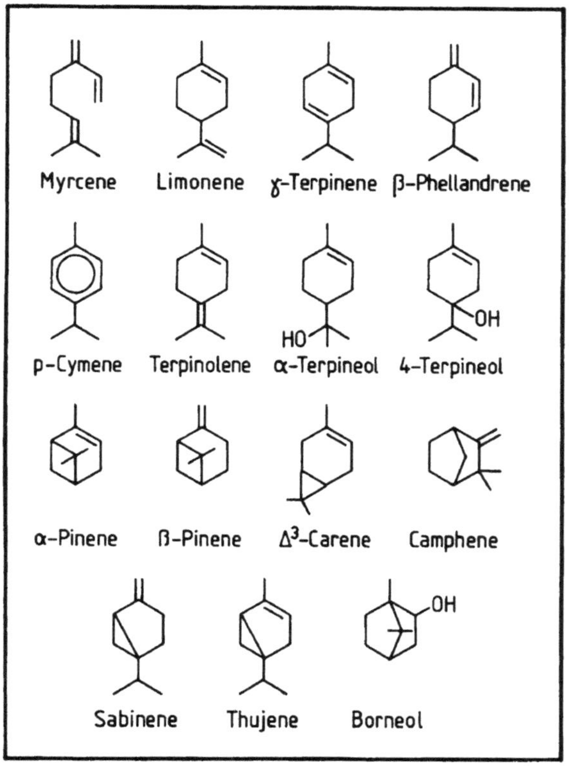

Fig. 7–2. Monoterpenes from softwoods.

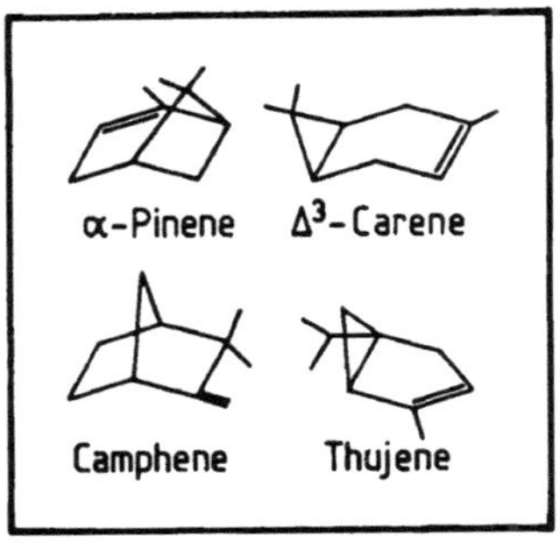

Fig. 7–3. Sterical structures of softwood monoterpenes.

Table 7–2: Composition of the monoterpene fraction from various softwoods

Wood species	α-Pinene %	β-Pinene %	Δ^3-Carene %	Camphene %	Limonene %	p-Cymene %	Terpinene %	Terpinolene %	Myrcene %	Sabinene %	β-Phellandrene %	References
Abies alba	39.0	3.0	4.5	tr.	53.5	–	–	–	tr.	–	tr.	Zavarin, Snajberk 1965
Abies amabilis	15.5	16.5	38.0	0.5	1.0	–	–	tr.	1.0	–	26.5	
Abies grandis	13.0	18.5	0.5	34.5	5.0	–	–	–	0.5	–	26.0	
Picea abies	58	24	2.1	3.0	4.5	0.5	–	0.4	–	–	–	Kimland, Norin 1972
Pinus aristata	8.8	11.2	64.4	0.1	7.0	–	–	3.4	3.6	–	2.3	Zavarin et al. 1976
Pinus balfouriana	81.0	1.9	tr.	0.9	14.1	0.3	–	1.0	0.6	–	0.2	Snajberk et al. 1979
Pinus elliottii	62.6	20.6	–	1.4	1.7	–	–	0.4	1.7	–	8.1	Drews, Pylant 1966
Pinus heldreichii	11.0	0.5	–	0.2	82.8	–	–	–	0.6	–	–	Weissmann 1973
Pinus monophylla												
sapwood	46	2	3	1	7	–	5	4	22	–	8	Anderson et al. 1970
heartwood	67	2	–	1	3	–	1	1	20	–	3	
Pinus occidentalis	63.8	22.2	7.7	0.2	–	–	–	–	1.1	–	–	Mirov et al. 1962
Pinus ponderosa	10.2	16.5	36.3	1.4	12.1	1.0	1.4	1.9	–	–	1.4	Drews, Pylant 1966
Pinus strobus	67.0	18.0	–	2.9	0.9	–	–	0.5	0.9	–	0.5	
Pinus taeda	64.0	28.0	–	1.3	1.5	–	–	–	2.0	–	0.5	
Pinus tropicalis	93.2	3.1	–	1.2	0.8	–	–	0.1	0.8	–	0.6	Weissmann, Vorher 1975
Pseudotsuga menziesii	31.0	36.0	10.0	0.5	5.0	–	–	5.5	1.0	9.5	1.5	Zavarin, Snajberk 1965

and between single trees must be considered (Zavarin, Snajberk 1965a; Mirov et al. 1966; Snajberk, Zavarin 1976; Snajberk et al. 1979). The variability of composition is attributed to environmental factors as well as to heredity. Thus the α-pinene content of *Pinus balfouriana* varied between 96.6% and 81.0% for different localities in the USA; the balsams of 7 trees of *Abies amabilis* contained Δ^3-carene in concentrations between 19.2% and 62.5%, based on total monoterpenes (Zavarin, Snajberk 1965a; Snajberk et al. 1979).

More rare are those monoterpenes consisting of a 7-ring which can be defined as tropolone derivatives. Compounds such as thujic acid, thujaplicin and thujaplicinol were found in species of the *Cupressaceae* family (Fig. 7–4) (Zavarin et al. 1959; Buchanan 1963).

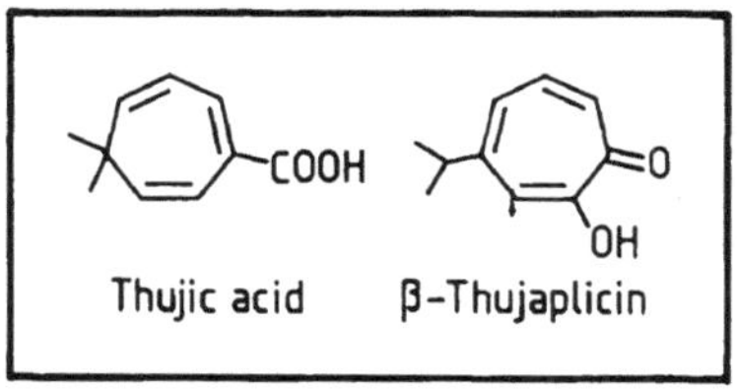

Fig. 7–4. Monoterpenoidic tropolone derivatives.

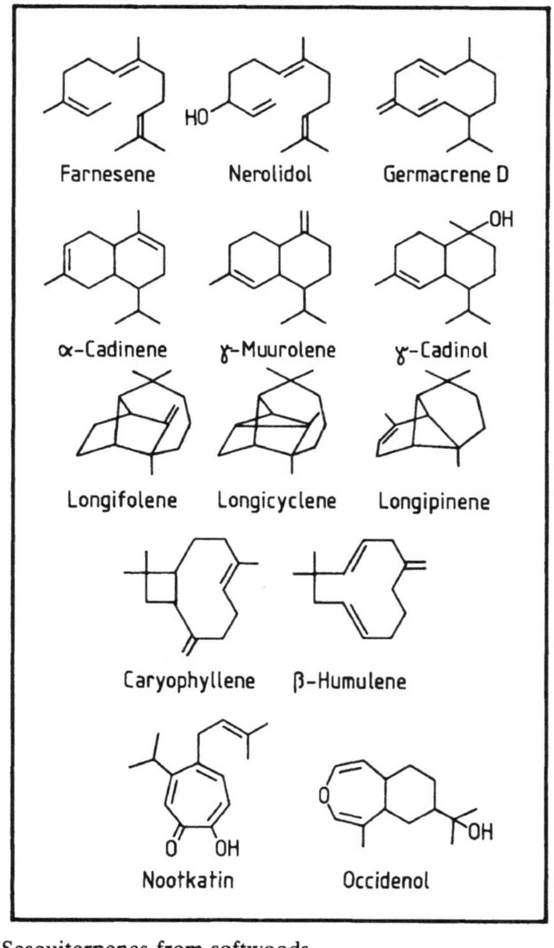

Fig. 7–5. Sesquiterpenes from softwoods.

The acyclic citronellic acid detected in the wood species *Thujopsis dolabrata* and *Chamaecyparis taiwanensis* is responsible for the termite resistance of these woods (Weissmann, Dietrichs 1975b).

Further components of the volatile softwood oils are compounds belonging to the sesquiterpenes. Among them are the acyclic compounds farnesene and nerolidol, the monocyclic germacrene, the bicyclic cadinene, cadinol and muurolene, and the more complicated longifolene, longipinene and longicyclene (Fig. 7–5). These compounds are present in pine and spruce species (Mirov et al. 1962; Kimland, Norin 1972; Zavarin et al. 1976; Zavarin, Snajberk 1980). In *Pinus* and *Pseudotsuga* species additional sesquiterpenes such as caryophyllene, humulene and sibirene have been found (Snajberk, Zavarin 1976; Zavarin et al. 1976). Sesquiterpenic tropolone derivatives called nootkatin and hydronootkatinol were isolated from extractives of *Cupressus, Chamaecyparis* and *Juniperus* species (Fig. 7–5) (Zavarin et al. 1959; Bicho et al. 1963). An unusual sesquiterpenoid, occidenol, from *Thuja koraiensis,* was identified by Tomita and Hirose (1970). The amount of sesquiterpenes is about 1–5% of the monoterpene portion in softwoods.

The oleoresin of softwood species contains a relatively high percentage of diterpenes and diterpenoidic acids apart from fats, fatty acids and alcohols. The neutral diterpenes consist of hydrocarbons (thunbergene, pimaradiene), oxides (manoyloxide), alcohols (abienol, pimarinol, larixol), and aldehydes (levopimaral) (Fig. 7–6). The resin acids are mostly tricyclic compounds; those present in most softwoods of temperate zones are shown in Fig. 7–7. Their total amount is in the range of 0.2–0.8% (based on dry wood) as determined for various pine species and

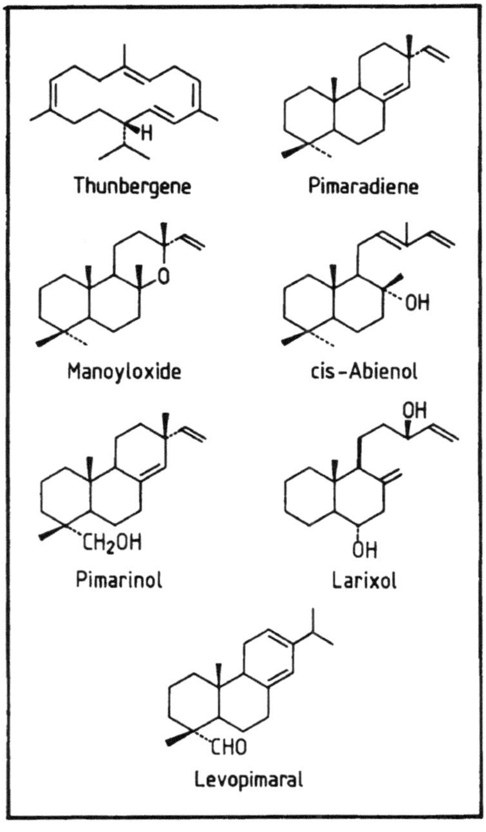

Fig. 7–6. Neutral diterpenes from softwoods.

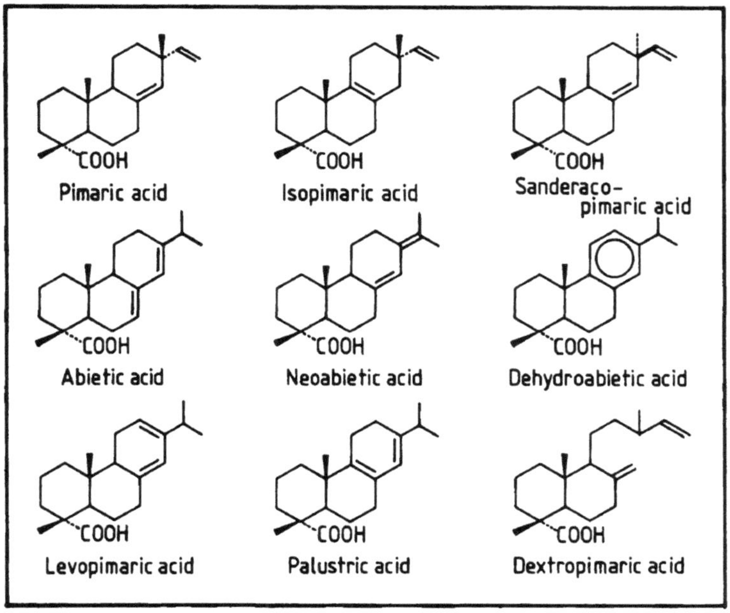

Fig. 7–7. Acidic diterpenes (resin acids) from softwoods.

spruce (Holmbom, Ekman 1978; Conner et al. 1980). In the heartwood of pinyons *(Pinus edulis, P. monophylla, P. quadrifolia)* a resin acid content of 2.4–4.6% was determined (Zavarin, Snajberk 1980). The composition of the resin acid fraction of various softwoods is summarized in Table 7–3. It can be seen that there are also variations in the composition of the resin acids of sapwood and heartwood.

Ekman et al. (1978) traced the radial distribution of the resin acids in *Picea abies.* They found a decrease of the sum of levopimaric, palustric, abietic, and neoabietic acid from the outer sapwood to the inner heartwood, and a reverse distribution of dehydroabietic acid, whereas the concentration of isopimaric, sandaracopimaric, and pimaric acid remained approximately constant. A slight increase of the percentage of resin acids and diterpene alcohols in the upward direction of the sapwood was observed.

Some of the more rare diterpenoids were isolated from certain softwoods. Among them are lambertianic acid from *Pinus lambertiana,* taxusin from *Taxus baccata,* anticopalic acid from *Pinus monticola,* strobic acid fom *Pinus strobus* (cortex) and *Pinus quadrifolia,* and merkusinic acid from *Pinus merkusii* (Fig. 7–8) (Dauben, German 1966; Della Casa de Marcano, Halsall 1969; Zinkel, Spalding 1971; Zinkel et al. 1971; Zinkel, Conner 1973; Weissmann 1974).

There are also diterpenes containing phenolic rings, e.g. ferruginol and podocarpic acid (Fig. 7–9). Ferruginol is a constituent of the heartwood extractives of *Cryptomeria japonica* and is obviously responsible for the colour reversion of bleached sulfate pulps from this wood (Akimoto, Sumimoto 1980). Podocarpic acid is the

Table 7-3: Composition of the resin acid fraction of various softwoods

Wood species	Pimaric %	Sandaraco-pimaric %	Palustric %	Levopimaric %	Isopimaric %	Anticopalic %	Abietic %	Dehydroabietic %	Neoabietic %	Secodehydroabietic %	% References
Picea abies	6.2	6.4	13.5	16.2	13.3	–	11.2	22.6	10.2	0.4	Holmbom, Ekman 1978
Picea abies											
sapwood	3.3	5.6	17.5	28.6	11.3	–	9.3	13.2	11.0	–	Ekman 1979a
heartwood	4.4	6.4	16.4	10.5	11.8	–	8.2	35.7	6.5	–	
Pinus banksiana											
sapwood	6	2	13	1	6	–	15	51	3	–	Conner et al. 1980
heartwood	7	2	11	tr.	8	–	29	35	4	–	
Pinus contorta											
sapwood	4	1	16	7	17	–	8	38	4	–	Conner et al. 1980
heartwood	7	2	14	tr.	18	–	15	40	2	–	
Pinus elliottii	7.4	2.0	19.4	9.6	23.7	–	12.0	4.8	18.4	–	Zinkel, Foster 1980
Pinus palustris	6.0	1.8	15.2	27.9	5.0	–	15.1	15.3	13.7	–	
Pinus pinaster	10	1.0	50		4.1	–	15	4.2	16	–	Hemingway et al. 1973
Pinus ponderosa											
sapwood	9	2	10	5	15	–	17	31	8	–	Conner et al. 1980
heartwood	9	2	20	2	10	–	28	15	13	–	
Pinus strobus											
sapwood	1	2	8	–	34	14	29	5	5	–	Conner et al. 1980
heartwood	1	2	8	–	36	11	32	5	3	–	
Pinus sylvestris	8.1	1.6	15.1	30.0	3.5	–	15.8	14.4	11.1	0.4	Holmbom, Ekman 1978
Pseudotsuga menziesii											
sapwood	–	3.2	23.5	6.2	27.8	–	16.1	7.8	13.8	–	Foster et al. 1980
heartwood	–	3.6	22.8	2.3	27.0	–	18.9	10.1	12.2	–	

fungicidic component in the wood of some *Podocarpaceae* (Bauch et al. 1977). The heartshakes of some *Dacrydium* species contain almost pure podocarpic acid which is found to be synthesized only in association with shakes in the living tree (Hillis et al. 1976).

A small portion of the resin acids exists in the form of methyl esters (Kimland, Norin 1972; Conner, Rowe 1975). Ekman (1980a) succeeded in detecting esters of the acyclic geranylgeranol with higher fatty acids in *Picea abies*.

After sulfate pulping the resin acids are recovered in tall oil which is separated from black liquor (→ 16.4.3.). During the pulping process some of the resin acids are isomerized (Zinkel 1975; Holmbom, Ekman 1978; Holmbom 1981). In particular levopimaric acid is converted to abietic acid.

The tricyclic ring system is obviously a very stable one as can be seen from studies in which diterpenes and similar compounds were found to be present in fossil woods, soil and amber (Swan 1965; Beck 1972).

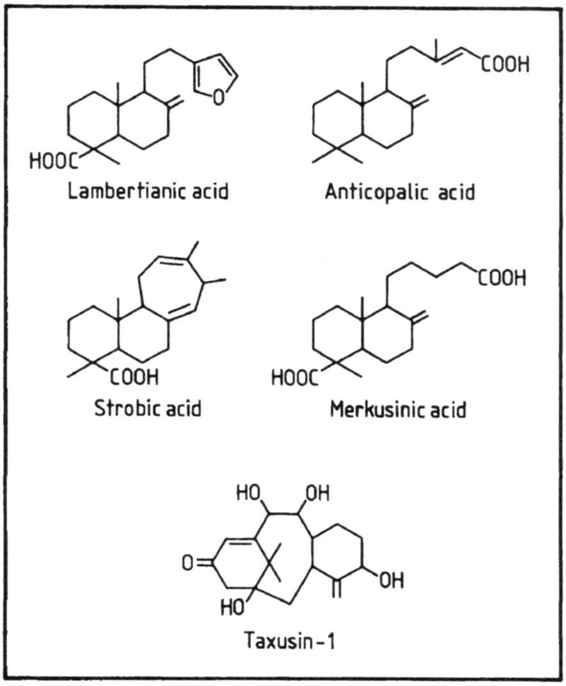

Fig. 7–8. Some rare diterpenoids from softwoods.

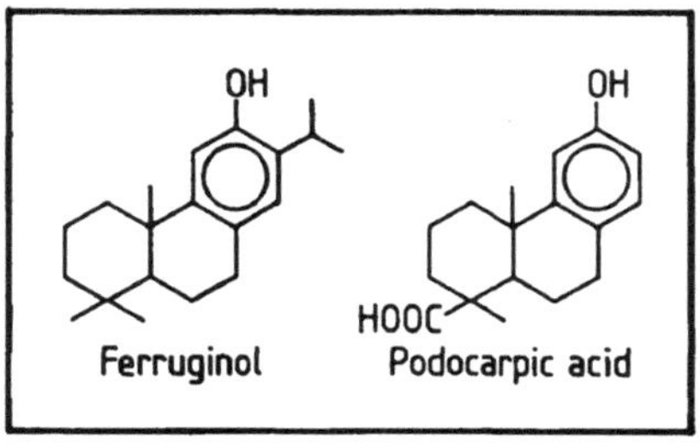

Fig. 7–9. Aromatic diterpenoids from softwoods.

Sesterterpenes were unknown before 1965. They have been isolated from insect protective waxes, from fungal sources and marine sponges but not from higher plants (Faulkner 1973; Cordell 1974).

In the extractives of softwood triterpenes could also be identified. Most of them have a sterane structure and must be assigned to the steroids. The acyclic squalene is the precursor for the cyclic triterpenes, and can be found in very small amounts in the wood of conifers. Another triterpene found in softwoods is serratenediol which consists of 5 rings (Fig. 7–10). The main component of the steroid group in softwood is β-sitosterol; minor members are campesterol, sitostanol, cycloarthenol and citrostadienol (Fig. 7–10) (Conner, Rowe 1975; Holmbom, Ekman 1978; Ekman 1979). Most of the triterpenols and sterols are esterified with fatty acids. Kimland

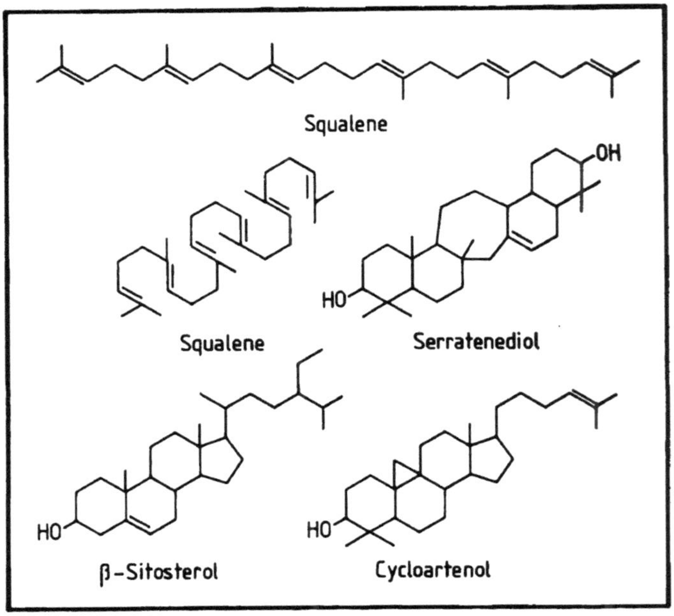

Fig. 7–10. Triterpenes and sterols from softwoods.

and Norin (1972) assume that the sterols are mainly concentrated in the parenchyma cells of rays as oleoresin from the resin canals contains only negligible amounts of these compounds.

7.2.2. Fats, Waxes and Their Components

Fats are defined as esters of higher carbonic acids (fatty acids) with glycerol whereas waxes are esters of fatty acids with higher alcohols (Fig. 7–11). Fats and waxes are extractable from wood with organic solvents (diethyl ether, petroleum ether, acetone etc.). The content of fats is about 0.3–0.4%, that of waxes about 0.08–0.09% (based on dry wood), as determined for *Picea abies* and *Pinus sylvestris* (Assarsson, Åkerlund 1966). Apart from fats and waxes free fatty acids and alcohols are also components of the extractives. Nevertheless, most of the fatty acids present in wood extractives are combined; an esterification with glycerol is dominant (Table 7–4). Among the glycerides (fats) the triglycerides are dominant as compared to mono- and diglycerides. The percentage of free fatty acids is considerably higher in heartwood than in sapwood.

More than 20 different fatty acids have been identified in various softwoods (Assarsson, Åkerlund 1966; Holmbom, Ekman 1978; Ekman 1979a, 1980b; Conner et al. 1980; Foster et al. 1980). In general they are saturated, monoenic, dienic and trienic acids with 16–22 carbon atoms, but also lower acids (C_{10}–C_{14}), higher ones (C_{24}–C_{30}) and tetraenic ones were detected. Comparing the composition of the total fatty acids of various conifers the bulk of them are composed of oleic (C_{18}:1(9)),

Table 7–4: Neutrals and acids in the extractives of various pines and Douglas fir (Foster et al. 1980; Zinkel, Foster 1980)

Extractive type	*P. echinata* sapwood %	*P. elliottii* sapwood %	*P. palustris* sapwood %	*P. virginiana* sapwood %	*Pseudotsuga menziesii* sapwood %	heartwood %
Neutrals	20.1	20.1	22.7	10.6	5.5	7.0
Fatty acids	17.6	18.3	20.3	9.5	3.5	1.6
Non-saponifiables	2.5	1.8	2.4	1.1	1.7	1.6
Free acids	5.9	6.6	6.0	3.4	2.1	2.8
Resin acids	5.7	6.5	5.8	3.3	2.2	2.7
Fatty acids	0.2	0.1	0.2	0.1	0.1	0.1
Total fatty acids	17.8	18.4	20.5	9.6	3.5	1.7
Strong acids	+	+	+	+	0.1	0.1

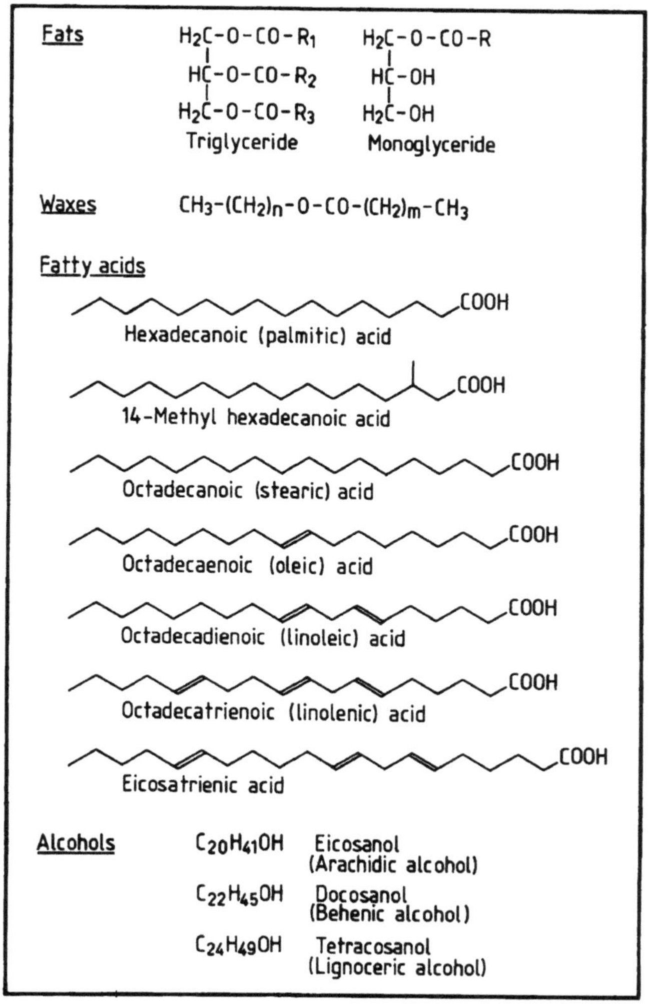

Fig. 7–11. Fats, waxes and their components isolated from wood.

linoleic (C_{18}:2(9,12)) and 5,9,12-octadecatrienic acid (C_{18}:3(5,9,12)). Smaller amounts were determined for palmitic (C_{16}:0), stearic (C_{18}:0), 14-methyl hexadecanoic (C_{17}ai) and 5,11,14-eicosatrienic acid (C_{20}:3(5,11,14)) (Fig. 7–11). A relatively high amount of triacontatrienoic acid (C_{30}:3(5,11,14) was found in the fatty acid fraction of *Pseudotsuga menziesii* (Foster et al. 1980).

Those alcohols which are free or combined in waxes are saturated straight-chain compounds with 16–28 carbon atoms. The main alcohols determined in *Picea abies* and *Pinus sylvestris* are 1-docosanol (behenic alcohol, C_{22}) and 1-tetracosanol (lignoceric alcohol, C_{24}) (Fig. 7–11) (Holmbom, Ekman 1978). Higher amounts of 1-hexadecanol (cetyl alcohol, C_{16}), 1-pentacosanol (C_{25}) and 1-octacosanol (montanyl alcohol, C_{28}) were identified in Southern pine tall oil (Conner, Rowe 1975; Traitler, Kratzl 1980b). Studying the extractives of *Picea abies* Ekman (1979a) found that most of the fatty alcohols are esterified with ferulic acid.

Swan (1968) observed a seasonal variation of the composition of the total fatty acids in spruce wood with the presence of short-chain acids during the early summer and an increase of linolenic acid (C_{18}:3) during winter time. By way of contrast Ekman et al. (1979) detected no variation of fatty acids in the same wood species during different seasons. They interpreted the preceding study as a result of between-tree variation.

7.2.3. Phenolic Compounds

The extractives of wood also contain a large number of various phenolic compounds, and thus a further subdivision is necessary for them as well. Some of these compounds are obviously residues and by-products of lignin biosynthesis (→ 6.2.1.).

Among the simple phenols which can be isolated from spruce extractives *(Picea abies)* are vanillin, p-hydroxybenzaldehyde, coniferyl aldehyde, guaiacylglycerol, p-ethylphenol as well as coniferin and syringin (Fig. 7–12) (Kimland, Norin 1972; Ekman 1976). Some of these simple phenols were also found in Western hemlock *(Tsuga heterophylla)* and Southern pines (*Pinus* spp.) (Barton 1968; Traitler, Kratzl 1980c). By fractionating the extractives obtained by a supercritical gas extraction of spruce wood guaiacol, eugenol, isoeugenol, cresols and further phenols were also identified (Calimli, Olcay 1978). But because of the high temperatures applied in this process these compounds probably derive from a beginning lignin degradation.

The presence of traces of quinol and hydroquinone, respectively, in the heartwood of *Pinus resinosa* and *P. radiata* was proved by von Rudloff (1965) and Hillis and Inoue (1968).

A second group are the lignans which are compounds consisting of two phenylpropane units linked in a different manner. Some of these compounds represent dimeric structures which are also present in the lignin molecule (→ 6.3.2.). Many of the

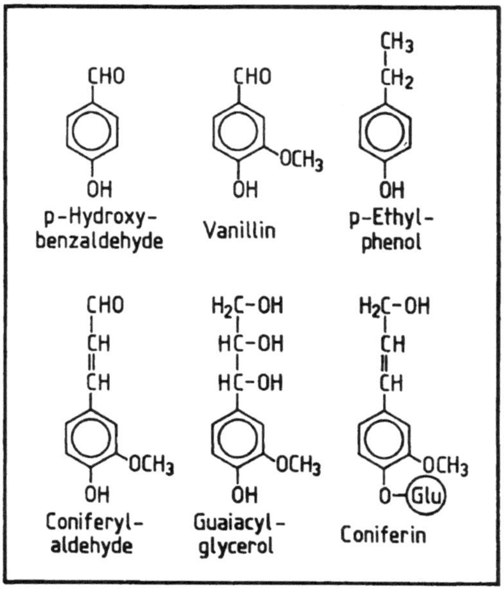

Fig. 7–12. Some simple phenols from softwoods.

lignans as identified in the extracts of *Picea, Pinus, Larix, Abies* and *Tsuga* species contain a tetrahydrofuran ring, such as pinoresinol, lariciresinol, matairesinol, conidendrin and liovil (Fig. 7–13) (Freudenberg, Knof 1957; Freudenberg, Weinges 1959; Erdman, Tsuno 1969; Barton 1973; Ekman 1976). But simple β-β linkages between two units (secoisolariciresinol) and 6-rings (isolariciresinol, conidendrin, plicatin) also exist. In two trees of *Picea abies* Ekman (1979a) found 1.8 and 0.9 mg lignans/g dry wood with a main portion of hydroxymatairesinol (Table 7–5). The lignans thujaplicatin, plicatin, plicatic acid and plicatinaphthol were found in the heartwood of western red cedar *(Thuja plicata)*, and are assumed to be connected by a biochemical pathway as indicated in Fig. 7–14 (Swan et al. 1969; Swan, Jiang 1970).

Table 7–5: Composition of lignans in heartwood of two spruce trees *(Picea abies)* (Ekman 1979a)

Compound	Tree A %	Tree B %
Isolariciresinol	0.4	1.4
Secoisolariciresinol	2.0	0.9
Liovil	7.0	8.1
α-Conidendric acid	0.6	1.1
Picearesinol	3.1	2.6
Lariciresinol	3.9	6.0
Hydroxymatairesinol isomers	17.3	13.8
	41.2	47.7
Pinoresinol	1.0	0.5
α-Conidendrin	4.1	2.6
Other	16.6	13.5

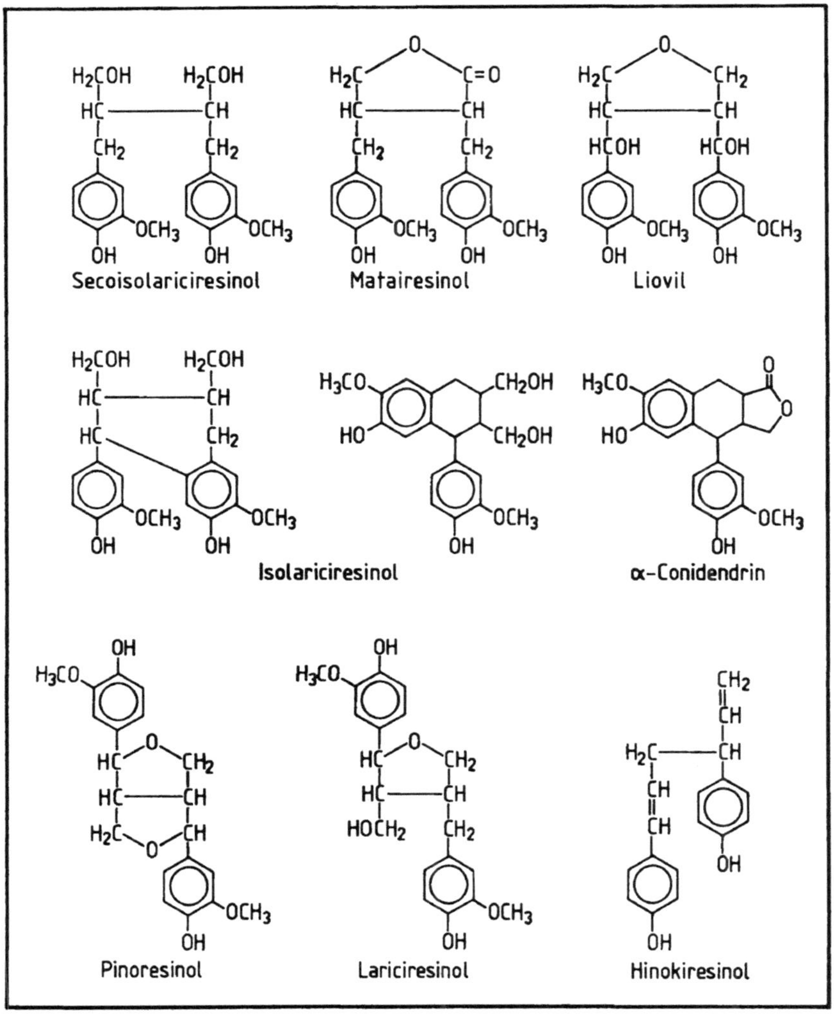

Fig. 7–13. Lignans from softwoods.

In addition to lariciresinol, secoisolariciresinol and pinoresinol Takehara and Sasaya (1979) isolated three lignans with dihydrobenzofuran structure from *Larix leptolepsis* (Fig. 7–15). A lignan with dihydrobenzofuran structure was previously identified in *Tsuga heterophylla* (Barton 1970). The ethanol extract of knots of Parana pine *(Araucaria angustifolia)* amounts to 30% (based on knots) and contains approximately 90% lignans, predominantly consisting of secoisolariciresinol relatives (Anderegg, Rowe 1974). Another lignan of Parana pine was hinokiresinol, a compound with an α-γ linkage which was also discovered in *Chamaecyparis obtusa* and *Agathis australis* (Hirose et al. 1965; Enzell et al. 1967) (Fig. 7–13).

Various studies have shown that the lignans are typical heartwood components, and their amount in the sapwood is small or negligible (Barton, Gardner 1966; Erdman, Tsuno 1969; Krahmer et al. 1970; Ekman 1979b). Hydroxysugiresinol and

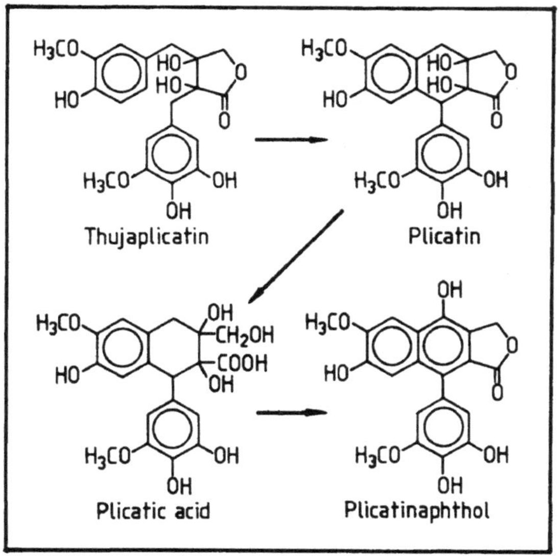

Fig. 7–14. Probable biochemical pathway of four lignans from *Thuja plicata*.

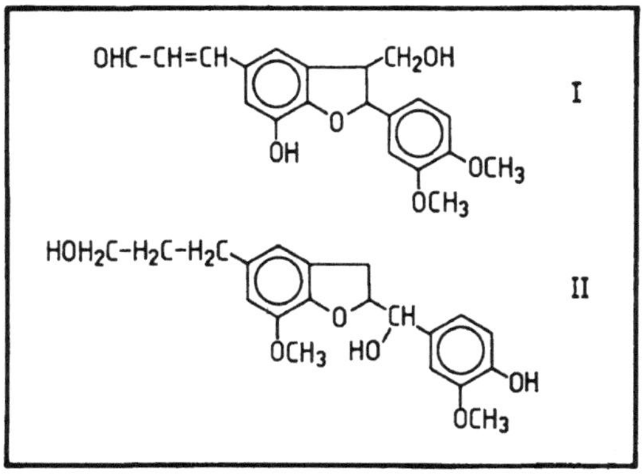

Fig. 7–15. Softwood lignans with dihydrobenzofuran structures, I from *Larix leptolepsis*, II from *Tsuga heterophylla*.

sequirin-C are found to be responsible for the colouration of the heartwood of sugi *(Cryptomeria japonica)* (Kai et al. 1972; Takahashi 1981).

But the callus resin of spruce is also rich in lignans (Weinges 1960). In the heartwood of *Tsuga heterophylla* it could be shown that lignans lined tracheid walls as surface films and often encrusted the bordered pits. The location and nature of the lignan deposits indicated that their biosynthesis probably takes place at the heartwood periphery in the vicinity of half-bordered pits (Krahmer et al. 1970).

A further group of aromatic compounds are the stilbenes, members of which are particularly present in the heartwood of pines. These compounds, mainly 4-hydroxystilbene, 4-methoxystilbene, pinosylvin, and pinosylvin mono- and diether, are responsible for the light-induced darkening of the wood as well as for difficulties during acidic pulping (Fig. 7–16) (Morgan, Orsler 1968; Hemingway et al. 1973; Manell, Pensar 1975). A stilbene glycoside called piceid (resveratrol-3-glucoside) was detected in *Picea glehnii* (Sandermann 1966).

A large group of various vegetable extractives are summarized as flavonoids, which comprises sub-groups such as flavones, flavanes, flavanones, isoflavones (→ 7.3.4.). In softwoods several flavonoids have been identified such as chrysin, taxifolin, pinocembrin and pinobanksin (Fig. 7–17) (Buchanan 1963; Sandermann 1966; Hemingway et al. 1973). Studies with *Pseudotsuga menziesii* and *Larix occidentalis* showed a considerable variation of the content of taxifolin within and between trees (Gardner, Barton 1960). The values ranged from 0 to 1.7% in the heartwood of Douglas fir, and from 0 to 1.8% in western larch. Lower concentrations of taxifolin were found in the sapwood of both species (0–0.6%). Within trees the concentration increased with increasing distance from pith, with highest values near the heartwood-sapwood boundary. The presence of catechin in *Tsuga heterophylla* is seen to be responsible for discolourations of groundwood as this compound is able to form chromophoric polymers with quinone structures (Barton 1973).

Catechin is also the structural unit of condensed tannins or phlobaphenes (→ 7.3.3.). Phlobaphenes were extracted from the heartwood of various conifers in yields of 0.2–6% (Zavarin, Snajberk 1965b). These extractives, however, showed a high content of methoxyl, which can be explained by the presence of a considerable amount of lignin.

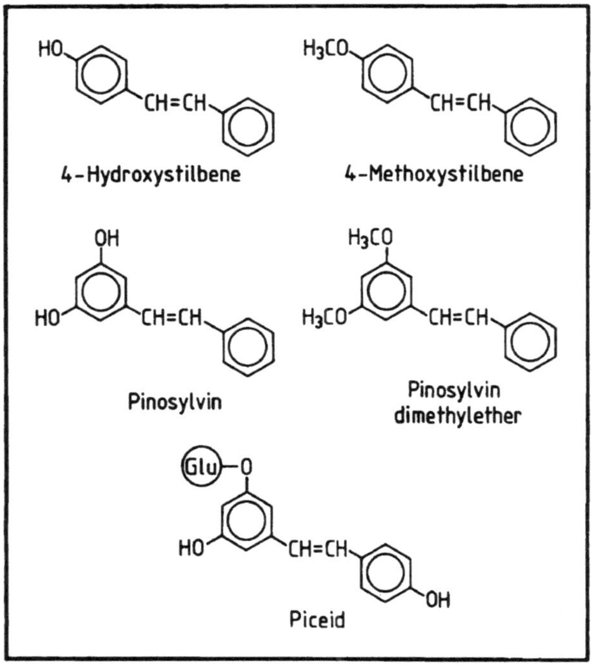

Fig. 7–16. Stilbene derivatives from softwoods.

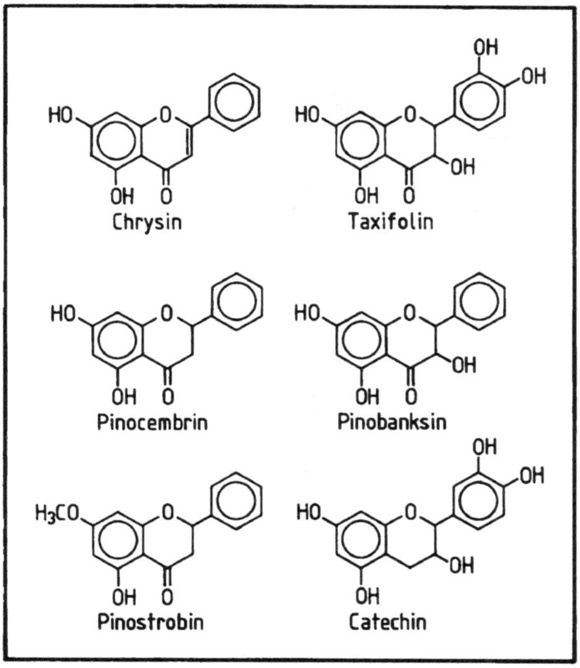

Fig. 7–17. Flavonoids from softwoods.

7.2.4. Various Compounds

Apart from terpenes other hydrocarbons, particularly n-alkanes, can also be found in the extractives of softwood. In the petroleum ether solubles of *Picea abies* and *Pinus sylvestris* saturated hydrocarbons in amounts of 0.2 and 0.6% (based on the petroleum ether extract) were determined (Assarsson, Åkerlund 1966). A homologous series from undecane (C11) to tritriacontane (C33) was identified. The volatiles of *Pinus jeffreyi* contain 10% n-heptane in addition to n-pentane and n-nonane (Zavarin et al. 1978).

Studies with *Pinus radiata* established a production of ethene in the sapwood (Shain, Hillis 1973). Because of a relatively high production at the sapwood-heartwood boundary particularly during the winter a connection with the heartwood formation is assumed. Furthermore a stimulation of the biosynthesis of pinosylvin could be proved.

Ethine derivatives were discovered in the wood of *Chamaecyparis pisifera* (Saeki et al. 1973). Chamaecynone and isochamaecynone are seen to be responsible for the termiticidal behaviour of this wood (Fig. 7–18).

The water and acetone-water extracts also contain mono- and disaccharides besides water-soluble polyoses (→ 5.3.2., 5.5.). The bulk of the sugars are glucose, fructose and sucrose. In *Picea abies* their amount is about 3–4 mg/g dry wood (Ekman 1979a). Dietrichs (1964) found the highest percentage of simple sugars in the

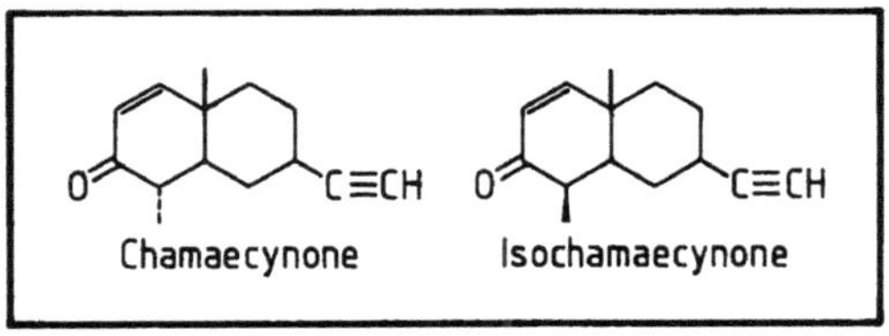

Fig. 7–18. Ethine derivatives from *Chamaecyparis pisifera.*

phloem and at the phloem-xylem boundary, and a rapid decrease of the sugar content towards the sapwood-heartwood boundary. Within the heartwood of spruce a low content of merely mannose was determined.

The protein content of wood influences the development of wood-destroying organisms, and has therefore a certain importance. Softwoods *(Picea, Pinus, Abies* spp.) have a protein content of 0.2–0.8% with a distribution over the stem cross-section resembling sugars; the heartwood, however, contains 0.2–0.4% protein (Becker 1962). The proteins in *Pinus sylvestris* consist of 16 amino acids; aspargin, glutamin, tryptophane and cysteine, which make up the remaining universal amino acids, are lacking (Adelsberger, Petrowitz 1976). With aging of wood the protein content decreases, but even after 300 years proteins were traceable in considerable amounts.

7.3. Extractives of Hardwoods

7.3.1. Terpenes and Terpenoids

The structural system of the terpenes has been described in one of the preceding sections (→ 7.2.1.). Some monoterpenes are constituents of the oleoresins of tropical woods. One of the most prominent representatives is camphor from *Cinnamomum camphora.*

Rare components of hardwoods from temperate zones are the sesquiterpenes. Derivatives of cadalene and calamene were isolated from several elms (*Ulmus, Celtis, Zelkova* spp.) (Fig. 7–19) (Suzuki et al. 1971; Rowe et al. 1972; Hayashi, Takahashi 1980). Furthermore, mansones were detected in elmwood. These are quinoid compounds which are characteristic components in the heartwood of *Mansonia altissima* (Marini Bettòlo et al. 1965). Sesquiterpenes are found in many tropical woods (Sandermann 1960). Other examples are α- and β-santalol (Fig. 7–19) which amount to more than 90% of the oil from sandalwood *(Santalum album),* and the tropolone derivative apitonene which is found in *Dipterocarpus gracilis* (Ikeda, Kitao 1972).

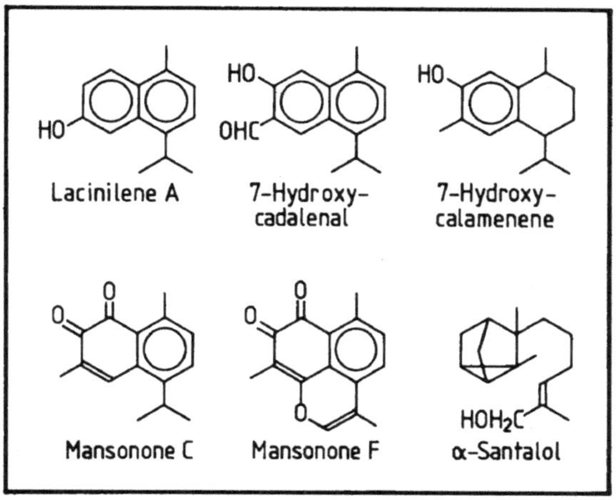

Fig. 7–19. Sesquiterpenes from hardwoods.

The diperpenes seem to be restricted to softwood species mainly in the form of resin acids. Sandermann (1960) mentions only two diterpenoid compounds from tropical hardwoods.

In contrast the triterpenes are present in a great variety in many hardwoods of tropical and temperate zones. In birch wood (*Betula* spp.) squalene and acetylated methyl betulinate were found (Figs. 7–10, 7–20) (Bergman et al. 1965; Paasonen 1967). In the extracts of various *Quercus* species triterpenes such as betulin, friedelin, taraxerol, α-, β-amyrin etc. have been identified (Fig. 7–20) (Ribo, Raventos 1972; Hui, Li 1977). Some of the compounds seem to be specific to a species, e.g. gilvanol for *Quercus gilva* (Itokawa et al. 1978).

The number of triterpenes detected in resins from tropical trees is large (Sandermann 1960). Among them are amyrins in the latex sap of *Hevea* species, sumaresinolic acid in Sumatra benzoin, and elemolic acid in elemi resin (Fig. 7–20).

Like softwoods, hardwoods also contain steroids. In species of *Betula, Populus, Quercus* and *Ulmus* steroids, mainly β-sitosterol, were found (Selleby 1960; Abramovitch et al. 1963; Chen 1970; Ribo, Raventos 1972; Rowe et al. 1972). The composition of the steroid fraction of *Quercus alba* is given as 85% β-sitosterol, 7% stigmasterol, 3% campesterol and a trace of dihydro-β-sitosterol (Chen 1970) (Fig. 7–10). *Ulmus rubra* contains 0.02% free and 0.005% esterified sterols; the free sterols consist of 88% β-sitosterol, 7% stigmasterol, 3% campesterol, 2% citrostadienol and 1% cholesterol (Rowe et al. 1972). Some of the steroids present in wood are esterified with fatty acids (Table 7–6).

Triterpenes and steroids may survive the pulping process and are varied and degraded during bleaching. The consequent products are seen to be jointly responsible for the yellowing of pulps (Bergman 1965; Lindgren 1967).

Several tropical woods contain glycosides of triterpenes and steroids which produce lathering solutions in water, and which are called saponins. The aglycones of the

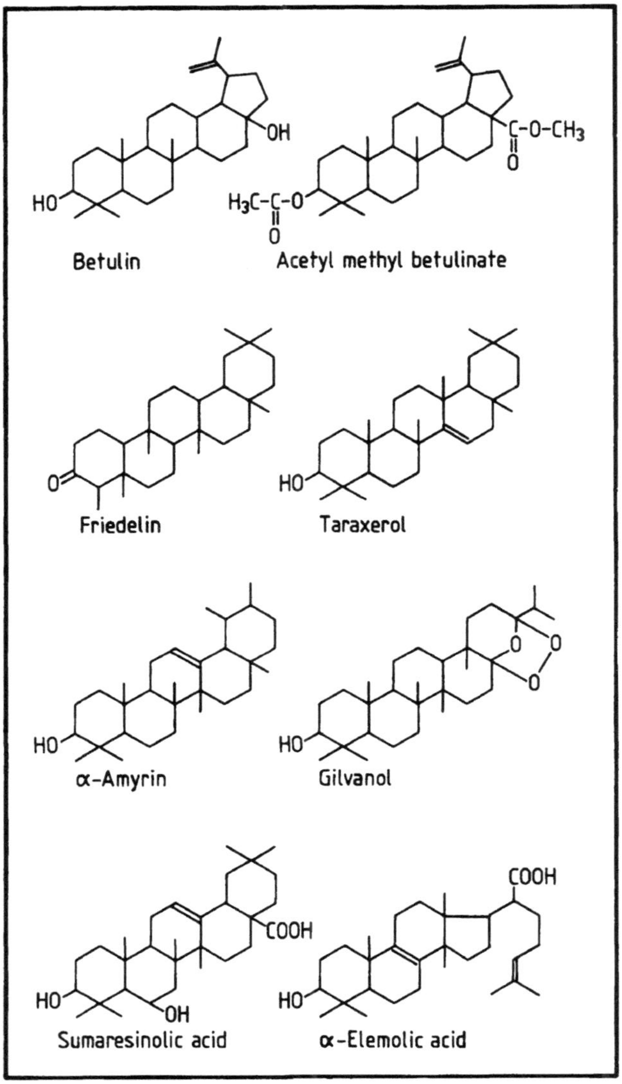

Fig. 7–20. Triterpenoids from hardwoods.

saponins are called sapogenins. The triterpenoic sapogenins are pentacyclic compounds whereas the steroid sapogenins contain an additional heterocyclic 6-ring at the fifth ring in most cases. Examples of triterpenoic and steroid saponins are shown in Fig. 7–21. Aescin is the saponin of chestnuts *(Aesculus hippocastanum),* bassiaic acid is the sapogenin of Makoré saponins *(Thieghemella heckelii),* and diosgenin is the sapogenin of saponins found in several plants (Sandermann, Barghoorn 1955; Tsukamoto et al. 1955; Hoppe et al. 1968). Saponins with aglycones of hydroxy- and dihydroxy-oleanolic acid are seen to be responsible for the resistance of zapote wood *(Manilkara zapota)* against fungi and termites (Sandermann, Funke 1970). Some of the saponins act as a narcotic for fish (fish poison); others cause dermatitis or other deseases during processing (Hausen 1981).

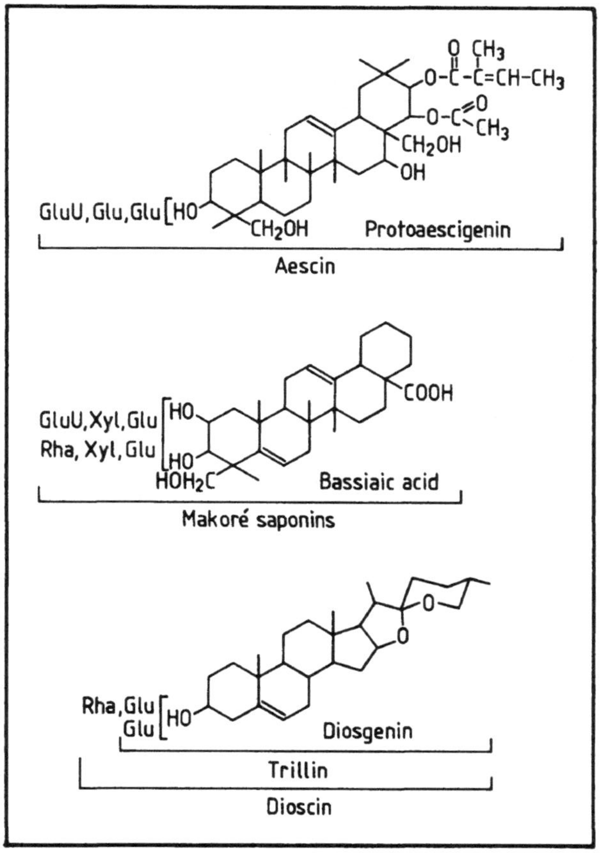

Fig. 7–21. Various sapogenins and saponins.

Compounds containing more than 6 isoprene units are known as polyprenes. Polyprenes are present in many plants in the form of gutta and caoutchouc, both being polymers differing in their chain conformation. While the isoprene units in gutta are arranged in trans-form, those in caoutchouc are arranged in cis-form. In both compounds the isoprene units are linked mainly by 1,4-bonds; only a small percentage (1.3 and 2.2%, respectively) is linked by 3,4-bonds (Sandermann et al. 1963). An investigation of 150 wood species showed that only eight had caoutchouc in the xylem, only four of them with a content of more than 1%: *Dyera costulata, Tabebuia serratifolia, Terminalia superba, Adina microcephala, Mitragyna stipulosa, Nauclea trillesii, Tectona grandis* and *Guaiacum officinale.*

Alcohols consisting of 6–9 isoprene units (betulaprenols) and esterified with various saturated fatty acids were found in *Betula verrucosa* (Table 7–6) (Bergman et al. 1965; Paasonen 1967a).

Table 7–6: Composition of the light petroleum soluble portion of the extractives from *Betula verrucosa* (Bergman et al. 1965)

Component	Percentage of the extractives %
Squalene	3
Fatty acid esters of	
betulaprenols	12
cycloartenol	2
methylene-cycloartenol	4
methylstigmastadienol	5
β-sitosterol	6
Triglycerides	48
Acetate of methyl betulinate	+
β-Sitosterol	4

7.3.2. Fats, Waxes and Their Components

Hardwoods contain fats, waxes, fatty acids and alcohols similar to those found in softwoods (→ 7.2.2.). Most of the fatty acids isolated from *Betula, Populus* and *Quercus* were bound as triglycerides (Selleby 1960; Abramovitch et al. 1963; Assarsson, Åkerlund 1966; Ribo, Raventos 1972). The ether extract of *Betula verrucosa* contained about 30–45% triglycerides, mainly consisting of linoleic (C_{18}:2(9,12), 62%) and lignoceric acid (C_{24}:0, 22%) (Fig. 7–11) (Paasonen 1967a).

The saponification of the triglycerides of *Quercus alba* yielded 75% linoleic, 10% stearic (C_{18}:0) and 10% palmitic acid (C_{16}:0) (Chen 1970). A similar composition was evaluated for *Tilia cordata:* 50% linolic, 15% oleic, 6% palmitic and 3% linoleic acid (Weissmann 1976). The latter study also proved the presence of cyclopropenic acids (malvalic, sterulic acid) in several *Tilia* species (Fig. 7–22).

The portion of free fatty acids in *Betula* species *(B. verrucosa, B. pubescens)* was about 5–10% of the combined fatty acids, half of them being linoleic acid, followed by palmitic and lignoceric acid with amounts of 7–17% (Paasonen 1967a). The foremost of the free and combined fatty acids of *Eucalyptus globulus* is linoleic acid, followed by palmitic acid (Swan, Åkerblom 1967).

Linoleic acid is also the main acid in the waxes of *Betula verrucosa, Populus tremula* and *P. tremuloides* (Abramovitch et al. 1963; Assarsson, Åkerlund 1966). Normal

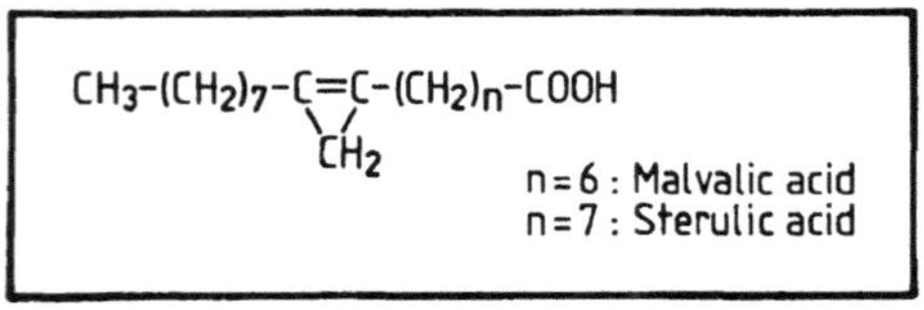

Fig. 7–22. Cyclopropenic acids from *Tilia* species.

saturated (C_{24}, C_{26}, C_{28}) and unsaturated alcohols, sterols and polyprenic alcohols are the alcoholic components of the waxes. In *Quercus alba* ferulic acid esters with tetracosyl alcohol were also found to be the major component (Chen 1970).

7.3.3. Phenols, Lignans, Quinones

The extracts of hardwoods also contain low molecular phenols, some of them are probably degradation products of compounds which can easily be hydrolyzed during extraction or steam distillation (e.g. glycosides). Buchanan (1963) reports on earlier studies in which p-hydroxybenzoic, vanillic, syringic, ferulic acid, vanillin and syringaldehyde were found in *Populus* and *Salix* species. According to Paasonen (1967a) ferulic acid is esterified with fatty alcohols in *Betula* species. Seikel et al. (1971) identified several phenols in the extracts of *Quercus alba* wood. Among them were sinapaldehyde, coniferylaldehyde, syringaldehyde, vanillin, p-hydroxybenzaldehyde and propioguaiacone (Figs. 7–12, 7–23). Egorov et al. (1976) isolated a volatile fraction from oak wood which contained phenol, cresols, guaiacol, p-ethylphenol, eugenol, hexenal, furfural etc. (Figs. 7–12, 7–23). Sinapaldehyde was also present in the low-molecular fraction of the hot-water extract of eucalyptus wood (Lundquist, Nelson 1971).

A group of compounds always mentioned with the softwood extractives are the lignans. From hardwoods, particularly from *Alnus, Quercus* and *Ulmus* species, lignans such as syringaresinol, lyoniresinol, thomasic and thomasidioic acid are known (Fig. 7–24) (Freudenberg, Weinges 1959; Seikel et al. 1971; Rowe et al. 1972). Some of them are linked with rhamnose or xylose as glycosides. Earlier studies also revealed the presence of lignans in *Populus* species as well as in a

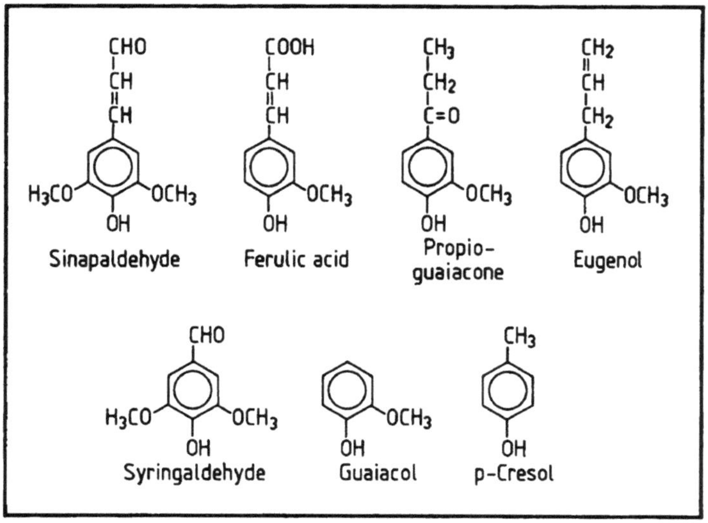

Fig. 7–23. Simple phenols from hardwoods.

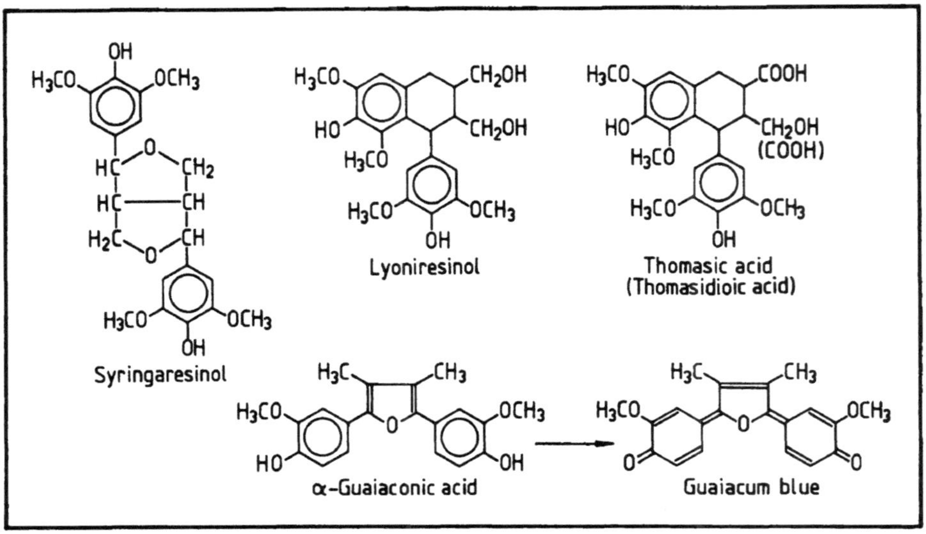

Fig. 7–24. Lignans from hardwoods.

number of subtropical and tropical woods (Hathway 1962). One of the lignans from tropical woods is α-guaiaconic acid from *Guaiacum* species which can easily be converted to a characteristic blue water-insoluble pigment called guaiacum blue (Fig. 7–24) (Kratochvil et al. 1971). Guaiacum blue has a quinoid structure.

Quinoid sesquiterpenes (mansonones) were described above as being present in *Ulmus* and *Mansonia* species (→ 7.3.1.) (Fig. 7–19). Dimethoxybenzoquinone was found in extracts from *Quercus rubra* (Seikel et al. 1971). The quinones from teak wood *(Tectona grandis)* and from *Dalbergia* species are well-known (Sandermann, Simatupang 1966; Dietrichs, Hansen 1971). In teak wood various quinones

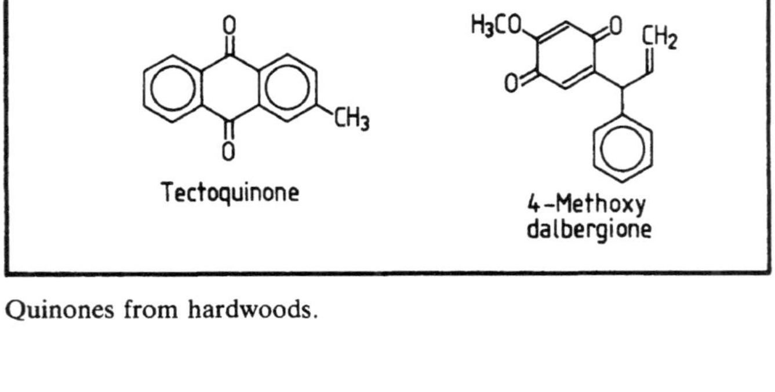

Fig. 7–25. Quinones from hardwoods.

are present belonging to the groups of naphthoquinones (lapachol, dehydrolapachol) and anthraquinones (tectoquinone) (Fig. 7–25). The quinones from *Dalbergia* species are called dalbergiones and must be assigned to the group of neoflavonoids (Fig. 7–25). Quinones have also been detected in other tropical woods, e.g. anthraquinones in esenge *(Maesopsis eminii)* (Cumming, Thomson 1970).

7.3.4. Tannins and Flavonoids

Compounds referred to by the term tannins are determined more by their tanning action on the proteins of hides than by a common chemical structure. Nevertheless all vegetable tannins are phenolic compounds ranging from simple phenols to condensed flavonoid systems. The tannins are subdivided into hydrolyzable tannins and non-hydrolyzable or condensed tannins (phlobaphenes).

The hydrolyzable tannins are esters of gallic acid and its dimers (digallic, ellagic acid) (Fig. 7–26) with monosaccharides, mainly glucose. In the extracts of various woods *(Castanea, Eucalyptus, Liquidambar, Quercus* spp.) the phenolic aglycones are also found, however in low percentages (Mayer et al. 1967a; Swan, Ackerblom 1967; Chen 1970; Spencer, Choong 1977). The hydrolyzable tannins are often subdivided into gallotannins yielding gallic acid after hydrolysis, and ellagitannins yielding ellagic acid after hydrolysis (Jurd 1962).

The tannins present in *Castanea sativa* and *Quercus petraea* were separated into four main compounds. 78% of the tannin fraction of *Castanea* wood consisted of vescalagin and its anomer castalagin (Mayer et al. 1967a). In oak wood these two tannins amounted to 44%. The minor anomeric tannins were called vescalin and castalin. Mayer et al. (1967b, 1969, 1971a, b) also determined the structure of these compounds which belong to the group of ellagitannins (Fig. 7–27). Ellagic acid, methyl derivatives of ellagic acid, and glycosides of both are the components of the tannin extractives of *Eucalyptus* species (Lowry 1968; Seikel, Hillis 1970; Hillis, Yazaki 1973). Further structures proposed for *Eucalyptus* tannins are shown in Fig. 7–27. Gallotannins (e. g. hamamelitannin, Fig. 7–27) were detected in the sap-

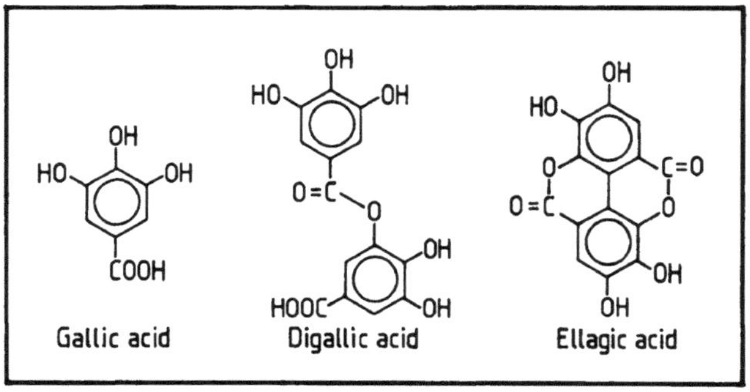

Fig. 7–26. Gallic acid and its dimers.

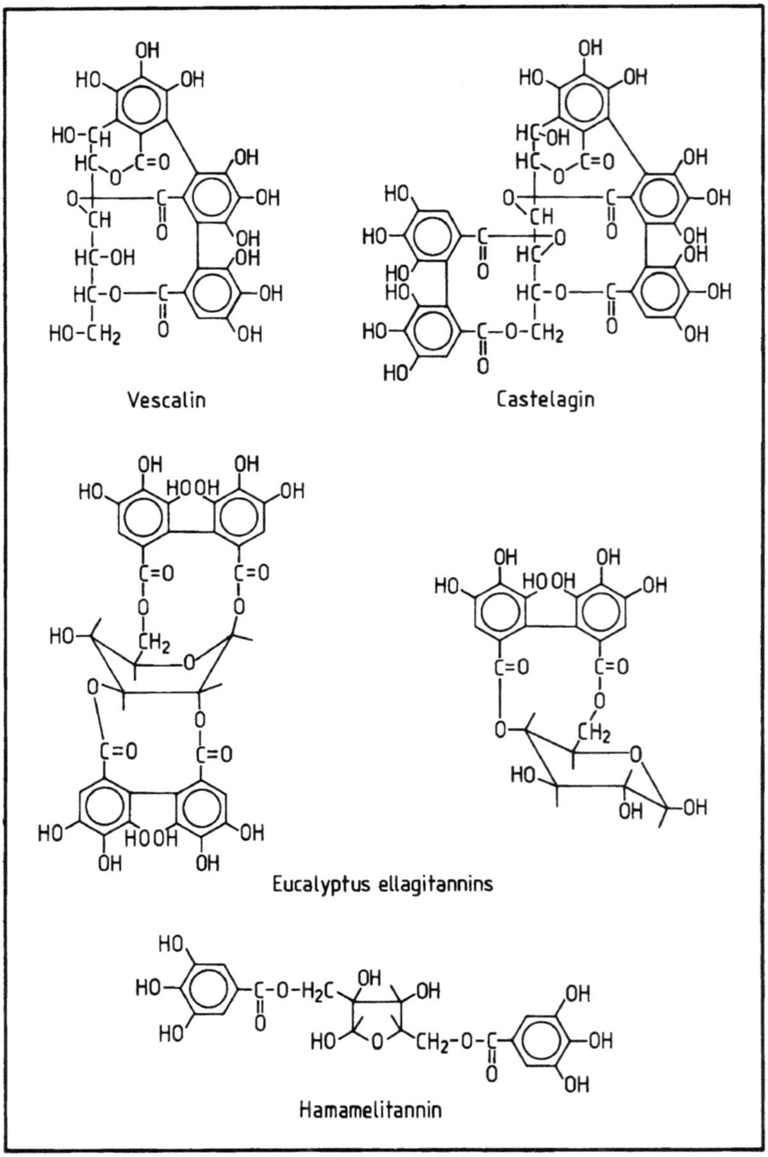

Fig. 7–27. Structure of some hydrolyzable tannins.

wood and heartwood of *Quercus alba* and *Q. rubra* (Chen 1970; Seikel 1971).

The study of the behaviour of ellagitannins with alkali showed that under the conditions of cold soda and alkaline groundwood pulping the tannins are notable resistant (Hemingway, Hillis 1971). Under conditions comparable to kraft and soda pulping the decarboxylation of gallic and ellagic acid was the major reaction.

Compared to the condensed tannins the hydrolyzable tannins occur less frequently in woods. Though apart from *Quercus, Castanea* and *Eucalyptus* species hydrolyzable tannins have also been found in *Terminalia, Phyllantus* and *Caesalpinia* species most of the wood species, particularly tropical woods, contain condensed tannins (Pizzi 1980). The study of the tannins of some *Quercus* species *(Q. petraea, Q. suber, Q. ilex)* and of several tropical woods showed the presence of hydrolyzable tannins (gallic acid type) only for the oaks, all other woods containing merely condensed tannins (catechin type) (Doat 1978).

The main components of the condensed tannins are the catechins (flavan-3-ols) and the leucoanthocyanidins (flavan-3,4-diols). These compounds belong to the group

Table 7–7: Flavonoids as isolated from various hardwood species

Basic structure	$OH-(OCH_3)$-position	Name	Occurrence
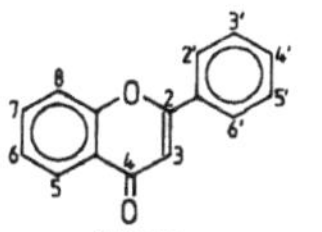 Flavones	3, 7, 3', 4'	Fisetin	*Acacia, Rhus, Schinopsis*
	3, 5, 7, 4'	Kaempferol	*Afzelia*
	3, 7, 3', 4', 5'	Robinetin	*Acacia, Robinia, Schinopsis*
	3, 5, 7, 3', 4'	Quercetin	*Acacia, Aesculus, Quercus*
	3, 5, 7, 2', 4'	Morin	*Chlorophora*
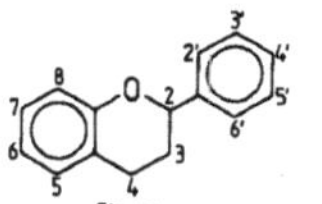 Flavanes	3, 7, 3', 4'	Fisetinidol	*Acacia*
	3, 4, 7, 3', 4'	Mollisacacidin	*Acacia, Gleditsia*
	3, 5, 7, 3', 4'	Catechin	*Acacia, Schinopsis*
	3, 4, 5, 7, 3', 4'	Leucocyanidin	*Schinopsis*
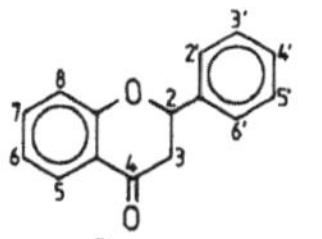 Flavanones	7, 3', 4'	Butin	*Acacia*
	3, 7, 3', 4'	Fustin	*Acacia, Schinopsis*
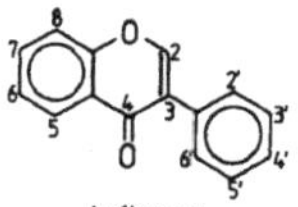 Isoflavones	5, 4', (7)	Prunetin	*Prunus, Pterocarpus*
	5, 3', 4', (7)	Santal	*Pterocarpus, Santalum*
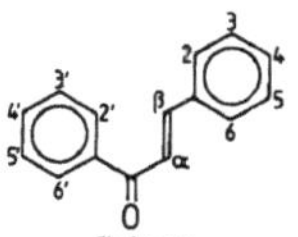 Chalcones	3, 4, 2', 4'	Butein	*Acacia, Pseudosindora*
	3, 4, 2', 3', 4'	Okanin	*Cyclicodiscus*
	α, 3, 4, 2', 4'	Pentahydroxy-chalcone	*Peltogyne, Trachylobium*
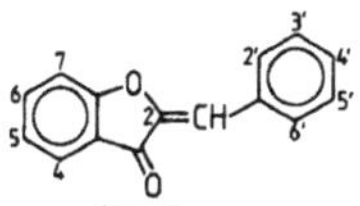 Aurones	6, 3', 4'	Sulfuretin	*Pseudosindora*
	6, 3', 4', (4)	Rengasin	*Melanorrhoea, Pseudosindora*
	2, 6, 3', 4'	Tetrahydroxybenzylcoumaranone	*Schinopsis*
	2, 6, 3', (4')	Methoxy trihydroxy-benzylcoumaranone	*Schinopsis*

of flavonoids, the members of which are wide-spread in the plant kingdom. Many of them are also present in the extracts of woods, and are called 'phenolic nontannins'.

The flavonoids can be derived from flavone which itself is interpreted as a 2-phenyl benzopyrone. Its derivatives containing a hydrated pyrane ring are called flavanes. Other derivatives of the basic flavone structure are the flavanones and the isoflavones. Structures containing an opened pyrone ring are called chalcones, and those with a furanone ring aurones. Representatives of most of these flavone derivatives are found in various woods. The nontannin phenols in wattle heartwood (*Acacia* spp.) and quebracho (*Schinopsis* spp.) have been intensively studied (Roux, Paulus 1960, 1961; Roux et al. 1975, 1976; King et al. 1961). A compilation of the basic structures and of various flavonoids as isolated and identified from woods is shown in Table 7–7.

A number of varieties of the above-mentioned flavonoids are known. Thus isoflavones and chalcones with additional ring formations were isolated from woods. For example pterocarpin from *Pterocarpus santalinus* and mopanol from *Goniorrhachis marginata* should be mentioned (Fig. 7–28) (Puth 1962; Gottlieb, de Sousa 1972). Some of the flavonoids determine the colour of the respective wood, e.g. fisetin, morin, santal. Others (butein, sulfuretin, rengasin) are responsible for coloured specks in pulps from tropical woods (Tachibana, Sumimoto 1980, 1982; Ohtani et al. 1982).

Most of the extractive components of the so-called colour woods, e.g. red woods *(Pterocarpus, Baphia, Casesalpina* spp., *Haematoxylon brasiletto),* blue wood *(Haematoxylon campechianum),* yellow wood *(Chlorophora tinctoria),* are flavonoids and related compounds. These compounds were used as natural dyes in former times (Puth 1962)(→18.7.). They are often present in the wood as colourless leuco-compounds and the colour must be developed. Well-known examples are the oxidation of haematoxylin to the blue haematein, and of brasilin to the red brasilein (Fig. 7–29). The colour and its graduation can be influenced by further treatments with acids, alkali or metal salts.

There seems to be a close interrelation between the various flavonoids, e.g. the chalcones are in an equilibrium with the corresponding flavanones. Roux and Fer-

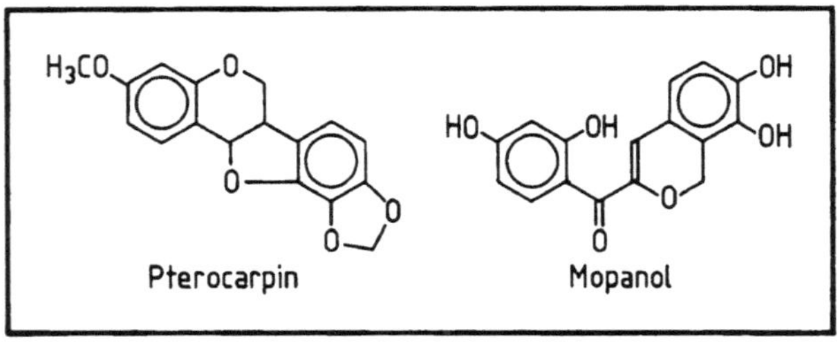

Fig. 7–28. Flavonoids with an additional ring formation.

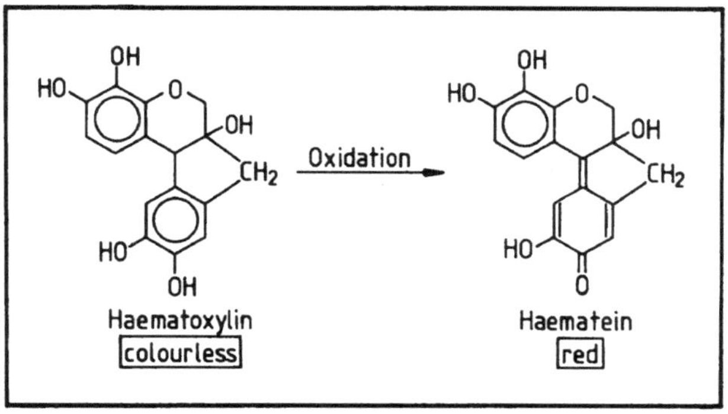

Fig. 7–29. Haematoxylin and haematein from *Haematoxylon campechianum.*

reira (1974) regard the α-hydroxychalcones as prominent intermediates in the biogenesis of flavonoids. They outline a metabolic pathway leading to the various flavonoid types and to the condensed tannins.

According to Roux et al. (1975) only flavonoids of the type flavan-3-ol and flavan-3,4-diol participate in the formation of condensed tannins. The carbonyl group in 4-position of the other types of flavonoids reduces the nucleophilic character and occupies one of the positions of condensation. The meta-substitution of the resorcinol A-ring of the flavan-3,4-diols by hydroxyl and heterocyclic oxygen groups creates strong nucleophilic centres at the 6- and 8-position. Thus the units of the condensed tannins are predominantly linked by 4,6- and 4,8-bonds.

The first step of this condensing reaction is the formation of biflavonoids (proanthocyanidins). This group of natural compounds is wide-spread in various plant tissues. Proanthocyanidins have been described for fruits, leaves, bark, phloem and xylem from many plant families (Thompson et al. 1972; Bate-Smith 1975; Foo, Porter 1980) (→ 7.4.). In the heartwood of several species (*Acacia, Colophospermum, Eucalyptus, Schinopsis* spp.) various types of biflavonoids occur differing in their components (catechin, gallocatechin, leucofisetinidin, leucorobinetin), in their linkages (4,6, 4,8), and in their sterical arrangements (cis-, trans-conformation (Nisi 1966; Roux 1972; Roux et al. 1975, 1976). Some examples of biflavonoids as isolated from hardwoods are shown in Fig. 7–30. In addition to 4,6-(I,II) and 4,8-linked proanthocyanidins (III, IV), those linked by ether bonds are also known (V, VI).

The true condensed tannins consist of 3 to 8 flavonoid units with a basic structure as can be seen in Fig. 7–31. Variations in the composition and the sterical arrangement are the same as described for the biflavonoids. But tannins with higher molecular weight (>3 000) corresponding to deca- and undecaflavonoids have also been isolated (Roux et al. 1975). The transition to the polyphenols which are mainly present in bark (→ 9.2.5.) is fluent.

The comparison of tannins from bark and heartwood of black wattle *(Acacia mearnsii)* demonstrated that bark tannins consist of leucodelphinidin, leucorobine-

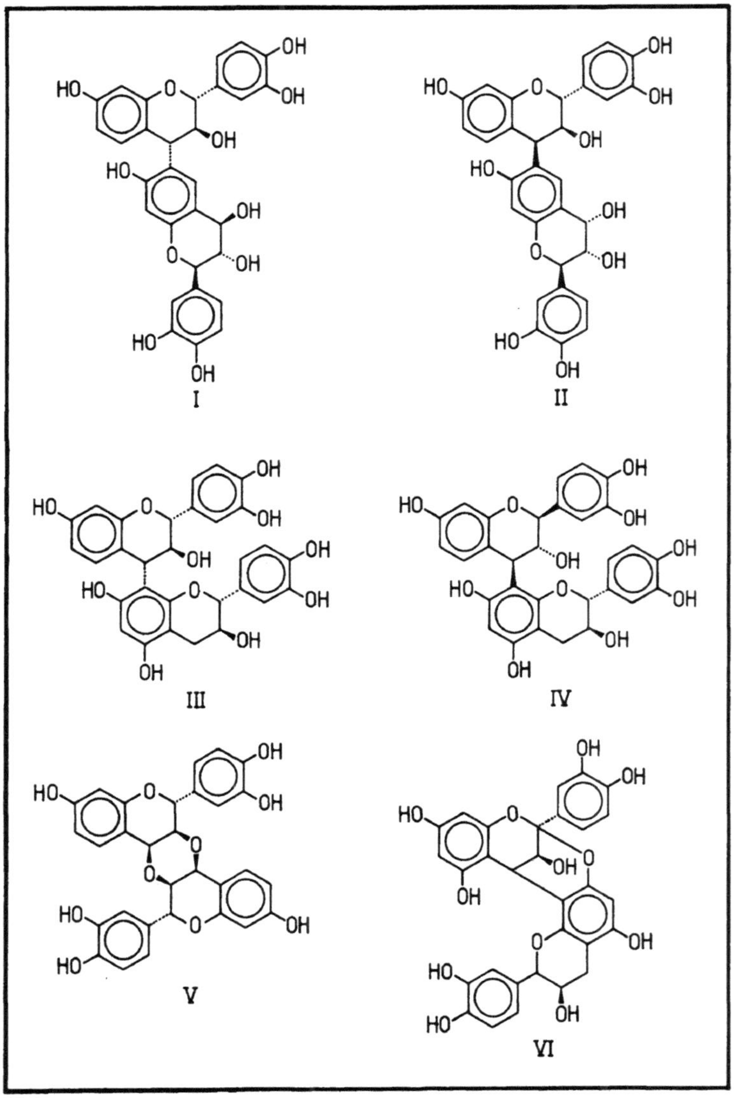

Fig. 7–30. Biflavonoids (proanthocyanidins) from hardwoods.

tin and leucofisetinidin (→ 9.2.5.) whereas heartwood tannins consist only of leucofisetinidin units (Saayman, Roux 1965).

Various studies are concerned with the utilization of condensed tannins (Roux 1972; Roux et al. 1975, 1976; Pizzi 1980). Experiments have been performed yielding simple phenols by fusion processes, and adhesives for plywood and particleboard by autocondensation, condensation with formaldehyde, or sulfitation (→ 18.7.).

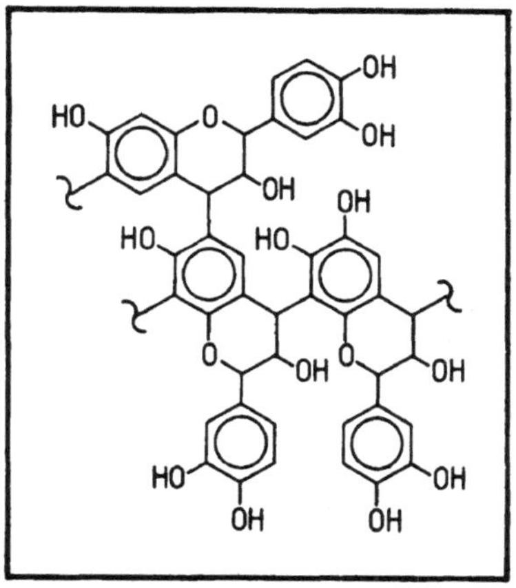

Fig. 7–31. Structure of condensed tannins.

7.3.5. Various Compounds

The portion of saturated hydrocarbons amounts to 0.4 and 1.7% of the petroleum ether solubles from *Betula verrucosa* and *Populus tremula*, respectively (Assarson, Åkerlund 1966). As already described above for spruce and pine (→ 7.2.4.) a homologous series from C11 to C33 has been determined for these species as well, with main components in the range of C22- to C30-alkanes.

The production of ethene as a product of heartwood formation was shown for black walnut *(Juglans nigra)* and black cherry *(Prunus serotina)* (Nelson 1978). (→ 7.2.4.).

The incidence of methane in the heartwood of *Ulmus, Salix* and *Populus* species could be attributed to an infestation by anaerobic bacteria (Zeikus, Ward 1974) (→ 14.8.).

As shown for several hardwoods (*Betula alba, Fagus sylvatica, Quercus* spec., *Terminalia superba*) heartwood formation is combined with a sharp decrease of the soluble sugars (Dietrichs 1964; Dietrichs, Schaich 1964). The main sugars in the sapwood are glucose, fructose and sucrose. Stachyose and raffinose were additionally found in buds.

Amino acids which are free or linked as proteins were also identified in the xylem and other tissues (leaves, phloem) of *Fagus sylvatica, Quercus robur, Betula alba* and *Eucalyptus* spp. (Dietrichs 1969; Scurfield, Nicholls 1970). According to these studies the qualitative composition of the wood proteins is the same as in other plant proteins with the exception of hydroxyprolin which occurred additionally. Though the proportion of proteins in wood is very low – the nitrogen content of most woods is lower than 0.1% – there are some indications that the proteins participate in the cell wall structure (Lüdtke, Lerch 1968; Lüdtke, Röpsch 1971).

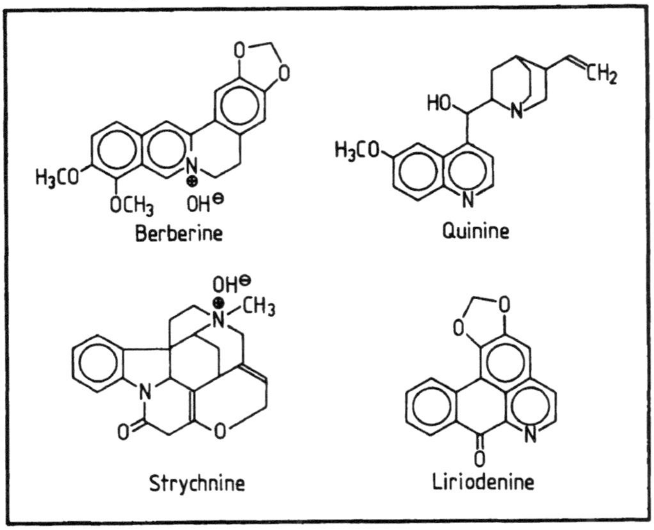

Fig. 7–32. Alkaloids from hardwoods.

The nitrogen content particularly of many tropical woods cannot be attributed merely to the proteins as they may also contain alkaloids. These are nitrogen-containing compounds of differing chemical structures. Among them are compounds such as berberine present in *Berberis* spp., quinine in *Cinchona* spp., liriodenine in *Magnoliaceae,* and strychnine in *Strychnos* spp. (Fig. 7–32). Lange (1976b) compiled 125 wood species in the xylem of which alkaloids have been detected. Except for *Taxus baccata* all are hardwoods. There are 67 different alkaloids from these species.

7.4. Extractives from Foliage, Buds and Fruits

Various studies are concerned with the extractives of tree tissues other than xylem. In particular foliage is seen to be a potential resource for silvichemicals and fodder. Investigations of needles showed that in addition to compounds also present in wood (mono-, diterpenes, fatty acids, simple phenols, lignans, flavonoids, sugars, proteins), some particular resin acids (e.g. imbricataloic acid, pinifolic acid), cyclic acids (quinic acid, shikimic acid), and various cyclitols (myoinositol, pinitol, sequoyitol) are also found (Fig. 7–33) (Spalding et al. 1971; Popoff, Theander 1976; Barton, McDonald 1978; Cranswick, Zabkiewicz 1979; Bardyshev et al. 1981; Theander 1981). Analyses of green needles are summarized in Table 7–8.

Apart from proteins, chlorophyll and carotenes, studies of leaves of angiospermous woods concentrate on phenols, flavonoids and tannins (Barton 1976). From leaves and nuts of beech *(Fagus sylvatica)* several phenolcarbonic acids (p-hydroxyben-

Table 7–8: Composition of green needles from pine and spruce (percentages based on dry weight) (Theander 1978)

Components	*Pinus sylvestris* %	*Picea abies* %
Neutral polysaccharide constituents, calculated as:		
Glucans	25.5	16.3
Mannans	4.4	7.5
Galactans	1.8	1.5
Arabinans	2.6	3.1
Xylans	1.3	2.1
Acidic polysaccharide constituents	3.3	3.2
Klason lignin	14.8	14.4
Ash	2.3	3.0
Crude protein	5.3	5.9
Hydrophilic components		
Glucose	2.3	
Fructose	1.6	
Sucrose	0.7	
Mannitol	1.8	
Pinitol	2.2	
Myoinositol	0.3	
Glycoside 2	0.3	
Taxifolin glucoside	0.3	
Lipophilic components		
Hydrocarbons	0.3	
Steryl esters	1.2	
Triglycerides	2.8	
Free acids	1.1	
Diterpene alcohols and sterols	0.9	

zoic, vanillic, p-coumaric, ferulic, sinapic, coffeic, chlorogenic acid) were separated (Beckmann, Volkmann 1965). The investigation of leaves from *Populus* and *Salix* species revealed that compounds such as benzoic, salicylic and coumaric acid are linked as glycosides which are called populin, salicin, salicortin, tremuloidin etc. (Fig. 9–12) (Pearl, Darling 1965, 1971a, b; Pearl et al. 1966).

The proanthocyanidins in leaves, fruits, barks and phloem of 14 widely distributed plant families were described by Foo and Porter (1980). All are based on a common 4,6-and 4,8-linked flavan-3-ol structure. In several plant tissues epicatechin-3-O-gallate was detected (Fig. 7–34, I). Most of the flavonoids seem to be linked as glycosides (Dietrichs, Schaich 1963; Pearl, Darling 1971a; Santamour 1977). In leaves of *Fagus* and *Populus* spp. glycosides of chrysin, quercetin and kaempferol were identified. Buds of *Fagus sylvatica* contain a kaempferol-4'-p-coumaryl-3-diglucoside (Fig. 7–34, II).

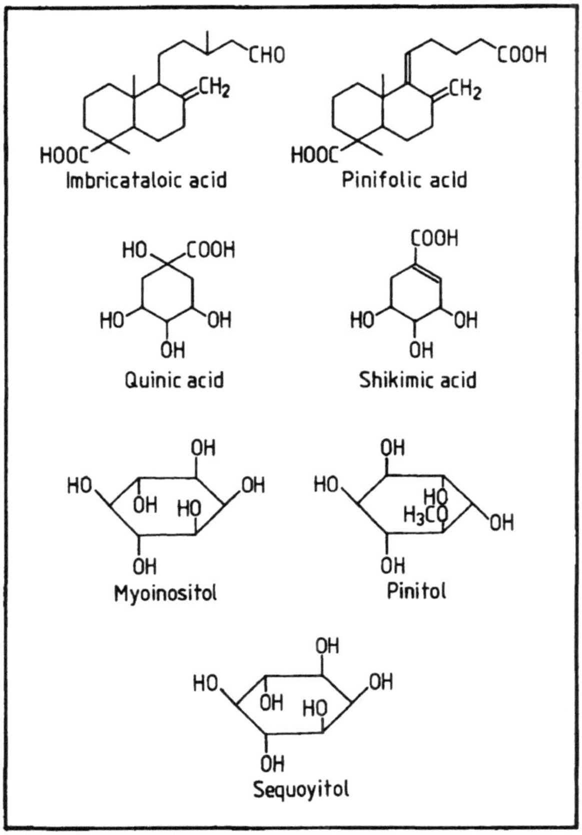

Fig. 7–33. Extractives from needles.

In the leaves of *Eucalyptus* species the presence of glycosides of ellagic acid was proved (Hillis, Yazaki 1973). The fruitcups of *Quercus* species contain castavalonic acid (Fig. 7–34, III) in addition to castalagin and vescalagin (Mayer et al. 1971c, 1976) (→ 7.3.4.).

Amino acids are mainly linked as proteins. The percentage of free amino acids is about ten to the third or fourth lower than that of combined amino acids in leaves (Dietrichs 1969). For *Pinus sylvestris* a decrease of free amino acids from 1.03 mg/g in first-year needles to 0.54 mg/g in the fourth-year needles was observed (Repyakh, Levin 1977).

More rare compounds from leaves and fruits are anthraquinones and anthracene derivatives as identified in *Cassia* species (Rai 1977, 1978).

The fruits of some angiospermous trees contain saponins among other compounds. Aescin, for example, is known in horse-chestnuts *(Aesculus hippocastanum)* (Fig. 7–21) and a glycoside of albigenic acid in the beans of kokko *(Albizzia lebbeck)* (Barua, Raman 1959; Hoppe et al. 1968).

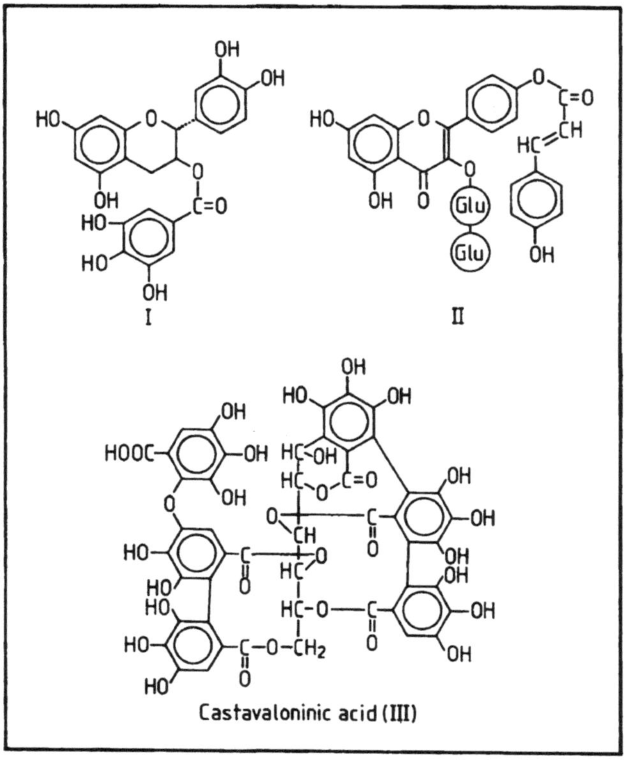

Fig. 7–34. Extractives from leaves and buds.

7.5. Inorganic Components

The inorganic constituents of wood are entirely contained in the ash, the residue after burning the organic matter. Though the percentage of the mineral components is low (woods of temperate zones contain 0.1–1.0%, those of tropical and subtropical zones up to 5% ash) (Tables 3–4, 3–5, 3–6), many of them are essential for the trees' growth. There is a certain dependence of the mineral content and composition on environmental conditions (site, climate) and on the location within the tree. Regarding the whole tree, the highest content of inorganic components is in the needles or leaves, respectively. The sequence of decreasing ash content is subsequently bark, tiny roots, twigs, roots, branches and stem (Yoshida 1961; Young, Guinn 1966). Several methods have been applied for the determination of the elemental composition of wood ashes (→ 3.2.5.).

The main ash components of wood are calcium, potassium and magnesium (Table 7–9). In many woods Ca amounts to 50% and more of the total elements in wood ashes. K and Mg are in second and third place, respectively, followed by Mn, Na, P

Table 7–9: Inorganic composition of several softwoods and hardwoods (ppm values based on dry wood) (according to Young and Guinn 1966, and Cutter et al. 1980)

Species	Ca	K	Mg	Mn	Na	Cl	P	Al	Fe	Zn
Abies balsamea	830	770	270	127					13	11
Picea rubens	820	200	70	144			50		14	8
Pinus strobus	210	290	70	28					10	11
Pinus spec.										
Earlywood	743		132	134	49	102		10		
Latewood	589	117	138	111	34	23		17		
Wholewood	764	39	110	97	28	48		6		
Pseudotsuga menziesii	295		41	25	44	67		13		
Thuja plicata	1013	229	76	1	10	12		2		
Tsuga spec.	750	400	110	145					6	2
Acer rubrum	820	690	120	72			30	2	11	29
Acer saccharum	1064	990	140	36	29	82		1		
Betula papyrifera	740	270	180	34			150	23	10	28
Populus spec.	1130	1230	270	29			100		12	17
Quercus alba	674	780	11	2	3	8		6		
Tilia americana	1125	543	117	11	74	93		15		

and Cl (Ellis 1962, 1965; Young, Guinn 1966; Choong et al. 1976; Cutter, et al. 1980). Fig. 7–35 shows these elements in an X-ray spectrum of spruce ash *(Picea abies)*.

Many other elements are present in wood in concentrations of less than 50 ppm; they must be defined as trace elements. The 12 trace elements Ba, Al, Fe, Zn, Cu, Ti, Pb, Ni, V, Co, Ag and Mo were detected in 34 wood species in varying concentrations (Ważny, Ważny 1964). More than 50 trace elements were detected in woods by neutron activation analysis. Young and Guinn (1966) give a compilation of the elements found in red spruce *(Picea rubens)* (Table 7–10). Because of the variations between various woods and even between wood from various trunks of the same species, it is difficult to set limits to the frequency of the elements, particularly of the minor ones. Nevertheless the upper-limit ranges in Table 7–10 may be seen as approximate values for many wood species.

Compared to woods of temperate zones many tropical woods stand out by having a high percentage of silicon, which may exceed the content of calcium in certain species (Hillis, de Silva 1979). A correlation between the silica content and the resistance against marine borers of certain woods *(Dicorynia guaianensis, Syncarpia glomulifera)* could not be verified (Sandermann, Lange 1967; de Silva, Hillis 1980).

The variation in the content of Ca, Mg and Mn of pine wood *(Pinus sylvestris)* of different origin (chips from 13 pulp mills in Finland and Sweden) was studied by Fossum et al. (1972). The values varied as follows: Ca: 557–805 ppm, Mg: 85–188 ppm, Mn: 51–112 ppm. An enhanced manganese content in wood may adversely affect the bleachability of the resulting pulp (Ginzel 1971).

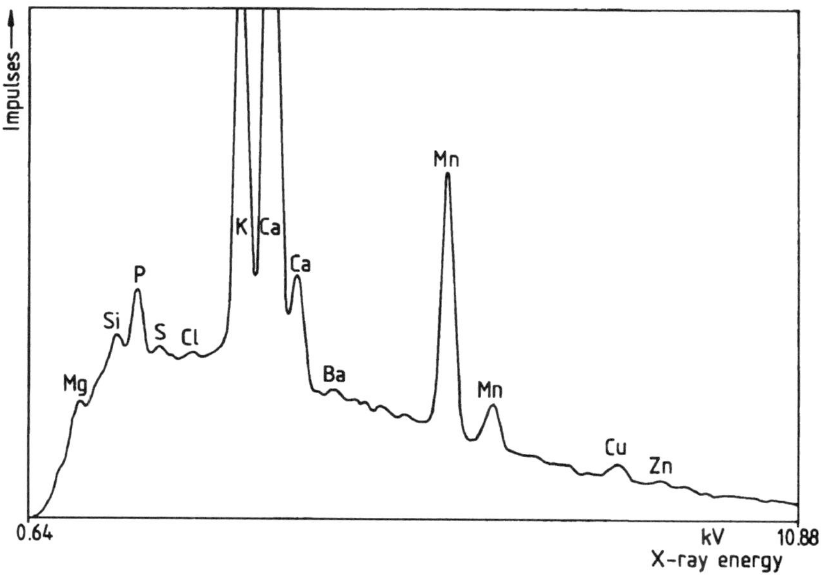

Fig. 7–35. X-ray spectrum of spruce ash *(Picea abies)*.

Table 7–10: Upper-limit values of elements in a wood sample *(Picea rubens)* (Young, Guinn 1966)

Range of upper-limit values, ppm	Elements
1000–100	Fe, Mg, P, Zr
100–10	Ba, F, Ni, Si, Sn, Sr, Ti, Y, Zn
10–1	Cd, Ce, Cr, Ga, Gd, Ge, Hg, I, Nb, Nd, Pd, Pr, Pt, Rb, Ru, Se, Te, Tm
1–0.1	Ag, Al, As, Br, Co, Cs, Cu, Er, Hf, Ho, La, Os, Rh, Sb, Ta, Tb, W, Yb
0.1–0.01	Eu, In, Re, Sc, Sm, V
0.01–0.001	Au, Dy, Ir, Lu

Analyses of the elemental distribution across the stem of pine *(Pinus sylvestris, P. taeda)* showed an increase of the mineral content from the outer edge of the sapwood to the heartwood and to the pith (Galligan et al. 1965; McMillin 1969b, 1970; Fossum et al. 1972). Highest concentration was found in the cambium as compared to the adjacent wood and bark (→ 9.2.8.). The content of inorganics was lower in latewood than in earlywood.

A careful ashing of wood cross-sections yields ash skeletons which can be observed in the light and electron microscopes (Zicherman, Thomas 1971; Fossum et al. 1972). It was found that there is an accumulation of inorganic substances in resin canals and ray cells. In the tracheid walls the minerals are concentrated in the compound middle lamella and in the tertiary wall layer. But pits and spiral thickenings in hardwood vessels also contain mineral components (Chatters 1963).

The minerals are not only incorporated in the cell walls but are also deposited in the lumina of parenchyma cells and libriform fibres. The mineral deposits consist mostly of calcium carbonate, calcium oxalate or silicate and have different shapes (Mia 1969; Scurfield et al. 1973, 1974; Hillis, de Silva 1979). Silica mostly appears as grains or grain aggregations; the other inorganic inclusions occur as needles, prismatic or polyedric crystals. Such crystals were detected not only in various tropical woods but also in *Abies, Acer* and *Populus* species. Crystals which appear in wood after an attack by fungi or bacteria are attributed to metabolic products of the microorganisms (Kühne et al. 1970; Scurfield et al. 1973; Aho et al. 1979) (→ 14.7.).

7.6. pH Values

The pH value is a measure of the concentration of H-ions (or OH-ions, respectively) in solutions, and is used for the determination of their acidic, neutral or alkaline behaviour. The pH value of wood, or more exactly of the aqueous solution within humid wood, is highly important for various ranges of its utilization. Metals in contact with wood may corrode, the adhesive power of glues and the fixation of wood protectives may be influenced, and attention to the pH of wood is also called for in connection with pulping, fibre- and particleboard production, and plastification (Sandermann, Rothkamm 1959; Kehr, Schilling 1965; Lábsky 1974).

Several methods have been developed for measuring the pH value of wood, as a direct measurement of solid wood is only possible with some uncertainties if the water content of the wood is higher than the fibre saturation point. The extraction of wood with a certain quantity of water is only applicable for comparative determinations. For an absolute measurement two methods are mainly applied:

- the determination of the moisture-pH (MpH) by assimilating the pH of an acidic or basic aqueous solution to the pH of the suspended disintegrated wood (Sandermann, Rothkamm 1959);
- the slope-intercept method, a graphic evaluation of the pH value of wood from measurements after immersion in dilute NaOH and HCl solution (Stamm 1961; McNamara et al. 1970).

A more direct method is the measurement of the fluid which is press-extracted from steam-treated wood (Kubinsky, Ifju 1973).

In Table 7–11 the pH values determined for various softwoods and hardwoods of temperate zones as well as for some tropical woods are summarized. While the pH values of woods from temperate zones are in the weak-to-moderate acidic range (3.3–6.4), those of tropical woods range from weak acidic to weak alkaline (3.7–8.2). A pH value of 7.5 to 8.0 was also determined for *Populus cathayana,* a species growing in China (Li, Hsiang 1963).

Table 7–11: pH Values of various woods (Sandermann, Rothkamm 1959; McNamara et al. 1970)

Wood species	pH-Value
Larix decidua	4.3
Picea abies	5.3
Pinus resinosa	6.0
Pinus sylvestris	5.1
Pinus strobus	4.9
Pseudotsuga menziesii	3.3
Tsuga canadensis	5.5
Acer saccharum	5.1
Betula papyrifera	5.5
Betula verrucosa	4.8
Carpinus betulus	5.2
Fagus grandifolia	5.5
Fagus sylvatica	5.4
Fraxinus excelsior	5.8
Populus spec.	5.8
Quercus alba	4.1
Quercus petraea	3.9
Quercus rubra	4.2
Tilia spec.	5.2
Ulmus spec.	6.4
Dalbergia melanoxylon	8.0
Gossypiospermum praecox	6.9
Lophira procera	4.7
Mansonia altissima	4.3
Ochroma lagopus	6.7
Pterocarpus soyauxii	3.7
Schinopsis balansae	4.3
Shorea spec.	4.7
Tectona grandis	5.1
Terminalia superba	8.2

The acidic reaction of most woods is caused by free acids and acidic groups which are easy to split off, i.e. predominantly acetic acid and acetyl groups, respectively. But other acids, particularly in tropical woods, may also attribute to the pH value.

There are only slight differences in pH between sapwood and heartwood, and furthermore the season of cutting seems to have only a small effect on the pH value (Sandermann, Rothkamm 1959; McNamara et al. 1970). The acidity of wood increases during storage under damp conditions and with a rise in temperature (Packman 1960; Volz 1971) (Table 9–15).

Literature

Abramovitch, R. A., Micetich, R. G. and Smith, S. J. 1963, Tappi *46,* 37–40
Adelsberger, U. and Petrowitz, H. J. 1976, Holzforschung *30,* 109–113
Aho, P. E., Cromack, K. and Li, C. Y. 1979, USDA For. Serv., Res. Note PNW-328
Akimoto, H. and Sumimoto, M. 1980, Mokuzai Gakkaishi *26,* 347–357
Anderegg, R. J. and Rowe, J. W. 1974, Holzforschung *28,* 172–175
Anderson, A. B., Riffer, R. and Wong, A. 1970, Holzforschung *24,* 182–184
Assarsson, A. 1966, Svensk Papperstid. *69,* 291–299
Assarsson, A. and Åkerlund, G. 1966, Svensk Papperstid. *69,* 517–525
Assarsson, A. and Åkerlund, G. 1967, Svensk Papperstid. *70,* 205–212
Back, E. 1960, Svensk Papperstid. *63,* 647–651
Bardyshev, I. I., Degtyarenko, A. S., Pertsovkij, A. L. and Kryuk, S. I. 1981, Khim. Drev. No. 3, 102–104
Barton, G. M. 1968, For. Prod. J. *18,* No. 5, 76–80
Barton, G. M. 1970, Wood Fiber *2,* 144–150
Barton, G. M. 1973, Tappi 56, No. 5, 115–118
Barton, G. M. 1976, Appl. Polym. Symp. No. 28, 465–484
Barton, G. M. and Gardner, J. A. F. 1966, Dept. For. Publ. No. 1147, Ottawa
Barton, G. M. and McDonald, B. F. 1978, Tappi *61,* No. 1, 45–48
Barua, A. K. and Raman, S. P. 1959, Tetrahedron *7,* 19–23
Bate-Smith, E. C. 1975, Phytochemistry *14,* 1107–1113
Bauch, J., Schmidt, O., Hillis, W. E. and Yazaki, Y. 1977, Holzforschung *31,* 1–7
Beck, C. W. 1972, Naturwissenschaften *59,* 294–298
Becker, G. 1962, Holz Roh-Werkst. *20,* 368–375
Becker, G. 1966, Holz Roh-Werkst. *24,* 429–432
Beckmann, S. and Volkmann, D. 1965, Naturwissenschaften *52,* 208
Bergman, J. 1965, Svensk Papperstid. *68,* 339–347
Bergman, J., Lindgren, B. O. and Svahn, C. M. 1965, Acta Chem. Scand. *19,* 1661–1666
Bicho, J. G., Zavarin, E. and Bhacca, N. S. 1963, J. Org. Chem. *28,* 2927–2929
Buchanan, M. A. 1963, Extraneous Components of Wood. In: The Chemistry of Wood (Browning, B. L., Ed.). Intersci. Publ., New York, London, pp. 313–367
Calimli, A. and Olcay, A. 1978, Holzforschung *32,* 7–10
Chatters, R. M. 1963, For. Prod. J. *13,* 368–372
Chen, C. L. 1970, Phytochemistry *9,* 1149
Choong, E. T., Abdullah, G. and Kowalczuk, J. 1976, LSU Wood Utiliz. Notes No. 29
Collier, T. J. 1977, Lightwood Res. Coord. Council, Proc. Ann. Meeting, Atlantic Beach, Fla., pp. 172–191
Conner, A. H. and Rowe, J. W. 1975, J. Am. Oil Chem. Soc. *52,* 334–338
Conner, A. H., Diehl, M. A., Wroblewska, H. and Rowe, J. W. 1977, Lightwood Res. Coord. Council, Proc. Ann. Meeting Atlantic Beach, Fla., pp. 34–56
Conner, A. H., Diehl, M. A. and Rowe, J. W. 1980, Wood Sci. *12,* 183–191, 194–198
Cordell, G. A. 1974, Phytochemistry *13,* 2343–2364
Cranswick, A. M. and Zabkiewicz, J. 1979, J. Chromatogr. *171,* 233–242
Croon, I. 1965, Papier *19,* 711–719
Cumming, A. M. and Thomson, R. H. 1970, Phytochemistry *9,* 2399–2400
Cutter, B. E., McGinnes, E. A. and McKown, D. H., 1980, Wood Fiber *12,* 72–79
Dahm, H. P. 1970, Svensk Papperstid. *73,* 613–618
Dauben, W. G. and German, V. F. 1966, Tetrahedron *22,* 679–683
Della Casa de Marcano, D. P. and Halsall, T. G. 1969, Chem. Comm. 1282–1283
Dietrichs, H. H. 1964, Holzforschung *18,* 14–24

Dietrichs, H. H. 1969, Holzforschung *23,* 177–181
Dietrichs, H. H. and Hansen, B. M. 1971, Holzforschung *25,* 183–187
Dietrichs, H. H. and Schaich, E. 1963, Naturwissenschaften *50,* 478
Dietrichs, H. H. and Schaich, E. 1964, Forstwiss. Centralbl. *83,* 212–222
Doat, J. 1978, Bois For. Trop. *182,* 37–54
Donetzhuber, A. and Swan, B. 1965, Svensk Papperstid. *68,* 419–429
Drew, J. and Pylant, G. D. 1966, Tappi *49,* 430–438
Egorov, I. A., Pisarnitskii, A. F., Zinkewich, E. P. and Gavrilov, A. I. 1976, Prikl. Biokhim. Mikrobiol. *12,* 108–112
Ekman, R. 1976, Holzforschung *30,* 79–85
Ekman, R. 1979a, Acta Acad. Abo., Ser. B *39,* No. 4, 1–15
Ekman, R. 1979b, Acta Acad. Abo., Ser. B *39,* No. 3, 1–6
Ekman, R. 1980a, Phytochemistry *19,* 321–322
Ekman, R. 1980b, Phytochemistry *19,* 147–148
Ekman, R., Peltonen, C., Hirvonen, P., Pensar, G. and von Weissenberg, K. 1979, Acta. Acad. Abo., Ser. B *39,* No. 8, 1–26
Ellis, E. L. 1962, For. Prod. J. *12,* 271–274
Ellis, E. L. 1965, Inorganic Elements in Wood. In: Cellular Ultrastructure of Woody Plants (Côté, W. A., Ed.). Syracuse Univ. Press, Syracuse N. Y., pp. 181–189
Enzell, C. R., Hirose, Y. and Thomas, B. R. 1967, Tetrahedron Lett. 793–798
Erdman, H. and Tunso, K. 1969, Phytochemistry *8,* 931–932
Faulkner, D. J. 1973, Tetrahedron Lett. 3821–3822
Foo, L. Y. and Porter, L. J. 1980, Phytochemistry *19,* 1747–1754
Fossum, T., Hartler, N. and Libert, J. 1972, Svensk Papperstid. *75,* 305–309
Foster, D. O., Zinkel, D. F. and Conner, A. H. 1980, Tappi *63,* No. 12, 103–105
Freudenberg, K. and Knof, L. 1957, Chem. Ber. *90,* 2957–2969
Freudenberg, K. and Weinges, K. 1959, Tetrahedron Lett. 19–22
Galligan, W. L., Stern, H. and Hohenschuh, P. 1965, For. Prod. J. *15,* No. 5, 185–189
Gardner, J. A. F. 1965, J. Paint Technol. Eng. *37,* 698–706
Gardner, J. A. F. and Barton, G. M. 1960, For. Prod. J. *10,* No. 3, 171–173
Ginzel, W. 1971, Papier *25,* 305–311
Gottlieb, O. R. and de Sousa, J. R. 1972, Phytochemistry *11,* 2841–2846
Grosser, D., Fengel, D. and Schmidt, H. 1974, Forstwiss. Centralbl. *93,* 191–207
Hathway, D. E. 1962, The Lignans. In: Wood Extractives and Their Significance to the Pulp and Paper Industries (Hillis, W. E., Ed.). Academic Press, New York, London, 1962, pp. 159–190
Hausen, B. M. 1981, Woods Injurious to Human Health; A Manual. W. de Gruyter, Berlin, New York
Hayashi, Y. Takahashi, T. 1980, Mokuzai Gakkaishi 26, 54–55
Hemingway, R. W. and Hillis, W. E. 1971, Tappi *54,* 933–936
Hemingway, R. W., Hillis, W. E. and Lau, L. S. 1973, Svensk Papperstid. *76,* 371–376
Hillis, W. E. 1962, Wood Extractives and Their Significance to the Pulp and Paper Industries. Academic Press, New York, London
Hillis, W. E. and Inoue, T. 1968, Phytochemistry *7,* 13–22
Hillis, W. E. and de Silva, D. 1979, Holzforschung *33,* 47–53
Hillis, W. E. and Yazaki, Y. 1973, Phytochemistry *12,* 2963–2968
Hillis, W. E., Yazaki, Y. and Bauch, J. 1976, Wood Sci. Technol. *10,* 60–69
Hirose, Y., Oishi, N., Nagaki, H. and Nakatsuka, T. 1965, Tetrahedron Lett. 3665–3668
Holmbom, B. 1981, Some Recent Advances in Tall Oil Chemistry. In: The Ekman Days 1981, Int. Symp. Wood Pulp. Chem., Stockholm, Vol. 4, pp. 53–58
Holmbom, B. and Ekman, R. 1978, Acta Acad. Abo., Ser. B *38,* No. 3, 1–11

Hoppe, W., Gieren, A., Brodherr, N., Tschesche, R. and Wulff, G. 1968, Angew. Chem. *80,* 563–564
Hui, W. H. and Li, M. M. 1977, J. Chem. Soc., Perkin Trans. *1,* 297–304
Ikeda, T. and Kitao, K. 1972, Mokuzai Gakkaishi *18,* 323–324
Itokawa, H., Tachi, Y., Kamaho, Y. and Iitaka, Y. 1978, Chem. Pharm. Bull. *26,* 331–333
Jurd, L. 1962, The Hydrolyzable Tannins. In: Wood Extractives and Their Significance to the Pulp and Paper Industries (Hillis, W. E., Ed.). Academic Press, New York, London, pp. 229–260
Kai, Y., Kuroda, H. and Teratani, F. 1972, Mokuzai Gakkaishi *18,* 315–321
Kehr, E. and Schilling, W. 1965, Holztechnologie *6,* 225–232
Kimland, B. and Norin, T. 1972, Svensk Papperstid. *75,* 403–409
King, H. G. C., White, T. and Hughes, R. B. 1961, J. Chem. Soc. 3234–3239
Krahmer, R. L., Hemingway, R. W. and Hillis, W. E. 1970, Wood Sci. Technol. *4,* 122–139
Kratochvil, J. F., Burris, R. H., Seikel, M. K. and Harkin, J. M. 1971, Phytochemistry *10,* 2529–2531
Kubinsky, E. and Ifju, G. 1973, For. Prod. J. *23,* No. 2, 54–56
Kühne, H., Leukens, U., Sell, J. and Wälchli, O. 1970, Holz Roh-Werkst. *28,* 223–229
Lábky, O. 1974, Drev. Vysk. *19,* 125–135
Lange, W. 1976a, Holz Roh-Werkst. *34,* 23–30, 101–105
Lange, W. 1976b, Holz Roh-Werkst. *34,* 213–218
Li, H. H. and Hsiang, Y. M. 1963, Sci. Silvae (Peking) *8,* 263–266
Lindgren, B. O. 1967, Svensk Papperstid. *70,* 532–536
Lowry, J. B. 1968, Phytochemistry *7,* 1803–1813
Lüdtke, M. and Lerch, B. 1968, Holzforschung *22,* 1–8
Lüdtke, M. and Röpsch, A. 1971, Holzforschung *25,* 187–194
Lundquist, K. and Nelson, P. F. 1971, Holzforschung *25,* 117–119
Manell, D. and Pensar, G. 1975, Pap. Puu *57,* 117–128
Marini Bettòlo, G. B., Casinovi, C. G. and Galeffi, C. 1965, Tetrahedron Lett. 4857–4864
Mayer, W., Gabler, W., Riester, A. and Korger, H. 1967a, Liebigs Ann. Chem. *707,* 177–181
Mayer, W., Einwiller, A. and Jochims, J. C. 1967b, Liebigs Ann. Chem. *707,* 182–189
Mayer, W., Seitz, H. and Jochims, J. C. 1969, Liebigs Ann. Chem. *721,* 186–193
Mayer, W., Kuhlmann, F. and Schilling, G. 1971a, Liebigs Ann. Chem. *747,* 51–59
Mayer, W., Seitz, H., Jochims, J. C., Schauerte, K. and Schilling, G. 1971b, Liebigs Ann. Chem. *751,* 60–68
Mayer, W., Bilzer, W. and Schauerte, K. 1971c, Liebigs Ann. Chem. *754,* 149–152
Mayer, W., Bilzer, W. and Schilling, G. 1976, Liebigs Ann. Chem. 876–881
McMillin, C. W. 1969a, Wood Sci. Technol. *3,* 232–238
McMillin, C. W. 1969b, Wood Sci. *2,* 26–30
McMillin, C. W. 1970, Holzforschung *24,* 152–157
McNamara, W. S., Sullivan, C. E. and Higgins, J. C. 1970, Wood Sci. *3,* 48–51
Meyer, R. W. and Barton, G. M. 1971, For. Prod. J. *21,* No. 4, 58–60
Mia, A. J. 1969, Wood Sci. *2,* 120–124
Mirov, N. T., Zavarin, E. and Bicho, J. G. 1962, J. Pharm. Sci. *51,* 1131–1135
Mirov, N. T., Zavarin, E., Snajberk, K. and Costello, K. 1966, Phytochemistry *5,* 343–355
Morgan, J. W. W. and Orsler, R. J. 1968, Holzforschung *22,* 11–16
Nelson, N. D. 1978, Can. J. Bot. *56,* 626–634
Newman, A. A. 1972, Chemistry of Terpenes and Terpenoids. Academic Press, London, New York
Nisi, D. 1966, Cellul. Carta *17,* No. 6., 18–23
Ohtani, Y., Shutoh, K. and Sumimoto, M. 1982, Mokuzai Gakkaishi *28,* 59–66
Paasonen, P. K. 1967a, Pap. Puu *49,* 3–15
Paasonen, P. K. 1967b, Pap. Puu *49,* 503–508
Packman, D. F. 1960, Holzforschung *14,* 178–183

Pearl, I. A. and Darling, S. F. 165, Tappi *48,* 506–508, 607–608
Pearl, I. A. and Darling, S. F. 1971a, Can. J. Chem. *49,* 49–55
Pearl, I. A. and Darling, S. F. 1971b, Phytochemistry *10,* 483–484, 3161–3166
Pearl, I. A., Darling, S. F. and Heller, S. F. 1966, Tappi *49,* 278–280
Pizzi, A. 1980, J. Macromol. Sci.-Rev. Macromol. Chem. *C 18,* 247–315
Popoff, T. and Theander, O. 1976, Appl. Polym. Symp. No. 28, 1341–1347
Popper, R. 1975, SAH-Bull. *3*/1, 2–11
Puth, M. 1962, Holz Zentralbl. *88,* 743–744, 787–788
Rai, P. P. 1977, Curr. Sci. *46,* 814–815
Rai, P. P. 1978, Curr. Sci. *47,* 271–272
Repyakh, S. M. and Levin, E. D. 1977, Khim. Drev. No. 2, 102–103
Ribo, J. M. and Raventos, J. 1972, Phytochemistry *11,* 3089
Roffael, E. and Rauch, W. 1974, Holz Roh-Werkst. *32,* 182–187
Roux, D. G. 1972, Phytochemistry *11,* 1219–1230
Roux, D. G. and Ferreira, D. 1974, Phytochemistry *13,* 2039–2048
Roux, D. G. and Paulus, E. 1960, Biochem. J. *77,* 315–320
Roux, D. G. and Paulus, E. 1961, Biochem. J. *78,* 785–789, *80,* 62–63
Roux, D. G., Ferreira, D., Hundt, H. K. L. and Malan, E. 1975, Appl. Polym. Symp. No. 28, 335–353
Roux, D. G., Ferreira, D., Botha, J. J. and Garbutt, D. C. F. 1976, Appl. Polym. Symp. No. 28, 1365–1376
Rowe, J. W., Seikel, M. K., Roy, D. N. and Jorgensen, E. 1972, Phytochemistry *11,* 2513–2517
von Rudloff, E. 1965, Chem. Ind. 180–181
Saayman, H. M. and Roux, D. G. 1965, Biochem. J. *97,* 794–801
Saeki, I., Sumimoto, M. and Kondo, T. 1973, Holzforschung *27,* 93–96
Sandermann, W. 1960, Naturharze, Terpentinöl, Tallöl, Chemie und Technologie. Springer Verlag, Berlin, Göttingen, Heidelberg
Sandermann, W. 1966, Naturwissenschaften *53,* 513–525
Sandermann, W. and Barghoorn, A. W. 1955, Holzforschung *9,* 112–117
Sandermann, W. and Funke, H. 1970, Naturwissenschaften *57,* 407–414
Sandermann, W. and Lange, W. 1967, Holzforschung *21,* 154–157
Sandermann, W. and Puth, M. 1965, Farbe Lack *71,* 13–25
Sandermann, W. and Rothkamm, M. 1959, Holz Roh-Werkst. *17,* 433–441
Sandermann, W. and Simatupang, M. H. 1966, Holz Roh-Werkst. *24,* 190–204
Sandermann, W., Dietrichs, H. H., Simatupang, M. H. and Puth, M. 1963, Holzforschung *17,* 161–168
Santamour, F. S. 1977, J. Arboricult. *3,* 14–18
Scurfield, G. and Nicholls, P. W. 1970, J. Exp. Bot. *21,* 857–886
Scurfield, G., Michell, A. J. and Silva, S. R. 1973, Bot. J. Linnean Soc. *66,* 277–289
Scurfield, G., Anderson, C. A. and Segnit, E. R. 1974, Aust. J. Bot. *22,* 211–229
Seikel, M. K. and Hillis, W. E. 1970, Phytochemistry *9,* 1115–1128
Seikel, M. K., Hall, S. S., Feldman, L. C. and Koeppen, R. C. 1965, Am. J. Bot. *52,* 1046–1049
Seikel, M. K., Hosteller, F. D. and Niemann, G. J. 1971, Phytochemistry *10,* 2249–2251
Selleby, L. 1960, Svensk Papperstid. *63,* 81–85
Shain, L. and Hillis, W. E. 1973, Can. J. Bot. *51,* 1331–1335
de Silva, D. and Hillis, W. E. 1980, Holzforschung *34,* 95–97
Snajberk, K. and Zavarin, E. 1976, Biochem. Syst. Ecol. *4,* 159–163
Snajberk, K., Zavarin, E. and Bailey, D. 1979, Biochem. Syst. Ecol. *7,* 269–279
Spalding, B. P., Zinkel, D. F. and Roberts, D. R. 1971, Phytochemistry *10,* 3289–3292
Spencer, R. W. and Choong, E. T. 1977, Holzforschung *31,* 25–31
Stamm, A. J. 1961, For. Prod. J. *11,* 310–312

Su, Z. A., Zhai, Q. H., Liang, Z. Q. and Guo, C. T. 1981, Chem. Ind. For. Prod. (China) *1,* 1–11
Suzuki, H., Yasuda, S. and Hanzawa, M. 1971, Mokuzai Gakkaishi *17,* 221–222
Swan, E. P. 1965, For. Prod. J. *15,* 272
Swan, B. 1968, Svensk Papperstid. *71,* 436–440
Swan, B. and Åkerblom, I. S. 1967, Svensk Papperstid. *70,* 239–244
Swan, E. P. and Jiang, K. S. 1970, Tappi *53,* 844–846
Swan, E. P., Jiang, K. S. and Gardner, J. A. F. 1969, Phytochemistry *8,* 345–351
Tachibana, S. and Sumimoto, M. 1980, Holzforschung *34,* 131–137
Tachibana, S. and Sumimoto, M. 1982, Mokuzai Gakkaishi *28,* 45–58
Takahashi, K. 1981, Mokuzai Gakkaishi *27,* 654–657
Takehara, T. and Sasaya, T. 1979, Mokuzai Gakkaishi *25,* 516–517, 660–664
Theander, O. 1978, Tappi *61,* No. 4, 69–72
Theander, O. 1981, Hydrophilic Extractives from Scots Pine and Norway Spruce. In: The Ekman Days 1981, Int. Symp. Wood Pulp. Chem., Stockholm, Vol. 1, pp. 89–93
Thompson, R. S., Jacques, D., Haslam, E. and Tailor, R. J. 1972, J. Chem. Soc., Perkin Trans. *1,* 1387
Tomita, B. and Hirose, Y. 1970, Tetrahedron Lett. 235–238
Traitler, H. and Kratzl, K. 1980a, J. Am. Oil Chem. Soc. *57,* 153–156
Traitler, H. and Kratzl, K. 1980b, Wood Sci. Technol. *14,* 101–106
Traitler, H. and Kratzl, K. 1980c, Wood Sci. Technol. *14,* 9–20
Tsoumis, G. 1965, Tappi *48,* 451–455
Tsukamoto, T., Kawasaki, T., Naraki, A. and Yamauchi, T. 1955, Chem. Abstr. *49,* 2032
Volz, K. R. 1971, Holz Zentralbl. *97,* 1783
Ważny, H. and Ważny, J. 1964, Holz Roh-Werkst. *22,* 299–304
Weinges, K. 1960, Tetrahedron Lett. 1–2
Weissmann, G. 1973, Holz Roh-Werkst. *31,* 341–343
Weissmann, G. 1974, Holzforschung *28,* 186–188
Weissmann, G. 1976, Holz Roh-Werkst. *34,* 171–174
Weissmann, G. and Dietrichs, H. H. 1975a, Holz Roh-Werkst. *33,* 54–56
Weissmann, G. and Dietrichs, H. H. 1975b, Holzforschung *29,* 68–71
Weissmann, G. and Vorher, W. 1975, Holz Roh-Werkst. *33,* 21–25
Yoshida, E. 1961, J. Fac. Agric. Iwate Univ. *5,* 87–95
Young, H. E. and Guinn, V. P. 1966, Tappi *49,* 190–197
Zavarin, E. and Snajberk, K. 1965a, Phytochemistry *4,* 141–148
Zavarin, E. and Snajberk, K. 1965b, Tappi *48,* 612–616
Zavarin, E. and Snajberk, K. 1980, Agric. Food Chem. *28,* 829–834
Zavarin, E., Smith, R. M. and Anderson, A. B. 1959, J. Org. Chem. *24,* 1318–1321
Zavarin, E., Snajberk, K. and Bailey, D. 1976, Biochem. Syst. Ecol. *4,* 81–92
Zavarin, E., Wong, Y. and Zinkel, D. F. 1978, Lightwood Res. Coord. Council, Proc. Ann. Meeting, pp. 19–30
Zeikus, J. G. and Ward, J. C. 1974, Science *184,* 1181–1183
Zichermann, J. B. and Thomas, R. J. 1971, Tappi *54,* 1727–1730
Zinkel, D. F. 1975, Appl. Polym. Symp. No. 28, 309–327
Zinkel, D. F. and Conner, A. H. 1973, Phytochemistry *12,* 938–939
Zinkel, D. F. and Foster, D. O. 1980, Tappi *63,* No. 5, 137–139
Zinkel, D. F. and McKibben, C. R. 1978, Lightwood Res. Coord. Council, Proc. Ann. Meeting, pp. 133–156
Zinkel, D. F. and Spalding, B. P. 1971, Tetrahedron Lett. 2459–2462
Zinkel, D. F., Toda, J. K. and Rowe, J. W. 1971, Phytochemistry *10,* 1161–1163

8. Distribution of the Components within the Wood Cell Wall

8.1 Estimation and Calculation

Wood is heterogeneous not only in its structure but also in the distribution of its components within the cell walls. As shown in the chapter on the ultrastructure of wood cells (→ 2.2.) cellulose determines the structure of the wall layers. From the package of the cellulosic fibrils a difference in the concentration of cellulose in the compound middle lamella and the secondary walls can be seen (Fig. 2–13).

After a careful treatment of wood sections with acids (H_2SO_4, HCl, HF) the lignin skeleton of the cell wall remains, which can be cut to ultrathin sections (Jayme, Fengel 1961; Sachs et al. 1963; Côté et al. 1968a). In the electron microscope the sections show a high lignin concentration in the compound middle lamella, and an approximately equal distribution in the secondary walls (Fig. 8–1). Additionally the orientation of the lignin particles according to the run of the previously present cellulose fibrils is visible. A form-dichroism of lignin in ultraviolet light caused by the cellulose texture was observed in sections of spruce wood and jute fibres (Frey 1959; Ruch, Hengartner 1960).

Because of the small dimensions of the wall layers (Tables 2–3, 2–4) a chemical analysis of parts of the cell walls is very difficult if not impossible. Therefore in most cases indirect methods have been used for the quantitative or half-quantitative determination of the components of the cell wall layers.

Early studies were made by Lange (1954a, b; Asunmaa, Lange 1954) using micro-spectroscopic measurements of thin wood sections. The UV absorption of lignin,

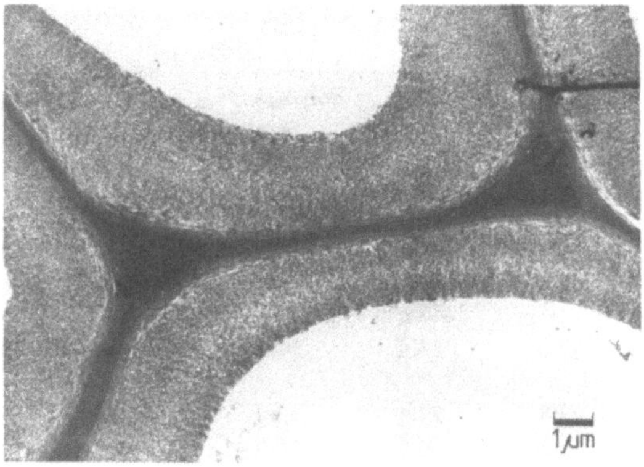

Fig. 8–1. Lignin skeleton of tracheid walls of spruce *(Picea abies)* produced by acid hydrolysis of the polysaccharides. TEM micrograph.

the absorption in the visible range of esterified polysaccharides, and the X-ray absorption of the cell wall substance gave evidence that the proportion of lignin is about 85% of the compound middle lamella and about 12 to 20% of the lumen-side part of the cell wall. The polyoses take about one half of the polysaccharides in the outer part and about 10 to 20% of the lumen-side part of the cell wall.

A special staining method made it possible to estimate the distribution of polyoses in spruce tracheids (Hoffmann, Parameswaran 1976; Parameswaran, Liese 1981). The highest concentration of polyoses was in the S1, but the content in the S1–S2 border region and in the outermost part of the S2 was higher than in the middle of the S1. The lower polyose content of the middle part of S2 increases slightly toward the border of the tertiary wall.

The findings of Lange (1954b), and the results of wood analysis had previously served as the basis for a calculation of the distribution of cellulose, polyoses and lignin in the cell walls of earlywood and latewood of spruce (Fengel 1969, 1970a) (Table 8–1). According to the calculated data nearly 60% of the substance of the compound middle lamella consists of lignin, the content of cellulose being only about 14%. About 60% of the secondary wall 2 is cellulose; about 27% is lignin, but the bulk of all three components is concentrated in the secondary wall 2 as this layer is the thickest one. Its volume percentage is about 73% in earlywood and 82% in latewood. Thus the higher proportion of cellulose in latewood is determined by the larger thickness of the S2. On the other hand the higher proportion of the compound middle lamella in earlywood is the reason for the higher percentage of lignin in this part of an annual ring.

A similar calculation was made by Cave (1976). He proceeded from an equal volume ratio of polyoses to cellulose of 1:2 throughout all wall layers, and an equal lignin content of 22 vol.% in the S1, S2 and T.

Various studies show that the extractives are not only present in the resin canals and parenchyma cells but also deposited within the fibre and tracheid cell walls. Micro-

Table 8–1: Calculated distribution of the components in the cell wall layers of spruce tracheids (Fengel 1969, 1970a)

Region	Wall layer	Cellulose		Polyoses		Lignin	
		% of the wall layer	% of total cellulose	% of the wall layer	% of total polyoses	% of the wall layer	% of total lignin
Earlywood	Compound middle lamella	13.9	4.1	27.1	20.6	59.0	26.8
	Secondary wall 1	36.4	8.9	36.4	23.2	27.2	10.4
	Secondary wall 2 + tertiary wall	58.5	87.0	14.4	56.1	27.1	62.8
Latewood	Compound middle lamella	13.7	2.5	27.4	15.0	58.9	18.4
	Secondary wall 1	34.6	5.2	34.6	15.6	30.8	7.9
	Secondary wall 2 + tertiary wall	58.4	92.3	14.5	69.4	27.1	73.7

spectroscopic methods were used to detect polyphenols in the cell walls of *Eucalyptus* heartwood and extractives in larchwood (Bland, Hillis 1969; Imagawa, Fukazawa 1978). The deposition of extractives within the secondary walls of tamrit cypress tracheids *(Cupressus dupreziana)* could be visualized in the electron microscope (Grosser et al. 1974). Substantial amounts of extractives were also found to be located in the cell walls of redwood *(Sequoia sempervirens)* and incense cedar *(Libocedrus decurrens)* heartwood (Kuo, Arganbright 1980).

8.2. Determination of the Lignin Distribution

The distribution of lignin within the cell walls can be made visible by ultraviolet microscopy because of its absorption at 280 nm (→ 6.4.2.). Attempts at a quantitative determination of lignin content in the cell wall layers were made by Lange (1954b), Frey (1959), and Ruch and Hengartner (1960). A successful method of quantitative evaluation of the lignin content using ultrathin sections was developed

Table 8–2: Distribution of lignin in softwood tracheids (Fergus et al. 1969; Wood, Goring 1971)

Wood	Wall layers of tracheids	Tissue volume % of total	Lignin % of total	Lignin concentration g/g
Picea mariana				
Earlywood	Compound middle lamella r, t	8.7	15.8	0.497
	Compound middle lamella cc	3.9	12.1	0.848
	Secondary walls	87.4	72.1	0.225
Latewood	Compound middle lamella r, t	4.1	9.7	0.60
	Compound middle lamella cc	2.2	8.6	1.00
	Secondary walls	93.7	81.7	0.222
Pseudotsuga menziesii				
Earlywood	Compound middle lamella r, t	10.1	17.9	0.56
	Compound middle lamella cc	4.1	10.7	0.83
	Secondary walls	74.4	58.3	0.248
Latewood	Compound middle lamella r, t	4.6	10.3	(0.60)
	Compound middle lamella cc	1.9	6.2	(0.88)
	Secondary walls	89.7	77.9	0.228

r = radial, t = tangential, cc = cell corner

Table 8–3: Distribution of lignin in the cells of birch wood *(Betula papyrifera)* (Fergus, Goring 1970)

Cell element	Wall layers	Type of lignin	Tissue volume % of total	Lignin % of total	Lignin concentration g/g
Fibre	Compound middle lamella r, t	Syringyl-guaiacyl 1:1	5.2	8.9	0.34–0.40
	Compound middle lamella cc	Syringyl-guaiacyl 1:1	2.4	8.8	0.72–0.85
	Secondary walls	Syringyl	73.4	59.9	0.16–0.19
Vessel	Compound middle lamella	Guaiacyl	0.8	1.5	0.35–0.42
	Secondary walls	Guaiacyl	8.2	9.4	0.22–0.27
Ray cells	Secondary walls	Syringyl	10.0	11.4	0.22–0.27

r = radial, t = tangential, cc= cell corner
The variation in lignin concentration is caused by a calculation using different lignin contents of xylem (0.199 g/g and 0.231 g/g)

by Goring and co-workers (Scott et al. 1969). The micrographs were taken with monochromatic ultraviolet light and subsequently measured with a microdensitometer.

By means of this method the lignin content of the compound middle lamella and the secondary walls was determined separately in tracheids of black spruce *(Picea mariana)* and Douglas fir *(Pseudotsuga menziesii)* as well as in fibres, vessels and ray cells of white birch *(Betula papyrifera)* (Fergus et al. 1969; Fergus, Goring 1970a; Wood, Goring 1971) (Tables 8–2, 8–3). The lignin values determined for spruce earlywood and latewood coincide well with the values calculated by Fengel (1970a) (Table 8–1). The densitometer curves show the lignin distribution within the cell walls as well as differences in the lignin distribution of various cell types (Fig. 8–2). Similar studies with woods of the genera *Pinus, Picea, Cryptomeria, Fagus* and *Abies* were made by Fukuda and Haraguchi (1971), Evdokimov (1974), and Fukazawa and Imagawa (1981).

A systematic study of several softwood species showed that a high UV-absorption at the lumen border of the cell walls is much more common in the branchwood than in the stemwood (Scott, Goring 1970). Thus it can be concluded that in general the stemwood tracheids have a rather uniform lignin distribution across their secondary and tertiary walls.

A study of compression wood by this method shows a high lignin concentration in the middle region of secondary walls, which is in the same range as the lignin concentration of the compound middle lamella (Wood, Goring 1971; Fukuzawa 1974). From lignin skeletons obtained by hydrolysis with HF, Parham and Côté (1971) concluded that the highest lignin concentration is found in the outer region of the S2 of compression wood tracheids.

The UV-micrographs make it possible to determine not only the lignin content in the cell wall layers but also the type of lignin. Because of the different absorption

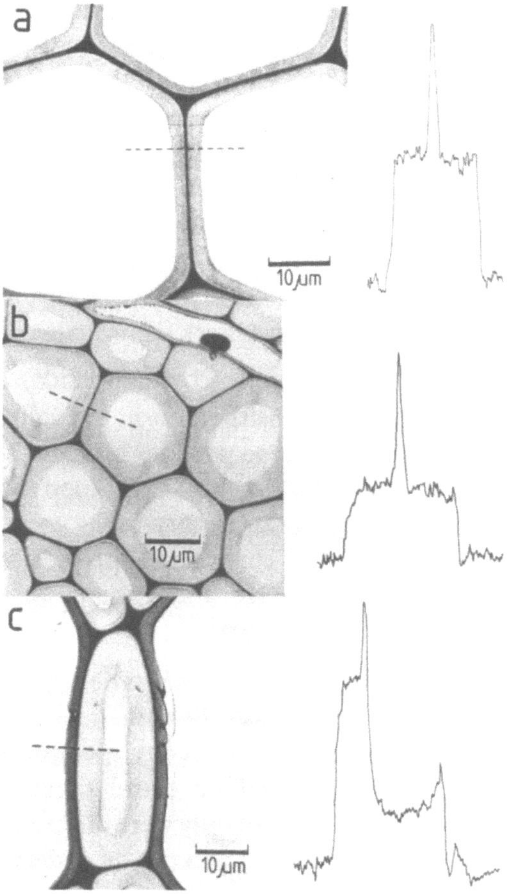

Fig. 8–2. Ultraviolet micrographs of ultrathin sections with the microdensitometer curves of the lignin distribution in the cell walls obtained by measuring along the dotted lines (by courtesy of D. A. I. Goring).

a) Spruce tracheids *(Picea mariana);*

b) Birch fibres *(Betula papyrifera);*

c) Birch vessels (left and right hand side) and fibre (centre) *(Betula papyrifera).*

maxima of guaiacyl (280 nm) and syringyl compounds (270 nm) the presence of guaiacyl and syringyl lignin in various cells and wall layers can be detected (Fergus, Goring 1970b; Musha, Goring 1975a; Gromov et al. 1977) (Table 8–3). Studying various fibre fractions from high-yield pulp of birch *(Betula ermanii)* Cho et al. (1980) found that the lignin of the compound middle lamella contains both guaiacyl and syringyl units, the lignin of the secondary wall layers consisting predominantly of syringyl units.

Musha and Goring (1975b) showed that syringyl content increases with increasing wall thickness of fibres and vessels in hardwoods. They suggested that this trend is related to the availability of oxygen during the lignification process.

Variations in the lignin composition of the wall layers were also found by Hardell et al. (1980). They obtained fractions containing different wall layers by decantation of a suspension of desintegrated wood in water. The analysis of these fractions gave evidence of a high portion of syringyl units in the secondary walls of birchwood fibres and ray cells *(Betula papyrifera),* while the lignin of the compound middle lamella and the vessel walls are rich in guaiacyl units. In spruce tracheids *(Picea abies)* the lignin of the secondary walls contains twice as many phenolic OH-groups as the lignin in the compound middle lamella. The latter coincides with the results of Yang and Goring (1978) who found more phenolic hydroxyl groups in the secondary wall lignin than in the middle lamella lignin of black spruce *(Picea mariana)* and Douglas fir *(Pseudotsuga menziesii)* by UV spectroscopic methods.

Another method for quantification of the lignin distribution in the cell walls was applied by Boutelje (1972). He calculated the volume portions of lignin in the wall layers from the refractive indices. In the secondary walls of earlywood tracheids of spruce he found a slightly higher portion of lignin (20%) than in the corresponding wall parts of latewood tracheids (18%). In the cell corners the volume portion was very high: 84% for earlywood and latewood of spruce *(Picea abies),* 73% for earlywood and 87% for latewood of pine *(Pinus sylvestris).*

A relative method using energy dispersive analysis of X-rays was developed by Saka et al. (1978). Wood and pulp fibres were treated with bromine and the bromine distribution in the fibre walls was determined with an X-ray spectrometer. The bromine reacts specifically with lignin, and there is a direct correlation between the amount of reacted bromine and the lignin content. Not only were differences shown in the lignin content of the compound middle lamella and the secondary walls but also slight variations of the lignin content within the secondary walls.

8.3. Determination of the Polysaccharide Distribution

The determination of the polysaccharide distribution is more difficult than that of the lignin distribution. Several previous studies are concerned with the distribution of pentosans. Thus Shimizu and Yamada (1959) determined the pentosans in isolated fibres from softwoods and hardwoods by partial degradation. They found 50–60% pentosans at the surface and 2–4% at the lumen border of softwood tracheids; the respective data for hardwood fibres were nearly 100% and 8–10%.

The problem of subdividing cells into their wall layers was solved by using radial sections of developing tissue from which cells of different maturation were separated with the aid of a micromanipulator (Meier, Wilkie 1959; Meier 1961). By means of this procedure four fibre fractions were obtained: the first fraction consisting of middle lamellae and primary walls (M+P), the second containing the

Table 8–4: Polysaccharide composition of the cell wall layers of fibres and tracheids (Meier 1961; Côté et al. 1968)

Polysaccharide	M+P*	S1	S2 outer part	S2 inner part +T
Betula verrucosa				
Galactan	16.9	1.2	0.7	0.0
Cellulose	41.4	49.8	48.0	60.0
Glucomannan	3.1	2.8	2.1	5.1
Arabinan	13.4	1.0	1.5	0.0
4-O-Methyl-glucuronoxylan	25.2	44.1	47.7	35.1
Picea abies				
Galactan	16.4	8.0	0.0	0.0
Cellulose	33.4	55.2	64.3	63.6
Glucomannan	7.9	18.1	24.4	23.7
Arabinan	29.3	1.1	0.8	0.0
Arabino-4-O-methyl-glucuronoxylan	13.0	17.6	10.7	12.7
Pinus sylvestris				
Galactan	20.1	5.2	1.6	3.2
Cellulose	35.5	61.5	66.5	47.5
Glucomannan	7.7	16.9	24.6	27.2
Arabinan	29.4	0.6	0.0	2.4
Arabino-4-O-methyl-glucuronoxylan	7.3	15.7	7.4	19.4
Abies balsamea Normal wood				
Galactan	17.0	14.8	19.9	
Cellulose	40.8	53.5	48.0	
Galactoglucomannan	7.4	22.7	23.7	
Arabinan	20.7	1.1	0.6	
Arabino-4-0-methyl-glucuronoxylan	14.1	7.9	7.8	
Abies balsamea Compression wood**				
Galactan	19.6	26.0	25.0	8.9
Cellulose	34.5	45.2	47.7	61.6
Galactoglucomannan	10.9	17.9	19.2	19.3
Arabinan	23.3	0.4	0.6	1.9
Arabino-4-O-methyl-glucuronoxylan	11.7	10.5	7.5	8.4

* Contains also a high percentage of galacturonan

** Has no tertiary wall (T)

same plus secondary walls 1 (M+P+S1), the third containing additionally the outer part of secondary walls 2 (M+P+S1+$S2_{part}$), and, finally, the complete cell walls (M+P+S1+S2+T). After acid hydrolysis of the fractions their sugar compositions were quantitatively determined, and the distribution of the polysaccharides in the wall layers was calculated (Table 8–4). Côté et al. (1968b) used this method to determine the polysaccharide composition in the wall layers of normal and compression wood of balsam fir *(Abies balsamea)*. As it was unknown at that time Meier (1961) did not consider that some of the galactose residues in softwood are part of the galactoglucomannan. The percentages of the single polysaccharides obtained by this method must be viewed as approximate. The sum of the percentages in the wall layers does not coincide with the percentages of cellulose and polyoses known from wood analysis. The reason for this discrepancy is a change of the polyose proportion during maturation of the cells. By incorporating $^{14}CO_2$ in the developing walls of pine tracheids *(Pinus resinosa)* an increase of xylose and a decrease of arabinose and galactose residues with cell maturity was observed (Larson 1969). The percentages of mannose residues remained relatively constant, as did the cellulose-polyoses ratio of 60:40 within the secondary wall.

Pursuing the change in the composition of cotton fibre cell walls during the development Huwyler et al. (1979) found a maximum amount of fucose, galactose, mannose, rhamnose, arabinose, uronic acid and non-cellulosic glucose residues at the end of the primary wall formation or at the beginning of the secondary wall formation. Only the absolute amounts of xylose and cellulosic glucose residues increased until the end of the fibre development.

8.4. Models of the Internal Cell Wall Structure

Studies using chemical and physical methods result in a different composition of the cell wall layers and a different distribution of cellulose, polyoses and lignin in the cell wall. But even in the submicroscopic and supramolecular range there are variations in the distribution of the components. In several studies a lamellation of lignin concentrically to the lumen was observed in the secondary wall by electron

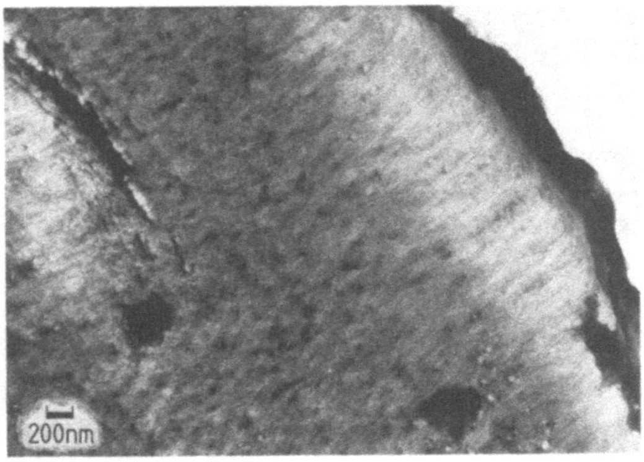

Fig. 8–3. Ultrathin cryo-section of a tracheid wall of spruce *(Picea abies)* in swollen state after delignification and alkali extraction. Places of previously high polyose concentration are more intensively stained. TEM micrograph.

microscopy (Hepler et al. 1970; Stone et al. 1971; Ruel et al. 1978). The microdensitometric measuring resulted in a mean distance of the lignin lamellae of 7–8.5 nm determined for softwoods *(Picea mariana, Abies alba)*, and about 10 nm for hardwoods *(Betula papyrifera, Fagus sylvatica)* and a reed *(Arundo donax)* (Ruel et al. 1978, 1979).

Polarized IR-spectra and birefringence give evidence that the polyoses are arranged parallel to the cellulose fibrils in the cell wall (Liang et al. 1960; Page et al. 1976). Electron microscopic observations also make the longitudinal orientation of xylan molecules in the parenchyma cells of birch *(Betula verrucosa)* and of *Arundo donax* very probable (Meier, Welck 1965; Ruel et al. 1976). From the shrinkage behaviour of cell walls after removal of the polyoses Kerr and Goring (1975, 1977) concluded that a lamellar arrangement exists for the polyoses.

If the swelling state of the fibre walls is maintained after the removal of polyoses by alkali extraction the places of previously high polyose concentration are visible in the form of intensively stained spots approximately in lamellar orientation (Fengel 1980) (Fig. 8–3).

On the other hand parts of polyoses resist oxidative degradation under mild conditions, which is explained by an intimate interpenetration of the polysaccharides within the relatively ordered regions of the native plant cell wall (Moore et al. 1966). The change of pore size distribution in holocellulose after a careful removal of polyoses also gives evidence of a penetration of the lignin network by polyoses (Sawabe 1980).

After an intensive fibrillation of wood meal relatively thick fibrils are visible in the electron microscope in addition to more or less amorphous looking longitudinal particles (Fengel 1978) (Fig. 8–4). These fibrils are probably units consisting of a cellulosic core enclosed by polyoses. As a linkage between polyoses and lignin is very probable, and the polyoses are orientated essentially parallel to cellulose mo-

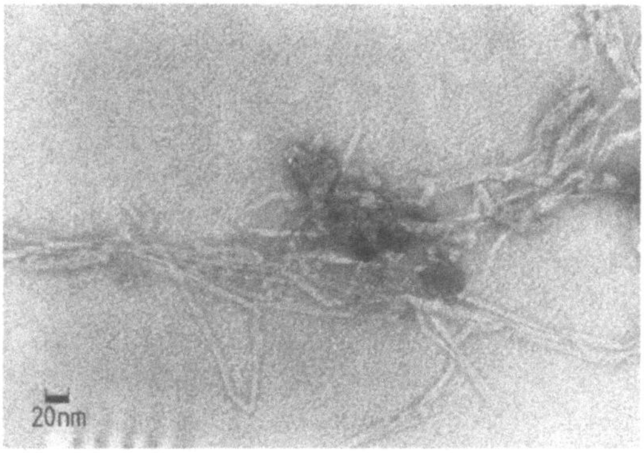

Fig. 8–4. Fibrillar and coiled structure units from intensively disintegrated wood meal of spruce *(Picea abies)*. TEM micrograph.

lecules, polyoses should be considered as a coupling agent between cellulose and lignin in the supramolecular range (Page 1976).

The various findings regarding the distribution of cellulose, polyoses and lignin within the cell wall layers provide the basis for the development of models on the supramolecular structure of the wall components. Models considering only the association of the polysaccharides include those Preston (1962) and Marchessault (1964). In Preston's model (1962) the cellulose fibrils are enclosed by less ordered cellulose molecules and polyose molecules (Fig. 8–5a). Additionally polyose molecules are incorporated in the fibrillar structure of cellulose. Marchessault (1964) arranged the polyose molecules as slack sheaves between the cellulose fibrils (Fig. 8–5b).

A model which also takes into account the presence of lignin was proposed by Fengel (1970b) (Fig. 8–5d). It explains the existence of various fibrillar units of cellulose by layers of polyoses differing in thickness. The smallest units (3 nm in diameter) are separated from each other by monomolecular layers of polyoses, and the largest units (25 nm in diameter) are enclosed by polyoses and lignin.

From their observation of a lamellar arrangement of the polyoses Kerr and Goring (1975) developed a model consisting of layers of cellulose-polyoses blocks interrupted in the radial and tangential direction by lignin-polyoses blocks (Fig. 8–5c). The realization of such a structure can be explained by Scallan's model (1974) which starts from an internal fibrillation of cellulose during swelling (Fig. 8–6). The in-

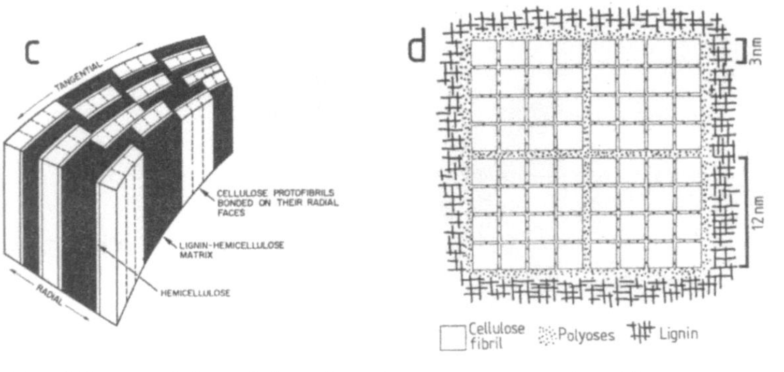

Fig. 8–5. Models of the association of the cell wall components developed by various authors. a) Preston (1962); b) Marchessault (1964); c) Kerr and Goring (1975); d) Fengel (1970).

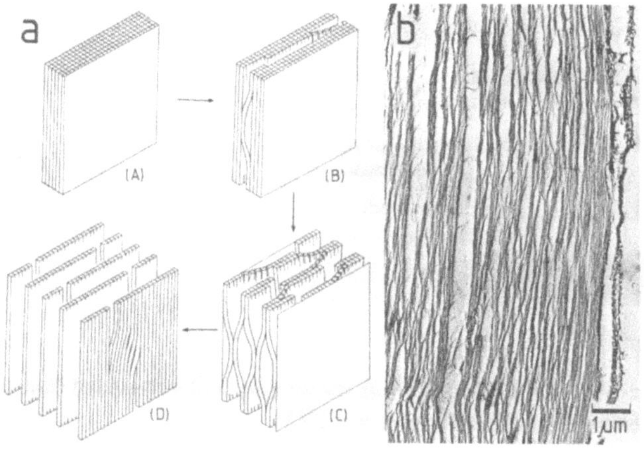

Fig. 8–6. Internal fibrillation of cellulose during swelling.
a) Model proposed by Scallan (1974);
b) Honeycomb structure in the wall of a spruce sulphite fibre according to (C).

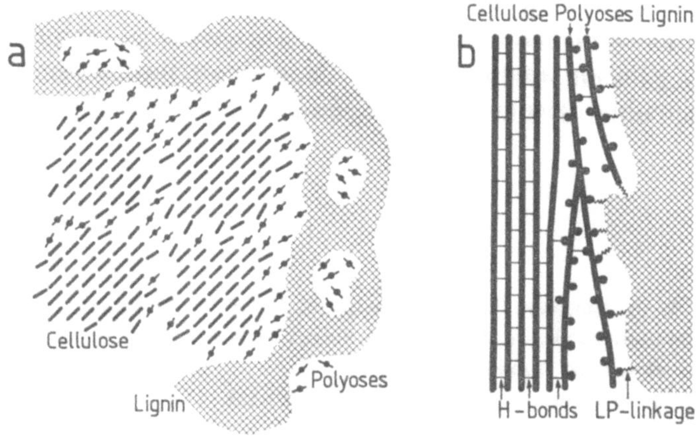

Fig. 8–7. Detailed model of the association of cellulose, polyoses and lignin in the wood cell wall with reference to the presence of less ordered cellulose molecules and the linkages at the interfaces in transverse (a) and longitudinal view (b).

corporation of lignin in a highly swollen cell wall of that type would result in the deposition of lignin in disk-shaped, tangentially orientated platelets.

Both Fengel's model (1970b) and that of Kerr and Goring (1975) consider the intimate association of cellulose and polyoses, and of lignin and polyoses at the interfaces. In the latter model polyoses are additionally incorporated in the lignin matrix.

A more detailed model which also treats the intermolecular linkages at the interfaces is presented in Fig. 8–7. It shows a separation of the fibrillar units not only by polyose molecules but also by less ordered cellulose molecules. The diameter of the

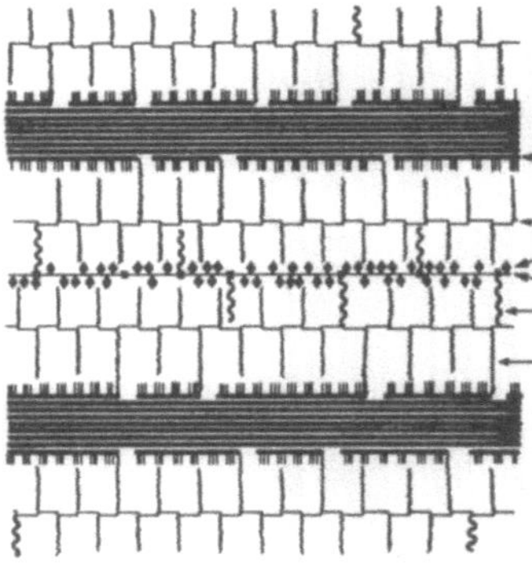

Fig. 8–8. Model of the strict order of the components in the primary wall as revealed from cell cultures of sycamore *(Acer pseudoplatanus)* by Albersheim et al. (1973).

fibrils varies within certain limits (→ 4.3.4.). Cords of polyose molecules are additionally deposited within the lignin layer.

A very strict order of the components in the primary wall can be seen in the model by Albersheim et al. (1973; Keegstra et al. 1973). From their studies on cell cultures of sycamore *(Acer pseudoplatanus)* they deduced a sequence of orientated molecules of xyloglucan, galactan, arabinogalactan, rhamnogalacturonan and glycoprotein linked with each other by covalent and hydrogen bonds between the cellulose fibrils (Fig. 8–8).

Literature

Albersheim, P., Bauer, W. D., Keegstra, K. and Talmadge, K. W. 1973, The Structure of the Wall of Suspension-cultured Sycamore Cells. In: Biogenesis of Plant Cell Wall Polysaccharides (Loewus, F., Ed.). Academic Press, New York, London, pp. 117–147

Asunmaa, S. and Lange, P. W. 1954, Svensk Papperstid. *57,* 501–516

Bland, D. E. and Hillis, W. E. 1969, Appita *23,* 204–210

Boutelje, B. 1972, Svensk Papperstid. *75,* 683–686

Cave, L. D. 1976, Wood Sci. Technol. *10,* 19–28

Cho, N. S., Lee, J. Y., Meshitsuka, G. and Nakano, J. 1980, Mokuzai Gakkaishi *26,* 527–533

Côté, W. A., Day, A. C. and Timell, T. E. 1968a, Wood Sci. Technol. *2,* 13–37

Côté, W. A., Kutscha, N. P., Simson, B. W. and Timell, T. E. 1968b, Tappi *51,* 33–40

Evdokimov, A. M. 1974, Khim. Drev. No. 1, 26–30

Fengel, D. 1969, Wood Sci. Technol. *3,* 203–217

Fengel, D. 1970a, Wood Sci. Technol. *4,* 15–35

Fengel, D. 1970b, Tappi *53,* 497–503

Fengel, D. 1978, Unpublished results

Fengel, D. 1980, Papier *34,* 428–432

Fergus, B. J. and Goring, D. A. I. 1970a, Holzforschung *24,* 118–124

Fergus, B. J. and Goring, D. A. I. 1970b, Holzforschung *24,* 113–117

Fergus, B. J., Procter, A. R., Scott, J. A. N. and Goring, D. A. I. 1969, Wood Sci. Technol. *3,* 117–138

Frey, H. P. 1959, Holz Roh-Werkst. *17,* 313–318

Fukazawa, K. 1974, Res. Bull. Coll. Exp. For., Hokkaido Univ. *31,* 87–114

Fukazawa, K. and Imagawa, H. 1981, Wood Sci. Technol. *15,* 45–55
Fukuda, K. and Haraguchi, T. 1971, Bull. Exp. For., Tokyo Univ. Agr. Technol. No. 9, 25–29
Gromov, V. S., Evdokimov, A. M. Abramovich, T. L. and Khrol, Y. S. 1977, Khim. Drev. No. 6, 73–79
Grosser, D., Fengel, D. and Schmidt, H. 1974, Forstwiss. Centralbl. *93,* 191–207
Hardell, H. L., Leary, G. J., Stoll, M. and Westermark, U. 1980, Svensk Papperstid. *83,* 44–49, 71–74
Hepler, P. K., Fosket, D. F. and Newcombe, E. H. 1970, Am. J. Bot. *57,* 85–96
Hoffmann, P. and Parameswaran, N. 1976, Holzforschung *30,* 62–70
Huwyler, H. R., Franz, G. and Meier, H. 1979, Planta *146,* 635–642
Imagawa, H. and Fukazawa, K. 1978, Mokuzai Gakkaishi *24,* 583–586
Jayme, G. and Fengel, D. 1961, Holzforschung *15,* 97–102
Keegstra, K., Talmadge, K. W., Bauer, W. D. and Albersheim, P. 1973, Plant Physiol. *51,* 188–196
Kerr, A. J. and Goring, D. A. I. 1975, Cell. Chem. Technol. *9,* 563–573
Kerr, A. J. and Goring, D. A. I. 1977, Wood Sci. *9,* 136–139
Kuo, M. L. and Arganbright, D. G. 1980, Holzforschung *34,* 17–22
Lange, P. W. 1954a, Svensk Papperstid. *57,* 533–537
Lange, P. W. 1954b, Svensk Papperstid. *57,* 563–567
Larson, P. R. 1969, Holzforschung *23,* 17–26
Liang, C. Y., Bassett, K. H., McGinnes, E. A. and Marchessault, R. H. 1960, Tappi *43,* 1017–1024
Marchessault, R. H. 1964, Texture and Morphology of Xylan. In: Chimie et Biochimie de la Lignine, de la Cellulose et des Hémicelluloses. Actes du Symposium International de Grenoble. Les Imprimeries réunies, Chambéry, pp. 287–301
Meier, H. 1961, J. Polym. Sci. *51,* 11–18
Meier, H. and Welck, A. 1965, Svensk Papperstid. *68,* 878–881
Meier, H. and Wilkie, K. C. B. 1959, Holzforschung *13,* 177–182
Moore, W. E., Effland, M., Sinha, B., Burdick, M. P. and Schuerch, C. 1966, Tappi *49,* 206–209
Musha, Y. and Goring, D. A. I. 1975a, Wood Sci. Technol. *9,* 45–58
Musha, Y. and Goring, D. A. I. 1975b, Can. J. For. Res. *5,* 259–268
Page, D. H. 1976, Wood Fiber *7,* 246–248
Page, D. H., El-Hosseiny, F., Bidmade, M. L. and Binet, R. 1976, Appl. Polym. Symp. No. 28, 923–929
Parameswaran, N. and Liese, W. 1981, Ultrastructural Localization of Wall Components in the Wood Cells. In: The Eckman Days 1981, Int. Symp. Wood Pulp. Chem., Stockholm, Vol. 1, pp. 16–31
Parham, R. A. and Côté, W. A. 1971, Wood Sci. Technol. *5,* 49–62
Preston, R. D. 1962, Symp. Int. Soc. Cell. Biol. *1,* 325, acc. to Preston, R. D. 1974, The Physical Biology of Plant Cell Walls. Chapman and Hall, London, pp. 169–170
Ruch, F. and Hengartner, H. 1960, Beih. Schweiz. Forstverein No. 30, 75–90
Ruel, K., Joseleau, J. P., Comtat, J. and Barnoud, F. 1976 Appl. Polym. Symp. No 28, 971–981
Ruel, K., Barnoud, F. and Goring, D. A. I. 1978, Wood Sci. Technol. *12,* 287–291
Ruel, K., Barnoud, F. and Goring, D. A. I. 1979, Cell. Chem. Technol. *13,* 429–432
Sachs, I. B., Clark, I. T. and Pew, J. C. 1963, J. Polym. Sci., Pt. C, No. 2, 203–212
Saka, S., Thomas, R. J. and Gratzl, J. S. 1978, Tappi *61,* No. 1, 73–76
Sawabe, O. 1980, Mokuzai Gakkaishi *26,* 641–646
Scallan, A. M. 1974, Wood Sci. *6,* 266–271
Scott, J. A. N. and Goring, D. A. I. 1970, Cell. Chem. Technol. *4,* 83–93
Scott, J. A. N., Procter, A. R., Fergus, B. J. and Goring, D. A. I. 1969, Wood Sci. Technol. *3,* 73–92
Shimizu, M. and Yamada, H. 1959, J. Soc. Text. Cell. Ind., Tokyo *15,* 540–544
Stone, J. E., Scallan, A. M. and Ahlgren, P. A. V. 1971, Tappi *54,* 1527–1530
Wood, J. R. and Goring, D. A. I. 1971, Pulp Pap. Mag. Can. *72,* No. 3, 61–68
Yang, J. M. and Goring, D. A. I. 1978, Trans. Techn. Sect. Can. Pulp Pap. Assoc. *4* TR2–TR5

9. Constituents of Bark

9.1. Anatomical Feature

After wood the bark is the second most important tissue of a trunk. It amounts to about 10–20% of a stem depending on the species and on growing conditions. As regards the whole tree the proportion of bark is highest for branches and the top with values of 20–35%; the bark portion of the stump and the roots is also higher than that of the stem (Young 1971). Bark yields a high amount of material during the processing of wood. It was long considered an annoying waste product, and was burned or stored. Only the barks of a few species were utilized, e.g. barks of oak and chestnut for extraction of tanning substances. In recent years bark has moved increasingly into the centre of interest. Numerous studies of its structure and composition as well as experiments of utilization were made. There are only a few surveys of the anatomical structure of bark, mostly as contributions to books on wood and plant anatomy, or on microscopy (Holdheide 1951; Esau 1964, 1965; Srivastava 1964; Parameswaran, Liese 1970; Nakano, Côté 1980).

The border between wood and bark is the cambium. This living cell layer produces xylem cells towards the inside of the stem and phloem cells towards the outside (Figs. 9–1, 9–2). The phloem or inner bark consists of conducting, sclerenchymatous and parenchymatous cells similar to the xylem. In the phloem of coniferous trees the conducting elements are sieve cells, relatively narrow cells with tapering ends arranged in longitudinal rows. In deciduous trees sieve tubes are formed, consisting of single elements connected end to end. The walls of both sieve cells and

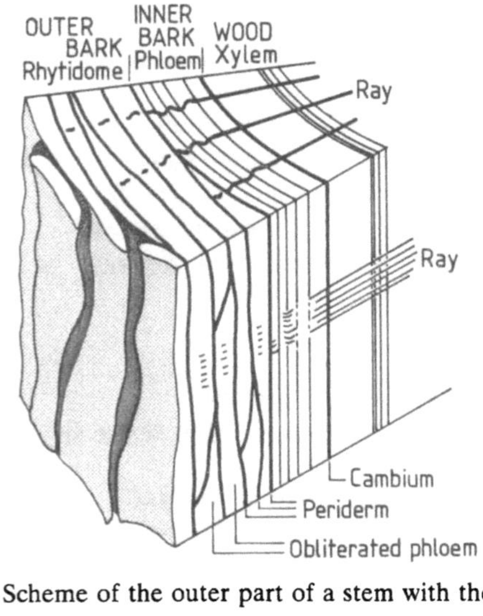

Fig. 9–1. Scheme of the outer part of a stem with the various tissues.

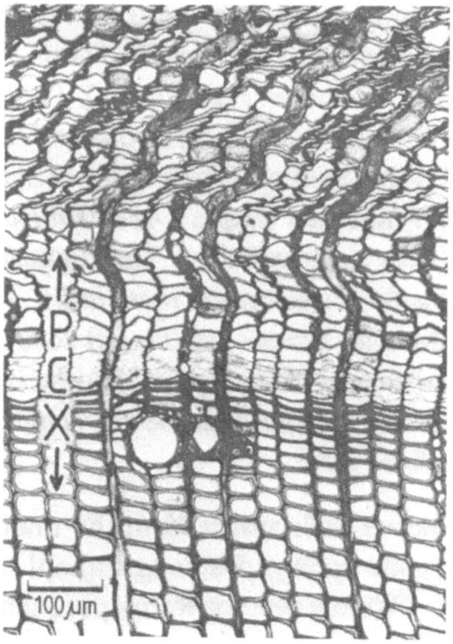

Fig. 9–2. Cross section of spruce *(Picea abies)* showing xylem (X), cambium (C) and inner bark (phloem, P). Light micrograph, by courtesy of D. Grosser.

sieve tube elements are perforated by numerous small pores arranged in varying areas (sieve fields). The degree of differentiation and the arrangement of the sieve fields are different in sieve cells and sieve tubes. In particular the cross walls of the sieve tube elements have relatively large pores, and that is why they are called sieve plates. Sieve cells and sieve tube elements are living cells with plasmatic content, but without a nucleus. The cell contents are interconnected by thin plasma strands (plasmodesmata) which pass through the pores of the sieve fields. The sieve cells and tubes serve to transport assimilates from the leaves downward.

The bast fibres and the stone cells are sclerenchymatous cells. Bast fibres are oblong, thick-walled cells with tapered overlapping ends usually arranged in tan-

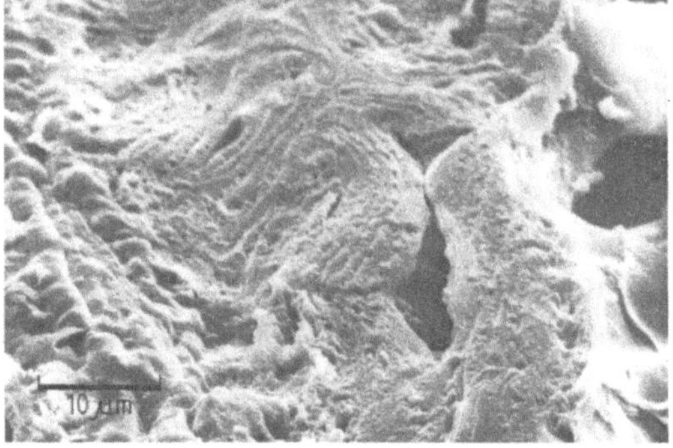

Fig. 9–3. Transverse plane of the inner bark of spruce *(Picea abies)* with thick-walled sclereids, the walls of which are layered. SEM micrograph.

gential rows. Stone cells (sclereids) have a polygonal shape and derive from parenchyma cells, the walls of which have been thickened and lignified (Fig. 9–3). The content of fibres and stone cells varies in a wide range. Harder et al. (1978) determined the percentages of both cell types in the barks of 42 coniferous and deciduous trees. Highest fibre proportions are found in the barks of black willow (*Salix nigra*) 23%, white ash *(Fraxinus americana)* 18% and red maple (*Acer rubrum*) 17%, whereas the barks of coniferous trees are very poor in fibres. The highest proportion of 5% was found in the bark of Douglas fir *(Pseudotsuga menziesii)*. The barks of conifers are more rich in sclereids: Virginia pine *(Pinus virginiana)* 26%, red pine *(Pinus resinosa)* 23%, balsam fir *(Abies balsamea)* 22%. Barks of some deciduous trees also have a high proportion of sclereids, e.g. American beech *(Fagus grandifolia)* 24% and white birch *(Betula papyrifera)* 22%.

The third group comprises the parenchymatous cells which form the main portion of the phloem tissue. The parenchyma cells are arranged in longitudinal strands or bands, or dispersed among the sieve cells. Phloem also contains parenchymatous rays, the structure of which is similar to that of the corresponding wood. Phloem rays are continuous with xylem rays, and are responsible for the conduction of metabolic products in a radial direction. At a certain distance from the cambium the radial alignment of rays becomes distorted; the rays assume an undulating pattern in the cross sectional view. The rays may also dilate by multiplication of the cells which results in an arch-like pattern in the outer part of the phloem (Howard 1971, 1978; Kramár et al. 1977).

Barks of certain species, particularly of conifers, are also interlaced by resin canals which are lined by epithelial cells. A further type of parenchyma cells is the companion cell intimately associated with sieve cells and sieve tubes.

The seasonal changes in cell dimensions resulting in earlywood and latewood are also visible in the phloem (Fig. 9–1), and an aging comparable to heartwood formation also occurs. The sieve cells and tubes loose their function in the outer part of the inner bark; the sieve plates are closed by deposition of callose (→ 9.2.3.). At the outermost end of the phloem a drastic change and the beginning of a new tissue, called outer bark or rhytidome, can be observed (Fig. 9–1). The rhytidome is mainly dead tissue with several depositions in the cells, and its function is to protect the stem against attack by microorganisms and loss of water.

The rhytidome is characterized by a collapse of the sieve cells and tubes, and an enlargement of the vertical parenchyma cells, a process which is called obliteration. The obliterated phloem is interrupted by irregularly shaped layers of periderm, which contains newly formed cambial cells. The periderm consists of three layers (Fig. 9–4):

- the cork cambium (phellogen), a uniserate layer of meristematic initials;
- the phellem, a layer of cork cells procuded by the phellogen to the outside;
- the phelloderm, a layer of cells produced by the phellogen to the inside.

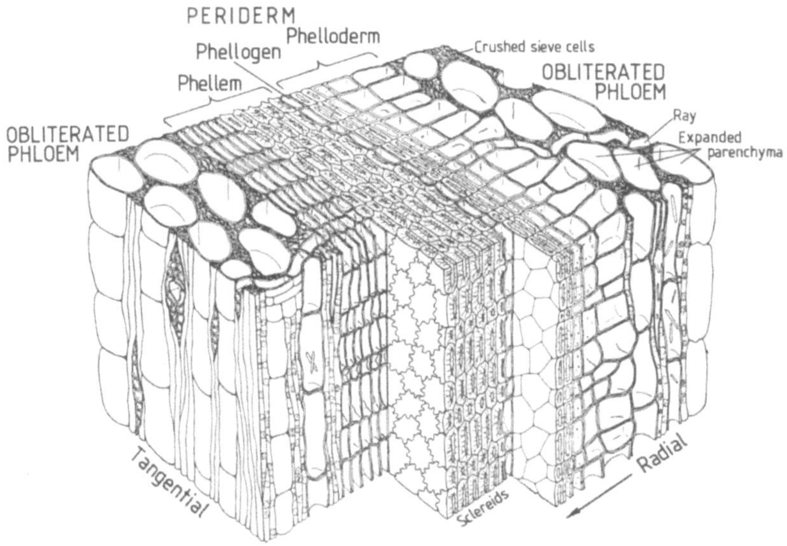

Fig. 9–4. Model of details of the outer bark of conifers (according to Howard 1971).

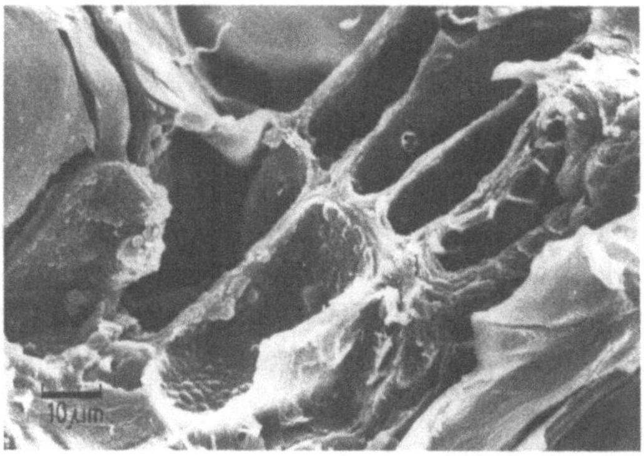

Fig. 9–5. Transverse plane of the outer bark of pine *(Pinus sylvestris)* with cork cells. SEM micrograph.

The cork cells are usually thin-walled and contain several layers of suberin incorporated into the cell walls (Fig. 9–5) (→ 9.2.6.). The tangential walls directed to the outside are often thickened.

Depending on the species the phelloderm cells adjacent to the phellogen may be thin-walled, and may develop into sclereids (Parameswaran, Liese 1970). In species such as *Pinus* the phelloderm cells have thickened and lignified walls, and with aging, i.e. increasing distance from phellogen, these cells become progressively larger and their walls thinner (Howard 1971).

9.2. Chemical Composition

9.2.1. General Analysis

In addition to its anatomical structure the chemical composition of bark determines the properties which are important in view of its utilization. Barks have a different swelling behaviour, are less anisotropic, possess slightly lower heat transfer coefficients, and are much weaker in all mechanical properties than wood (Martin 1969; Cassens 1974).

In its chemical composition bark differs from wood by the presence of polyphenols and suberin, by a lower percentage of polysaccharides, and a higher percentage of extractives. Analyses have been made of several barks, but their comparison is restricted because of the different extraction steps. Studying the bark of loblolly pine *(Pinus taeda)* McGinnis and Parikh (1975) determined the total extractive content as 19.9%, using a sequence of petroleum ether, benzene, ethanol, cold and hot water, whereas Labosky (1979) received a value of 27.5% after extraction with hexane, benzene, ethyl ether, ethanol, water and 1% NaOH. Extraction with ethanol-benzene removed 18.3% of loblolly pine bark (Pearl, Buchanan 1976) (Table 9–1). A sequence of ether, ethanol and hot water dissolved about 20% of European spruce and pine barks, and 12 and 16%, respectively, of beech and oak barks. By additional extraction with hot 1% NaOH the value for the extractives increased to 26.5–47.4% (Dietrichs et al. 1978) (Table 9–2).

The extraction procedure influences the content of polysaccharides and lignin in the residue. Comparing the values of Tables 9–1 and 9–2 it can be seen that the hot alkali extraction removes phenolic compounds which are insoluble in other solvents; the values for holocelluloses increase whereas the contents of lignin are lower after this treatment.

The anatomical differences between phloem and rhytidome are reflected in the chemical composition. Generally the contents of extractives and polysaccharides decrease, the contents of lignin and polyphenolic substances increase from the inner to the outer bark (Table 9–3). Sano and Tanaka (1976) subdivided the bark of Hokkaido spruce *(Picea glehnii)* into four layers: newly formed inner bark (I_1),

Table 9–1: Chemical analyses of the barks of some American conifers

Species	Extractives	Holocellulose	Aromatics	Lignin	Suberin	Ash	References
Pinus taeda	19.9	41.7		46.0		0.7	McGinnis, Parikh (1975)
	18.3	29.9		50	1.8		Pearl, Buchanan (1976)
	27.5	37.0	59.1	52.2	1.7	0.9	Labosky (1979)
Pinus echinata	29.4	35.9	61.2	51.2	1.8	0.9	Labosky (1979)
Pinus virginiana	23.0	37.4	62.2	57.7	0.9	1.0	Labosky (1979)
Pinus elliottii	35.8	37.9	60.8	49.7	2.0	0.5	Labosky (1979)
Tsuga mertensiana		45.9		52.9		2.2	Wilson (1961)

Table 9–2: Chemical analyses of the barks of some European trees

Species	Extractives	Holocellulose	Cellulose	Polyoses	Lignin	Suberin	Ash	References
Picea abies	21.0	65.3	47.9	17.4	37.8		2.1	Dietrichs et al. (1978)
Pinus sylvestris	20.7	54.6	37.0	15.6	44.7		1.1	Dietrichs et al. (1978)
Larix sibirica		51.6	24.6	18.3*	39.9	2.7		Gvozdeva et al. (1979)
Fagus sylvatica	11.4	61.6	38.1	23.1	39.0		7.3	Dietrichs et al. (1978)
	20.3	83.4	23.8	16.9**	43.0	4.3	6.3	Kramár, Ebringerová (1976)
Quercus robur	15.7	63.2	53.9	9.3	38.1		2.2	Dietrichs et al. (1978)

* Hydrolyzable carbohydrates
** Pentosans only

Table 9–3: Chemical analysis of the phloem and the rhytidome of chestnut oak *(Quercus prinus)* (Binotto, Murphey 1975)

Components	Phloem %	Rhytidome %
Extractives*	31.4	23.0
Polyphenolic acids	21.6	32.8
Lignin	18.2	28.9
Cellulose	43.2	24.9
Polyoses	16.9	13.1

* Based on ovendry matter
All other values based on extracted ovendry matter

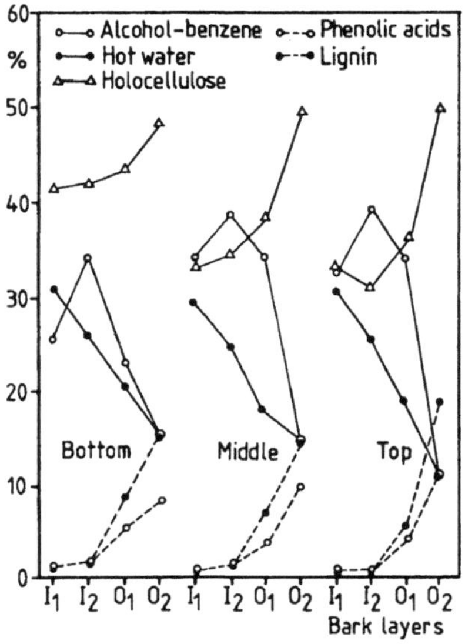

Fig. 9–6. Distribution of the various components in the inner and outer bark of the bottom, the middle and the top of Hokkaido spruce *(Picea glehnii)*. The bark is subdivided into four layers: newly formed inner bark (I_1), mature inner bark (I_2), outer bark (O_1) and outermost bark (O_2) (according to Sano and Tanaka 1976).

mature inner bark (I_2), outer bark (O_1) and outermost bark (O_2); additional investigations were made at various heights of a stem. According to Fig. 9–6 the percentages of the polysaccharides increase from the innermost to the outermost bark layer as these analyses derive from the fibre fraction only, which was separated by screening and amounted to more than one third (34–38%) of the whole bark substance.

9.2.2. Cellulose

As in wood, the main sugar in the hydrolyzates of bark is glucose, the content of which is in the range of 16–41% (Table 9–4). The outer bark contains fewer glucose units than the inner bark. Based on unextracted bark the cellulose percentages evaluated by Dietrichs et al. (1978) are between 20.2% (pine) and 32.6% (oak). After the successive extraction including NaOH treatment there is a relatively high cellulose amount in the residue (Table 9–2).

Apart from the studies of Dietrichs et al. (1978) the cellulose contents of various barks were determined by Timell (1961a; Mian, Timell 1960a) and Haas and Kremers (1961). The celluloses isolated during these studies had number-average degrees of polymerization ($\bar{P}_n$) of 125–700 and weight-average degrees of polymerization ($\bar{P}_w$) of 4 000–6 900 (Table 9–5). The values of $\bar{P}_n$ and $\bar{P}_w$ show a lower degree of polymerization for bark cellulose than for wood cellulose and a high polydispersity.

Table 9–4: Sugar composition of the hydrolyzates of barks of various trees

Species	Glu %	Man %	Gal %	Xyl %	Ara %	Rha %	UrA %	Ac %	References
Abies amabilis	37.4	8.0	1.6	3.2	3.2		5.6	0.8	Timell (1961a)
Ginkgo biloba									
inner bark	38.3	3.4	4.5	4.0	6.2		11.5	0.2	
outer bark	23.8	2.1	1.7	3.5	3.5		9.4	0.3	Timell (1961a)
Picea abies	36.6	6.5	1.3	4.8	1.8	0.3			Dietrichs et al. (1978)
Picea engelmannii	35.7	2.9	2.4	3.8	3.3		8.0	0.5	Timell (1961a)
Pinus contorta									
inner bark	40.9	2.5	4.3	3.7	10.6		9.9	0.2	
outer bark	26.8	2.5	4.2	3.4	5.5		7.7	0.8	Timell (1961a)
Pinus sylvestris	30.2	5.4	2.4	5.8	2.1	0.3			Dietrichs et al. (1978)
Pinus taeda	26.0	4.1	2.8	3.9	1.7				McGinnis, Parikh (1975)
Pinus taeda									
inner bark	21.3	2.5	3.1	2.1	5.6	0.3	4.6		
outer bark	15.8	2.6	2.5	3.8	1.8	0.1	2.1		Pearl, Buchanan (1976)
Betula papyrifera									
inner bark	28.0	0.2	1.0	21.0	2.7		2.2		Mian, Timell (1960a)
Fagus sylvatica	29.7	0.2	3.1	20.1	3.1	1.2			Dietrichs et al. (1978)
Quercus robur	32.3	0.5	1.3	16.4	2.0	0.5			Dietrichs et al. (1978)

Table 9–5: Yield and degree of polymerization of celluloses from barks of various trees

Species	Yield %	$\bar{P}_n$	$\bar{P}_w$	References
Abies amabilis	38.1	216	4 000	Timell (1961a)
Gingko biloba	37.6	500	6 200	Timell (1961a)
Picea engelmannii	30.9	412	5 300	Timell (1961a)
Pinus contorta	30.4	702	6 900	Timell (1961a)
Betula papyrifera	28.4	125	4 300	Mian, Timell (1960a)
*Populus grandidentata**	32.3		4 000	Haas, Kremers (1961)
*Populus tremuloides**	27.2		3 900	Haas, Kremers (1961)

* From sclereids only

Table 9–6: Degree of polymerization of celluloses from barks isolated by various procedures (Garves 1976)

Procedures	*Picea abies* $\bar{P}_w$	*Pinus sylvestris* $\bar{P}_w$	*Fagus sylvatica* $\bar{P}_w$
Pre-extraction 1% NaOH Delignification 70 °C Alkali extraction	630	530	840
Pre-extraction 1% NaOH Delignification 20 °C Alkali extraction	1 300		
Delignification 20 °C Alkali extraction	1 800 1 900*		 1 300*
Direct nitration	2 800*		3 300*

* From cellulose nitrate in butylacetate, all other values from cellulose in cadoxene solution

Garves (1976) found a high sensibility of bark celluloses against the isolation procedure. He obtained low $\bar{P}_w$-values after delignification at 70 °C and alkali extraction. The values were higher after delignification at room temperature and highest after a direct nitration of the bark and extraction of the cellulose nitrate (Table 9–6).

Bark cellulose has the same crystalline lattice (cellulose I) as is known from wood cellulose, but its degree of crystallinity is lower (Binotto et al. 1971; Parameswaran 1973). Whereas in bark of chestnut oak *(Quercus prinus)* the crystallinity of cellulose from phloem and rhytidome was approximately the same (Fig. 9–7), in barks of Central American cedar *(Cedrela odorata)* and makoré *(Tieghemella heckelii)* a somewhat higher crystallinity was found for cellulose from phloem than for cellulose from rhytidome.

9.2.3. Polyoses

Table 9–4 shows that in barks from conifers the percentage of sugars present in the hydrolyzates apart from glucose is less than 10% except for arabinose in the inner

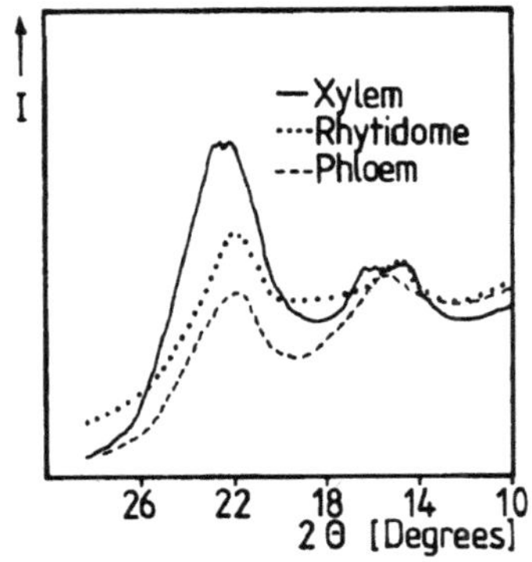

Fig. 9–7. X-ray diffraction diagram of the xylem, phloem and rhytidome of chestnut oak *(Quercus prinus)* (according to Binotto et al. 1971).

bark of lodgepole pine *(Pinus contorta)* and uronic acids in the inner bark of maidenhair tree *(Ginkgo biloba)*. Dietrichs et al. (1978) calculated the content of the various polysaccharides in the barks of four European trees (Table 9–7). From this calculation it can be seen that, as in the corresponding wood, the main polyose in conifer barks is galactoglucomannan and in deciduous barks arabino-4-O-methylglucuronoxylan. The combination of the sugars to these types of polyoses is based on several studies which verify the presence of 4-O-methylglucuronoxylans, arabino-4-O-methylglucuronoxylans, glucomannans and galactoglucomannans in barks, with differences of varying degree in their composition (Timell 1965; Dietrichs 1975) (Table 9–8).

From the barks of white birch *(Betula papyrifera)*, trembling aspen *(Populus tremuloides)* and white willow *(Salix alba)* 4-O-methylglucuronoxylans with very similar structures were isolated (Mian, Timell 1960b; Jiang, Timell 1972a; Toman 1973). The xylose units are β-(1→4)-linked and the uronic acid groups are attached to the backbone by α-(1→2)-linkages. The ratio of about 10:1 (Xyl:Me-GluU) is in the same range as in xylans from wood. The determination of the degree of polymerization ($\overline{P}_n$) resulted in values between 171 and 234.

Table 9–7: Carbohydrate composition of the barks of four European trees, calculated from the sugar analytical data (Dietrichs et al. 1978)

Carbohydrates	*Picea abies* %	*Pinus sylvestris* %	*Fagus sylvatica* %	*Quercus robur* %
Cellulose	31.7	23.1	26.2	25.4
Glucomannan			0.3	0.9
O-acetyl-galactoglucomannan	10.1	8.4		
O-acetyl-4-O-methylglucuronoxylan			24.6	20.1
Arabino-4-O-methylglucuronoxylan	5.4	6.6		
Pectin (without galacturonic acid)	2.5	3.6	7.4	3.8
Starch	0.6	0.2	0.4	0.6
Combined glucose	1.9	5.1	2.9	6.2

Table 9–8: Polyoses isolated from barks

Types	Species	Yield %	Sugar units	Ratio	Linkage	$[\alpha]_D$	$\bar{P}_n$	References
Xylans	*Betula papyrifera*	26.6	β-D-Xylp	10	1→4			
			α-D-Me-GluU	1	1→2	−68	234	Mian, Timell (1960b)
	Populus tremuloides	18–20	β-D-Xylp	12	1→4			
			α-D-Me-GluU	1	1→2	−84.6	216	Jiang, Timell (1972a)
	Salix alba	3.7	Xyl	9				
			Me-GluU	1		−57.7	171	Toman (1973)
	Abies amabilis	2.1	β-D-Xylp	6	1→4			
			α-D-Me-GluU	1	1→2			
			L-Araf	0.6	1→3	−57	124	Timell (1961c)
	Picea engelmannii	3.7	β-D-Xylp	9	1→4			
			α-D-Me-GluU	1	1→2			
			L-Araf	1.4	1→3	−35		Ramalingam, Timell (1964)
Mannans	*Salix alba*	1.6	Man	1.4				
			Glu	1		−6.6	32	Toman, Karácsonyi (1972)
	Salix alba	2	Man	1.1				
			Glu	1.2				
			Gal	0.5				Pavlovova et al. (1970)
	Populus tremuloides	low	β-D-Man	1.3	1→4			
			β-D-Glu	1	1→4			
			α-D-Gal	0.5	1→6	+10	50	Jiang, Timell (1972b)
	Picea glauca	low	Man	4.5				
			Glu	1				
			Gal	0.5		−12	45	Painter, Purves (1960)
	Abies amabilis	3	β-D-Man	2.5	1→4			
			β-D-Glu	1	1→4			
			α-D-Gal	0.1	1→6	−34	70	Timell (1961c)
	Abies amabilis	2.8	β-D-Man	2.7	1→4			
			β-D-Glu	1	1→4			
			α-D-Gal	0.4	1→6	−14.4	80	Timell (1962)
	Picea engelmannii	2	β-D-Manp	3	1→4			
			β-D-Glup	1	1→4			
			α-D-Galp	0.2	1→6	−43	86	Ramalingam, Timell (1964)
	Pinus sylvestris	2.6	β-D-Manp	1.7	1→4			
			β-D-Glup	1	1→4			
			α-D-Galp	0.08	1→6	−33	60	Fu,Timell (1972a)
Glucans	*Pinus sylvestris*	2.5	β-D-Glup		1→3			Fu et al. (1972)
	Picea engelmannii	1.2	β-D-Glup	4	1→4			
			β-D-Xylp	3	1→4			
			D-Galp	1	1→6	+35		Ramalingam, Timell (1964)
Galactans	*Salix alba*		β-D-Gal		1→4			
			Ara		1→6	+49.5	33	Toman et al. (1972)
	Picea glauca	low	Gal	10				
			Ara	1		−22		Painter, Purves (1960)
Arabinans	*Populus tremuloides*		α-L-Araf		1→5			
					1→2,3	−150,3	45	Jiang, Timell (1972c)
	Pinus sylvestris	2.5	α-L-Araf		1→5			
					1→2,3		95	Fu, Timell (1972b)

Types	Species	Yield %	Sugar units	Ratio	Link-age	$[\alpha]_D$	$\bar{P}_n$	References
Galacturonans	*Betula papyrifera*	3–4	α-D-GalU	9	1→4			
			L-Araf	1				
			α-D-GalU	7	1→4			
			L-Araf	3				
			D-Galp	1			130	Timell, Mian (1961)
	Picea glauca	7	GalU					
			Ara					
			Gal			+210		Painter, Purves (1960)
	Abies amabilis	1	α-D-GalU		1→4	+246	450	
		0.6	α-D-GalU	10	1→4			
			L-Araf	1	1→2,3			
			D-Galp	2	1→2,3	+225		Bhattacharjee, Timell (1965)

Xylans isolated from coniferous barks consist of a β-(1→4)-linked backbone of xylose with side groups of 4-O-methylglucuronic acid and L-arabinose (Timell 1961b; Ramalingam, Timell 1964). The ratio of Xyl:Me-GluU:Ara is 6:1:0.6 for xylan from bark of amabilis fir *(Abies amabilis)* and 9:1:1.4 for xylan from bark of Engelmann spruce *(Picea engelmannii)*.

In the bark of Asian camphorwood *(Cinnamomum iners)* a water-soluble arabinoxylan was detected consisting of arabinose and xylose in a molar ratio of 1.45:1. The polysaccharide contains a backbone of β-(1→4)-linked xylose units, each of which is substituted both at C2 and C3 with arabinofuranose and 3-O-α-xylopyranosyl-arabinofuranose groups (Gowda et al. 1980).

Glucomannans from deciduous barks *(Populus tremuloides, Salix alba)* contain mannose and glucose units in a ratio of about 1:1 to 1.4:1 forming a heteropolymer backbone which is essentially linear (Pavlovová et al. 1970; Toman, Karácsonyi 1972; Jiang, Timell 1972b). In the mannans from the barks of aspen and willow twigs galactose units were additionally found in a relatively high percentage as compared to wood. The ratios are 1.3:1:0.5 (Man:Glu:Gal) and 0.9:1:0.5 respectively. The $\bar{P}_n$ is relatively low: 30–50.

Timell (1961c, 1962) isolated a water-soluble and an alkali-soluble galactoglucomannan from the bark of amabilis fir *(Abies amabilis)* which differed mainly in the proportion of galactose units. The ratio of mannose, glucose and galactose was 2.5:1:0.1 for the alkali-soluble mannan, and 2.7:1:0.4 for the water-soluble mannan. Somewhat differing ratios of the sugar composition were found in galactoglucomannans from barks of spruce *(Picea engelmannii, P. glauca)* and pine *(Pinus sylvestris)* (Painter, Purves 1960; Ramalingam, Timell 1964; Fu, Timell 1972a) (Table 9–8).

Callose has already been mentioned in connection with the wood polyoses (→ 5.4.). It is also found in phloem particularly as a plugging substance of the sieve plates. The walls of cork cells also contain callose (Litvay, Krahmer 1976). Callose consists of β-(1→3)-linked glucose units. Fu et al. (1972) isolated callose in a yield of 2.5%

from Scots pine bark *(Pinus sylvestris)*. A branched galactoxyloglucan was found in the bark of Engelmann spruce (Ramalingam, Timell 1964). It consists of a backbone of β-(1→4)-glycosidic linked glucose residues, which is branched by β-(1→6)-linkages. Xylose and galactose units form side chains. The Glu:Xyl:Gal-ratio is 4:3:1.

Polysaccharides belonging to the pectic substances are also present in barks. Galactans have been isolated from the bark of white willow *(Salix alba)* consisting of galactose only, or of galactose and arabinose units (Pavlovová et al. 1970; Toman et al. 1972). The pure galactan was water-soluble; the molecular chain consisted of 33 β-(1→4)-linked galactose units with few side chains and branching points at C6 of the backbone. A highly branched arabinogalactan with a ratio of 10:1 (Gal:Ara) was detected in white spruce bark *(Picea glauca)* (Painter, Purves 1960).

A certain importance seems to attach to arabinans in bark. Polysaccharides of this type were isolated from barks of *Populus tremuloides, Picea glauca* and *Pinus sylvestris* (Painter, Purves 1960; Fu, Timell 1972b; Jiang, Timell 1972c; Pavlová et al. 1972). Arabinans consist of α-(1→5)-linked arabinofuranose units. In the case of aspen the backbone has 45 units and is laced with 25 final arabinose residues linked to C2 and C3 of backbone units. The pine bark arabinan has a similar structure with a $\bar{P}_n$ of 95.

Timell and Mian (1961) isolated polygalacturonans in a yield of 3–4% from the inner bark of white birch *(Betula papyrifera)*. The substance could be separated into two uniform compounds, one of them consisting of galacturonic acid and arabinose units in a ratio of 9:1, the other one consisting of galacturonic acid, arabinose and galactose in a ratio of 7:3:1. Additionally traces of glucose, xylose and rhamnose were present. The galacturonic acid units are α-(1→4)-linked. Similar galacturonans were also found in coniferous barks. The pectin isolated in a yield of 7% from spruce *(Picea glauca)* by Painter and Purves (1960) consisted mainly of galacturonic acid with about 10% of galactose and arabinose units. Polygalacturonans from fir *(Abies amabilis)* also consist of galacturonic acid, galactose and arabinose, and additionally of a few rhamnose residues (Bhattacharjee, Timell 1965). One of the fractions has a ratio of about 10:1:2 (GalU:Gal:Ara).

All studies show that the polyoses from barks have the same chemical structure as those from wood though there are some variations in the composition.

9.2.4. Lignin

Histochemical studies have shown that the cell walls of fibres and sclereids are lignified and also the cells of periderm and rhytidome give a lignin reaction (Srivastava 1966). Extracts of coniferous barks contain compounds deriving from the lignin metabolism (shikimic acid, ferulic acid, coniferylaldehyde, vanillin etc.) (Hergert 1960; Holmes, Kurth 1961). The lignin content of barks can only be determined after alkali extraction, as that part of polyphenols which is insoluble in the usual solvents is likewise resistant to hydrolysis, and thus contributes to the lignin value.

Studies of the structure of bark lignin were made mainly with sclereids as they are free of polyphenols and extractives, and can easily be separated by screening and floating. The lignin content of these stone cells is about 23% for aspen bark *(Populus grandidentata, P. tremuloides)* and about 25% for kosipo bark *(Entandophragma candollei)* (Haas, Kremers 1961; Parameswaran et al. 1975).
A study of bark lignins of three Japanese conifers *(Pinus thunbergii, Abies firma, Cryptomeria japonica)* and three Japanese broadleaved trees *(Fagus crenata, Magnolia obovata, Quercus crispula)* by nitrobenzene oxidation and ethanolysis showed that the same degradation products occur as are found in wood lignin, but their ratio is different; in particular the ratio of syringyl to guaiacyl units is moderately lower for deciduous bark than for the corresponding wood (Higuchi et al. 1967).

The determination of functional groups and elementary composition of sclereid lignin from aspen bark *(Populus tremuloides)* resulted in a lower OCH_3 content and a higher ratio of phenolic OH to OCH_3 as compared to wood lignin (Clermont 1970). The alkaline nitrobenzene oxidation gave a vanillin-syringaldehyde ratio of 1:1, while that of wood lignin was 1:3. The NMR spectra indicated a higher degree of condensation in at least two bark lignin fractions.

A methoxyl content of 18.1% was determined in bark lignin of beech *(Fagus sylvatica),* a value which is higher in the corresponding wood lignin (21%) (Paulínyová et al. 1978). According to the reaction products of nitrobenzene oxidation the ratio of vanillin:syringaldehyde:p-hydroxybenzaldehyde was 46:48:6.

Degradation studies of sulfate lignins from barks of spruce *(Picea abies),* yew *(Taxus baccata),* birch *(Betula verrucosa),* ash *(Fraxinus excelsior)* and vine *(Vitis vinifera)* confirm the higher frequency of guaiacyl units for deciduous bark lignin as compared to wood (Andersson et al. 1973). The lignins from the coniferous barks have a much higher proportion of p-hydroxyphenyl units.

No structural differences have been found between lignins from wood and from bark sclereids of kosipo *(Entandophragma candollei);* the latter however was richer in syringyl units (Parameswaran et al. 1975). The ratio of p-hydroxyphenyl:guaiacyl:syringyl units was 0.4:1:1.4 for sclereid lignin and 0.5:1:1.2 for wood lignin of kosipo. Thus the bark lignin of this species has a higher methoxyl content (20.1%) than the corresponding wood lignin (18.05%).

All studies show a similar structure of wood and bark lignins though there are some differences in the ratio of the components.

9.2.5. Polyphenols

The term polyphenols refers to a large number of related compounds deriving mainly from flavane derivatives. The polyphenols present in barks are classified according to their molecular weight and solubility.

Lowest in molecular weight are the procyanidins (proanthocyanidins), which are di- and trimeric flavanols (Porter 1974; Yazaki, Hillis 1977; Samejima, Yoshimoto 1979). These compounds are soluble in methanol, hot water and ethyl acetate. Alkaline fusion of oligomeric flavanols from hemlock bark *(Tsuga heterophylla)* yielded protocatechuic acid, catechin and phloroglucinol (Hergert et al. 1965). Studies of water-extracted procyanidins from barks of pines *(Pinus radiata, P. taeda, P. echinata)* showed that the pentahydroxyflavanol (catechin) units are joined by C4–C8 or C4–C6 linkages (Porter 1974; Hemingway, McGraw 1976). Di- and triflavanols with a similar structure were also isolated from the inner bark of sugi *(Cryptomeria japonica)* (Samejima, Yoshimoto 1979) (Fig. 9–8). Samejima and Yoshimoto (1981) characterized the bark tannins of numerous conifer species by determining the tannin-flavanol ratio (T/F value). According to this study the bark tannins of many conifers consist mainly of proantocyanidins.

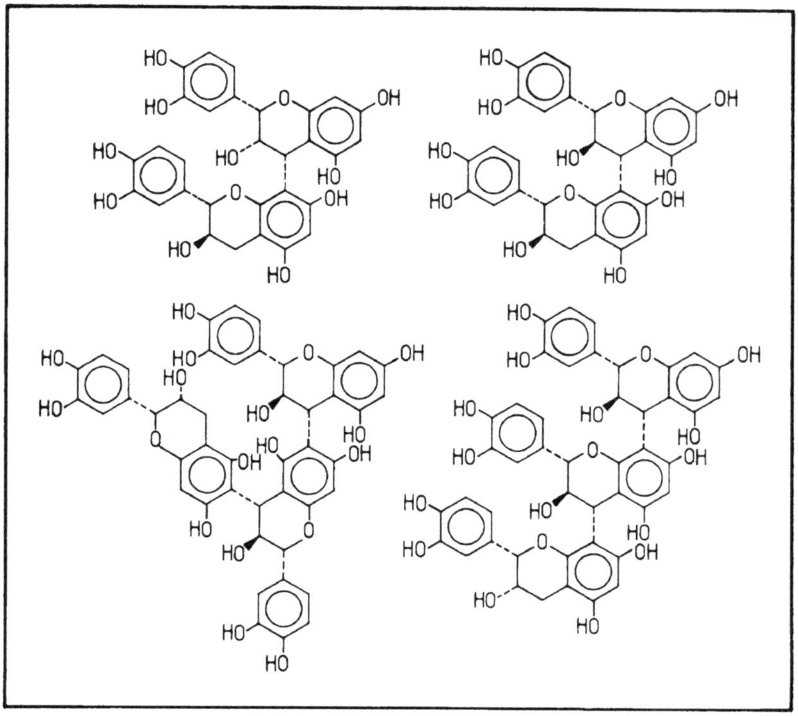

Fig. 9–8. Diflavanols and triflavanols as isolated from pine *(Pinus radiata)* and sugi *(Cryptomeria japonica)* barks.

Extracts from bark of black wattle *(Acacia mearnsii)* contain triflavanols consisting of catechin (I), gallocatechin (II), leucofisetinidin (III) and leucorobinetinidin (IV) which are C4–C6 and C4–C8 linked (Roux et al. 1975) (Fig. 9–9).

Similar in structure but higher in molecular weight are the condensed tannins (phlobaphenes). They are soluble in hot water and comprise compounds with molecular weights between about 1 000 (tetramers) 3 000 (undecamers). The content of condensed tannins in barks varies widely from 5–50% (Table 9–9). These polyphenols consist of catechin, gallocatechin and other flavanols, but flavonols and chalcones are also present (Hathway 1962). Hergert et al. (1965) distinguish between condensed tannins soluble in hot water and phlobaphenes soluble in 95% ethanol, and they attribute chemical structures to them differing in the linkages between the catechin units. In the literature several linkages have been proposed for the structure of phlobaphenes as well as of polyphenolic acids (Hemingway, McGraw 1976) (Fig. 9–10).

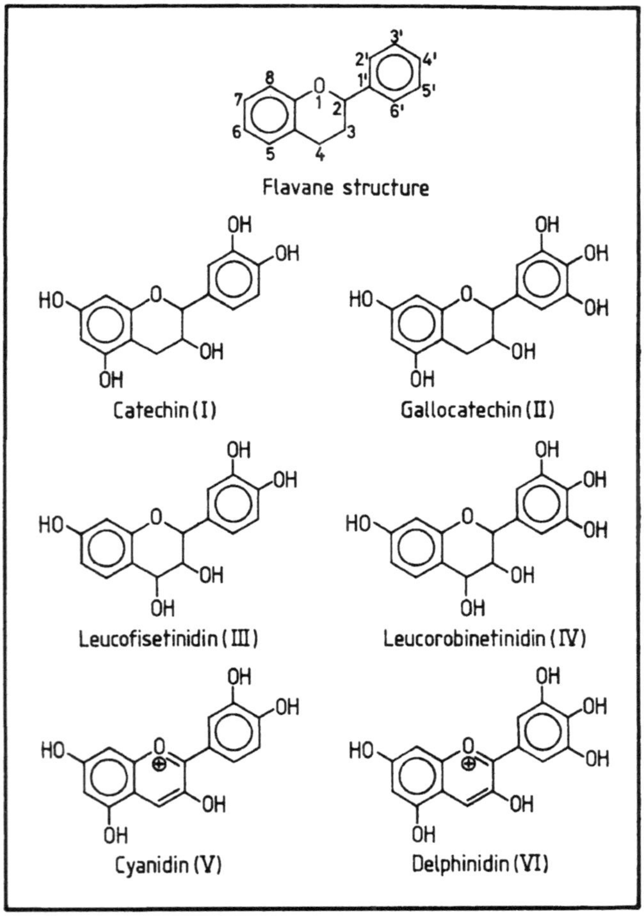

Fig. 9–9. Flavanols, units of procyanidins from the bark of black wattle *(Acacia mearnsii)* and of polyphenolic acids from conifer barks.

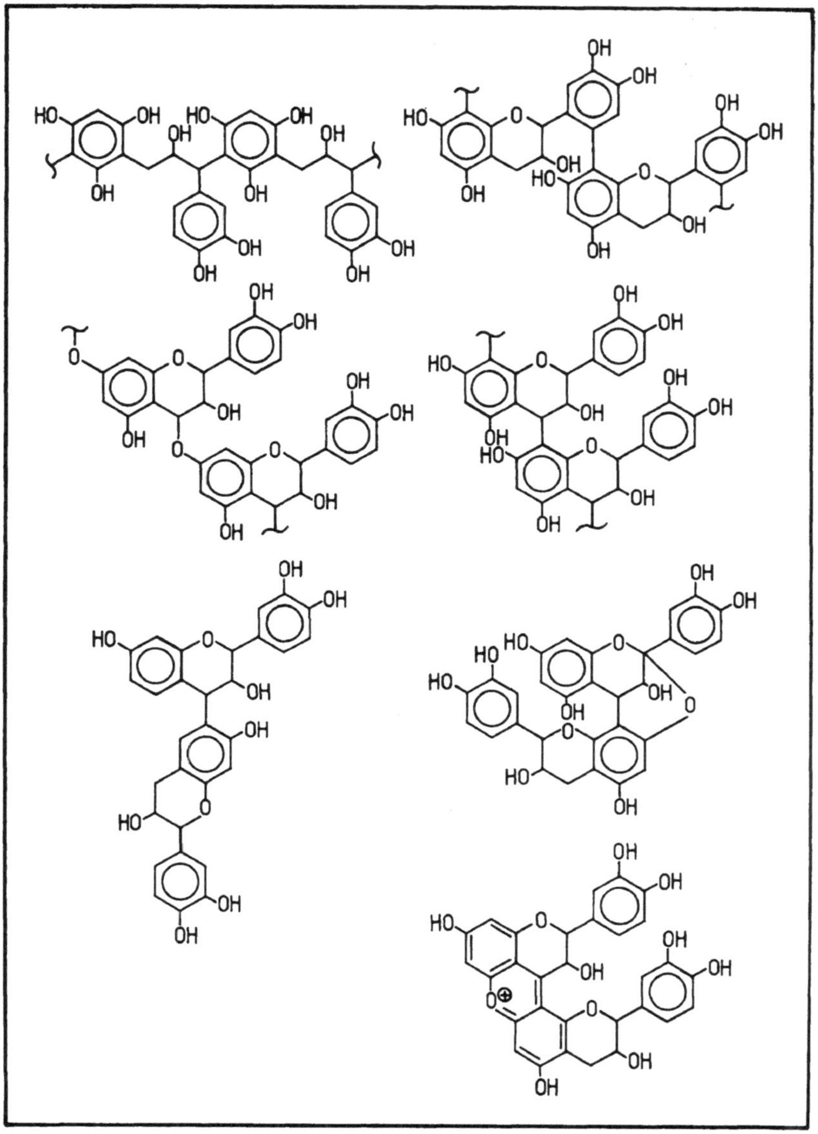

Fig. 9–10. Polyflavonoid structures as proposed in the literature (according to Hemingway and McGraw 1976).

The third group of compounds belonging to the polyphenols are the polyphenolic acids, which differ from the above-mentioned compounds by the fact that they can be extracted only with 1% NaOH at 100 °C. After precipitation with mineral acids the compounds are partially soluble in water and polar organic solvents. In general they are characterized by 1–4% aliphatic hydroxyl groups and a methoxyl content of less than 2%. The molecular weight of polyphenolic acids from loblolly pine bark *(Pinus taeda)* is in the range of 1 500–6 700 (Fang, McGinnes 1975). The polyphenols in the hot-water extracts of barks from *Picea abies* and *Pinus brutia* have a mo-

Table 9–9: Yields of water-soluble condensed tannins from barks of various trees (according to Hathway 1962)

Species	Yield %
Betula alba	10–15
Castanea sativa	8–14
Eucalyptus adstringens	40–54
Eucalyptus wandoo	13–15
Larix decidua	5–20
Larix leptolepis	10–25
Picea abies	5–18
Picea sitchensis	11–37
Pinus densiflora	6
Pinus nigra var. *calabrica*	13–25
Pinus ponderosa	5–11
Pinus radiata	17–18
Pinus sylvestris	16
Pseudotsuga menziesii	5–25
Quercus robur	12–16
Robinia pseudoacacia	7
Sequoia sempervirens	2–8
Tsuga canadensis	10–11
Tsuga heterophylla	15–16

lecular weight ($\bar{M}_w$) of 2 700 and 3 800, respectively, with a dispersity ($\bar{M}_w/\bar{M}_n$) of about 2 (Weissmann 1981). For polyphenolic acids from slash pine bark *(Pinus elliottii)* Erman and Lyness (1965) calculated an empirical formula of $(C_{15}H_{11}O_4(OCH_3)_3)_{17}$.

Pyrolysis yielded mainly phenols (phenol, cresols, guaiacol, catechin) and only small amounts of aliphatic compounds (methanol, acetic acid, acetone). By nitrobenzene oxidation a small portion of typical lignin degradation products were found, e.g. vanillin, vanillic acid, p-hydroxybenzaldehyde and a significant portion of protocatechualdehyde (Fahey, Kurth 1957; Sarkanen, Hergert 1971). Ethanolysis of fines from Douglas fir bark *(Pseudotsuga menziesii)*, rich in polyphenolic acids, resulted in ethyl esters of ferulic acid, vanillic acid and protocatechuic acid apart from the free acids (Fujii, Kurth 1966). Higuchi et al. (1967) believe that degradation products such as vanillin, syringaldehyde and p-hydroxybenzaldehyde derive from lignin which in small amounts may be dissolved with 1% NaOH.

Degradation products of mild acid treatment of conifer barks are catechin (I), cyanidin (V) and delphinidin (VI) depending on source and method (Fig. 9–9). From these studies the polyphenolic acids in coniferous barks would be expected to have the same structure as phlobaphenes, possibly with an additional C–C-linkage in the phloroglucinol ring to give a three-dimensional network (Hergert et al. 1965; Sarkanen, Hergert 1971). The change in solubility after alkali extraction is probably caused by a molecular rearrangement under alkaline condition.

9.2.6. Suberin

Suberin is an insoluble constistuent of the outer bark, and is concentrated in the cork cells. The cork of cork oak *(Quercus suber)* contains 40–45% suberin apart from 12% polysaccharides, 27% lignin, waxes, tannins and ash constituents (Holloway 1972; Pereira 1979). But cork cells of other barks also contain high amounts of suberin, e.g. Swedish birch *(Betula verrucosa):* 43.3%, Pao Santo *(Kielmeyera coriacea):* 46% (Guillemonat, Triaca 1968; Holloway 1972). The suberin is intimately associated with soluble waxes and phenolics forming altering lamellae within the cell walls (Sitte 1962; Wattendorff 1973; Litvay, Krahmer 1977). Suberin cannot be isolated in whole from cork tissue, it has to be saponified by alkali treatment. The constituents are extracted and separated by chromatographic methods.

Jensen et al. (1957, 1963) isolated from the saponification mixtures of suberin from birch *(Betula verrucosa)* and oak *(Quercus suber)* C_{18} and C_{22} mono- and dibasic acids and hydroxy acids, among them:

- docosane-1,22-dioic (phellogenic) acid
  $$HOOC{-}(CH_2)_{20}{-}COOH$$
- 22-hydroxydocosanoic (phellonic) acid
  $$HOCH_2{-}(CH_2)_{20}{-}COOH$$
- octadec-9-ene-1,18-dioic acid
  $$HOOC{-}(CH_2)_7{-}CH{=}CH{-}(CH_2)_7{-}COOH$$
- 18-hydroxyoctadec-9-enoic acid
  $$HOCH_2{-}(CH_2)_7{-}CH{=}CH{-}(CH_2)_7{-}COOH$$
- 9,10-dihydroxyoctadecane-1,18-dioic (phloionic) acid
  $$HOOC{-}(CH_2)_7{-}(CH_2OH)_2{-}(CH_2)_7{-}COOH$$
- 9,10,18-trihydroxyoctadecanoic (phloionolic) acid
  $$HOCH_2{-}(CH_2)_7{-}(CH_2OH)_2{-}(CH_2)_7{-}COOH$$

The analysis of suberin from birch and oak barks shows specific differences in the composition (Table 9–10). Thus the content of phloionic acid is higher in oak suberin, whereas the content of phloionolic acid is higher in birch suberin; there are also minor differences in the content of the other components.

Even carbon numbered C_{16}–C_{24} aliphatic acids and hydroxy acids were found in suberins from barks of white fir *(Abies concolor)*, Douglas fir *(Pseudotsuga menziesii)* and western red cedar *(Thuja plicata)* (Kurth 1967; Swan 1968). The main components in suberin from white fir are C_{16} and C_{22} hydroxy acids, from Douglas fir lignoceric acid $C_{23}H_{47}COOH$, and from western red cedar C_{18} acids.

A tropical tree whose cork is used industrially is Pao Santo *(Kielmeyera coriacea)* growing in Brazil. The suberin of its bark is characterized by the presence of 9,10-dihydroxyoctadecane-1,18-dioic, hexadec-8-ene-1,16-dioic, octacosanoi acids, and C_{15}–C_{30} hydrocarbons (Guillemonat, Triaca 1968).

In the saponification mixture of suberin a considerable amount (25–35%) of phenolic compounds is found (Hergert 1958; Swan 1968; Holloway 1972). Among them

Table 9–10: Composition of suberin from the cork of *Quercus suber* and *Betula verrucosa* (Holloway 1972)

Constituents	*Quercus suber* %	*Betula verrucosa* %
Neutral constituents:		
Alkan-1-ols (C_{22}–C_{28})	2.7	
Unidentified	3.4	10.3
Acid constituents:		
Monobasic (C_{16}–C_{26})	1.9	
α,ω-Dibasic (C_{16}–C_{26})	7.6	8.5
ω-Hydroxymonobasic (C_{16}–C_{26})	47.4	21.3
Dihydroxyoctadecanoic		1.3
Dihydroxyhexadecanoic		3.6
9,10-Dihydroxyoctadecane-1,18-dioic	15.4	1.3
9,10,18-Trihydroxyoctadecanoic	7.7	42.7
9,10-Epoxy-18-hydroxyoctadecanoic		1.8
Unidentified	13.9	9.2

are ferulic and sinapic acid in amounts of 5–6%, apart from high-molecular phenolic acids.

From the nature of the constituents it is concluded that suberin has a polyester structure composed mainly of long chain fatty and hydroxy fatty acids with variations of the composition depending on the species. The presence of phenolic compounds indicates that these are probably also incorporated in the high-molecular suberin complex.

9.2.7. Extractives

The content of extractives is higher in bark than in wood. It depends not only on the species but also on the solvents used. The diversity of the extractable compounds usually requires a sequential extraction, the yields from which give a preliminary characterization of the composition. The variation of this composition can be very large even within one genus. Table 9–11 summarizes the results of the sequential extraction of the barks of American southern pines, the European Scots pine and the Aleppo pine from Asia Minor. The latter has an extremely high content of extractives: more than two thirds of the ovendry substance.

Some of the bark extractives were already described in one of the preceding sections (→ 9.2.5.). The alkali extract as well as the ethanol extract contain higher polymer flavanoids including the polyphenolic acids. Apart from di-, tri- and oligoflavanoids in the hot-water extract there are also monomeric flavanes and flavones, i.e. compounds such as catechin, gallocatechin, myricetin, quercetin, taxifolin, cyanidin etc., most of them also present in coniferous wood (→ 7.2.3.; Figs. 7–17, 9–9) (Hergert 1960). Catechin and the cis-isomeric epicatechin are the main com-

Table 9–11: Yields from sequential solvent extraction of various pine barks (Labosky 1979; Weissmann, Ayla 1980)

Solvent	*Pinus echinata* %	*Pinus elliottii* %	*Pinus taeda* %	*Pinus virginiana* %	*Pinus sylvestris* %	*Pinus brutia* %
Hexane	2.6	2.1	1.7	1.5		
Benzene	1.2	2.0	1.3	1.0		5.0
Ethyl ether	1.1	1.2	1.3	1.0	4.6	
95% ethanol	4.4	7.3	2.0	3.5	1.2	25.7
Hot water	2.9	3.3	1.9	1.9	4.8	17.8
1% NaOH	17.2	19.9	19.3	19.3	39.1	19.7
Total	29.4	35.8	27.5	28.2	49.7	68.3

ponents in the methanol extractives of the bark of radiata pine *(Pinus radiata)* and sugi *(Cryptomeria japonica)* (Yazaki, Hillis 1977; Samejima, Yoshimoto 1979). Quercetin and dihydroquercetin (taxifolin) are mainly found in the outer bark as shown by studies of radiata pine, Virginia pine and Aleppo pine *(Pinus halepensis* var. *brutia*) (Yazaki, Hillis 1977; Laboski, Sellers 1980; Weissmann, Ayla 1980).

As observed in Douglas-fir bark dihydroquercetin may be deposited in the form of white crystals in the lumina of cork cells (Krahmer, Wellons 1973). Tannins and other phenolic extractives also seem to be located in the cell lumina whereas waxes are located within the cell walls.

The extractives soluble in hexane, benzene and ether are compounds belonging to the group of fats, waxes and their components as well as of terpenes. A very

Table 9–12: Composition of loblolly pine bark *(Pinus taeda)* (Pearl, Buchanan 1976)

Components	Inner bark %	Outer bark %	Composite bark %
Total fatty acids	3.6	0.65	0.99
Free resin acids	0.44	0.23	0.25
Glycerol (from saponification)	0.39	0.15	0.17
Fatty alcohols	0.07	0.20	0.19
Sterols	0.40	0.26	0.27
Other benzene-soluble components*	1.8	1.9	1.9
Sugars and polygalacturonic acid in extractives	8.0		0.88
Other aqueous-ethanol solubles*	17.9	6.3	7.6
Additional extractives**		6.8	6.1
Suberin		2.0	1.8
Polysaccharides	39.5	28.7	29.9
Phenolic acids, lignin etc.*	27.9	52.8	50.0

* By difference

** With water and with methanol

Table 9–13: Amounts of single compounds in the extractives of loblolly pine bark *(Pinus taeda)* (Pearl, Buchanan 1976)

Compounds	Inner bark %	Outer bark %	Composite bark %
C_{16} and C_{18} free acids			
Palmitic acid	0.026	0.0062	0.0084
Stearic acid	0.003	0.0031	0.0031
Oleic acid	0.067	0.021	0.026
Linoleic acid	0.044	0.014	0.017
Linolenic acid	0.012	0.0039	0.0048
Total	0.15	0.048	0.059
Higher satured free acids			
Arachidic acid	0.009	0.020	0.019
Behenic acid	0.011	0.036	0.033
Lignoceric acid	0.009	0.12	0.11
Cerotic acid	0.004	0.096	0.085
Total	0.033	0.27	0.25
C_{16} and C_{18} combined acids			
Palmitic acid	0.39	0.013	0.055
Stearic acid	0.03	0.004	0.007
Oleic acid	1.39	0.046	0.19
Linoleic acid	1.21	0.041	0.17
Linolenic acid	0.12	0.004	0.017
Total	3.1	0.11	0.44
Higher satured combined acids			
Arachidic acid	0.16	0.016	0.032
Behenic acid	0.18	0.041	0.056
Lignoceric acid	0.06	0.10	0.097
Cerotic acid		0.058	0.052
Total	0.40	0.22	0.24
Resin acids			
Palustric acid	0.024	0.022	0.022
Isopimaric acid	0.018	0.005	0.006
Abietic acid	0.24	0.072	0.092
Dehydroabietic acid	0.15	0.11	0.11
Neoabietic acid	0.012	0.021	0.020
Total	0.44	0.23	0.25
Sterols			
β-Sitosterol	0.29	0.060	0.085
Campesterol	0.04	0.004	0.008
Total	0.33	0.064	0.093
Fatty alcohols			
Eicosanol	0.014	0.007	0.008
Docosanol	0.004	0.005	0.005
Tetracosanol	0.048	0.13	0.12
Hexacosanol	0.008	0.053	0.048
Total	0.074	0.20	0.18

detailed analysis of the compounds of loblolly pine bark *(Pinus taeda)* soluble in non-polar organic solvents was achieved by Pearl and Buchanan (1976). They analysed the extracts from inner and outer bark separately and calculated the amounts of the components for the composite bark from the percentages of both parts of the bark (11.1% and 88.9%) (Table 9–12). The single components could be determined even in amounts of 0.004% based on unextracted ovendry bark solids (Table 9–13). Similar compositions concerning the free and combined acids and the fatty alcohols were evaluated in the extractives of barks from slash pine *(Pinus elliottii)*, white fir *(Abies concolor)*, Douglas fir *(Pseudotsuga menziesii)* and mountain hemlock *(Tsuga mertensiana)* (Kurth 1967; Loveland, Laver 1972; Pearl 1975; Laver et al. 1977).

Sterols present in the benzene extracts from barks of jack pine *(Pinus banksiana)*, lodgepole pine *(P. contorta)*, sugar pine *(P. lambertiana)* and loblolly pine *(P. taeda)* were analysed by Rowe (1965). In all cases β-sitosterol was the predominant component, followed by campesterol with amounts of about one tenth of the β-sitosterol portion. Other sterols such as cholesterol, α-sitosterol etc. were found only in traces.

As shown in Table 9–13 the amount of resin acids (diterpenes) is about 0.25% for bark of loblolly pine. Detailed studies of the terpenes in pine barks were made by Rowe (1964; Rowe, Shaffer 1965; Rowe et al. 1971, 1972). During these studies more rare terpenes were identified such as nordehydroabietane (VIII) derivatives in jack pine *(Pinus banksiana)* and western white pine *(P. monticola)*, oplopanone (VII), epimanool (IX) and agathadiol (X) in lodgepole pine *(P. contorta)*, and serratenediol (XI) in jack pine *(P. banksiana)*, sugar pine *(P. lambertiana)*, loblolly pine *(P. taeda)* and longleaf pine *(P. palustris)* (Fig. 9–11).

A further group of compounds present in bark extractives are the glycosides, the most important of them being the hydrolyzable tannins (→ 7.3.4.). Tannins and their aglycones gallic and ellagic acid were isolated e.g. from barks of European and sessile oak *(Quercus robur, Q. petraea)*, sweetgum *(Liquidambar styraciflua)* and western hemlock *(Tsuga heterophylla)* (Hathway 1958, 1959; Hergert et al. 1965; Spencer, Choong 1977).

Stilbenes (piceatannol, astringenin etc.) have been isolated as aglycones from the barks of various spruce species *(Picea abies, P. engelmannii, P. glauca, P. mariana)* (Grassmann, Endres 1965; Manners, Swan 1971; Pearson et al. 1977). The hot-water extract from cork of cork oak *(Quercus suber)* contains catechin, orcinol, trihydroxybenzoic acid and gallic acid (Soares, Teodoro 1973).

In a series of investigations Pearl (1969; Pearl, Darling 1965; Pearl, Estes 1965; Estes, Pearl 1967; Erickson et al. 1970) was engaged with water-soluble extractives and particularly with the glycosides from the barks of various American *Populus* species *(P. balsamifera, P. grandidentata, P. tremuloides, P. trichocarpa)*. The glycosides consist of glucose which is combined e.g. with salicyl alcohol (salicin XII),

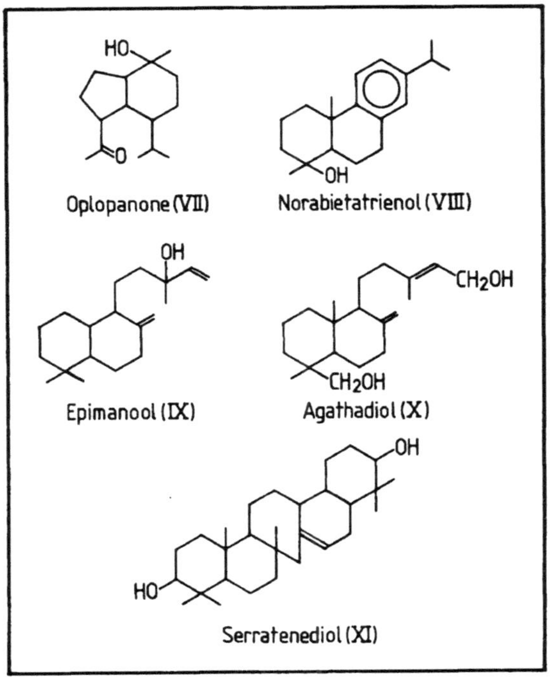

Fig. 9–11. Sesquiterpene, diterpenes and triterpene from the barks of various pines.

benzoic acid (tremuloidin XIII, populin XIV), gentisic acid and its benzylester (salireposide XV, trichocarpin XVI), p-coumaric acid, and 1,2-cyclohexanediol (grandidentin XVII) (Fig. 9–12). The amount of the predominants of these glycosides is less than 1% of the dry bark, e.g. salicin in bigtooth aspen *(P. grandidentata)* 0.6%, trichocarpin in black cottonwood *(P. trichocarpa)* 0.75%. Glycosides with a sugar component of glucose or rhamnose and an aglycon consisting of derivatives of glycerol or dehydrobenzofuran (XVIII) were isolated from the inner bark of Japanese larch (*Larix leptolepis,* Fig. 9–12) (Miki, Sasaya 1979).

9.2.8. Inorganic Components and pH Value

Bark is generally richer in minerals than the corresponding wood. The frequency of the elements is also different from that of wood. The analyses of the elements in the barks and woods of American broad-leaved trees show that the ash content is usually more than 10% of the bark and more than ten times higher than in wood (Table 9–14) (Choong et al. 1976). The predominant element is calcium (82–95%); potassium and magnesium occur in secondary quantities. In most cases the amount of the other elements is less than 1% of the sum of all elements.

The variation in the content of Ca, Mg and Mn in wood and bark of pine (*Pinus* spec.) was studied by Fossum et al. (1972). Except for the cambium they found a decrease of the Mg- and Mn-content and an increase of the Ca-content from wood

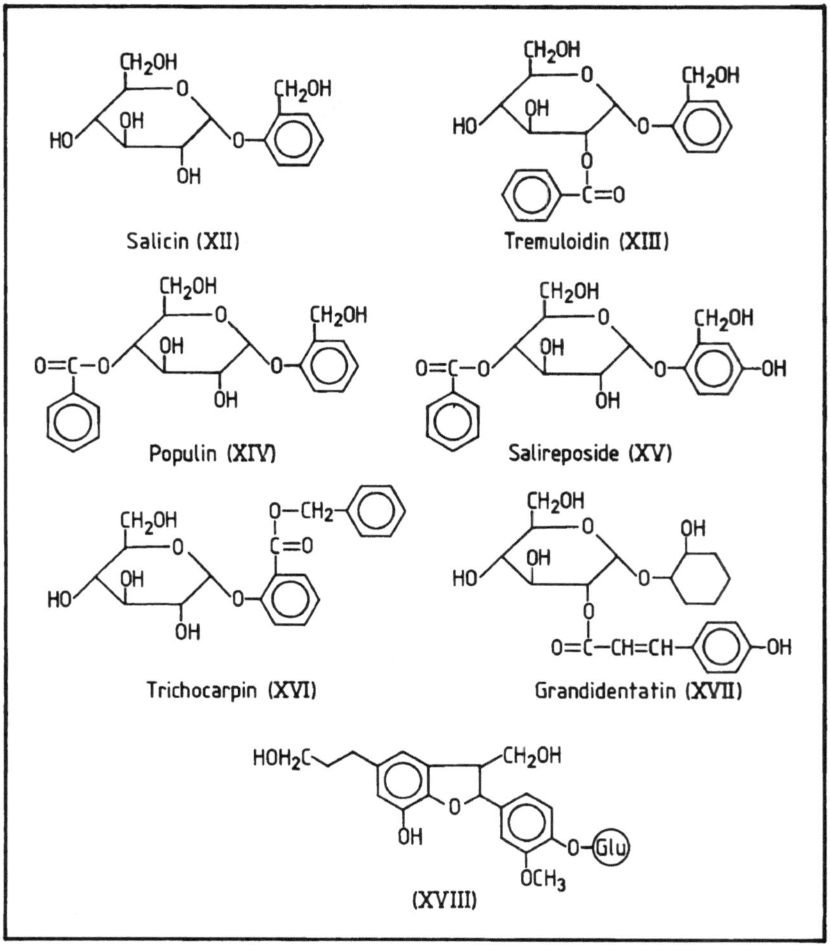

Fig. 9–12. Various glycosides from the barks of *Populus* species (XII–XVII) and of *Larix leptolepis* (XVIII).

to bark. The mineral compounds are distributed over the whole cell walls with a certain concentration in the rays and resin canals. The high calcium content is caused by the presence of calcium oxalate crystals, which are found in sieve cells and longitudinal parenchyma, where they may occupy a large part of the lumina (Wattendorff 1969, 1978; Howard 1971).

Bark is more acidic than wood because of the higher content of acidic compounds. Martin and Gray (1971) determined the pH values of 9 southern pines ranging from about 3.1 to 3.8, averaging from 3.4 to 3.5 with very little differences between most species. They measured a hot-water extract of 4 g of ground 40-mesh bark in 25 g of distilled water. In the cold-water extract of ground 40-mesh barks from 7 deciduous trees Murphy et al. (1970) determined pH values between 4.9 (red maple, *Acer rubrum*) and 6.0 (American elm, *Ulmus americana;* American ash, *Fraxinus americana*). pH values of 5.5 were measured in the extracts of sugar maple *(Acer sac-*

Table 9–14: Mineral content of bark and wood of various American deciduous trees (Choong et al. 1976)

Species Tissue	Ash* %	Na %	K %	Ca %	Mg %	Mn %	Zn %	P %
Willow								
Outer bark	11.5	0.82	6.41	88.80	2.24	0.79	0.39	0.53
Inner bark	13.1	0.91	14.01	81.85	1.31	0.53	0.30	1.08
Sapwood	0.9	4.89	52.17	26.42	2.85	2.24	0.88	10.53
Sweetgum								
Outer bark	10.4	0.45	2.07	94.42	1.98	0.38	0.18	0.51
Inner bark	12.8	0.46	5.29	90.64	2.45	0.22	0.16	0.77
Sapwood	0.5	3.38	43.80	27.98	10.63	0.63	4.71	8.88
Red oak								
Outer bark	8.9	0.31	3.45	92.06	2.58	0.97	0.09	0.54
Inner bark	11.1	0.26	2.58	95.00	1.24	0.34	0.07	0.51
Sapwood	0.9	2.22	42.14	40.34	6.12	0.53	0.57	8.07
Ash								
Outer bark	12.3	0.22	5.37	90.38	3.08		0.50	0.45
Inner bark	12.1	0.23	12.90	82.44	2.81		0.40	1.21
Sapwood	0.9	5.09	49.34	30.13	3.43		0.45	11.51

* Based on ovendry matter, other values based on the sum of analyzed elements

charum), American white oak *(Quercus alba)* and black walnut *(Juglans nigra),* a pH value of 5.9 in beech bark *(Fagus grandifolia).*

A comparison of hot- and cold-water extracts from bark and wood of European trees was made by Volz (1971) (Table 9–15). In both tissues acidic compounds are likewise liberated by hot-water treatment thus these extracts have a lower pH than the cold-water extracts. During this study it was determined that the outer bark is more acidic than the inner bark, and that there is a slight decrease in the pH of bark with the age of a tree.

Table 9–15: pH Values in cold and hot water extracts from bark and wood of European trees (Volz 1971)

	Bark		Wood	
Species	cold water pH	hot water pH	cold water pH	hot water pH
Picea abies	3.9	3.6	4.9	4.6
Pinus sylvestris	3.8	3.5	4.4	4.2
Fagus sylvatica	5.4	5.0	5.5	5.3
Quercus robur	4.2	3.9	4.9	4.8

Literature

Andersson, A., Erickson, M., Fridh, H. and Miksche, G. E. 1973, Holzforschung *27,* 189–193
Bhattacharjee, S. S. and Timell, T. E. 1965, Can. J. Chem. *43,* 758–765
Binotto, A. P. and Murphy, W. K. 1975, Wood Sci. *7,* 185–190
Binotto, A. P., Murphy, W. K. and Cutter, B. E. 1971, Wood Fiber *3,* 179–181
Cassens, D. L. 1974, For. Prod. J. *24,* No. 4, 40–44
Choong, E. T., Abdullah, G. and Kowalczuk, J. 1976, LSU Wood Utilization Notes No. 29
Clermont, L. P. 1970, Tappi *53,* 52–57
Dietrichs, H. H. 1975, Holz Roh-Werkst. *33,* 13–20
Dietrichs, H. H., Garves, K., Behrensdorf, D. and Sinner, M. 1978, Holzforschung *32,* 60–67
Erickson, R. L., Pearl, I. A. and Darling, S. F. 1970, Tappi *53,* 240–244
Erman, W. F. and Lyness, W. I. 1965, Tappi *48,* 249–256
Esau, K. 1964, Structure and Development of the Bark in Dicotyledons. In: Formation of Wood in Forest Trees (Zimmermann, M. H., Ed.). Academic Press, New York, pp. 37–50
Esau, K. 1965, Plant Anatomy, 2nd Ed.. J. Wiley, New York, London
Estes, T. K. and Pearl, I. A. 1967, Tappi *50,* 318–323
Fahey, M. D. and Kurth, E. F. 1957, Tappi *40,* 506–512
Fang, P. and McGinnis, G. D. 1975, Appl. Polym. Symp. No. 28, 263–376
Fossum, T., Hartler, N. and Libert, J. 1972, Svensk Papperstid. *75,* 305–309
Fu, Y. L. and Timell, T. E. 1972a, Cell. Chem. Technol. *6,* 517–519
Fu, Y. L. and Timell, T. E. 1972b, Cell. Chem. Technol. *6,* 513–515
Fu, Y. L., Gutmann, P. J. and Timell, T. E. 1972, Cell. Chem. Technol. *6,* 507–512
Fujii, M. and Kurth, E. F. 1966, Tappi *49,* 92–96
Garves, K. 1976, Cell. Chem. Technol. *10,* 249–259
Gowda, J. P., Gowda, D. C. and Anjaneyala, Y. V. 1980, Carbohydr. Res. *87,* 241–248
Grassmann, W. and Endres, H. 1959, Leder *10,* 237–244
Guillemonat, A. and Triaca, M. 1968, Bull. Soc. Chim. France No. 3, 950–952
Gvozdeva, E. N., Artemeva, I. S. and Levanova, V. P. 1979, Khim. Drev. No. 3, 100–102
Haas, B. R. and Kremers, R. E. 1961, Tappi *44,* 747–748
Harder, M. L., Einspahr, D. W. and Parham, R. A. 1978, Tappi *61,* No. 11, 121–122
Hathway, D. E. 1958, Biochem. J. *70,* 34–42
Hathway, D. E. 1959, Biochem. J. *71,* 533–537
Hathway, D. E. 1962, The Condensed Tannins. In: Wood Extractives and Their Significance to the Pulp and Paper Industry (Hillis, W. E., Ed.). Academic Press, New York, London, pp. 191–228
Hemingway, R. W. and McGraw, G. W. 1976, Appl. Polym. Symp. No. 28, 1349–1364
Hergert, H. L 1958, For. Prod. J. *8,* 335–339
Hergert, H. L 1960, For. Prod. J. *10,* 610–617
Hergert, H. L., Van Blaricom, L. E., Steinberg, J. C. and Gray, K. R. 1965, For. Prod. J. *15,* 485–491
Higuchi, T., Ito, Y., Shimada, M. and Kawamura, I. 1967, Cell. Chem. Technol. *1,* 585–595
Holdheide, W. 1951, Anatomie mitteleuropäischer Gehölzrinden. In: Handbuch der Mikroskopie in der Technik (Freund, H., Ed.). Umschau-Verlag Frankfurt/M., Vol. V/1, pp. 193–367
Holloway, P. J. 1972, Chem. Phys. Lipids *9,* 158–170
Holmes, G. W. and Kurth, E. F. 1961, Tappi *44,* 893–898
Howard, E. T. 1971, Wood Sci. *3,* 134–148
Howard, E. T. 1977, Wood Fiber *9,* 171–183
Jensen, W., Ihalo, P. and Varsa, K. 1957, Pap. Puu *39,* 237–242
Jensen, W., Fremer, K. E., Sierilä, P. and Wartiovaara, V. 1963, The Chemistry of Bark. In: The Chemistry of Wood (Browning, B. L., Ed.). Intersci. Publ., New York, London, pp. 587–666
Jiang, K. S. and Timell, T. E. 1972a, Cell. Chem. Technol. *6,* 493–498

Jiang, K. S. and Timell, T. E. 1972b, Cell. Chem. Technol. *6,* 503–505

Jiang, K. S. and Timell, T. E. 1972c, Cell. Chem. Technol. *6,* 499–502

Krahmer, R. L. and Wellons, J. D. 1973, Wood Sci. *6,* 97–105

Kramár, A. and Ebringerová, A. 1976, Vysk. Prac. Odbora Pap. Celul. *21,* V38–V40

Kramár, A., Zakutná, L. and Ebringerová, A. 1977, Pap. Celul. *11,* V57–V62

Kurth, E. F. 1967, Tappi *50,* 253–258

Labosky, P. 1979, Wood Sci. *12,* 80–85

Labosky, P. and Sellers, J. A. 1980, Wood Sci, *13,* 32–35

Laver, N. L., Copeland, P. M., Chen, C. H., Fang, H. H. L., Liu, Y. C. L. and Zerrudo, J. V. 1977, Wood Sci. *10,* 85–92

Litvay, J. D. and Krahmer, R. L. 1976, Wood Fiber *8,* 146–152

Litvay, J. D. and Krahmer, R. L. 1977, Wood Sci. *9,* 167–173

Loveland, P. M. and Laver, M. L. 1972, Phytochemistry *11,* 430–432, 3080–3081

Manners, G. D. and Swan, E. P. 1971, Phytochemistry *10,* 607–610

Martin, R. E. 1969, For. Prod. J. *19,* No. 8, 23–30

Martin, R. E. and Gray, G. R. 1971, For. Prod. J. *21,* No. 3, 49–52

McGinnis, G. D. and Parikh, S. 1975, Wood Sci. *7,* 295–297

Mian, A. J. and Timell, T. E. 1960a, Can. J. Chem. *38,* 1191–1198

Mian, A. J. and Timell, T. E. 1960b, Tappi *43,* 775–781

Miki, K. and Sasaya, T. 1979, Mokuzai Gakkaishi *25,* 361–366, 437–441

Murphy, W. K., Beall, F. C., Cutter, B. E. and Baldwin, R. C. 1970, For. Prod. J. *20,* No. 2, 58–59

Nakano, H. and Côté, W. A. 1980, Bark Structure of Hardwoods Grown on Southern Pine Sites. Syracuse University Press, Syracuse N. Y.

Painter, T. J. and Purves, C. B. 1960, Tappi *43,* 729–736

Parameswaran, N. 1973, Holzforschung *27,* 151–153

Parameswaran, N. and Liese, W. 1970, Mikroskopie der Rinde tropischer Holzarten. In: Handbuch der Mikroskopie (Freund, H., Ed.). Umschau-Verlag, Frankfurt/M. Vol. V/1, pp. 227–306

Parameswaran, N., Faix, O. and Schweers, W. 1975, Holzforschung *29,* 1–4

Paulínyová, E., Rybárik, I. and Šutý, L. 1978, Drev. Vysk. *23,* 15–24

Pavlová, T. A., Tsvetkova, L. Y. and Sharkov, V. I. 1972, Khim. Drev. No. 12, 49–53

Pavlovová, E., Rendoš, F. and Kováč, P. 1970, Cell. Chem. Technol. *4,* 255–260

Pearl, I. A. 1969, Tappi *52,* 428–431

Pearl, I. A. 1975, Tappi, *58,* No. 9, 135–137

Pearl, I. A. and Buchanan, M. A. 1976, Tappi *59,* 136–139

Pearl, I. A. and Darling, S. F. 1965, Tappi *48,* 506–508

Pearl, I. A. and Estes, T. K. 1965, Tappi *48,* 532–535

Pearson, T. W., Kriz, G. S. and Taylor, R. J. 1977, Wood Sci. *10,* 93–98

Pereira, H. 1979, Bol. Inst. Prod. Flor., Cortiça No. 483, 259–264

Porter, L. J. 1974, N.Z.J. Sci. *17,* 213–218

Ramalingam, K. V. and Timell, T. E. 1964, Svensk Papperstid. *67,* 512–521

Roux, D. G., Ferreira, D., Hundt, H. K. L. and Malan, E. 1975, Appl. Polym. Symp. No. 28, 335–353

Rowe, J. W. 1964, Tetrahedron Lett. No. 34, 2347–2353

Rowe, J. W. 1965, Phytochemistry *4,* 1–10

Rowe, J. W. and Shaffer, G. W. 1965, Tetrahedron Lett. No. 30, 2633–2637

Rowe, J. W., Nagasampagi, B. A., Burgstahler, A. W. and Fitzsimmons, J. W. 1971, Phytochemistry *10,* 1647–1651

Rowe, J. W., Ronald, R. C. and Nagasampagi, B. A. 1972, Phytochemistry *11,* 365–369

Samejima, M. and Yoshimoto, T. 1979, Mokuzai Gakkaishi *25,* 671–677

Samejima, M. and Yoshimoto, T. 1981, Mokuzai Gakkaishi *27,* 491–497, 606–609

Sano, Y. and Tanaka, K. 1976, Res. Bull. Coll. Exp. For. Hokkaido Univ. *33,* 223–233

Sarkanen, K. V. and Hergert, H. L. 1971, Lignins and Phenolic Polymers in Tree Barks. In: Lignins, Occurrance, Formation, Structure and Reactions (Sarkanen, K. V. and Ludwig, C. H., Eds.). Wiley-Intersci., New York, London, Sydney, Toronto, pp. 81–89
Sitte, P. 1962, Protoplasma *54,* 555–559
Soares, J. M. A. and Teodoro, J. R. 1973, Bol. Inst. Prod. Flor., Cortiça *35,* 192–196
Spencer, R. W. and Choong, E. T. 1977, Holzforschung *31,* 25–31
Srivastava, L. M. 1964, Int. Rev. For. Res. *1,* 203–277
Srivastava, L. M. 1966, Tappi *49,* 173–183
Swan, E. P. 1968, Tappi *51,* 301–304
Timell, T. E. 1961a, Svensk Papperstid. *64,* 651–661, 685–688
Timell, T. E. 1961b, Svensk Papperstid. *64,* 748–750
Timell, T. E. 1961c, Svensk Papperstid. *64,* 744–747
Timell, T. E. 1962, Svensk Papperstid. *65,* 843–846
Timell, T. E. 1965, Wood and Bark Polysaccharides. In: Cellular Ultrastructure of Woody Plants (Côté, W. A., Ed.). Syracuse University Press, Syracuse, N. Y., pp. 127–156
Timell, T. E. and Mian, A. J. 1961, Tappi *44,* 788–793
Toman, R. 1973, Cell. Chem. Technol. *7,* 351–357
Toman, R. and Karásonyi, S. 1972, Collect. Czech. Chem. Commun. *37,* 3640–3645
Toman, R., Karásonyi, S. and Kováčik, V. 1972, Carbohyd. Res. *25,* 371–378
Volz, K. R. 1971, Holz-Zentralbl. *97,* 1783
Wattendorff, J. 1969, Z. Pflanzenphysiol. *60,* 307–347
Wattendorff, J. 1973, Verh. Schweiz, Naturforsch. Ges. 71–73
Wattendorff, J. 1978, Protoplasma *95,* 193–206
Weissmann, G. 1981, Holz Roh-Werkst. *39,* 457–461
Weissmann, G. and Ayla, C. 1980, Holz Roh-Werkst. *38,* 307–312
Wilson, J. D. 1961, For. Prod. J. *11,* 260–263
Yazaki, Y. and Hillis, W. E. 1977, Holzforschung *31,* 20–24
Young, H. E. 1971, For. Prod. J. *21,* No. 5, 56–59

10. Reactions in Acidic Medium

10.1. General Aspects of Wood Reactions

Like other materials wood is subjected to very different conditions and media in its wide field of utilization. Many reactions are unavoidable, others are desired for special analytical purposes or for obtaining technical products.

Changes of wood composition and properties may be caused by solutions or gases of different pH values as described in this and the following chapter, by the influence of temperature (→ 12.) and radiation (→13.) as well as by enzymatic degradation (→14.) and aging over long periods (→15.). Derivation reactions of cellulose are important both in basic research and in the production of technical products (→17.).

Reaction mechanisms and kinetic considerations are mainly evaluated by use of low-molecular-weight model compounds reacting in homogeneous phase.

The reactions of wood are predominantly reactions of the main macromolecular cell wall components cellulose, polyoses and lignin, though extractives may also be involved, e.g. in pulping.

From a practical standpoint the most important reactions of lignin and wood polysaccharides take place during pulping and bleaching, mainly as delignification reactions of wood and pulp, respectively.

10.2. Reactions of Polysaccharides

10.2.1. Mechanism of Acidic Hydrolysis

Besides degradation by enzymatic attack, chemical acidic hydrolysis is the essential and most typical degradation reaction of glycosidically linked glycosides, di-, oligo- and polysaccharides.

With regard to wood polysaccharides, hydrolysis deals with cellulose and the different polyoses in analytical and technical procedures. While hydrolytic effects are undesired e.g. in acidic pulping and bleaching (e.g. Pfister, Sjöström 1977; Teder, Tormund 1978), a total hydrolysis of cellulose is necessary in wood saccharification to obtain high yields of glucose (→ 18.3.).

The principal molecular mechanism of acidic hydrolysis, which was confirmed by model and tracer studies, is outlined in Fig. 10–1. Acidic hydrolysis, leading finally

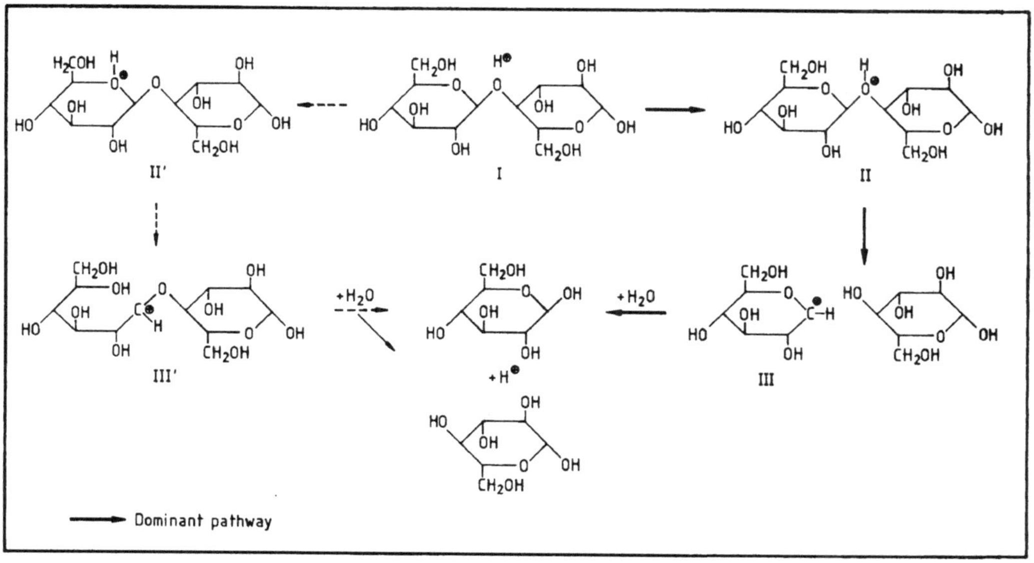

Fig. 10–1. Mechanism of acidic hydrolysis of glycosidic linkages.

to a fission of glycosidic bonds, proceeds in three steps. In the first step the proton of the catalyzing acid interacts rapidly with the glycosidic oxygen linking two sugar units (I), forming a so-called conjugate acid (II). This step is followed by a slow cleavage of the C–O bond yielding an intermediate cyclic carbonium cation (III). The protonation may also occur at the ring oxygen (II'), resulting in a ring opening and a non-cyclic carbonium cation (III'). It is uncertain which type of carbonium ion is most likely to be formed. Probably both modifications of protonation take place with a greater likelihood of the cyclic cation in the most cases. The tautomeric carbonium-oxonium ion occurs in a half-chair conformation as demonstrated in Fig. 10–2. The carbonium cation finally initiates rapid addition of a water molecule, resulting in the stable end product and release of the proton (Shafizadeh 1963; Timell 1964a; Harris 1975; Szejtli 1976; Philipp et al. 1979).

In the case of solvolysis with acid-containing solvents in the presence or absence of water the cleavage of glycosidic linkages is mainly governed by the type of solvent.

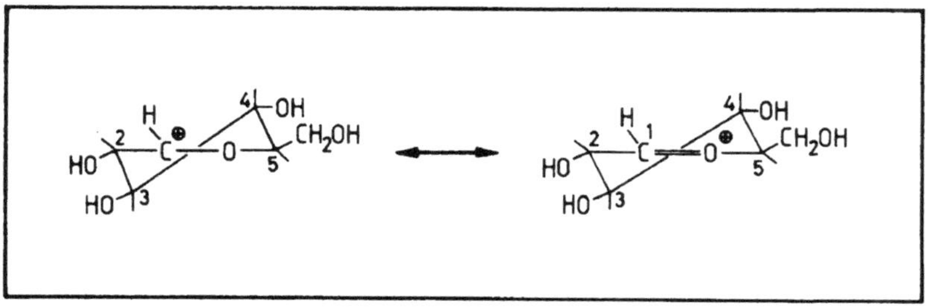

Fig. 10–2. Half-chair conformation of the tautomeric carbonium-oxonium ion pair.

The non-aqueous solvents which are miscible with water, such as ethanol or dioxane, may act as reaction partners (e.g. ethanol) or not (e.g. dioxane). The addition of benzene is only used in pure alcoholic solutions, resulting in exclusively solvolytic reactions.

Generally, solvolytic effects improve the cleavage rate of glycosidic bonds. It was shown e.g. that in alcoholysis of cellulose in the presence of water (96% ethanol) the reaction rate was substantially increased as compared to hydrolysis under comparable conditions, probably due to higher activation energy of the solvolysis system (Sharkov 1961; Szejtli 1976).

Solvolysis of wood polysaccharides in the system dioxane/H_2O/HCl was also found to be much faster than without dioxane (Poller, Hohmuth 1979). Methanolysis in hydrochloric acid medium was used to yield methyl mannoside from wood prehydrolysis material (Herrick et al. 1975). A special system for homogeneous solvolysis of cellulose, suggested by Garves (1979), calls for dissolving cellulose in a mixture of acetic anhydride/DMF/H_2SO_4.

10.2.2. Factors Influencing Hydrolysis

Though the principles of cleaving glycosidic linkages by acid-catalyzed reactions are generally valid, the special kinetic data and the overall course of degradation are influenced both by the applied acidic medium and the characteristics of the sample.

One of the basic factors influencing hydrolysis is the phase state of the reaction partners, covering 4 main possibilities (Table 10–1). The most important and frequent hydrolytic reactions of cellulose and partly of polyoses take place in heterogeneous phase, with the polysaccharide component occurring in solid state in a more or less acidic aqueous medium. If the sample to be hydrolyzed dissolves easily and totally in the acidic solution, both reaction partners exist in homogeneous phase. Less frequently the combinations of a dissolved or solid sample with an acidic component in a solid state (e.g. a cation-exchange resin) are used for special hydrolysis problems such as the selective cleavage of certain linkages in partial hydrolysis (Browning 1967; Szejtli 1976).

Table 10–1: Different phase conditions of acid hydrolysis

Phase of sample	Phase of acid	Examples
solid	liquid	Cellulose/dilute acid
dissolved	liquid	Cellulose/conc. acid Sucrose/dilute acid
dissolved	solid	Sucrose/ion-exchange resin
solid	solid	Cellulose/ion-exchange resin (pulverized)

The hydrolyzing medium can be characterized by:
- kind of acid
- acid concentration
- pH value
- acid strength
- temperature and pressure.

With regard to the sample, important factors are:
- phase state
- physical structure and accessibility in the case of heterogeneous hydrolysis
- conformation effects
- ring structure and substituents.

Hydrolysis of glycosidic bonds usually follows a first-order reaction (Springer 1966; Daruwalla, Shet 1962; Szejtli 1976). The individual hydrolysis rate depends on the parameters of the acid and the sample, as mentioned above.

The choice of acid and its concentration depends on the kind of sample and the aim of hydrolysis. In wood analysis (→ 3.2.7.–3.2.9.) and wood saccharification (→ 18.3.) mostly mineral acids are applied at various concentrations, including sulfuric acid, hydrochloric acid, but also phosphoric acid, nitric acid, and more recently trifluoroacetic acid (TFA). Acids are characterized by their acid strength values (pK_a-values) as listed for some important acids in Table 10–2. Apart from the pK_a value the hydrolytic activity (expressed by the pH value) and the activation coefficient of the hydronium ion (expressed by the Hammett acidity function) also influence the overall hydrolysis rate of glycosidic linkages. Additional parameters are temperature and pressure. Generally, increased temperature and pressure accelerate the hydrolysis velocity to some extent, depending on the special acid characteristics mentioned above (Vink 1966; Szejtli 1976).

Kinetic equations are generally used to elucidate reactions in homogeneous phase. Therefore they apply exactly only to glycosides, di-, oligo- and those polysaccharides which are readily and totally soluble in the hydrolyzing medium.

For cellulose generally an initial heterogeneous reaction can be stated, even if cellulose dissolves after some time in strong acids such as 72% H_2SO_4, 41% HCl or

Table 10–2: pK_a values of several acids

Acid	pK_a
HCl	−6
H_2SO_4	−3
HNO_3	−1.32
CF_3COOH	0.23
H_3PO_4	1.96
HCOOH	3.7
CH_3COOH	4.8

100% TFA. In this case, bulk hydrolysis is accomplished as a homogeneous reaction with cellulose occurring intermediately as an ester or addition compound (→ 4.2.2.) (Valtsaar, Dunlap 1952; Geddes 1956; Rydholm 1965; Fengel et al. 1978).

A totally heterogeneous reaction takes place in the case of partial hydrolysis with highly diluted acids, yielding so-called hydrocellulose, a product with reduced DP (LDP, → 4.3.3.) but higher crystallinity (Nelson 1960). In the case of classical acetolysis, applying acetic anhydride and conc. H_2SO_4 on cellulose, cellobiose octaacetate is formed by combined hydrolysis and substitution (Browning 1967).

The rates of heterogeneous hydrolysis of celluloses, which are 1–2 orders of magnitude less than those of homogeneous hydrolysis of model compounds, are mainly influenced by the degree of crystallinity and the swelling state of cellulose. Both characteristics can be influenced by mechanical disintegration and/or decrystallization procedures (→ 18.3.) (Ant-Wuorinen, Visapää 1969, 1970; Visapää 1971, 1972; Philipp et al. 1979; Millett et al. 1979).

The theoretical and fundamental relations of the molecular structure, conformation and interunit linkages of polysaccharides to hydrolysis data were evaluated in numerous model experiments (Szejtli 1976). The hydrolytic behaviour of glycosidic bonds is substantially influenced by two facts effective by themselves or much more frequently overlapping each other. The first one is the conformation of the sugar units, the other one is the so-called inductive effect, caused by certain substituents.

The half-chair conformation occurring intermediately during the hydrolytic attack (Fig. 10–2) is caused by a slight rotation of the substituents around the bonds between the carbon atoms 2 and 3, and C4 and C5 respectively. The possibility and extent of this rotation depends on the size and steric position of the ring substituents. Generally hydrolysis is supported if the axial substituents change to an equatorial position (→ 4.2.1., Fig. 4–4). For example the hydroxyl group at the C2 in β-D-glucose changes from the equatorial to the axial position, while the opposite applies to β-D-mannose; thus mannose is hydrolyzed more rapidly than glucose.

As the hydrolysis rate increases with the number of axial groups, the β-anomers are generally hydrolyzed more rapidly than the corresponding α-forms, with exception of L-arabinose (Table 10–3) (Nakano, Rånby 1962; Shafizadeh 1963).

The relations between the hydrolysis rates of the important sugars occurring in wood polysaccharides were found to be similar to β-D-glucose: β-D-mannose: β-D-galactose: β-D-xylose = 1:3:4–5:4–6 (Shafizadeh 1963; Harris 1975). Other conformation effects are the faster hydrolysis of the linkage to the non-reducing end group as compared to the reducing end, and the impeding effect of large C5-substituents on hydrolysis (Timell 1964b; Harris 1975).

Table 10–3: Hydrolysis of methylpyranosides (according to Shafizadeh 1963)

Hexose/Pentose	Relative rate of hydrolysis
α-D-Glucose	0.4
β-D-Glucose	0.8
α-D-Mannose	1.0
β-D-Mannose	2.4
α-D-Galactose	2.2
β-D-Galactose	3.9
α-D-Xylose	1.9
β-D-Xylose	3.8
α-L-Arabinose	5.5
β-L-Arabinose	3.8

Furanosidic ring structures are hydrolyzed more rapidly than pyranosidic rings, due to higher structural angle strains in the conformation of furanosidic sugar units as compared to the strain-free pyranose ring. Thus, for example, α-D-galactofuranosides are hydrolyzed about 100 times more quickly than α-D-galactopyranosides (Shafizadeh 1963).

The term inductive effect describes the fact that different ring substituents cause an alteration of the electron density of the ring oxygen. Electrophilic substituents such as carboxyl or carbonyl groups reduce protonation and impede C–O fission, thus having a stabilizing effect on the glycosidic bond. This was even demonstrated for hydroxyl groups, as the relative hydrolysis velocities of model compounds with none, 1 or 2 hydroxyl groups are related by 3 100:2.6:1. This explains why, for example, deoxyglycosidic units are hydrolyzed more rapidly than the corresponding glycosides (Marchessault, Rånby 1959; Shafizadeh 1963).

Though hydroxyl groups are the only principal functional groups in the idealized cellulose chain, isolated celluloses from wood or other plant materials contain a greater or lesser number of other functional groups (mainly carboxyl groups), randomly distributed along the chains. By inductive effects these modified units cause β-(1→4)-linkages, which are more sensitive than the normal ones.

The hydrolysis velocity of these 'weak linkages' is about twice as fast as that of normal bonds. Fig. 10–3 demonstrates the inductive effect caused by the chemically

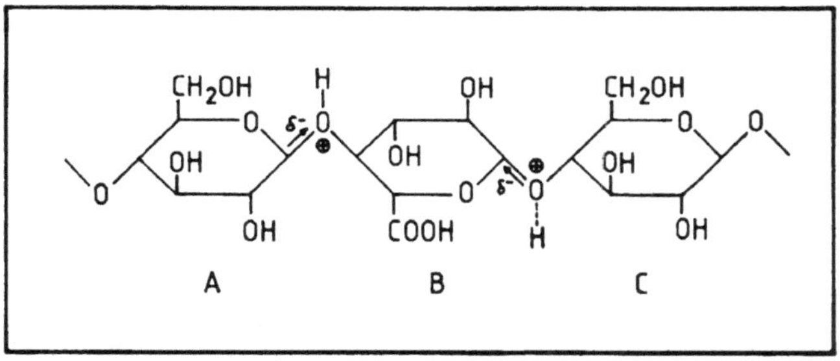

Fig. 10–3. Inductive effect of the carboxyl group on acidic hydrolysis of polysaccharides.

modified unit (glucuronic acid group B). The carboxyl group induces different electron densities at the glycosidic oxygen atoms between A–B and B–C, respectively. The higher nucleophility at the oxygen atom between B and C impedes protonation. Thus this linkage is stabilized while the glycosidic bond between A and B is activated by the same effect, acting controversially. This fact was proved in principle for both heterogeneous and homogeneous hydrolysis of wood and cotton cellulose (Marchessault, Rånby 1959; Rånby 1961; Daruwalla, Narsian 1966).

While the fact that glucuronides are more resistant to hydrolysis is well established, the formation of weak linkages and its explanation were occasionally questioned (Timell 1964a; Semke et al. 1964).

The stabilizing effect of a modified sugar unit can also be demonstrated by partial hydrolysis of xylan bearing 4-O-methylglucuronic acid side-groups. The uronic acid group stabilizes the linkage to the chain, favouring the formation of aldobiuronic acid. Thus, the generally simpler hydrolysis of xylans as compared to cellulose, and especially the higher hydrolytic resistance of softwood xylans (Xyl:4-O-MGluU ~ 5:1) as compared to hardwood xylans (Xyl:4-O-MGluU ~ 10:1) can be explained (Czirnich, Patt 1976; Hoffmann, Patt 1976). Combined conformational, sterical and inductive effects are also responsible for the rapid cleavage of arabinose side-groups from xylans and galactose side-groups from mannans during acidic sulfite pulping (Makkonen 1967; Pereira et al. 1974; Pfister, Sjöström 1977).

Model experiments on the cleavage of glycosidic bonds by ozone in acidic aqueous solution (pH ~ 3) indicate a direct attack by ozone and the formation of polyoxide intermediates, which react further according to ionic or radical mechanisms (Pan et al. 1981).

10.2.3. Dehydration

Dehydration reactions typically occur during thermal treatments of polysaccharides (→ 12.4.2., 12.4.3.). In addition they are also rather unavoidable side reactions under acidic hydrolysis conditions, causing a decomposition of hydrolyzed sugars. Depending on the acid concentration and the applied temperature, numerous reaction products are possible, most of them being rather unstable or occurring only in very low concentrations (Popoff, Theander 1972, 1976; Harris 1975).

The acid-catalyzed dehydration under mild conditions leads to the formation of anhydro sugars with intramolecular glycosidic linkages, resulting from the elimination of a water molecule from two hydroxyl groups (e.g. 1,6-anhydroglucose (levoglucosan), Fig. 12–14). As these glycosidic linkages can easily be hydrolyzed, a series of further degradation products may be formed, part of them being aromatic and condensed compounds (Fig. 10–4). The most important degradation products with regard to yield and potential utilization (→ 18.4., 18.5.) are the cyclic compounds furfural (2-furaldehyde) formed from pentoses and uronic acids, and hydroxymethylfurfural (5-(hydroxymethyl)-2-furaldehyde) (HMF) from hexose sugars, mainly glucose (Fig. 10–5). High yields of these compounds are only obtained in concentrated acids at elevated temperatures. If the temperature is additionally increased the cyclic HMF molecule is converted to levulinic acid and formic acid.

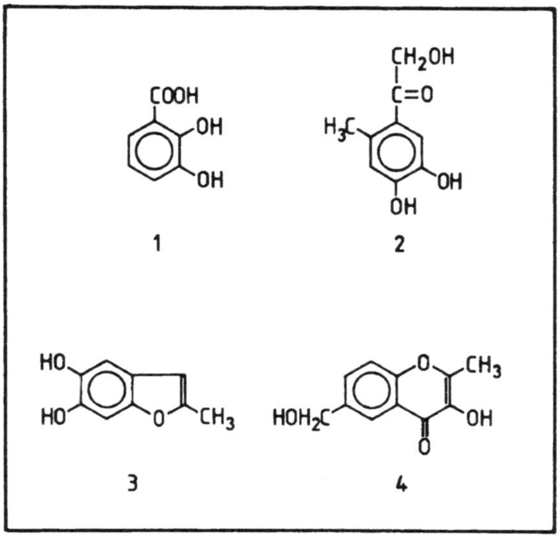

Fig. 10–4. Aromatic compounds resulting from dehydration of sugars in acidic medium
1: 2,3-dihydroxybenzoic acid
2: 1-(3,4-dihydroxy-6-methylphenyl)-2-hydroxyethanone
3: 5,6-dihydroxy-2-methyl-benzofuran
4: 3-hydroxy-6-hydroxymethyl-2-methylchromone.

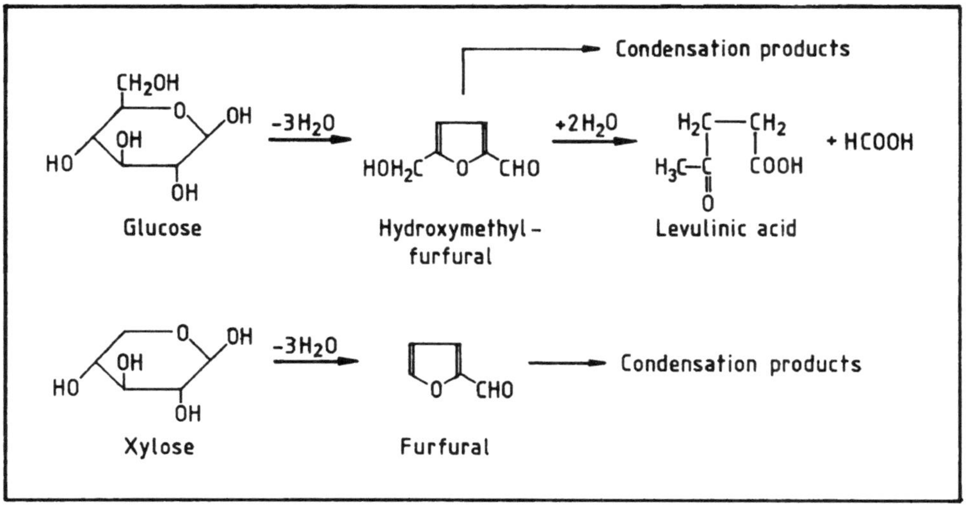

Fig. 10–5. Formation of furfural, hydroxymethylfurfural, levulinic acid and formic acid from monosaccharides in acidic medium.

Furfural is also rather unstable in the dehydrative medium and follows a typical growth-and-decay curve (Root et al. 1959; Harris 1975).

10.2.4. Oxidation

Oxidation reactions involving polysaccharides and sugars in an acidic medium are important, occurring more or less simultaneously with hydrolytic degradation during acidic pulping and bleaching processes.

In principle the hydroxyl groups of the sugar units and the reducing end-groups of di-, oligo- or polysaccharides may be subject to oxidative attack. The converted groups are aldehyde, keto and carboxyl groups. The ring structure can be preserved or may be destroyed by cleavage of the ring oxygen bond or C–C bonds.

Important final degradation acids are uronic acids, e.g. glucuronic acid (I, Fig. 10–6), aldonic acids, e.g. gluconic acid (II), and aldaric acids, e.g. glucaric acid (III). Pentonic acids, e.g. xylonic acid (IV) are formed by decarboxylation of hexuronic acids. Severe oxidation yields keto groups at C2 and C3, finally resulting in a

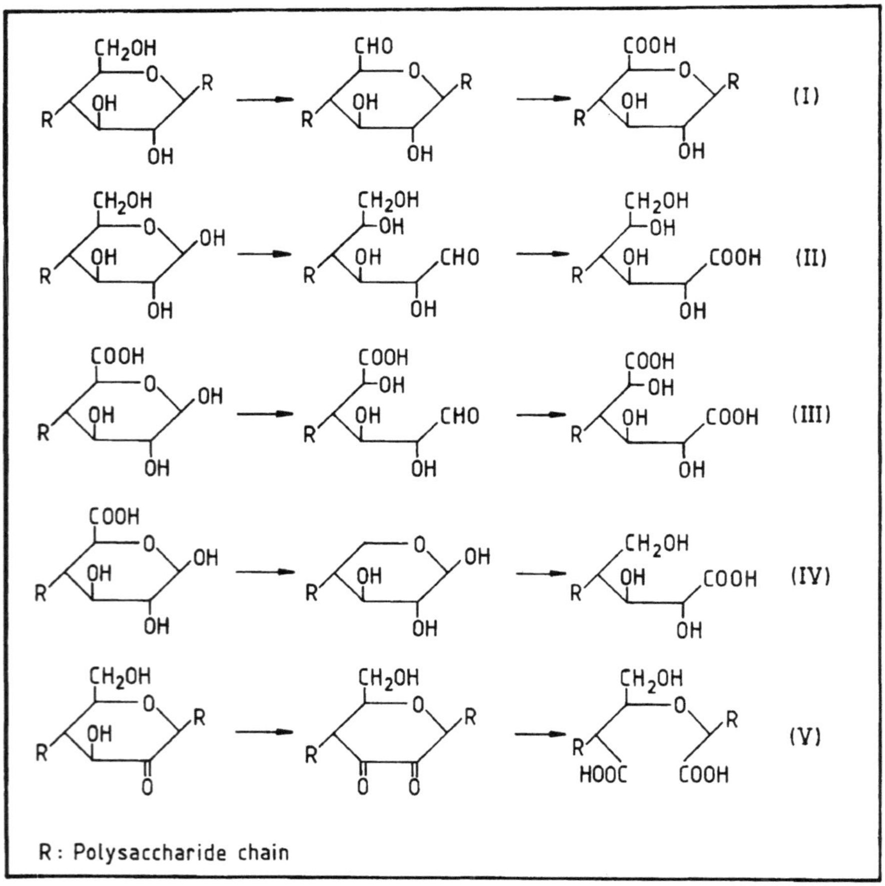

Fig. 10–6. Oxidative formation of different sugar acid units from polysaccharides.

dicarboxylic acid (V). The free aldonic and aldaric acids occur in acidic solutions mainly in the form of their lactones, which show different sensitivities to fission (Slavik et al. 1967).

Fundamental results on the oxidative behaviour of polysaccharides were obtained by selective analytical oxidation procedures. Oxidized preparations of cellulose are generally called oxycelluloses, independent of the kind and degree of oxidative changes. Mild and specific oxidation in an acidic medium is performed e.g. with periodic acid, chlorous acid, aqueous bromine solutions, chromic acid, nitrogen dioxide, lead tetraacetate or potassium permanganate (Mutton 1964; Browning 1967).

The formation of uronic acids is a common fact in acidic pulping. Glucuronic acid occurs both in sulfite pulps and in sulfite spent liquors, partly also deriving from hydrolytic demethylation of 4-O-methylglucuronic acid groups from xylans. Furthermore small amounts of galacturonic acid, cellobiuronic and iduronic acid were found in bisulfite and neutral sulfite spent liquors (Sjöström, Enström 1967; Nelson 1968; Larsson, Samuelson 1969; Pfister, Sjöström 1977).

The action of acidified sodium chlorite on cellulose causes both conversion of primary hydroxyl groups to uronic acid groups along the chain, and oxidation of reducing end-groups to gluconic acid (Alfredsson et al. 1961; Petterson, Samuelson 1968; Smelstorius 1972; Miyazaki et al. 1974; Fengel et al. 1979).

Table 10–4: Monocarboxylic acids in hydrolyzates from cellulose treated with aqueous chlorine solution (Alfredsson, Samuelson 1974)

Acid	Yield mg/100 g cellulose
D-Gluconic	83
Cellobionic	tr.
D-Arabinonic	16
D-Erythronic	9
Glyoxylic	tr.
Cellobiuronic	47
D-Glucuronic	26
2-Deoxy-erythro-pentonic	13
3-Deoxy-ribo-hexonic	tr.
3-Deoxy-arabino-hexonic	tr.
3-Deoxy-erythro-pentonic	22
3-Deoxy-threo-pentonic	10
Glycolic	29
2-Hydroxypropionic	tr.
Glyceric	13
3,4-Dihydroxybutyric	6

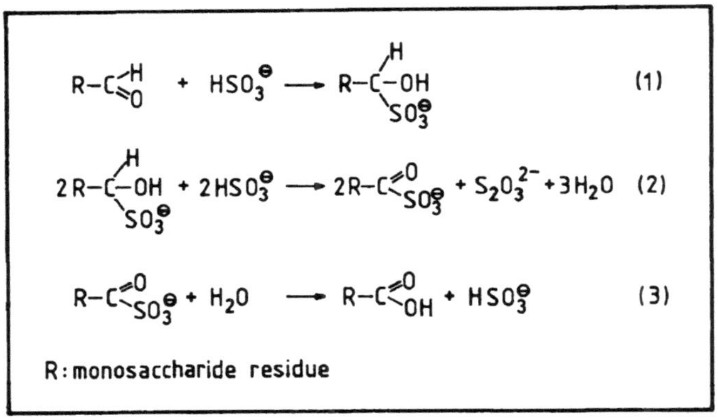

Fig. 10–7. Reactions of hydrogen sulfite with monosaccharides, yielding α-hydroxysulfonic and aldonic acids.

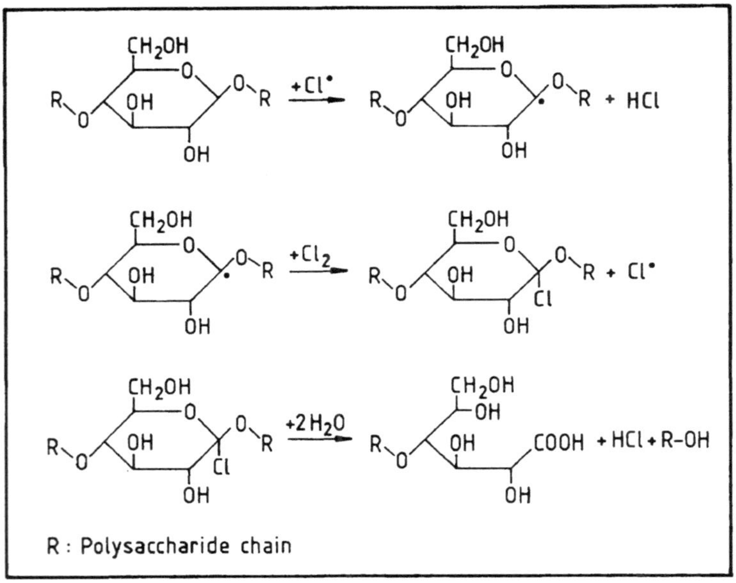

Fig. 10–8. Oxidation of cellulose by chlorine.

Aldonic acids are the predominant oxidized compounds in spent liquors from acidic and neutral sulfite pulping (Nelson 1968; Pfister, Sjöström 1977). Apart from gluconic acid mannonic, xylonic, arabinonic, ribonic and galactonic acids were also identified in smaller amounts in sulfite pulps from spruce (Larsson, Samuelson 1969).

Some of the monosaccharides in sulfite spent liquors react with hydrogen sulfite ions (Fig. 10–7). After addition of a hydrogen sulfite ion, yielding an α-hydroxysulfonic acid (1), oxidation takes place with a resultant reduction of hydrogen sulfite to thiosulfate (2) and formation of an aldonic acid (3) (Herrick et al. 1975).

Oxidation of polysaccharides in aqueous chlorine solutions, as occurs e.g. in lignin-removing bleaching (→ 16.7.3.), causes cleavage of glycosidic linkages governed by a radical mechanism (Fig. 10–8). Mainly gluconic acid is formed, but also numerous other monocarboxylic acids, such as arabinonic, erythronic, deoxy-hexonic and -pentonic acids were identified in chlorine-treated cellulose (Table 10–4). The presence of glucuronic acid also indicates oxidative reactions without chain cleavage (Alfredsson, Samuelson 1974; Andersson, Samuelson 1976).

The oxidative action of gaseous sulfur dioxide and gaseous chlorine on spruce wood is also expected to occur mainly according to the reactions with acidic sulfite and aqueous chlorine solutions (Besold 1982). Oxidation of wood polysaccharides, catalyzed by rusting iron under acidic conditions, yields oxycellulose with reducing properties (Emery, Schroeder 1974).

10.3. Reactions of Lignin

10.3.1. Reactive Sites of Lignin

Lignin reactions can be most simply classified into reactions occurring at the carbon atoms of the propane side-chain, at the aromatic nucleus, at the methoxyl group, and at the less frequent phenolic hydroxyl group (Fig. 10–9).

In principle, lignin or lignin fragments represent p-hydroxy or p-alkoxy arylalkane and arylpropene units with lone electron pairs at the phenolic oxygen atom, which can overlap with the π-electron cloud of the aromatic ring, creating high electron densities (δ^-) at different sites. In arylalkane units these centres are located in ortho- and para-positions, while in units with an aliphatic double bond in conjunction with the ring an additional site of high electron density is the β-C atom (Fig. 10–10). Centres of electron deficiencies (δ^+) are formed by the transition of arylalkane and arylpropene units to intermediate quinone methides by elimination

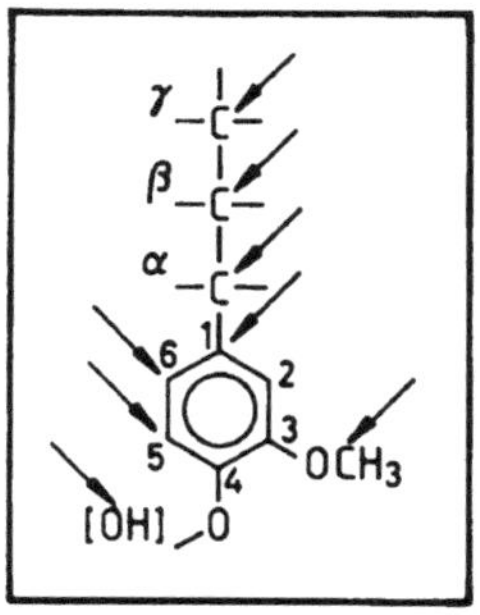

Fig. 10–9. Principal sites of reactions in phenylpropane units.

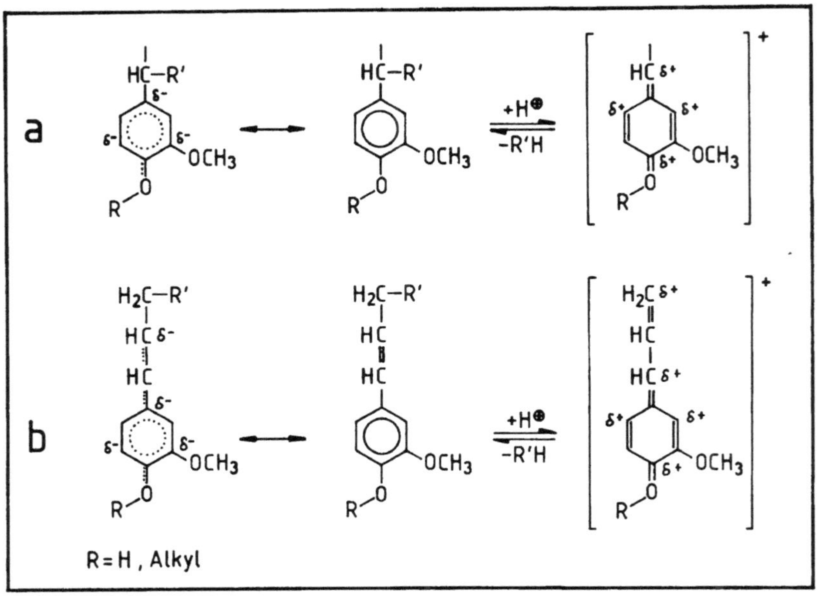

Fig. 10–10. Sites of high (δ^-) and low (δ^+) electron densities in arylalkane (a) and arylpropene (b) units in acidic medium (according to Gierer 1982).

of α- and γ-C-substituents (Fig. 10–10). Positions with high electron densities are the sites of electrophilic attacks while nucleophiles attack the δ^+-positions.

The presence of protons causes the formation of intermediate carbonium ions of different types (Fig. 10–11) which have a high affinity for any nucleophilic reaction partner. The protonation of the frequently occurring hydroxyl or ether groups at the benzylic carbon atom (α-C) of units with free or etherified phenolic hydroxyl groups generates a carbonium-oxonium system (Fig. 10–12). This system ist primarily involved as a reaction partner in lignin fragmentation or condensation reactions, depending on the type of the active nucleophile (Gierer 1970, 1981, 1982; Glennie 1971).

Typical reactions involving carbonium ions can be visualized e.g. by the colour transformation from yellow to green (green resulting from the superposition of yellow and blue), if wood is treated with conc. sulfuric acid, or by the well-known reaction of lignin with phloroglucinol/HCl (Wiesner reaction), yielding a red condensation product (Fig. 10–13). The reactions are restricted to lignins containing coniferyl aldehyde end-groups (Srivastava 1966; Allan 1971).

$$R{-}CH{=}CH{-}CH_2OH + H^{\oplus} \rightarrow R{-}CH{=}CH{-}C^{\oplus}H_2 + HOH$$

$$R{-}CH{=}CH{-}CHO + H^{\oplus} \rightarrow R{-}CH{=}CH{-}C^{\oplus}HOH$$

$$R{-}CH(OR''){-}R' + H^{\oplus} \rightarrow R{-}C^{\oplus}H{-}R' + R''OH$$

Fig. 10–11. Different types of carbonium ions.

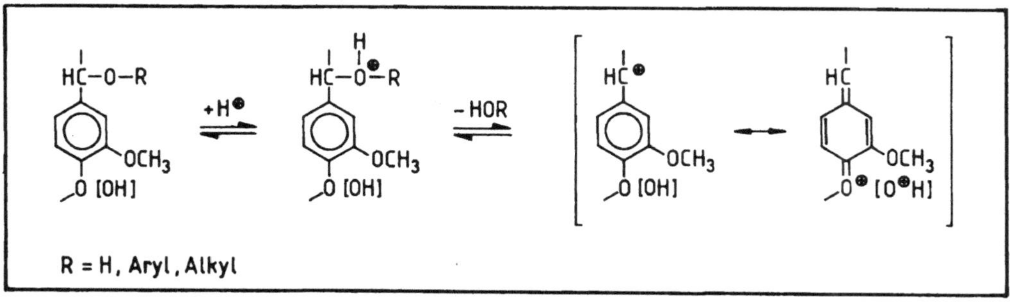

Fig. 10–12. Conversion of α-ether groups to carbonium and oxonium ions, respectively (according to Gierer 1982).

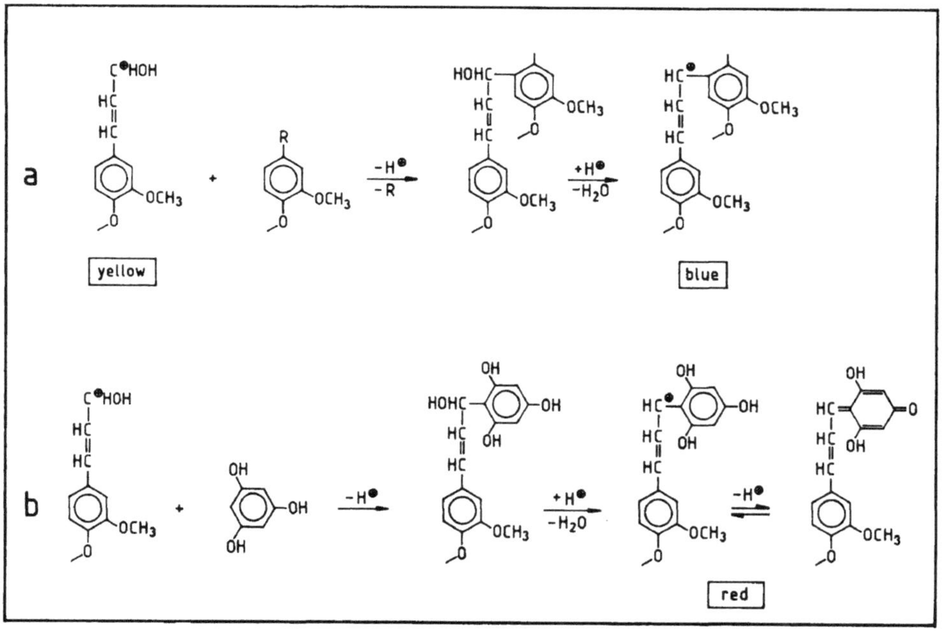

Fig. 10–13. Condensation reactions of lignin with concentrated mineral acids
a: conc. H_2SO_4
b: Phloroglucinol/HCl.

Treatments of wood with concentrated mineral acids do not cause dissolution but condensation reactions of lignin, a fact which is used in analytical lignin determination (acid lignins, → 3.2.9.).

The carbonium ions evidently also react with the acid anions, as sulfuric acid lignin and hydrochloric acid lignin contain HSO_4^- and Cl^- groups, respectively (Yasuda et al. 1980, 1981a, b). Condensation reactions of lignin model compounds with phenol or resorcinol with catalysis of hydrochloric acid, occurring e.g. during phenol pulping (→ 16.6.), support the fact of a dominant reactivity of the benzylic carbon atom (Nimz 1969; Schweers, Rechy 1972; Kratzl, Oburger 1980).

10.3.2. Sulfite Pulping

The main aim in pulping is the removal of lignin from wood to obtain a more or less delignified pulp. Therefore the desired reactions are degradation and solubilization of lignin, which are, however, accompanied by competing reactions.

According to a general concept of delignification, the reactions of lignin in pulping are limited to nucleophilic addition and displacement reactions (Gierer 1982).

Sulfite pulping is carried out at different pH values leading to different nucleophiles in the cooking liquor (→ 16.5.1.). Aqueous sulfur dioxide ($SO_2 \cdot H_2O$) is effective in acidic sulfite pulping, while in neutral sulfite pulping hydrogen sulfite and sulfite ions (HSO_3^-, SO_3^{2-}) promote the delignification.

Under acidic sulfite conditions the initial cleavage reaction at the α-C atom of phenolic and non-phenolic units (Fig. 10–12) is followed by the addition of a SO_3H-group to the intermediary carbonium ion (Fig. 10–14). This sulfonation of the ben-

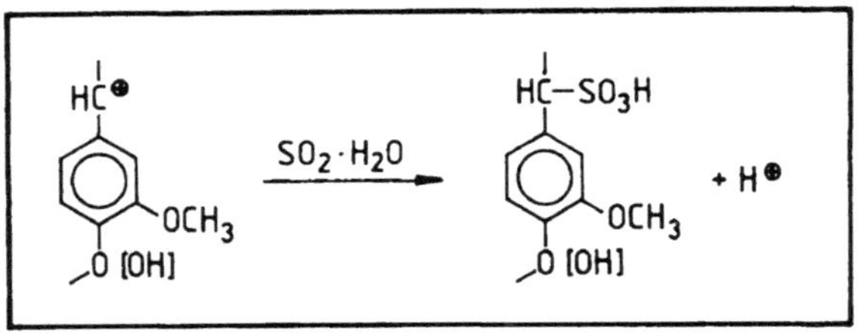

Fig. 10–14. Sulfonation of phenolic or non-phenolic lignin units in acidic sulfite pulping.

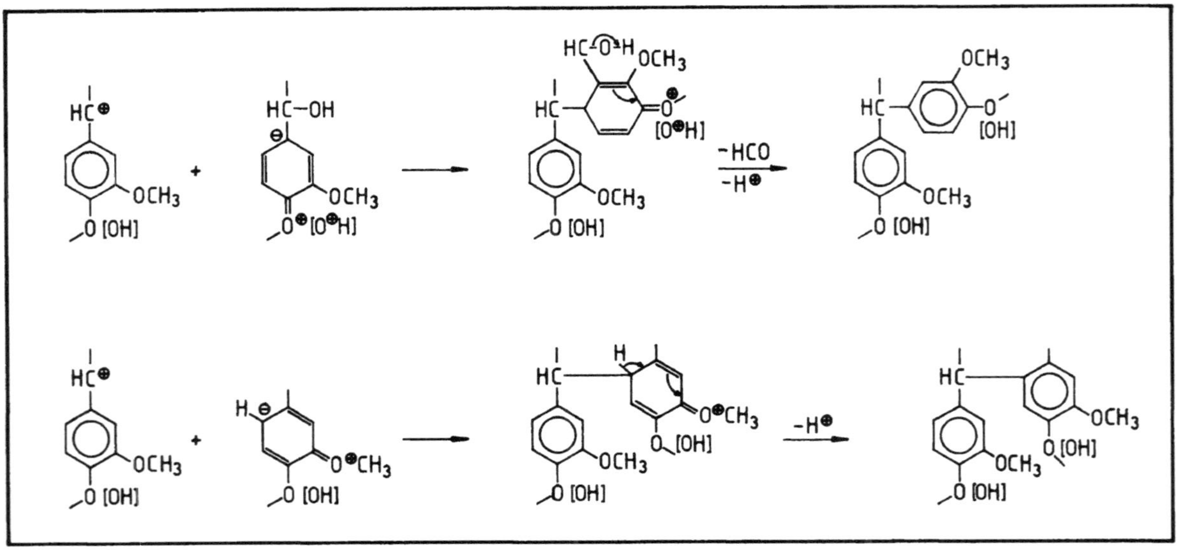

Fig. 10–15. Condensation reactions with weak nucleophilic lignin fragments (according to Gierer 1970).

zylium ion increases the hydrophility of lignin, thus increasing the solubility in the aqueous liquor. The solubility is additionally increased by sulfonation of γ-C aldehyde end-groups and α-C carbonyl groups (Gellerstedt 1976). In contrast to neutral sulfite pulping β-aryl ether linkages are not attacked, even if the α-position is sulfonated. The cleavage of methoxyl groups is negligible.

The desired sulfonation reactions compete with condensation reactions, especially at low pH values. Lignin condensation is also caused by a nucleophilic addition at the α-C atom. But instead of the external nucleophilic pulping reagent, internal lignin fragments with weak nucleophilic 1-, 6- or 5-positions are added (Fig. 10–15). This leads to new C–C bonds, increased C/O ratio, reduction of hydrophility and an increase in molecular weight.

Reactive phenolic extractives (e.g. pinosylvin or its methyl ethers in pine heartwood, → 7.2.3.) may also act as nucleophilic agents, preventing pulping of this wood with acidic sulfite liquors. Retardation or even inhibition of the delignification may also be caused by thiosulfate ions in the pulping liquor (Fig. 10–7), resulting in interunit cross-links (Fig. 10–16).

The principal initial reaction in neutral sulfite pulping is the addition of sulfite or hydrogen sulfite ions to quinone methide intermediates of exclusively phenolic lignin units. The resulting α-sulfonic acid structures undergo sulfitolytic cleavage of β-aryl ethers, the essential lignin degradation reaction in NSSC pulping (Fig. 10–17). The type and number of additional side reactions are mainly dependent on the current pH level. Due to the higher nucleophility of sulfite and hydrogen sulfite as compared to aqueous sulfur dioxide, cleavage of some methyl-aryl ether bonds takes place (Fig. 10–18).

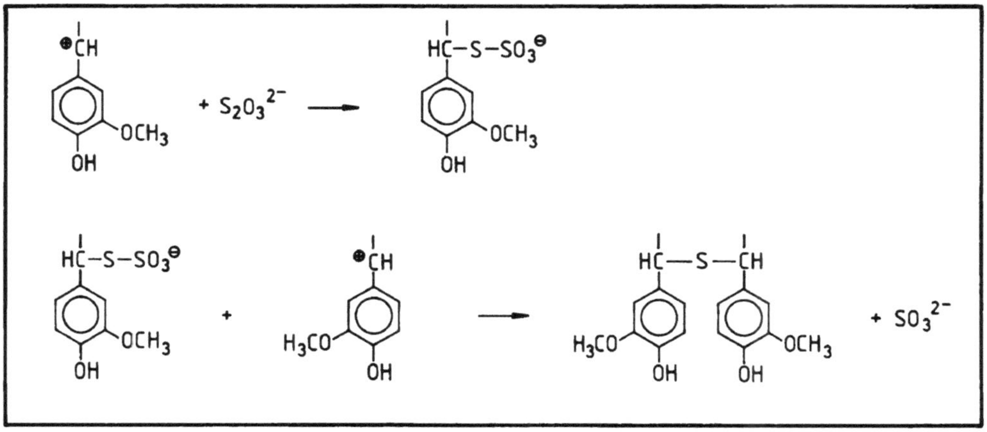

Fig. 10–16. Cross-linking of phenolic lignin units by action of thiosulfate ions (according to Goliath, Lindgren 1961).

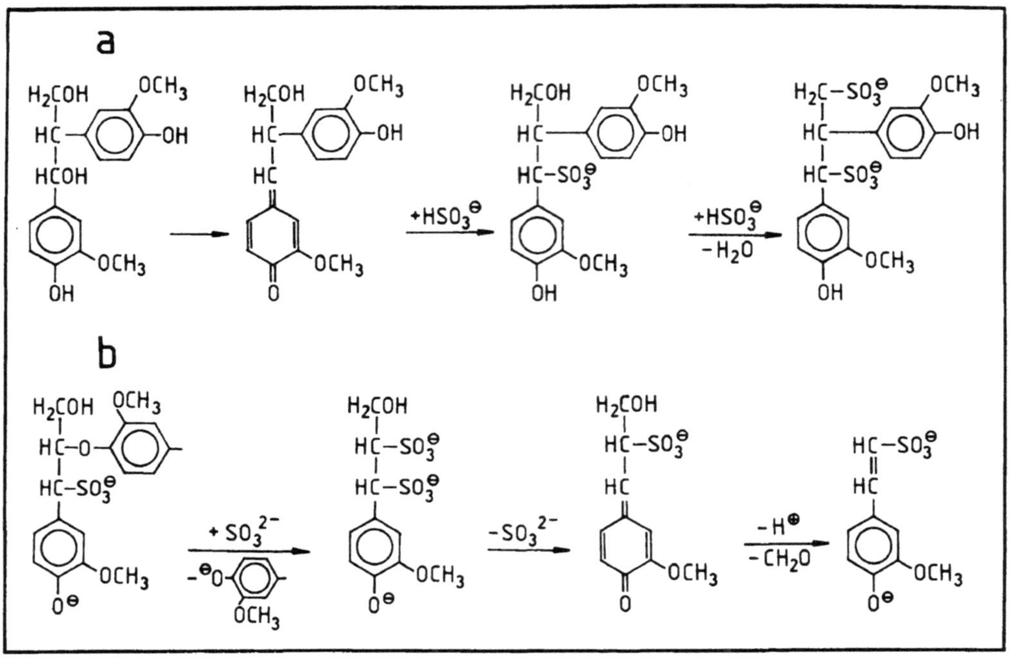

Fig. 10–17. Reactions in neutral sulfite pulping.
a: Sulfonation of 1,2-diarylpropane-1,3-diol structures
b: Cleavage of β-aryl ether bonds followed by elimination of the α-sulfonic acid group (according to Gierer 1982).

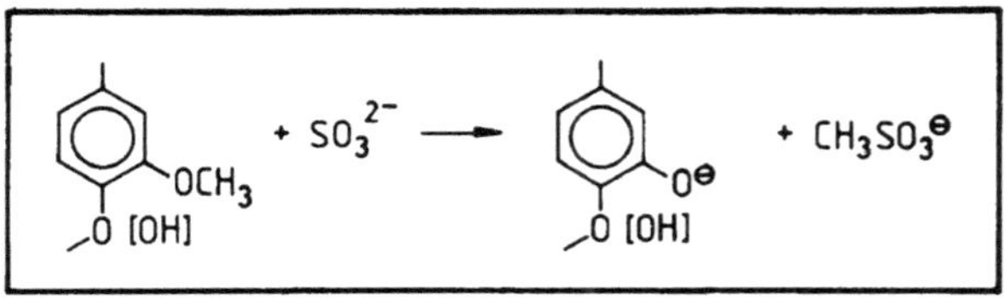

Fig. 10–18. Demethylation by sulfite ions.

10.3.3. Acidic Bleaching Reactions

Acidic bleaching procedures include chlorination and treatments with chlorine dioxide, hydrogen peroxide, peroxyacetic acid and ozone (→ 16.7.3.).

The initial step of lignin-removing bleaching reactions is generally an electrophilic attack on sites of high electron densities (Fig. 10–10), followed by nucleophilic reactions.

In bleaching with acidic aqueous solutions of chlorine (<pH 5.5) chlorine predominates in the liquor, while in the pH region between 5.5 and about 7 hypochlorous acid is the main component. In both cases the reactive chlorination species is the chloronium ion (Cl^+), deriving from heterolysis of Cl_2, or the protonated hypo-

chlorous acid ($ClOH_2^+$). Smaller proportions of chlorine radicals may be involved in chlorination, which are formed in the gaseous state under the influence of light.

The main reactions taking place between chlorine and lignin (Fig. 10–19) are:
- electrophilic substitution (1)
- electrophilic side-chain displacement (2)
- oxidative cleavage of aryl-alkyl ether linkages and oxidation of displaced side-chain structures (2)
- oxidative decomposition of aromatic rings to dicarbonic acid derivatives via quinoid structures (3)

(Dence 1971; Chang, Allan 1971; Lindgren 1974; Gierer, Sundholm 1971; Gierer 1981, 1982).

The substitution occurs mainly at C5- or C6-positions. Substitution at the C1-position causes side-chain displacement, representing an important lignin fragmentation, in addition to the oxidative fragmentation by the cleavage of β-aryl ether linkages. Minor amounts of chlorine are introduced to the lignin molecule by being added to less frequent olefinic or ring-conjugated structures.

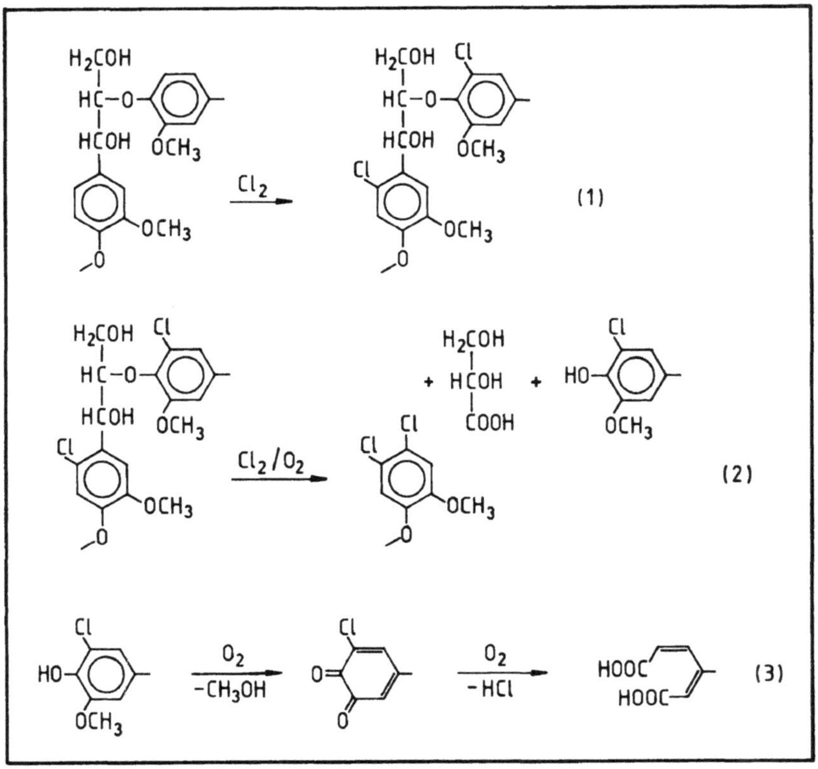

Fig. 10–19. Degradation of lignin aqueous chlorine solutions.

In the alkaline extraction stage, following the chlorine treatment, the chlorinated lignin fragments are converted to the respective hydroxy compounds, which dissolve easily in the alkaline liquor.

Chlorine dioxide, which is often used in combination with chlorine (C+D, D/C, C_D; Table 16–13), causes a more efficient delignification than chlorine, as related to the applied active chlorine. This is mainly due to the fact that in contrast to chlorine the reactions with lignin are exclusively oxidative reactions of CO_2 phenolic lignin units with chlorine dioxide radicals ($ClO_2^{\cdot}$), initiated by hydrogen abstraction. The final oxidation products are muconic acid derivatives (without liberation of methanol) or quinoid structures (Fig. 10–20). Chlorine-substituted fragments in chlorine dioxide bleaching liquors must derive from the action of chlorine generated by partial decomposition of chlorine dioxide (Lindgren 1971, Gierer 1982).

By oxidation and reduction of ClO_2 chlorate, chlorite and ClO-radicals are also formed, and may influence the overall reaction of lignin with chlorine dioxide (Murphy et al. 1961; Lindgren 1971, 1974; Lindgren, Nilsson 1972, 1975; Kolar et al. 1981).

Peroxyacetic acid is known as a specific oxidant for lignin, e.g. it is used in acidic solution for the preparation of holocellulose (Haas et al. 1955; → 3.2.6.). In neutral and alkaline media it is recommended as a lignin-preserving bleaching agent (Wayman et al. 1965; Nimz, Schwind 1979, 1981).

Since the hydroxonium ion (HO^+) is the reactive species in the reactions of both peroxyacetic acid and hydrogen peroxide in acidic medium, the principle mechanisms are valid for both chemicals (Johnson 1980). Hydroxy radicals may also be partly involved as reactive agents (Gierer 1982).

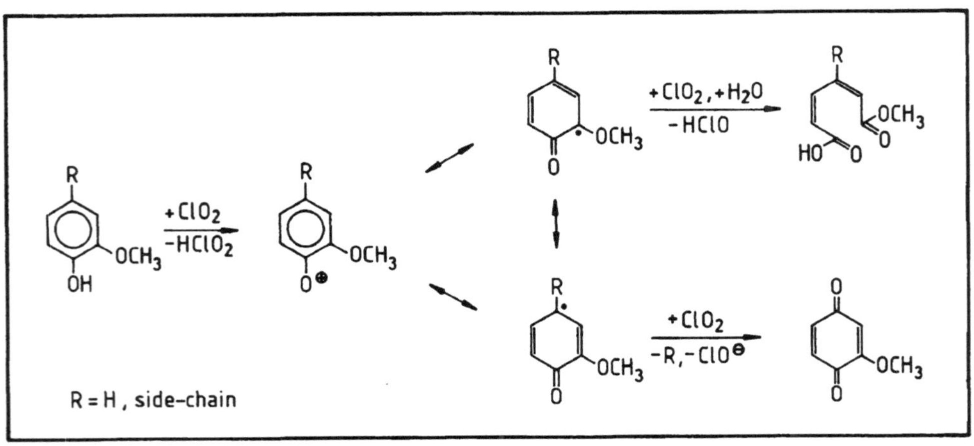

Fig. 10–20. Oxidative degradation of phenolic lignin units by chlorine dioxide (according to Gierer 1982).

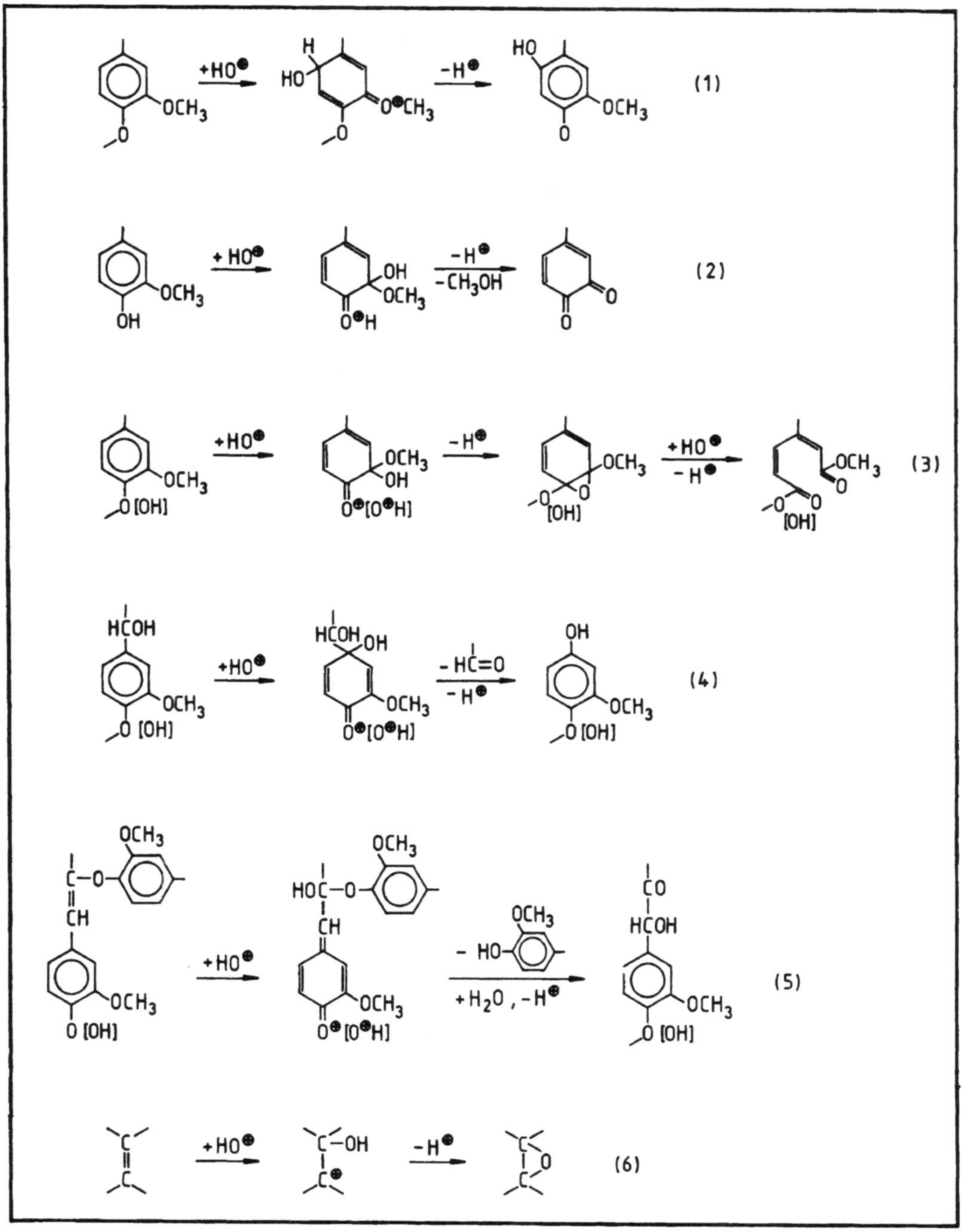

Fig. 10–21. Principal reactions of lignin units with hydroxonium ions (according to Gierer 1982).

Studies with model compounds revealed that the initial reactions are electrophilic additions of hydroxonium ions to the δ^--positions of lignin units. Fig. 10–21 shows the main types of reactions:

- addition of hydroxyl groups to the ring (1)
- oxidative demethylation (2)
- oxidative ring opening (3)
- side-chain displacement (4)
- cleavage of β-aryl ether linkages (5)
- epoxidation of olefinic structures (6)

(Sakai et al. 1972; Oki et al. 1972, 1974; Sakai, Kondo 1975; Lawrence et al. 1980).

Ozone may have a future importance as a non-polluting and effective bleaching agent (→ 16.7.3.). The mechanisms of lignin degradation by ozone (O_3) (Fig. 10–22) have not yet entirely been solved, but some essential principles were established by means of model studies (Fig. 10–23). Important reactions are elec-

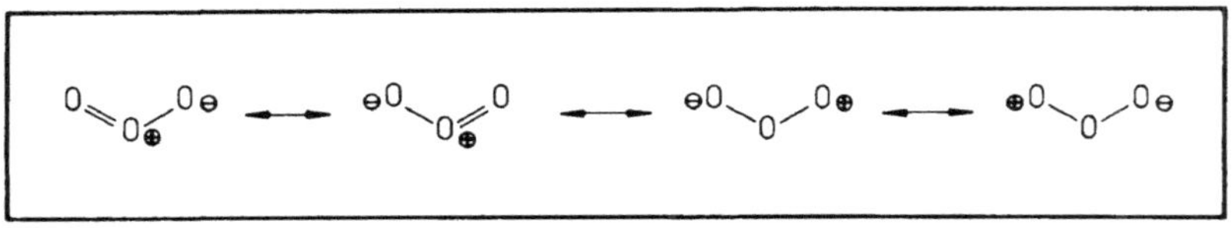

Fig. 10–22. Mesomeric nature of the ozone molecule.

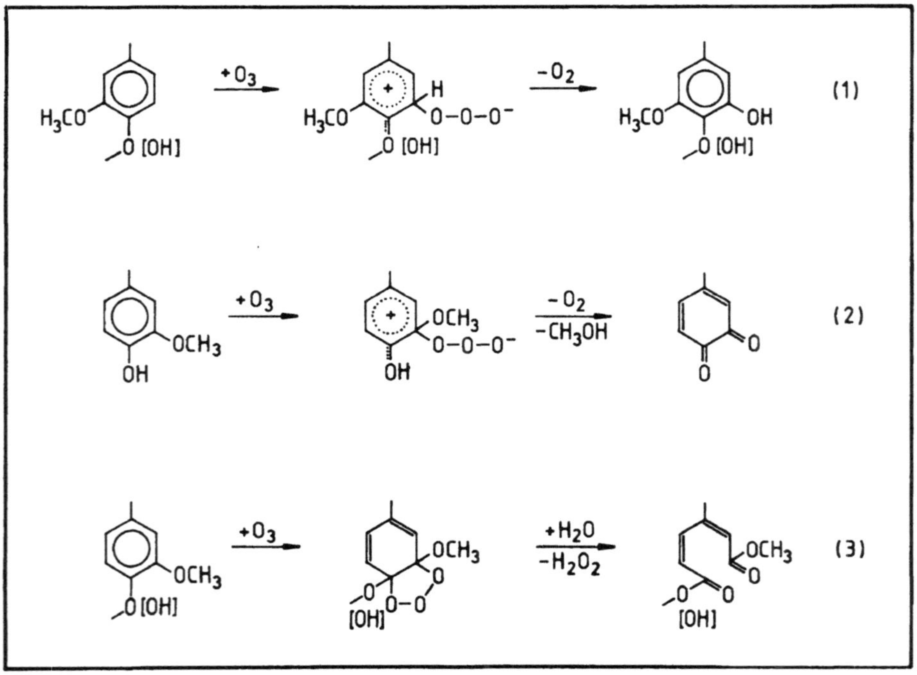

Fig. 10–23. Mechanisms of lignin degradation by ozone (according to Gierer 1982).

trophilic substitution yielding hydroxylated rings (1), oxidative cleavage of methoxyl groups (2), and ring cleavage (3) after a cyclo-addition of the ozone molecule (Kratzl et al. 1976; Kojima et al. 1978; Eckert, Singh 1980; Balousek et al. 1981; Kaneko et al. 1979, 1981).

10.3.4. Mild Solvolysis

The degradation of lignin by means of mild solvolytic reactions is the most valuable tool for the elucidation of lignin structure (→ 6.3.1.).

The most important procedures for obtaining low-molecular-weight degradation compounds for analytical purposes are:

- percolation with boiling water, partly in the presence of 2% acetic acid (mild hydrolysis)
- treatment with dioxane-water mixtures in or without the presence of hydrochloric acid (acidolysis)
- treatment with alcohol in the presence of hydrochloric acid (alcoholysis)
- treatment with thioacetic acid (thioacetolysis).

A mild depolymerization of lignin can be achieved by percolation of wood at 100 °C for several weeks, partly also in the presence of 2% acetic acid (Nimz 1966, 1974). Thus 40% of beech lignin and 20% of spruce lignin could be dissolved. Apart from high-molecular-weight lignin fragments eight dimers (I–VIII), a diastereomeric trimer (IX) and a tetrameric lignol (X) were isolated and identified (Fig. 10–24). The mild conditions are effective enough to cleave the labile benzyl ether bonds (α-O-4) and preserve the side-chains and β-O-4 linkages without condensation under formation of new C–C bonds.

A mild acidolysis is achieved by refluxing lignin with HCl (0.2 moles/l) in a dioxane/water-mixture (9:1) (Lundquist 1976; Adler 1977). The general mechanism of lignin degradation follows the scheme shown in Fig. 10–25. The predominant depolymerization reaction is the cleavage of arylglycerol-β-aryl ether linkages, preceded by the formation of a benzyl-carbonium ion. The primary monomeric C-9 unit resulting from this cleavage is ω-hydroxyguaiacylketone. This is further converted to the ketones I–IV. Minor side, conversion, and some condensation reactions yield numerous compounds from spruce and birch lignin, some of them shown in Fig. 10–26.

The treatment of wood or lignin with a dioxane/water-mixture (1:1) at 180 °C for 20 min also yields mono-, di- and oligomers, mostly similar to those produced by mild hydrolysis. Additional structures arise from condensation and recombination reactions taking place under these less mild conditions (Sakakibara 1980).

If alcohols are involved in mild solvolysis of lignin the term alcoholysis or, more specifically, methanolysis or ethanolysis is used. While methoxylation of lignin with methanolic hydrogen chloride at ambient temperature does not solubilize lignin,

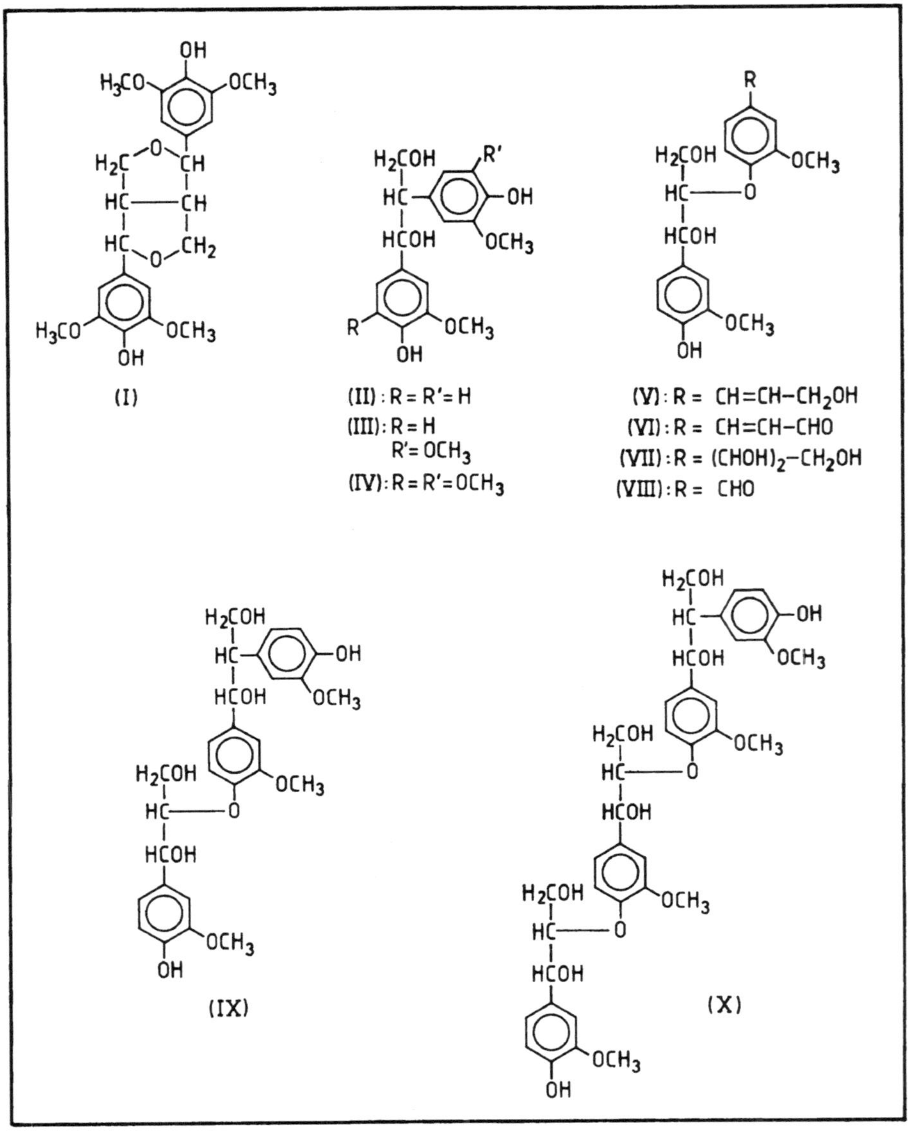

Fig. 10–24. Lignols obtained by mild hydrolysis of beech and spruce lignin (according to Nimz 1974).

the treatment of wood or lignin with boiling ethanol containing 2% hydrochloric acid causes lignin to dissolve. As in acidolysis (Fig. 10–25) the essential depolymerization reaction is the cleavage of β-aryl ether linkages. In addition to ketones (I–IV) ethoxylated compounds are formed (V, VI), known collectively as Hibbert's ketones and amounting to about 10% of lignin. In addition to guaiacyl compounds syringyl compounds are formed in ethanolysis of hardwoods (Kratzl 1961; Töppel 1960; Kratzl, Claus 1962; Adler 1977).

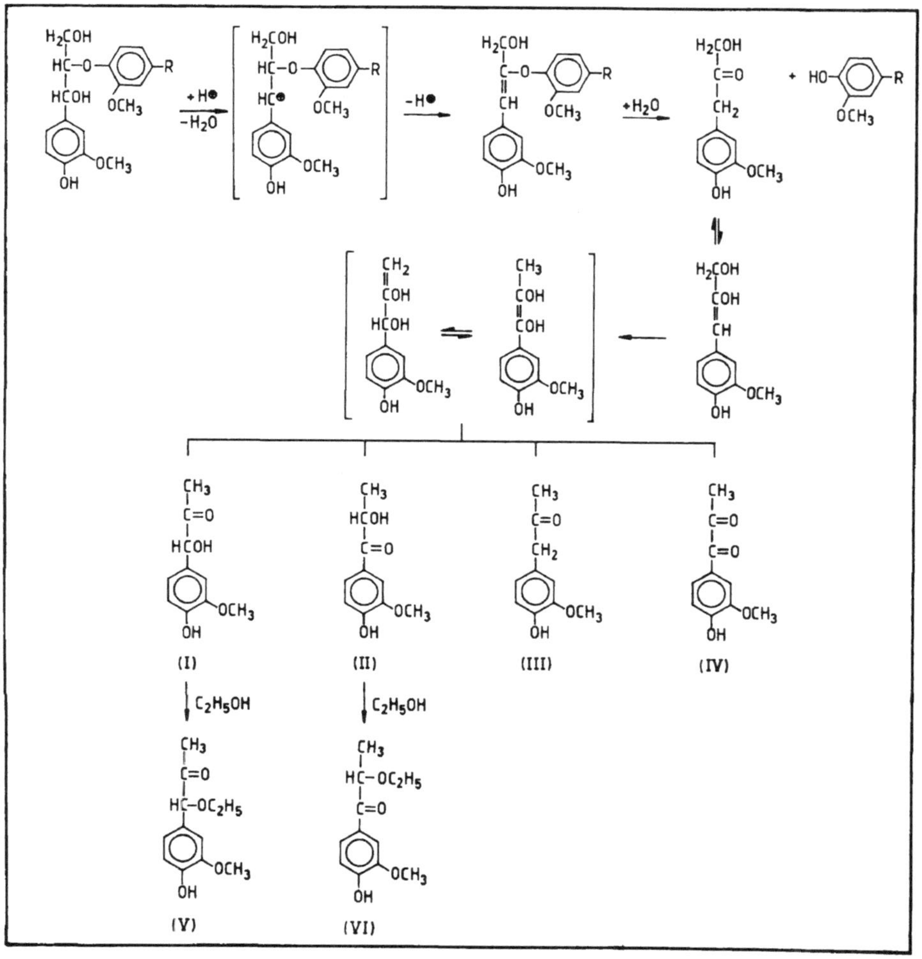

Fig. 10–25. Degradation mechanism in acidolysis and ethanolysis.

In the case of thioacetolysis wood is treated with thioacetic acid in the presence of borontrifluoride, followed by alkaline hydrolysis (Nimz 1969, 1974; Nimz et al. 1971; Skamla, Rybárik 1975). Mixtures of mono- to tetrameric lignols were obtained in high yields of 77 and 91% from spruce and beech lignin. The simplified degradation mechanism is outlined in Fig. 10–27.

Autohydrolysis of milled wood lignin and acidic solvolysis ('organosolv hydrolysis') of hydrochloric acid lignin under varying conditions yield much less clearly defined low-molecular-weight degradation products (Wayman, Lora 1980; Lora, Wayman 1980; Goldstein et al. 1981).

Acidic solvolytic reactions are also involved in isolation of numerous analytical lignin preparations (e.g. ethanol lignin, dioxane lignin, thioglycolic acid lignin) (→ 3.2.9.).

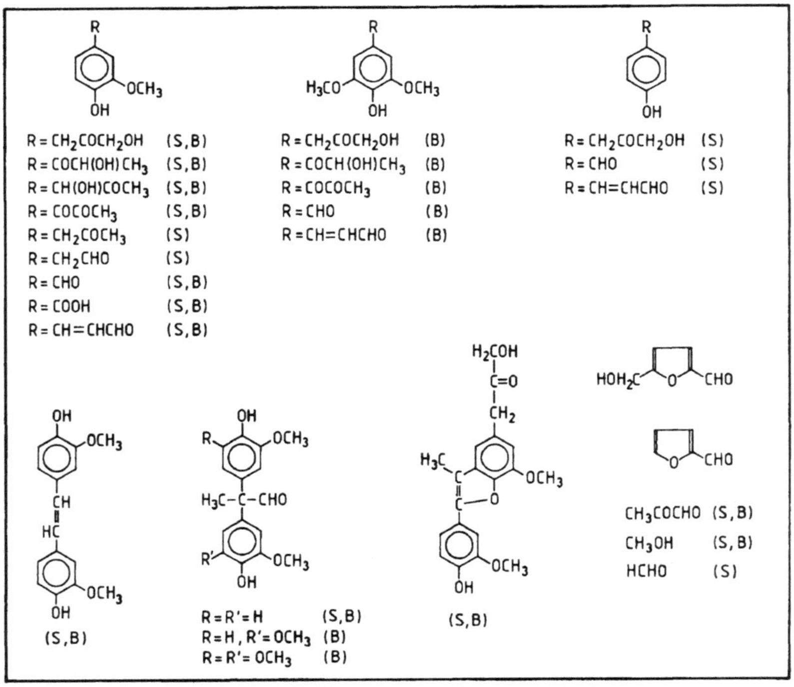

Fig. 10–26. Compounds obtained by acidolysis of milled wood lignin from spruce (S) and birch (B) (Lundquist 1976).

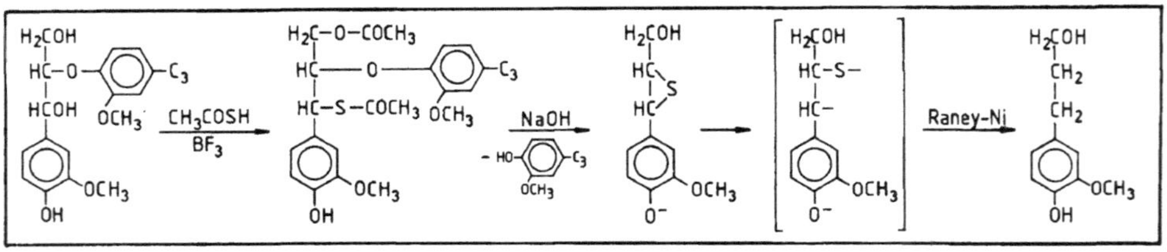

Fig. 10–27. Cleavage of β-O-4 linkages in thioacetolysis (according to Nimz 1974).

Literature

Adler, E. 1977, Wood Sci. Technol. *11,* 169–218

Alfredsson, B. and Samuelson, O. 1974, Svensk Papperstid. *77,* 449–452

Alfredsson, B., Czerwinsky, W. and Samuelson, O. 1961, Svensk Papperstid. *64,* 812–814

Allan, G. G. 1971, Modification Reactions. In: Lignins. Occurrence, Formation, Structure and Reactions (Sarkanen, K. V. and Ludwig, C. H., Eds.). Wiley-Intersci., New York, pp. 511–573

Andersson, S. I. and Samuelson, O. 1976, Cell. Chem. Technol. *10,* 209–218

Ant-Wuorinen, O. and Visapää, A. 1969, Pap. Puu *51,* 107–113

Ant-Wuorinen, O. and Visapää, A. 1970, Pap. Puu *52,* 765–779

Balousek, P. J., McDonough, T. J., McKelvey, R. D. and Johnson, D. C. 1981, Svensk Papperstid. *84,* R49–R54

Besold, G. 1982, Systematische Untersuchungen der Wirkung aggressiver Gase auf Fichtenholz. Doctor Thesis, Universität München

Browning, B. L. 1967, Methods of Wood Chemistry. Vol. II., Intersci. Publ. New York
Chang, H. M. and Allan, G. G. 1971, Oxidation. In: Lignins. Occurrence, Formation, Structure and Reactions (Sarkanen, K. V. and Ludwig, C. H., Eds.). Wiley-Intersci., New York, pp. 433–485
Czirnich, W. and Patt, R. 1976, Holzforschung *30,* 19–27
Daruwalla, E. H. and Narsian, M. G. 1966, Tappi *49,* 106–111
Daruwalla, E. H. and Shet, R. T. 1962, Text. Res. J. *32,* 942–954
Dence, C. W. 1971, Halogenation and Nitration. In: Lignins. Occurrence, Formation, Structure and Reactions (Sarkanen, K. V. and Ludwig, C. H., Eds.). Wiley-Intersci., New York, pp. 373–432
Eckert, R. C. and Singh, R. P. 1980, Ozone Reactions in Relation to the Aromatic Structure of Lignin: A Review of Selected Topics in Ozone Chemistry. In: Chemistry of Delignification with Oxygen, Ozone, and Peroxides (Gratzl, J. S., Nakano, J. and Singh, R. P., Eds.). Uni Publ. Co., Ltd., Tokyo, pp. 229–243
Emery, J. A. and Schroeder, H. A. 1974, Wood Sci. Technol. *8,* 123–137
Fengel, D., Uçar, H. and Wegener, G. 1979, Papier *33,* 233–239
Fengel, D., Wegener, G., Heizmann, A. and Przyklenk, M. 1978, Cell. Chem. Technol. *12,* 31–37
Garves, K. 1979, Rapid Hydrolysis of Cellulose in Homogeneous Solution. In: Hydrolysis of Cellulose: Mechanisms of Enzymatic and Acid Catalysis (Brown, Jr. R. D. and Jurasek, L., Eds.). Adv. Chem. Ser. 181, ACS, Washington, D. C., pp. 159–165
Geddes, A. L. 1956, J. Polym. Sci., No. 22, 31–39
Gellerstedt, G. 1976, Svensk Papperstid. *79,* 537–543
Gierer, J. 1970, Svensk Papperstid. *73,* 571–596
Gierer, J. 1981, Chemical Aspects of Delignification. In: The Ekman Days 1981. Int. Symp. Wood Pulp. Chem. Stockholm, Vol. 2, pp. 12–17
Gierer, J. 1982, Holzforschung *36,* 43–51; 55–65
Gierer, J. and Sundholm, L. 1971, Svensk Papperstid. *74,* 345–351
Glennie, D. W. 1971, Reactions in Sulfite Pulping. In: Lignins. Occurrence, Formation, Structure and Reactions (Sarkanen, K. V. and Ludwig, C. H., Eds.). Wiley-Intersci., New York, pp. 597–637
Goldstein, I. S., Preto, R. J., Pittman, J. L. and Schultz, T. P. 1981, Hydrolytic Depolymerization of HCl Hydrolysis Lignin. In: The Ekman Days 1981. Int. Symp. Wood Pulp. Chem. Stockholm, Vol. 4, pp. 9–13
Goliath, M. and Lindgren, B. O. 1961, Svensk Papperstid. *64,* 469–471
Haas, H., Schoch, W. and Ströle, U. 1955, Papier *9,* 469–475
Harris, J. F. 1975, Appl. Polym. Symp. No. 28, 131–144
Herrick, F. W., Casebier, R. L., Hamilton, J. K. and Wilson, J. D. 1975, Appl. Polym. Symp. No. 28, 93–108
Hoffmann, P. and Patt, R. 1976, Holzforschung *30,* 124–132
Johnson, D. C. 1980, Lignin Reactions in Delignification with Peroxyacetic Acid. In: Chemistry of Delignification with Oxygen, Ozone, and Peroxides (Gratzl, J. S., Nakano, J. and Singh, R. P., Eds.). Uni Publ. Co., Ltd., Tokyo, pp. 217–228
Kaneko, H., Hosoya, S. and Nakano, J. 1979, Mokuzai Gakkaishi *25,* 503–509
Kaneko, H., Hosoya, S. and Nakano, J. 1981, Mokuzai Gakkaishi *27,* 678–683
Kojima, Y., Miura, K. and Kayama, Y. 1978, Res. Bull. Coll. Exp. For. Hokkaido Univ. XXXV, 1, 165–184
Kolar, J. J., Lindgren, B. O. and Pettersson, B. 1981, The Reaction of Chlorine Dioxide with Lignin. In: The Ekman Days 1981. Int. Symp. Wood Pulp. Chem. Stockholm, Vol. 2, pp. 94–96
Kratzl, K. 1961, Holz, Roh-Werkst. *19,* 219–232
Kratzl, K. and Claus, P. 1962, Monatsh. Chem. *93,* 219–229
Kratzl, K. and Oburger, M. 1980, Holzforschung *34,* 11–16; 191–196
Kratzl, K., Claus, P. and Reichel, G. 1976, Tappi *59,* No. 11, 86–87
Larsson, K. and Samuelson, O. 1969, Svensk Papperstid. *72,* 97–100
Lawrence, W., McKelvey, R. D. and Johnson, D. C. 1980, Svensk Papperstid. *83,* 11–18

Lindgren, B. O. 1971, Svensk Papperstid. *74,* 57–63
Lindgren, B. O. 1974, Papier *28,* No. 10A, V29–V32
Lindgren, B. O. and Nilsson, T. 1972, Svensk Papperstid. *75,* 161–168
Lindgren, B. O. and Nilsson, T. 1975, Svensk Papperstid. *78,* 66–68
Lora, J. H. and Wayman, M. 1980, Can. J. Chem. *58,* 669–676
Lundquist, K. 1976, Appl. Polym. Symp. No. 28, 1393–1407
Makkonen, H. 1967, Pap. Puu *49,* 383–390; 437–456
Marchessault, R. H. and Rånby, B. G. 1959, Svensk Papperstid. *62,* 230–240
Millett, M. A., Effland, M. J. and Caulfield, D. F. 1979, Influence of Fine Grinding on the Hydrolysis of Cellulosic Materials – Acid vs. Enzymatic. In: Hydrolysis of Cellulose: Mechanisms of Enzymatic and Acid Catalysis (Brown, Jr. R. D. and Jurasek, L., Eds.). Adv. Chem. Ser. 181, ACS, Washington, D. C., pp. 71–89
Miyazaki, K., Smelstorius, J. A., Harwood, B. J. and Stewart, C. M. 1971, Appita *24,* 452–454
Murphy, R. A., Kakehi, K. and Sarkanen, K. V. 1961, Tappi *44,* 465–467
Mutton, D. B. 1964, Pulp Pap. Mag. Can. *65,* No. 2, T41–T51
Nakano, J. and Rånby, B. G. 1962, Svensk Papperstid. *65,* 29–33
Nelson, M. L. 1960, J. Polym. Sci. *43,* 351–371
Nelson, P. F. 1968, Svensk Papperstid. *71,* 369–372
Nimz, H. 1966, Holzforschung *20,* 105–109
Nimz, H. 1969a, Holzforschung *23,* 84–88
Nimz, H. 1969b, Chem. Ber. *102,* 799–810; 3803–3817
Nimz, H. 1974, Angew. Chem. Int. Ed. Engl. *13,* 313–321
Nimz, H. H. and Schwind, H. 1979, Cell. Chem. Technol. *13,* 35–46
Nimz, H. H. and Schwind, H. 1981, Oxidation of Lignin and Lignin Model Compounds with Peracetic Acid. In: The Ekman Days 1981. Int. Symp. Wood Pulp. Chem., Stockholm, Vol. 2, pp. 105–111
Nimz, H., Das, K. and Minemura, N. 1971, Chem. Ber. *104,* 1871–1876
Oki, T., Okubo, K. and Ishikawa, H. 1972, Mokuzai Gakkaishi *18,* 601–610
Oki, T., Okubo, K. and Ishikawa, H. 1974, Mokuzai Gakkaishi *20,* 549–557
Pan, G., Chen, C. L., Chang, H. and Gratzl, J. S. 1981, Model Experiments on the Splitting of Glycosidic Bonds by Ozone. In: The Ekman Days 1981. Int. Symp. Wood Pulp. Chem. Stockholm, Vol. 2, pp. 132–144
Pereira, H., Patt, R. and Sinner, M. 1974, Pap. Puu *56,* 321–336
Pettersson, S. and Samuelson, O. 1968, Svensk Papperstid. *71,* 429–431
Pfister, K. and Sjöström, E. 1977, Pap. Puu *59,* 711–720
Philipp, B., Jacopian, V., Loth, F., Hirte, W. and Schulz, G. 1979, Influence of Cellulose Physical Structure on Thermohydrolytic, Hydrolytic and Enzymatic Degradation of Cellulose. In: Hydrolysis of Cellulose: Mechanisms of Enzymatic and Acid Catalysis (Brown, Jr. R. D. and Jurasek, L., Eds.). Adv. Chem. Ser. 181, ACS, Washington, D. C. pp. 127–143
Poller, S. and Hohmuth, K. 1979, Holztechnologie *20,* 161–164
Popoff, T. and Theander, O. 1972, Carbohydr. Res. *22,* 135–149
Popoff, T. and Theander, O. 1976, Acta Chem. Scand. *30,* 397–402
Rånby, B. G. 1961, J. Polym. Sci. *53,* 131–140
Root, D. F., Harris, J. F., Saeman, J. F. and Neill, W. K. 1959, For. Prod. J. *9,* No. 5, 158–165
Rydholm, S. A. 1965, Pulping Processes. Intersci. Publ., New York
Sakai, K. and Kondo, T. 1975, Mokuzai Gakkaishi *21,* 87–92
Sakai, K., Kuroda, K. and Kishimoto, S. 1972, Tappi *55,* 1702–1706
Sakakibara, A. 1980, Wood Sci. Technol. *14,* 89–100
Schweers, W. and Rechy, M. 1972, Papier *26,* 585–590
Semke, L. K., Thompson, N. S. and Williams, D. G. 1964, J. Org. Chem. *29,* 1041–1047
Shafizadeh, F. 1963, Tappi *46,* 381–383
Sharkov, V. I. 1961, Celul. Hirtie 1961, No. 10, 297–303

Sjöström, E. 1981, Wood Chemistry. Fundamentals and Applications. Academic Press, New York, London
Sjöström, E. and Enström, B. 1967, Tappi *50,* 32–36
Skamla, J. and Rybárik, I. 1975, Drev. Výsk. *20,* 107–112
Slavik, I., Pästeka, M. and Kučerová, M. 1967, Svensk Papperstid. *70,* 365–370
Smelstorius, J. A. 1972, Holzforschung *26,* 92–96
Springer, E. L. 1966, Tappi *49,* 102–106
Srivastava, L. M. 1966, Tappi *49,* 173–183
Szejtli, J. 1976, Säurehydrolyse glycosidischer Bindungen. VEB Fachbuchverlag Leipzig
Teder, A. and Tormund, D. 1978, Tappi *61,* 59–62
Timell, T. E. 1964a, Can. J. Chem. *42,* 1456–1472
Timell, T. E. 1964b, Chem. Ind. (London), 503–504
Töppel, O. 1960, Holzforschung *14,* 139–146
Valtsaar, H. and Dunlap, R. 1952, Some Observations on the Behavior of Cellulose in Trifluoroacetic and Heptafluorobutyric Acids. In: Abstracts of Papers, 122 nd Meeting, A.C.S., 3D
Vink, H. 1966, Makromol. Chem. *94,* 1–14
Visapää, A. 1971, Pap. Puu *53,* 397–408
Visapää, A. 1972, Pap. Puu *54,* 353–364
Wayman, M. and Lora, J. H. 1980, J. Appl. Polym. Sci. *25,* 2187–2194
Wayman, M., Anderson, C. B. and Rapson, W. H. 1965, Tappi *47,* 270–275
Yasuda, S., Terashima, N. and Ito, T. 1980, Mokuzai Gakkaishi *26,* 552–557
Yasuda, S., Terashima, N. and Ito, T. 1981, Mokuzai Gakkaishi *27,* 879–884
Yasuda, S., Hayashi, K., Ito, T. and Terashima, N. 1981, Mokuzai Gakkaishi *27,* 478–483

11. Reactions in Alkaline Medium

11.1. Reactions of Polysaccharides

11.1.1. Endwise Degradation and Alkaline Hydrolysis

The alkaline degradation of cellulose and polyoses is an essential factor in kraft and soda pulping (→ 16.4.), oxygen delignification (→ 16.7.) and hot-alkali purification in dissolving pulp manufacture.

Initial reactions are solvation of hydroxyl groups by hydroxyl ions causing a swollen state. At elevated temperatures the polysaccharides are attacked by strong alkali solutions, with a large number of reactions taking place. The most important ones are:
- dissolution of undegraded polysaccharides
- peeling of end-groups with formation of alkalistable end-groups
- alkaline hydrolysis of glycosidic bonds and acetyl groups
- degradation and decomposition of dissolved polysaccharides, hydrolyzed fragments, and peeled monosaccharides

(Rydholm 1965; Sjöström 1977, 1981a; Samuelson 1981).

The most important reactions responsible for the loss of polysaccharides and reduction of the chain length of cellulose in alkaline pulping are peeling and hydrolytic reactions.

At temperatures of about 100 °C, occurring during the heating period in pulping, the degradation of polysaccharide chains starts from the existing reducing end-groups by the so-called peeling reaction (primary peeling). At temperatures above 150 °C chains are split by alkaline hydrolysis. Thus new reducing end-groups are formed, which are also subjected to endwise degradation (secondary peeling). The peeling reaction of polysaccharides involves the elimination of reducing end-groups by a β-alkoxy elimination to various carboxylic acids, thus reducing the chains by one monomeric unit at a time (Lindberg 1956; Mutton 1964). The reaction mechanism of endwise degradation is outlined in Fig. 11–1 a. The initial step is an isomerization of the reducing end-group (I) to a ketose (II), which is in equilibrium with the corresponding 2,3- enediol. These structures are alkali-labile, and the C4-substituent is cleaved, leading to a new reducing end of the shortened polysaccharide chain and to the eliminated end-group (III). The latter is tautomerized to a dicarbonyl compound (IV).

The main reaction of the deliberated end-group corresponds to a benzilic acid rearrangement in the alkaline medium to yield isosaccharinic acid (3-deoxy-2-C-hydroxymethyl aldonic acid) (V) (gluco-isosaccharinic acid in the case of cellulose

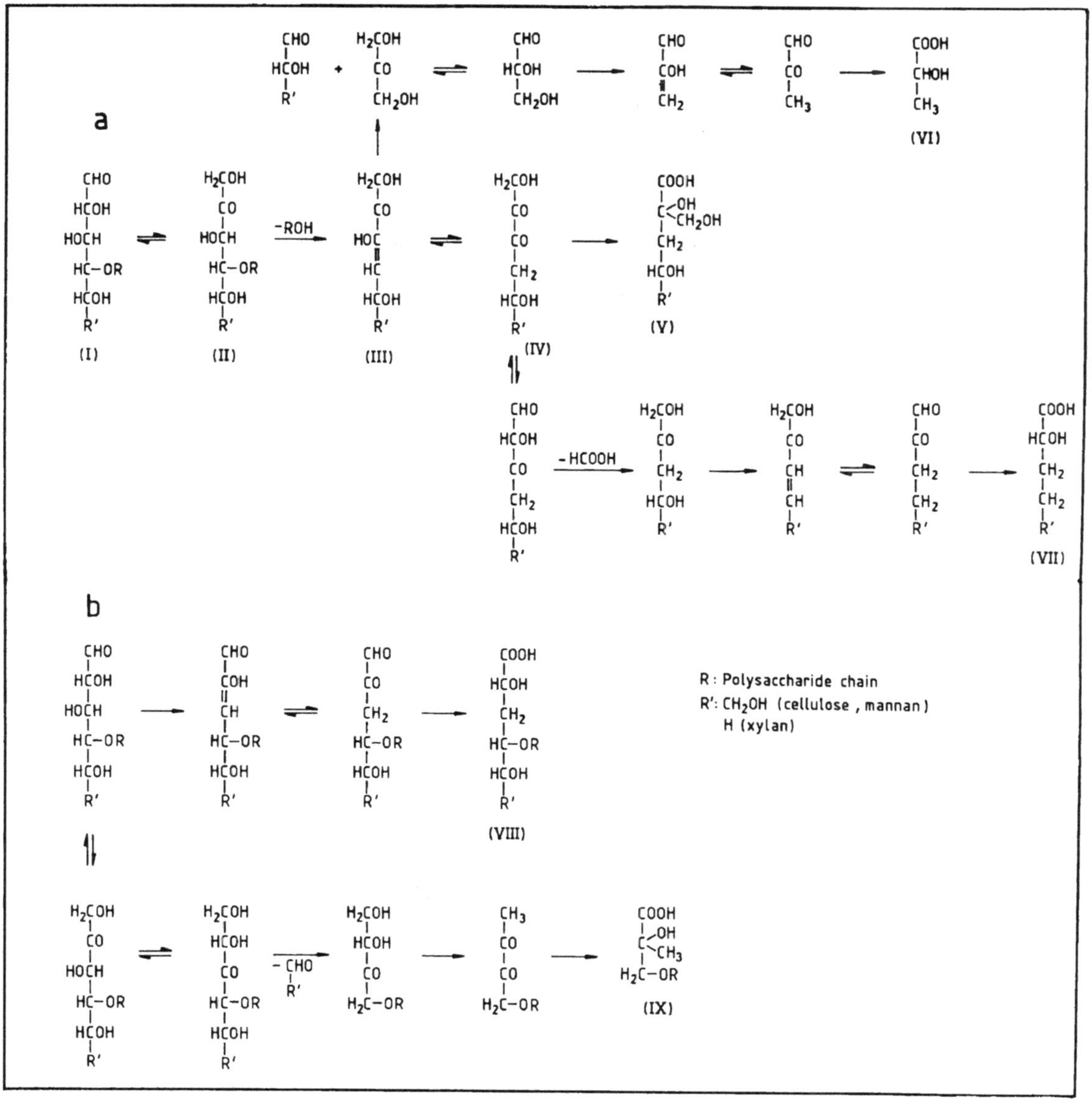

Fig. 11–1. Endwise degradation of polysaccharides
a: Reactions involved in peeling
b: Main stopping reactions
(according to Sjöström 1977).

and mannan; xylo-isosaccharinic acid in the case of xylan). Other possible final degradation products are lactic acid (VI) or 2-hydroxy-butanoic acid and 2,5-dihydroxypentanoic acid, respectively (VII) (Sjöström 1977; Green et al. 1977). The polyoses are peeled much more than cellulose. The individual reaction rates depend on the type of polyoses (Casebier, Hamilton 1965; Lindberg et al. 1966; Malinen, Sjöström 1975). Xylans are more stable than glucomannans, which was attributed in the case of birch xylan to the stabilizing effect of the galacturonic acid side-groups adjacent to the reducing end of the xylan chain (Rydholm 1965; Johansson, Samuelson 1977a, b). The easy cleavage of arabinose side-groups in softwood xylans also has a stabilizing effect against alkaline peeling, since with the loss of the side-group an alkali-stable metasaccharinic acid end-group is formed.

In the case of cellulose about 50–60 glucose units are expected to be cleaved on an average by endwise peeling until a competing reaction takes place which terminates the degradation (Franzon, Samuelson 1957; Alfredsson et al. 1961) (Fig. 11–1 b). The highly important, so-called stopping reaction is initiated by a β-hydroxy elimination at the C4-position to a tautomeric intermediate, which is converted to an alkali-stable metasaccharinic acid end-group (3-deoxyaldonic acid) (VIII). Other possible end-group formations are e.g. C2-methylglyceric acid (IX) and to some extent aldonic acid, indicating the participation of some oxidative reactions (Johansson, Samuelson 1974) (→ 11.1.2.). Without the stopping reaction even a whole molecule may be destroyed by peeling (Johansson, Samuelson 1975).

In addition to the above-mentioned acids formic acid, acetic acid and small amounts of dicarboxylic acids are also among the alkaline degradation products of cellulose, hexosans and pentosans (Malinen, Sjöström 1974, 1975a; Löwendahl et al. 1976a, b). While formic acid is liberated during the peeling reaction, acetic acid results from the cleavage of acetyl groups of hardwood xylans and softwood mannans, occurring at very early stages of alkaline pulping procedures. The dissolved, deacetylated xylan chains are known to be redeposited on the fibres (Yllner, Enström 1956, 1957; Croon, Enström 1961; Axelsson et al. 1962). Glucuronic acid groups of xylans are mostly lost at maximum cooking temperature of alkaline pulping, whether by dissolution of xylan portions bearing acid side-groups or by alkaline hydrolysis of these groups.

The treatment of glucose and xylose with sodium hydroxide at 96 °C under nitrogen revealed that in addition to the aliphatic acids small amounts of cyclic enols and phenolic compounds are also formed from monosaccharides (Fig. 11–2). (Forsskåhl et al. 1976). As those structures may also be formed to some extent under milder conditions, the UV absorption of lignin-polysaccharide fractions obtained from alkaline isolation procedures can be strongly influenced (→ 6.4.2.).

To avoid or at least to diminish peeling reactions, the reducing aldehyde end-groups can be converted by reduction or oxidation to alcohol or carboxyl groups, respectively, or substituted to yield other alkali-stable end-groups.

A polysaccharide stabilization, meaning increased pulp yields, can be reached e.g. by the presence of polysulfides, causing oxidation to aldonic and metasaccharinic acids. Reduction to alditol and thioalditol groups is performed by treatments with sodium borohydride and hydrogen sulfide, respectively (Fig. 11–3) (Hartler 1959, 1967; Rydholm 1965; Vinje, Worster 1969; Procter, Wiekenkamp 1969; Kleppe 1970; Hartler, Olsson 1972; Bryce 1980; Fleming et al. 1980; Samuelson 1981).

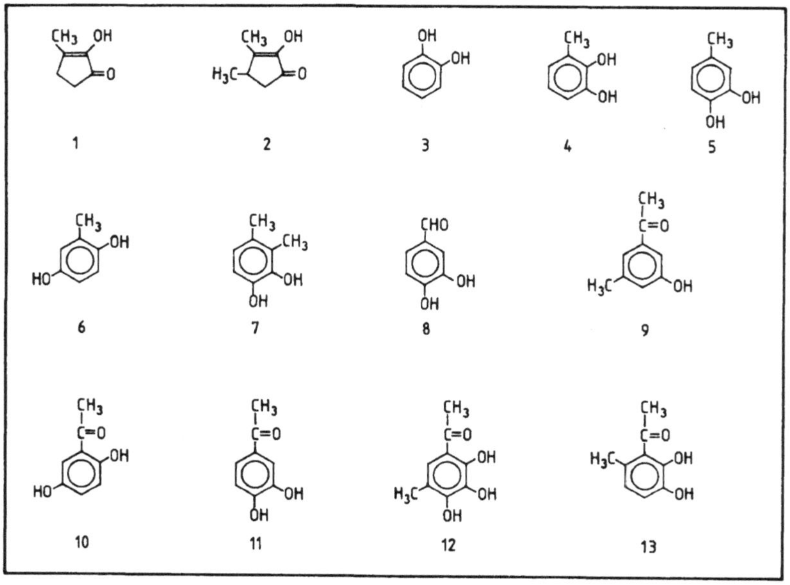

Fig. 11–2: Cyclic enols and phenolic compounds obtained by alkali treatment of xylose and glucose (Forsskåhl et al. 1976).
1: 2-hydroxy-3-methyl-2-cyclopenten-1-one
2: 2-hydroxy-3,4-dimethyl-2-cyclopenten-1-one
3: pyrocatechol
4: 3-methyl-1,2-benzenediol
5: 4-methyl-1,2-benzenediol
6: 2-methyl-1,4-benzenediol
7: 3,4-dimethyl-1,2-benzenediol
8: 3,4-dihydroxybenzaldehyde
9: 3-hydroxy-5-methylacetophenone
10: 2,5-dihydroxyacetophenone
11: 3,4-dihydroxyacetophenone
12: 2,3,4-trihydroxy-5-methylacetophenone
13: 2,3-dihydroxy-6-methylacetophenone.

Another type of additive which stabilizes polysaccharides against alkaline peeling is anthraquinone (AQ) or related compounds such as anthraquinone-2-sulfonic acid (→ 16.4.5.). AQ causes an oxidation of aldehyde end-groups to alkali-stable aldonic acids of different types and amounts, itself being reduced to anthrahydroquinone (AHQ) (Löwendahl, Samuelson 1977, 1978). This compound was proved to react with lignin, resulting in improved delignification (→ 11.2.1.), itself being oxidized to AQ (Fleming et al. 1978; Kubes et al. 1980). This reduction-oxidation conversion is a prerequisite for an effective polysaccharide stabilization in alkaline pulping, as large portions of the involved catalytic AQ amounts (about 0.1% based on wood) are consumed by solubilized degradation compounds in the liquor (Carlson, Samuelson 1979; Bryce 1980).

In addition to the peeling of end-groups alkaline hydrolysis (depolymerization) of polysaccharides becomes important at high temperatures of about 170 °C, which

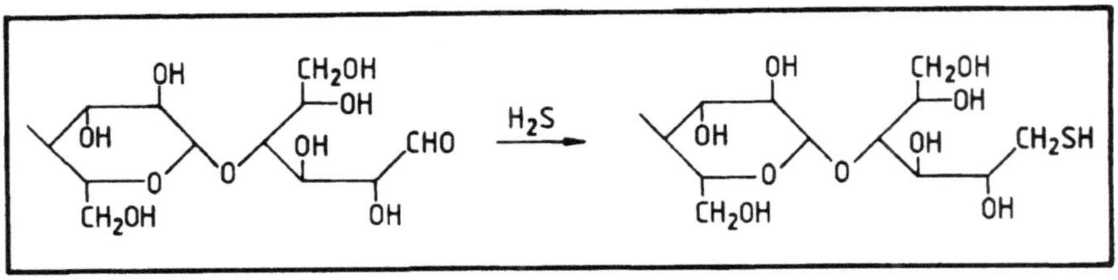

Fig. 11–3. End-group stabilization by hydrogen sulfide.

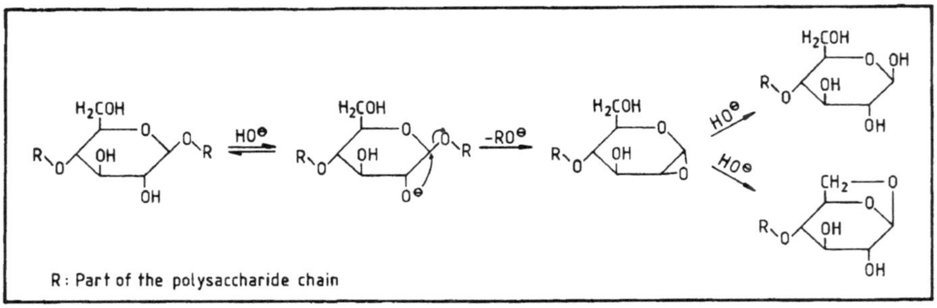

Fig. 11–4. Hydrolysis of glycosidic bonds in alkaline medium.

give rise to further peeling reactions. The probable mechanism of alkaline hydrolysis, occurring much more slowly than acidic hydrolysis, is outlined in Fig. 11–4 (Franzon, Samuelson 1957; Brooks, Thompson 1966; Matthews 1974; Lai 1981). The glycosidic linkage is cleaved by elimination of the alkoxy group (β-elimination) after ionization of the hydroxyl group at C2 and formation of an 1,2-epoxide (oxirane structure).

The hydrolytic cleavage results in a new reducing end-group, or possibly an 1,6-anhydride end is formed, which is easily degraded. Gierer (1980) has pointed out parallels between lignin degradation reactions and polysaccharide hydrolysis in alkaline medium with regard to the β-elimination step and the oxirane intermediate (→ 11.2.1.).

11.1.2. Oxidative Degradation

The oxidative attack on polysaccharides is an undesirable but unavoidable reaction in alkaline bleaching steps, and especially in oxygen bleaching and pulping, respectively (→ 16.7.2., 16.7.3.). As these procedures are performed in alkaline solutions the oxidative reactions are only part of the overall alkaline degradation, involving peeling and hydrolysis of polysaccharides as well.

In principle oxygen can be reduced to water by a one-electron transfer mechanism in four steps, yielding peroxy radicals ($HO_2^{\cdot}$), hydrogen peroxide (H_2O_2) and hydroxyl radicals ($HO^{\cdot}$) as reactive intermediates. In the practical case of delignification of wood or pulp, lignin is expected to be the main source for the formation of organic radical species (Fig. 11–5), but also heavy metal impurities catalyze the decomposition of hydrogen peroxide to reactive radicals (Lindgren, Sundin 1978; Sjöström 1981b).

Our knowledge of polysaccharide reactions in oxygen delignifications was increased by numerous experiments with low-molecular-weight compounds (Malinen, Sjöström 1972, 1973, 1974, 1975b). The most important reaction induced by oxygen radicals is the formation of a carbonyl group at the C2-position of a monomeric sugar unit, leading to a cleavage of the glycosidic bond by β-alkoxy elimination (Fig. 11–6) (Lai 1981; Sjöström 1981b). A comparable reaction occurs if the C3-position is oxidized, and also the oxidation at the C6-hydroxyl can induce a chain cleavage. If the C2- and C3-positions are oxidized simultaneously a 2,3-diketo structure is formed, which may be converted to a carboxy furanosidic group without chain cleavage or easily degraded in the alkaline medium (Malinen 1975; McCloskey et al. 1975; Malinen, Sjöström 1975c, d).

Oxidative depolymerization forms reducing end-groups which are subjected to the typical peeling reaction. Due to the preceding oxidation other acids are formed than under non-oxidative alkaline conditions, e.g. glycolic and 3,4-dihydroxybutanoic acids.

a $$O_2 \xrightarrow{e^\ominus, H^\oplus} HO_2^\bullet \xrightarrow{e^\ominus, H^\oplus} H_2O_2 \xrightarrow{e^\ominus} HO^\bullet (+HO^\ominus) \xrightarrow{e^\ominus, H^\oplus} H_2O$$

b $$R^\bullet + O_2 \longrightarrow RO_2^\bullet \xrightarrow{e^\ominus, H^\oplus} RO_2H \xrightarrow{e^\ominus} RO^\bullet (+HO^\ominus) \xrightarrow{e^\ominus, H^\oplus} ROH$$

$R^\bullet$: Organic radical

Fig. 11–5. Formation of reactive intermediates during reduction of oxygen to water
a: Principal mechanism of reduction
b: Participation of organic radicals.

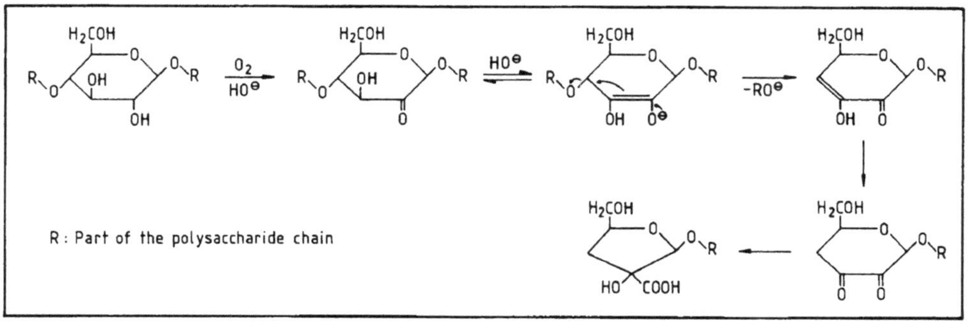

Fig. 11–6. Cleavage of glycosidic bonds by oxidative attack.

It is an interesting fact that oxidation causes not only a degradation of polysaccharides but also some stabilization, as has been shown for different carbohydrate materials, especially in the presence of anthraquinone (Malinen et al. 1973; Heikkilä, Sjöström 1975; Ruoho, Sjöström 1978; Sjöström 1981b). This stabilizing effect is due to the conversion of reducing end-groups via aldosuloses to aldonic acid end-groups, mainly mannonic acid in the case of cellulose and mannan (Samuelson, Thede 1969; Samuelson, Stolpe 1969; Sjöström 1981a). The relative importance of the oxidative reactions in oxygen-alkali systems depends on various factors such as alkali concentration, type and amount of active oxygen, temperature, and additives reducing degradation reactions. The volatile substances liberated in alkali-oxygen bleaching (CO, CO_2, formic acid etc.) are formed from polysaccharides and lignin decomposition fragments, as shown by tracer studies (Kratzl et al. 1974a; Kratzl, Schwarz 1975).

To inhibite the oxidative degradation of polysaccharides, a large number of additive compounds have been suggested, only some of them being in practical use, like triethanolamine and magnesium salts. Their functions are the stabilization of peroxides, decomposition of reactive species or the scavenging of radicals. The stabilizing effect of magnesium salts results mainly from the deactivating (adsorbing) properties of precipitated magnesium hydroxide in alkaline medium (Sjöström, Välttilä 1972; Löwendahl, Samuelson 1974; Christensen 1975).

The action of alkaline hypochlorite is rather unspecific. Apart from oxidation of aldehyde end-groups to aldonic end-groups, C–C linkages between C2 and C3 are cleaved, resulting in a ring-opening and a formation of dicarboxylic acids. Oxidation of the C6-hydroxyl leads to carboxyl groups along the chains of cellulose and mannans (Norstedt, Samuelson 1965). Additionally a multitude of degradation products were identified in the bleaching liquors (Lorås 1980).

11.1.3. Hydrogenation of Sugars

The well-known hydrogenation reaction of sugars yields sugar alcohols (polyols). The polyols of greatest importance to technology are sorbitol, xylitol and mannitol,

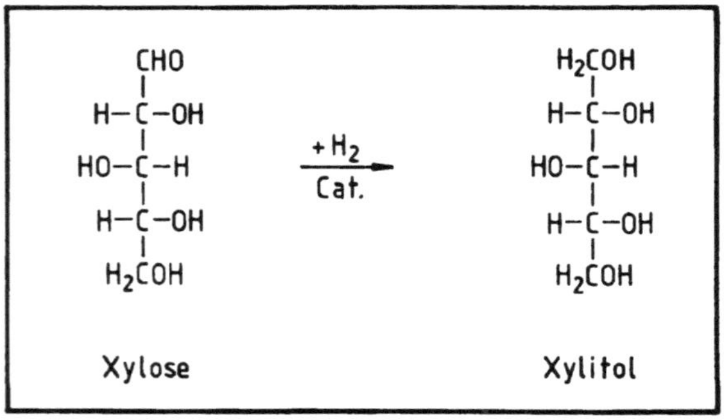

Fig. 11-7. Hydrogenation of xylose to xylitol.

which have attracted increasing interest in recent years for the production of vitamin C, of surface-active agents, and as diabetic sweeteners or basic material for tablets (→ 18.5.) (Herrick et al. 1975; Herrick, Hergert 1977; Haidegger 1977).

The principal reaction is a catalytic reduction in alkaline medium (Fig. 11–7). The reaction parameters influencing the course of the reaction and the final yields of polyols are temperature (130–170 °C), pressure (about 20 N/mm^2) and pH value (7–9). Table 11–1 shows different possibilities of polyol formations. Due to the isomeric function of the keto-group in fructose, both sorbitol and mannitol are formed from fructose. For the same reason, and because of some alkaline epimerization of glucose to mannose, the hydrogenation of sucrose also yields mannitol.

Hydrogenolysis (hydrogenation combined with hydrolysis) is applied if plant hydrolyzates (e.g. from wood, straw, bagasse) are used as raw materials, which still contain di-, oligo-, and polysaccharide moieties.

Hydrogenation under suitable cracking conditions (Ling et al. 1967; Ling, Vlugter 1969) yields lower polyols as main reaction products, as demonstrated with a wood hydrolyzate (Table 11–2).

Table 11–1: Production of polyols from sugars

Sugar	Reaction	Sugar alcohol (Polyol)
D-Glucose	Hydrogenation	D-Sorbitol (D-Glucitol)
D-Mannose	Hydrogenation	D-Mannitol
D-Xylose	Hydrogenation	D-Xylitol
Fructose	Hydrogenation	D-Sorbitol D-Mannitol
Sucrose	Hydrogenation	D-Sorbitol D-Mannitol
Glucose syrup	Hydrogenolysis	Hydrogenated glucose syrup
Sugar mixture	Hydrogenation/ Cracking	Mixed sugar alcohols Lower polyols (e.g. glycerine, ethylene glycol)

Table 11–2: Polyols obtained by hydrocracking of a wood hydrolyzate (according to Haidegger 1977)

Hydrolyzate	%	Distillate	%	Distillate residue	%
Xylose	32	Glycerine	54	Sorbitol	14
Mannose	29	Methyl glycerine	9	Xylitol	4
Glucose	28	Propylene glycol	8	Mannitol	2
Galactose	5	Ethylene glycol	7	Erythrite	1
Arabinose	5			Other polyalcohols	1
Polymeric material	1				

11.2. Reactions of Lignin

11.2.1. Alkaline Pulping

The most important field of alkaline lignin reactions is sulfate (kraft) and soda pulping (→ 16.4.).

Hydrogen sulfide and hydroxyl ions are involved in sulfate pulping, while in soda pulping only hydroxyl ions cause delignification. As in the case of sulfite pulping (→ 10.3.2.) the alkaline pulping reactions with lignin are also nucleophilic reactions (Gierer 1982). They can be grouped into fragmentations, contributing to the degradation and dissolution of lignin, and condensations of lignin units to fragments with increased molecular weight and reduced solubility. In both types of reactions common intermediates are involved.

An essential aspect in alkaline pulping is the different behaviour and stability of the various types of linkages and structural elements in lignin. While all types of aryl ether bonds are typically cleaved in alkaline medium, and also aryl-alkyl or alkyl-alkyl carbon-carbon bonds are destroyed to some extent, diaryl ether and C–C bonds are stable under the same conditions. As α-and β-aryl ether linkages are the dominant types of linkages in both softwood and hardwood lignins (Table 6–2), the cleavage of these bonds contributes essentially to lignin degradation (Rydholm 1965; Gierer 1970, 1980, 1981, 1982; Marton 1971; Gierer, Norén 1981).

The most easily cleaved linkages are α-aryl ether bonds in phenolic arylpropane units (Gierer, Norén 1962). The cleavage leads to the formation of a quinone methide structure and an elimination of the α-substituent (Fig. 11–8). The fragmentation into two separate units is only possible in the absence of an additional β-aryl ether linkage. The only prerequisite for this reaction is a sufficient ionization of the phenolic group in the alkaline medium as it is independent of the action of hydrogen sulfide ions.

Moreover, β-aryl ether bonds in phenolic lignin units are also rapidly cleaved in a first-order reaction, but only in the presence of hydrogen sulfide ions, which are

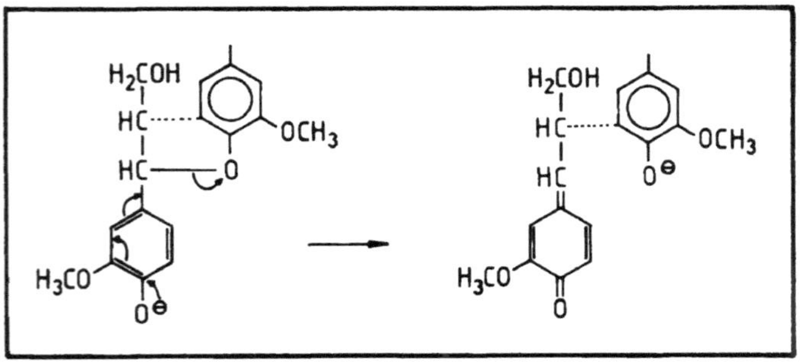

Fig. 11–8. Formation of a quinone methide by alkaline cleavage of the α-aryl ether bond in a phenolic lignin unit (Dotted line: possible β-C linkage to another unit; according to Gierer 1980).

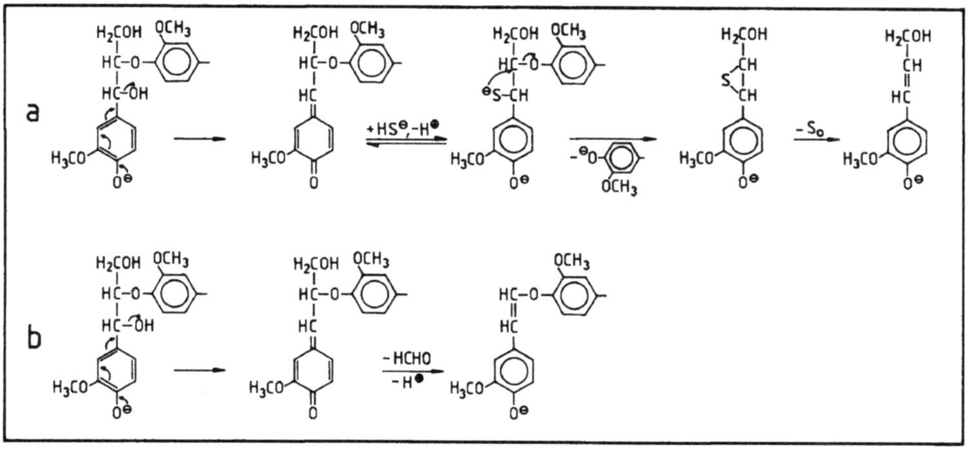

Fig. 11–9. Reactions of β-aryl ether bonds in phenolic lignin units.
a: Sulfidolytic cleavage in sulfate pulping
b: Formation of β-aroxy styrene units in soda pulping
(according to Gierer 1980).

sufficiently nucleophilic to cause a cleávage. This sulfidolytic cleavage of β-aryl ether bonds is outlined in Fig. 11–9 a. After the elimination of the α-substituent (Fig. 11–8) hydrogen sulfide ions are added, resulting in a benzyl mercaptide structure. The anions induce the β-aryl ether cleavage, and the bound sulfur is released via a thiirane intermediate (Gierer, Smedman 1965; Gierer et al. 1973; Gierer, Ljunggren 1979a). In the case of soda pulping without sulfur, alkali-stable structures are formed without cleavage at the β-C atom (Fig. 11–9 b). As α- and β-aryl ether bonds are cleaved relatively easy and quickly, independent of the alkali concentration, they are the main reactions during the initial phase of alkaline pulping procedures (Gierer, Norén 1980; Ljunggren 1980).

β-Aryl ether bonds in non-phenolic units are cleaved much more slowly. The reaction velocity depends on the hydroxide concentration, but is independent of the presence of hydrogen sulfide ions. Fig. 11–10 demonstrates the cleavage at the β-C atom via an oxirane intermediate, which is additionally promoted by anions at vicinal C-atoms (Gierer, Norén 1962; Gierer, Ljunggren 1979b).

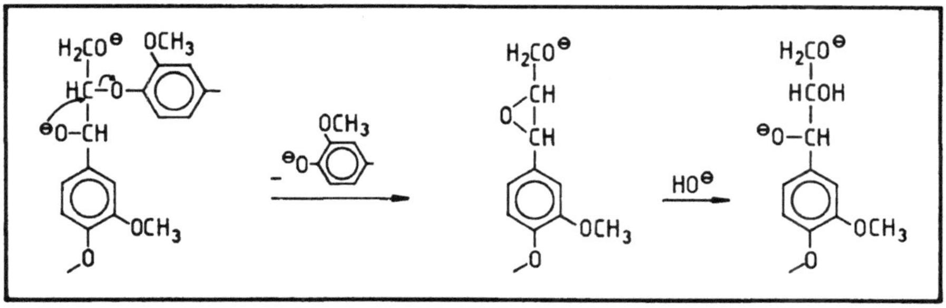

Fig. 11–10. Cleavage of β-aryl ether bonds in non-phenolic lignin units (according to Gierer 1980).

Small portions of lignin are degraded by cleavage of less frequent linkages such as C–C bonds, leading to a reduction or total elimination of side-chains. For example, the γ-C may be split from quinone intermediates, releasing formaldehyde (Fig. 11–11). Thereby no lignin fragmentation is reached. The same is true of the cleavage of methyl-aryl ether bonds. The methoxyl groups are predominantly cleaved by hydrogen sulfide ions and only to a small extent by the less nucleophilic hydroxyl ions (Fig. 11–12). Thus only minor amounts of methanol are formed. The demethylation deliberates methyl mercaptane (CH_3SH), which may be converted to dimethylsulfide (CH_3SCH_3) or dimethyldisulfide ($CH_3S_2CH_3$), all compounds causing the typical odour problems in kraft pulping (→ 16.4.2.).

As in sulfite pulping, condensation reactions compete with reactions leading to a dissolution of lignin. Condensation in alkaline medium can be generally described as an addition of internal nucleophiles (carbanions from phenolic or enolic structures) to conjugated carbonyl structures such as quinone methides or other enones (Gierer 1970, 1980, 1982; Marton 1971).

If a phenolate unit is added to a quinone methide intermediate a new α-5 linkage is formed by an irreversible release of a proton (primary condensation) (Fig. 11–13a). In reaction b (Fig. 11–13) the same principal mechanism is demonstrated, if fragments with conjugated side-chain structures, resulting e.g. from sulfidolytic ether cleavage (Fig. 11–9), liberate a proton and formaldehyde (secondary condensation). A condensation of phenolic units with formaldehyde, resulting also from side-chain reduction (Fig. 11–11), yields diaryl methane structures (Fig. 11–13c) (Ekman 1965; Gierer 1970; Marton 1971).

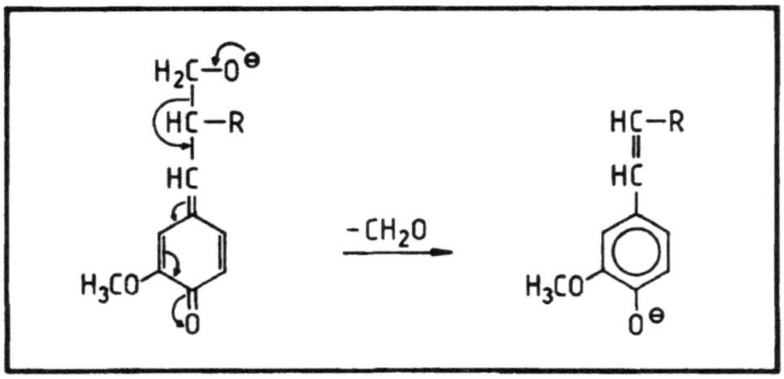

Fig. 11–11. Cleavage of C–C bonds, liberating formaldehyde (according to Gierer 1970).

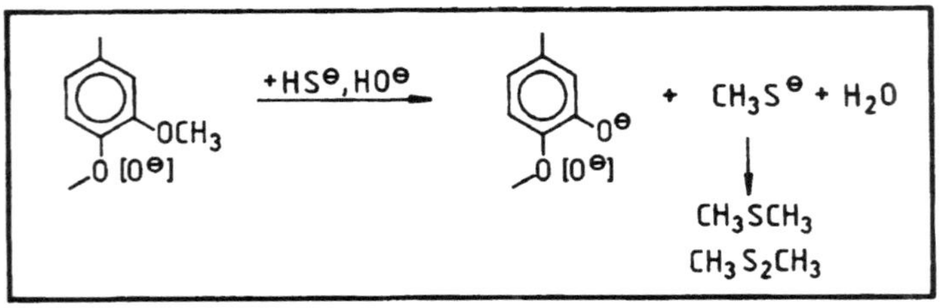

Fig. 11–12. Alkaline cleavage of methoxyl groups.

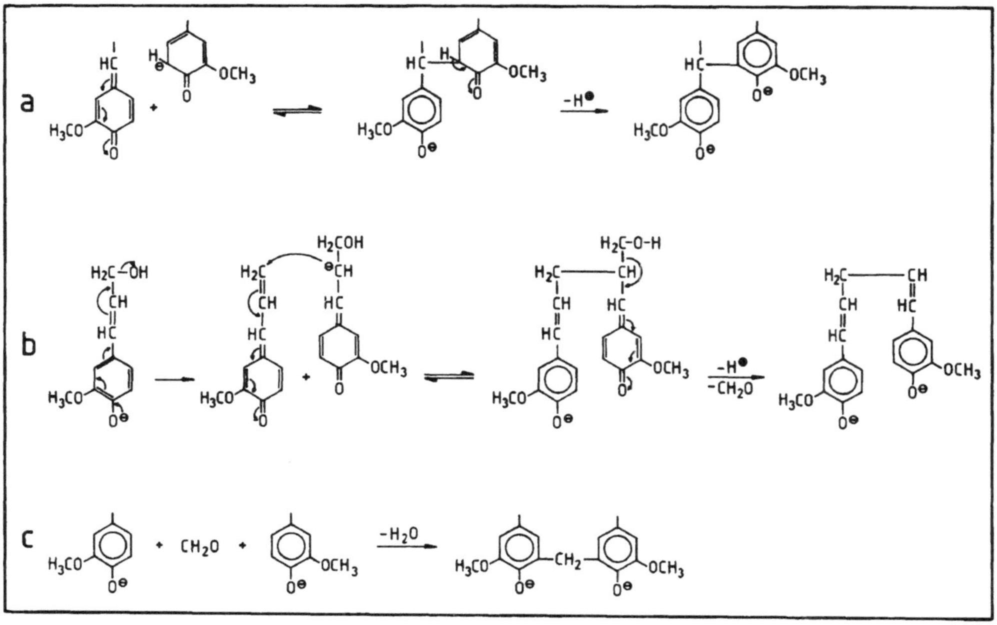

Fig. 11–13: Condensation reactions of lignin in alkaline medium (according to Gierer 1980)
a: primary condensation
b: secondary condensation
c: condensation with formaldehyde.

As already mentioned in connection with the stabilization of polysaccharides against peeling (→ 11.1.1.) the reduced form of anthraquinone (AHQ) reacts with lignin and is itself oxidized to AQ. Numerous other additives are also known to represent comparable redox systems (→ 16.4.5.) (Fleming et al. 1978; Kubes et al. 1980; Algar et al. 1980; Eckert, Amos 1981).

The reaction of AHQ with phenolic lignin units leads to a cleavage of β-aryl ether bonds after the addition of AHQ anions to quinone intermediates, thus accelerating the delignification (Fig. 11–14).

Recently the presence of AQ radicals ($AQ^{\cdot -}$) was proved in soda-AQ and kraft-AQ cooking liquors, which may possibly be involved in some way in polysaccharide oxidation and lignin degradation (Mattar, Fleming 1981).

11.2.2. Alkaline Bleaching Reactions

Bleaching in alkaline medium comprises the hypochlorite stage and oxidation with oxygen and peroxides, respectively (→ 16.7.2., 16.7.3.).

Hypochlorite is usually applied in later stages of lignin-removing bleaching sequences (e.g. CEH), but sometimes even within the extraction step to obtain an substantial increase in brightness. This practical aspect becomes understandable if the reactions of hypochlorite anions (ClO^-) in alkaline medium are taken into consid-

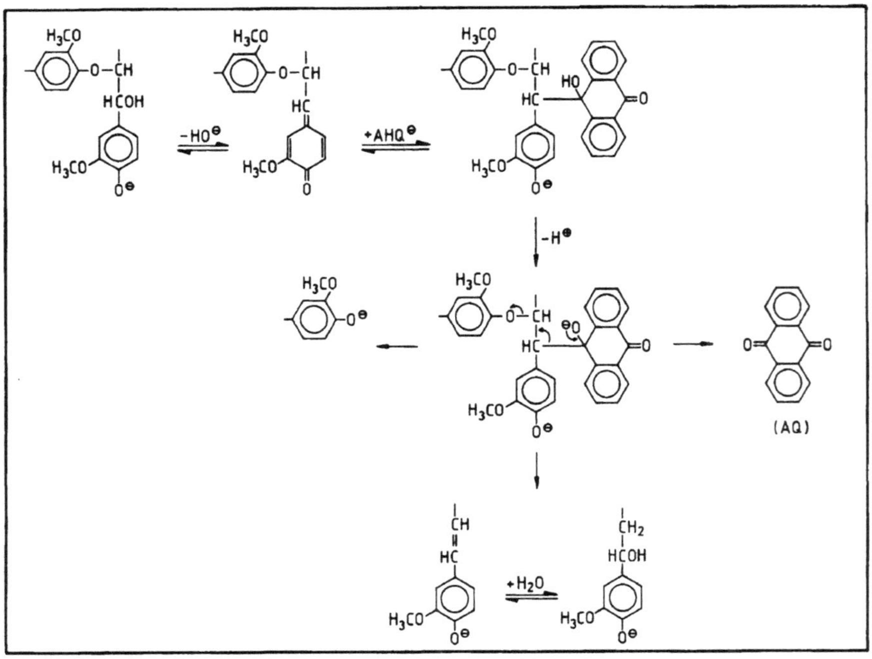

Fig. 11–14. Reaction of anthrahydroquinone with phenolic arylpropane units (according to Gierer 1980).

eration. In contrast to chlorine and chlorine dioxide they act as strong nucleophiles, which are easily added to positively charged sites of lignin, present in quinoid and other enone structures formed in the preceding oxidation reactions (Fig. 10–10). By this addition those structures are either converted to carbonyl acid compounds or other alkali-soluble degradation products via oxirane intermediates (Fig. 11–15).

By a partial decomposition of hypochlorite electrophilic radicals (Cl˙, ClO˙) may also be formed (Gierer 1982).

The oxidative action of oxygen and peroxides in alkaline medium is closely related because both are involved in oxygen bleaching and peroxide bleaching as well. On

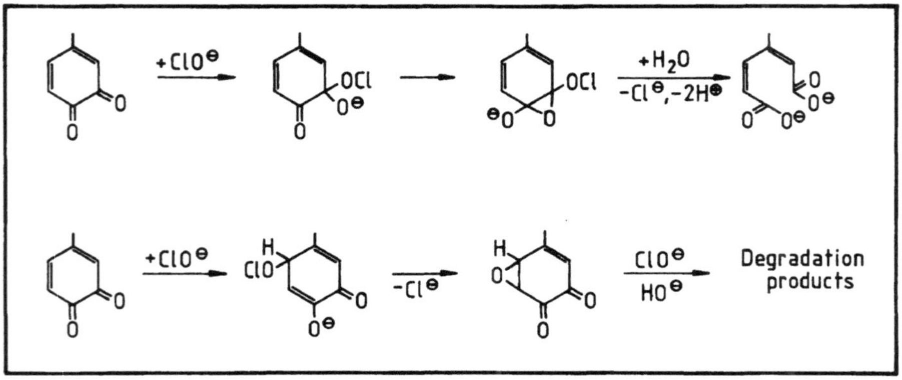

Fig. 11–15. Action of hypochlorite ions on quinoid lignin structures (according to Gierer 1982).

the other hand the initial reactions with lignin units are quite different. Molecular oxygen causes an electrophilic attack on carbanions, which are formed from phenolic and enolic lignin units by autoxidation in alkaline medium (Fig. 11–16a). Fig. 11–16b shows the formation of carbonyl and conjugated carbonyl structures. Hydrogen peroxide anions (HOO^-), which are introduced in peroxide bleaching or formed by autoxidation of lignin structures in oxygen bleaching (Fig. 11–17), act as nucleophiles, however, and are added to carbonyl and conjugated carbonyl structures (Kratzl et al. 1974b, 1978; Gierer, Imsgard 1977; Sindhwani et al. 1979; Omori, Dence 1981).

The reaction of oxygen with carbanions and the addition of hydrogen peroxide anions to carbonylic structures both involve hydrogen peroxide intermediates of lignin. Fig. 11–18 shows an example of the oxygenation of carbanions (I) to a hydrogen peroxide intermediate (II) which is converted via a dioxetane compound (III) to a muconic acid structure (IV). Two examples of the action of hydrogen peroxide anions are given in Fig. 11–19. In the first reaction (a) the quinoid structure is degraded to a dicarboxylic acid via the peroxide and dioxetane intermediates. The second example (b) shows the formation of degradation products with oxirane structures.

Highly reactive radicals are also partly involved in alkaline bleaching with oxygen and peroxides, because hydrogen peroxide decomposes to hydroxy, hydrogen peroxide and peroxide ion radicals ($HO^{\bullet}$, $HOO^{\bullet}$, $O_2^{\bullet -}$) (Roberts et al. 1978).

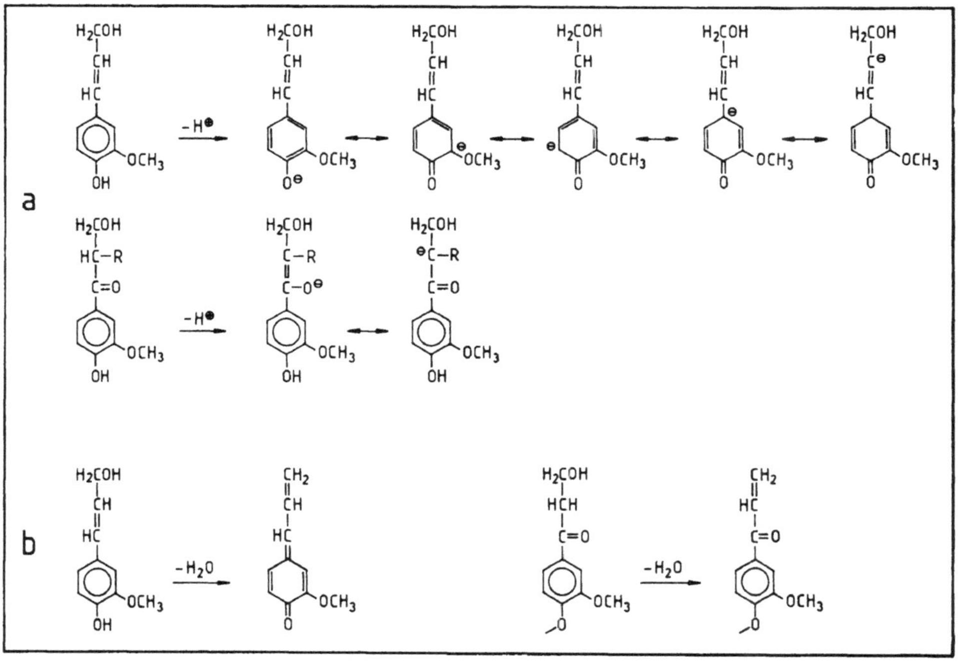

Fig. 11–16. Autoxidative formation of carbanions (a) and conjugated carbonyl structures (b) (according to Gierer, Imsgard 1977).

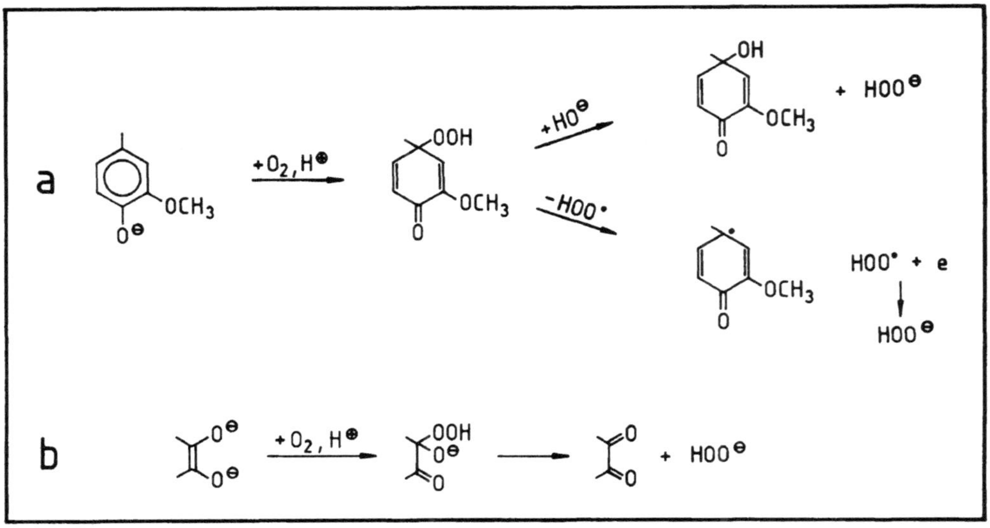

Fig. 11–17. Formation of hydrogen peroxide from phenolic (a) and enediol structures (b) (according to Gierer 1982).

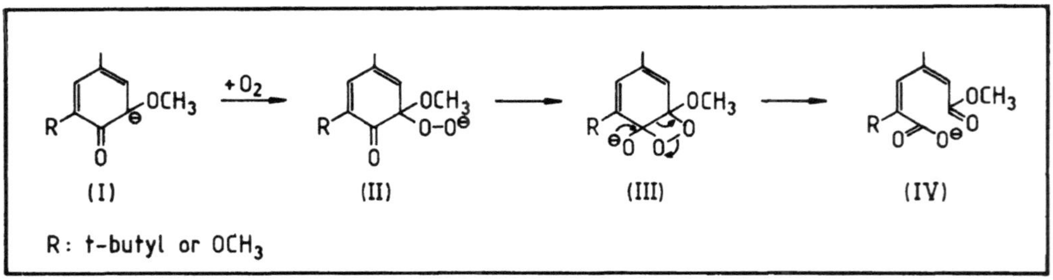

Fig. 11–18. Action of molecular oxygen on carbanionic lignin structures (according to Gierer, Imsgard 1977).

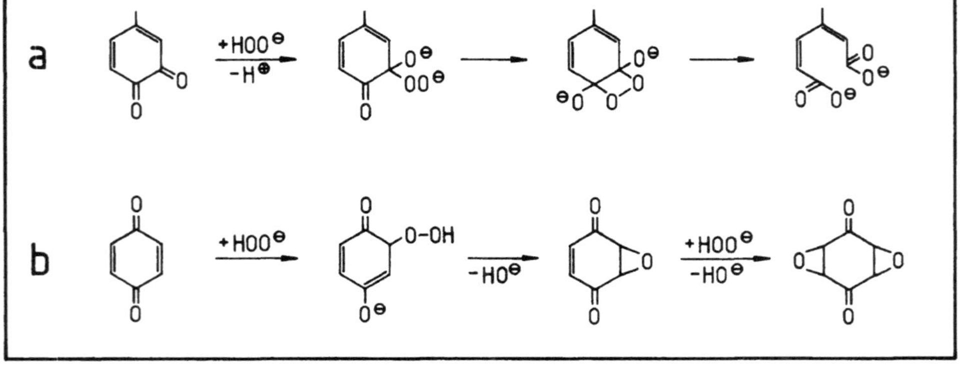

Fig. 11–19. Action of hydrogen peroxide ions on quinoid lignin structures (according to Gierer, Imsgard 1977)

a: Formation of a dicarboxylic acid

b: Formation of oxirane structures.

Comparing the overall effects of oxygen and hydrogen peroxide on lignin, it is obvious that molecular oxygen tends to form chromophoric structures accompanying lignin degradation, while the advantage of peroxides is the decomposition of chromophoric structures. This advantage is used in lignin-preserving bleaching with hydrogen peroxide (Spittler, Dence 1977; Krüger 1979).

11.2.3. Mild and Selective Oxidative Degradation

Apart from the oxidation reactions of lignin in bleaching procedures a series of analytical oxidative degradation methods are known, and mainly applied to the characterization of lignins (→ 6.3.1., 6.3.3.). Common to these reactions is the preservation of the aromatic rings in lignin. A first group of oxidative reactions degrades lignin into aromatic carbonyl compounds and carboxylic acids, while in a second group only special functional groups are involved (Chang, Allan 1971).

The most important techniques of the first group, excluding alkali/oxygen oxidation discussed in the previous section, are the following:
- alkaline nitrobenzene oxidation
- metal oxide oxidations
- permanganate oxidation.

Nitrobenzene in aqueous alkaline solutions (mostly sodium hydroxide at elevated temperatures of 170–180 °C) degrades lignin typically to aromatic aldehydes. In the case of softwood lignin vanillin is the major reaction product. From hardwood lignin a mixture of vanillin and syringaldehyde is obtained, while grass lignins additionally yield p-hydroxybenzaldehyde. Minor amounts of vanillic acid and other compounds were identified e.g. in spruce wood lignin *(Picea abies)* (Table 11–3).

The principal mechanism of nitrobenzene oxidation is based on the action of nitrobenzene as a two-electron acceptor and the presence of hydroxyl ions, as demonstrated in Fig. 11–20 with isoeugenol as a model substance. Numerous degradation products from various model compounds and different lignins are discussed elsewhere (Brauns, Brauns 1960; Freudenberg, Neish 1968; Chang, Allan 1971).

In the case of oxides of various metals (Cu, Ag, Hg, Co) in alkaline medium some similarities with the nitrobenzene oxidative degradation are obvious with regard to the action of hydroxyl ions, the intermediate formation of quinone methides, and the final oxidation products. Also aromatic aldehydes and aromatic carboxylic acids are formed. The yields of acids depend on the oxidative power of the applied metals, cupric oxide yielding the most comparable results with the nitrobenzene oxidation. In contrast to the reaction of nitrobenzene with lignin the metal oxides involve a one-electron transfer, and the reaction probably starts with a resonance-stabilized phenoxy radical (Kratzl et al. 1966).

Table 11–3: Compounds obtained by nitrobenzene oxidation of spruce wood *(Picea abies)* (according to Leopold 1952)

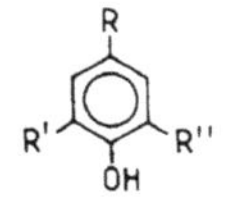

Oxidation Compounds	R	R'	R''	Yield % on Klason lignin
Vanillin	CHO	OCH_3	H	27.5
Syringaldehyde	CHO	OCH_3	OCH_3	0.06
p-Hydroxybenzaldehyde	H	H	H	0.25
5-Formylvanillin	CHO	OCH_3	CHO	0.23
Dehydrodivanillin	CHO	OCH_3	*	0.80
Vanillic acid	COOH	OCH_3	H	4.8
Syringic acid	COOH	OCH_3	OCH_3	0.02
5-Formylvanillic acid	COOH	OCH_3	CHO	0.1
5-Carboxyvanillin	CHO	OCH_3	COOH	1.2
Dehydrodivanillic acid	COOH	OCH_3	**	0.03
Acetoguaiacone	$COCH_3$	OCH_3	H	0.05

* Dehydrovanillin residue
** Dehydrovanillic acid residue

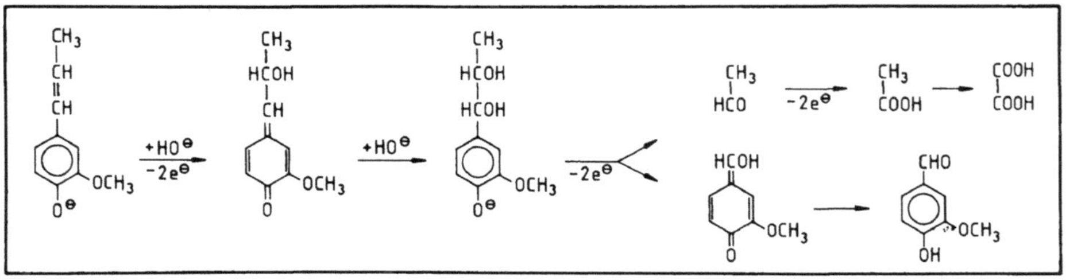

Fig. 11–20: Nitrobenzene oxidation of isoeugenol to vanillin.

The oxidation of lignin with potassium permanganate in neutral medium at temperatures between 50 and 100 °C is also a typical and useful tool for the elucidation of lignin structure. The treatment with permanganate is usually preceded by an alkaline hydrolysis step and a methylation for stabilizing free hydroxyl groups against oxidation. The resulting aromatic acids, of which 40 were obtained e.g. by oxidation of Björkman lignin from spruce, are formed only in very low yields with an average total of 10–30% based on lignin (Freudenberg, Neish 1968; Chang, Allan 1971; Erickson et al. 1973; Adler 1977).

Some prominent acids deriving from softwood and hardwood lignins are shown in Fig. 11–21. Modified procedures include e.g. preheating with sodium hydroxide and cupric oxide, or a brief treatment with alkaline hydrogen peroxide after a permanganate oxidation in aqueous sodium carbonate (Erickson et al. 1973).

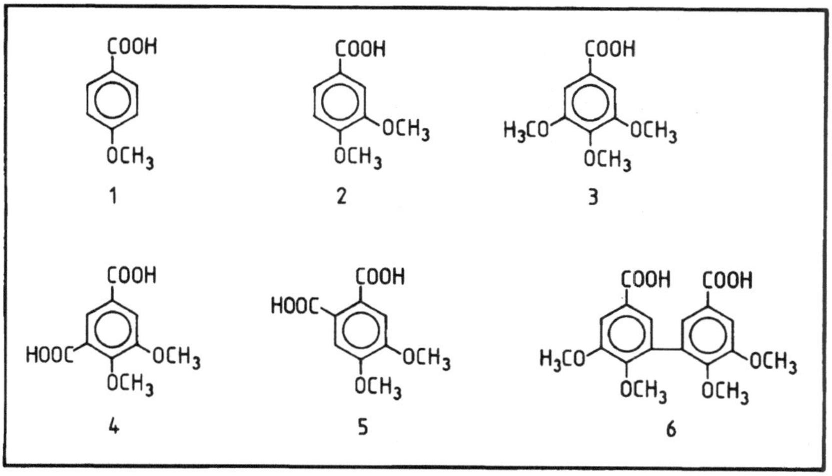

Fig. 11–21. Prominent degradation acids resulting from permanganate oxidation
1: p-Anisic acid
2: Veratric acid
3: Trimethyl gallic acid
4: Isohemipinic acid
5: Metahemipinic acid
6: Dehydrodiveratric acid.

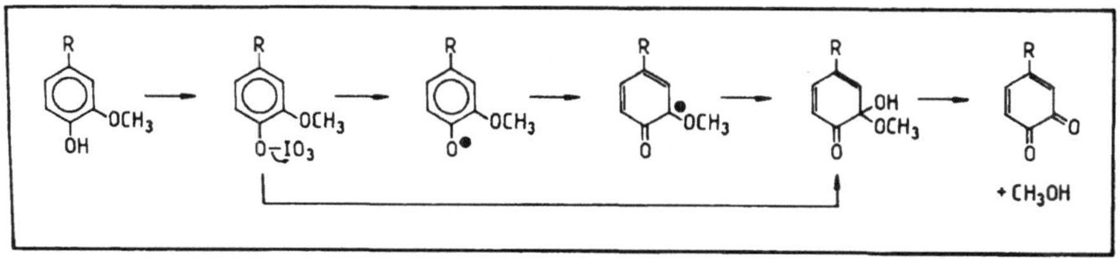

Fig. 11–22. Conversion of a phenolic lignin unit to a quinoid structure by periodate oxidation (according to Adler et al. 1962).

The most typical oxidation directed to a special functional group takes place with sodium periodate. Free phenolic hydroxyl groups are oxidized to quinones, from which methanol is liberated by splitting the methoxyl group or groups, respectively (Fig. 11–22). The periodate ester thus formed may be converted directly to a quinone half ketal or via its aroxy cation (Adler et al. 1962).

11.2.4. Hydrogenolysis

Hydrogenolysis generally means a hydrogen treatment causing additional hydrolytic effects. Hydrogenolysis of lignin is of interest both with regard to structural studies (→ 6.3.1.) and the production of phenols from technical lignins (→ 18.6.2.).

For the hydrogenolytic degradation of lignin a number of procedures are known, including treatments in different alkaline and neutral solvents, in the presence of suitable catalysts, nascent hydrogen or hydrogen donators (Brauns, Brauns 1960; Schweers 1966; Schweers 1969; Hrutfiord 1971; Hoffmann, Schweers 1975a). By means of these procedures a multitude of degradation products can be isolated. Generally hydrogenolysis of lignins yields complex mixtures of mono-, di-, tri-, oligo- and polymeric degradation compounds. Types and amounts of the low-molecular-weight portions (mainly mono- and dicompounds) depend on the hydrolytic effects involved in the preceding lignin isolation procedure (e.g. pulping or mild solvolysis) or on the hydrolytic action of the hydrogenation process itself. Under mild conditions, e.g. ethanol/water or dilute NaOH as solvent and at 160–170 °C under pressure, typical phenylpropane mono- and dimers are obtained in total yields of up to 25%, based on lignin (Hrutfiord 1971; Sudo et al. 1978, 1979; Hwang, Sakakibara 1979). A comprehensive scheme for the hydrogenolysis reactions of phenylpropane model compounds was given by Hoffmann and Schweers (1975b). Some probable

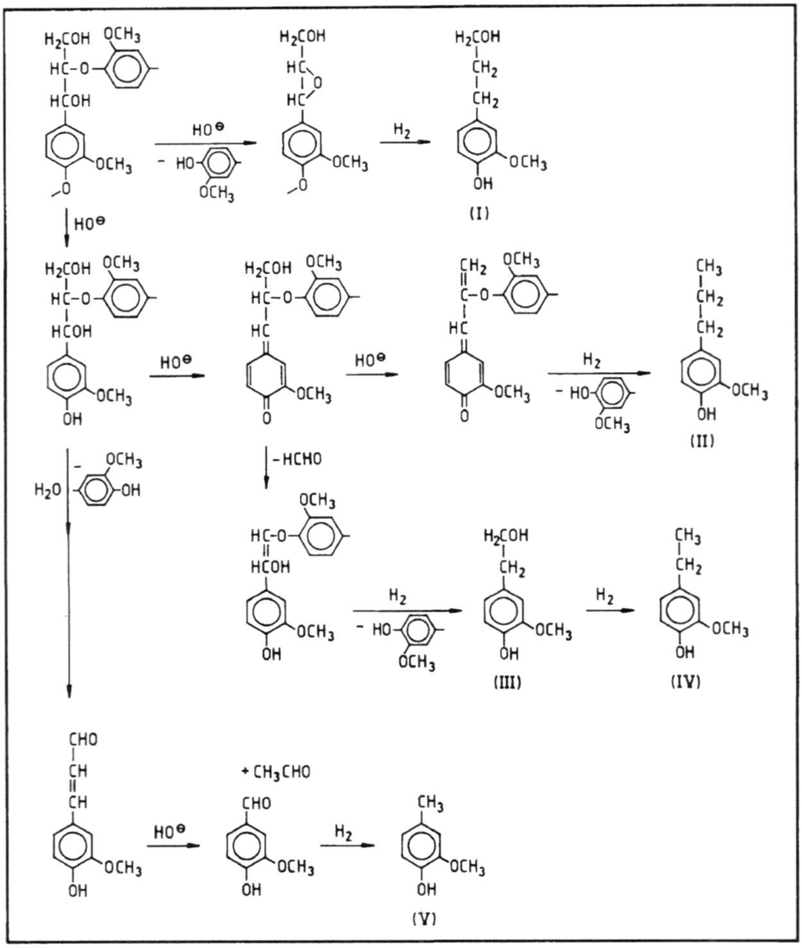

Fig. 11–23. Hydrogenolysis reactions (according to Hrutfiord 1971).

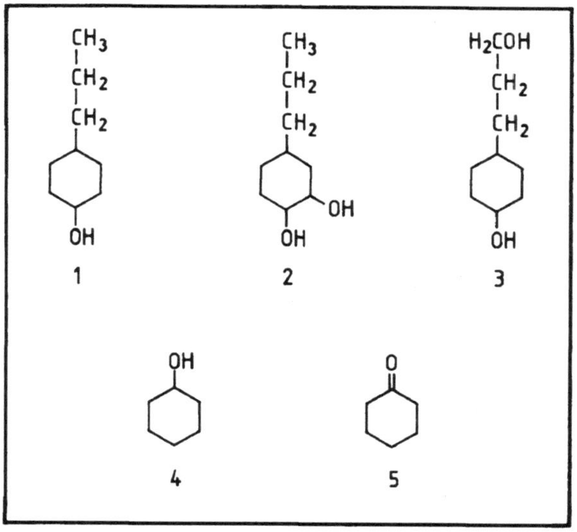

Fig. 11–24. Cyclohexane derivatives obtained under strong hydrogenolysis conditions
1: 4-n-propylcyclohexanol-1
2: 4-n-propylcyclohexandiol-1,2
3: 3-(4-hydroxycyclohexyl)-propanol-1
4: Cyclohexanol
5: Cyclohexanone.

reactions are outlined in Fig. 11–23, showing the formation of guaiacyl propanol (I), guaiacyl propane (II), guaiacyl ethanol (III), guaiacyl ethane (IV) and methyl guaiacol (V) (Hrutfiord 1971).

Under more severe hydrogenolysis conditions with temperatures of 200–350 °C or by hydrogenation of hydrogenolysis lignin the aromatic character of lignin is lost and cyclohexane derivatives are obtained, as shown e.g. in Fig. 11–24.

Literature

Adler, E. 1977, Wood Sci. Technol. *11,* 169–218

Adler, E., Falkehag, I. and Smith, B. 1962, Acta Chem. Scand. *16,* 529–540

Alfredsson, B., Gedda, L. and Samuelson, O. 1961, Svensk Papperstid. *64,* 694–698

Algar, W. H., Farrington, A., Jessup, B., Nelson, P. F. and Vanderhoek, N. 1980, Appita *33,* 33–37

Axelsson, S., Croon, I. and Enström, B. 1962, Svensk Papperstid. *65,* 693–697

Brauns, F. E. and Brauns, D. A. 1960, The Chemistry of Lignin. Supplement Volume. Academic Press, New York, London

Brocks, R. D. and Thompson, N. S. 1966, Tappi *49,* 362–366

Bryce, J. R. G. 1980, Alkaline Pulping. In: Pulp and Paper. Chemistry and Chemical Technology (Casey, J. P., Ed.), 3rd Ed., Vol. I. Wiley-Intersci., New York, pp. 377–492

Carlson, U. and Samuelson, O. 1979, Svensk Papperstid. *82,* 48–52

Casebier, R. L., Hamilton, J. K. 1965, Tappi *48,* 664–669

Chang, H. M. and Allan, G. G. 1971, Oxidation. In: Lignins. Occurrence, Formation, Structure and Reactions (Sarkanen, K. V. and Ludwig, C. H., Eds.). Wiley-Intersci., New York, pp. 433–485

Christensen, P. K. 1975, Norsk Skogind. *29,* 84–88
Croon, I. and Enström, B. F. 1961, Tappi *44,* 870–874
Eckert, R. C. and Amos, L. W. 1981, Tappi *64,* No. 6, 123–124
Ekman, K. H. 1965, Tappi *48,* 398–402
Erickson, M., Larsson, S. and Miksche, G. E. 1973, Acta Chem. Scand. *27,* 127–140
Fleming, B. I., Kubes, G. J., MacLeod, J. M. and Bolker, H. I. 1978, Tappi *61,* No. 6, 43–46
Fleming, B. I., Bolker, H. I., Kubes, G. J., MacLeod, J. M., Werthemann, D. B. 1980, Tappi *63,* No. 11, 73–77
Forsskåhl, I., Popoff, T. and Theander, O. 1976, Carbohydr. Res. *48,* 13–21
Franzon, O. and Samuelson, O. 1957, Svensk Papperstid. *60,* 872–877
Freudenberg, K. and Neish, A. C. 1968, Constitution and Biosynthesis of Lignin. Springer-Verlag, Berlin
Gierer, J. 1970, Svensk Papperstid. *73,* 571–596
Gierer, J. 1980, Wood Sci. Technol. *14,* 241–266
Gierer, J. 1981, Chemical Aspects of Delignification. In: The Ekman Days 1981. Int. Symp. Wood Pulp. Chem. Stockholm, Vol. 2, pp. 12–17
Gierer, J. 1982, Holzforschung *36,* 43–51; 55–65
Gierer, J. and Imsgard, F. 1977, Svensk Papperstid. *80,* 510–518
Gierer, J. and Ljunggren, S. 1979a, Svensk Papperstid. *82,* 503–508
Gierer, J. and Ljunggren, S. 1979b, Svensk Papperstid. *82,* 71–81
Gierer, J. and Norén, I. 1962, Acta Chem. Scand. *16,* 1713–1729
Gierer, J. and Norén, I. 1980, Holzforschung *34,* 197–200
Gierer, J. and Norén, I. 1981, Papier *35,* No. 10A, V18–V24
Gierer, J. and Smedman, L. Å. 1965, Acta Chem. Scand. *19,* 1103–1112
Gierer, J., Pettersson, I., Smedman, L. Å. and Wennberg, I. 1973, Acta Chem. Scand. *27,* 2082–2094
Green, J. W., Pearl, I. A., Hardacker, K. W., Andrews, B. D. and Haigh, F. C. 1977, Tappi *60,* No. 10, 120–125
Haidegger, E. 1977, Stärke *29,* 430–435
Hartler, N. 1959, Svensk Papperstid. *62,* 467–470
Hartler, N. 1967, Tappi *50,* 156–160
Hartler, N. and Olsson, L. A. 1972, Svensk Papperstid. *75,* 559–565
Heikkilä, H. and Sjöström, E. 1975, Cell. Chem. Technol. *9,* 3–11
Herrick, F. W. and Hergert, H. L. 1977, Utilization of Chemicals from Wood: Retrospect and Prospect. In: The Structure, Biosynthesis, and Degradation of Wood; Rec. Adv. in Phytochem., Vol. 11 (Loewus, F. A. and Runeckles, V. C., Eds.). Plenum Press, New York, pp. 443–515
Herrick, F. W., Casebier, R. L., Hamilton, J. K. and Wilson, J. D. 1975, Appl. Polym. Symp. No. 28, 93–108
Hoffmann, P. and Schweers, W. 1975a, Holzforschung *29,* 73–79
Hoffmann, P. and Schweers, W. 1975b, Pap. Puu *57,* 581–592
Hrutfjord, B. F. 1971, Reduction and Hydrogenolysis. In: Lignins. Occurence, Formation, Structure and Reactions (Sarkanen, K. V. and Ludwig, C. H., Eds.). Wiley-Intersci., New York, pp. 487-507
Hwang, B. H. and Sakakibara, A. 1970, Mokuzai Gakkaishi *25,* 647–652
Johansson, M. H. and Samuelson, O. 1974, Carbohydr. Res. *34,* 33–43
Johansson, M. H. and Samuelson, O. 1975, J. Appl. Polym. Sci. *19,* 3007–3013
Johansson, M. H. and Samuelson, O. 1977a, Wood Sci. Technol. *11,* 251–263
Johansson, M. H. and Samuelson, O. 1977b, Svensk Papperstid. *80,* 519–524
Kleppe, P. J. 1970, Tappi *53,* No. 1, 35–47
Kratzl, K. and Schwarz, H. A. 1975, Holzforschung *29,* 29–31
Kratzl, K., Gratzl, J. S. and Claus, P. 1966, Adv. Chem. Ser. No. 59, 157–176

Kratzl, K., Silbernagel, H. and Vierhapper, F. W. 1974a, Holzforschung *28,* 77–81
Kratzl, K., Claus, P., Lonsky, W. and Gratzl, J. S. 1974b, Wood Sci. Technol. *8,* 35–49
Kratzl, K., Claus, P. K., Hruscha, A. and Vierhapper, F. W. 1978, Cell.Chem.Technol. *12,* 445–462
Krüger, H. 1970, Papier *33,* No. 10A, V1–V9
Kubes, G. I., Fleming, B. I., MacLeod, J. M. and Bolker, H. I. 1980, Wood Sci. Technol. *14,* 207–228
Lai, Y. Z. 1981, Kinetics of Base-Catalyzed Cleavage of Glycosidic Linkages. In: The Ekman Days 1981. Int. Symp. Wood Pulp. Chem. Stockholm, Vol. 2, pp. 26–33
Leopold, B. 1952, Acta Chem. Scand. *6,* 38–48
Lindberg, B. 1956, Svensk Papperstid. *59,* 531–534
Lindberg, B., Theander, O. and Uddegard, J. E. 1966, Svensk Papperstid. *69,* 360–363
Lindgren, B. O. and Sundin, S. 1978, Svensk Papperstid. *81,* 485–488
Ling, G., van and Vlugter, J. 1969, J. Appl. Chem. *19,* 43–45
Ling, G., van, Ruijterman, C. and Vlugter, J. C. Carbohydr. Res. *4,* 380–386
Ljunggren, S. 1980, Svensk Papperstid. *83,* 363–369
Lorås, V. 1980, Bleaching. In: Pulp and Paper. Chemistry and Chemical Technology (Casey, J. P., Ed.), 3rd Ed., Vol. I. Wiley-Intersci., New York, pp. 633–764
Löwendahl, L. and Samuelson, O. 1974, Svensk Papperstid. *77,* 593–602
Löwendahl, L. and Samuelson, O. 1977, Svensk Papperstid. *80,* 549–551
Löwendahl, L. and Samuelson, O. 1978, Tappi *61,* No. 2, 19–21
Löwendahl, L., Petersson, G. and Samuelson, O. 1976a, Cell. Chem. Technol. *10,* 471–477
Löwendahl, L., Petersson, G. and Samuelson, O. 1976b, Tappi *59,* No. 9, 118–120
Malinen, R. 1975, Pap. Puu *57,* 193–204
Malinen, R. and Sjöström, E. 1972, Pap. Puu *54,* 451–468
Malinen, R. and Sjöström, E. 1973, Pap. Puu *55,* 547–556
Malinen, R. and Sjöström, E. 1974, Pap. Puu *56,* 895–909
Malinen, R. and Sjöström, E. 1975a, Pap. Puu *57,* 728–736
Malinen, R. and Sjöström, E. 1975b, Cell. Chem. Technol. *9,* 231–238
Malinen, R. and Sjöström, E. 1975c, Cell. Chem. Technol. *9,* 651–655
Malinen, R. and Sjöström, E. 1975d, Pap. Puu *57,* 101–114
Malinen, R., Sjöström, E. and Ylijoki, J. 1973, Pap. Puu *55,* 5–13
Marton, J. 1971, Reactions in Alkaline Pulping. In: Lignins. Occurrence, Formation, Structure and Reactions (Sarkanen, K. V. and Ludwig, C. H., Eds.). Wiley-Intersci., New York, pp. 639–694
Mattar, S. M. and Fleming, B. I. 1981, Tappi *64,* No. 4, 136–137
Matthews, C. H. 1974, Svensk Papperstid. *77,* 629–635
McCloskey, J. T., Schroeder, L. R., Sinkey, J. D. and Thompson, N. S. 1975, Pap. Puu *57,* 131–145
Mutton, D. B. 1964, Pulp Pap. Mag. Can. *65,* No. 2, T41–T51
Norstedt, I. and Samuelson, O. 1965, Svensk Papperstid. *68,* 565–571
Omori, S. and Dence, C. W. 1981, Wood Sci. Technol. *15,* 113–123
Procter, A. R. and Wiedenkamp, R. H. 1969, J. Polym. Sci., Part C, *28,* 1–13
Roberts, Jr. J. L., Morrison, M. M. and Sawyer, D. T. 1978, J. Am. Chem. Soc. *100,* 329–330
Ruoho, K. and Sjöström, E. 1978, Tappi *61,* No. 7, 87–88
Rydholm, S. A. 1965, Pulping Processes. Intersci. Publ., New York
Samuelson, O. 1981, Some Undesirable Carbohydrate Reactions During Alkaline Cooking and Bleaching. In: The Ekman Days 1981. Int. Symp. Wood Pulp. Chem., Stockhom, Vol. 2, pp. 78–82
Samuelson, O. and Stolpe, L. 1969, Tappi *52,* 1709–1711
Samuelson, O. and Thede, L. 1969, Tappi *52,* 99–104
Schweers, W. 1966, Pap. Puu *48,* 161–173
Schweers, W. 1969, Holzforschung *23,* 5–9
Sindhwani, K. L., Singh, M. M. and Guha, S. R. D. 1979, Holzforsch. Holzverwert. *31,* 37–40

Sjöström, E. 1977, Tappi *60,* No. 9, 151–154
Sjöström, E. 1981a, Wood Chemistry. Fundamentals and Applications. Academic Press, New York, London
Sjöström, E. 1981b, Pap. Puu *63,* 438–442
Sjöström, E. and Välttilä, O. 1972, Pap. Puu *54,* 695–705
Spittler, T. D. and Dence, C. W. 1977, Svensk Papperstid. *80,* 275–284
Sudo, K., Hwang, B. H. and Sakakibara, A. 1978, Mokuzai Gakkaishi *24,* 424–425
Sudo, K., Hwang, B. H. and Sakakibara, A. 1979, Mokuzai Gakkaishi *25,* 61–66
Vinje, M. G. and Worster, H. E. 1969, Tappi *52,* 1341–1345
Yllner, S. and Enström, B. 1956, Svensk Papperstid. *59,* 229–232
Yllner, S. and Enström, B. 1957, Svensk Papperstid. *60,* 549–554

12. Influence of Temperature

12.1. General Effects

In many kinds of processing, wood is subjected to a treatment at elevated temperatures, e.g. drying, size stabilization, pulping, production of particle- and fibreboard. As temperature influences the physical, structural and chemical properties of wood, numerous publications are concerned with this subject. Summaries of earlier studies can be read in the publications of Sandermann and Augustin (1963), Košik et al. (1968a), and Beal and Eickner (1970).

The processes mentioned above require temperatures which usually do not exceed 200 °C as a thermal degradation is not intended. But it is not the temperature alone which causes changes in the wood substance; there are additional factors which influence thermal degradation processes, e.g. time of treatment, atmosphere, pressure, water content and state of distribution. Thus under certain conditions changes in wood can be observed even from 100 °C up.

A change of physical properties, e.g. the reduction of sorption, dry weight, and dry dimensions of wood, was established by Kollmann and Schneider (1963, 1964). During the heating of ground spruce wood *(Picea abies)* for 24 hours, a loss of weight began at 120 °C with 0.8%, and increased to 15.5% at 200 °C (Fengel 1966) (Fig. 12–1). The thermal treatment of beech wood *(Fagus sylvatica)* with an increasing temperature of 5° per minute resulted in a loss of weight of 8.1% at 150 °C and of 9.8% at 200 °C (Košík et al. 1969a) (Fig. 12–2). At the same time the carbon content of the wood increased, an indication of the chemical conversion of the residue.

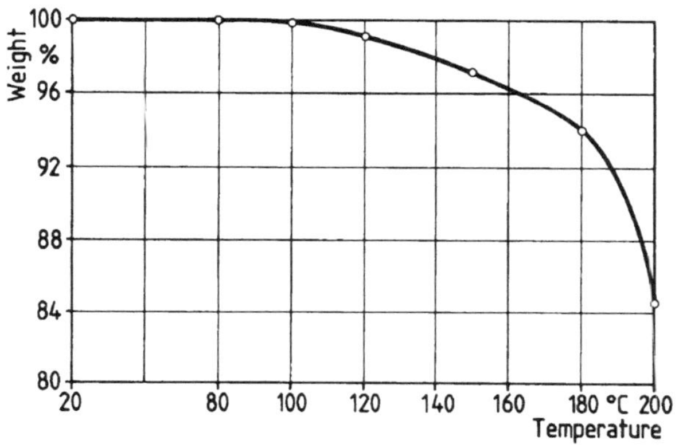

Fig. 12–1. Loss by weight of spruce wood *(Picea abies)* after heating at various temperatures for 24 hours (Fengel 1966).

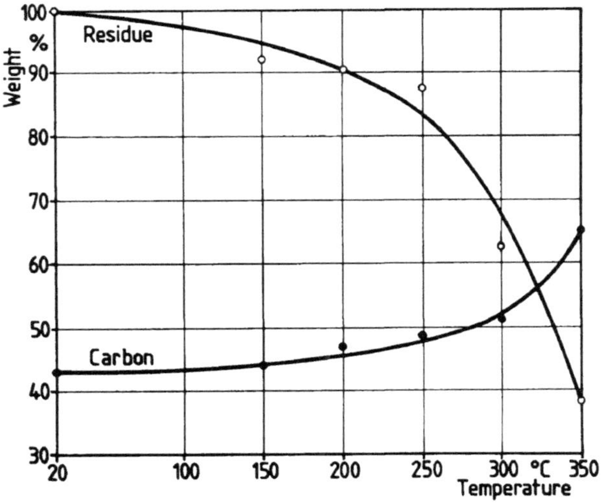

Fig. 12–2. Loss by weight and carbon content of beech wood *(Fagus sylvatica)* after heating with an increasing temperature of 5° per minute (according to Košík et al. 1969a).

The influence of water content can be realized e.g. from the thermal softening of wood. Softwoods and hardwoods soften near 180 °C with a maximum at 380 °C (Chow, Pickles 1971). In the presence of moisture another softening peak appears below 180 °C. The higher the moisture content the greater is the absolute softening and the lower the softening temperature. The changes caused during the heating of wood in water (hydrothermal treatment) are predominantly determined by hydrolytical reactions. These reactions are described in a preceding chapter (→ 10.).

Damages of the structure, conversion of the components, occurrence of gaseous degradation products are observed at temperatures above 200 °C; the pyrolysis of wood begins at temperatures of more than 270 °C. Increasing temperature causes an increase of gaseous products and temperatures of more than 500 °C are applied for the gasification of wood.

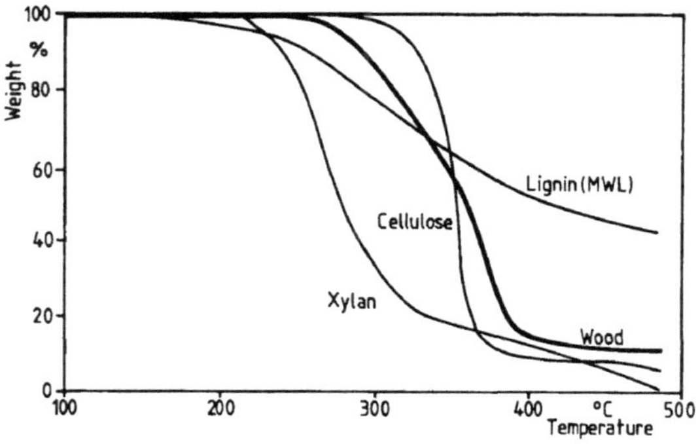

Fig. 12–3. Thermogravimetric diagram of cottonwood (*Populus* spec.) and its components (Shafizadeh, DeGroot 1976).

The components of wood may show other changes during thermal treatment if they are in an isolated condition than if they are incorporated in the cell walls. The degradation of cellulose, polyoses and lignin, as indicated by the loss of weight, is considerably different as compared to each other or to wood (Fig. 12–3). The thermally conditioned changes in the wood can be interpreted only within certain limits as an additional effect of the changes of the components. This fact was established by means of differential thermal analyses.

12.2. Thermal Reactions

The method of differential thermal analysis (DTA) allows the measurement of thermally caused reactions within a wood sample which is heated with constantly increasing temperature. This method was applied on several wood species, e.g. spruce *(Picea abies)*, balsam fir *(Abies balsamea)*, Douglas fir *(Pseudotsuga menziesii)*, beech *(Fagus sylvatica)*, red alder *(Alnus rubra)*, cottonwood *(Populus trichocarpa)* (Arseneau 1961; Sandermann, Augustin 1963; Košík et al. 1968b; Chow, Pickles 1971; Shafizadeh, McGinnis 1971). Thermograms beginning only at 220 °C were also made with black locust *(Robinia pseudoacacia)*, red ironwood *(Lophira alata)* and teak *(Tectona grandis)* by Keylwerth and Christoph (1960).

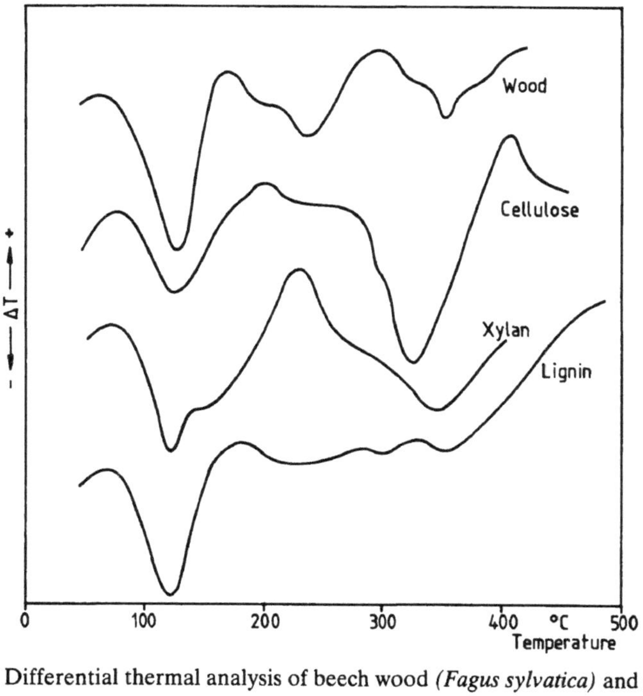

Fig. 12–4. Differential thermal analysis of beech wood *(Fagus sylvatica)* and its components (according to Košík et al. 1968b).

Generally regarded the thermograms show at first an endothermal maximum in the range of 120–150 °C which is attributed to the evaporation of water being more strongly absorbed (Fig. 12–4). The exothermal peaks at 200–250 °C and 280–320 °C and those higher than 400 °C are due to a degradation of the wood components. In Fig. 12–4 the thermograms of cellulose, xylan and lignin are shown, which are different from each other and from that of wood. The endothermal and exothermal peaks visible in the thermogram of wood can only be deduced within certain limits to the peaks of the isolated components. A mixture of isolated components in the ratio as present in wood results at best in a thermogram similar to that of the corresponding wood (Fig. 12–5) (Domanský, Rendoš 1962; Sandermann, Augustin 1963).

With the aid of a similar method (differential calorimetric analysis, DCA) Beall (1971) found individual thermograms for several polyoses in the temperature range of 20–800 °C; the thermograms of cellulose, polyoses and lignin are additionally influenced by the kind of isolation of the respective component. During thermal analyses with oxygen excluded the exothermal peaks are shifted to higher temperatures. The oxidation of the components seems to be an important factor for the position of the thermal peaks. Sandermann and Augustin (1963) showed that oxidized wood components have their endo- and exothermal peaks at lower temperatures than the corresponding non-oxidized compounds. Furthermore the conditions of heating and the size of a sample may influence the thermal reactions. The appearance of a microexothermal point in wood at temperatures lower than 150 °C under certain conditions was assumed by Kollmann (1960).

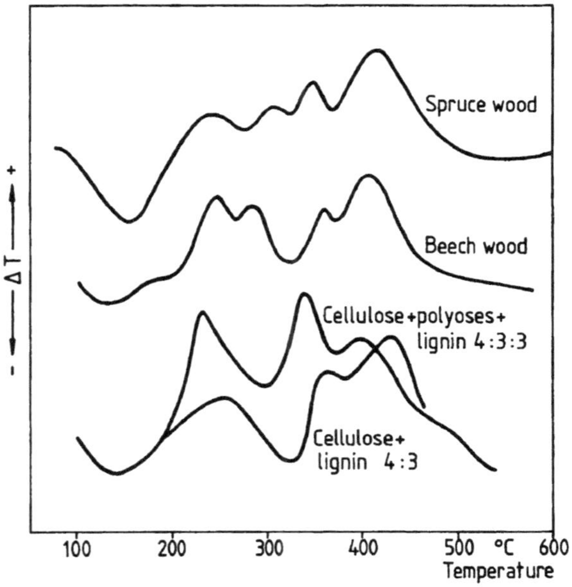

Fig. 12–5. Differential thermal analysis of spruce wood *(Picea abies)*, beech wood *(Fagus sylvatica)*, and mixtures of cellulose, polyoses and lignin (according to Sandermann and Augustin 1963).

Table 12–1: Weight loss and decomposition temperature of wood components during pyrolysis in nitrogen (P) and combustion in oxygen (C) at 6 °C/min (Beall 1969)

Compound	Reaction	Maximum rate of weight loss mg/min	Temperature at maximum rate of weight loss °C	Decomposition temperature °C 10%	50%
Brownell lignin	P	0.6	303	241	366
	C	1.5	353	254	356
Björkman lignin	P	0.8	363	228	358
	C	1.2	429	273	407
Sulfuric acid lignin	P	0.8	381	331	428
	C	1.9	432	293	392
Cellulase lignin	P	0.7	297	266	367
	C	1.4	414	261	372
O-acetyl-4-O-methylglucurono-	P	4.5	254	216	265
xylan *(Betula papyrifera)*	C	5.3	238	233	264
4-O-methylglucuronoxylan	P	3.4	283	248	292
(Betula alleghaniensis)	C	9.2	236	242	254
Galactoglucomannan, alkali-soluble	P	1.9	291	223	277
(Pinus strobus)	C	3.0	238	218	258
Galactoglucomannan, water-	P	1.6	279	238	293
soluble *(Tsuga canadensis)*	C	2.5	239	231	271
Arabino-4-O-methylglucurono-	P	1.4	249	218	265
xylan *(Tsuga canadensis)*	C	2.2	213	211	248
Compression wood galactan	P	2.3	302	249	303
(Abies balsamea)	C	–	200	–	–
Arabinogalactan	P	3.6	305	260	305
(Larix laricina)	C	5.1	277	255	284

Table 12–2: Weight loss and activation energy during thermal treatment of beech wood *(Fagus sylvatica)* and its components (Košík et al. 1968b)

Sample	Zero order reaction			First order reaction		
	Temperature °C	Weight loss %	Activation energy kJ	Temperature °C	Weight loss %	Activation energy kJ
Wood	170–220	5.5	63	248–310	47.3	130
Holocellulose	120–200	5.4	54	–	–	–
Cellulose	220–300	5.0	78	300–380	55.1	243
Lignin	200–320	3.9	34	–	–	–
4-O-methyl-glucuronoxylan	100–160	9.5	46	180–290	43.2	100

Experiments with beech wood gave evidence that the thermal decomposition follows an overall first order reaction (Roberts, Clough 1963). It was concluded that in the lower temperature range an initial reaction of activation energy of 105 kJ/mole predominates, whereas in the higher temperature range this reaction is rapidly completed and a further reaction with an activation energy of 63 kJ/mole becomes important. Similar studies with lignins and polyoses resulted in specific decomposition temperatures for each compound (Beall 1969) (Table 12–1). For most compounds the degradations are zero-order reactions in the lower temperature range, and first order reactions in the higher temperature range. The activation energy is, on the average, lower for the lignin samples (20–100 kJ/mole) than for the polyose samples (50–200 kJ/mole). A similar study of beech wood *(Fagus sylvatica)* was made by Košík et al. (1968b) (Table 12–2).

12.3. Structural Changes

The shrinkage of wood during drying has its origin in a shrinkage of the cell walls. The dimensions of the cell walls are reduced on a remarkable scale. Boutelje (1962) determined the volumetric shrinkage of the cell walls in earlywood of spruce as 26.5% and in latewood as 29.5%. This shrinkage results in a decrease of the pore volume in earlywood and an increase in latewood. From elevated temperatures ensue additional shrinking forces as the volume shrinkage increases because of substance losses by thermal decomposition. These losses are different for the various wall layers. Different shrinkages produce fissures and compressions within the cell walls. Fissures can be observed mainly in the weakest region, i.e. the border

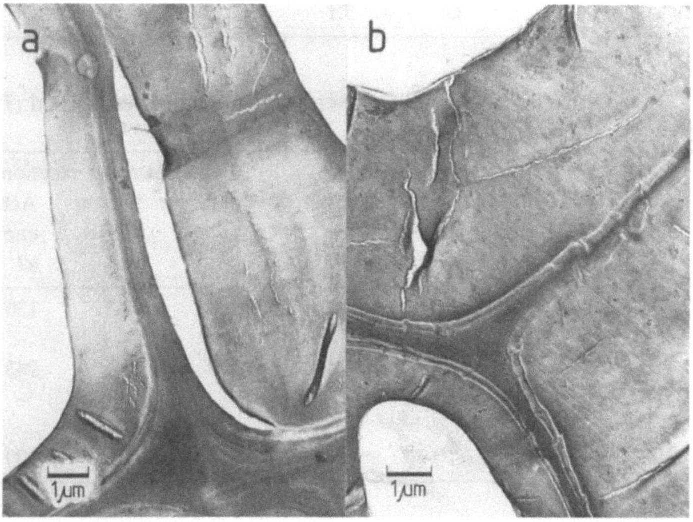

Fig. 12–6. Thermally caused damages in the tracheidal cell walls of spruce *(Picea abies)*. Ultrathin sections, TEM micrographs.

a) Fissures between S1 and S2 after heating at 150 °C (24 hrs.)

b) Compression zones in the compound middle lamella and S1 of latewood tracheids after heating at 200 °C (24 hrs.)

between S1 and S2 originating obviously in a cell corner (Fig. 12–6a). In the latewood tracheids of spruce *(Picea abies)* heated at 180 and 200 °C compression effects also occurred in the compound middle lamella and the adjacent secondary wall 1 (Fengel 1966) (Fig. 12–6b).

Koran (1968) studied the tangential tensile failure at the surface of black spruce *(Picea mariana)* through the wide range of −190 to +250 °C. The fibre faces exposed at +100 °C and below mainly reveal the S1 surface structure. In contrast the fibre faces produced at 150 °C and above expose predominantly the primary wall surface structure, heavily embedded in, or covered by, an amorphous matrix of lignin and polyoses. This softening of the amorphous cell wall components continues with elevation of temperature.

A hydrothermal treatment applied to beech wood *(Fagus sylvatica)* and birch wood *(Betula pubescens)* loosened the cell wall structure mainly at the interface of the S1 and S2 layers even at temperatures of 120 and 160 °C (Filló, Peres 1970; Gromov et al. 1972).

A softening and disappearance of warts at the lumen side of vessels in heated beech wood *(Fagus grandifolia, F. sylvatica)* was observed by Kollmann and Sachs (1967). During a thermal treatment of spruce wood the amorphous substances deposited in the tori of bordered pits soften and flow along the margo and the pit border (Fengel 1966) (Fig. 12–7a). A chemical change of the torus substance is indicated by a change in its solubility. At temperatures of 180 and 200 °C the torus substance becomes soluble in ethanol-benzene, and after extraction with this solvent the cellulose membranes of the tori remain as empty wraps (Fig. 12–7b).

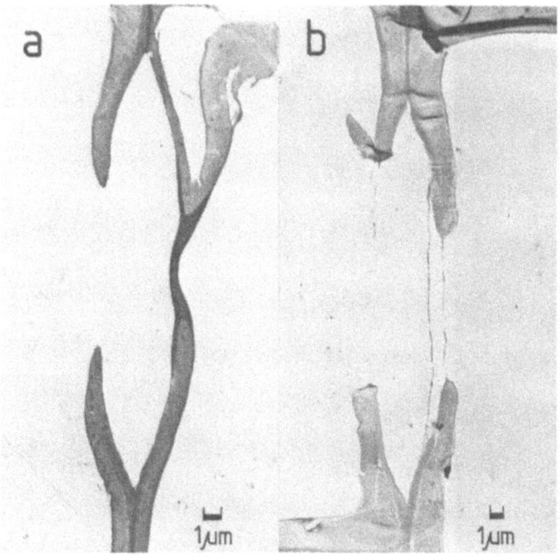

Fig. 12–7. Changes of the bordered pits in earlywood of spruce *(Picea abies)* after heating at 200 °C for 24 hours. Ultrathin sections, TEM micrographs.
a) The dark torus substance has flowed along the pit border.
b) After extraction with ethanol-benzene for 4 hours the amorphous torus substance has been dissolved.

Knudson and Williamson (1971) report on the formation of crystals covering the surface of the tori in Douglas fir after a treatment at 550 °C. At the same temperature the spiral thickenings of the tracheid walls are destroyed. Furthermore the rips of the ladder-like perforations in beech wood vessels show smoothening and flowing effects after heating at 250 °C (Pajtík, Chovanec 1974).

After cooling a heated wood sample the plasticity of the amorphous components, particularly in the compound middle lamella, is maintained until a temperature of about 60 °C is reached. Below this temperature the components consolidate again (Nečesaný 1965).

Cellulose and polyoses isolated from thermally treated spruce wood showed structures which were different from those of the polysaccharides isolated from unheated wood (Fengel 1967, 1969). The original fibrillar structures had been destroyed after isolation from wood which was heated at 200 °C. The degraded cellulose was able to form new fibrillar systems (Fig. 12–8).

12.4. Changes of the Chemical Components

12.4.1. Wood Analysis

The chemical analysis of woods after thermal treatment at various temperatures shows relatively good stability of the components up to 100 °C and up to a treating time of 48 hours (Fig. 12–9). At higher temperatures the content of polysaccharides

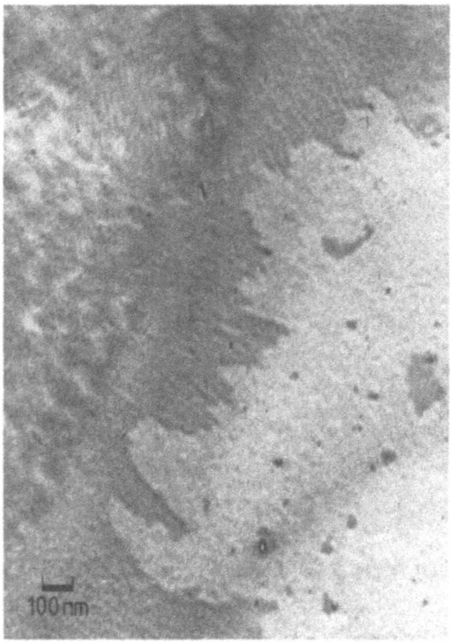

Fig. 12–8. Cellulose isolated from spruce wood heated at 200 °C for 24 hours. Suspension from buffer solution, TEM micrograph.

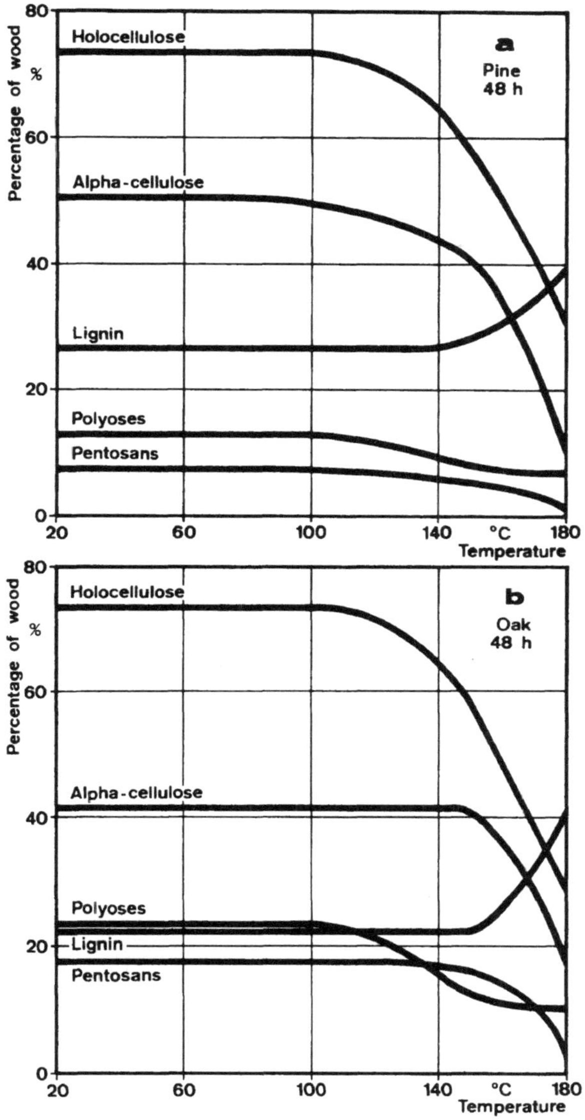

Fig. 12–9. Variation of the composition of pine wood *(Pinus sylvestris)* (a) and oak wood *(Quercus robur)* (b) after heating at various temperatures for 48 hours, as evaluated by wood analysis (according to Kollmann and Fengel 1965).

(holocellulose) decreases more and more, the polyoses reacting obviously more sensitively than cellulose.

Alpha-cellulose of softwoods contains a relatively high amount of polyoses and residual lignin; the decrease of alpha-cellulose which begins at 100 °C can be reduced to the loss of these accompanying compounds, so that the content of pure cellulose remains constant up to 150 °C even in softwoods (Fengel 1967). The lignin content, too, remains constant in a wide temperature range; above 140–150 °C the lignin content increases.

12.4.2. Polyoses

The results of wood analysis give only a rough impression of the thermally caused changes of the wood components, changes which were identified more exactly by detailed studies. The content of the total polyoses in softwood is characterized by only a small decrease up to 180 °C or by an increase particularly above 180 °C as determined with spruce wood (Fengel 1966). The determination of the various fractions of polyoses (polyoses A, polyoses B, pentosans) shows that the polyoses content results from decomposition especially of pentosans and an addition of alkali-soluble cellulose produced by thermally caused chain-splitting (Fig. 12–10). During treatments of beech wood *(Fagus sylvatica)* with continuously increasing temperature (7°/min, 5°/min) the loss of pentosans was 8% at 170 °C and 100% at 350 °C (Sandermann, Augustin 1964; Košík et al. 1969a). The acetyl content was reduced from 4.6% at the beginning (20 °C) to 0% at 350 °C.

The O-acetyl groups have an important influence on the thermal stability of the polyoses. A comparison of the thermal degradation of O-acetyl-galactoglucomannan and deacetylated galactoglucomannan from larch *(Larix leptolepis)* showed that the latter is more stable (Shimizu et al. 1972). The first-order reaction range at 182–210 °C with an activation energy of 160 kJ/mole for the O-acetyl compound was attributed to the hydrolysis of the acetyl group. Ramiah and Goring (1967) determined the beginning of degradation of birch xylan *(Betula papyrifera)* to be at 117 °C with a first-order activation energy of 193 kJ/mole, and of pine glucomannan *(Pinus strobus)* at 127 °C with 207 kJ/mole.

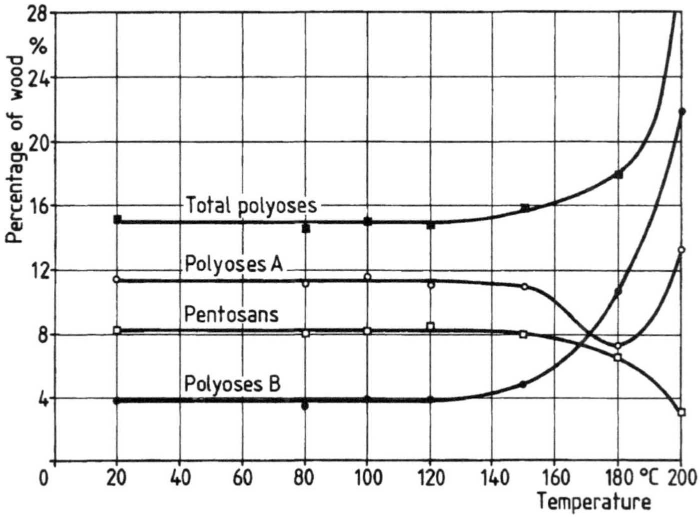

Fig. 12–10. Variation of the percentages of the polyose fractions isolated from spruce wood *(Picea abies)* after heating at various temperatures for 24 hours (based on absolutely dry and extractive-free wood) (Fengel 1966).

Košík et al. (1969b) studied the oxygen consumption of beech wood at elevated temperatures (150–230 °C). 4-O-methylglucuronoxylan consumed more than double the amount of oxygen than did cellulose. The total polyoses, which contained a large portion of the xylan, were oxidized more slowly, and consumed less oxygen than did the pure xylan.

According to the studies of Domburg et al. (1966) the decomposition of hardwood xylan begins near 200 °C in normal atmosphere. The glycosidic linkages rupture as well as some C–C bonds of the pyranose rings. Treatments at 225 °C entail a complete obliteration of the molecular structure. In the temperature range of 275–290 °C the molecular fragments are dehydrated, forming furfural. Under vacuum a rapid decomposition occurs from about 220 °C with the formation of anhydrate compounds.

The thermal treatment in nitrogen atmosphere of 4-O-methylglucuronoxylan isolated from hornbeam *(Carpinus betulus)* resulted in depolymerization and dehydration at temperatures of more than 150 °C (Ebringerová 1976). Keto groups, γ-

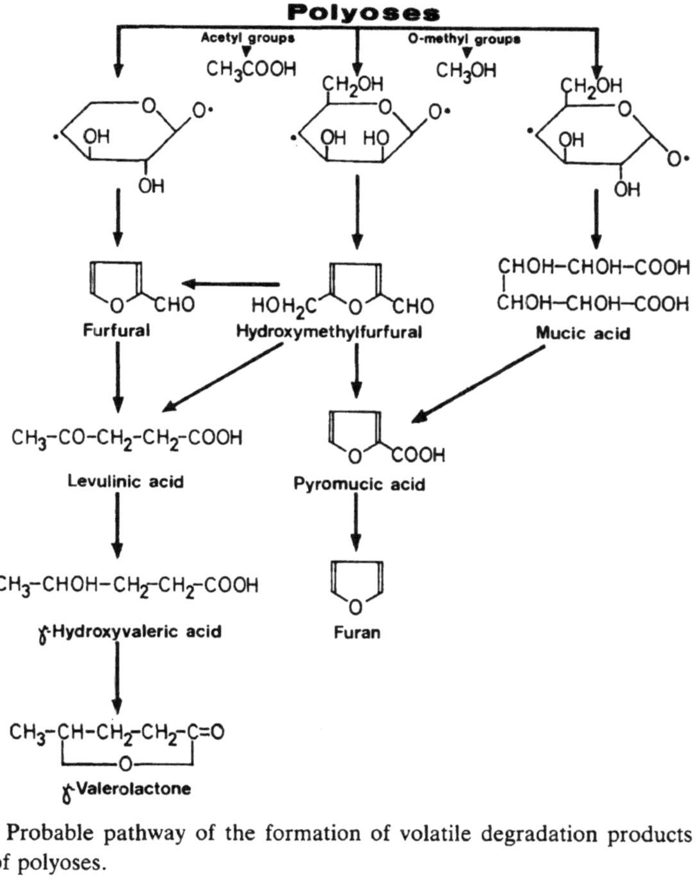

Fig. 12–11. Probable pathway of the formation of volatile degradation products during the thermal treatment of polyoses.

lactones and furfural are formed by intramolecular dehydration, and ester linkages by intermolecular dehydration.

A scheme of probable pathways for the thermal degradation of polyoses was proposed by Sandermann and Augustin (1963). The variation of this scheme, as shown in Fig. 12–11, takes into account that no monosaccharides are produced during the thermal cleavage of the polysaccharidic chains. In cold-water extracts of spruce wood which was pretreated at temperatures between 80 and 200 °C for 24 hours, no monosaccharides could be detected (Fengel 1966). Thus it has to be concluded that the intermediates for the formation of low molecular degradation products are radicals resulting from a non-hydrolytical cleavage of glycosidic bonds. According to the proposed pathway the final products of thermal degradation are methanol, acetic acid and volatile heterocyclic compounds (furan, γ-valerolactone).

Košík et al. (1968c), after a thermal treatment of 4-O-methylglucuronoxylan from beech wood at 180 °C, also detected acetaldehyde, methylacetate, propylaldehyde and methylketone in addition to acetic acid, methanol and furfural. At temperatures of 200–300 °C carbon dioxide is produced in addition to the compounds mentioned above, showing that the degradation proceeds rapidly with increasing temperature (Shimizu et al. 1971).

The yield of furfural and acetic acid can be influenced by impregnating the wood samples with acidic compounds. Reiser et al. (1968) obtained the highest yield in furfural by impregnating beech wood with H_2SO_4, $ZnCl_2$ or $SnCl_2$. The highest yield of acetic acid was achieved with $AlCl_3$. Also at higher temperatures (500 °C) the yield of furan and furan derivatives can be improved in the presence of $ZnCl_2$ as shown with cellulose, xylan and O-acetylxylan (Shafizadeh, DeGroot 1976).

12.4.3. Cellulose

The increase of the polyoses content after a thermal treatment of wood is caused by short-chain cellulose, as mentioned above. This result means that the first step in the thermal degradation of cellulose is a cleavage of the macromolecules producing alkali-soluble products. But the degree of polymerization of the residual cellulose is also reduced. In cellulose isolated from thermally pretreated spruce wood the DP remained constant up to 120 °C, followed by a rapid decrease with increasing temperature (Fengel 1967). The thermal treatment of isolated cellulose (bleached sulfite pulp) resulted in a decrease of DP even at 100 °C, though the heating time was only 20 minutes (Roffael, Schaller 1971). The degree of degradation was influenced by the water content of the sample (Fig. 12–12). After a treatment at 200 °C a cellulose sample with a high initial moisture content (60%) had a DP about 200 units higher than a sample with a low initial moisture content (7%). It is assumed that the de-swelling of the highly swollen sample with 60% water results in recrystallization and an inhibition of the chain splitting during a thermal treatment. Cotton celluloses with an initial DP of 1 140 had a final DP of 200 after heating at

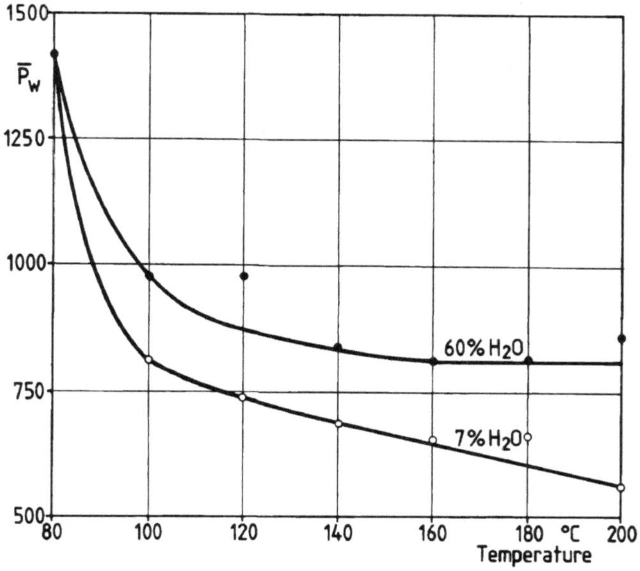

Fig. 12–12. Variation of the degree of polymerization ($\bar{P}_w$) of cellulose (bleached sulfite pulp) containing different amounts of water, after heating at various temperatures for 20 minutes (Roffael, Schaller 1971).

200 °C, which is in the range of the leveling-off DP (→ 4.3.3.) (Dmitrijeva et al. 1964).

The crystalline structure of cellulose is not changed or even improved up to a certain temperature, which may be as high as 200 °C depending on the conditions involved. The crystallinity of the alkali-resistant cellulose from thermally treated spruce wood increased up to a temperature of 200 °C because of a preferred degradation of the less ordered molecules (Fengel 1967). From the behaviour of cotton heated at various temperatures up to 160 °C, during a subsequent hydrolysis with 10% H_2SO_4, Levanova et al. (1976) concluded that the amorphous regions in cellulose grow with increasing temperature and time of heating. The annealing of stretched rayon fibres at 200 °C initiates a growing of the cellulose crystallites and an increase of the degree of crystallinity (Basch, Lewin 1975).

Roffael and Schaller (1971) observed an increase of crystallinity in thermally treated cellulose up to temperatures of 120–160 °C followed by a decrease (Fig. 12–13). The temperature at which the maximum value was reached depended on the water content of the cellulose sample.

During the heating of pine wood *(Pinus densiflora)* Taniguchi and Nakato (1966) found no change of X-ray diffraction up to 210 °C. Beyond this temperature the supramolecular structure is destroyed and a completely amorphous state was reached at about 270 °C. A similar result was obtained with sulfite pulp from slash pine *(Pinus elliottii)* by Millett and Goedken (1965). At 240 °C the crystalline struc-

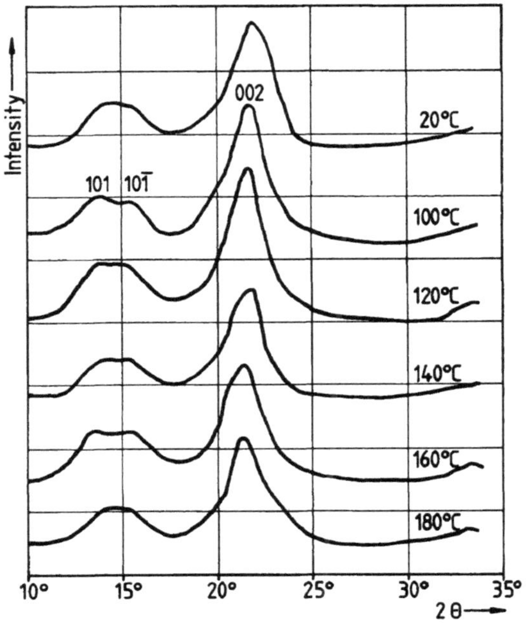

Fig. 12–13. X-ray diffraction diagrams of cellulose (bleached sulfite pulp, containing 7% water) after heating at various temperatures for 20 minutes (Roffeal, Schaller 1971).

ture of cellulose obviously breaks down as the DP decreases within two hours below 200 and within 8 hours below 100.

The thermal degradation of cellulose is not restricted to the cleavage of molecular chains, but there are additional dehydration and oxidation reactions. Chain cleavage and dehydration follow a zero-order reaction whereas oxidation is a first-order reaction (Hernádi 1976; Bolkunevich 1979). Heating in air causes oxidation of the hydroxyl groups resulting in an increase of carbonyl and subsequently of carboxyl groups. The ratio of both reactions depends mainly on the temperature (Hernádi 1977). Hernádi (1977) additionally found a correlation between the tendency of yellowing and the content of aldehydic groups. In a preceding study with rayon fibres Sihtola and Neimo (1965) determined the heat of formation of carbonyl groups as 17 kJ/mole and the heat of formation of the yellowing compounds as 92 kJ/mole.

By raising the temperature above 200 °C the thermal degradation of cellulose and the formation of volatile products proceeds rapidly. Levoglucosan is frequently said to be the most important primary degradation product, but other anhydroglucoses (1,2-, 1,4-anhydroglucose, 1,6-anhydroglucofuranose, enones, Fig. 12–14), furan and furan derivatives are also produced.

Shafizadeh and DeGroot (1976) classified the reactions taking place during the thermal degradation of cellulose and other polysaccharides into the following categories:

- Depolymerization of the polysaccharides by transglycosylation at about 300 °C to provide a mixture of levoglucosan, other monosaccharide derivatives, and a variety of randomly linked oligosaccharides. This mixture is generally referred to as the tar fraction.
- The above reactions are accompanied by dehydration of sugar units in cellulose. These give unsaturated compounds, including 3-deoxyglucosenone, levoglucosenone, furfural, and a variety of furan derivatives which are found partly in the tar fraction and partly among the volatiles.
- At somewhat higher temperatures, fission of sugar units provides a variety of carbonyl compounds, such as acetaldehyde, glyoxal and acrolein, which readily evaporate.
- Condensation of the unsaturate products and cleavage of the side chains by means of free radical mechanisms leave a highly reactive carbonaceous residue containing trapped free radicals.

Byrne et al. (1966) proposed various pathways for the formation of anhydroglucoses, furan and furan derivatives as well as low molecular compounds (glycolaldehyde, glyoxal, acrolein etc.) deriving from the rupture of the pyranose rings. A more simplified scheme of the formation of anhydroglucoses, furan and low molecular compounds is presented in Fig. 12–15.

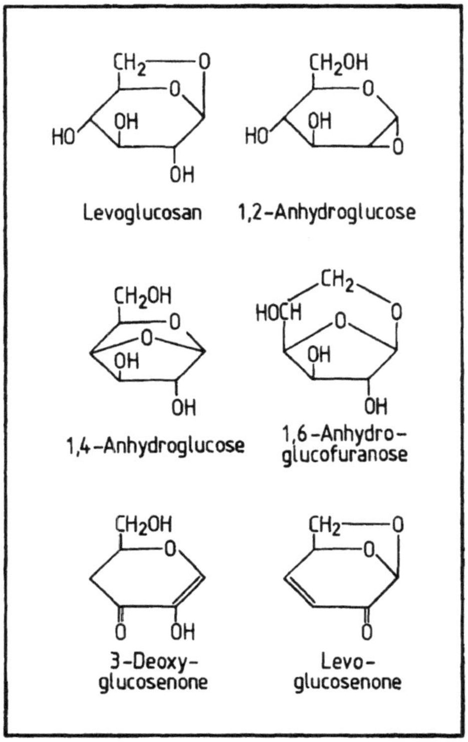

Fig. 12–14. Anhydroglucoses and enones as thermal degradation products of cellulose.

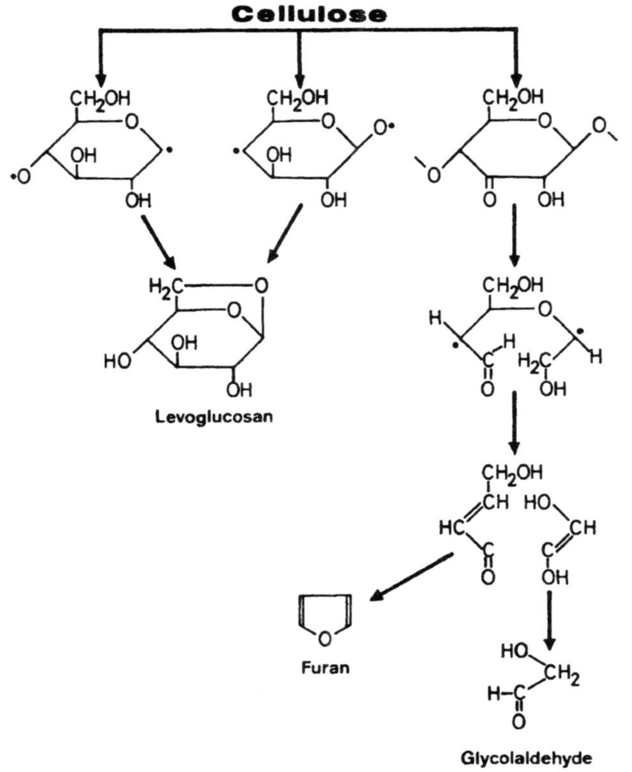

Fig. 12–15. Probable pathway of the formation of volatile degradation products during the thermal treatment of cellulose.

Vacuum pyrolysis of cotton cellulose and wood lignocellulose yielded tars containing 38–39% levoglucosan (Shafizadeh et al. 1979; Alsup et al. 1980). Additionally the residue can be easily hydrolyzed, an effect promising an improved glucose yield in utilizing waste cellulose (Millett, Goedken 1965).

The possibility of recombinating monomeric degradation products of cellulose can be derived from a study by Sugisawa and Edo (1964). The heating of dry glucose resulted in several di- and oligosaccharides apart from dehydration products.

A study comparing the thermal properties of the various crystalline forms of cellulose was made by Cabradilla and Zeronian (1978). The authors produced cellulose II, III and IV from cotton in fibrous as well as in microcrystalline form. It appeared that the thermal stability of cellulose in the cellulose III form is less than that of cellulose I, II or IV. The levoglucosan producing reactions occur primarily in the crystalline areas. Heating of the microcrystalline samples showed that the levoglucosan yield is less for cellulose II than for the other three crystalline forms. Comparing the levoglucosan yield of fibrous and microcrystalline samples it is obvious that the amorphous part in the fibrous state inhibits the levoglucosan producing reactions. Heat treatments which increased the crystallinity of fibrous cellulose increased the levoglucosan yield.

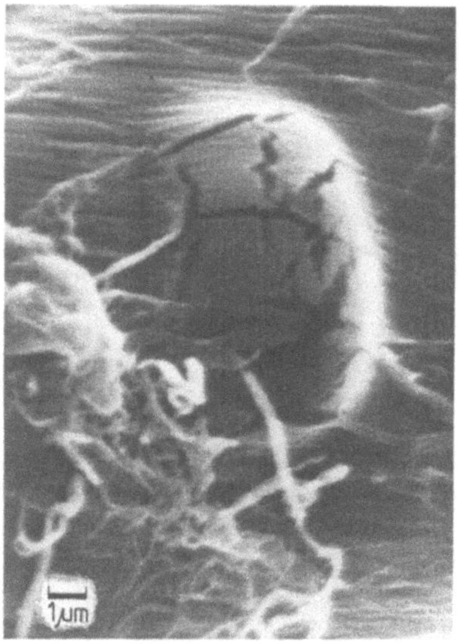

Fig. 12–16. Formation of a bubble and disappearance of the fibrillar structure on the surface of a cotton fibre after exposure to a high-energy electron beam. REM micrograph.

Because of the high reactivity at elevated temperatures it is probably impossible to determine a melting point for crystalline cellulose. Nordin et al. (1973) estimated the melting point (T_m) of cellulose from its glass transition temperature (T_g = 230 °C) to be roughly 450 °C using an empirical correlation $T_g = 0.7 \cdot T_m$ [K]. They succeeded in producing melting effects on the surface of cellulose fibres by applying a rapid heating of the sample with a continuous carbon dioxide laser which permits a heating time as short as 0.1 ms to 500 °C. This caused the formation of heating bubbles and the disappearance of the fibrillar structure. Similar effects can be produced in the scanning electron microscope if a small area of a cellulosic fibre wall is exposed to a high-energy electron beam (Fig. 12–16).

12.4.4. Lignin

Though lignin is seen to be the thermally most stable component of wood various changes were observed even at temperatures below 200 °C. The determination of the lignin content in thermally treated woods *(Picea abies, Pinus sylvestris, Quercus robur)* gave evidence of an increase in non-hydrolyzable residue with increasing temperature up to 200 °C (Kollmann, Fengel 1965; Fengel, Przyklenk 1970) (Fig. 12–17). The yield of ethanolysis products of spruce wood increased also, whereas the methoxyl content decreased after heating the wood at temperatures of 180 °C and 200 °C for 24 hours.

Using beech wood *(Fagus sylvatica)* Kürschner and Melcerová (1965) observed a decrease in lignin content at temperatures above 100 °C, and with prolongation of the heating time up to 28 days. At 160 °C and 28 days the acid-insoluble lignin

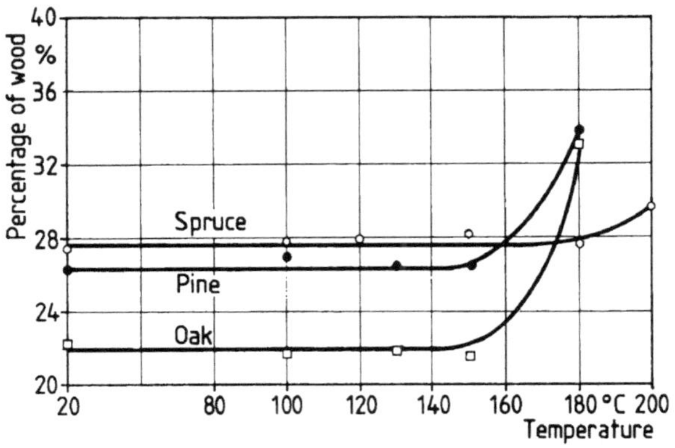

Fig. 12–17. Variation of the percentages of acid-insoluble lignin isolated from wood of spruce, pine and oak after heating at various temperatures for 24 hours (based on absolutely dry and extractive-free wood) (according to Kollman and Fengel 1965, and Fengel and Przyklenk 1970).

content was less than 1%. The thermogravimetric analysis of isolated beech wood lignin showed a slight decrease beginning at 100 °C, and a loss by weight of about 15% at 400 °C (Košík et al. 1968b).

Regarding the degree and shape of fibre swelling as criteria of thermally caused changes, Sergeeva and Miljutina (1960) found no change in lignin up to 155 °C. Heating at 175 °C caused a lignin condensation which increased with the heating temperature up to 240 °C. At 260–280 °C the lignin condensation was accompanied by other changes in the lignin molecules which led to a reduction of hydrophilic capacity.

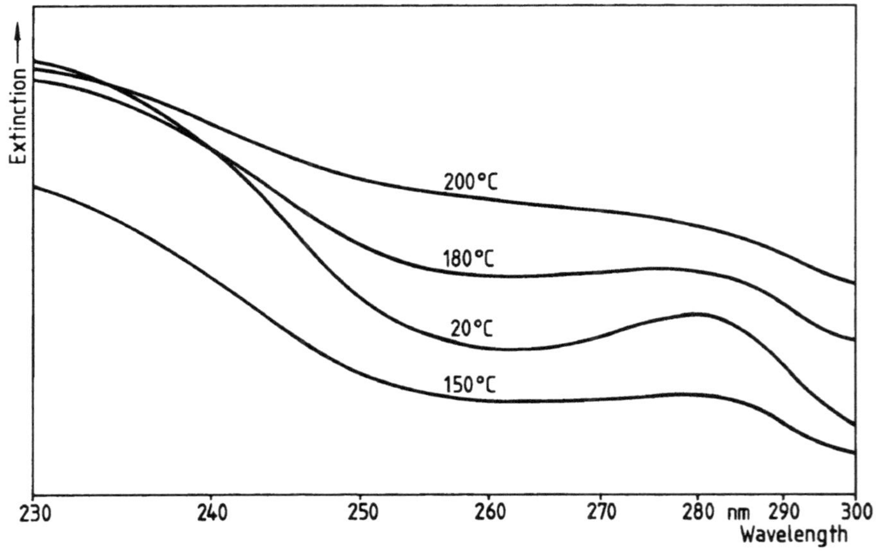

Fig. 12–18. Ultraviolet spectra of milled wood lignins isolated from spruce wood after heating at various temperatures for 24 hours (Fengel, Przyklenk 1970).

Milled wood lignin isolated from spruce wood which was heated at various temperatures showed a change in the ultraviolet spectrum beginning at 150 °C (Fengel, Przyklenk 1970). The absorption maximum at 280 nm sloped down and disappeared after treatment at 200 °C (Fig. 12–18).

An infrared spectroscopic study of various lignin samples isolated from spruce *(Picea jezoensis)* and cypress *(Chamaecyparis obtusa)*, and heated in the range of 20–250 °C gave evidence of a cleavage of hydrogen bonds at 60–80 °C, as evaluated from changes in the C–O streching and C–O deformation bands (Hatakeyama et al. 1969). A change in the bands representing the aromatic skeletal vibration was found between 100 and 180 °C. These changes observed at relatively low temperatures may have their explanation in the very fine distribution state of the lignin samples as they had been heated in the KBr pellets as used for infrared spectroscopy. Dioxan lignin from spruce heated at 140 and 180° also showed some changes in the infrared spectrum which, however, could not be interpreted as a result of distinct decomposition (Potutkin 1969).

The softening temperature of lignin depends on the kind of isolation which determines the changes of the chemical structure. Further criteria influencing the softening temperature are the molecular size and the water content of the sample (Goring 1963). Periodate lignin had a softening temperature of 195 °C in the dry state, and 90 °C with 27.1% water. The softening temperatures of various lignin preparations are compiled in Table 12–3. No softening was observed in hydrolytically isolated lignins and lignins from spent liquors of sulfite and sulfate processes (Fengel et al. 1981). The lowest softening temperature (165–170 °C) was determined for milled wood lignin of beech.

Ramiah and Goring (1967) found an incipient thermal decomposition of dioxane lignin at 130 °C, and of periodate lignin at 145 °C by dilatometric measurements. According to a study by Domburg et al. (1974) with HCl-lignin from aspen, alkyl-

Table 12–3: Softening temperatures of various lignin preparations

Lignin	Softening temperature °C	References
MWL spruce	180–185	Fengel et al. 1981
MWL beech	165–180	Fengel et al. 1981
MWL bamboo	162	Tanahashi et al. 1982
Ethanol lignin spruce	183–187	Fengel et al. 1981
Periodate lignin spruce	193	Goring 1963
Periodate lignin birch	179	Goring 1963
Dioxane lignin spruce	146	Goring 1963
Dioxane lignin aspen	134	Goring 1963
Ethanol/H_2O lignin spruce	185–195	Fengel et al. 1981
DHP (high mol. weight)	175	Tanahashi et al. 1982
DHP (low mol. weight)	134	Tanahashi et al. 1982

aryl ether linkages are broken up to 270 °C, and in the range of 270–300 °C an attack on C–C linkages begins.

The thermal stability of probable linkages between lignin and polysaccharides was investigated by Košíková et al. (1978). They heated model compounds (3-O-benzyl ethers of D-glucose and 4-O-benzyl ethers of methyl α-D-glucoside). The splitting of the ether linkages occurred at temperatures above 200 °C, and was influenced by the methoxyl groups at the aromatic body, and by the hydroxyl group in para-position to the benzyl ether bond. Both reduced the thermal stability of the aryl ethers.

12.5. Carbonization and Gasification

As known from the thermal analysis the main reaction of wood degradation starts at 270–280 °C, i.e. with the beginning of exothermal reactions. Up to 380 °C large amounts of distillates are produced, mainly acetic acid and methanol, and later increasing amounts of tar and gaseous products. The remaining residue is charcoal.

Experiments with various wood species heated at 400 °C yielded about 33–41% charcoal, 3–7% acetic acid, 1.5–2.5% methanol, 11–19% tar, and 15–17% non-condensable gases (Brocksiepe 1976) (Table 12–4). A carbonization of some Philippine woods at 482 °C produced a higher percentage of non-condensable gases (22.4–28.8%) (Laxamana 1971).

The separation of the volatile components occurs by distillation resulting in crude products. Thus the wood alcohol fraction consists of water with 45% methanol, 7% acetone, 5% methyl acetate, 3% acetaldehyde, and lower amounts of allyl alcohol, methyl formiate, furan and furfural. The wood vinegar fraction contains propionic, butyric and other acids in addition to acetic acid. Cresol, guaiacol, other phenols, and phenol ethers are the main components of the tar fraction (Brocksiepe 1976).

Table 12–4: Carbonization products of various woods at 400 °C (Brocksiepe 1976)

Wood	Charcoal %	Acetic acid %	Methanol %	Tar %	Gases %
Pine	32.8	3.9	1.5	18.9	15.4
Spruce	34.2	3.6	1.7	15.6	15.2
Alder	35.5	6.5	1.9	16.2	16.8
Beech	32.5	7.7	2.1	14.0	16.0
Eucalyptus	36.5	4.1	2.1	12.3	16.3
Maple	35.0	6.6	1.8	15.5	15.5
Oak	35.7	5.6	1.6	13.6	14.9
Red ironwood	41.4	3.1	2.4	11.0	17.2

Hileman et al. (1976) studied the thermal degradation of Douglas fir *(Pseudotsuga menziesii)* under various conditions, and analysed the degradation products. At a fast heating rate the distribution of the products was essentially identical in inert or in oxidative atmosphere, except for the percentage of carbon dioxide (Table 12–5). Under this condition the oxygen does not have sufficient time to enter into the degradation process.

The phenolic products which are formed during the thermal degradation of wood result almost completely from lignin and other aromatic compounds. Zavarin and Snajberk (1963) found only traces of phenols in the pyrolysate of cellulose powder, whereas the pyrolysates of natural phlobaphenes, tannins and extractive-free wood of redwood *(Sequoia sempervirens)* contained catechin, guaiacol, 1,2-dimethoxybenzene and phenol, with their para-methyl, -ethyl and -n-propyl homologues, as well as o- and m-cresol, 2,4-xylenol and 2,6-dimethoxyphenol. A similar result was obtained by Kratzl et al. (1965) who compared gas chromatograms of the pyrolysates of cellulose, milled wood lignin, and wood of spruce. Cellulose yielded only

Table 12–5: Inert and oxidative degradation products of Douglas fir at 400 °C using a fast heating rate (Hileman et al. 1976)

Product	Yield (%)	
	Inert	Oxidative
Carbon dioxide	5.7	11.8
Ethene	Trace	–
Ethane	Trace	–
Water	17.9	14.2
Propene	0.1	Trace
Methanal	0.1	0.1
Methanol	0.3	0.3
Ethanal	0.3	0.3
Propenal	0.5	0.3
Furan	0.1	Trace
2-Oxopropanal	0.7	0.6
Hydroxyethanal	0.7	0.9
Ethanoic acid	0.6	0.8
2,3-Butanedione	0.2	–
2-Hydroxypropanone	0.8	0.4
Furfural	0.5	0.5
Methylfurfural	0.1	0.1
Furfuryl alcohol	0.1	0.1
O-Methoxyphenol	0.3	0.3
2-Methoxy-4-methylphenol	0.8	0.9
2-Methoxy-4-methylanisole	0.3	0.2
4-Oxopentanoic acid	0.1	0.1
4-Hydroxypentanoic acid	0.2	0.2
p-Methoxyacetophenone	0.4	0.4
2-Methoxy-4-propenylphenol	0.3	0.3
5-Hydroxymethyl-2-furaldehyde	0.3	0.3

traces of phenols, while wood and lignin yielded several phenolics, mainly guaiacol and its derivatives (creosol, ethyl-, vinyl-, propylguaiacol, eugenol, vanillin) (Fig. 12–19).

Zavarin and Snajberk (1965, Zavarin et al. 1965) pyrolysed isolated phlobaphenes from various softwoods, and identified a large number of aromatic compounds (phenol, guaiacol, pyrogallol, anisole and corresponding derivatives).

An increase of temperature reduces the amount of monomeric organic degradation products. In an experiment with Douglas fir only traces of few compounds could be detected after heating at 550 °C (Hileman et al. 1976) (Table 12–6). At the same time the amount of carbon dioxide increased. A further rise of temperature increases the portion of gaseous products while the charcoal amount decreases (Fig. 12–20).

An analysis of the non-condensable gases shows that the major gases are hydrogen, methane, carbon monoxide and carbon dioxide. C_2-, C_3- and C_4-hydrocarbons are produced in relatively small amounts (Knight 1976). Carbon monoxide and carbon dioxide decrease significantly in yield with increasing pyrolysis temperature (Fig. 12–21). Methane shows an increase at higher temperatures. An enormous increase with increasing temperature can be observed with hydrogen. The yield of this gas is additionally influenced by the presence of water. Brink (1976) found a much higher proportion of hydrogen using wood with 52.5% water than using wood

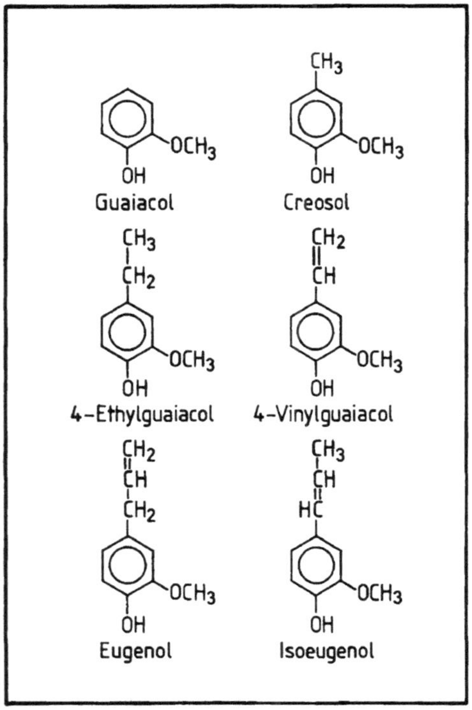

Fig. 12–19. Thermal degradation products of lignin.

Table 12–6: Oxidative degradation products of Douglas fir at various temperatures (Hileman et al.)

Product	Yield (%)		
	340 °C	400 °C	550 °C
Carbon dioxide	3.7	7.6	21.5
Ethene	–	Trace	Trace
Ethane	–	Trace	Trace
Water	10.4	7.5	7.9
Methanal	0.2	0.1	–
Methanol	0.1	0.3	0.1
Ethanal	Trace	0.1	–
Propenal	Trace	0.1	–
2-Oxopropanal	0.3	0.1	–
Hydroxyethanal	Trace	–	–
Ethanoic acid	0.2	0.1	–
2-Hydroxypropanone	Trace	Trace	–
Furfural	0.3	0.1	–
Furfuryl alcohol	Trace	Trace	–
O-Methoxyphenol	Trace	0.2	–
2-Methoxy-4-methylphenol	0.1	0.2	–
2-Methoxy-4-methylanisole	Trace	0.1	–
4-Oxopentanoic acid	–	–	Trace
4-Hydroxypentanoic acid	0.1	Trace	–
p-Methoxyacetophenone	0.1	Trace	–
2-Methoxy-4-propenylphenol	0.1	–	–
5-Hydroxymethyl-2-furaldehyde	Trace	–	–

with 6.5% water during gasification in the temperature range of 500–1 000 °C. The yield of CO and CO_2 also increased with increasing temperature during this experiment.

The yield and the composition of the gas mixture can be controlled by using catalysts during the combustion. With steel wool as a catalyst, Stern et al. (1965) reached an equimolar mixture of hydrogen and carbon monoxide at 1 000 °C. The percentages of carbon dioxide and methane were less than 0.5%, and the total gas yield was 103 m^3 per 100 kg wood.

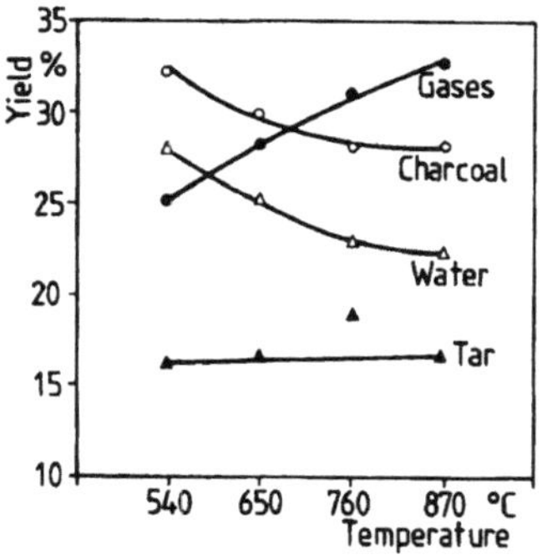

Fig. 12–20. Yield of the major pyrolytic products of wood depending on temperature (Knight 1976).

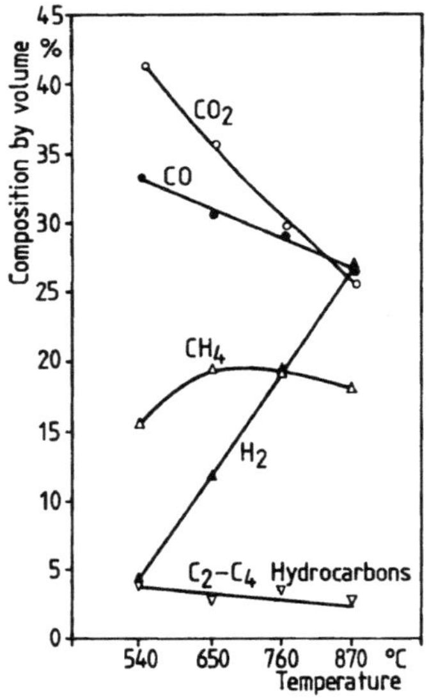

Fig. 12–21. Variation of the gas composition depending on the pyrolysis temperature (Knight 1976).

Literature

Alsup, I. A., Medne, B. A. and Berzinya, I. Y. 1980, Khim. Drev. No. 4, 88–92

Arseneau, D. F. 1961, Can. J. Chem. *39,* 1915–1919

Basch, A. and Lewin, M. 1975, J. Polym. Sci. Polym. Lett. Ed. *13,* 493–499

Beall, F. C. 1969, Wood Fiber *1,* 215–226

Beall, F. C. 1971, Wood Sci. Technol. *5,* 159–175

Beall, F. C. and Eickner, H. W. 1970, Thermal Degradation of Wood Components: A Review of Literature. USDA For. Service Res. Paper FPL 130

Bolkunevich, P. D. 1979, Khim. Drev. No. 3, 78–80

Boutelje, J. 1962, Svensk Papperstid. *65,* 209–215

Brink, D. L. 1976, Appl. Polym. Symp. No. 28, 1377–1391

Brocksiepe, H. G. 1976, Holzverkohlung. In: Ullmanns Encyklopädie der technischen Chemie (4th Ed.). Verlag Chemie, Weinheim, pp. 703–708

Byrne, G. A., Gardiner, D. and Holmes, F. H. 1966, J. Appl. Chem. *16,* No. 3, 81–88

Cabradilla, K. E. and Zeronian, S. H. 1978, Effect on Changes in Supramolecular Structure on the Thermal Properties and Pyrolysis of Cellulose. In: Modified Cellulosics (Rowell, R. M. and Young, R. A., Eds.). Academic Press, New York, pp. 321–340

Chow, S. Z. and Pickles, K. J. 1971, Wood Fiber *3,* 166–178

Dmitrijeva, O. A., Potapova, N. P. and Sharkov, V. I. 1964, Zhur. Prikl. Khim. *37,* 1583–1589

Domanský, R. and Rendoš, F. 1962, Holz Roh-Werkst. *20,* 473–476

Domburg, G. E., Sergejeva, V. N., Kalnins, A. I. and Kiselis, O. 1966, Latv. PSR Zinat. Akad. Vestis *12,* 52–57

Domburg, G. E., Skripcenko, T. N., Dobele, G. V., Sarapova, T. E., Sergejeva, V. N. and Gavars, M. P. 1974, Khim. Drev. No. 1, 51–57

Ebringerová, A. 1976, Cell. Chem. Technol. *10,* 121–129

Fengel, D. 1966, Holz Roh-Werkst. *24,* 9–14, 98–109, 529–536

Fengel, D. 1967, Holz Roh-Werkst. *25,* 102–111

Fengel, D. 1969, Svensk Papperstid. *72,* 209–215

Fengel, D. and Przyklenk, M. 1970, Holz Roh-Werkst. *28,* 254–263

Fengel, D., Wegener, G. and Feckl, J. 1981, Holzforschung *35,* 111–118
Filló, Z. and Peres, T. 1970, Holztechnologie *11,* 270–273
Goring, D. A. I. 1963, Pulp Pap. Mag. Can. *64,* T517–T527
Gromov, V. S., Khrol, Y. S. and Degina, R. A. 1972, Zb. Ved. Pr. Drev. Fak. Vys. Sk. Les. Drev. Zvolene 7–12
Hatakeyama, H., Nakano, J., Hatano, A. and Migita, N. 1969, Tappi *52,* 1724–1728
Hernádi, S. 1976, Svensk Papperstid. *79,* 418–423
Hernádi, S. 1977, Pap. Puu *59,* 566–573
Hileman, F. D., Wojcik, L. H., Futrell, J. H. and Einhorn, I. N. 1976, Comparison of the Thermal Degradation Products of α-Cellulose and Douglas Fir under Inert and Oxidative Environments. In: Thermal Uses and Properties of Carbohydrates and Lignins (Shafizadeh, F., Sarkanen, K. V. and Tillman, D. A., Eds.). Academic Press, New York, San Francisco, London, pp. 49–71
Keylwerth, R. and Christoph, N. 1960, Materialprüfung *2,* 281–288
Knight, J. A. 1976, Pyrolysis of Pine Sawdust. In: Thermal Uses and Properties of Carbohydrates and Lignins (Shafizadeh, F., Sarkanen, K. V. and Tillman, D. A., Eds.). Academic Press, New York, San Francisco, London, pp. 159–173
Knudson, R. M. and Williamson, R. B. 1971, Wood Sci. Technol. *5,* 176–189
Kollmann, F. 1960, Holz Roh-Werkst. *18,* 193–200
Kollmann, F. and Fengel, D. 1965, Holz Roh-Werkst. *23,* 461–468
Kollmann, F. and Sachs, I. B. 1967, Wood Sci. Technol. *1,* 14–25
Kollmann, F. and Schneider, A. 1963, Holz Roh-Werkst. *21,* 77–85
Kollmann, F. and Schneider, A. 1964, Untersuchungen über den Einfluß von Wärmebehandlungen im Temperaturbereich bis 200 °C und von Wasserlagerung bis 100 °C auf wichtige physikalische und physikalisch-chemische Eigenschaften des Holzes. Forschungsber. Nordrhein-Westfalen No. 1399. Westdeutscher Verlag, Köln, Opladen
Koran, Z. 1968, Svensk Papperstid. *71,* 567–576
Košík, M., Kozmál, F., Reiser, V. and Domanský, R. 1968a, Holzforsch. Holzverw. *20,* 11–15
Košík, M., Gerátová, L., Rendoš, F. and Domanský, R. 1968b, Holzforsch. Holzverw. *20,* 15–19
Košík, M., Herain, J. and Domanský, R. 1968c, Holzforsch. Holzverw. *20,* 56–59
Košík, M., Dandárová, M., and Domanský, R. 1969a, Holzforsch. Holzverw. *21,* 40–43
Košík, M., Micko, M. and Domanský, R. 1969b, Wood Sci. *1,* 167–171
Košíková, B., Kosáková, L., Joniak, D. and Košík, M. 1978, Cell. Chem. Technol. *12,* 657–663, 665–669
Kratzl, K., Czepel, H. and Gratzl, J. 1965, Holz Roh-Werkst. *23,* 237–240
Kürschner, K. and Melcerová, A. 1965, Holzforschung *19,* 161–178
Laxamana, N. B. 1971, Destructive Destillation of Wood. Forpridecom Techn. Note No. 110
Levanova, V. P., Smirnova, L. G. and Sharkov, V. I. 1976, Khim. Drev. No. 1, 3–8
Millett, M. A. and Goedken, V. L. 1965, Tappi *48,* 367–371
Nečesaný, V. 1965, Drev. Výsk. *10,* 149–154
Nordin, S. B., Nyrén, J. O. and Back, E. L. 1973, Svensk Papperstid. *76,* 609–610
Pajtík, J. and Chovanec, D. 1974, Drev. Výsk. *19,* 137–142
Potutkin, G. F. 1969, Lesn. Zhur. *13,* 169–171
Ramiah, M. V. and Goring, D. A. I. 1967, Cell. Chem. Technol. *1,* 277–285
Reiser, V., Košík, M., Durdovič, V. and Domanský, R. 1968, Holzforsch. Holzverw. *20,* 148–150
Roberts, A. F. and Clough, G. 1963, Thermal Decomposition of Wood in an Inert Atmosphere. In: Ninth Int. Symp. Combustion, Proceed. Academic Press, New York, pp. 158–166
Roffael, E. and Schaller, K. 1971, Holz Roh-Werkst. *29,* 275–278
Sandermann, W. and Augustin, H. 1963, Holz Roh-Werkst. *21,* 256–265, 305–315
Sandermann, W. and Augustin, H. 1964, Holz Roh-Werkst. *22,* 377–386
Sergejeva, V. N. and Miljutina, S. V. 1960, Trud. Inst. Lesohoz. Probl. Riga No. 21, 101–106
Shafizadeh, V. and DeGroot, W. F. 1976, Combustion Characteristics of Cellulosic Fuels. In: Thermal

Uses and Properties of Carbohydrates and Lignins (Shafizadeh, F., Sarkanen, K. V. and Tillman, D. A., Eds.). Academic Press, New York, San Francisco, London, pp. 1–17

Shafizadeh, F. and McGinnis, G. D. 1971, Carbohydr. Res. *16,* 273–277

Shafizadeh, F., Furneaux, R. H., Cochran, T. G., Scholl, J. P. and Sakai, Y. 1979, J. Appl. Polym. Sci. *23,* 3525–3539

Shimizu, K., Teratani, F. and Miyazaki, K. 1971, Mokuzai Gakkaishi *17,* 456–463

Shimizu, K., Teratani, F., Hashi, M. and Miyazaki, K. 1972, Mokuzai Gakkaishi *18,* 79–84

Sihtola, H. and Neimo, L. 1965, Faserforsch. Textiltechn. *16,* 481–487

Stern, E. W., Longindice, A. S. and Heinemann H. 1965, Ind. Eng. Chem. Process Des. Dev. *4,* 171–173

Sugisawa, H. and Edo, H. 1964, Chem. Ind. 892–893

Tanahashi, M., Aoki, T. and Higuchi, T. 1982, Holzforschung *36,* 117–122

Taniguchi, T. and Nakato, K. 1966, Bull. Kyoto Univ. For. No. 38, 192–199

Zavarin, E. and Snajberk, K. 1963, Tappi *46,* 320–323

Zavarin, E. and Snajberk, K. 1965, Tappi *48,* 612–616

Zavarin, E., Snajberk, K. and Smith, R. M. 1965, Tappi *48,* 574–577

13. Degradation by Light and Ionizing Rays

13.1. Changes Initiated by Light

13.1.1. Importance and General Effects

The wide range of the electromagnetic spectrum has a measurable influence on wood only in its short-waved and energy-rich part. In particular, changes caused by light and ionizing rays are of technological interest; that is the reason why various studies have been made in this field. Articles summarizing previous results were published by Kenaga and Cowling (1959), Desai (1968), Kringstad (1969) and Hon and Glasser (1979).

Light, for our purposes, denotes a range of electromagnetic waves comprising ultraviolet, visible and infrared rays. The latter, which transfer heat, were treated in the preceding chapter as heat is not only transferred by infrared rays but also by conduction.

Visible and the adjacent ultraviolet light changes the colour of wood to a lighter or a darker tone depending on the wood species. Some woods become bleached or grey, others yellow, red-orange or brown, colours which are influenced by the wood composition, in many cases particularly by extractive components (Sandermann, Schlumbom 1962). These colour changes may improve the impression of a wood surface; surfaces of other wood species may become unsightly.

The colour change is also influenced by additional factors, such as temperature, water and atmosphere. Thus spruce and larch wood weathered by sunlight and rain become grey; if the woods are protected against rain they become dark red-brown (Kleinert 1970a).

An additional visible effect during weathering is a roughening of the wood surfaces which indicates macroscopically a maceration and degradation of the wood structure.

Wood samples of birch and redwood exposed to ultraviolet light darkened during the first several hours in atmospheres of air, oxygen, nitrogen and argon (Wengert 1966). In continuing the UV exposure the samples in air and oxygen stopped darkening and became lighter, while those in nitrogen and argon continued to darken. In particular, chemically treated wood yellows during UV-irradiation, and is bleached again by irradiation with visible light (Claesson et al. 1968).

The penetration depth of light into wood was determined by various photometric methods and by electron spin resonance techniques (Browne, Simonson 1957; Hon, Ifju 1978). The latter method was used for the detection of photo-induced free radicals (→ 13.1.5.). In several softwoods UV rays did not penetrate any more deeply than 75 μm. Visible light penetrated up to 200 μm, the depth depending on the colour of the original wood and the colour change during exposure. Infrared rays penetrated much more deeply into the wood surface (1–1.5 mm), and the variation in transmission among different woods was comparatively small.

13.1.2. Effects Visible in Structure

In addition to chemical studies several microscopic observations were made to investigate the structural changes of wood after light exposure and weathering. By electron optical methods, however, degradation effects can be seen only after long exposure times, high-energy irradiation, or the influence of additional factors (rain, elevated temperature etc.).

Wood samples *(Pinus sylvestris)* from Norwegian stave churches and other wooden constructions several hundred years old were studied by Borgin (1970, 1971). He observed a slow disintegration of the outer layer of the fibres. The primary walls and the secondary walls 1 were often found to be partly flaked off or completely

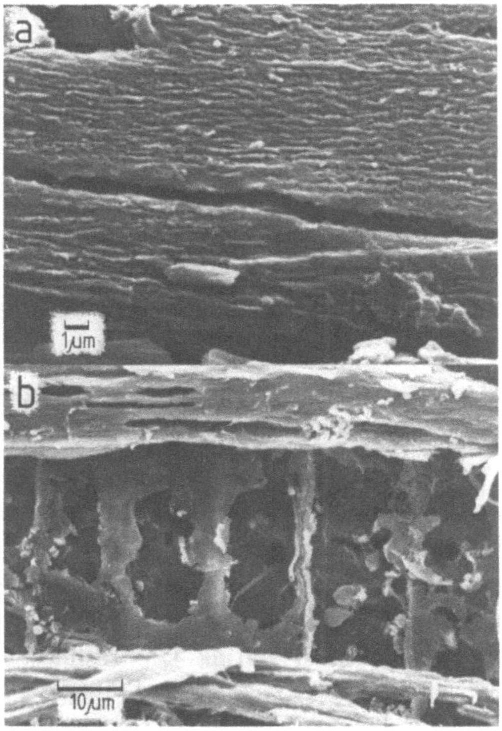

Fig. 13–1. Weathered pine wood *(Pinus sylvestris)*. SEM micrograph.
a) Surface of a tracheid with released S2-structure and fissures in the cell wall.
b) Degraded fenestriform pit membranes and parenchyma cell walls.

missing on the exposed side of the fibres. It was evident that the most stable and apparently uneffected part of the wood structure were the fibrils in the form of aggregates or bundles (Fig. 13–1). The destructive processes, however, were limited to a thin surface layer of 2–3 mm, and the very slow deterioration of this layer protected the bulk of the interior structure very efficiently.

Futó (1974) observed that the degradation begins at a relatively low irradiation intensity with an attack on the compound middle lamella. Higher intensities and longer exposure also degrade the secondary walls, as is made visible by the formation of cavernes (Fig. 13–2). Elevated temperature intensifies the photolytic degradation process (Futó 1976).

The degradation of the wall substance during UV irradiation effects a contraction of the cell walls, resulting in microchecks along the compound middle lamellae and, particularly in latewood, along the border between S1 and S2 (Miniutti 1973). Diagonal fissures that follow the fibril orientation of the S2 layer have also been observed. The apertures of the bordered pits in softwood were enlarged or ruptured by microchecks.

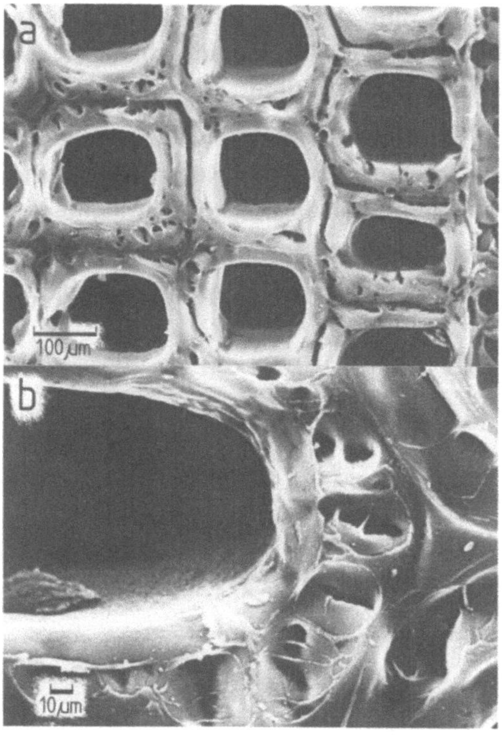

Fig. 13–2. Fir wood *(Abies alba)* irradiated with ultraviolet light. SEM micrograph, by courtesy of L. P. Futó.

a) Degradation effects in the compound middle lamella and the cell walls of the tracheids.

b) Radial orientation of cavernes in the secondary wall of a tracheid.

13.1.3. Chemical Changes of Wood and Lignin

From microscopic observations it could be seen that the surface of weathered wood becomes rough during the photodegradation of the wood components and washing-out of the degraded products. A study of the loss of weight by UV irradiation was carried out by Futó (1974, 1976). He used a lamp with a high energy portion at wavelengths of 240–310 nm. He found that the loss of weight is highly influenced by the temperature and the irradiation energy (Fig. 13–3). The loss of weight is much higher even at low temperatures if the wood samples are irradiated under water, which indicates that water-soluble products are formed in addition to volatile products.

Experiments in which wood samples and lignocellulosic materials were irradiated with UV lamps confirm that all wood components are attacked though the presence of lignin retards the photolytic degradation of cellulose. This phenomenon is explained by the high UV absorption and the strong capability of autoxidation of lignin (Kleinert 1970b). That is why the absorption of ultraviolet light by wood is primarily determined by the lignin content. According to Norrström (1969) lignin contributes 80–95%, the carbohydrates 5–20%, and the extractives about 2% to the absorption coefficient. The same also applies to visible light which is e.g. used for the determination of the brightness of pulp. Using wood sections and mechanical pulp of loblolly pine *(Pinus taeda)* Wilcox (1975) found a linear relation of the absorption coefficient at 457 nm to the lignin content.

Lignin is a good UV absorber and therefore the energy which is transferred in the range of 200–400 nm initiates degradation processes. The absorption occurs at chromophoric structural elements within the molecular network of lignin (Fig. 13–4).

Hon and Glasser (1979) classified the potential chromophoric groups as follows:
- Chromophoric functional groups: phenolic hydroxyl groups, double bonds, carbonyl groups etc.

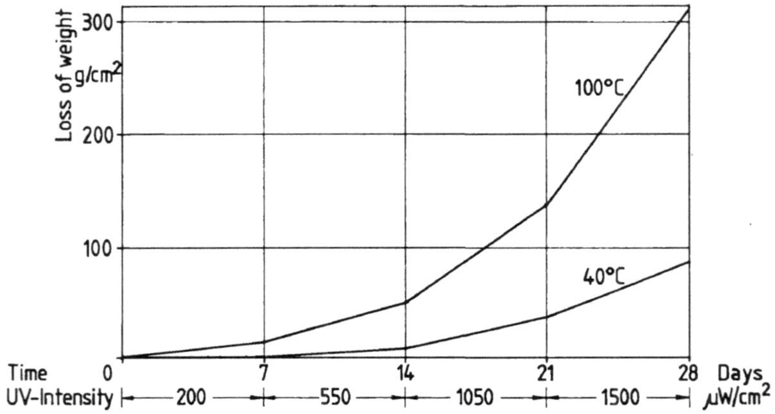

Fig. 13–3. Loss of weight in beech wood *(Fagus sylvatica)* after UV irradiation of various intensities at temperatures of 40 °C and 100 °C (Futó 1976).

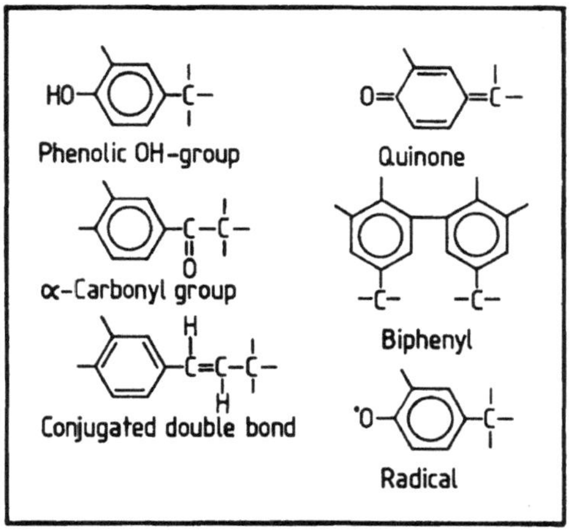

Fig. 13–4. Various chromophoric groups present in lignin molecules.

- Chromophoric systems: quinones, quinone methides, biphenyls etc.
- Leucochromophoric systems: methylenequinones, phenanthrenequinones, phenylnaphthalenediones, bimethylenequinones etc.
- Intermediates: free radicals
- Complexes: chelate structures with metal ions.

The photodegradation of lignin is visible in its UV spectrum after a short period of irradiation. The irradiation of milled wood lignin, lignin sulfonates and kraft lignin for 1–3 hours changed the absorption spectra; the characteristic peaks dropped drastically (Kleinert 1970c; Lin, Kringstad 1970a). During the study of wood microsections of Douglas fir *(Pseudotsuga menziesii)* in an ultraviolet microscope Scott and Goring (1970) observed a continuous reduction of the UV absorbancy at 280 nm to less than 25% of its original value after an exposure time of 2 hours. The irradiation of milled wood lignin dissolved in NaOH (1 mole/l) by UV light of wavelengths between 210 and 276 nm decreased the absorbancy at 240 nm, but new chromophores were built up which absorbed in the range of 250–380 nm (Gizetdinov, Chupka 1979). These changes in the absorption spectrum are attributed to redox reactions within the lignin molecule.

Various steps of lignin degradation have been observed. Among them are the reduction of the methoxyl content and the splitting of monomeric units (Sandermann, Schlumbom 1962; Leary 1967, 1968; Kleinert 1970b). Leary (1968) discovered that traces of oxygen are necessary for the cleavage of methoxyl groups. A complete degradation of lignin was found in spruce and larch wood from huts and houses in the Alps (1 200–1 600 m above SL) which had been exposed to the sun for 50 to 120 years but not leached by rain (Kleinert 1970a). The yield of extractives was high; about 55–75% of the wood samples were soluble in boiling 1% NaOH.

13.1.4. Chemical Changes of Cellulose

The degradation of cellulose under the influence of UV rays is indicated by loss of weight, and a reduction of the α-cellulose content and the degree of polymerization. The irradiation of filter paper (Whatman No. 1) resulted in a linear increase of the weight loss with the time of exposure up to 5.6% after 16 hours (Desai, Shields 1969). The DP dropped very rapidly within the first hour from 2 000 to 600, followed by a slower decrease to 300 after 16 hours of irradiation (Fig. 13–5). No or only small differences were observed between 40 and 60 °C. The results indicate that the degradation of cellulose is random and follows first order kinetics.

Similar results were obtained by LeNest and Silvy (1970) who irradiated paper consisting of regenerated pure cellulose at wavelengths less than 300 nm. The loss of weight versus irradiation time is a linear function with a slope increasing with elevating temperatures. The activation energy was determined to be 15.5 kJ/mole. A relatively high percentage of hydrocarbons (CH_4, C_2H_6, C_3H_8) of 9.3% was found in the gaseous degradation products after an irradiation of 300 h, and at a weight loss of 15%.

Bleached softwood pulp *(Pseudotsuga menziesii)* was irradiated by Hon (1979). The α-cellulose content decreased from 88% to about 50 and 40%, respectively, after an exposure of 10 hours. The first value applies to irradiation in vacuum, the second to that in oxygen. During the same time the DP was reduced from 850 to about 380 and 260, respectively.

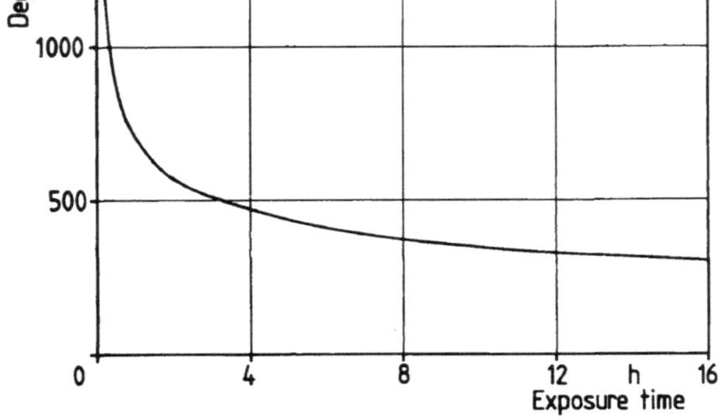

Fig. 13–5. Decrease of the degree of polymerization in cellulose with the exposure time of UV irradiation (550 W high-pressure mercury-vapor lamp, Desai and Shields 1969).

An influence of temperature was observed by Hernádi (1975). Softwood sulfate pulp had a DP value of about 800 after irradiation for 4 hours at 50 °C, and one of 550 at 100 °C. It was concluded that during the photodegradation of pulp the cleavage of the glycosidic bonds is the main reaction, while the oxidation plays only a secondary role. But, according to Hon (1979), the chain scission is accelerated by oxygen.

Because of the low penetration of ultraviolet rays photodegradation is more or less a surface reaction. In paper sheets, effects of ultraviolet exposure were found to be confined to a surface layer of approximately 0.15 mm (MacClaren et al. 1962). In another study the DP was strongly reduced at the irradiated surface where a DP of 350 was determined, while the DP of the opposite side was 1 400; the original sample had a DP of 1 900 (Desai 1970). The wood samples of the huts and houses in the Alps exposed to the sun for up to 120 years, as mentioned above, showed a DP of cellulose of less than 100 in the outermost part, and an increase of DP to about 1 600 at a distance of 28 mm from the surface (Kleinert 1970a).

During an UV irradiation of cellulose at 185 and 253.7 nm Kujirai (1965, 1966) found that the wavelength influences the degradation mechanism. Shorter wavelengths produced aldehyde groups indicating a hydrolytic chain cleavage whereas longer wavelengths produced peroxide groups, indicating a degradation by participation of oxygen. The investigation of the soluble degradation products of irradiated cellulose (pulp, cotton, model compounds) resulted in the occurrence of glucose, cellobiose and cellotriose, as well as xylose, xylo-oligomers, arabinose and 3-β-D-glucosido-D-arabinose (Beélik, Hamilton 1959, 1961; Gingras et al. 1963). A probable pathway of the formation of pentoses from glucose units in cellulose is presented in Fig. 13–6.

Among the volatile degradation products of cellulose are acetaldehyde, propionaldehyde, methyl formiate, acetone, methanol, ethanol, methane and ethane (Desai, Shields 1969). For the photolytic degradation of cellulose the cleavage of carbon-

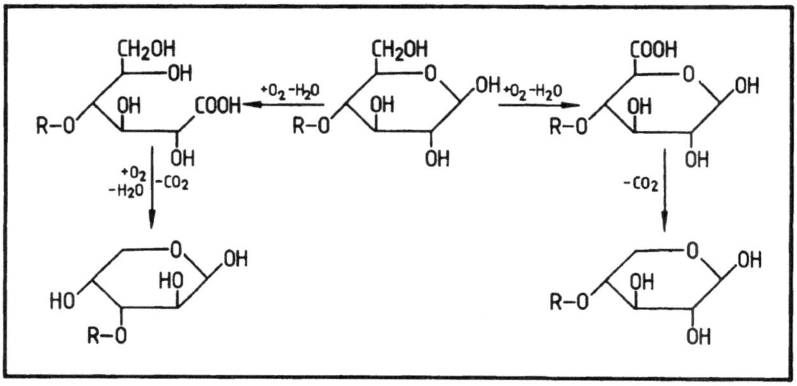

Fig. 13–6. Probable pathways of formation of pentose units during UV irradiation of cellulose.

Table 13–1: Energy and approximate bond dissociation energies of chemical bonds in the cellulose molecule (Hon 1976a)

Wavelength nm	Energy kJ/mole	Bond dissociation
200	599	
253.7	473	
280	427	O – H
286	419	
288.9	414	C – H
300	398	
307.5	389	C – O
325	368	
340	352	
345	347	C – O
360	331	
400	301	

oxygen or carbon-carbon bonds will require an energy level of 340–390 kJ/mole (Desai 1968; Hon 1976a). This energy level will be met by ultraviolet rays of wavelengths of 340 nm or shorter, which correspond to 350 kJ/mole (Table 13–1).

An explanation of the absorption of rays and the photolytic degradation in these wavelengths by cellulose and other polysaccharides is not easy to find. Solutions of cellulose do not show a distinct absorption in the range of 200–400 nm. Colourless solutions of cellulose in LiBr or H_2SO_4 have a broad shoulder at 260 and 290 nm, respectively; only a solution in aqueous cadoxene has a distinct absorption peak at 235 nm which, however, has to be attributed to the cellulose-cadoxene complex (Donetzhuber 1961; Schurz et al. 1963; Williams 1968). It has been suggested that the absorbing chromophores are the hydroxyl, carbonyl or carboxyl groups or the acetal group at the C1 position of non-reducing glucose units (Beélik, Hamilton 1959; Zapolskii 1964; Williams 1968) (Fig. 13–7).

Bos (1972) compared the reflectance spectra of cellulose and model compounds. The study indicated that the acetal linkage does not contribute significantly to the absorption peak at 260 nm. The spectra of photolyzed and photo-oxidized cellulose suggested that ketonic carbonyl groups are a more likely cause of this peak. The photolysis of cellulose and amylose under vacuum increased the concentration of

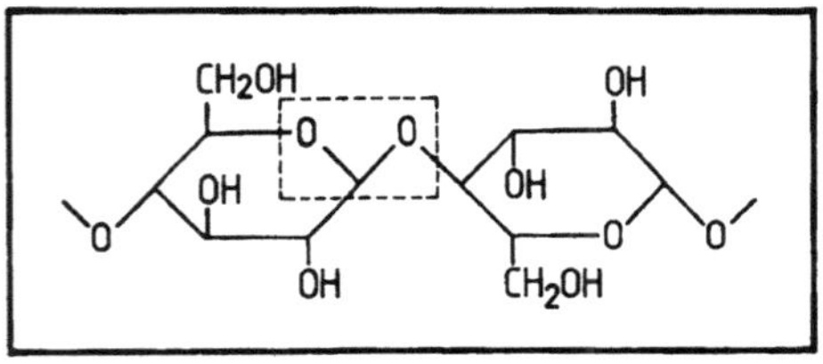

Fig. 13–7. Probable chromophoric group in cellulose molecules.

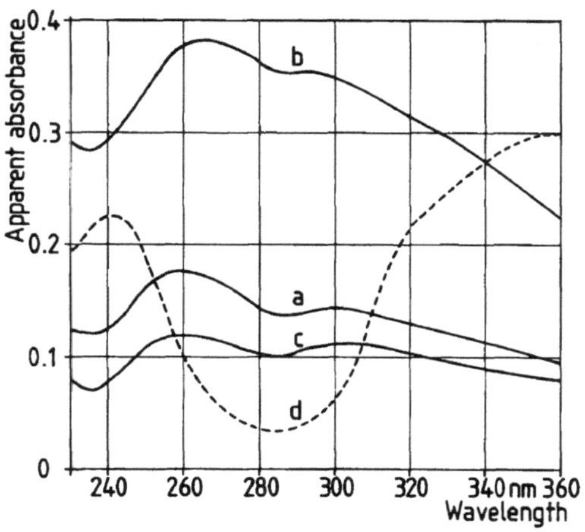

Fig. 13–8. UV reflectance spectra of cellulose, non-irradiated (a), UV irradiated in vacuum (b), UV irradiated in air (c), thermally oxidized at 140 °C (d) (Bos 1972).

ketonic carbonyls and thus promoted yellowing. A photo-oxidation with the same wavelength irradiation (254 nm) bleached both polysaccharides owing to the formation of carboxylic groups (Fig. 13–8).

13.1.5. Formation of Free Radicals

The radiation energy absorbed by wood and its components cause a generation of free radicals which can be proved by electron spin resonance spectrometry (ESR). Formation, lifetime and reactions of these radicals are greatly dependent on the light source and the exposure atmosphere. The ESR signals as observed with cellulose at various conditions are summarized in Table 13–2.

Various studies showed that green wood and wood stored in the dark contain no or only a few radicals (Kalnins et al. 1966; Rånby et al. 1969; Hon et al. 1980). A high amount of free radicals was found in fresh wood samples treated with sunlight, but

Table 13–2: ESR spectra of celluloses, observed at 77 K (Hon 1975d)

System	Wavelength of radiation nm	Spectra	Peak location G*
Untreated	>253.7	(1)seven-line	±8, ±12, ±34, ±44
		(2)doublet	64.5
		(3)doublet	254.0
	>280	(1)seven-line	±8, ±12, ±34, ±44
		(2)doublet	±254.0
	>340	–	
Fe^{3+}-sensitized	>253.7	(1)five-line	±3.5, ±13.5, ±34
	>280	(1)five-line	±7, ±13.5, ±34
	>340	(1)five-line	±7, ±13.5, ±34

most of the free radicals were relatively unstable. Artificial fluorescent light produced a low amount of free radicals which, however, were relatively stable at ambient temperature.

Air-dried wood specimens readily interacted with all kinds of ultraviolet irradiation to produce free radicals, either in the presence of air or under vacuum. Higher amounts of radicals and longer living radicals are generated under vacuum (Fig. 13–9). The addition of oxygen to samples treated under vacuum seems to be able to promote the formation of free radicals, but the lifetime of these radicals is relatively short (Fig. 13–10).

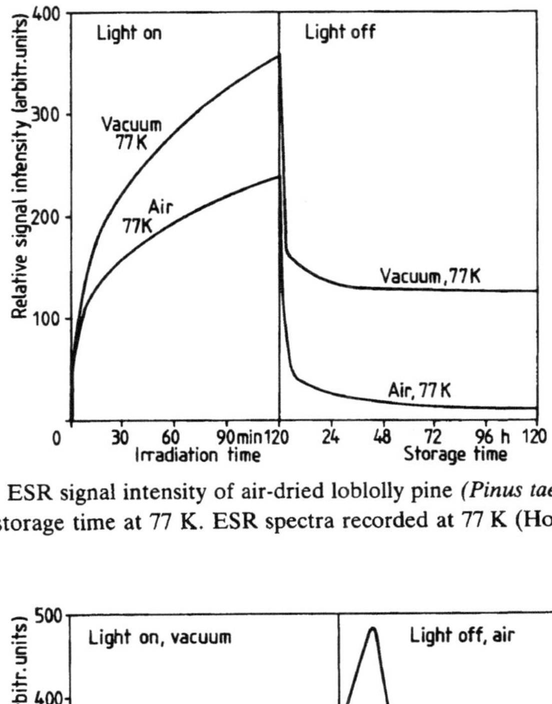

Fig. 13–9. ESR signal intensity of air-dried loblolly pine *(Pinus taeda)* as a function of UV irradiation time and storage time at 77 K. ESR spectra recorded at 77 K (Hon et al. 1980).

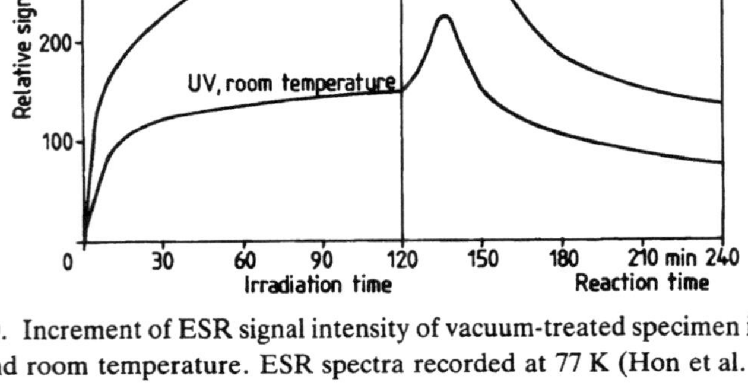

Fig. 13–10. Increment of ESR signal intensity of vacuum-treated specimen interacted with oxygen in air at 77 K and room temperature. ESR spectra recorded at 77 K (Hon et al. 1980).

During UV irradiation radicals are produced from polysaccharides as well as from lignin, but the formation of radicals from polysaccharides is greatly influenced by the presence of lignin. In several studies the formation and behaviour of free radicals in polysaccharides, particularly in cellulose, and in lignin were investigated.

The early investigations of Kleinert (1964a, b; Kleinert, Morton 1962) on UV irratiated cellulose using ESR spectrometry showed that there are approximately linear relations between the ESR signal area and the chain splitting, the formation of peroxide groups, and the solubility in boiling 1% NaOH.

Extended studies on the formation and behaviour of photoirradiated cellulose were made by Hon (1975a, b, c, d; 1976a, b; 1979). These studies show that a radiation of wavelengths longer than 330–340 nm does not form radicals, but wavelengths beneath this range are able to produce radicals in cellulose and other polysaccharides. The relative signal intensity of the ESR spectrum generated by cellulose irradiated with light of $\lambda > 253.7$ nm was observed to be greater than that of the spectrum generated with light of $\lambda > 280$ nm.

The ESR spectra of cellulose irradiated at various temperatures are different because of the different life-times of the several kinds of radicals generated during UV irradiation. Only the long-life components are trapped at 20 °C and 45 °C. At 45 °C a singlet spectrum, at 20 °C a poorly resolved three line spectrum was received. An irradiation at −80 °C and −196 °C yielded spectra with seven and eleven lines, respectively (Fig. 13–11). Also the moisture content of the sample greatly influences the formation of free radicals when irradiated with ultraviolet rays. A

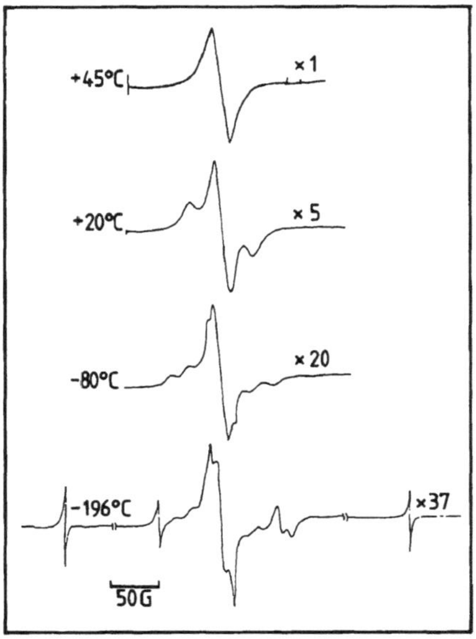

Fig. 13–11. ESR spectra of cellulose irradiated with wavelengths > 253.7 nm for 120 min at various temperatures. All spectra recorded at 77 k. Multiplication number represents the ESR relative signal intensity (Hon 1975a).

moisture content of 5–7% leads to a significant decrease of the radical formation; moisture contents below and above this range cause an increase of free radicals (Hon 1975a).

The presence of photosensitizers in a cellulose sample promotes the radical formation during irradiation with ultraviolet light, especially at wavelengths longer than 340 nm. Among several photosensitizers Fe^{3+}-ions have the greatest sensitizing effect on the formation of free radicals in cellulose (Hon 1975b).

The comparison of ESR spectra of various celluloses irradiated with UV light indicate that generally the same types of radical species are generated, but that the radical formation depends upon the degree of crystallinity, the lattice type and the arrangement of the cellulose molecules. The formation of free radicals was approximately related to the ratio of amorphous to crystalline portions (Table 13–3) (Hon 1976b).

The presence of lignin effects a protection of cellulose, and radicals are produced predominantly in the lignin molecules by which the species of radicals are changed as can be seen in the ESR spectra (Fig. 13–12). While in pure cellulose radicals are only formed in air when irradiated with wavelengths longer than 340 nm, in samples containing lignin, and in pure lignin, radicals are also formed in vacuum or nitrogen atmosphere (Hon 1975c). This yield of free-radical sites is lower in lignin-rich pulps as compared to those in wood cellulose if irradiated with wavelengths longer than 253.7 nm. It is concluded that the lignin in pulps effects energy transfer, particularly localization of energy. This would suggest that energy deposited within the cellulose could be rapidly localized to an essentially inert group, such as an aromatic group in lignin from which it could be dissipated.

The study of radical formation in various woods revealed that more free radicals are formed in earlywood than in latewood, presumably due to the higher lignin content in earlywood. (Hon, Feist 1981). On the other hand the formation of free radicals also depends on the wood species. An increase of the water content in wood from

Table 13–3: Correlation of degree of crystallinity and ESR signals of various celluloses (Hon 1976b)

Sample	Degree of crystallinity %*	ESR lines	Relative signal intensity**
Cellulose I	62	7	1.0
Cellulose II	45	7	4.5
Cellulose III	40	7	3.5
Cellulose IV	43	7	2.7
Rayon cellulose	35	7	4.0
Cotton linters	72	7	0.6
Amorphous cellulose	0	7	5.0
Absorbent cotton	65	7	1.1

* X-Ray diffraction method; amorphous cellulose as the standard material.

** Calculations based on the value of cellulose I in arbitrary units.

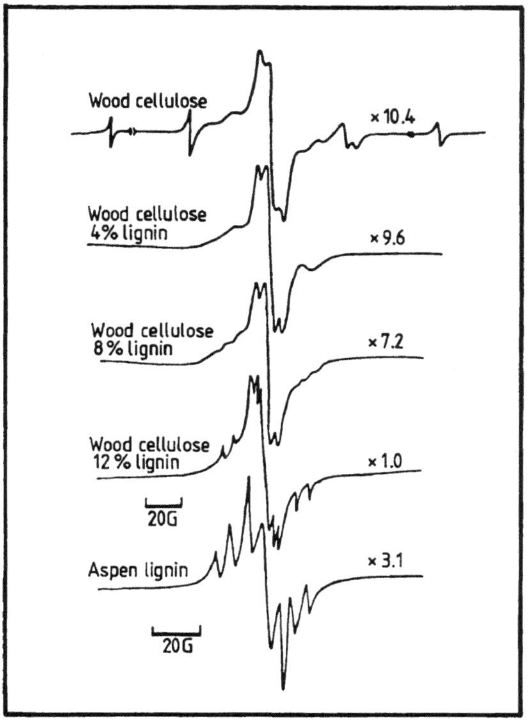

Fig. 13–12. ESR spectra of cellulose, lignin-containing cellulose, and lignin irradiated with wavelengths > 253.7 nm for 60 min at 77 K. All spectra recorded at 77 K. Multiplication number represents the ESR relative signal intensity (Hon 1975c).

0 to 6.3% enhanced the free radical formation, but beyond this water content the rate of radical decay increased. It is suggested that the excess water above 6.3% traps free water by forming phenoxy radical-water complexes.

Other studies show that the radicals generated in lignin and lignin degradation products are relatively stable, and that there is a continuous change in the chemical structure of such lignins on storage in daylight (Steelink 1964; Rånby et al. 1969). The chemical conversion of lignin e.g. into sodium salts increases the amount of radicals remarkably (Steelink et al. 1963; Steelink 1964).

13.1.6 The Mechanisms of Degradation

From the preceding description it can be seen that the photolysis of the wood components is initiated by the absorption of radiation energy. Energy-transfer effects lead to localization of the energy within the molecules resulting in splitting reactions such as depolymerization, dehydrogenation, dehydroxymethylation.

In cellulose two kinds of splitting are observed, and the ESR signals could be attributed to the corresponding radicals (Hon 1976a). The localization of energy in the vinicity of the glycosidic bond leads to a chain scission producing alkoxy radicals at the C1 and C4 position (Fig. 13–13). Signals which are attributed to hydrogen and formyl indicate a cleavage of these elements from the glycosidic rings. Some of the possible radicals could be attributed to corresponding signals in the ESR spectrum (Fig. 13–14).

The cleavage of the bonds between C5 and C6 results in a hydroxymethyl radical, which, however, could not be detected. Thus this radical is assumed to be rapidly split into a formyl radical and hydrogen. An additional proof that the formyl radical derives from the C6-group is the fact that after irradiation of xylan no CHO-signal could be detected in the ESR spectrum (Hon 1975a). A further reaction with light energy produces CO and hydrogen from CHO.

The primary reactions of photolytic degradation of cellulose can be formulated by means of the following equations (Hon 1975c):

$$\text{Cell-O-Cell} \xrightarrow{h\nu} \text{Cell-O}^{\cdot} + {}^{\cdot}\text{Cell}$$
$$\text{Cell-OH} \xrightarrow{h\nu} \text{Cell-O}^{\cdot} + {}^{\cdot}\text{H}$$
$$\text{Cell-H} \xrightarrow{h\nu} \text{Cell}^{\cdot} + {}^{\cdot}\text{H}$$
$$\text{Cell-OH} \xrightarrow{h\nu} \text{Cell}^{\cdot} + {}^{\cdot}\text{OH}$$
$$\text{Cell-CH}_2\text{OH} \xrightarrow{h\nu} \text{Cell}^{\cdot} + {}^{\cdot}\text{CH}_2\text{OH}$$
$${}^{\cdot}\text{CH}_2\text{OH} \xrightarrow{h\nu} {}^{\cdot}\text{CHO} + 2\text{H}$$
$$2{}^{\cdot}\text{CHO} \xrightarrow{h\nu} 2{}^{\cdot}\text{CO} + \text{H}_2$$

During the irradiation of lignin only a few reactions are assumed to produce phenoxy radicals:

$$\text{Lignin-OH} \xrightarrow{h\nu} \text{Lignin-O}^{\cdot} + {}^{\cdot}\text{H}$$

$$\text{Lignin-O-Lignin} \xrightarrow{h\nu} \text{Lignin-O}^{\cdot} + {}^{\cdot}\text{Lignin}$$

In lignin the radiation energy seems to be absorbed predominantly by the α-carbonyl group which is transferred into an excited state. The excited α-carbonyl group initiates an abstraction of hydrogen in an adjacent phenolic structure (Fig. 13–15) (Kringstad, Lin 1970).

Light-induced reactions under normal conditions, i.e. in presence of air, are greatly influenced by oxygen. Thus the radicals formed during irradiation under vacuum react immediately with oxygen. In cellulose peroxide radicals are produced which, however, are stable only at low temperatures (Figs. 13–13, 13–14) (Kleinert 1964b; Hon 1979). The formation of peroxide radicals is terminated when the sample temperature is elevated to 0 °C. A probable pathway is the abstraction of hydrogen from C–H and O–H bonds in the cellulose molecules to produce polymeric hydroperoxides.

Oxygen in its ground state is not able to react directly with unsaturated or other electron-rich organic compounds under photoirradiation. A photosensitizer to be

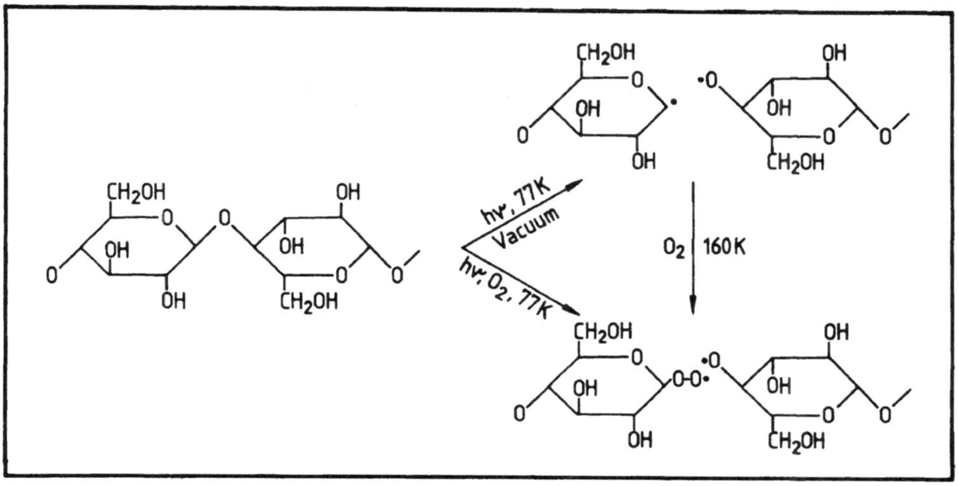

Fig. 13–13. Reaction mechanisms of the formation of C1 and C4 radicals in cellulose during UV irradiation.

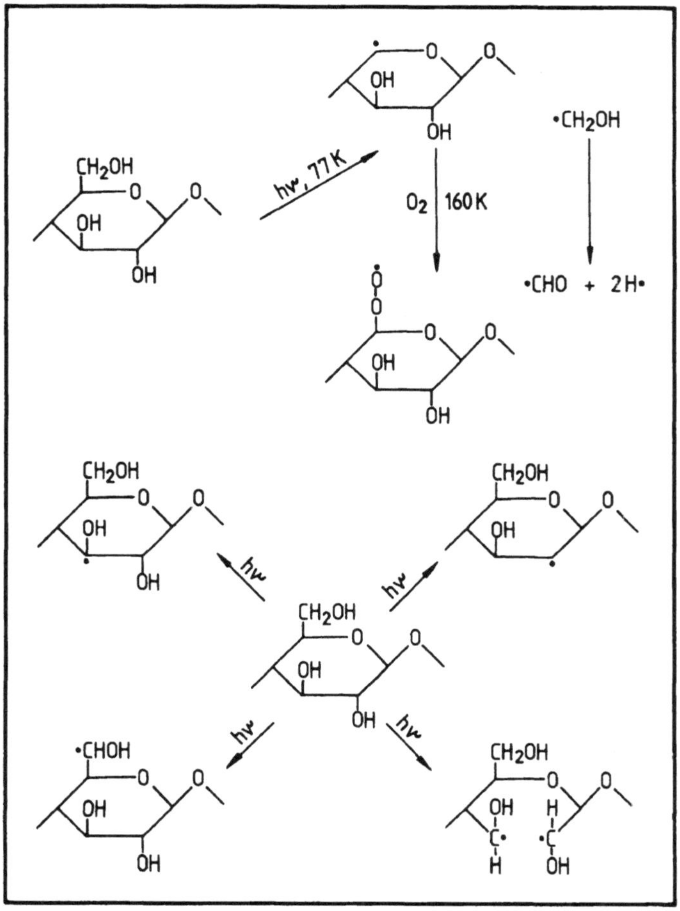

Fig. 13–14. Formation of various C2, C3 and C6 radicals in cellulose during UV irradiation.

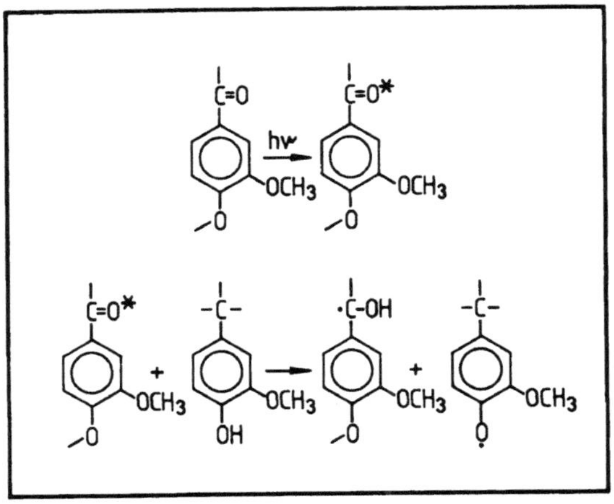

Fig. 13–15. Excitation of the α-carbonyl group and radical formation in the lignin molecule during UV irradiation.

excited by the radiation energy must be present. The excited (triplet) sensitizer transfers the energy onto the ground state (triplet, $^3\Delta^-{}_g$) oxygen which is converted to a reactive singlet molecule ($^1\Delta_g$) with an internal energy of 92 kJ/mole above ground state (Cotton, Wilkinson 1974):

$$
\begin{aligned}
{}^1\text{sens} &\xrightarrow{h\nu} {}^1\text{sens}^* \\
{}^1\text{sens}^* &\longrightarrow {}^3\text{sens}^* \\
{}^3\text{sens}^* + {}^3O_2 &\longrightarrow {}^1\text{sens} + {}^1O_2
\end{aligned}
$$

Matsuura et al. (1972) assume that possibly a charge-transfer interaction between singlet oxygen and an electron-rich phenol occurs, followed by an electron transfer giving a cation radical which then releases a proton, to form a phenoxy radical according to the equation:

$$\text{ArOH} + {}^1O_2 \rightarrow [\text{ArOH}\cdots O_2] \rightarrow O_2^{\overline{\cdot}} + [\text{ArOH}]^{\dot{+}} \rightarrow \text{ArO}^{\cdot} + H^+$$

In the oxidative photodegradation of lignin the α-carbonyl group seems to function as a photosensitizer which activates oxygen and converts it into the singlet state. The probable reaction mechanism for the phenoxy radical formation in lignin with

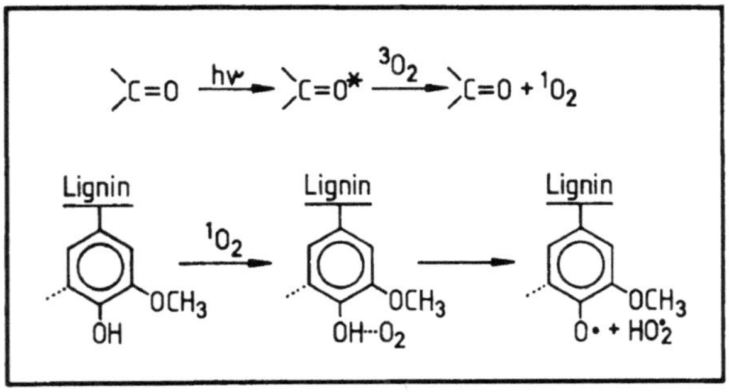

Fig. 13–16. Reaction mechanism of the formation of phenoxy radicals effected by singlet oxygen during UV irradiation.

participation of oxygen was formulated by Brunow and Sivonen (1975) (Fig. 13–16). Their experimental comparison of the reaction rates under nitrogen with those in air showed that oxygen quenches the excited α-carbonyl group and that phenol is dehydrogenated by excited oxygen. The presence of metal ions can accelerate the formation of phenoxy radicals (Landucci 1978).

13.1.7. Formation and Nature of Yellowing Products

The problem of yellowing is closely connected with pulp and paper. The colouring substances in these products are generated by lignin and lignin derivatives, and can be attributed to structures such as quinones, quinone methides and stilbenes (Hon, Glasser 1979).

Nevertheless cellulose also yellows under irradiation. The yellowing effect is attributed to the production of oxygen-containing groups, e.g. carbonyl, carboxyl and hydroperoxide groups (Kleinert, Marraccini 1966; Kleinert 1969; Hon 1979). The amount of these groups increases with the time of irradiation. The concentration of hydroperoxide groups is, however, rather low as compared to that of carbonyl and carboxyl groups. This may be due to the instability of hydroperoxides in UV light (Table 13–4).

Most of the colouring compounds generated in lignin during daylight irradiation come from further reactions between the intermediary phenoxy radicals and oxygen (Leary 1968; Kringstad, Lin 1970).

Important steps for the formation of chromophoric structures are the elimination of side chains and the formation of quinones as intermediate compounds. Model compounds in which β-arylether linkages were easily photolyzed are very rapidly discoloured by near ultraviolet light, whereas compounds in which the ether bonds are not cleaved under irradiation do not form chromophores (Gierer, Lin 1972). The initiation of the cleavage of the β-arylether linkage is also attributed to the excited α-carbonyl group (Fig. 13–17). The intermediate keto radical may probably abstract hydrogen or combine with hydroxyl radicals to yield an α-ketone or α-ketol, respectively. Additionally the formation of hydroperoxides and hydroperoxide ra-

Table 13–4: Formation of chromophoric groups in photooxidated cellulose, irradiated with UV light (λ >253.7 nm) (Hon 1979)

Irradiation time h	Carbonyl groups mmole/100 g	Carboxyl groups mmole/100 g	Hydroperoxide groups mmole/100 g
0	0.2	0.0	0.0
3	5.6	1.1	0.0
5	10.1	1.9	0.1
10	15.9	4.2	0.3
15	18.5	6.5	0.5
20	22.1	9.7	0.6

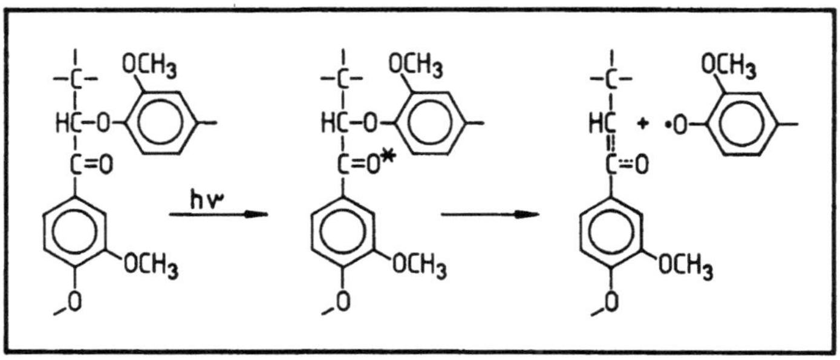

Fig. 13–17. Cleavage of the β-aryl ether linkage under the influence of the excited α-carbonyl group.

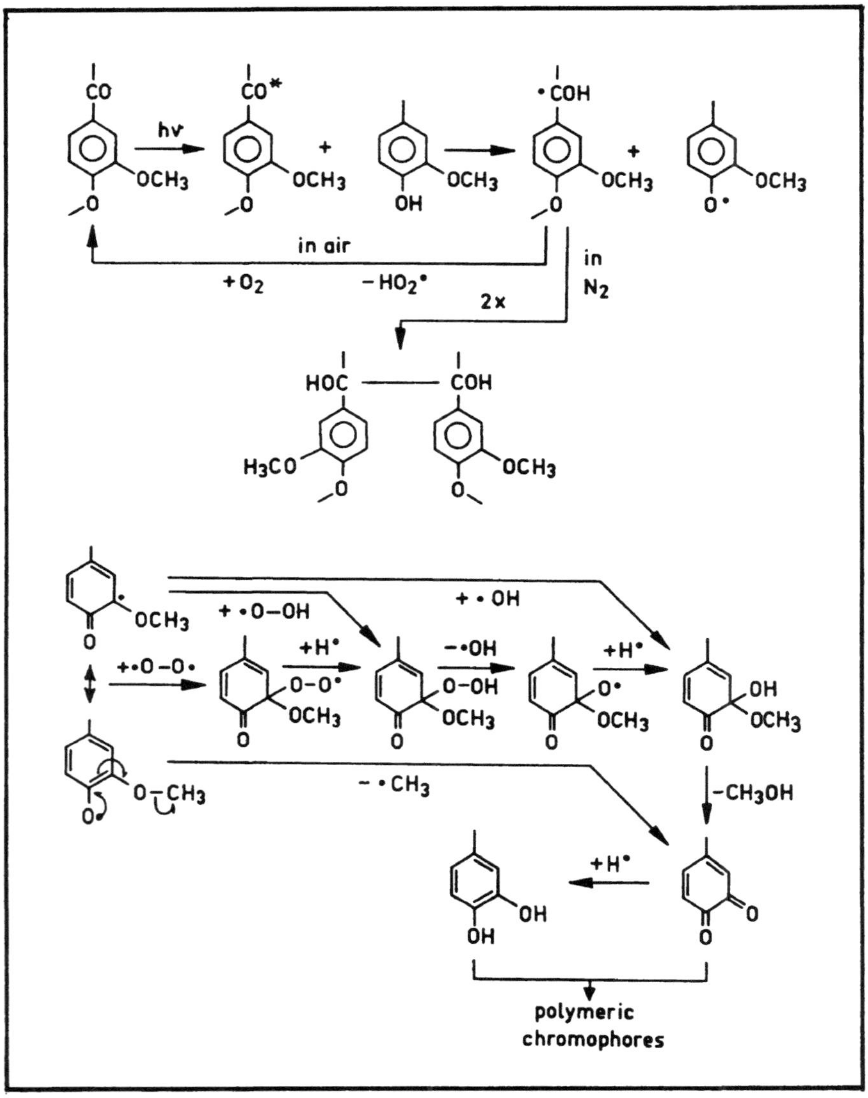

Fig. 13–18. Probable pathway of the formation of quinoid and dimer structures during UV irradiation of lignin, as proposed by Gierer and Lin (1972).

dicals by the interaction of oxygen and hydrogen is possible. Hydroperoxides and hydroperoxide radicals are probably important intermediates in the formation of quinoid compounds (Fig. 13–18).

Further reactions are the oxidative abstraction of the methoxyl group and the elimination of the propyl chain, both resulting in quinone structures (Gellerstedt, Pettersson 1977). The rate of the side chain elimination depends on the pH value with a minimum at about pH 6. The formation of 5,5′-dimers through radical coupling of 2,5-quinones has been found by Lin and Kringstad (1970b). Figs. 13–18 and 13–19

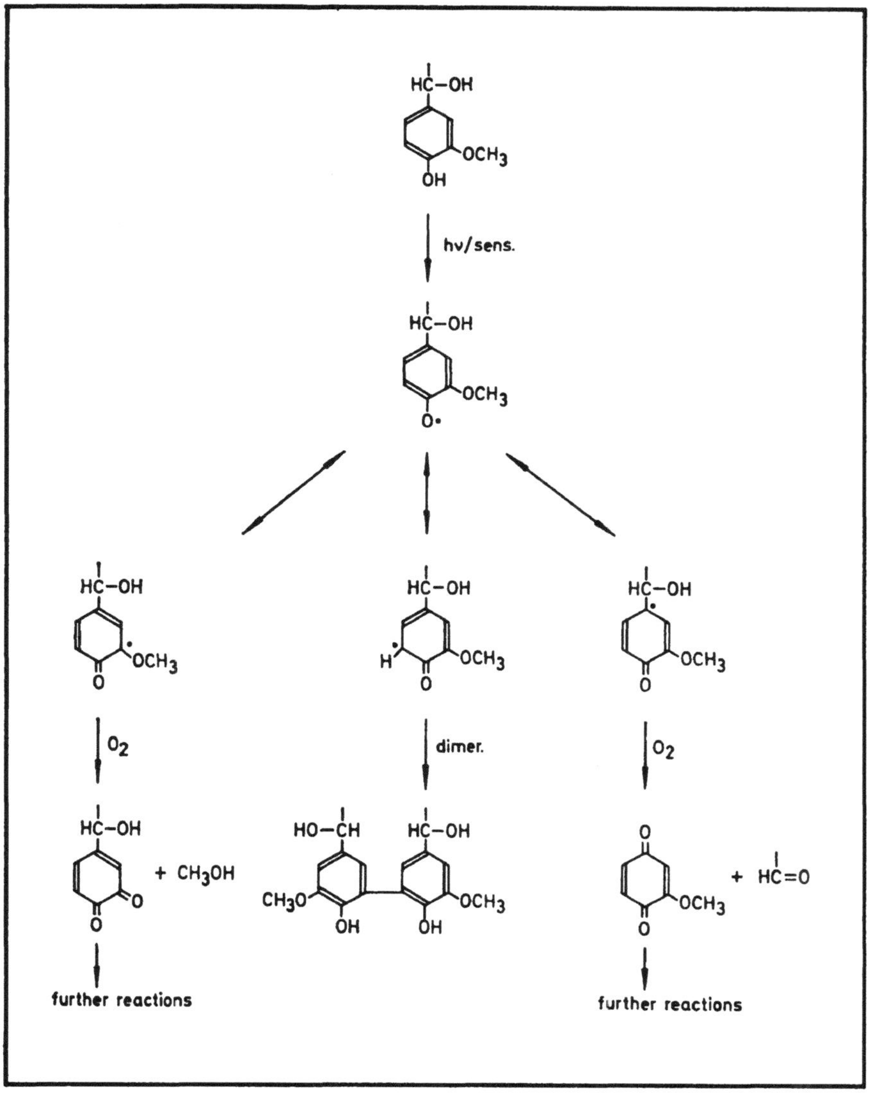

Fig. 13–19. Probable pathway of the formation of quinoid and dimer structures during UV irradiation of lignin, as proposed by Gellerstedt and Pettersson (1977).

show various pathways for the light-induced formation of quinoid structures in lignin units; Fig. 13–18 contains a more detailed mechanism of the probable formation of o-quinoid structures.

Various methods were proposed for stabilizing lignin against photolytic degradation with the aim of reducing the yellowing tendency of lignin-containing pulps (Lin, Kringstad 1970b; Gierer, Lin 1972; Gellerstedt, Pettersson 1977). All these methods aim to prevent the formation of phenoxy radicals. The chemical modification of the lignin molecule, such as the reduction or epoxidation of the α-carbonyl groups and hydrogenation of α, β-double bonds, is used to eliminate the photosensitive centres; by means of esterification the phenolic hydroxyls can be blocked. Another approach is the addition of antioxidants, singlet oxygen quenchers and ultraviolet absorbers (e.g. β-carotene, 2,4,6-trimethyl-phenol, 2,6-di-1',1'-dimethylethyl-4-methylphenol).

13.2. Changes Initiated by Ionizing Rays

13.2.1. Importance and General Effects

The application of ionizing rays, predominantly gamma-rays, has its main importance in the production of wood-plastic compounds. These high-energy rays may penetrate thick wood samples and initiate polymerization reactions within the samples. A radiolytic treatment of wood, cellulose and lignin has been taken into consideration for pulp processing and waste water cleaning. The gamma-irradiation changes the properties of wood in structural and chemical as well as in physical and mechanical behaviour. These variations depend mainly on dosage and wood species.

After irradiation with low doses a small increase in strength properties and balanced moisture content was observed (Weichert 1963; Burmester 1964, 1966). By increasing the radiation dosage strength and balanced moisture content are continuously reduced, while the capacity for water vapour sorption increases. Also a greater swelling and an increasing accessiblity for fungal decay was found (Kenaga, Cowling 1959; Becker, Burmester 1962).

13.2.2. Effects Visible in Structure

Low doses of gamma-irradiation up to about 1 MJ/kg (10^8 rad) do not effect a visible change in the wood structure as observed under the electron microscope. This result was found for various wood species: *Picea abies, Pinus sylvestris, Fagus sylvatica, Populus alba* (Polčin, Karhánek 1964; Burmester 1966; Antoine, Van Eyseren 1971). Variations, however, become apparent after disintegration by grinding or ultrasonic treatment. The fibrillar elements of the cell wall break perpendicularly to the fibrillar axes. Similar observations were made by Chiaverina et al. (1973) who irradiated cotton and pulp with gamma-rays of 1 MJ/kg. A change in the fibrillar structure appeared after mild acidic or alkaline treatments. Bacteria cellulose, too, shows a similar behaviour after beta- and gamma-irradiation (Purz, Schwarz 1976).

Antoine and Van Eyseren (1971) found an increase in the friability of wood and cellulose of spruce after a radiation dose starting at 4 MJ/kg. This explains breakages in any direction during the preparation.

After an exposure of oakwood *(Quercus alba)* to a gamma-irradiation up to 19 MJ/kg a change in the cell dimensions was observed (Tabirih et al. 1978). The tangential vessel diameter, the ray cell length and the width of intervessel pits increased, while the tangential vessel-wall thickness and the thickness of the ray-cell and latewood-fibre double walls decreased.

Goto et al. (1974) investigated spruce wood *(Picea abies)* which was irradiated with 6.55 MJ/kg. After a normal embedding procedure including immersion in water, the ultrathin sections showed a loosened structure which is explained by the re-

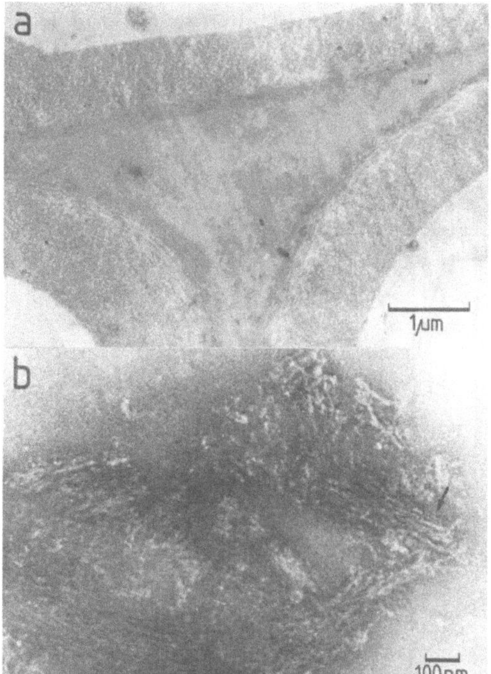

Fig. 13–20. Gamma-irradiated (6.55 MJ/kg) spruce wood *(Picea abies)*. TEM micrographs, by courtesy of H. Harada.

a) Ultrathin section showing a somewhat loosened structure.

b) Disintegrated sample showing fibrils broken down to short fragments.

moval of a high amount of water-soluble products (Fig. 13–20a). After a few minutes' treatment with ultrasonics the cellulose fibrils and the amorphous components of the gamma-irradiated wood were easily broken down to short fragments (Fig. 13–20b). The same ultrasonication was ineffective with non-irradiated wood.

13.2.3. Chemical Changes of Wood and Lignin

The chemical analysis of irradiated wood shows that all components, though in various intensities, are attacked by the high-energy rays. At low irradiation doses (0.1–100 J/kg) even a small increase of the cellulose and lignin content was determined which can be explained by a cross-linking of these components. The wood species *Picea abies, Pinus sylvestris* and *Fagus sylvatica* were investigated after gamma-irradiation with doses of 0.1 J/kg – 1.8 MJ/kg by Seifert (1964). The losses of weight were relatively low with values of 2.3–2.7% at the highest radiation doses, and are probably caused by decarboxylation and demethoxylation. With increasing radiation doses the cellulose content dropped rather constantly whereas on the average only 15% of the lignin was degraded (Fig. 13–21). On the other hand a high loss of methoxyl groups was observed. The pentosan content increased slightly, which may be attributed to an average value resulting from a loss by degradation and a formation of pentose units by oxidation of glucose-C6 and a subsequent decarboxylation.

Fig. 13–21. Loss of weight of cellulose and lignin in pine wood *(Pinus sylvestris)* as a function of the dose of irradiation (based on non-irradiated wood = 100%; according to Seifert 1964).

Higher gamma-doses (5–6.5 MJ/kg) reduce the content of alkali-resistant cellulose and increase the amount of extractives remarkably (Table 13–5) (Freijdin et al. 1959; Hachihama, Takamuku 1960; Antoine et al. 1971). The solubility of lignin in ethanol and dioxane was improved.

The influence of gamma-irradiation of wood on delignification was studied by Polčin and Kapustová (1963) and Crook et al. (1970). There was no influence on the calcium-bisulfite cooking of wood of *Picea abies, Fagus sylvatica* and *Populus alba* up to an irradiation dose of 10^4 J/kg. Doses of 10^5 J/kg, and particularly of $5 \cdot 10^5$ J/kg, resulted in a more intensive delignification with a reduction of the kappa-number, as compared to non-irradiated wood. At the same time the yield dropped, and the DP of cellulose was reduced. Also the kraft cooking of *Eucalyptus regnans* wood resulted in pulps of lower yield. Additionally these pulps had a dark colour which is explained by a higher condensation degree of the residual lignin. After an irradiation of wood chips with 10^3 J/kg Mori et al. (1979) obtained kraft and neutral sulfite pulps with a better defiberability.

A higher yield of lignin-polysaccharide complexes was obtained after an irradiation of wood of *Pinus densiflora* with 1.1 MJ/kg (Takamuku, Hachihama 1961). The percentage of the extracted complexes increased from 0.3% (untreated wood) up to 8.6%. Irradiation doses up to 0.5 MJ/kg do not cleave the linkages between lignin and polysaccharides whereby a protective effect is attributed to the lignin (Sergeeva et al. 1978). The irradiation of lignin-polysaccharide complexes with doses higher than 0.5 MJ/kg causes condensations.

In order to clarify the degradation mechanism of lignin various lignin model compounds were irradiated with 0.5 MJ/kg in the presence and absence of air by Meshitsuka and Nakano (1976) (Fig. 13–22). They found that dissolved oxygen is indispensible to the radiolytic degradation of lignin. Starting points of radiolytic degradation are the phenolic hydroxyl groups, the β-carbon adjacent to the α-carbo-

Table 13–5: Chemical analysis of untreated and γ-irradiated spruce wood *(Picea abies)* (Antoine et al. 1971)

Constituents	Untreated %	γ-irradiated 6.55 MJ/kg %
Ethanol-benzene extractives*	3.60	25.50
Water extractives	0.19	44.22
Cellulose	47.94	0.00
Lignin	29.95	27.34
Pentosans	6.94	2.31
Mannans	1.74	0.58
Ash	0.30	0.39

* Calculated on basis of oven-dry wood; all other values determined after ethanol-benzene extraction, and calculated on basis of unextracted oven-dry wood.

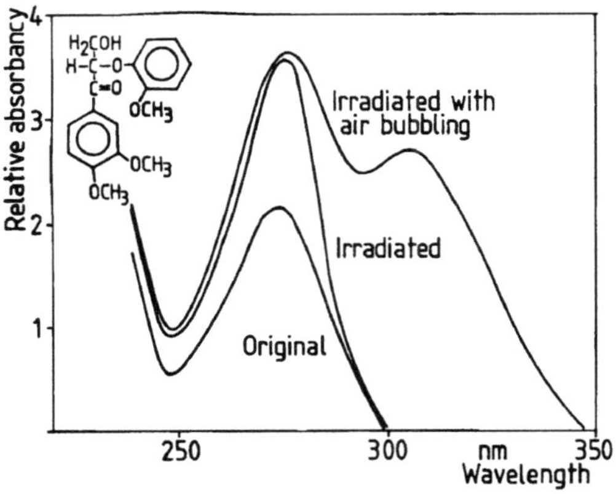

Fig. 13–22. Change of the UV absorption by gamma-irradiation of a β-aryl ether model compound (Meshitsuka, Nakano 1976).

nyl group, and structures containing conjugated double bonds. The degradation is initiated by the formation of radicals (phenoxy radical, β-carbon radical).

13.2.4. Chemical Changes of Cellulose

As described in the preceding chapter, cellulose and polyoses are more sensitive to gamma-irradiation than lignin. An irradiation with 6.55 MJ/kg degraded the cellulose in spruce wood completely without leaving an alkali-insoluble residue (Table 13–5). According to Polčin and Karhánek (1963) there is a critical irradiation dose in the range of 10^4–10^5 J/kg. Up to doses of 10^3 J/kg there are only slight increases of alkali solubility, copper number, and content of carboxyl groups, and a slight drop of the DP.

Berkolde et al. (1974) found a linear relation between the formation of carbonyl groups and the radiation dose of sulfite pulp. Up to 0.3 MJ/kg the formation rate of cold-water soluble products was higher than at higher doses.

The study of the distribution of the polymerization degrees shows that the maximum is shifted to lower DPs and the shorter fragments become more and more uniform in length (Fig. 13–23) (Polčin 1966; Simionescu et al. 1973; Kusama et al. 1976). The higher the DP of the original sample the more intensive is the cellulose degradation. The degree of crystallinity seems to have no influence on the kinetics of radiolytic cellulose depolymerization. Paszner (1968) degraded cotton linters and decrystallized cellulose with doses up to 0.5 MJ/kg, and obtained nearly identical degradation curves of DP.

Simionescu et al. (1973) found no influence on the crystallinity of cellulose I up to an irradiation dose of 0.5 MJ/kg. Doses of 1 MJ/kg slightly increased the crystalline portion. The irradiation of cellulose II caused a decrease of crystallinity. The crystallinity as evaluated by X-ray diffraction obviously indicates only an apparently undamaged structure of cellulose. A mild mechanical treatment breaks down the

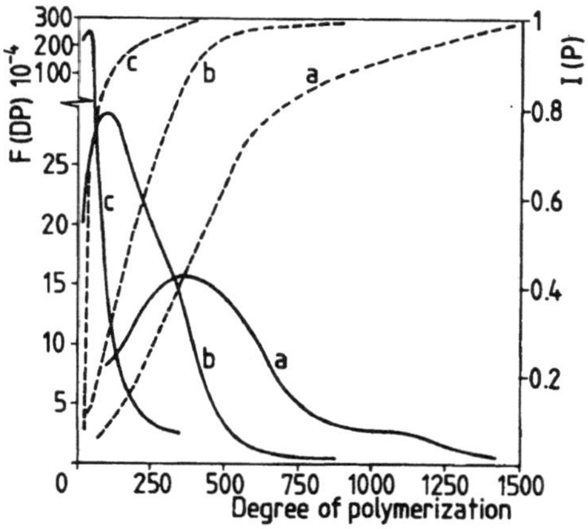

Fig. 13–23. Change of the DP distribution of cellulose (beech pulp) after γ-irradiation; original sample (a), irradiated with 4.6 · 10^4J/kg (b), irradiated with 4.6 · 10^5J/kg (c), (Polčin 1966).

crystalline structure, and the diffraction diagrams show amorphous cellulose. According to Goto et al. (1974) cellulose is attacked uniformly both in crystalline and amorphous regions if irradiated with high doses of gamma-rays. This attack produces many crystalline defects throughout the whole fibrils, but they retain their original shape. Even small external forces, such as a few minutes of ultrasonication, can reveal that the fibrils are only arrays of short fragments which can be easily disordered (Fig. 13–24).

The defects in the crystalline regions improve the accessibility for acids, and the hydrolysis of irradiated cellulose samples is facilitated (Millett, Goedken 1965; Klimentov et al. 1978, 1980). An increase of the doses increased the concentration

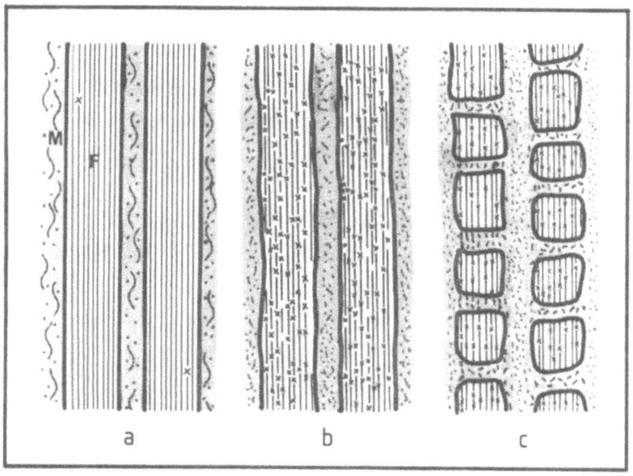

Fig. 13–24. Effect of gamma-irradiation on cellulose fibrils. F = cellulose fibril, M = matrix substances, X = crystalline defect (Goto et al. 1974).

a) Original sample;

b) after γ-irradiation;

c) after γ-irradiation and mild ultrasonic treatment.

of easily hydrolyzable and unsaturated substances. A complete decrystallization of the cellulose in oak wood was reached by a radiation dose of 19 MJ/kg (Cutter et al. 1980).

The degradation process in a cellulose sample also proceeds after irradiation has terminated. This "after-effect" could be traced up to 100 days (Glegg, Kertesz 1957; Christensen, Tolbert 1965). The after-effect degradation of cotton cellulose in air during 100 days after irradiation with $2.2 \cdot 10^4$ J/kg in vacuum corresponded to a decrease of DP caused by an additional irradiation for 7.5 hours. These studies indicate that about one-third of the chain breaks originate from carbonyl groups which occur during exposure to air. The remaining two-thirds of the breaks must in some way be caused by the radicals generated during irradiation.

ESR measurements of irradiated wood and cellulose show that the number of radicals decreases rapidly during 5–10 hours; the remainder are relatively stable, and their number decreases only slightly over more than 70 hours (Ramalingam et al. 1963). The types of radicals formed during irradiation are different for cellulose I and cellulose II, as can be seen in the ESR spectra (Park, Ward 1964; Simionescu et al. 1973). This finding is explained by the formation of radicals at the C1, C4 and C5-atom in cellulose-I molecules, and by the formation of radicals mainly by splitting of H or OH at the C6-atom in cellulose-II molecules, originating from the different hydrogen bonds in both modifications.

The radicals, as formed by irradiation of wood and cellulose impregnated with plastic monomers (styrene, diisocyanate etc.) or formaldehyde, provide the connection points for the grafting of plastics onto wood components or cellulose, respectively (Ramalingam et al. 1963; Burmester 1967).

Literature

Antoine, R. and Van Eyseren, J. C. 1971, C. R. Acad. Sci., Ser. B, *272,* 308–309

Antoine, R. C., Avella, T. and Van Eyseren, J. C. 1971, IAWA Bull. No. 4, 11–16

Becker, G. and Burmester, A. 1962, Materialprüfung *4,* 416–426

Beélik, A. and Hamilton, J. K. 1959, Papier *13,* 77–85

Beélik, A. and Hamilton, J. K. 1961, J. Organ. Chem. *26,* 5074–5080

Berkolde, V. A., Mujznieks, A. A., Rayavee, E. L. and Kalkis, V. J. 1974, Khim. Drev. No. 1, 101–103

Borgin, K. 1970, J. Microscopy *92,* Pt. I, 47–55

Borgin, K. 1971, J. Inst. Wood Sci. *5,* No. 4, 26–30

Bos, A. 1972, J. Appl. Polym. Sci. *16,* 2567–2576

Browne, F. L. and Simonson, H. C. 1957, For. Prod. J. *7,* 308–314

Brunow, G. and Sivonen, M. 1975, Pap. Puu *57,* 215–220

Burmester, A. 1964, Materialprüfung *6,* 95–99

Burmester, A. 1966, Materialprüfung *8,* 205–211

Burmester, A. 1967, Holz Roh-Werkst. *25,* 11–25

Chiaverina, J., Lazaro, R., Simionescu, C., Butnaru, R. and Rozmarin, G. 1973, Cell. Chem. Technol. *7,* 3–21

Christensen, P. K. and Tolbert, B. M. 1965, Norsk Skogind. *19,* 12–17
Claesson, S., Olson, E. and Wennerblom, A. 1968, Svensk Papperstid. *71,* 335–340
Cotton, F. A. and Wilkinson, G. 1974, Anorganische Chemie, 3rd Ed.. Verlag Chemie, Intersci. Publishers Weinheim, New York, p. 432
Crook, F. M., Nelson, P. F. and Thompson, R. G. 1970, Holzforschung *24,* 184–190
Cutter, B. E., McGinnes Jr., E. A. and Schmidt, P. W. 1980, Wood Fiber *11,* 228–232
Desai, R. L. 1968, Pulp Pap. Mag. Can. *69,* T322–T330
Desai, R. L. 1970, Pulp Pap. Mag. Can. *71,* 51–52
Desai, R. L. and Shields, J. A. 1969, Makromol. Chem. *122,* 134–144
Donetzhuber, A. 1961, Svensk Papperstid. *64,* 898–901
Frejdin, A. S., Malinskij, J. M. and Karpov, V. L. 1959, Gidroliz. Lesokhim. Promst. *12,* No. 4, 4–7
Futó, L. 1974, Holz Roh-Werkst. *32,* 303–311
Futó, L. P. 1976, Holz Roh-Werkst. *34,* 31–36, 49–54
Gellerstedt, G. and Pettersson, E. L. 1977, Svensk Papperstid. *80,* 15–21
Gierer, J. and Lin, S. Y. 1972, Svensk Papperstid. *75,* 233–239
Gingras, B. A., Cooney, D., Jackson, K. A. and Bayley, C. H. 1963, Text. Res. J. *33,* 1000–1004
Gizetdinov, F. M. and Chupka, E. I. 1979, Khim. Drev. No. 3, 62–65
Glegg, R. E. and Kertesz, Z. I. 1957, J. Polym. Sci. *26,* 289–297
Goto, T., Harada, H. and Saiki, H. 1974, Bull. Kyoto Univ. For. No. 46, 153–161
Hachihama, Y. and Takamuku, Y. 1960, Kogyo Kagaku Zasshi *63,* 1043–1046
Hernádi, S. 1975, Cell. Chem. Technol. *9,* 31–39
Hon, N. S. 1975a, J. Polym. Sci., Polym. Chem. Ed. *13,* 955–959, 1347–1361, 2363–2374
Hon, N. S. 1975b, J. Polym. Sci., Polym. Chem. Ed. *13,* 1933–1941
Hon, N. S. 1975c, J. Polym. Sci., Polym. Chem. Ed. *13,* 2641–2652
Hon, N. S. 1975d, J. Polym. Sci., Polym. Chem. Ed. *13,* 2653–2669
Hon, N. S. 1976a, J. Polym. Sci., Polym. Chem. Ed. *14,* 2497–2512
Hon, N. S. 1976b, J. Polym. Sci., Polym. Chem. Ed. *14,* 2513–2525
Hon, D. N. S. 1979, J. Polym. Sci., Polym. Chem. Ed. *17,* 441–454
Hon, D. N. S. and Feist, W. C. 1981, Wood Sci. *14,* 41–48
Hon, D. N. S. and Glasser, W. 1979, Polym. Plast. Technol. Eng. *12,* 159–179
Hon, D. N. S. and Ifju, G. 1978, Wood Sci. *11,* 118–127
Hon, D. N. S., Ifju, G. and Feist, W. C. 1980, Wood Fiber *12,* 121–130
Kalnins, M. A., Steelink, C. and Tarkow, H. 1966, US For. Serv. Res. Paper FPL No. 58
Kenaga, D. L. and Cowling, E. B. 1959, For. Prod. J. *9,* 112–116
Kleinert, T. N. 1964a, Holzforschung *18,* 24–28
Kleinert, T. N. 1964b, Mh. Chemie *95,* 387–389
Kleinert, T. N. 1969, Holzforsch. Holzverwert. *21,* 133–134
Kleinert, T. N. 1970a, Holzforsch. Holzverwert. *22,* 21–24
Kleinert, T. N. 1970b, Papier *24,* 207–208
Kleinert, T. N. 1970c, Papier *24,* 563–566
Kleinert, T. N. and Marraccini, L. M. 1966, Svensk Papperstid. *69,* 69–71, 159–160
Kleinert, T. N. and Morton, J. R. 1962, Nature *196,* 334–336
Klimentov, A. S., Ershov, B. G., Kraev, L. N. and Vysotskaya, I. F. 1978, Khim. Drev. No. 1, 68–71
Klimentov, A. S., Shakhanova, R. K., Vysotskaya, I. F. and Kraev, L. N. 1980, Khim. Drev. No. 3, 47–48
Kringstad, K. 1969, Tappi *52,* 1070–1074
Kringstad, K. P. and Lin, S. Y. 1970, Tappi *53,* 2296–2301
Kujirai, C. 1965, Sen-i Gakkaishi *21,* 183–196, 260–266
Kujirai, C. 1966, Sen-i Gakkaishi *22,* 20–23

Kusama, Y., Kageyama, E., Shimida, M. and Nakamura, Y. 1976, J. Appl. Polym. Sci. *20,* 1679–1688
Landucci, L. L. 1978, Trans. Techn. Sect. *4,* 25–29
Leary, G. J. 1967, Tappi *50,* 17–19
Leary, G. J. 1968, Tappi *51,* 257–260
Le Nest, J. F. and Silvy, J. 1970, C. R. Acad. Sci., Ser. C *271,* 102–105
Lin, S. Y. and Kringstad, K. P. 1970a, Tappi *53,* 658–663
Lin, S. N. and Kringstad, K. P. 1970b, Tappi *53,* 1675–1677
MacClaren, R. H., Wells, F. L., Rosequist, J. V. and Ingerick, D. F. 1962, Tappi *45,* 789–793
Matsuura, T., Yoshimura, N., Nishinaga, A. and Saito, I. 1972, Tetrahedron *28,* 4933–4938
Meshitsuka, G. and Nakano, J. 1976, Tappi *59,* No. 11, 123–125
Millett, M. A. and Goedken, V. L. 1965, Tappi *48,* 367–371
Miniutti, V. P. 1973, J. Paint Technol. *45,* 27–34
Mori, K., Sasaki, T. and Hanamura, N. 1979, Kami Pa Gikyoshi *33,* 480–482
Norrström, H. 1969, Svensk Papperstid. *72,* 25–38
Park, G. S. and Ward, J. C. 1964, Nature *202,* 389
Paszner, L. 1968, Svensk Papperstid. *71,* 822–828
Polčin, J. 1966, Faserforsch. Textiltechn. *17,* 128–132
Polčin, J. and Kapustová, J. 1963, Sb. Vys. Pr. Odboru Celul. Pap. *8,* 9–33
Polčin, J. and Karhánek, M. 1963, Faserforsch. Textiltechn. *14,* 357–363
Polčin, J. and Karhánek, M. 1964, Holzforschung *18,* 102–108
Purz, H. J. and Schwarz, H. H. 1976, Faserforsch. Textiltechn. *27,* 261–270
Ramalingam, K. V., Werezak, G. N. and Hodgins, J. W. 1963, Polym. Symp. No. 2, 153–167
Rånby, B., Kringstad, K., Cowling, E. B. and Lin, S. Y. 1969, Acta Chem. Scand. *23,* 3257–3259
Sandermann, W. and Schlumbom, F. 1962, Holz Roh-Werkst. *20,* 245–252, 285–291
Schurz, J., Kaempgen, D., Schlor, M. and Windisch, K. 1963, Papier *17,* 556–568
Scott, J. A. N. and Goring, D. A. I. 1970, Wood Sci. Technol. *4,* 237–239
Seifert, K. 1964, Holz Roh-Werkst. *22,* 267–275
Sergeeva, V. N., Kreicberga, Z., Ekabsome, M., Rajavee, E. and Muiznieks, A. 1978, Khim. Drev. No. 5, 58–67
Simionescu, C., Butuara, R. and Rozmarin, G. 1973, Cell. Chem. Technol. *7,* 153–169
Steelink, C. 1964, Geochim. Cosmochim. Acta *28,* 1615–1622
Steelink, C., Reid, T. and Tollin, G. 1963, J. Am. Chem. Soc. *85,* 4048–4049
Tabirih, P. K., McGinnes, E. A., Kay, M. A. and Harlow, C. A. 1978, Wood Fiber *9,* 211–215
Takamuku, S. and Hachihama, Y. 1961, Kogyo Kagaku Zasshi *64,* 1662–1665
Weichert, L. 1963, Holz Roh-Werkst. *21,* 290–300
Wengert, E. M. 1966, J. Paint. Techn. *38,* 71–76
Wilcox, M. D. 1975, Svensk Papperstid. *78,* 22–26
Williams, P. 1968, Arkiv Kemi *29,* 485–502
Zapolskii, O. B. 1964, Dokl. Akad. Nauk BSSR *8,* 234–236

14. Microbial and Enzymatic Degradation

14.1. Importance of Enzymatic Reactions

Enzymes play an important role in the lives of all organisms: in their development, their growth and their disappearance. In trees, too, the biosynthesis of wood components and the building-up of the cell walls are controlled by enzymes. Enzymes regulate the metabolism in the parenchyma cells. The natural degradation of wood also occurs by the action of enzymes supplied by wood-destroying organisms. They degrade the insoluble wood components to soluble products, and finally to simple chemical compounds which are introduced to the metabolism of the organisms.

A knowledge of the mechanisms of enzymatic degradation of wood and its components may be helpful for taking precautionary measures to prevent the attack of wood-destroying organisms. This knowledge is also important for the utilization of enzymatic reactions by which wood components can be converted into chemicals and foodstuff. Additionally enzymatic reactions are used for the isolation of wood components, and for clarifying linkages, e.g. in lignin-polysaccharide complexes.

In recent years a great many studies have appeared concerning the various aspects of enzymatic reactions of wood and its components; the results of these studies alone would fill more than one book. Summary descriptions of biological, physical and chemical aspects are given Cartwright and Findley (1958), Rypáček (1966), Côté (1968), Seifert (1968), Liese (1970), Bavendamm (1974) and Crawford (1981). Several symposia demonstrate the world-wide activity in this field (Becker, Liese 1965, 1976; Liese 1975; Kirk et al. 1980).

14.2. Wood-Destroying Organisms

Wood can be attacked by animal or vegetable parasites. The animal parasites include: insects, such as boring beatles (*Anobium, Lyctus, Hylotrupes* etc.), wasps, ants and termites; crustaceans such as gribble *(Limnoria);* and molusks such as shipworms (*Teredo, Banksia* etc.) (Schmidt 1962; Hickin 1963; Bletchly 1967; Côté 1968). All these animals or their larvae, respectively, destroy wood primarily by a mechanical process of boring and feeding.

Apart from hemiparasitic mistletoe *(Viscum album)* the vegetable pests belong to the groups of fungi and bacteria (Wagenführ, Steiger 1966; Côté 1968; Seifert 1968; Liese 1970).

Under favourable conditions fungi expand rapidly within a wood sample by hyphae growth. The simplest path for the expansion of the hyphae are the lumina of parenchyma and vascular cells (Radtke et al. 1981). The transition from one cell to another occurs by penetration of the pits or by passage of the cell walls. The hyphae of several fungi are also able to grow within the compound middle lamellae or within the secondary walls (Fig. 14–1). The hyphae excrete enzymes which decompose the components of the wood cell wall.

The fungal pests are subdivided into the following groups:

- Brown-rot: the fungi, belonging to the subdivision of Basidiomycetes, degrade mainly the polysaccharides of the wood. But there is also a certain change and degradation of lignin. The wood becomes brown and brittle. Most brown-rot fungi attack softwood. The mechanical strength is reduced after a short period of incubation. The degradation is accompanied by abnormal longitudinal shrinkage and a deformation of the cell walls.
- White-rot: these fungi also belong to the Basidiomycetes and degrade lignin as well as polysaccharides. The degraded wood is white and soft. Most of the white-rot fungi prefer hardwoods. An attack by white-rot effects a decrease of strength properties and an increase of swelling.
- Soft-rot: a group of fungi belonging to the Ascomycetes and Fungi imperfecti, which are able to degrade polysaccharides and lignin. The rate of degradation for the individual wood components differs among the various soft-rot fungi. Soft-rot is found in softwoods and hardwoods, and results in various rates of reduction of the strength properties.

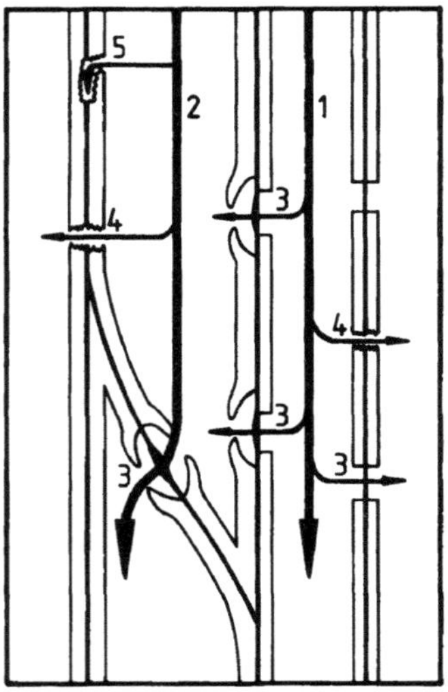

Fig. 14–1. Scheme of the possible paths of hyphae growth: (1) within parenchyma cell lumina, (2) within vascular cell lumina, (3) through simple, half-bordered and bordered pits, (4) through the cell walls, and (5) within the compound middle lamellae and the cell walls.

- Blue-stain: fungi which live mainly from residual proteins in the parenchyma cells predominantly of softwoods. They are Acomycetes and Fungi imperfecti which are able to degrade polysaccharides in a restricted manner. Their main damage to wood is a blue or black discolouration which is caused by dark deposits within the hyphae.

In addition to these groups of fungi numerous other fungi (e.g. mould fungi) are able to live in wood. But they do not, or only slightly, discolour the wood or cause negligible changes of weight and strength, though some of them show comparatively high cellulolytic and xylolytic activities (Fukuda et al. 1981). Their importance, which is known in considerable detail, is due to their influence on the growth behaviour of the wood-destroying fungi (antagonism, stimulation).

The degradation of wood by bacteria is limited as they propagate by cell division but cannot move within wood unless the wood is stored under water. Bacteria tend to form colonies in parenchyma cells using proteins as their source, and in pit chambers where they dissolve the pit membranes. They may also attack cell walls as they are able to degrade polysaccharides and lignin though on a restricted scale.

The mechanical properties of softwoods and hardwoods decrease rapidly in the first stage of decay by wood-destroying fungi. At weight losses of 5–10% the toughness and impact bending are reduced by about 60–80% (Wilcox 1978). Losses of 50–70% can be expected in the case of bending. Losses in the same range are probable for the modulus of rupture and the modulus of elasticity.

14.3. Characterization of Enzymes

The group of enzymes comprises a large number of various biocatalysts which accelerate and control biochemical reactions. Many of them show a high specificity, i.e. their action is restricted to certain molecules or molecular structures. This property is never attained by artificial catalysts.

Enzymes are often called according to the molecule on which they act or according to their action, whereby the suffix -ase is used, e.g. the cellulose-splitting enzyme is called cellulase and enzymes effecting a hydrolytic cleavage are called hydrolases. Apart from systematic names many enzymes bear trivial names unrelated to their function (e.g. catalase, pepsin). An international system of classification divides enzymes according to their function:

- Oxidoreductases, effect redox-reactions and act on alcoholic hydroxyl groups, keto groups, double bonds, C–N linkages etc.
- Transferases, transfer functional groups such as C_1-groups, aldehyde, keto, acyl and glycosyl groups etc.

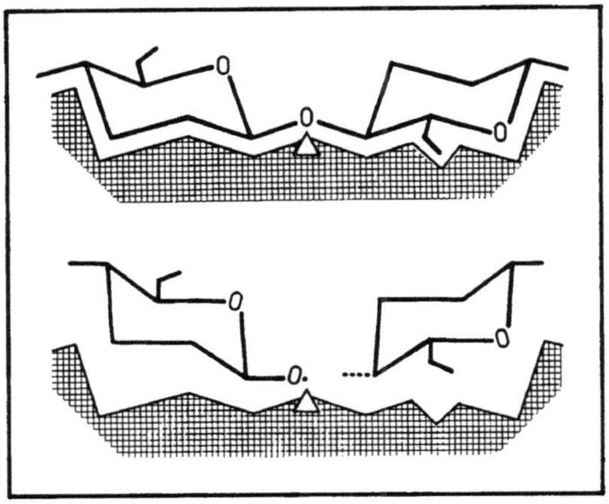

Fig. 14–2. Splitting of a β-1,4-glycosidic linkage at the active centre of an enzyme. Triangle = cofactor, checked area = apo-enzyme.

- Hydrolases, cleave esters, glycosides, peptides etc. hydrolytically.
- Lyases, effect additions at C = C, C = O and C = N double bonds.
- Isomerases, catalyse isomeration reactions.
- Ligases, effect the formation of new bonds during ATP-cleavage.

Enzymes are produced within the cells but they may also act outside the cell, and they can be separated from the cells without losing their reactivity. Thus enzymatic reactions can be carried out in a cell-free medium. The biocatalysts are macromolecular proteins with a defined supramolecular structure containing an active centre which is generally situated in a hole (Fig. 14–2). In many cases the active centre consists of a complex organic molecule or a metal-ion (cofactor), and may or may not be linked to the protein by an atom-bond. The complex of protein (apo-enzyme) and cofactor is called holo-enzyme:

apo-enzyme + cofactor → holo-enzyme

Depending on their place of reaction the enzymes are subdivided into endo-enzymes, which act within the cells and control processes of the internal metabolism, and exo-enzymes, which are excreted by the cell for the degradation of an insoluble substrate to soluble products which are able to diffuse through the cell membrane.

For more detailed information on the characteristics and reactions the reader is referred to books on biochemistry (e.g. Mahler, Cordes 1971; Kindl, Wöber 1975; Stryer 1975; Karlson 1977; Lehninger 1979).

14.4. Decay by Brown-rot Fungi and Action of Glycanhydrolases

Decay caused by brown-rot fungi results in a rather selective degradation of the polysaccharides, but the lignin is also subjected to a certain degradation process. In heavily brown-rotted wood a lignin skeleton remains which allows the study of lignin distribution in the cell walls (Cowling 1965; Côté et al. 1966; Ozolinja et al. 1974). The penetration of wood by the hyphae occurs through the rays. From there they spread over the pits and by penetrating the cell walls with bore- or microhyphae. The hyphae growing in the cell lumina are in close contact with the tertiary wall, but lysis zones as seen with white-rot fungi are not visible (Fig. 14–3). Only fissures, loosening of fibrils and a lamellation at the inner surface indicate a local degradation (Nečesaný 1963; Liese 1965, 1970).

Though there are various symptoms which indicate a degradation of the cell walls beginning at the lumenside, probably by decay pockets (Courtois 1965; Liese 1970), Chou and Levi (1971) found a lysis of carbohydrates within the secondary walls starting from microhyphae growing through the walls. According to these results

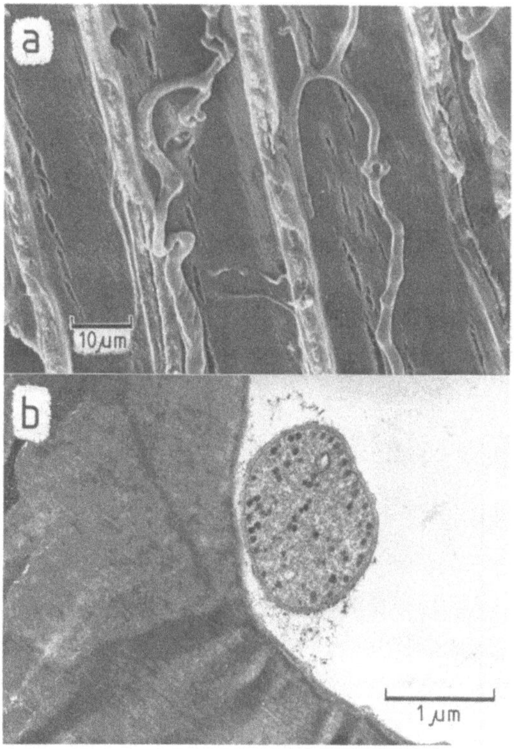

Fig. 14–3. Decay of spruce wood *(Picea abies)* by *Fomes marginatus.*
a) Hyphae in the lumina of tracheids. The cell walls show longitudinal fissures. SEM micrograph.
b) Cross view of a hypha which is in contact with a tracheid wall. Ultrathin section, TEM micrograph.

Table 14–1: Loss of the major components in conifer woods after decay by various brown-rot fungi (Kirk, Highley 1973)

Wood	Fungus	Loss in weight %	Loss in lignin %	Loss in glucan %	Loss in mannan %	Loss in xylan %
Tsuga heterophylla	*Poria placenta*	6	8	4	11	16
		26	4	26	38	39
		46	− 2	79	88	75
Picea engelmannii	*Poria placenta*	12	6	13	26	22
		26	10	29	41	69
		49	−13	81	93	79
Picea sitchensis	*Poria placenta*	6	3	9	11	11
		22	7	26	40	34
		47	12	62	81	62
		64	8	93	99	97
Picea sitchensis	*Gloeophyllum trabea*	10	9	12	14	18
		19	4	22	47	37
		43	11	55	80	65
Picea sitchensis	*Lentinus lepideus*	6	3	12	6	19
		27	− 4	37	68	49
		45	6	57	78	64
Pinus taeda	*Poria placenta*	9	− 4	13	25	1
		24	2	29	58	26
		45	4	68	81	69

the decay originates in the S1 and the inner S2 layer by diffusion of enzymes from the hyphae in the bore holes. The tertiary wall is more resistant to the attack of brown-rot fungi, whereas the G-layer of tension-wood fibres can easily be degraded (Zenker 1962). Numerous holes in the inner surface of decayed cell walls which do not contain microhyphae were attributed to collapse effects (Chou, Levi 1971). But according to new observations these holes probably result from the action of enzymes excreted by the hyphae (von Aufsess, Fengel 1982).

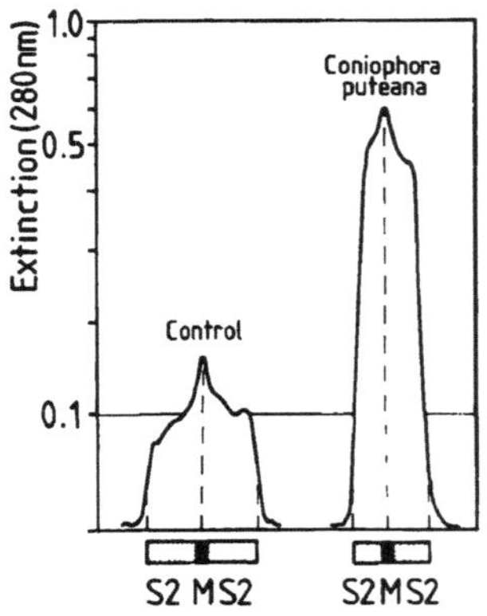

Fig. 14–4. UV absorption of cross sections of pine sapwood *(Pinus sylvestris)*, sound (left side) and decayed by *Coniophora puteana* (average weight loss 30%) (by courtesy of J. Bauch).

Table 14–2: Characteristics of cellulases isolated from various fungi (Keilich et al. 1970)

Source of enzyme	pH		Temperature	
	optimum	stability	optimum	stability
Coniophora puteana (Brown-rot)	3.5–4.0		50 °C	progressively inactivated from 40–70 °C
Phaeolus schweinitzii (Brown-rot)	4.0	2.5–7.5	60 °C	inactivated above 60 °C
Stereum sanguinolentum (White-rot)	3.7–4.2	2.0–7.5	45 °C	inactivated above 50 °C
Chaetomium globosum (Soft-rot)	4.5–5.5	3–8	35 °C	inactivated above 40 °C
Trichoderma viride (Soft-rot)	4.0	3–7	50 °C	activity decreases above 50 °C
Myrothecium verrucaria (Mould)	5–6	3–8	50–55 °C	inactivated above 60 °C

UV microscopic and UV spectroscopic studies of decayed wood show a high relative increase of the lignin content in the secondary wall layers as well as in the compound middle lamellae (Bauch et al. 1976) (Fig. 14–4).

Chemical analyses of wood after various amounts of decay by brown-rot fungi result in an increasing loss of polysaccharides (Table 14–1). During the fungal degradation of softwood, mannan is removed more quickly than cellulose. The variation in xylan removal depends more on the wood than on the fungus. Cellulose and possibly the polyoses undergo an extensive depolymerization before much weight loss has occurred (Kirk, Highley 1973). There is also a slight loss in lignin, but the analytical values are irregular. A graphic representation of the progressive loss of the wood components is shown in Fig. 14–5.

The degradation of the polysaccharides is caused by the action of enzymes which are specialized for the different carbohydrates present in the wood cell wall. En-

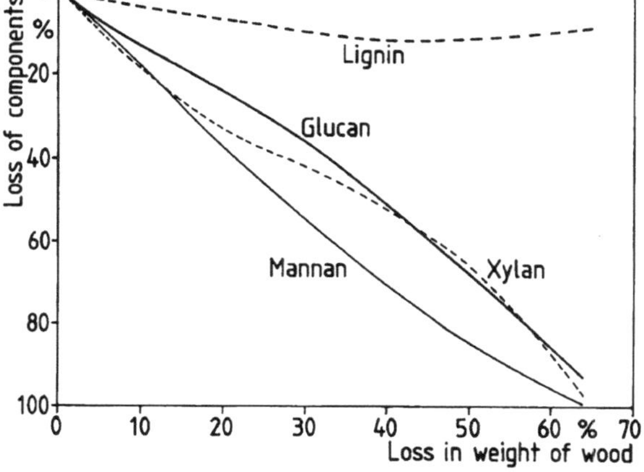

Fig. 14–5. Loss of wood components in *Picea sitchensis* during decay by *Poria monticola* (according to Kirk and Highley 1973).

zymes of the same type may have varying properties depending on the fungal source (Table 14–2) (Keilich et al. 1970; Highley 1975).

Various β-glycanhydrolases are summarized in Table 14–3 including components of the cellulase enzyme complex. The molecular weight of most of the enzymes ranges from 20 000–60 000. Higher values are found with some β-glucosidases but there is no uniform tendency: the molecular weight of β-glucosidases in the soft-rotter *Trichoderma koningii* was found to be 39 800 (Wood 1981), that of a β-glucosidase in *Sporotrichum thermophile* (soft-rot) was 400 000 (Meyer 1980). Very low (13 000) is the molecular weight of an endoglucanase in *Trichoderma koningii.*

Isolated glycanhydrolases can be fractionated into several components with different properties, e.g. molecular weight, activity, stability and isoelectric point (Bailey et al. 1969; Keilich et al. 1969; Sinner, Dietrichs 1975). Studies of the cellulase complex isolated from *Trichoderma* fungi showed the presence of at least two cellobiohydrolases, two β-glucosidases and six endoglucanases (Pettersson et al. 1981; Wood 1981). The hydrolytic effect of individual enzymes is low whereas the combination of exoenzymes (cellobiohydrolases) and endoenzymes increases the

Table 14–3: Molecular weight of β-glycanhydrolases isolated from various fungi (Keilich et al. 1970)

Source of enzyme	Hydrolases	Molecular weight (dalton)	Method of determination
Stereum sanguinolentum	Cellulase	21 200	Ultracentrifuge
Chaetomium globosum	Cellulase (endo-)	30 000 ± 3 000	Bio-Gel P-60*
Trichoderma viridi	Cellulase		Ultracentrifuge
	(endo-)	42 000	
	(exo-)	61 000	
	(C_1-)	60 000	
Phaeolus schweinitzii	Cellulase	45 000 ± 4 500	Sephadex G-100*
	(endo-)	47 000 ± 4 700	Bio-Gel P-60*
		42 000 ± 4 200	Bio-Gel P-100*
	Mannanase	37 000 ± 3 700	Bio-Gel P-60*
		35 000 ± 3 500	Bio-Gel P-100*
	Xylanase	~ 35 000	Bio-Gel P-100*
	β-Glucosidase	~ 120 000	Sephadex G-150*
Aspergillus wentii	Cellulase	40 000 ± 4 000	Bio-Gel P-100*
		37 000 ± 3 700	Bio-Gel P-60*
	Mannanase	42 000 ± 4 200	Bio-Gel P-100*
		47 000 ± 4 700	Bio-Gel P-60*
	Xylanase	25 000 ± 2 500	Bio-Gel P-100*
		28 000 ± 2 800	Bio-Gel P-60*
	β-Glucosidase		
	(A_3)	171 000	Sephadex G-200*
	(B_2)	145 000	

* Gel permeation chromatography

production of glucose. Thus various enzymes are seen to cooperate in the degradation of cellulose.

Reese (1975, 1977) developed a scheme in which he differentiates between C_1 and C_x enzymes:

$$\text{Crystalline cellulose} \xrightarrow{C_1} \text{swollen cellulose, alkali-soluble cellulose} \quad \overset{C_x}{\begin{cases} \xrightarrow{\text{endoglucanases}} \text{oligomers} \\ \xrightarrow{\text{cellobiohydrolase}} \text{cellobiose} \\ \xrightarrow{\text{glucohydrolase}} \text{glucose} \end{cases}}$$

The C_1 enzymes are thought to destroy the surface of the crystalline structure of cellulose by some swelling und cleavage of some covalent linkages. According to this scheme the C_1 enzymes prepare the substrate for the attack of the C_x enzymes. The studies of Wood and McCrae (1979) resulted in a more complex action of the C_1 and C_x components. They assume a formation of enzyme-enzyme complexes at the surface of the cellulose. These complexes, consisting of endo- and exoenzymes, solubilize the cellulose by a rapid sequential action.

The endoglucanases act more randomly on the cellulose chains, producing new reaction places for the endwise-acting cellobiohydrolases. This synergism seems to be a very important mechanism particularly for the highly ordered regions of cellulose, as there is no possibility for the large macromolecules of enzymes to penetrate the cellulose structure. In this connection we should note Wood's (1981) important observation that the endoglucanase with the lowest molecular weight (13 000 dalton) is the one which acts most randomly. He also explains the need for at least two enzymes of the same type by arguing that during the topochemical attack the enzymes meet two types of glycosidic linkages which differ with regard to steric conditions (Fig. 14–6).

The X-ray diffraction study of cellulosic material treated with cellulase shows an increase of crystallinity and a reduction of the crystallite width of cellulose, indicating an attack beginning at the paracrystalline surface (Betrabet et al. 1974; Betrabet, Paralikar 1978). Additionally a procession of the cellulase action along the 101 and 002 plane of the crystalline areas was observed.

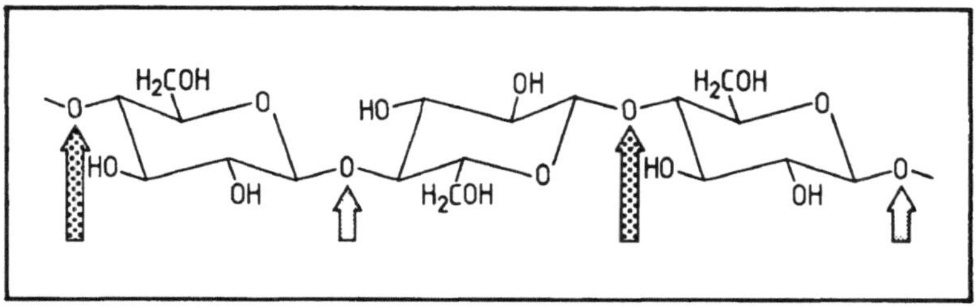

Fig. 14–6. Two types of glycosidic linkages due to the steric conditions which require two enzymes for hydrolytic cleavage.

There is evidence that the enzymatic degradation of cellulose is influenced by additional factors such as the presence of monosaccharides, of polyoses, or of an H_2O_2-Fe system which may act as a starter-substance or aid in decomposing the crystalline system (Koenigs 1974a; Highley 1976, 1977b, 1980). The production of small amounts of H_2O_2 and a partial oxidative degradation of cellulose by brown-rot fungi were reported by Koenigs (1972, 1974b). Most brown-rot fungi show no C_1 activity if they are growing on a pure cellulose substrate, whereas with white-rot fungi this activity is present. Nilsson (1974c) concludes that either the production of C_1-cellulase is induced by the presence of lignin or the brown-rot cellulase is most active against lignified cellulose.

Polyoses-splitting enzymes also consist of various components. Sinner and Dietrichs (1975) isolated three individual xylanases from a commercial product. Various β-xylanases, β-mannanases and β-mannosidases were separated from the brown-rot fungus *Tyromyces palustris* and those enzymes with the highest activity were purified (Ishihara et al. 1978; Ishihara, Shimizu 1980). Yamazaki et al. (1976) succeeded in purifying an individual endomannanase from a crude product of the mould fungus *Aspergillus niger,* which contained α-galactosidase, β-mannosidase and other enzymes.

The mode of action of a highly purified endo-1,4-β-mannanase on a galactomannan isolated from guar-seeds *(Cyamopsis tetragonoloba)* was studied by Yamazaki and Dietrichs (1979). They observed a tendency to attack the non-branched areas with a random splitting of the molecular chains. The mannanase could not degrade areas with alternating side-chain units. A similar action was found with xylanases and isolated xylans and xylans within delignified cell walls (Sinner, Dietrichs 1976 a, b). Apart from xylose and xylobiose, 4-O-methylglucuronoxylobiose and 4-O-methylglucuronoxylotriose were identified as degradation products.

The composition and activity of the polyoses-degrading enzymes seem to determine the preference of the fungi for hardwoods or softwoods (Lewis 1976; Rypácek 1977; Highley 1977 a). Species which are specialized in nature in the wood of conifers grow better in solutions containing galactose and mannose than in solutions containing cellulose, and mannose is utilized more advantageously than xylose. The xylanase complex is obviously more highly developed in species preferring angiosperms. There is also evidence that certain cell wall components may induce the production of the necessary enzymes. Thus cellulose induced the formation of carbohydrolases in the white-rotting fungus *Coriolus versicolor (= Polyporus versicolor),* whereas mannan was the best carbohydrolases inducer in the brown-rot *Poria placenta (= Poria monticola)* (Highley 1977a).

An apparently unique enzyme complex capable of degrading various polysaccharides and glycosides was isolated from *Poria placenta* by Wolter et al. (1980; Highley et al. 1981). This complex with a molecular weight of 185 000 is apparently an aggregate of various polypeptides which under the right conditions break down into small sub-units. The enzyme was found to be active for xylan, glucomannan, carboxymethylcellulose and some glycosides.

A synergism also with lignin-degrading enzymes is obviously necessary for the degradation of lignified cell walls. These cell walls are accessible for cellulolytical and polyolytical enzymes, but not before the wood has been physically or chemically treated, particularly with lignin-removing effects (Garves, Dietrichs 1975; Sinner et al. 1978; Katkevich et al. 1978).

The reduction of the lignin content of sugi *(Cryptomeria japonica)* by decay of a brown-rotter *(Pleurotus ostreatus)* from 34.5% in sound wood to 27.1% at a weight loss of 47.9% increased the saccharification of the cell wall polysaccharides by cellulase from 12.1% to 63.4% (Hiroi 1981).

After partial degradation by brown-rotters the remaining lignin becomes soluble in dioxane. At weight losses of about 40% only lignin from the S2 wall layer is extractable, whereas at higher weight losses lignin of the compound middle lamella can also be removed by dioxane extraction (Krejcberg et al. 1974 a). The lignin however has undergone chemical changes.

Brown-rotted lignin is characterized by a loss of methoxyl groups and increased carbonyl and carboxyl groups (Tables 14–4, 14–5). The main reactions are therefore seen to be an oxidative depolymerization and a demethylation of the lignin (Kirk, Adler 1970; Kirk 1975b). The oxidation results in a significant production of CO_2 with a loss of carbon particularly from the side chains and methoxyl groups. This loss in weight may be compensated by extensive incorporation of molecular oxygen into the residual polymer (Crawford 1981).

Table 14–4: Elemental and methoxyl analyses and C_9-formulae for sound spruce lignin and lignin decayed by *Gloeophyllum trabea* (Kirk 1975b)

Lignin sample	C %	H %	O %	OCH_3 %	C_9-formula	Molecular weight of C_9-unit
MWL, sound	62.85	6.08	31.07	15.11	$C_9H_{8.66}O_{2.75}(OCH_3)_{0.92}$	189.2
Extractive lignin, sound	61.95	6.08	31.87	15.34	$C_9H_{8.79}O_{2.89}(OCH_3)_{0.96}$	192.8
MWL, brown-rotted	58.98	5.31	35.73	9.67	$C_9H_{8.44}O_{3.75}(OCH_3)_{0.61}$	195.4

Table 14–5: Estimation of the functional groups in sound and brown-rotted lignin from spruce *(Picea abies)* (values in equivalents per C_9-unit) (Kirk 1975b)

Lignin sample	Conjugated carbonyl	Carboxyl group	Total hydroxyl	Phenolic hydroxyl	Aliphatic hydroxyl
MWL, sound	0.07	0.10	1.16	0.24	0.92
Extractive lignin, sound	0.08	0.04	(1.24)	0.44	(0.86)
MWL, brown-rotted	0.14	0.23	1.36	0.58	0.78

From the liberation of phenol from glycolmonophenyl ether by culture filtrates of *Coniophora puteana* Rösch and Dietrichs (1971) concluded that the brown-rot fungus is able to split phenylether linkages or to oxidize the glycolic residue. The crude enzyme extract from *Poria cocos* produced phenoxy radicals in substrates of milled wood lignin and syringaldehyde as evaluated by electron-spin resonance spectroscopy (Ferm et al. 1972).

From brown-rotted wood higher yields of lignin-polysaccharide complexes are obtained which, however, show certain changes in the lignin portion and in the sugar composition as compared to complexes from sound wood (Doi et al. 1974; Arončik et al. 1974; Krejcberg et al. 1974b).

Polčin and Bezuch (1978) treated ball-milled softwood *(Picea abies)* and hardwoods *(Betula verrucosa, Populus monilifera)* with a multicomponent enzyme complex containing cellulase and polyases, and succeeded in completely dissolving lignin in dioxane. A relatively high amount of polysaccharides in the extract indicated the presence of lignin-polysaccharide complexes.

14.5. Decay by White-rot Fungi and Action of Ligninolytic Enzymes

The white-rot fungi attack softwood and mainly hardwood with a preference for lignin. The lignin-degrading enzymes are well developed but there are also enzymes for the degradation of pectin, polyoses and even of cellulose. The hyphae of these fungi penetrate the wood tissue through the pit membranes and through the cell walls by forming bore holes (Schmid, Liese 1964; von Aufsess 1974; Radtke et al. 1981). In root-wood of spruce it was observed that the hyphae of *Heterobasidion annosum (= Fomes annosus)* tend to grow from the phloem rays into the wood rays and from there laterally into the adjoining tracheids (Peek et al. 1972).

The hyphae grow predominantly at the inner surface of the cell walls and degrade the walls by the action of exoenzymes which produce lysis zones in the vinicity of a hypha. By this process the hyphae burrow into the cell walls (Liese, Schmid 1966; von Aufsess 1972, 1974). But there are also lysis zones surrounding the hyphae growing within the cell walls (Fig. 14–7a). According to Ruel et al. (1981) the attack begins with the disturbance of the parallel lignin lamellae by swelling of the interlamellar spaces. The lamellae are then progressively disrupted and modified into chains of dark granules which agglomerate to form larger clusters.

A progressing attack makes the cell walls more and more porous and produces a honeycomb structure. In the lysis zones the cellulose fibrils are uncovered and loosened. Later the residual cell walls become thinner and the fibrils are also de-

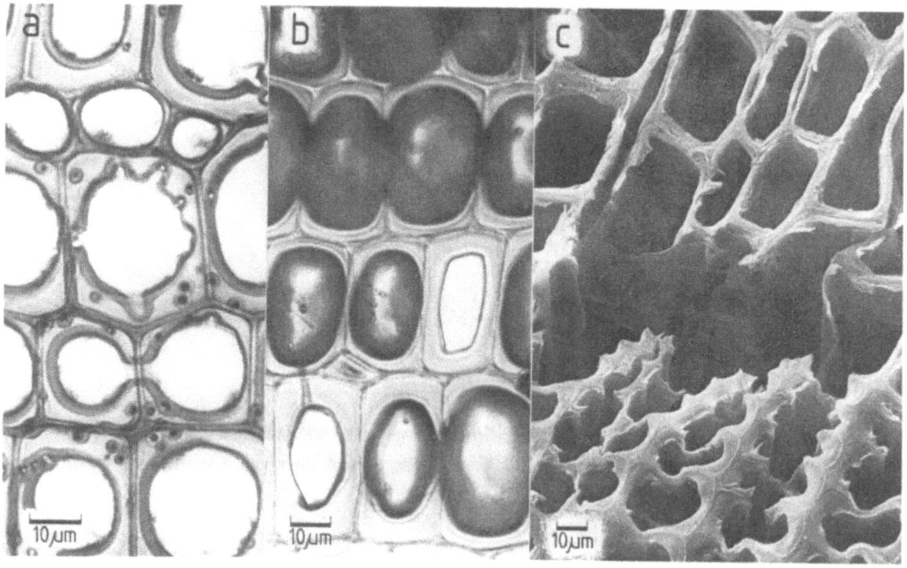

Fig. 14–7. Decay of spruce wood *(Picea abies)* by white-rot fungi (by courtesy of H. von Aufsess).
a, b) Different stages of degradation by *Heterobasidion annosum;* light micrograph.
c) Degradation of total cell walls by *Peniophora gigantea;* REM micrograph.

stroyed (Fig. 14–7b, c) (Nečesaný 1963; Schmid, Liese 1964). Cellulase-less mutants of white-rot fungi are not able to do this (Eriksson et al. 1980).

The degradation of the secondary walls proceeds more rapidly whereas at the compound middle lamella the fungal degradation is retarded to certain extent depending on the fungus species (Nečesaný 1965; Nečesaný, Sterka 1977). Microspectrometrical measurements on spruce wood *(Picea abies)* decayed by *Heterobasidion annosum* revealed that the compound middle lamella still shows a UV-absorption whereas the secondary wall does not contain any UV absorbing substance at an average weight loss of 10% (Bauch et al. 1976) (Fig. 14–8).

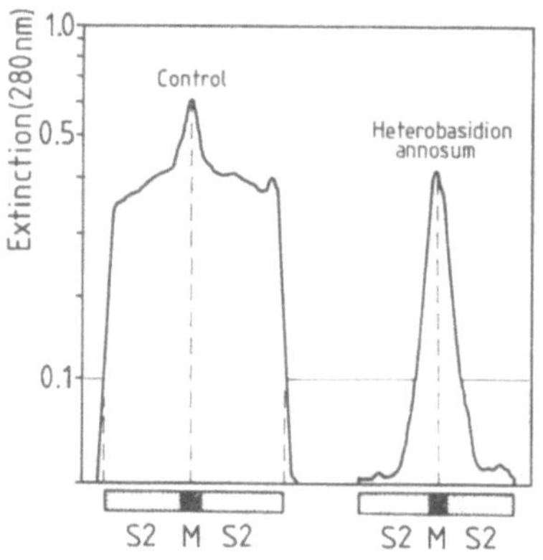

Fig. 14–8. UV absorption of cross sections of spruce sapwood *(Picea abies)*, sound (left side) and decayed by *Heterobasidion annosum* (average weight loss 10%) (by courtesy of J. Bauch).

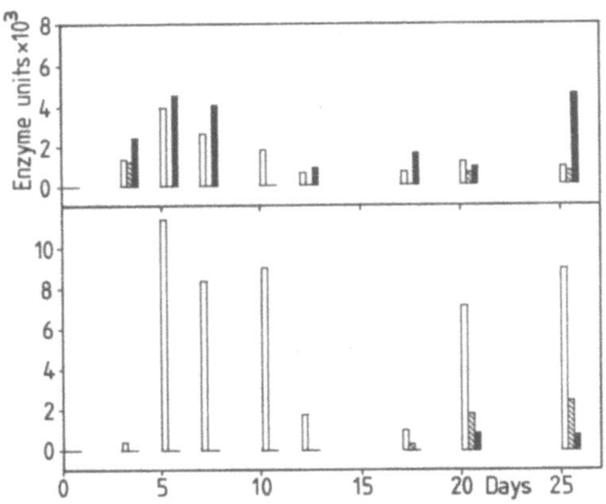

Fig. 14–9. Time course for the intracellular formation (above) and extracellular secretion (below) of laccase (open bars), tyrosinase (striped bars) and peroxidase (filled bars) by *Coriolus versicolor* (according to Ferm and Cowling 1972).

Apart from the compound middle lamella the tertiary wall is also found to have a particular resistance to fungal attack (Ruel et al. 1981). There is also a slower degradation of earlywood cells as compared to latewood cells, this being explained by a high density and a higher lignification of the former (Liese 1970).

Staining and growth tests of the fungal mycelium gave evidence of the presence of various enzymes which are able to metabolize lignin (von Aufsess et al. 1968; Ferm, Cowling 1972; Rösch 1972). Some of the white-rot fungi produce predominantly laccase, others peroxidase, and tyrosinase is also among the lignin degrading enzymes. The production of these enzymes differs according to whether they are applied inside or outside the hyphae, and may vary with the time of action (Fig. 14–9). In addition to ligninolytical enzymes β-glycanhydrolases, particularly cellulase and xylanase, were detected in white-rot fungi (Paice, Jurasek 1979; Schmidt, Liese 1980).

Eriksson and Hamp (1978) succeeded in clarifying which enzymatic system is responsible for the hydrolytic degradation of cellulose by *Phanerochaete chrysosporium (= Sporotrichum pulverulentum)*. According to this study the cellulolytic degradation occurs by the combined action of following enzymes:

- five endo-1,4-β-glucanases which split off β-glycosidic bonds at any place;
- one exo-1,4-β-glucanase which splits off cellobiose and glucose units from the non-reducing end;
- two 1,4-β-glucosidases which hydrolyze cellobiose to glucose, and cellobionic acid to glucose and gluconolactone.

Additionally an oxidating enzyme was detected which produces aldonic acids from cellobiose and cellodextrin.

Lignin and glycan degrading enzymes cause a progressive degradation of the cell wall components (Fig. 14–10). The decomposition of polysaccharides and lignin

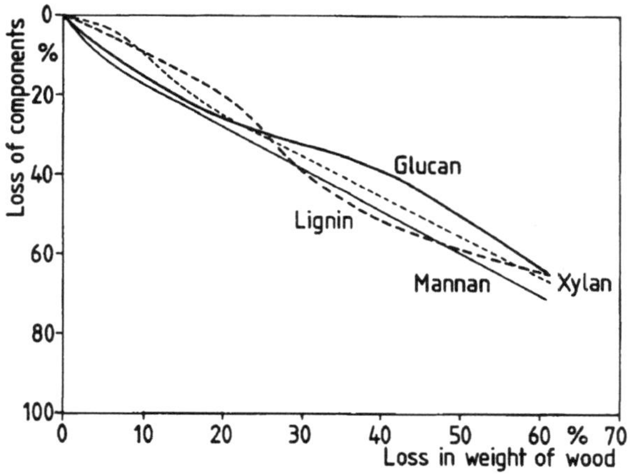

Fig. 14–10. Loss of wood components in *Picea sitchensis* by *Coriolus versicolor* (according to Kirk and Highley 1973).

varies for different fungi because of varying activities of the enzyme systems (Tables 14–6, 14–7). Varying activities may also occur between strains of the same fungus. In general mannan and xylan are removed faster than cellulose (Kirk, Moore 1972; Kirk, Highley 1973). From the analytical data a correlation between the removal of lignin and that of a particular polysaccharide cannot be deduced. The decomposition of the wood components also depends on the wood species. Thus e.g. *Fomes ulmarius (= Rigidoporus ulmarius)* degrades lignin of *Pinus taeda* and *Picea sitchensis* more quickly than lignin of *Pseudotsuga menziesii.*

With the aid of X-ray diffraction Cutter and Murphey (1970) found a linear relation between cellulose degradation and weight loss of yellow poplar sapwood *(Liriodendron tulipifera)* decayed by *Coriolus versicolor (= Polyporus versicolor)*. The diminution of the 101 and $10\bar{1}$ intensity possibly indicated a unit-cell transformation.

Table 14–6: Loss of the major components in birchwood *(Betula alleghaniensis)* after decay by white-rot fungi (Kirk, Moore 1972)

Fungus	Loss in weight %	Loss in lignin %	Loss in glucan %	Loss in mannan %	Loss in xylan %
Fomes ultramarius	15	41.59	3.06	15.93	28.53
Ganoderma applanatum	17	18.17	28.27	27.03	13.52
	32	36.90	38.07	49.80	35.61
Polyporus berkeleyi	8	30.89	15.58	8.45	3.58
	22	42.06	31.01	33.35	30.25
	39	63.15	43.90	50.97	39.54
Ischnoderma resinosum	11	34.88	17.12	25.38	10.03
	22	43.98	30.12	26.72	20.63
Coriolus versicolor	21	31.21	19.64	25.63	25.54
	36	38.96	38.89	54.25	39.08

Table 14–7: Loss of the major components in western white pine *(Pinus monticola)* after decay by white-rot fungi (Kirk, Highley 1973)

Fungus	Loss in weight %	Loss in lignin %	Loss in glucan %	Loss in mannan %	Loss in xylan %
Coriolus versicolor	13	27	4	13	13
	22	33	17	22	21
	43	52	43	47	47
	61	62	65	68	67
	83	86	85	89	89
Ganoderma applanatum	16	26	12	16	19
	43	57	42	52	51
	52	68	49	63	64
Peniophora "G"	11	15	6	3	6
	30	34	29	39	41
	46	51	51	59	59

The combined degradation of all wood components seems to be common to all wild-type white-rot fungi. The detection of an enzyme which needs cellobiose, a product of the cellulose degradation, for the decomposition of lignin in combination with the action of laccase indicates a certain correlation in the decay of polysaccharides and lignin (Westermark, Eriksson 1974). The enzyme was called cellobiose-quinone oxidoreductase. The probable joint action of both enzymes is shown in Fig. 14–11. Further studies with the white-rot fungus *Phanerochaete chrysosporium (= Sporotrichum pulverulentum)* indicated that the cellobiose-quinone oxidoreductase is important but not necessary for the decay of lignin (Ander, Eriksson 1975, 1976; Eriksson, Hamp 1978). Absolutely necessary is the presence of laccase. A fungus mutant lacking this phenol oxidase was not able to degrade lignin or wood, whereas a cellulase-less mutant was able to do so. A depletion of lignin seems not to be necessary for its degradation as polysaccharides apparently do not form protective barrier for the fungal enzymes (Kirk 1973).

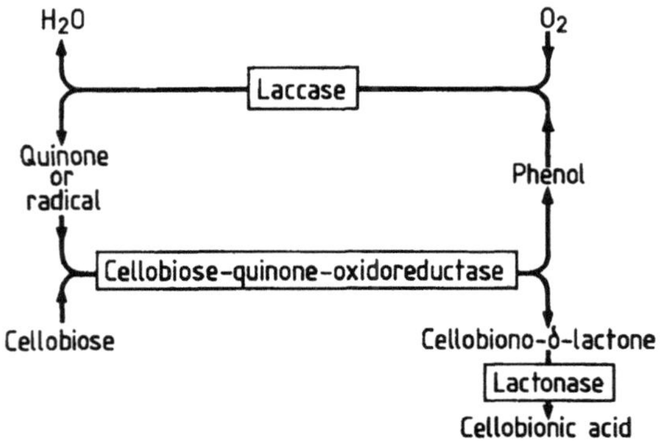

Fig. 14–11. Probable mechanism of action for cellobiose-quinone oxidoreductase.

According to Kirk (1975a) the lignin-decaying enzymes must act extracellularly as they have to degrade macromolecular substance. Judging from their kind of action the enzymes are probably bound at the surface of the hyphae in a way that permits contact with the lignin at the cell walls. On the other hand a nearly uniform degradation of the whole cell wall can be observed though only one or two hyphae are present.

Various studies are concerned with changes taking place during the microbial degradation of lignin by white-rot fungi (Hata 1966; Kirk, Lundquist 1970; Doi et al. 1974; Kirk, Chang 1974, 1975). The results of these studies evaluated by chemical methods, ultraviolet, infrared and proton resonance spectroscopy have been summarized by Kirk et al. (1978). Table 14–8 indicates the qualitative changes of the chemical properties of lignin caused by white-rot fungi.

From the elemental analysis of isolated lignin it can be seen that after fungal attack the amount of oxygen is increased whereas the methoxyl content is decreased as compared to lignin from sound wood (Kirk 1971; Kirk, Chang 1974; Oki et al. 1981) (Table 14–9). The increase of oxygen in the lignin molecule derives from an oxidation of α-carbon atoms and an oxidative cleavage of the β- and γ-carbon atoms in terminal position (Fig. 14–12) (Hata 1966; Kirk, Chang 1975; Nakatsubo et al. 1981). Guaiacylglycerol-β-coniferyl ether was converted by *Fusarium solani* to guaiacylglycerol-β-vanillic acid ether without oxidation of the benzylic secondary alcohol group or cleavage of the β-aryl ether linkage (Katayama et al. 1980). The demethylation of the methoxyl groups with white-rot fungi was shown by Haider et al. (1964), who made tests with various methoxylated phenols. By the decomposition of ^{14}C-labelled lignin with two white-rotters *(Coriolus versicolor, Phane-*

Table 14–8: Chemical changes in lignins after decay by white-rot fungi (Kirk et al. 1978)

Properties	Change[a] Increase	Decrease
Carboxyl content	+	
Hydroxyl content, phenolic	+	+
aliphatic		+
Carbonyl content	+	
Hydrogen/carbon		+
Oxygen/carbon	+	
Methoxyl/carbon		+
Yield of methoxylated aromatic acids on oxidative degradation after methylation[b]		+
Yield of principle acidolysis products[c]		+
Yield of major products on nitrobenzene oxidation[d]		+

[a] Compared to sound lignins
[b] The major product is 3,4-dimethoxybenzoic (veratric) acid from gymnosperm lignin. From angiosperms, veratric acid and tri-O-methyl gallic acid are dominant
[c] The major product from sound lignin is 3-hydroxy-1-(4-hydroxy-3-methoxyphenyl)-2-propanone. Vanillic acid is the major product from white-rotted lignin
[d] Vanillin (gymnosperms) and vanillin + syringaldehyde (angiosperms) are the major products

Table 14–9: Elemental and methoxyl analyses and C_9-formulae for sound and white-rotted lignins (Kirk 1971; Kirk, Chang 1974)

Lignin	Decayed by	C %	H %	O %	OCH_3 %	C_9-formula	Molecular weight of C_9-unit
MWL, pine	sound	63.70	6.29	30.01	15.5	$C_9H_{8.86}O_{2.58}(OCH_3)_{0.94}$	187.3
MWL, pine	*Coriolus versicolor*[a]	61.41	6.11	32.48	10.1	$C_9H_{9.56}O_{3.20}(OCH_3)_{0.61}$	187.7
MWL, spruce	sound	62.81	5.88	31.31	15.24	$C_9H_{8.28}O_{2.78}(OCH_3)_{0.93}$	189.6
MWL, spruce	*Poria subacida*[b]	60.44	5.67	33.89	12.32	$C_9H_{8.61}O_{3.34}(OCH_3)_{0.77}$	193.9
MWL, spruce	sound	62.85	6.08	31.07	15.11	$C_9H_{8.66}O_{2.75}(OCH_3)_{0.92}$	189.2
MWL, spruce	*Coriolus versicolor*[b]	57.97	4.70	37.23	11.33	$C_9H_{7.26}O_{3.95}(OCH_3)_{0.74}$	199.4
MWL, spruce	*Polyporus anceps*[b]	58.71	4.99	36.30	11.21	$C_9H_{7.70}O_{3.80}(OCH_3)_{0.72}$	198.8

[a] Recovered after incubation in liquid culture with fungus
[b] Purified from extract of white-rottet wood

rochaete chrysosporium) it was shown that the greatest proportion of the metabolic final product CO_2 derives from the methoxyl group; the residual portion derives from carbon of the side chains and the aromatic rings (Kirk et al. 1975a).

A further step in the fungal degradation of lignin is the oxidative cleavage of β-O-4-linkages of terminal phenylpropan units (Fig. 14–13) (Ishikawa, Oki 1964, 1966; Higuchi 1981). The pathways for the splitting of α-O-4-, β-5-, β-1- and β-β-linkages could also be clarified (Ohta et al. 1979; Nakatsubo et al. 1981; Higuchi 1981). The splitting reactions result in monomeric and dimeric compounds, most of them carrying carboxylic groups. For these compounds to enter the internal meta-

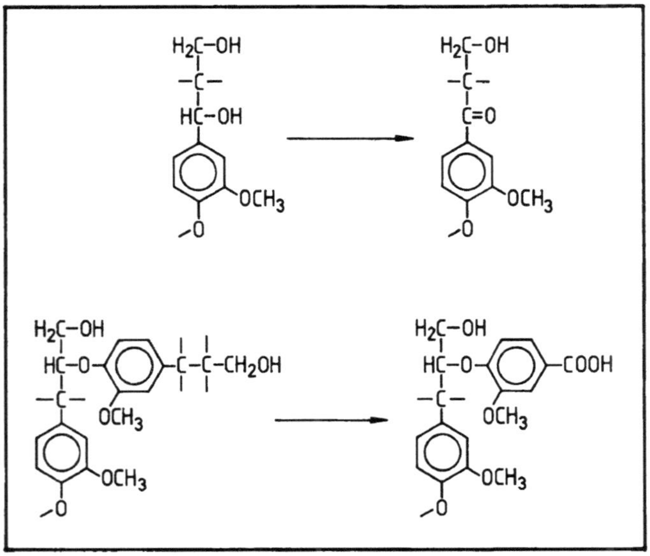

Fig. 14–12. Oxidative reactions at the lignin units during white-rot decay.

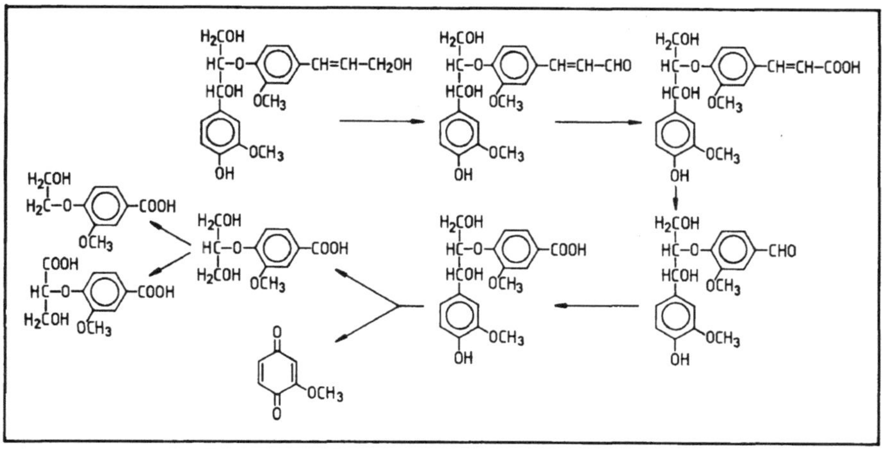

Fig. 14–13. Oxidative cleavage of terminal chains and β-aryl ether linkages by white-rot enzymes.

bolism of the fungus a cleavage of the aromatic rings seems to be necessary. Dioxygenases were identified in fungi which are able to catalyse an oxidative cleavage of protocatechuic acid (Fig. 14–14) (Kirk 1975a; Crawford, Crawford 1980). Protocatechuic acid derives from vanillic acid or veratric acid by demethylation, both monomers being found as fungal degradation products (Chen et al. 1981).

The cleavage of the aromatic rings needs not occur exclusively with monomeric degradation products. Aromatic rings incorporated in the lignin polymer are probably also cleaved by enzymes (Kirk et al. 1977; Chen et al. 1981). ^{13}C-NMR spectra of synthetic lignin (DHP) which was incubated with white-rot fungi showed a cleav-

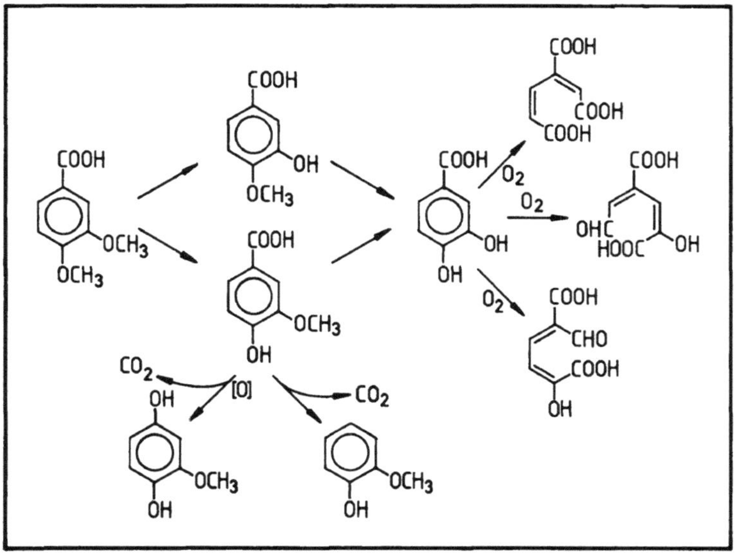

Fig. 14–14. Oxidative degradation of vanillic acid, including a cleavage of the aromatic ring by dioxygenases.

age of aromatic rings, which probably did not take place between oxygen-substituted carbon atoms, as well as a splitting of aryl ether linkages and a degradation of terminal chains (Ellwardt et al. 1981).

From the change in acidolysis products of white-rotted birch wood it was concluded that lignin is degraded from the exposed surface which becomes progressively larger (Kirk et al. 1975b). Changes occur primarily at the lignin surface, and the macromolecules are not fragmented. The degradation process proceeds by cleavage of terminal groups.

Carefully isolated lignin from decayed wood shows UV absorbance spectra which are not coincident with those of sound lignins (Fig. 14–15) (Doi et al. 1974; Kirk, Chang 1974). The lignin is obviously not completely incorporated into the metabolism of the fungi as a certain portion is converted to a high condensed product. Kuwahara et al. (1981) detected the formation of guaiacol from vanillic acid by *Fusarium* species. Some of the fungi strains oxidized guaiacol and converted it to polymer products. Similar reactions opposite to the enzymatic degradation were also observed using milled wood lignin, kraft lignin and lignin sulfonates as substrates for *Heterobasidion annosum* and *Coriolus versicolor* (Hüttermann et al. 1977, 1980; Brunow, Wallin 1978). Model experiments indicate the formation of biphenyl structures by enzymatic dehydration. Laccase seems to be responsible for this condensation as it is prevented if laccase inhibitors are added.

The condensing action of laccase was also tested by the addition of cellulose to a culture of *Pleurotus ostreatus* on lignin sulfonate (Hiroi, Eriksson 1976; Hiroi et al.

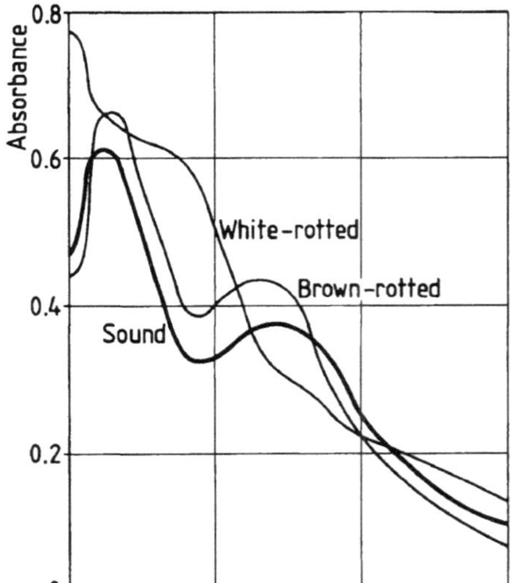

Fig. 14–15. UV absorption spectra of lignin (Braun's native lignin) from sound, brown-rotted *(Coriolellus palustris)* and white-rotted *(Pycnoporus coccineus)* Katsura wood *(Cercidiphyllum japonicum)* (according to Doi et al. 1974).

1976). The addition of cellulose inhibited the polymerization reactions by providing cellobiose as cosubstrate for cellobiose-quinone oxidoreductase. This enzyme reduces the phenol radicals and thus inhibits polymerization.

Condensed lignins containing biphenyl linkages are obviously highly inaccessible for fungal enzymes (Krisnangkura, Gold 1919). Also the fact that syringyl elements are degraded more rapidly than guaiacyl elements in hardwood lignin and synthetic guaiacyl-syringyl-lignin is explained by the higher amount of biphenyl structures in the guaiacyl portion, the phenolic groups of which are more resistant to the formation of phenoxy radicals (Kirk et al. 1975b; Noguchi et al. 1980; Oki et al. 1981).

The molecular weight distribution of decayed lignins is similar to that of sound lignins as has been shown by gel permeation chromatography (Doi et al. 1974; Kirk, Chang 1974; Kirk et al. 1978). Spruce lignin heavily degraded by *Coriolus versicolor* and *Polyporus anceps* consisted mainly of molecules greater than 1 700 dalton, which correspond to about 8–10 phenylpropan units.

During the fungal degradation of commercial lignin sulfonate the microorganisms preferred the low molecular portion (Ferm, Nilsson 1969, 1970). From a substrate consisting of a uniform high-molecular lignin sulfonate the fungus *Heterobasidion annosum* split off only small fragments with molecular weights below 1 000 dalton (Haars, Hüttermann 1980). Intermediate-size fragments were almost nonexistent. On the other hand Clayton and Srinivasan (1981) succeeded in obtaining a large spectrum of lignin fragments during the degradation of kraft lignin by *Candida* spec.

Experiments using white-rot fungi, particularly *Phanerochaete chrysosporium*, for a partial delignification of unbleached pulps showed the necessity of adding nutrient nitrogen and molecular oxygen to obtain a sufficient rate of degradation (Kirk, Yang 1979; Yang et al. 1980). The maximum rate of lignin degradation was 3% per day over a two-weeks incubation of thermomechanical pulp of red alder *(Alnus rubra)*.

14.6. Decay by Soft-rot Fungi

Soft-rot fungi contain enzymes which degrade all cell wall components. They differ from brown-rot and white-rot fungi by growing mainly within the cell walls (Fig. 14–16). The wood is invaded by the hyphae which grow primarily through the rays and vessels, from where they penetrate into the lumina of tracheids or fibres (Courtois 1963a; Liese 1964, 1970). Only few soft-rot species are able to attack the tertiary wall of softwood tracheids, whereas in general the tertiary walls are attacked easily. The degradation of the wall substance can be observed by the ap-

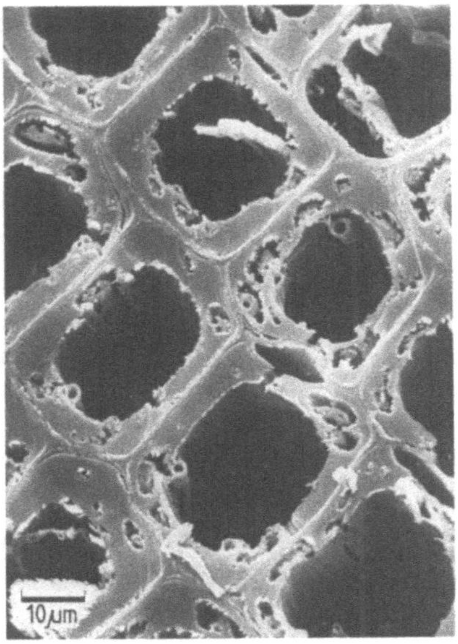

Fig. 14–16. Spruce wood *(Picea abies)* decayed by *Scytalidium lignicolum;* REM micrograph (by courtesy of H. von Aufsess).

pearance of lysis zones on both sides of the hyphae. The area of lysis enlarges and the hyphae burrow towards the inner part of the cell wall.

With regard to the invasion of the secondary walls, particularly of softwood tracheids, small perforation or bore hyphae are formed which perforate the cell walls in a lateral direction (Courtois 1963a; Levi, Preston 1965; Radtke et al. 1981). Within the cell walls the growing hyphae follow the direction of the S2-fibrils, producing characteristic caverns. These caverns have a hexagonal shape in the longitudinal direction, and run in an acute angle towards the fibre axis. Variations of the decay pattern depend more on structural and topochemical properties of the attacked cell wall than on the fungus species (Courtois 1963a; Nilsson 1976). As decay progresses the cell walls become interlaced with slit-like cavities until finally only the compound middle lamellae and the tertiary walls or parts of the latter remain (Liese 1970; Mennaga, Jutte 1972).

The aggressiveness of soft-rot fungi depends on the wood and the fungus species involved. A test with various fungi under identical experimental condition showed that beech wood *(Fagus sylvatica)* is decomposed more intensively than pine wood *(Pinus sylvestris)* (Courtois 1963b). While the weight loss of beech wood varied between 11 and 52% that of pine wood varied between 0.2 and 4.8% after an incubation period of 16 weeks.

Keilich et al. (1970) and Nilsson (1974a, 1976) evaluated the presence of several polysaccharide splitting enzymes in soft-rot fungi. The activity of cellulase, xylanase and mannanase could be demonstrated with several fungi. Though soft-rot fungi prefer hardwoods, experiments with isolated polyoses show that, e.g., *Chaetomium globosum* reveals a high level of activity for birch xylan as well as for larch xylan (Lewis 1976). From the same fungus an endo-cellulase could be isolated and characterized (Keilich et al. 1970). The enzyme had a molecular weight of 30 000 ± 3 000, a pH-optimum of 4.5–5.5 and an optimum temperature of 35 °C (Tables 14–2, 14–3).

Studies concerned with lignin degradation revealed that soft-rot fungi are provided with at least a weak enzyme system for this purpose. Specific tests for the presence of phenoloxidases took a negative course (von Aufsess et al. 1968). Nevertheless Haider and Trojanowski (1975) found several soft-rotters capable of converting DHPs to CO_2. The important structural elements, such as methoxyl groups, side chains and aromatic rings, of the ^{13}C-labelled DHPs were involved in the CO_2 production. The fungi also released CO_2 from maize lignin.

Delignified wood samples and pure cellulose fibres are much more accessible for soft-rot fungi, and the decay patterns are partly or totally changed (Courtois 1963b; Nilsson 1974b). The absence of typical cavities in delignified pine samples *(Pinus sylvestris)* after an attack of *Phialophora fastigiata* was given the following possible explanations by Zainal (1976):

- The fungus developed another kind of enzyme after delignification which caused an attack of all cell wall layers without differentiation.
- With removal of lignin from the other wall constituents, the enzyme system of the fungus has space in which to move freely through the wall. This clearly cannot be true in normal wood when cavities are formed.
- The removal of lignin opens up the cellulose structure and makes it readily available to an attack by the fungal enzyme.

The delignification of pine wood improved the decay by *Chaetomium globosum:* the weight loss was 7.4% after a 10 weeks' incubation as compared to 0.3% of untreated pine wood (Courtois 1963b). Nevertheless the weight loss of 30.5% as in the case of beech wood was not reached.

The chemical analysis of hardwoods after a degradation by soft-rot fungi results in a decomposition of all cell wall components though, at varying rates (Levi, Preston 1965; Seifert 1966; Eslyn et al. 1975). The lignin degradation in beech wood by *Chaetomium globosum* begins with a demethylation up to a total weight loss of 12%. With increase of weight loss a further decomposition of lignin also begins (Fig. 14–17). In this regard the fungus acts similarly to brown-rot fungi. With regard to the formation of alkali-soluble substances it is similar to white-rot fungi. Eslyn et al. (1975) concluded from their investigations that the action of various soft-rot fungi must not be regarded simplistically. Six soft-rot fungi utilized the carbohydrates of alder *(Alnus rubra)* and poplar *(Populus balsamifera)* before the lignin. But two of them (*Paecilomyces* spec., *Thielavia terrestris*) growing on pine *(Pinus monticola)* removed lignin more quickly than carbohydrates, as is more characteristic of white-rot fungi (Table 14–10).

Table 14–10: Loss of the major components in wood after decay by various soft-rot fungi (Eslyn et al. 1975)

Wood	Fungus	Loss in weight %	Loss in lignin %	Loss in glucan %	Loss in xylan %	Loss in mannan %
Alnus rubra	*Papulospora* spec.	10	9	15	6	14
		17	12	23	18	18
	Paecilomyces spec.	15	11	21	25	28
		25	13	37	23	20
		41	19	60	44	22
	Thielavia terrestris	7	9	10	1	19
		28	17	40	29	21
Populus balsamifera	*Papulospora* spec.	10	0	14	17	14
		21	4	27	29	25
	Paecilomyces spec.	14	10	15	35	28
		28	11	41	37	30
	Thielavia terrestris	10	6	11	12	29
		23	13	25	27	43
Pinus monticola	*Papulospora* spec.	15	12	18	18	13
	Paecilomyces spec.	10	14	7	8	6
	Thielavia terrestris	7	14	3	9	− 3

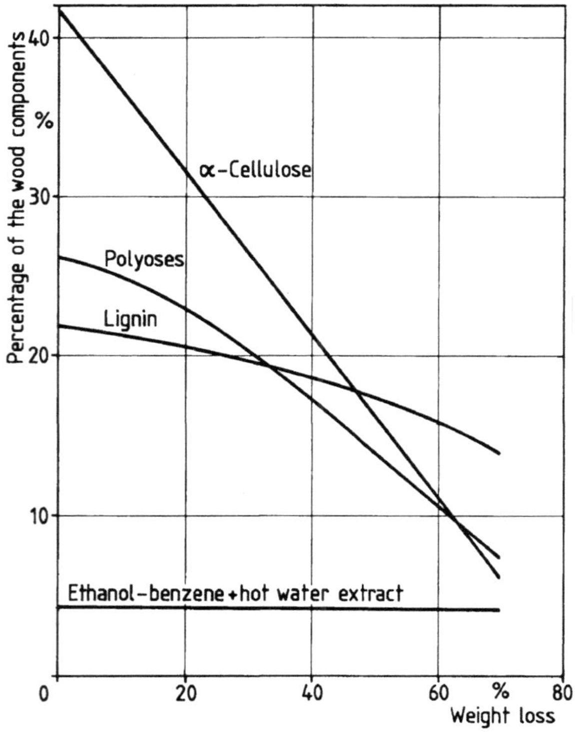

Fig. 14–17. Loss of wood components in *Fagus sylvatica* during decay by *Chaetomium globosum* (according to Levi and Preston 1965).

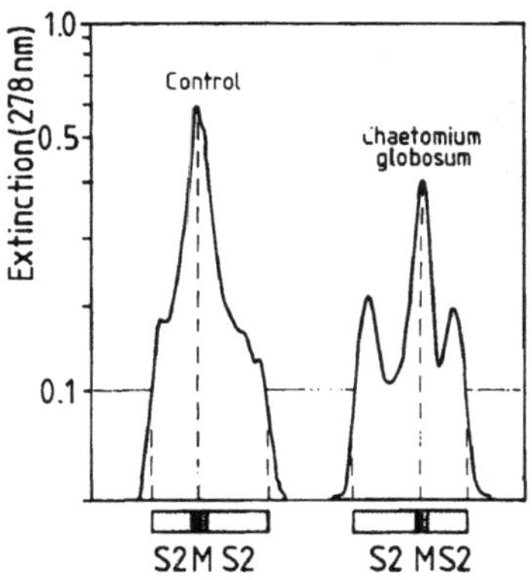

Fig. 14–18. UV absorption of cross sections of oak sapwood *(Quercus robur)*, sound (left side) and decayed by *Chaetomium globosum* (average weight loss 28%) (by courtesy of J. Bauch).

UV spectroscopic measurements of oak wood *(Quercus robur)* decayed by *Chaetomium globosum* showed a decrease of lignin content at least in the S2-layer (Bauch et al. 1976) (Fig. 14–18). It seemed that the aromatic rings of the lignin or of its degradation products are cleaved as no accumulation of aromatic compounds was observed.

The degradation of cellulose proceeds very quickly at an approximately constant rate during the attack by various soft-rotters (Levi, Preston 1965; Seifert 1966; Eslyn et al. 1975). *Chaetomium globosum* removes polyoses at an increasing rate as decay proceeds (Fig. 14–17). Other soft-rot fungi have similar effects on hardwoods (Table 14–10).

14.7. Effects of Blue-stain Fungi

Though belonging to the same group of fungi as the soft-rotters most of the blue-stain fungi are not wood-destroying in the common sense. But there are also blue-staining fungi which cause soft-rotting effects (e.g. *Scytalidium lignicolum, Alternaria tenuis*). Blue-stain fungi are typical softwood inhabitants but they may also occur in hardwoods. The discolouration of wood invaded by blue-stain fungi is caused by dark material which is deposited in vacuoles within the hyphae of these fungi (Fig. 14–19a).

The hyphae grow mainly in the parenchyma cells living on the proteinaceous content, but they are also found in tracheids, where they grow on the inner cell wall surface without an enzymatic alteration of the cell wall structure (Liese 1970). Blue-stain fungi may also grow on wood and coating surfaces where they are involved in the formation of grey stain (Kühne et al. 1970).

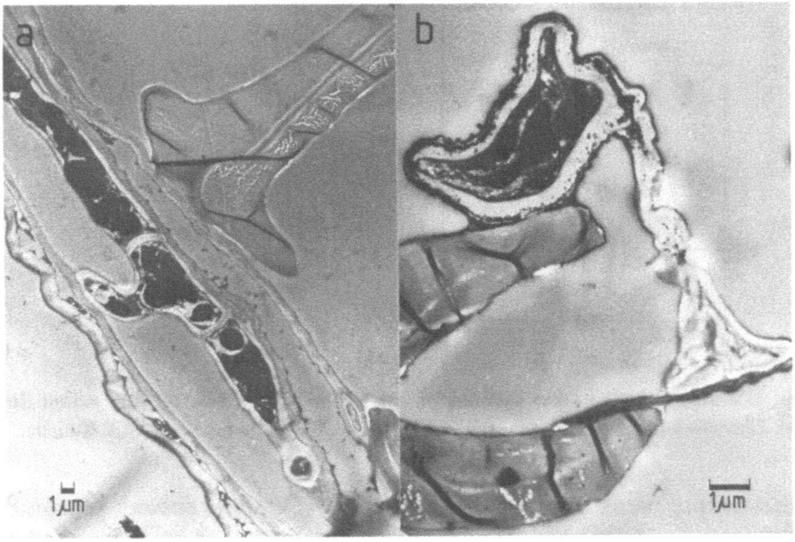

Fig. 14–19. Hyphae of a blue-stain fungus in pine wood *(Pinus sylvestris)*; ultrathin sections, TEM micrographs.
a) Hypha in a ray parenchyma cell.
b) Cross section of a hypha with transpressorium which is attached to the torus of a bordered pit.

The hyphae pass from one cell to another by penetrating the pit membranes or even the cell walls in its full width. For this purpose the end of a hypha is transformed into a boring instrument called a transpressorium which is able to penetrate the torus or the cell wall by localized enzymatic action and mechanical pressure (Liese, Schmid 1964; Liese 1970) (Fig. 14–19b).

The electron microscopic study of hyphae of blue-stain fungi shows various structures of the surface which seem not to be significant for the species (Schmid, Liese 1965). The smooth, granular or fibrillar appearance was assumed to be due to external factors. The hyphae can also be coated by a slime substance. The slime layer of the hyphae of the fungus *Aureobasidium pullulans* contains an α-glucan which is called pullulan (Bender et al. 1959). With a specific staining method Schmid and Balderman (1967) found additional acidic mucopolysaccharides in this extracellular layer.

The enzymes acting intra- and extracellularly of the hyphae were studied by Liese (Rösch et al. 1969; Rösch, Liese 1971; Berndt, Liese 1971). The blue-stain fungi produce enzymes for degrading polysaccharides and pectins, such as cellulase, polygalacturonase and mannanase. The hydrolysis of pectin in a wide pH range (3.5–7.5) is made possible by the presence of at least two enzymes (polygalacturonase and pectin-transeliminase) with optimum values at pH 4.5 and 6.5. The presence of lignin-splitting enzymes (phenoloxidases) was proved for 19 blue-stain species. In most species laccase was found, which, however, occurs predominantly as an intracellular enzyme.

The treatment of pine wood samples *(Pinus sylvestris)* with blue-stain fungi *(Aureobasidium pullulans)* for 10 weeks resulted in a weight loss of 1.7 and 2.1%, respectively, by dissimilation (Seifert 1964). The cellulose content was reduced by 6.9 and 3.5%, and the total pentosans were reduced by 3.9 and 3.1%, respectively. Alkali-insoluble pentosans, however, decreased up to 29%. The lignin content decreased only slightly with a maximum value of 1.3%. From these results it can be seen that, as far as the cell wall components are concerned, the main action of blue-stain fungi is a splitting of the polyose molecules, which become soluble in alkali.

14.8. Bacterial Attack

The bacterial attack of wood cell walls proceeds very slowly as compared to fungal attack. Bacteria are not able to grow in size and thus their expansion is determined by cell division. The initial colonization of a wood sample generally occurs by invasion through the ray parenchyma cells though additionally a random distribution at the walls of the other cells in wood can be observed. The colonization in the pit holes results in a degradation of the pit membranes by pectinolytic and cellulolytic enzymes (Fig. 14–20) (Greaves 1969; Johnson, Giovik 1970; Liese, Greaves 1975). Particularly at the outer edge of the pit borders circular or elliptic perfora-

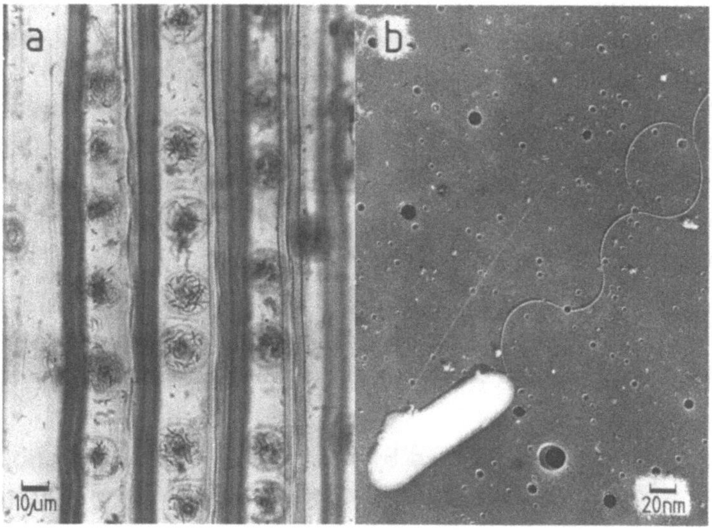

Fig. 14–20. Attack of bacteria on wood.
a) Colonies of bacteria in the bordered pits of spruce *(Picea abies)* after storage in water; light micrograph (by courtesy of H. von Aufsess).
b) Bacterium of *Pseudomonas* spec. from a culture filtrate; TEM micrograph (by courtesy of H. J. Preusser).

tions become visible. Levy (1975) observed that the parenchyma pit membranes are destroyed prior to the bordered pit membranes. The degradation of the cell walls begins with a lysis zone at the contact area of the bacteria. As the bacterial attack proceeds these patterns grow to erosion troughs and cavities which expand until finally the whole cell wall is destroyed (Courtois 1966; Greaves 1969; Holt, Jones 1978). In the first phase of cell wall degradation the loss of birefringence indicates an attack on the ordered regions of the cellulose (Liese 1970).

The bacterial attack is mainly restricted to the sapwood; heartwood constituents seem to impede such an attack. On the other hand bacterial colonies are found in the wet heartwood of several conifers. In wetwood cores of *Ulmus americana, Salix nigra, Populus alba* and *P. deltoides* an infestation of methanobacteria was observed (Zeikus, Ward 1974). The bacteria produce methane, which may reach a high pressure so that it can be ignited at the stem surface after boring a hole into the heartwood.

Bacterial decay was not found to be impeded by resin in the vicinity of resin canals (Courtois 1966). Various bacteria species are not able to attack the cell walls unless the lignin content is reduced by chemical treatment. The reduction of the lignin content in beech wood from 25.8% to 20.1% effected a weight loss of 71.3% after a 4-weeks period of incubation with bacteria under optimal conditions (Schmidt 1978). After 5 years of storage a natural bacteria population degraded beech *(Fagus sylvatica)* and oak samples *(Quercus robur)* minimally without technological depreciation (Courtois, Erasmy 1976).

From the large spectrum of various bacteria species which may settle on wood trunks under natural conditions, only a part is able to degrade wood components and only a few are able to degrade wood. Schmidt and Dietrichs (1976) tested 150 bacteria strains by feeding them various substrates. 23% of them were able to degrade pectin, 17% xylan, 10% carboxymethylcellulose, 9% holocellulose, 6% α-cellulose and only two of 80 bacterial strains were able to degrade untreated wood. The degradation of wood components by bacteria under aerobic conditions was more frequent than under anaerobic conditions.

Various bacteria living on sources containing cellulose and other carbohydrates produce larger amounts of extracellular slime (Martin et al. 1968). This slime forms a fibrillar envelope, and consists of heteropolysaccharides of varying composition depending on the species.

A natural bacteria population was used by Seifert (1967) for degradation experiments with pine wood *(Pinus sylvestris)* and beech wood *(Fagus sylvatica)*. An incubation of 240 days resulted in a weight loss of 7% for pine and 4.5% for beech. The highest loss was observed for polyoses with 13.3%; the decrease of cellulose was 5.2% while the lignin content remained constant. An attack of the lignin was indicated by a 13.6% loss of methoxyl groups. The increase of alkali solubility was caused by a reduction of the chain length of the cellulose molecules.

Experiments using pine wood *(Pinus sylvestris)* and defined bacteria strains belonging to the species of *Cellulomonas* spec. and *Bacillus polymyxa* were performed by Schmidt (1980; Schmidt, Bauch 1980). The incubation period was up to half a year (static) or 5 to 14 days with shaking. The loss of weight was 1.7% (*Cellulomonas* spec.) and 1.1% *(Bacillus polymyxa)*. *Cellulomonas* decreased the total polysaccharide content by 3.8%; this was apparently due to the polyoses as cellulose was obviously not attacked. The lignin content was reduced by 4.2% *(C.)* and 6.8% *(B. p.)*. A partial delignification before incubation facilitated the bacterial attack. A chemical reduction of the lignin content to 15% and 17%, respectively, effected a weight loss 36% *(C.)* and 25% *(B. p.)*, and a decrease of the lignin content of 54% *(C.)* and 42% *(B. p.)*. 50% of the carbohydrates were degraded by *Cellulomonas*, and in this case cellulose was also highly (47%) degraded.

A progressive degradation of beech wood *(Fagus sylvatica)* and pine wood *(Pinus sylvestris)* with an increasing degree of delignification was also found by Holt and Jones (1978) who used the bacterium *Cellvibrio vulgaris*. Untreated wood sections showed a weight loss of 5.1% (beech) and 1.2% (pine). With an increasing delignification period with sodium chlorite the weight loss increased up to 53% (beech) and 38% (pine) after a two-week period of incubation. During the action of a certain *Pseudomonas* strain on thin sections of poplar wood *(Populus euamericana)* the lignin content was reduced by 30% after 7 days and by 53% after 30 days of incubation (Monties et al. 1981).

There are various experiments which demonstrate the ability of various bacteria to degrade lignin, lignin model compounds and DHPs (Crawford et al. 1973, 1975; Kawakami 1976; Dagley 1978; Haider et al. 1978). The metabolism of bacteria involves a reaction sequence of oxidation of the α-carbon, oxidative splitting of the aryl ether linkage, and cleavage of two carbon atoms of the propyl side chain (Fig. 14–21). Nevertheless it is assumed that bacteria respire predominantly carbohydrates whereas lignin remains as "ballast" in the nutrient medium or becomes absorbed by the bacterial cells (Schmidt 1980).

Experiments with ^{35}S-labelled ligninsulfonic acids showed that some bacteria, particularly *Pseudomonas* species, are able to split off sulfonate groups if an easily digestible carbon source is present (Haider, Trojanowski 1981). The degradation of the lignin frame, however, was considerably reduced.

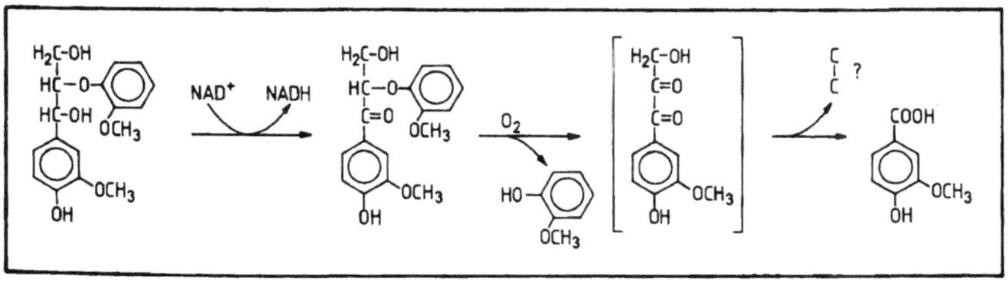

Fig. 14–21. Enzymatic cleavage of β-arylether linkages in a lignin model compound by bacteria (*Pseudomonas* spec.).

Literature

Ander, P. and Eriksson, K. E. 1975, Svensk Papperstid. *78,* 643–652
Ander, P. and Eriksson, K. E. 1976, Mater. Org. Beih. No. 3, 129–140
Arončik, B. M., Krejcberg, Z. N. and Sergeeva, V. N. 1974, Khim. Drev. No. 1, 78–82
von Aufsess, H. 1972, Forstwiss. Centralbl. *91,* 98–105
von Aufsess, H. 1974, Eur. J. For. Pathol. *4,* 193–203
von Aufsess, H. and Fengel, D. 1982, Unpublished results
von Aufsess, H., von Pechmann, H. and Graessle, E. 1968, Holz Roh-Werkst. *26,* 50–61
Bailey, P. J., Liese, W., Rösch, R., Keilich, G. and Afting, E. G. 1969, Biochim. Biophys. Acta *185,* 381–391
Bauch, J., Seehann, G. and Fitzner, H. 1976, Mater. Org. Beih. No. 3, 141–152
Bavendamm, W. 1974, Die Holzschäden und ihre Verhütung, Wiss. Verlagsges., Stuttgart
Becker, G. and Liese, W. 1965, Holz und Organismen, Mater. Org. Beih. No. 1
Becker, G. and Liese, W. 1976, Organismen und Holz, Mater. Org. Beih. No. 3
Bender, H., Lehmann, J. and Wallenfels, K. 1959, Biochem. Biophys, Acta *36,* 309–316
Berndt, H. and Liese, W. 1971, Arch. Mikrobiol. *79,* 140–146
Betrabet, S. M. and Paralikar, K. M. 1978, Cell. Chem. Technol. *12,* 241–252
Betrabet, S. M., Khandeparkar, V. G. and Patil, N. B. 1974, Cell. Chem. Technol. *8,* 339–344
Bletchly, J. D. 1967, Insect and Marine Borers Damage to Timber and Woodwork. Her Majesty's Stationary Office, London
Brunow, G. and Wallin, H. 1978, Holzforschung *32,* 189–192
Cartwright, K. S. G. and Findley, W. P. K. 1958, Decay of Timber and its Prevention. Her Majesty's Stationary Office, London
Chen, C. L., Chua, M. G. S., Evans, J. E., Chang, H. M. and Kirk, T. K. 1981, Chemistry of Lignin Biodegradation by *Phanerochaete chrysosporium.* In: The Ekman Days 1981, Int. Symp. Wood Pulp. Chem., Stockholm, Vol. 3, pp. 75–87
Chou, C. K. and Levi, M. P. 1971, Holzforschung *25,* 107–112
Clayton, N. E. and Srinivasan, V. R. 1981, Naturwissenschaften *68,* 97–98
Côté, W. A. 1968, Biological Deterioration of Wood. In: Principles of Wood Science and Technology, Vol. 1. Solid Wood (Kollmann, F. F. P. and Côté, W. A., Eds.). Springer, Berlin, Heidelberg, New York, pp. 97–135
Côté, W. A., Timell, T. E. and Zabel, R. A. 1966, Holz Roh-Werkst. *24,* 432–438
Courtois, H. 1963a, Holzforsch. Holzverwert. *15,* 88–101
Courtois, H. 1963b, Holzforschung *17,* 176–183
Courtois, H. 1965, Mater. Org. Beih. No. 1, 41–53
Courtois, H. 1966, Holzforschung *20,* 148–154
Courtois, H. and Erasmy, J. J. 1976, Holz Roh-Werkst. *34,* 181–184
Cowling, E. B. 1965, Microorganisms and Microbial Enzyme Systems as Selective Tools in Wood Anatomy. In: Cellular Ultrastructure of Woody Plants (Côté, W. A., Ed.). Syracuse University Press, Syracuse, N. Y., pp. 341–368
Crawford, R. L. 1981, Lignin Biodegradation and Transformation. J. Wiley, New York, Chichester, Brisbane, Toronto
Crawford, D. L. and Crawford, R. L. 1980, Enzyme Microb. Technol. *2,* 11–21
Crawford, R. L., Kirk, T. K., Harkin, J. M. and McCoy, E. 1973, Appl. Microbiol. *25,* 322–324
Crawford, R. L., Kirk, T. K. and McCoy, E. 1975, Can. J. Microbiol. *21,* 577–579
Cutter, B. E. and Murphey, W. K. 1970, Wood Sci. *3,* 54–58
Dagley, S. 1978, Naturwissenschaften *65,* 85–95
Doi, S., Morohoshi, N. and Haraguchi, T. 1974, Mokuzai Gakkaishi *20,* 230–237
Ellwardt, P. C., Haider, K. and Ernst, L. 1981, Holzforschung *35,* 103–109
Eriksson, K. E. and Hamp, S. 1978, Papier *32,* 545–550

Eriksson, K. E., Grünewald, A., Nilsson, T. and Vallander, L. 1980, Holzforschung *34,* 207–213
Eslyn, W. E., Kirk, T. K. and Effland, M. J. 1975, Phytopathology *65,* 473–476
Ferm, R. and Cowling, E. B. 1972, Svensk Papperstid. *75,* 767–772
Ferm, R. and Nilsson, A. C. 1969, Svensk Papperstid. *72,* 531–536
Ferm, R. and Nilsson, R. 1970, Svensk Papperstid. *73,* 283–286
Ferm, R., Kringstad, K. P. and Cowling, E. B. 1972, Svensk Papperstid. *75,* 859–865
Fukuda, K., Tabe, S. and Haraguchi, T. 1981, Mokuzai Gakkaishi *27,* 691–695
Garves, K. and Dietrichs, H. H. 1975, Holzforschung *29,* 39–40
Greaves, H. 1969, Wood Sci. Technol. *3,* 150–166
Haars, A. and Hüttermann, A. 1980, Naturwissenschaften *67,* 39–40
Haider, K. and Trojanowski, J. 1975, Arch. Microbiol. *105,* 33–42
Haider, K. and Trojanowski, J. 1981, Holzforschung *35,* 33–38
Haider, K., Lim, S. and Flaig, W. 1964, Holzforschung *18,* 81–88
Haider, K., Trojanowski, J. and Sundman, V. 1978, Arch. Microbiol. *119,* 103–106
Hata, K. 1966, Holzforschung *20,* 142–147
Hickin, N. E. 1963, The Insect Factor in Wood Decay. Hutchinson, London
Highley, T. L. 1975, Wood Fiber *6,* 275–281
Highley, T. L. 1976, Mater. Org. *11,* 33–46
Highley, T. L. 1977a, Mater. Org. *12,* 161–174
Highley, T. L. 1977b, Mater. Org. *12,* 25–36
Highley, T. L. 1980, Appl. Environ. Microbiol. *40,* 1145–1147
Highley, T. L., Wolter, K. E. and Evans, F. J. 1981, Wood Fiber *13,* 265–274
Higuchi, T. 1981, Biosynthesis and Biodegradation of Lignin. In: The Ekman Days 1981. Int. Symp. Wood Pulp. Chem., Stockholm, Vol. 3, pp. 16–24
Hiroi, T. 1981, Mokuzai Gakkaishi *27,* 684–690
Hiroi, T. and Eriksson, K. E. 1976, Svensk Papperstid. *79,* 157–161
Hiroi, T., Eriksson, K. E. and Stenlund, B. 1976, Svensk Papperstid. *79,* 162–166
Holt, D. M. and Jones, E. B. G. 1978, Mater. Org. *13,* 13–30
Hüttermann, A., Gebauer, M., Volger, C. and Rösger, C. 1977, Holzforschung *31,* 83–89
Hüttermann, A., Herche, C. and Haars, A. 1980, Holzforschung *34,* 64–66
Ishihara, M. and Shimizu, K. 1980, Mokuzai Gakkaishi *26,* 811–818
Ishihara, M., Shimizu, K. and Ishihara, T. 1978, Mokuzai Gakkaishi *24,* 108–115
Ishikawa, H. and Oki, T. 1964, Mokuzai Gakkaishi *10,* 207–213
Ishikawa, H. and Oki, T. 1966, Mokuzai Gakkaishi *12,* 101–107
Johnson, B. R. and Giovik, L. R. 1970, Proc. Am. Wood Preserv. Assoc. *66,* 234–242
Karlson, P. 1977, Kurzes Lehrbuch der Biochemie, Thieme, Stuttgart
Katayama, T., Nakatsubo, F. and Higuchi, T. 1980, Arch. Microbiol. *126,* 127–132
Katkevich, Y. Y., Katkevich, R. G., Perkone, S., Lipina, D. and Zeltins, R. 1978, Khim. Drev. No. 4, 82–87
Kawakami, H. 1976, Mokuzai Gakkaishi *22,* 252–257
Keilich, G., Bailey, P. J., Afting, E. G. and Liese, W. 1969, Biochim. Biophys. Acta *185,* 392–401
Keilich, G., Bailey, P. and Liese, W. 1970, Wood Sci. Technol. *4,* 273–283
Kindl, H. and Wöber, G. 1975, Biochemie der Pflanzen. Springer, Berlin, Heidelberg, New York
Kirk, T. K. 1971, Ann. Rev. Phytopathol. *9,* 185–210
Kirk, T. K. 1973, Phytopathology *63,* 1504–1507
Kirk, T. K. 1975a, Biotechnol. Bioeng. Symp. No. 5, 139–150
Kirk, T. K. 1975b, Holzforschung *29,* 99–107
Kirk, T. K. and Adler, E. 1970, Acta Chem. Scand. *24,* 3379–3390
Kirk, T. K. and Chang, H. 1974, Holzforschung *28,* 217–222
Kirk, T. K. and Chang, H. 1975, Holzforschung *29,* 56–64
Kirk, T. K. and Highley, T. L. 1973, Phytopathology *63,* 1338–1342

Kirk, T. K. and Lundquist, K. 1970, Svensk Papperstid. *73,* 294–306
Kirk, T. K. and Moore, W. E. 1972, Wood Fiber *4,* 72–79
Kirk, T. K. and Yang, H. H. 1979, Biotechnol. Lett. *1,* 347–352
Kirk, T. K., Connors, W. J., Bleam, R. D., Hackett, W. F. and Zeikus, J. G. 1975a, Proc. Nat. Acad. Sci. USA *72,* 2515–2519
Kirk, T. K., Chang, H. and Lorenz, L. F. 1975b, Wood Sci. Technol. *9,* 81–86
Kirk, T. K., Connors, W. J. and Zeikus, J. G. 1977, Advances in Understanding the Microbiological Degradation of Lignin. In: The Structure, Biosynthesis and Degradation of Wood (Loewus, F. A. and Runeckles, V. C., Eds.). Plenum Press, New York, pp. 369–394
Kirk, T. K., Yang, H. H. and Keyser, K. 1978, Developm. Industr. Microbiol. *19,* 51–61
Kirk, T. K., Higuchi, T. and Chang H. M. 1980, Lignin Biodegradation: Microbiology, Chemistry and Potential Applications. CRC Press, Boca Raton, Florida
Koenigs, J. W. 1972, Phytopathology *62,* 100–110
Koenigs, J. W. 1974a, Wood Fiber *6,* 66–80
Koenigs, J. W. 1974b, Arch. Mikrobiol. *99,* 129–145
Krejcberg, Z. N., Ozolinja, N. R. and Sergeeva, V. N. 1974a, Khim. Drev. No. 2, 99–102
Krejcberg, Z. N., Arončik, B. M. and Sergeeva, V. N. 1974b, Khim. Drev. No. 1, 74–77
Krisnangkura, K. and Gold, M. H. 1979, Holzforschung *33,* 174–176
Kühne, H., Leukens, U., Sell, J. and Wälchli, O. 1970, Holz Roh-Werkst. *28,* 223–229
Kuwahara, M., Takegami, H., Yonehana, M., Sato, T. and Iwahara, S. 1981, Mokuzai Gakkaishi *27,* 885–892
Lehninger, A. L. 1979, Biochemie. Verlag Chemie, Weinheim, New York
Levi, M. P. and Preston, R. D. 1965, Holzforschung *19,* 183–190
Levy, J. F. 1975, Bacteria Associated with Wood in Ground Contact. In: Biological Transformation of Wood by Microorganisms (Liese, W., Ed.). Springer, Berlin, Heidelberg, New York, pp. 64–73
Lewis, P. F. 1976, Mater. Org. Beih. No. 3, 113–119
Liese, W. 1964, Holz Roh-Werkst. *22,* 289–295
Liese, W. 1965, Mater. Org. Beih. No. 1, 13–26
Liese, W. 1970, Ann. Rev. Phytopathol. *8,* 231–258
Liese, W. 1975, Biological Transformation of Wood by Microorganisms. Springer, Berlin, Heidelberg, New York
Liese, W. and Greaves, H. 1975, Micromorphological of Bacterial Attack. In: Biological Transformation of Wood by Microorganisms (Liese, W., Ed.). Springer, Berlin, Heidelberg, New York, pp. 74–88
Liese, W. and Schmid, R. 1964 Phytopathol. Z. *51,* 385–393
Liese, W. and Schmid, R. 1966, Holz Roh-Werkst. *24,* 454–460
Mahler, H. R. and Cordes, E. H. 1971, Biological Chemistry. Harper and Row Publ., New York, Evanston, San Francisco, London
Martin, H. H., Preusser, H. J. and Verma, J. P. 1968, Arch. Mikrobiol. *62,* 72–84
Mennega, A. M. W. and Jutte, S. M. 1972, Acta Bot. Neerl. *21,* 343–345
Meyer, H. P. 1980, Die β-Glucosidasen von *Sporotrichum thermophile* (Apinis): Regulation, Lokalisation und Charakterisierung. Doctor Thesis, Universität Freiburg, Switzerland
Monties, B., Odier, E., Janin, G. and Czaninski, Y. 1981, Holzforschung *35,* 217–222
Nakatsubo, F., Kirk, T. K., Shimada, M. and Higuchi, T. 1981, Arch. Microbiol. *128,* 416–420
Nečesaný, V. 1963, Holzforschung *17,* 57–60
Nečesaný, V. 1965, Mater. Org. Beih. No. 1, 27–39
Nečesaný, V. and Sterka, V. 1977, Drev. Vysk. *22,* 42–53
Nilsson T. 1974a, Studia For. Suec. No. 114
Nilsson, T. 1974b, Studia For. Suec. No. 117
Nilsson, T. 1974c, Mater. Org. *9,* 173–198
Nilsson, T. 1976, Mater. Org. Beih. No. 3, 103–112

Noguchi, A., Shimada, M. and Higuchi, T. 1980, Holzforschung *34,* 86–89

Ohta, M., Higuchi, T. and Iwahara, S. 1979, Arch. Microbiol. *121,* 23–28

Oki, T., Watanabe, H. and Ishikawa, H. 1981, Mokuzai Gakkaishi *27,* 696–702

Ozolinja, N. R., Krejcberg, Z. N. and Sergeeva, V. N. 1974, Khim. Drev. No. 2, 96–98

Paice, M. G. and Jurasek, L. 1979, Structural and Mechanistic Comparison of some β-(1–4) Glycoside Hydrolases. In: Hydrolysis of Cellulose: Mechanisms of Enzymatic and Acid Catalysis (Brown, R. D., Jr. and Jurasek, L., Eds.). Adv. Chem. Ser. 181, ACS, Washington, D. C., pp. 361–374

Peek, R. D., Liese, W. and Parameswaran, N. 1972, Eur. J. For. Pathol. *2,* 237–248

Pettersson, G., Fägerstam, L., Bhikhabhai, R., Leandoer, K. 1981 The Cellulase Complex of *Trichoderma reesei* QM 9414. In: The Ekman Days 1981. Int. Symp. Wood Pulp. Chem., Vol. 3, pp. 39–42

Polčin, J. and Bezúch, B. 1978, Wood Sci. Technol. *12,* 149–158

Radtke, D., Kilbertus, G. and Mangenot, F. 1981, Holzforschung *35,* 141–148

Reese, E. T. 1975, Polysaccharases and the Hydrolysis of Insoluble Substrates. In: Biological Transformation of Wood by Microorganisms (Liese, W. Ed.). Springer, Berlin, Heidelberg, New York, pp. 165–181

Reese, E. T. 1977, Degradation of Polymeric Carbohydrates by Microbial Enzymes. In: The Structure, Biosynthesis and Degradation of Wood (Loewus, F. A. and Runeckles, V. C., Eds.). Plenum Press, New York, London, pp. 311–367

Rösch, R. 1972, Zentralbl. Bakteriol. II *127,* 555–563

Rösch, R. and Dietrichs, H. H. 1971, Arch. Mikrobiol. *75,* 197–202

Rösch, R. and Liese, W. 1971, Arch. Mikrobiol. *76,* 212–218

Rösch, R., Liese, W. and Berndt, H. 1969, Arch. Mikrobiol. *67,* 28–50

Ruel, K., Barnoud, F. and Eriksson, K. E. 1981, Holzforschung *35,* 157–171

Rypáček, V. 1966, Biologie holzzerstörender Pilze. VEB G. Fischer, Jena

Rypáček, V. 1977, Wood Sci. Technol. *11,* 59–67

Schmid, R. and Baldermann, E. 1967, Naturwissenschaften *54,* 521

Schmid, R. and Liese, W. 1964, Arch. Mikrobiol. *47,* 260–276

Schmid, R. and Liese, W. 1965, Phytopathol. Z. *54,* 275–284

Schmidt, H. 1962, Tierische Schädlinge im Bau- und Werkholz. Parey, Hamburg, Berlin

Schmidt, O. 1978, Holzforschung *32,* 214–215

Schmidt, O. 1980, Mater. Org. *15,* 207–224

Schmidt, O. and Bauch, J. 1980, Wood Sci. Technol. *14,* 229–239

Schmidt, O. and Dietrichs, H. H. 1976, Mater. Org. Beih. No. 3, 91–102

Schmidt, O. and Liese, W. 1980, Holzforschung *34,* 67–72

Seifert, K. 1964, Holz Roh-Werkst. *22,* 405–409

Seifert, K. 1966, Holz Roh-Werkst. *24,* 185–189

Seifert, K. 1967, Holz Roh-Werkst. *25,* 377–379

Seifert, K. 1968, Holz Roh-Werkst. *26,* 208–215

Sinner, M. and Dietrichs, H. H. 1975, Holzforschung *29,* 123–130, 168–177, 207–214

Sinner, M. and Dietrichs, H. H. 1976a, Holzforschung *30,* 50–59

Sinner, M. and Dietrichs, H. H. 1976b, Holz Roh-Werkst. *34,* 63–66

Sinner, M., Parameswaran, N. and Dietrichs, H. H. 1978, Papier *32,* 530–538

Stryer, L. 1975, Biochemistry. Freeman, San Francisco

Wagenführ, R. and Steiger, A. 1966, Pilze auf Bauholz. Ziemsen, Wittenberg

Westermark, U. and Eriksson, K. E. 1974, Acta Chem. Scand. B *28,* 204–214

Wilcox, W. W. 1978, Wood Fiber *9,* 252–257

Wolter, K. E., Highley, T. L. and Evans, F. J. 1980, Biochem. Biophys. Res. Commun. *97,* 1499–1504

Wood, T. M. 1981, Enzyme Interactions Involved in Fungal Degradation of Cellulosic Materials. In: The Ekman Days 1981. Int. Symp. Wood Pulp. Chem. Stockholm, Vol. 3, pp. 31–38

Wood, T. M. and McCrae, S. I. 1979, Synergism between Enzymes Involved in the Solubilization of Native Cellulose. In: Hydrolysis of Cellulose: Mechanisms of Enzymatic and Acid Catalysis (Brown, R. D., Jr. and Jurasek, L., Eds.). Adv. Chem. Ser. 181, ACS, Washington, pp. 181–209

Yamazaki, N. and Dietrichs, H. H. 1979, Holzforschung *33,* 36–42

Yamazaki, N., Sinner, M. and Dietrichs, H. H. 1976, Holzforschung *30,* 101–109

Yang, H. H., Effland, M. J. and Kirk, T. K. 1980, Biotechnol. Bioeng. *22,* 65–77

Zainal, A. S. 1976, Mater. Org. Beih. No. 3, 121–127

Zeikus, I. G. and Ward, J. C. 1974, Science *184,* 1181–1183

Zenker, R. 1962, Der Abbau der Zugholzlamelle durch holzzerstörende Pilze: In: Holzzerstörung durch Pilze, Int. Symp. Eberswalde. Akademie-Verlag, Berlin, pp. 77–81, 379–380

15. Aging and Fossilization

15.1. Kinds of Conversion

Wood is primarily the strengthening tissue of a tree. After a tree has died its wood has no further function and is degraded as quickly as possible by microorganisms so that its basic components can return to the natural cycle. In prehistoric times wood was sometimes able, under favourable conditions, to age naturally without the application of chemical preservatives. Thus, for example, the hot, dry climate of Egypt has preserved wooden sculptures for 4 000 to 5 000 years, and wooden spears and arrows have been preserved for over 100 000 years in muddy soils which preclude the presence of oxygen. In addition to wooden artifacts made by humans, wood can also be preserved under certain natural conditions which cause a very, almost infinitely slow process of aging resulting in fossilized wood. Two different types of fossilization should be distinguished: silicification and carbonization.

Scientists have shown great interest in what happens during the aging of wood, and accordingly a large number of publications deal with this matter, beginning with excavated wood samples some hundreds or thousands of years old and ranging to fossilized samples 100 million or more years of age. The results of earlier studies were summarized by Sen (1956; Sen, Basak 1957).

Environmental conditions determine which kind of changes will take place in wood. Samples a few hundred years old may be more degraded than those which are several thousand years old. Thus in Alaska well preserved pieces of pine wood (probably *Pinus monticola*) about 40 000 years old were recovered 760 m below the surface of the earth (Smith et al. 1973).

Borgin et al. (1979) studied the physical properties, particularly density and shrinkage, of old woods (*Picea abies, Pinus sylvestris, Quercus* spec.) with ages of 300 to 100 000 years. In most cases density was reduced and shrinkage increased. These changes did not relate to time of exposure, but to species, state of wood (sapwood, heartwood) and condition of exposure.

Silicification means an impregnation of the cell walls with water rich in soluble minerals. These minerals crystallize slowly within the walls and replace the cell wall components. Minerals such as quartz, magnesite, pyrite, serpentin, chalcedony, opal etc. have been identified in silicified wood samples (Mitra, Sen 1956; Scurfield et al. 1974).

Various influences determine carbonization. It is assumed that the process resulting in peat and brown coal has its beginning in a partial degradation by aerobe microorganisms, whereas the formation of pit-coal, anthracite and graphite begins

Table 15–1: Elementary analyses of wood and various carbonized products

Source	Age Million years	Carbon %	Hydrogen %	Oxygen %
Wood		48–51	5–7	43–45
Peat	0.6	55–60	5–6	34–38
Brown coal	0.6–60	65–76	5–6	17–25
Pit-coal	60–280	78–91	4–5	4–15
Anthracite	280–350	92–98	1–4	1–3
Graphite	800–1 100	98–100	0–2	0

with an attack of anaerobe microorganisms in the absence of oxygen. In a later phase geochemical processes influenced by pressure and high temperatures come into effect. During carbonization the relative amount of carbon in the organic matter increases, while the amounts of oxygen and hydrogen decrease (Table 15–1).

15.2. Structural Studies

The microscopic investigation of fossilized wood may be used to

- identify the species and classify it within the plant system;
- study the degradation and conversion process in the cell walls.

Only studies of the second type are within the scope of this book.

Wood samples of a relatively young age, i.e. up to several thousand years, show no or only little change in cell wall structure provided that no attack by microorganisms took place.

Borgin et al. (1975a) studied several softwoods and hardwoods from excavations in Norway (*Pinus sylvestris,* 900 and 2 200 years old; *Quercus robur,* 1 100 years old) and from pyramids in Egypt (*Juniperus phoenicea,* 4 000 years old; *Pinus pinea,* 4 300 years old; *Acacia nilotica,* 4 400 years old). Though there was no visible change in the cell wall structure the authors found an increased sensitivity of the samples against mechanical influences during the microscopical preparation. The birefringence of the samples was reduced as compared to new wood indicating a reduction in the amount of crystalline cellulose. Similar results were obtained by Chowdhury et al. (1967) with wood of *Shorea robusta* with an age of about 2 000 years.

The cell walls may be preserved millions of years. Samples of conifers were found with ages between 10 and 180 million years, the walls of which were still recognizable though details such as pitting and microfibrillar structure have partly or com-

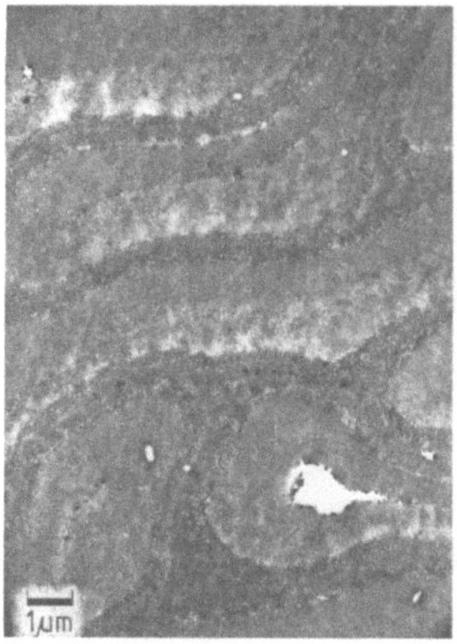

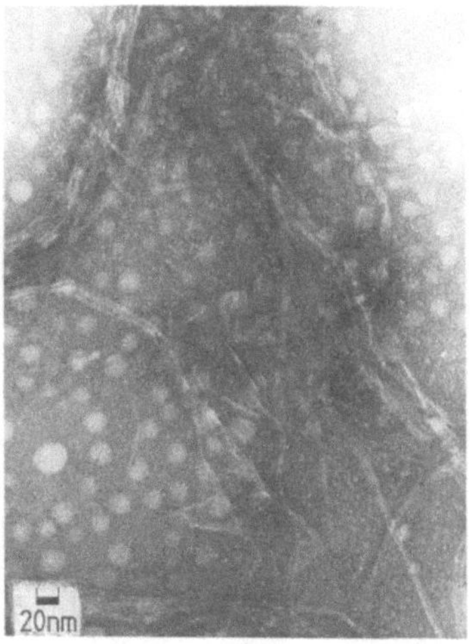

Fig. 15–1. Collapsed cells in a *Protopinacea* with an age of 180 million years. The compound middle lamella and the secondary wall layers are still visible. TEM micrograph.

Fig. 15–2. Cellulose fibrils isolated from a 20-million-year-old *Taxodioxylon*. TEM micrograph.

pletely disappeared (Wayman et al. 1971; Grosser et al. 1974; Fengel 1976) (Fig. 15–1). A fossil *Phyllocladus* wood from Oligocene (about 50 million years ago) yielded, after delignification, fibres which did not differ markedly from that of modern pulp fibres after a severe treatment (Crook et al. 1965). Cellulose fibrils from 20-million-year-old *Taxodioxylon* showed the same structure as those from recent woods even at high electron microscopic magnification (Fengel 1974) (Fig. 15–2). Purelis (1962) detected in coal from fossil ferns microstructures in the dimensions of cellulose fibrils.

Apart from the conversion processes which preserve the cell wall, in some cases even in its ultrastructure, other processes occur during which the cell walls are degraded. Fengel (1971; Fengel et al. 1973; Bednar, Fengel 1974) describes a degradation of the cell walls beginning at the lumenside and proceeding towards the middle lamella with increasing age. In the first stage layers of dark conversion products are deposited at the lumenside of the cell walls (Fig. 15–3a, b). Later on the dark conversion substance fills up the whole lumen (Fig. 15–3c), and finally only single lamellae of the cell wall and residues of the middle lamella are incorporated in the dark substance (Fig. 15–3d). The observation of differences in the degradation of earlywood and latewood cells, i.e. the latter being more resistant to conversion, gave rise to the assumption that the penetration of water influences this kind of degradation process.

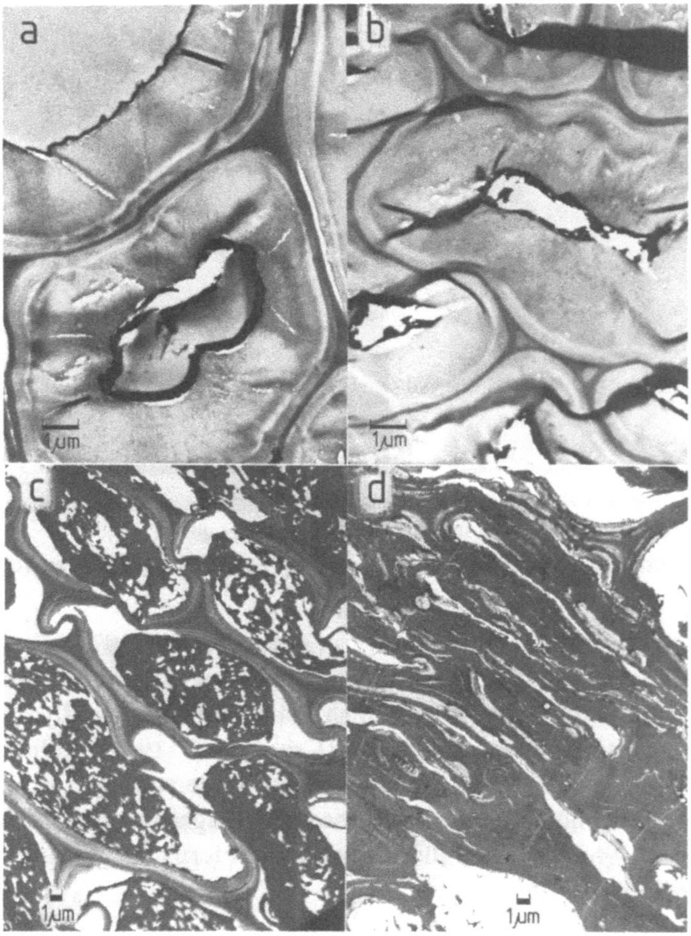

Fig. 15–3. Degradation and conversion of the cell wall substance in wood species of various ages. TEM micrographs.
a) Oak (*Quercus* spec.), 8 500 years old.
b) Buckthorn *(Hippophae rhamnoides)*, 20 000 years old.
c) Spruce *(Picea abies)*, 100 000 years old.
d) Fossil redwood *(Taxodioxylon gypsaceum)*, 20 million years old.

15.3. Chemical Investigations

Apart from the morphological view the change of physical properties indicates a chemical conversion of the cell wall components. Studies of oak samples *(Quercus robur, Q. petraea)* of ages between 400 and 8 500 years resulted in an increase of sorption and swelling and a decrease of mechanical properties depending on the degree of degradation (Dzbeński 1970a, b; Bednar, Fengel 1974; Scheiber, Wagenführ 1976; Kommert 1980).

The chemical analyses of old and fossilized woods generally show a decrease of polysaccharides and an increase of the non-hydrolyzable residue (NHR) depending, respectively, on age and on degree of degradation. In relatively young samples with a strong degradation of polysaccharides the presence of microorganisms was proved (Ważny 1976).

Several studies of ancient oak samples (Table 15–2) as well as ancient samples of birch, ash and pine (Kazanskaya et al. 1975) show that the degree of degradation is strongly influenced by environmental conditions, particularly in the early stages. Thus the oak samples of Lednickie and Wilgi (900 years old) had a relatively low amount of polysaccharides whereas the much older sample of Grossenzersdorf (8 500 years old) had a relatively high amount of polyoses as compared to the younger samples (Table 15–2).

From the relative increase of the cellulose content in most of the samples in Table 15–3 it can be seen that degradation and conversion begin predominantly at the polyoses. Among the polyoses the more highly soluble ones, particularly the pentosans, are degraded more quickly than the less soluble ones. The polyoses of recent oak heartwood soluble in 5% KOH amount to 22.4%. This value decreases to 12.9% in a 8 500-year-old sample, whereas the value for polyoses soluble in 24% KOH decreases from 6.1% to only 5.4% (Bednar, Fengel 1974).

Though the degradation of polysaccharides begins at an early stage of the fossilizing process parts of them may survive for many millions of years. Wayman et al. (1971) found about 2% sugar in a 100-million-year-old conifer. Brasch and Jones (1959) isolated 1.1% holocellulose from a wood sample from the Lower Cretaceous period (140 million years ago), the main sugars in the hydrolyzates being glucose and

Table 15–2: Analyses of recent and ancient oak samples

Age Years	Place of discovery	Holocellulose %	Cellulose %	Polyoses %	Lignin %	Ash %	Extract %	References
Recent	Sapwood	78.7	39.9	27.7	24.9	0.5	2.4	
	Heartwood	77.0	37.6	28.5	24.5	0.3	4.4	Bednar, Fengel (1974)
600	"Kogge" of Bremen (FRG)	46			36	5.3		Dietrichs (1964)
900	Ostrów Tumski Poznań (Poland)		46	13*	29	2.2		Grzeczyński, Surmiński (1962)
900	Sypniewo	45.1	45.1	19.7	33.2	2.0	1.0	
	Lednickie		29.7	15.8	47.1	4.0		
	Wilgi (Poland)	28.0			59.0–64.1	1.1	0.6	Dzbeński (1970b)
3 000	Kyffhäuser (GDR)				47.8	17.3		Kommert, Wienhaus (1970)
4 700	Bitterfeld (GDR)		43.2	17.5*	34.9	0.8	5.1	Scheiber, Wagenführ (1976)
8 500	Großenzersdorf (Austria)	60.0	43.4	18.3	29.6	1.5	3.5	Bednar, Fengel (1974)

* Pentosans only

Table 15–3: Analyses of various fossil wood samples and of some recent woods for comparison

Species	Age Years	Holo-cellulose %	Cellulose %	Polyoses %	NHR %	E-B-extract %	Cold water extract %	Hot water extract %	Ash %	References
Picea abies	Recent	81.0	56.5	15.3	27.3	2.0	1.0	2.0	0.3	
Picea abies	10^5	17.6	12.2	4.5	74.4	1.1	1.2	2.3	2.3	Fengel (1971)
Pinus strobus	10^7	38.5	22.3		80.6				3.1	Wayman et al. (1971)
Cedrus penhallowii	$2 \cdot 10^7$		29.2	17.4	44.3	4.5	3.3		3.4	Brasch, Jones (1959)
Sequoia sempervirens	Recent		49.9	16.7	37.0	13.5	3.9	8.7	0.2	
Taxodioxylon gypsaceum	$2 \cdot 10^7$		14.5	2.6	71.2	14.3	1.8	2.4	1.2	Fengel et al. (1973)
Softwood	$2 \cdot 10^7$	11.8	10.5	1.2	89.9	1.7	0.1	0.6	1.1	
Hardwood	$2 \cdot 10^7$	0.8			93.3	2.3	0.5	1.3	1.1	Fengel (1976)
Phyllocladus asplenifolius	Recent		51.2	7.2*	31.0	15.9		14.0	0.1	
Phyllocladus asplenifolius	$3 \cdot 10^7$		35.8	0.7*	68.6	8.4		1.4	3.4	
Araucaria spec.	$3 \cdot 10^7$		13.6	0.7*	79.1	3.7		4.2	5.3	Crook et al. (1965)

* Pentosans only

mannose. A *Protopinacea* contained very small amounts of cellulose and polyoses even after 180 million years as the hydrolyzates consisted of glucose, mannose and xylose (Fengel 1976).

Though lignin seems to persist for millions of years, it can change even within a relatively short timespan. Borgin et al. (1975b) found oxidative changes in the lignin molecules of samples with ages of 900–4 400 years. An oxidation of lignin was also detected in samples which were 30 million years old (Crook et al. 1965). A loss of two-thirds of the methoxyl content was determined by Wayman et al. (1971). The infrared spectrum indicates the presence of condensed aromatic ring systems, products mainly deriving from lignin.

The conversion of lignin to condensed aromatic rings proceeds with age; pit-coal contains a considerable number of these ring systems. The ring condensation is the reason for the increase of the carbon content during the carbonization process as shown in Table 15–1. At the end of this process stands graphite, a carbon modification with a hexagonal connection of the carbon atoms forming layers arranged one upon the other. Aromatic rings may also derive from degradation products of the polysaccharides. It is known that e.g. pyruvic aldehyde can easily be aromatized via quinones, a process occurring during the formation of humic acids (Wayman et al. 1971).

Literature

Bednar, H. and Fengel, D. 1974, Holz Roh-Werkst. *32,* 99–107
Borgin, K., Parameswaran, N. and Liese, W. 1975a, Wood Sci. Technol. *9,* 87–98
Borgin, K., Faix, O. and Schweers, W. 1975b, Wood Sci. Technol. *9,* 207–211
Borgin, K., Tsoumis, G. and Passialis, C. 1979, Wood Sci. Technol. *13,* 49–57
Brasch, D. J. and Jones, J. K. N. 1959, Tappi *42,* 913–920
Chowdhury, K. A., Preston, R. D. and White, R. K. 1967, Proc. Roy. Soc. London, Ser. B *168,* 148–157
Crook, F. M., Nelson, P. F. and Sharp, D. W. 1965, Holzforschung *19,* 153–156
Dietrichs, H. H. 1964, unpublished. According to Noack, D. 1969, Zur Verfahrenstechnik der Konservierung des Holzes der Bremer Kogge. In: Die Bremer Hanse-Kogge, Verlag F. Röver, Bremen, pp. 127–156
Dzbeński, W. 1970a, Holztechnologie *11,* 57–62
Dzbeński, W. 1970b, Sylvan *64,* 1–27
Fengel, D. 1971, Holz Roh-Werkst. *29,* 305–314
Fengel, D. 1974, Naturwissenschaften *61,* 450
Fengel, D. 1976, Holz Roh-Werkst. *34,* 459–463
Fengel, D., Grosser, D. and Wegener, U. 1973, Palaeontographica *B 144,* 31–43
Grosser, D., Fengel, D. and Selmeier, A. 1974, Forstwiss. Centralbl. *93,* 332-346
Grzeczyński, T. and Surmiński, J. 1962, Folia For. Polon. No. 4 B, 145–153
Kazanskaya, S. Y., Vikhrov, Y. V. and Golman, L. P. 1975, Khim. Drev. No. 2, 41–44
Kommert, R. 1980, Holztechnologie *21,* 230–233
Kommert, R. and Wienhaus, O. 1970, Holztechnologie *11,* 177–182
Mitra, G. B. and Sen, J. 1956, Am. J. Sci. *254,* 257–259
Purelis, G. A. 1962, Am. J. Bot. *49,* 663
Scheiber, C. and Wagenführer, R. 1976, Holztechnologie *17,* 133–139
Scurfield, G., Segnit, E. R. and Anderson, C. A. 1974, Proc. Workshop SEM & Plant Sci., ITT Res. Inst. Chicago, 385–396
Sen, J. 1956, Bot. Rev. *22,* 343–374
Sen, J. and Basak, K. 1957, Geolog. För. i. Stockholm Förhandl. *79,* 737–758
Smith, B. N., Rivera, E. R. and Keel, J. R. 1973, Naturwissenschaften *60,* 202
Wayman, M., Azhar, M. R. and Koran, Z. 1971, Wood Fiber *3,* 153–165
Ważny, J. 1976, Mater. Org. Beih. No. 3, 53–60

16. Pulping Processes

16.1. Introductory Survey

As pointed out in the introduction (→ 1.) the production of pulp is the most important technique for the chemical conversion of wood. This fact is illustrated by current figures of world-wide pulp and paper production and the dynamic growth predicted for it to the year 2000, with assumed annual increases between 2 and 4% (Fig. 1–2) (Keays 1975; Hagemeyer 1976; FAO 1977; Patt 1978).

The total wood mass required today for chemical and mechanical pulping is about 460 Mio. m^3, corresponding to more than one-third of the total harvested timber mass (2 600 Mio. m^3) after subtraction of fuel wood consumption (1 500 Mio. m^3).

Table 16–1 shows the 8 largest pulp producing countries, covering about 85% of the total world pulp production.

Apart from uncertainties concerning the future consumption figures for pulp and paper in the developing countries, it is evident that even in countries with a high standard of living and slowly increasing population paper consumption is still expanding (Table 1–2). Production and consumption data in the future will be influenced by, among other things, the following factors:

- general growth rates of the gross national product (GNP) in the industrial and developing countries which is known to have a strong influence on paper consumption;
- pulp production costs, resulting from:
  wood prices

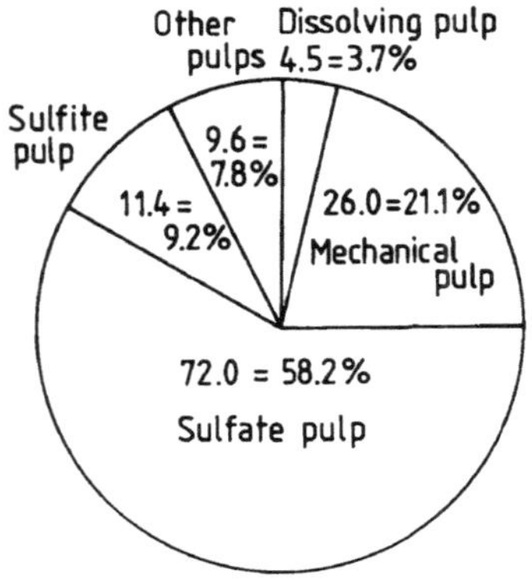

Fig. 16–1. Portions of different pulp types related to total world pulp production 1979 (million t) (according to VDP 1981).

Table 16–1: The most important pulp producing countries (VDP 1981)

	Production 1979 (1 000 t)	
	Chemical pulp	Mechanical pulp
U.S.A.	41 500	4 000
Canada	12 100	7 600
Japan	8 200	1 800
U.S.S.R.	7 600	1 900
Sweden	7 100	2 000
Finland	4 800	2 200
Brazil	2 300	150
France	1 500	450

labour costs
energy costs
costs for environmental protection
initial capital investment.

It is obvious that most of these factors will be quite different e.g. in Canada as compared to Brazil. Other factors influencing the future pulp and paper scene are the consumption development of different paper grades and the choice of pulp types to produce paper and paperboards of different types and qualities (Patt 1978; Gorsler 1980; Göttsching 1980a; Annergren 1981a).

With regard to the pulping processes some general trends may be outlined here in brief. The term pulp is used generically for stone groundwood, refiner mechanical pulps, semichemical and chemical pulps. The raw material basis is characterized by the increased use of hardwoods from the temperate zones and the tropics (Ekström 1976; Hergert 1976; Welte, Patt 1978; Kubes et al. 1979; Phillips et al. 1979; Foelkel, Zvinakevicius 1980). The use of residues from other wood manufacturing industries (e.g. saw mills, planing mills) is becoming more and more important. Of greater importance to the future may be chips from whole-tree or full-tree utilization, and the use of nonwood fibres, particularly in many developing countries (Eskillson 1972, 1974; Eskillson, Hartler 1973; Keays 1974; Nyholm, Auchter 1976; Virtanen 1976; Misra 1980). Wastepaper and waste paperboard are already indispensible fibre sources and will become even more important in the future due to improved techniques of secondary fibre pulping (Coletti, Buongiorno 1980; Koffinke 1980).

Besides optimizing pulp quality and quality control important aims of improved pulping technologies are:

- improved pulp yields
- reduction of energy consumption
- reduction of the amounts of chemicals required for pulping and bleaching, including improved recovery processes for the chemicals
- reduction of air and water pollution

- development of sulfur-free pulping processes and chlorine-free bleaching sequences
- high flexibility with regard to pulp yield, quality and bleachability
- process conditions allowing the suitable preparation of pulping by-products
- profitable smaller production units requiring lower set-up costs for new mills and reduced raw material demand.

These and other aims form the background to an increasing diversification of the pulping processes. Important process modifications are alkaline pulping with additives (e.g. soda-AQ, → 16. 4. 5.), modified refiner mechanical pulping (e.g. chemithermomechanical pulping, CTMP, → 16. 2. 2.) or sulfur-free pulping (e.g. soda-oxygen pulping, organosolv pulping, → 16. 4. 6., 16. 6.) (Jensen 1979a, b; Palenius 1980).

Table 16–2 gives a classification of pulping processes combined with a survey of important data.

In general it can be stated that pulping and bleaching procedures will be modified and combined in the future in order to face in particular the monstrous increase of costs and the severe pollution regulations in many countries. The importance of mechanical and high-yield pulps is increasing.

Though an increasingly more sophisticated pulping market is to be expected in the future the dominant role of the kraft pulp will remain. This type of pulp covers more than one-half of the total chemical and mechanical pulp production (58.2%) or nearly three-quarters of the chemical pulps (73.5%) (Fig. 16–1).

Within this chapter a short description is given of established processes but emphasis is directed to latest developments and trends. Due to limitations of space technical aspects of the processes are more or less excluded. More detailed information on pulping principles and processes is given in the large-scale monographs by Rydholm (1965), Lengyel and Morvay (1973) and Casey (1980).

16.2. Mechanical Pulping

16.2.1. Stone Grinding

The stone groundwood process is the oldest process for converting wood into pulp and its invention by Keller in 1843 marked a milestone in the history of paper making. Since that time wood has been the predominant raw material for paper.

Within the group of mechanical pulping processes (Table 16–2) the stone groundwood process, including some modifications, still holds a predominant position covering about 90% of world-wide mechanical pulp production in 1977 and about 75% in 1981 (Copeland 1977; Venter 1981; Collicutt et al. 1981).

Table 16–2: Survey of pulping processes (McGovern 1980; Annergren 1981b; Wegener 1981, 1983)

Process/Pulp type		Chemical treatment	Mechanical treatment	Wood+	Yield %
Mechanical pulping					80–99
Stone grinding*	Groundwood (SGW)	None	Grindstone	S	93–99
	Steamed groundwood	Steam	Grindstone	S	80–90
**	Pressure groundwood (PGW)	None	Grindstone (pressure)	S	
Refiner pulping	Refiner mechanical pulp (RMP)	None	Disk refiner	S	93–98
**	Pressurized refiner mechanical pulp (PRMP)	None	Disk refiner (pressure)	S	
	Thermomechanical pulp (TMP)	Steam	Disk refiner (pressure)	S	91–98
	Asplund pulp	Steam	Disk refiner	S	80–90
Chemimechanical and Chemi-thermomechanical pulping					65–97
Stone grinding*	Chemigroundwood (CGW)	Neutral sulfite	Grindstone	S/H	80–92
		or Acidic sulfite		S/H	80–90
		or $Na_2S + NaOH$		S/H	85–90
Refiner pulping	Chemi-refiner mechanical pulp (CRMP)	$NaOH$ or $NaHSO_3$ or Alkaline sulfite or Acidic sulfite	Disk refiner	S/H	80–90
	Chemi-thermo-mechanical pulp (CTMP)	Steam + Na_2SO_3 + $NaOH$	Disk refiner (pressure)	S/H	65–97
Semichemical pulping					65–92
	Neutral sulfite (NSSC-pulp)	$Na_2SO_3 + Na_2CO_3$ or $NaHCO_3$	Disk refiner	H	65–90
	Cold soda	$NaOH$	Disk refiner	H	85–92
	Alkaline sulfite	Na_2CO_3, Na_2S, $NaOH$	Disk refiner	H/S	80–90
	Sulfate	$NaOH + Na_2S$	Disk refiner	H/S	75–85
	Soda	$NaOH$	Disk refiner	H	65–85
	Green liquor	$Na_2S + Na_2CO_3$	Disk refiner	H	65–85
	Nonsulfur	$Na_2CO_3 + NaOH$	Disk refiner	H	65–85

Forts. Table 16–2

Process/Pulp type		Chemical treatment	Mechanical treatment	Wood⁺	Yield %
High-yield chemical pulping					55–70
	Kraft	Na_2S + NaOH	Disk refiner	S/H	55–65
	Sulfite	Acidic sulfite (Ca, Na, Mg) or Bisulfite (Na, Mg)	Disk refiner	S	55–70
Full chemical pulping					30–60
Alkaline pulping	Kraft (+ AQ)	NaOH + Na_2S (+ AQ)	Mild to none	S/H	40–55
	Kraft (Polysulfide)	$(NaOH + Na_2S)_x$	None	S/H	45–60
	Soda	NaOH	None	H	40–55
	Soda-AQ	NaOH + AQ	Mild to none	H	45–55
	Soda-oxygen, two-stage	NaOH, oxygen	Disk refiner	H	45–60
Sulfite pulping	Acidic sulfite	Acidic sulfite (Ca, Na, Mg, NH_3)	Mild to none	S (no pine)	45–55
	Bisulfite	Bisulfite (Na, Mg, NH_3)	Mild to none	S/H	45–60
	Neutral sulfite	Neutral sulfite	Mild to none	S/H	45–60
	Magnefite	Mg-bisulfite	Mild to none	S/H	45–60
	Multi-stage sulfite	Na_2SO_3 + $NaHSO_3/SO_2$ or $NaHSO_3$ + SO_2/Na_2CO_3	None	S/H	45–55
**	Alkaline sulfite	Na_2SO_3 + NaOH	None	S/H	
Dissolving pulping	Acidic sulfite	Acidic sulfite (Ca, Na)	None	H/S	35–42
	Prehydrolysis kraft	Na_2S + NaOH after prehydrolysis	None	H/S	30–35

* Wood used in the form of bolts. All the other processes use chips.
⁺ Wood mainly used (S: softwood, H: hardwood)
** Not commercially established

In principle logs of debarked wood (60–120 cm long, predominantly softwoods, but also suitable hardwoods) are pressed with their long side parallel to the surface of a rotating grindstone while water is sprayed on the grinding zone. Friction raises the temperature in the grinding zone to 150–190 °C thus plasticising the lignin component of wood. Fibre bundles, fibres and fibre fragments are torn out of the wood surface and transported to the grinder pit. A schematic illustration of the grinding operation is given in Fig. 16–2. An exact theory of the proceedings during grinding

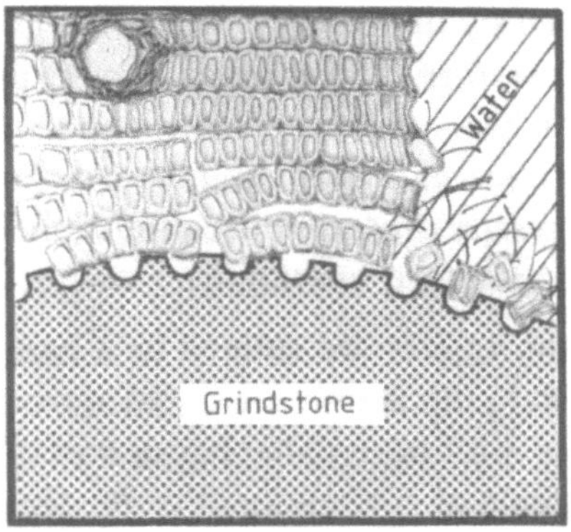

Fig. 16–2. Schematic illustration of the grinding operation.

has not yet been established, but it is generally accepted that the grinding procedure involves a defibration of the wood surface by the action of the grindstone grits and a simultaneous refining of fibres and fibre bundles to smaller units (Klemm 1957a; Atack, May 1962; Bruun et al. 1979a; Chang et al. 1979).

The groundwood properties can be influenced by several variables. Above all the wood feed value must be chosen in such a way that during the so-called initial grinding time of generally 50–60 ms (quotient of the grinding zone length and the stone surface velocity) a wood layer is ground corresponding to the average fibre diameter of the respective wood (e.g. about 40 μm for spruce). Otherwise fibres are split lengthwise or slivered, thus reducing the freeness value of the groundwood pulp (Klemm 1957a, 1978).

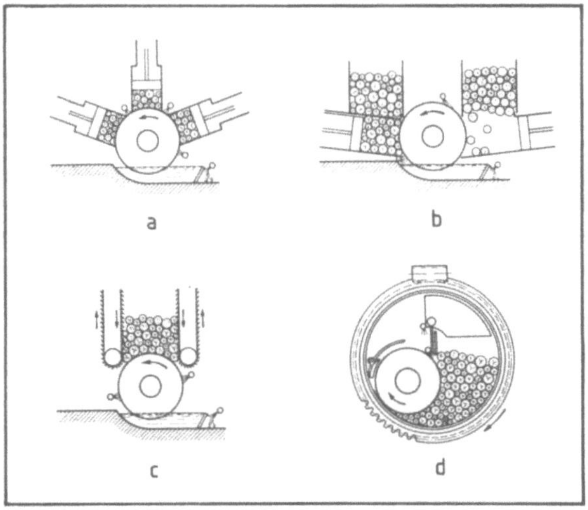

Fig. 16–3. Types of grinders (according to Klemm 1957b).
a: Pocket grinder (intermittent)
b: Magazine grinder (intermittent)
c: Chain grinder (continuous)
d: Ring grinder (continuous)

The grinding temperature can be influenced by the speed at which the grindstone revolves as well as by the initial water temperature. The grinders, having diameters up to 1.80 m and revolving speeds up to 356 rpm or stone velocities up to 40 $m \cdot s^{-1}$, are classified into discontinuous (intermittent) and continuous types (Fig. 16–3). Today continuous grinder systems are generally preferred because of their higher productivity due to the faster and easier wood feed operations.

After the screening and cleaning stages the rejects are refined in a hammermill, shredder or refiner. Finally the groundwood pulp is thickened and stored in slush form or in the form of wet laps. Overdrying must be avoided because of difficulties in redispersing the fibres in water. The yields of conventional stone groundwood are between 93–99% (based on wood), depending mainly on the wood species, wood quality and the grinding characteristics.

Apart from the important relations between wood feed and characteristic grinding data mentioned above, the total grinding process is influenced by numerous variables, which are highly dependent on each other:

– wood characteristics:	wood species (involving average fibre lengths, wood colour) wood density wood moisture content wood cleanness
– stone characteristics:	stone surface properties stone speed
– grinding characteristics:	grinding pressure grinding temperature grinding consistency grinding energy input and consumption
– pulp related characteristics:	pulp yield pulp production/time pulp properties

The effects of the individual variables are discussed in detail by Daniell (1980). In principle the wood quality required is higher for grinding procedures than for chemical pulping. Bark residues, which are generally restricted to 0.2–0.5% (Annergren 1981b), and discoloured or decayed wood portions have a strong influence on the groundwood quality. Concerning wood species, softwoods are generally preferred, especially spruce species because of their relatively low grinding energy consumption together with high yields and good pulp qualities (e.g. freeness and strength values). Fig. 16–4 demonstrates the approximate linear relationship of groundwood yields and wood densities. This relation is valid for both softwoods and hardwoods, the latter being increasingly used for groundwood production in North America

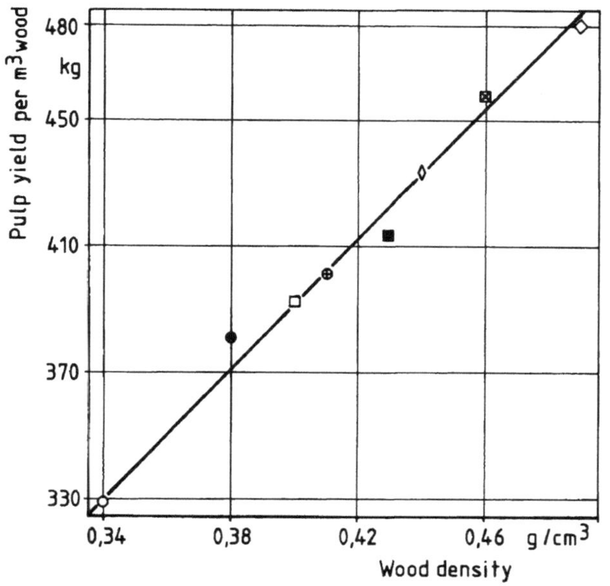

Fig. 16–4. Relationship between groundwood yield and wood density (data according to Daniell 1980).

◇	: Loblolly pine	⊕	: Jack pine
☒	: Shortleaf pine	□	: Eastern hemlock
◊	: Cottonwood	●	: Spruce
■	: Poplar	○	: Balsam fir

(e.g. aspen, cottonwood (*Populus* spec.)) (McGovern 1981). But while the yields increase with higher densities the strength properties of high-density hardwoods such as birch or maple are very poor. This is the main factor for the limited use of these species in groundwood production, and a stimulus for investigations of chemi-groundwood techniques to yield better hardwood groundwood characteristics.

The optimum moisture content of wood for grinding depends on the wood species and the special grinding conditions, but is generally about 40–50% to obtain long fibres and optimum strength values.

One of the most important factors in mechanical pulping is energy consumption. Industrial grinding and refiner processes are coming increasingly to be judged by that criterion. But energy consumption cannot be mentioned without considering the resulting pulp quality, including the optical and mechanical properties. In general higher strength properties and higher freeness values require more energy. Table 16–3 shows typical values of energy consumption for producing pulps with a given freeness of 100 ml CSF (Canadian Standard Freeness, Tappi Standard T 227 os-58). It is evident that stone grinding has the lowest energy consumption of all mechanical pulping procedures. As groundwood is used as single fibrous material only in low-quality papers and is otherwise mixed with about 15–20% of chemical

Table 16–3: Energy consumption of different mechanical pulping processes at constant freeness (CSF 100) (Arjas, Perälä 1980)

RMP MJ/t	TMP MJ/t	PRMP MJ/t	SGW MJ/t	PGW* MJ/t
6 480	7 920	7 200	5 220	5 580

* Pilot-plant data

pulp in newsprint to reach acceptable runnability and printability, the overall energy balance may be not better than when thermomechanical pulp (TMP) is used with a 5% furnish of chemical pulp in newsprint (Mannström 1972).

In addition to energy problems, wood consumption as related to paper quality is beginning to take on an increasingly interesting aspect. Though yields from stone grinding are in the same range as those of TMPs (Table 16–2) the above-mentioned lower chemical pulp furnish for TMP newsprint results in a considerably lower overall wood consumption for TMP newsprint (Hartler 1978).

Special properties of groundwood required for different paper grades can be obtained by the choice of wood species, variation of the grinding conditions, and particularly by the so-called post refining after screening and cleaning. The latter improves the printability properties of papers furnished with mechanical pulps by reducing the shive content together with a drop in freeness (Kurdin 1980). As groundwood fibre material is practically non-delignified ('wood containing' paper) and therefore yellows during aging, the application is generally restricted to more temporary papers such as newsprint, catalogs, magazines, toilet tissues and paperboards. Groundwood bleaching is described below (→ 16.7.2.).

From a morphological standpoint groundwood pulp consists of a mixture of intact fibres, fibre bundles, broken fibres and fibre fragments. The surfaces of the fibrillar elements are generally more uneven and less smooth than in the case of thermomechanical pulps, a prerequisite for the good fibre-fibre bond properties of soft-

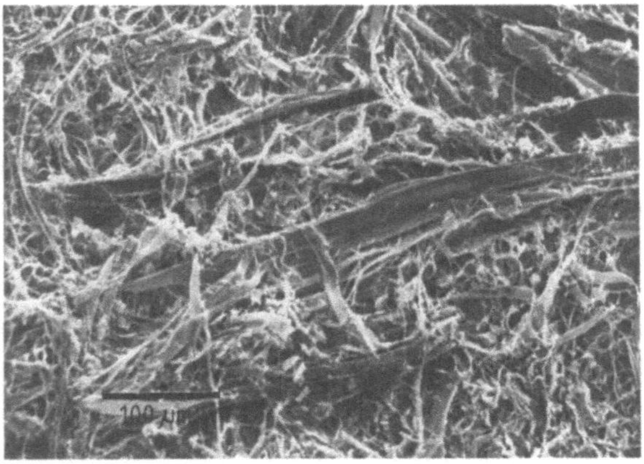

Fig. 16–5. Bleached groundwood pulp. SEM micrograph.

wood stone groundwood. Fig. 16–5 gives an impression of a screened, bleached spruce groundwood pulp.

Modifications of the traditional grinding process include groundwood production with a preceding steaming step, application of chemicals, and more recently the grinding process under pressure (Table 16–2). All processes aim at reducing power consumption and improving groundwood properties, especially of hardwoods.

So-called steamed groundwood or brown groundwood (in contrast to normal white groundwood) is produced by a two-step procedure and applied mainly to extractive-rich pine species and hardwoods. The steam pretreatment is performed in separate digesters resulting in an improved wood and especially lignin softening. By this, the required grinding energy is considerably reduced and the wood fibres are released much more easily and remain undamaged even at lower grinding temperature ('cold' grinding). The resultant material has better strength properties, but is obtained in lower yields (80–90%) demonstrating considerable wood losses. Though the term chemical treatment is generally restricted to an addition of chemicals some chemical reactions also take place under steaming conditions, e.g. formation of acetic acid, causing a degradation of the polysaccharides.

The processes which apply chemicals before or during grinding yield so-called chemigroundwood (CGW). The addition of sodium carbonate, sulfite, bisulfite or sulfate to the grinding water results in energy savings and increased brightness of the groundwood (Cochrane 1956; Brecht, Weiss 1957).

Though there have also been many attempts at pretreatment with very different chemicals (sulfite, bisulfite solutions, kraft-like liquors, sodium hydroxide, sodium bicarbonate) chemigroundwood has not attained the importance once predicted for it (Daniell 1980).

More recently experiments with pressure grinding processes (PGW) have shown acceptable results. The main differences as compared to atmospheric grinding are the use of a pressure-resistant grinder chamber and much higher temperatures. Intensive research work on this process was carried out in Finland, leading to the installation of the first full-scale plant in 1980 (Kärnä 1980).

Improved pressurized grinding of hardwoods was achieved in the presence of 5% caustic soda or 5% caustic soda + 2.5% H_2O_2 within the pressure grinder (chemi-pressurized groundwood, CPGW). The experimental process yielded excellent pulps from birch and aspen wood at low energy requirements, suitable for cardboard and corrugating medium (Anonymous 1980).

16.2.2. Refiner Mechanical Pulping

The second group of mechanical pulping covers the refiner mechanical processes. The principal characteristics of refiner mechanical pulping are the use of chips (but also wafers or even sawdust) predominantly from softwoods, and the application of disk refiners of various types for defibration and fibrillation of the raw material.

Refiner mechanical pulping processes yield the following types of mechanical pulps, depending on the process conditions (Table 16–2):

- refiner mechanical pulp (RMP)
- chemi-refiner mechanical pulp (CRMP or CMP)
- pressurized refiner mechanical pulp (PRMP)
- chemi-thermomechanical pulp (CTMP).

The Asplund process and the Masonite process can also be classified as refiner processes but they are principally more closely related to the production of fibreboard material. The Asplund process was the first thermomechanical process, applying steam at high temperature and disk refining under pressure; thus it forms the basic procedure of todays' thermomechanical pulping. In the so-called Masonite process wood is defibred by steam expansion ('explosion'), performed by a quick release of high pressure (up to 70 bar) after a steam treatment of chips in a digester ('Mason gun') at temperatures up to 285 °C for only a few seconds. The crude fibre product is further refined and screened (Sandermann 1956, 1963; Rydholm 1965).

As shown in Table 16–2 disk refiners are also applied in semichemical pulping and the production of high-yield chemical pulps. Therefore, with regard to the resulting pulps, the distinction between some semichemical and high-yield chemical pulping processes on the one hand and chemimechanical as well as chemi-thermomechanical processes on the other hand is beginning to blur. But concerning the process conditions in semichemical and high-yield chemical pulping it is obvious that the chemical treatment is the essential stage of the process (average yields 60–80%), while in all chemimechanical processes the mechanical treatment is the main step (average yields: 80–95%).

The original refiner mechanical pulping (RMP), which was a low-temperature, atmospheric disk refining process, has largely been replaced by processes using pressurized presteaming and refining (e.g. TMP) and applying chemicals (e.g. CTMP). This has produced improvements in pulp quality but generally calls for higher energy demands.

Common to all refiner mechanical processes are the two basic operations taking place during refining: defibration of the wood to single fibres and fibre bundles, and fibrillation involving the conversion of fibres to fibrillar elements. As the quality and properties of mechanical pulps mainly depend on their fibre characteristics it is obvious that the number of refining stages and especially the refiner design are highly important (Onisko 1980; Leask 1981). In principle, refiner mechanical pulping can be carried out in one stage (single-stage process) or two, three or four stages (multi-stage process) with a trend to two-stage refining because of advantages in

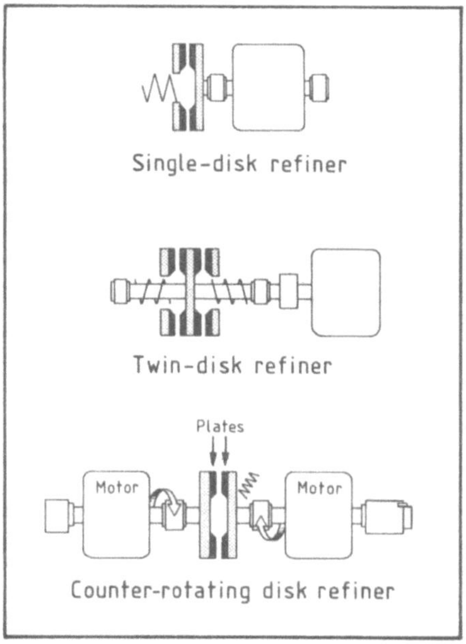

Fig. 16–6. Major types of disk refiners (according to Kurdin 1980).

wood feeding and energy consumption. The first stage is mostly pressurized today while the following stages generally take place under atmospheric conditions.

The disk refiners are of three principal types (Fig. 16–6). The first type has one stationary and one rotating disk (single-disk refiner), the second type is equipped with two counter-rotating disks (double-disk refiner), while the so-called twin-disk refiner combines two single-disk refiners in one operation unit.

With the increasing importance of refiner mechanical pulping among other things the refiner design has been intensively studied and has become more and more sophisticated. The sizes and capacities of the refiners were enlarged considerably (Kurdin 1980; Collicutt et al. 1981).

The most important industrial refiner mechanical pulping process today is thermomechanical pulping (TMP) with still enormous world-wide growth rates of about 25% per year. The process is established in North America, Scandinavia and Japan as well as in South Africa, West Germany and some other countries with a total operating capacity of 6.5 million tons per year in 1980, enabling a more intensive utilization of several pine species as a fibre source (Collicutt et al. 1981). This development has been accompanied by numerous publications concerned with different aspects of thermomechanical pulping (Hauan et al. 1975; Jensen 1976; Anonymous 1978; Peltonen, Rahkila 1979; Marton et al. 1979; Flowers et al. 1979; Gavelin, Lunden 1980; Venter 1981; Marton et al. 1981).

The principal process involves an impregnation and preheating step of washed wood chips with saturated steam under pressure. Thus pretreated chips are fed to the disk refiner for defibration at approximately the same temperature and pressure as in the preheating stage. The secondary refining stage is generally carried out at atmospheric pressure. Therefore the defibred material is expanded into a cyclone where the steam is removed, and refined in one or two stages to the desired freeness. The rejects from screening and cleaning are thickened and recycled to the refining step or separately refined.

Apart from technical details of the process design like optimal chip handling, steam transport or process computerization, the principal factors influencing the TMP properties are the following:

- steam temperature and pressure in the preheating and refining stages
- refining consistency
- refining energy.

Concerning the steam conditions in thermomechanical pulping there are several conflicting aspects. A high steam temperature and pressure would improve the application possibilities of the recovered steam. On the other hand temperatures higher than 130 °C during preheating and refining (these stages are generally closely coupled) have been shown to cause uncontrolled lignin softening in the middle lamellae, resulting in long fibres with poor bending and optical properties (Huusari, Syrjaenen 1980). At temperatures of 165–185 °C, which are applied e.g. in the Asplund process to produce fibreboard material, the glass transition point of lignin (→ 12.4.4.; Goring 1963) is passed and the softened lignin coats the surface of the separated fibres. Thus the fibrillation ability in the second step of refining is considerably reduced. Optimum refining temperatures in the production of TMP for paper furnish are between 115–130 °C. In this case the secondary walls are ruptured by the refining operation while the middle lamellae remain relatively unchanged. Softwood TMP contains larger amounts of long-fibre material with intact fibres than groundwood. The surfaces of the undamaged fibres are smooth, resulting in poor strength properties, especially of the coarse fraction (plus 30-mesh), usually

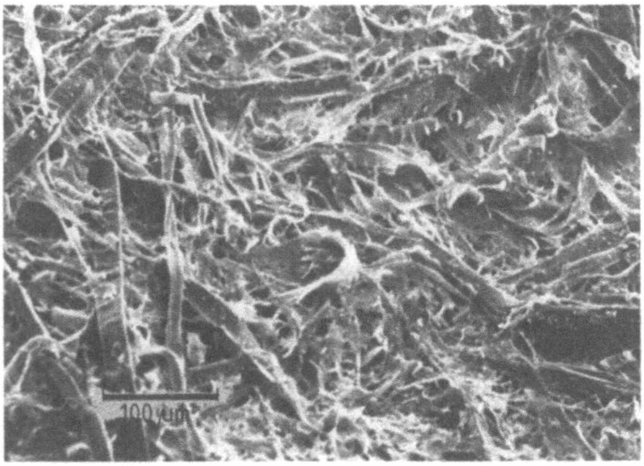

Fig. 16–7. Spruce thermomechanical pulp. SEM micrograph.

representing 30–40% of the total TMP. The smaller fractions (50–100 mesh, 100–200 mesh) consist mainly of ribbon-shaped, flexible lamellae, and determine the overall bonding strength of TMP (Anonymous 1977a; Kurdin 1980). Fig. 16–7 shows a typical unfractionated spruce TMP.

As the refining consistency affects strength properties as well as freeness of the mechanical pulp it is a very important factor in thermomechanical pulping. Fig. 16–8 demonstrates the dependence of strength values and freeness of unscreened TMP on the refiner consistency. It is obvious that consistencies between 20 and 30% yield high strength values while the freeness is reduced correspondingly. Low consistency (about 10%) in single-stage refining is necessary because of the high energy input. In two- or multistage processes the second and further refining stages require half as much energy as the primary stage, and consistencies up to 50% can be used (Holzer et al. 1962).

The refining energy is a factor which influences the pulp properties as well as the overall economy of TMP processes. Concerning pulp quality there is an increase in pulp strength (except tear strength) if the refining energy is increased, but a later drop in strength if the energy input is substantially higher, thus raising the refining temperature.

With regard to economics, TMP processes require somewhat more energy than atmospheric RMP processes but considerably more energy (up to 40–60% more) as compared to the production of stone groundwood of the same freeness (Table 16–3). In pressurized refining, the refining power is known to be nearly completely transferred to heat in the form of pressurized steam. As only about one-sixth of this steam is sufficient for chip preheating, intensive work was directed to economic steam recovery and further utilization. One of the latest advances in this field is a two-pressurized-stage process (Tandem-TMP) developed in Finland. In this proc-

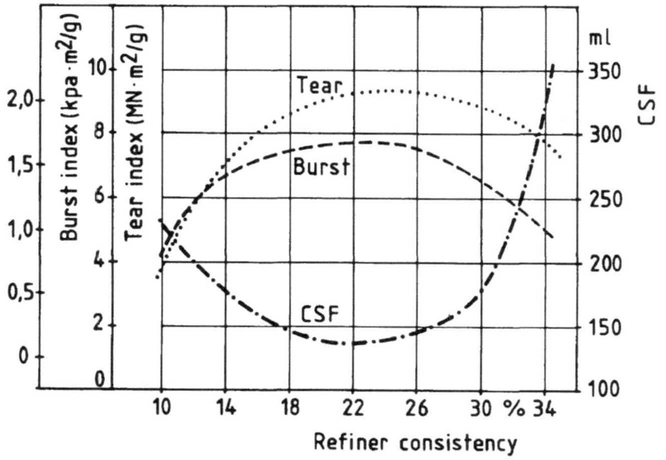

Fig. 16–8. Dependence of burst strength, tear strength and freeness on the refiner consistency (according to Kurdin 1980).

ess about 80% of the refining power is further used as pressurized steam (2.5 bar). The recovered steam is described as practically free of air, which is achieved by a small amount of steam flowing against the chip infeed at the top of the preheater. The steam is used for generating shaft power and especially for paper drying, thus reducing energy costs by about 20% (Huusari, Syrjaenen 1980). Even the total steam demand of a newsprint mill can be supplied by a TMP plant (Rydquist et al. 1981).

As already pointed out in connection with stone grinding, the energy demand of mechanical processes must be related to the respective pulp qualitites. Thermomechanical pulps are generally much stronger than groundwood pulp, thus enabling lower furnish of reinforcing chemical pulp for newsprint, catalogue and magazine papers. Besides the above-mentioned applications, TMP ist also used as a furnish e.g. in printing papers, paperboard, tissue paper and computer cards.

The thermomechanical pulps, obtained in yields of 91–98%, have on the average a lower brightness than groundwood of the respective wood species. This is caused by chip storage effects, higher bark portions of the starting material, and the high presteaming temperatures (Lorås 1981). To obtain bright and high-yield TMP a non-dissolving bleaching must be carried out, e.g. with dithionite or hydrogen peroxide (→ 16.7.2.).

Air pollution problems in mechanical pulping are less significant than in chemical pulping. But also the production of stone groundwood, RMP, TMP and particularly the chemi-modified CRMP and CTMP processes cause mill effluents with considerable amounts of softwood extractives which are highly toxic to aquatic life (Leach, Thakoré 1976; Rogers et al. 1979).

In the following a short overview is given of modifications to refiner mechanical processes. A very recently published modification of thermomechanical pulping omits the preheating stage, thus saving about 10% of the process energy demand. Results obtained from a single-stage pilot-plant indicate that the mechanical and optical properties of this type of pulp (called PRMP) are comparable to those of conventional TMP (Sundholm, Mannström 1982).

As already mentioned softwoods are the main fibre source for all types of mechanical pulping, because hardwoods give rather poor pulp strength properties. This is mainly caused by the fact that hardwood fibres do not tend to form fibrils during refining but break down to short rigid debris. This is even true of presteamed wood material in thermomechanical pulping, though low-density hardwood species such as aspen, poplar, cottonwood, birch, maple and eucalyptus are and can be used in TMP processes alone, in hardwood mixtures or in mixtures with softwoods (Leask 1968; Giertz 1974, 1977; Marton et al. 1979). The resulting hardwood TMP strengths are also very low but the pulps are characterized by high cleanness and good optical properties (high scattering coefficient), and are used mainly as filler grade pulp.

Based on the forecasts of increasing hardwood utilization for fibre production (→16.1.) many attempts have been made to obtain mechanical pulp qualities from hardwoods which are comparable to those of softwood pulps. This aim can be reached by the application of chemicals which cause a more intensive softening of the wood chips and especially the release of more or less undamaged fibres as is the case with softwoods. The chemicals, mostly applied at elevated temperatures, affect both lignin and polyoses in the cell walls, increasing the hydrophilicity of lignin in conjunction with a partial delignification as well as some dissolution of polyoses (e.g. Stevens, Marton 1966; Erinš et al. 1976; Rothenberg et al. 1981). With these reactions a considerable loss of substance takes place, reducing the yields of chemi-modified mechanical pulping processes generally to the range between 80 and 90%.

Though these processes focus mainly the application of hardwoods, they are used for softwoods as well. There are numerous chemicals which have been used or at least investigated for this purpose under different conditions with regard to their effect on refining energy savings, on lowering the refining temperature and particularly on improving pulp strengths (Hartler 1980). The chemical treatment can be performed in principle with sodium sulfite, sodium bisulfite (Leask 1968; Hauan et al. 1975; Frederiksson, Hoglund 1978; Vikström, Hammar 1979; Atack et al. 1980), sodium sulfite and sodium hydroxide (Marton et al. 1979; Vikström, Nelson 1980; Kobo et al. 1981), acid sulfite (Atack et al. 1978), hydrogen peroxide (Marton et al. 1979), sodium chlorite and peracetic acid (Stevens, Marton 1966; Sakai, Kondo 1976). Generally the chemicals are applied in the pretreatment step according to different principles (Peltonen, Rahkila 1979).

In a recently described chemimechanical process mechanically destructured chips are used to increase impregnation efficiency. The alkaline peroxide impregnation liquor ($NaOH/H_2O_2$) is applied at 40–60 °C at atmospheric pressure for 1.5–2 hours before low-consistency (5%) refining. High-yield (86–93%) light pulps (70% brightness) were produced from a French hardwood mixture containing 50% oak, 25% beech and 25% birch wood (Lachenal et al. 1979).

The treatment with hydrogen peroxide was also performed within the refining stage (Marton et al. 1979) while the delignifying action of peracetic acid was used in a pretreatment step as well as in posttreatment after refining (Stevens, Marton 1966).

In commercial CRMP and CTMP processes variations of sulfite (CTMP) and bisulfite (CRMP) pretreatments are mainly applied to softwoods and sodium hydroxide/sodium sulfite pretreatments to hardwoods. Sodium bisulfite (at pH 4–6) is a preferred pretreating chemical, which causes a sulfonation of lignin. Thereby at least part of the lignin becomes more hydrophilic and the fibres are capable of swelling. Additionally, sulfonation lowers the glass transition temperature of lignin considerably to the range of 70–90 °C, depending on the degree of sulfonation. To avoid the above-mentioned covering effect of softened lignin on the fibre surfaces, the refiner temperature must be kept below the glass transition range of the modified lignin. Generally hardwoods, but especially high-density hardwoods, require higher degrees of sulfonation than softwoods to obtain pulps with good strength properties.

Chemi-thermomechanical pulps are generally obtained in yields which are only a few percent lower than TMP yields. The properties of chemi-thermomechanical pulps can be varied by altering the main process parameters: preheating time and temperature, amount of sodium sulfite, and the refiner energy input.

The future of all mechanical pulps will depend on the development of energy costs and the trend in finding substitutes for more expensive chemical pulps as a paper furnish, especially in newsprint. The world-wide increase in the application of mechanical, thermomechanical and chemimechanical pulping processes to produce pulps from nonwood fibre sources such as straw, bagasse and bamboo is described extensively by Misra (1980).

16.3. Semichemical Pulping

16.3.1. The Neutral Sulfite Semichemical (NSSC) Process

Semichemical pulping processes are characterized in principle by a chemical treatment preceded by a mechanical refining step. As already mentioned this general definition also applies to the chemimechanical processes and the high-yield sulfite and kraft processes. In spite of this fact semichemical pulps (German: Halbzellstoffe) are a typical group of pulps which are obtained predominantly from hardwoods in yields between 65 and 85% but even up to 92% (Table 16–2).

The most important semichemical process is undoubtedly the NSSC process which has been widely established in the U.S.A. since 1926, and within the last 20 years has also taken hold in Europe and many other countries throughout the world (Crönert 1966; Marteny 1980).

The general advantages of the NSSC process are low requirements with regard to wood quality and species, high yields, relatively low consumption of chemicals at a given residual lignin content, low capital investment and profitable small production units as compared to full chemical pulping.

The principal process involves three main steps:

- impregnation with a sodium sulfite liquor
- cooking at temperatures between 160 and 190 °C
- defibration by disk refining.

The impregnation with the pulping liquor is generally carried out at about 125 °C for one hour under pressure after a short steaming of the chips at atmospheric pressure. The sodium sulfite solution is generally buffered at about pH 7 or somewhat higher (up to pH 10) with small amounts of sodium hydroxide, sodium carbonate, sodium bicarbonate or sodium bisulfite to avoid acidic conditions, which may result from organic acids set free by the cooking procedure. The pulping liquor can be prepared by any of several techniques, e.g. by dissolving sodium sulfite and sodium bicarbonate in water or gassing up a sodium carbonate solution with suitable amounts of sulfur dioxide. The composition of the pulping liquor can be characterized by the degree of sulfitation expressed as a quotient of molar amounts of SO_2 and Na_2O. Generally values of 0.75–1 are used (Lengyel, Morvay 1973).

The cooking temperature is mainly dependent on the cooking time, which may vary between 15 min and 8 hours, depending on the type of digester used and the desired pulp type and quality. Defibration can be carried out by means of one or two refining steps or as a progressive multi-stage refinement process. Refining can be done under digester conditions or at atmospheric pressure.

There are several variants of NSSC pulping and of the techniques of spent liquor recovery, and the total process is influenced by numerous variables. The main parameters of the cooking procedure are the amount and composition of the pulping chemicals, and cooking time and temperature. The proper overall design of the NSSC plant and the special process conditions will be mainly determined by:

- the wood species used for pulping
- the desired pulp type and quality.

With regard to the recovery process it is essential whether or not the NSSC pulp production is an integrated part of a kraft mill.

In principle spent NSSC pulping liquors are not suitable for energy purposes because of the low content of organic material. The general problem of chemical recovery in all sulfite-based processes is the conversion of sulfide to sulfite without the formation of considerable amounts of thiosulfate, which is known to cause problems in connection with bleachability and digester corrosion (Marteny 1980). The recovery process is mostly carried out by means of dry combustion of the concentrated black liquor in a recovery furnace by treating the sodium sulfide containing smelt with a sodium sulfite solution to produce a new cooking liquor. Several other recovery principles are in use. These include e.g. wet combustion to produce sodium sulfate, which can be converted to sodium sulfite by reacting with calcium sulfite and sulfur burner gas; direct oxidation of sodium sulfide to sodium sulfite; and pyrolysis of the evaporated spent liquor to soda and hydrogen sulfide for further processing (atomized suspension technique; Lee, Gauvin 1958). When NSSC pulping is integrated in a kraft mill, the spent liquor is recovered by so-called cross recovery, i.e. burning the spent liquor and producing fresh liquor within kraft liquor recovery (→ 16.4.3.). (Markant 1960; Boyer 1960; Sandermann 1963; Rydholm 1965; Lengyel, Morvay 1973; Marteny 1980).

As already mentioned, cooking conditions vary depending on the digester system used and especially on the grade of pulp to be produced. NSSC pulping is carried out either discontinuously or in continuously working digesters. Discontinuous rotary globe digesters or stationary vertical digesters are still in use, but continuous systems are becoming increasingly important and typical for NSSC pulping. Continuous digesters have the advantages of providing an uninterrupted process to feed the refiners with cooked chips at high temperatures while requiring little space and offering several other economic benefits. Numerous digester designs were developed including e.g. horizontal-tube, vertical down-flow and inclined-tube digesters (Sandermann 1963; Marteny 1980). Due to generally higher pulping temperatures applied in continuous pulping systems, the reaction rates are higher than in discontinuous pulping, thus considerably reducing cooking time.

In the NSSC process, and in semichemical pulping generally, hardwood species are the typical raw material. A large number of hardwoods are used for NSSC pulping including low-density, medium-density and high-density woods with low but also high lignin contents. Table 16–4 lists the most frequently used hardwoods.

The advantage offered by the NSSC process with regard to the raw material is the possibility of utilizing many different hardwood species, even in mixtures of up to 12 species in proportions according to their occurrence in natural mixed stands (Marteny 1980). Tropical hardwoods can also be used for semichemical pulping, but many high-density species have very high contents of lignin and extractives (e.g. Lophira alata, →Table 3–6), reducing their ability to produce suitable semichemical and even kraft pulps (Rahmani 1978). A comprehensive laboratory study of NSSC pulping properties of 31 wood species (softwoods and hardwoods from temperate

Table 16–4: Hardwoods used for NSSC pulping (Rydholm 1965; Morin 1974; Marteny 1980; Madan 1981)

General name	Botanical species/genus
Aspen Poplar Cottonwood	*Populus* spec.
Willow	*Salix* spec.
Black alder	*Alnus glutinosa*
Tupelo	*Nyssa* spec.
Black cherry	*Prunus* spec.
Maple	*Acer* spec.
Chestnut	*Aesculus hippocastanum*
Birch	*Betula* spec.
Ash	*Fraxinus* spec.
Oak	*Quercus* spec.
Beech	*Fagus* spec.
Elm	*Ulmus* spec.
Hornbeam	*Carpinus betulus*
Eucalyptus	*Eucalyptus* spec.

zones, tropical hardwoods) using discontinuous and continuous processes was carried out by Riese (1962), resulting in a scheme of suitability for semichemical pulping.

Besides single hardwoods and hardwood mixtures, mixtures of hardwoods and softwoods may also be pulped successfully with the NSSC process. Improved properties were reported for pulping birch wood with the addition of up to 30% spruce wood, or by adding poplar, birch and spruce wood to beech wood (Riese 1957; Dahm 1958).

The predominant use of hardwoods can be explained by the fact that most hardwood species generally have lower lignin contents and in particular are more easily delignified than softwoods, in addition to the above-mentioned economic considerations. Under the conditions of semichemical pulping softwoods cannot be delignified to suitable pulps without a considerably higher consumption of chemicals and refiner energy.

The principal aim in neutral sulfite semichemical pulping is to obtain a sufficient but highly selective delignification (removing up to 50% of the lignin) and a preservation of polyoses (removing up to 40% of the polyoses). The lignin contents of NSSC pulps are high as compared to those from chemical pulps, and range generally between 10 and 15% residual lignin. Figure 16–9 demonstrates that the course of pulping selectivity (expressed as the ratio of dissolved lignin and dissolved polyoses) is quite different as compared to kraft pulping. The optimum is already reached if 40–50% of the lignin is removed, corresponding to pulp yields of 75–80%. The chemical mechanism in NSSC cooking is primarily a sulfonation of mainly middle lamella lignin, causing a partial dissolution. By this and a partial solubilization of

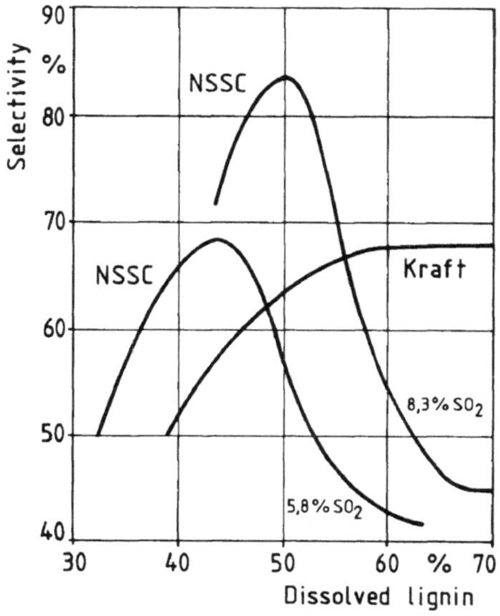

Fig. 16–9. Selectivity of lignin dissolution in NSSC and kraft pulping (Lengyel, Morvay 1973).

more soluble polyoses the fibres are weakened for the subsequent mechanical defibration.

Due to the high lignin and polyoses content the pulp has low conventional strength properties. Comparing hardwood and softwood NSSC pulps it is evident that softwood NSSC pulps are generally stronger than hardwood NSSC pulps but weaker than average chemical softwood pulps, while hardwood NSSC pulps are stronger than the corresponding chemical pulps. This is demonstrated e.g. by Table 16–5, which compares strength values of aspen pulps produced by different processes.

The effect of adding anthraquinone (AQ) in NSSC pulping of pine species (*Pinus radiata*, *Pinus elliottii*) and eucalyptus (*Eucalyptus* spec.) was recently investigated by Cameron et al. (1981), showing different responses of the wood species.

The typical NSSC pulp is normally much more rigid and stiff than e.g. kraft pulp. Therefore it is the most typical and suitable fibre material for the production of corrugating medium. For this purpose unbleached pulp in high yields of about 80% is produced. Otherwise bleached grade pulps can also be obtained by semichemical pulping in low yields even down to 65% after bleaching. In the first case usually unbarked wood is cooked at lower temperatures and corresponding pressures, lower chemicals-to-wood ratios and longer cooking times than is the case when pulps are produced for subsequent bleaching. The latter require debarked wood and considerably higher charges of chemicals, amounting up to 20%, based on dry wood, as compared to generally half as much in the case of high-yield unbleached grades.

Besides the predominant and typical utilization of NSSC pulps for corrugating medium production, other fields of NSSC pulp applications include furnish printing papers, grease-proof papers, bond papers and many other paper grades, depending on the special pulp properties obtained by varied pulping conditions (Marteny 1980).

Like kraft and soda chemimechanical pulping and some other processes NSSC pulping is also used for nonwood plants and residues because of the generally low lignin contents of these materials and the widely variable conditions offered by NSSC pulping (e.g. Lengyel, Morvay 1973; Misra 1980; Botes 1981).

Table 16–5: Strength properties of aspen pulps from different pulping processes (Marteny 1980)

Process	Burst index $kPa \cdot m^2/g$	Tear index $mN \cdot m^2/g$
NSSC, bleached	1.32	10.7
NSSC, unbleached	0.96	8.7
Sulfate, unbleached	1.08	7.6
Sulfate, bleached	0.96	9.2
Soda, unbleached	0.78	7.6
Soda, bleached	0.60	6.1
Sulfite, unbleached	0.60	6.1

16.3.2. The Cold Soda Process

The cold soda or cold caustic process, which is less important than the NSSC process, involves in principle the treatment of chips with a sodium hydroxide solution at temperatures generally between 20 and 30 °C and a final refiner defibration. The process is almost exclusively used for hardwoods, including high-density hardwoods such as oaks, but also for nonwood plants (Misra 1980; Gruber 1980). The cold soda process is usually integrated in a kraft mill to allow a combined recovery of the sodium hydroxide.

The most important step in cold soda pulping is the impregnation with alkaline liquor to reach a very fast but total penetration of the chips, causing the necessary swelling of the fibres and avoiding considerable losses of polyoses. The impregnation was originally performed by simple soaking or steeping, but was improved by several systems, including:

- pressure impregnation of smaller sized chips
- compression and decompression of the chips in roller mills
- continuous two-step pulping in fibre presses or screw-pressing after impregnating the chips by spraying, and subsequent soda treatment in an inclined reactor tube.

Impregnation times are between 15 and 120 min with generally short reaction times of 15–30 min in pressurized and continuous systems. The concentrations of NaOH are generally low (0.25–2.5%), but up to 10% in the case of the roller mill impregnation process. Cold soda pulping requires little installation capital, and despite the cost of the chemicals and the processing costs are actually lower than in stone grinding, because of reduced energy consumption. In some process modifications the liquor is reused up to 20 times.

The pulp yields lie between 85 and 92%, whereby the selectivity of lignin and polyoses dissolution is much lower than in NSSC pulping. The main disadvantage of cold soda pulps is a generally low brightness level (40–50%), which can be effectively increased by a two-stage peroxide-hypochlorite bleaching. The strength properties and freeness values of high-grade cold soda pulps are generally somewhat superior to those of softwood groundwood pulps.

As cold soda pulps have properties comparable to NSSC pulps they are used as unbleached coarse grades for corrugating medium production, and as bleached grades for printing papers and newsprint in combination with groundwood pulp and chemical pulp. Further details about cold soda pulping and various process modifications as well as earlier literature are given by Runkel and Patt (1958), Bryne, Voelker (1960), Runkel (1962), Sandermann (1963), Rydholm (1965) and Marteny (1980).

16.3.3. Other Semichemical Processes and High-Yield Chemical Pulping

Apart from the well-known NSSC and cold soda processes some other semichemical processes should briefly be mentioned. Semichemical pulps can be produced by using kraft liquor ($NaOH + Na_2S$) or green liquor ($Na_2S + Na_2CO_3$) and disk refining after cooking. Soda pulping at elevated temperatures combined with disk refining is used for hardwoods and particularly for pulping of nonwood fibres (Sandermann 1963; Rydholm 1965; Marteny 1980; Misra 1980).

A semichemical process applying sodium hydroxide and hydrogen peroxide at 70 °C was investigated by Ginzel et al. (1969). Pulps in high yields of more than 90% and with high brightness values (up to 80%) were produced from wood of poplar *(Populus marilandica)* and hornbeam *(Carpinus betulus)*. Another nonsulfur process yields semichemical pulp from hardwoods for corrugating medium production by means of continuous cooking with sodium hydroxide and sodium carbonate followed by pressurized refining (Dillard et al. 1976).

There is no clear borderline between semichemical pulping and so-called high-yield chemical pulping, or between the resulting pulps. From a practical aspect many high-yield chemical pulps are more or less semichemical pulps with regard to their yields (55–70% and even higher) and typical process design (disk refining after the chemical treatment).

High-yield chemical pulps are obtained in principle by modified sulfite and sulfate processes, i. e. by applying reduced charges of chemicals and/or reduced cooking time and temperature, and a refining step after cooking. High-yield sulfite pulps can be produced by an acidic sulfite, bisulfite or alkaline sulfite process (Table 16–2). In high-yield acidic sulfite pulping (with calcium, magnesium or sodium as bases) the reaction rate is usually decreased by cooking at lower temperatures (120–130 °C) and with lower acidity of the liquor, i. e. less sulfur dioxide, than in full chemical sulfite pulping. The resultant stuff can be defibred with low energy consumption, if the yield does not exceed the range of 70–80% (Wong 1980). Common yields of unbleached pulps are between 60 and 70%. High-yield acidic sulfite pulps are often produced in newsprint sulfite mills, saving up to 30% of wood compared with full chemical pulps (Pamén 1960). The pulps are mainly used as newsprint furnish in mixtures with groundwood and chemical pulp. Softwoods are used almost exclusively, but poplar and eucalyptus wood (*Populus* spec., *Eucalyptus saligna*) were also shown to give pulps with good strength properties (Jayme, Branscheid 1959; Jayme, Krause 1961). Generally the strength values (except burst strength) and the freeness are lower than those of comparable full chemical pulps.

In high-yield bisulfite pulping the cooking liquor has no excess of sulfur dioxide and the maximum cooking temperature is somewhat higher than in acidic sulfite pulping. The preferred bases are sodium and magnesium. High-yield bisulfite pulps are generally used in the same fields as acidic sulfite pulps, e.g. in newsprint. Also produced were hardwood high-yield pulps suitable for corrugating medium (Marteny 1980).

High-yield alkaline sulfite pulping is a modification of alkaline sulfite pulping, which is becoming an increasingly interesting alternative to the kraft process (→ 16.5.2., Wong 1980). The cooking liquor is buffered between pH 9 and 12, containing sodium sulfite and one or all of the chemicals sodium carbonate, sodium hydroxide and sodium sulfide. The yields lie around 60% and the pulp is brighter than normal kraft pulp. Due to its good strength properties the pulp is usable as linerboard (Marteny 1980).

In high-yield kraft or sulfate pulping the yields are generally between 55 and 65%, but even up to 80%. The typical kraft process is modified either by reducing the charge of chemicals by about one-half or by decreasing the cooking time and temperature. Usually high-yield kraft pulps have lower strength values and are darker than normal kraft pulps and NSSC pulps. The process is used for both hardwoods and softwoods , being especially suitable for oaks and pines, respectively (Rydholm 1965; Marteny 1980). Together with the numerous modifications of mechanical and chemimechanical pulping processes, semichemical and high-yield chemical pulping will make an increasingly significant contribution to optimum wood utilization and a diversified pulp supply in the future.

16.4. Alkaline Chemical Pulping

16.4.1. General Aspects

The sulfate or kraft process and the soda process are the two principal alkaline pulping techniques and the basis for several modified alkaline processes, including kraft pulping after a prehydrolysis step for the production of dissolving pulp (e.g. Marttala et al. 1981; Blom et al. 1981). Sodium hydroxide is the principal cooking chemical in both processes, while in sulfate pulping sodium sulfide is an additional active pulping component. Both processes received their names from the regeneration chemicals used to compensate for the loss of sodium hydroxide, namely sodium carbonate and sodium sulfate, respectively.

Today the sulfate process is not only the dominant alkaline pulping process for wood, but also the most important pulping process altogether. The first statement is mainly based on the fact that kraft pulps (kraft means strength or power in German and Swedish) are obtained in higher yields and with properties superior to soda pulps. Secondly, there are several well-known reasons why the world-wide trend in chemical pulping over the last 50 years has put the sulfate process ahead of sulfite pulping (→ 16.5.).

The main advantages of sulfate pulping, listed below, give a first characterization of the process and the resulting pulps:

- low demands on wood species and wood quality, including all types of softwoods and hardwoods, even in combination, and toleration of high amounts of extractives as well as considerable portions of decayed wood and bark residues
- short cooking times
- well established processing of the spent liquor, including the recovery of the pulping chemicals, generation of process heat, and the production of valuable by-products such as tall oil and turpentine from pine species
- excellent pulp strength properties.

The main drawbacks of sulfate pulping are the odour problems, lower yields than in sulfite pulping (usually 45–50%), the dark colour of the unbleached pulp, and not least the enormous costs for the installation of a new mill.

Though the classical soda process has been largely replaced by the sulfate process, especially in softwood pulping, it is still an important process for the production of nonwood fibre pulps (Lengyel, Morvay 1973; Saad, Fahmy 1979; Corradini 1979; Misra 1980). The application of additives such as anthraquinone (soda-AQ process) (→ 16.4.5.) and the development of soda-based oxygen delignification procedures (soda-oxygen processes) will increase the importance of these sulfur-free processes in the future (→ 16.4.6.).

16.4.2. Process Conditions and Variables

In contrast to sulfite pulping the sulfate process usually follows a relatively uniform process scheme though there are numerous variations in the special equipment and the cooking conditions. Fig. 16–10 shows a simplified flow-scheme of the kraft process. In principle the wood is charged to the digester in the form of chips together with fresh cooking liquor (white liquor) from the chemical recovery line. The heart of the pulping system is the digester.

The principal digester systems are discontinuous (batch system) or continuous. Though continuous pulping has many technical and economical advantages such as lower energy requirements or better emission control, a considerable number of kraft mills (e.g. about 60% in North America) (Helberg et al. 1976, 1977) still use the older discontinuous digester systems, which are characterized e.g. by higher flexibility, lower maintenance costs and higher turpentine yields. The batch digestion is sometimes carried out without liquor circulation, though more frequently liquor circulation is used to obtain a more uniform liquor distribution within the digester. The digesters can be heated by direct steaming or indirectly by heat exchange in the circulating liquor system. The digesters are generally smaller than in sulfite pulping (volumes of discontinuous digesters are between 60 and 180 m^3), but the total production capacities of pulp plants can be as much as 1 500 t. This is due

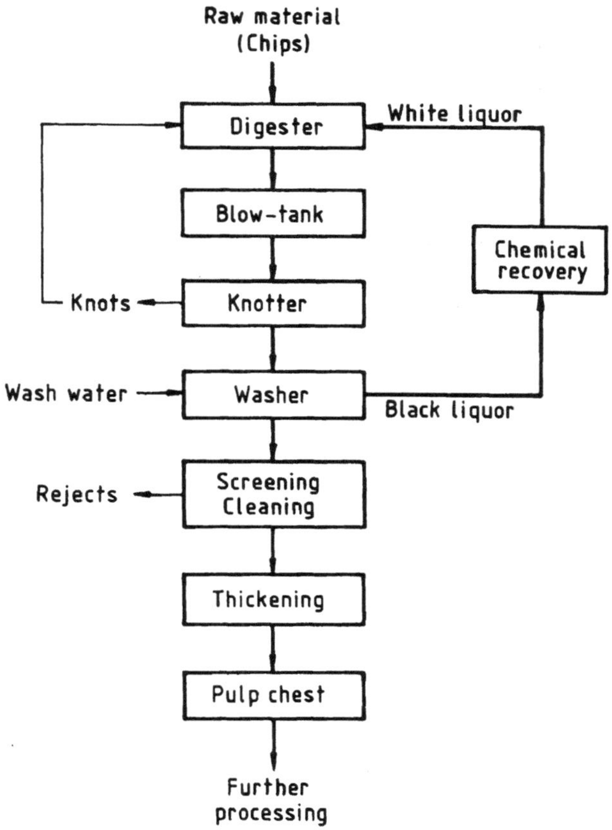

Fig. 16–10. Simplified flowchart of the kraft process.

to short cooking times of usually 4–6 hours and a large number of digesters in one plant. The normal kraft cooking is performed at temperatures between 160 and 180 °C at pressures between 7 and 11 bar.

Continuous rapid pulping processes, applying temperatures of 190–200 °C, require cooking times of only 15–30 min (Nolan 1957; Kleinert 1965).

The typical digester used in the kraft process is the vertical down-flow digester. In continuous kraft pulping the chips are pretreated with steam in a presteaming vessel before being charged to the digester.

Improved digester systems include:

- in-digester pulp washing
- low-temperature blowing (the ready-cooked pulp is cooled below 100 °C before blowing)
- countercurrent cooking (the white liquor is introduced in the middle of the digester and flows upwards)
- digester surging (boiling is induced at the bottom of the digester by reducing the pressure in the upper part).

After cooking, the pulp and the spent liquor (black liquor) are discharged at the bottom of the digester at reduced pressure into a blow tank with a capacity of several cooking volumes. Insufficiently cooked large-size rejects (knots) are screened on knotter screens and generally retransported to the digester for repeated cooking. The spent liquor is removed after countercurrent washing of the pulp and further processed within the recovery line (→ 16.5.3.). The pulp is further screened, cleaned, sometimes mildly refined, and finally thickened and stored for further processing. The by-product sulfate turpentine is removed during presteaming, and condensed from the relief gases. The tall oil soap is removed during evaporation of the black liquor by skimming.

During digester blowing and evaporation of the black liquor odourous and toxic components escape, mainly hydrogen sulfide (H_2S) and methyl mercaptan (CH_3SH) together with smaller amounts of dimethylsulfide (CH_3SCH_3) and dimethyl disulfide ($CH_3S_2CH_3$). Great efforts have been made to reduce these typical kraft pulping emissions by chlorination to less volatile components, incineration, oxidation, absorption in white liquor, or scrubbing in black liquor. But at present emission problems cannot be completely avoided even though large sums are invested in environmental protection equipment (Sarkanen et al. 1970; Louden, Weiner 1976; Vettenranta 1978; Kiiskilä, Virkola 1978; Ikeda et al. 1979; Ezzat 1979; Silander 1981).

As already mentioned, the essential pulping chemical in alkaline pulping is sodium hydroxide. But as other sodium compounds are also involved in both soda and sulfate pulping, it is necessary to have a common basis for all chemicals, which is either the equivalent amount of sodium oxide (Na_2O) (used in North America) or sodium hydroxide (NaOH) (used e.g. in Scandinavia). Widely used terms and definitions in soda and sulfate pulping are given in the Tappi Standards T 1202 os-72 and T 1203 os-61, respectively.

In soda pulping the cooking liquor is composed mainly of sodium hydroxide (80–85%) and smaller amounts of sodium carbonate (15–20%) deriving from the incomplete causticizing reaction to yield sodium hydroxide. The cooking liquor in sulfate pulping has more components. In addition to sodium hydroxide and sodium

Table 16–6: Composition of typical white and green liquors in sulfate pulping (Wenzl 1960)

Components	White liquor g/l	Green liquor g/l
Solids	12.5	15.0
NaOH	65.6	3.2
Na_2CO_3	25.6	83.3
Na_2S	30.4	33.6
Na_2SO_4	1.6	1.6
$Na_2S_2O_3$	0.1	0.1

Na-compounds calculated as Na_2O

carbonate, sodium sulfide is the essential pulping chemical. But sodium sulfate, sodium thiosulfate, polysulfides and sodium sulfite may also be present in small amounts. Table 16–6 gives an example of the composition of a sulfate white liquor. The charge of alkali used in kraft pulping, which is an important pulping factor, as discussed later on in this section, can be expressed as active alkali ($NaOH + Na_2S$) or as effective alkali ($NaOH + {}^1/_2Na_2S$). The latter is derived from the single equivalent point of the hydrolysis reaction of sodium sulfide in water according to the equation:

$$Na_2S + H_2O \rightleftharpoons NaOH + NaHS$$

Good chip impregnation is an essential prerequisite for a homogeneous delignification of wood. As alkaline solutions penetrate into wood much better than acidic solutions, the heating time to the maximum pulping temperature is much shorter than in acidic sulfite pulping. The impregnation of the chips is a combined result of penetration of the cooking liquor through the capillary system of wood, and diffusion through the totally impregnated chips. While the bulk penetration is mainly dependent on the size of the single capillaries, the diffusion rate depends on the total effective cross-sectional area of all capillaries (Stone 1957). The most critical dimension for the delignification is the chip thickness, which strongly influences the pulp yield and amount of screening rejects (Akhtaruzzaman, Virkola 1979). But the total size and shape of the chips also have effects, e.g. on digester filling and the liquor flow-rate. Generally the chip-thickness is restricted to 8–10 mm (Hatton 1977).

The delignification proceeds in three phases as a heterogeneous reaction. The initial delignification takes place below 140 °C, while the main delignification runs at temperatures above 140 °C until about 90% of the lignin is dissolved. The last stage of lignin removal is called residual delignification (Kleppe 1970). The chemical reactions of lignin and polysaccharides in alkaline pulping are described in chapter 11.

The kraft pulping process and the resulting pulps are influenced by several parameters:

- raw material (wood species and quality)
- liquor-to-wood ratio
- cooking time and temperature
- charge and concentration of the cooking chemicals
- composition of the cooking chemicals.

As already mentioned the kraft process is not very sensitive to the raw material and is highly suitable for softwoods and hardwoods of different densities and ages, even in combination and with considerable toleration of bark residues, reaching about 2% for bleachable pulp grades (Augustin, Rahmani 1975; Jett, Zobel 1975; Roffael, Parameswaran 1978; Hunt 1981; Annergren 1981a). There is no bark content limitation for the production of unbleached coarse pulp grades (Reside, Garvin

1978). An increased use of hardwood barks as a fibre source was recently recommended (Harder et al. 1978; Einspahr, Harder 1980). A listing of more than 70 softwood and hardwood species used and investigated in kraft pulping has been compiled by Rydholm (1965). Today most of the chips are stored outdoors in open piles. Different wood species, or at least different classes, are generally pulped separately (Helberg et al. 1977; Annergren 1981a).

Whole-tree and full-tree material, including the stump, roots, branches, twigs, bark and needles, has proved to be a suitable raw material for kraft pulping in principle. But the advantage of an increased fibre supply is diminished by considerably lower pulp yields, poorer strength properties and high costs for raw material cleaning and higher consumption of chemicals (Eskilsson, Hartler 1973; Eskilsson 1974; Keays 1974; Nyholm, Virtanen 1976; Horn, Auchter 1976; Stade 1976; Marton et al. 1976; Hartler et al. 1977; Chase 1978; Genco et al. 1978; Law, Koran 1979; Amidon 1981).

The liquor-to-wood ratio is mainly determined by the digester size and the chip density packing in the digester, and varies with different process conditions in batch or continuous digesting. Generally, higher liquor-to-wood ratios result in better chip impregnation.

Cooking time and temperature are closely related. In principle the cooking time can be decreased to some extent by increasing the cooking temperature, but in the range of normal pulping temperatures between 160 and 180 °C there is no distinct effect on the pulping rate. At higher temperatures the pulp yield and quality normally decrease.

The charge of chemicals used in alkaline pulping is expressed as effective alkali charge and depends on factors such as wood species, cooking conditions and required residual lignin in pulp. Effective alkali charges range between 11% (based on oven-dry wood) for unbleached coarse grades and 17% for bleachable paper grades, and are much higher for dissolving pulp qualities.

The alkali concentration is the main parameter of lignin and polysaccharide dissolution. The concentration of sodium hydroxide at the beginning of the cook can vary widely from 20 to 80 g/l.

In the initial period much alkali is consumed for the neutralization of acids deriving from the polysaccharides (acetic acid, uronic acids) (10% of the alkali) and for neutralizing lignin degradation products (25–30% of the alkali) (Sjöström 1981). Especially at the end of the cooking procedure too high alkali concentrations must be avoided. Otherwise overproportional degradation and dissolution of polyoses and cellulose take place, resulting in reduced yields and pulp strength properties. Softwoods generally require higher alkali charges and concentrations than hardwoods to reach comparable degrees of delignification.

The composition of the cooking liquor in sulfate pulping is expressed as so-called sulfidity, representing the ratio of Na_2S to active alkali, both of which are expressed

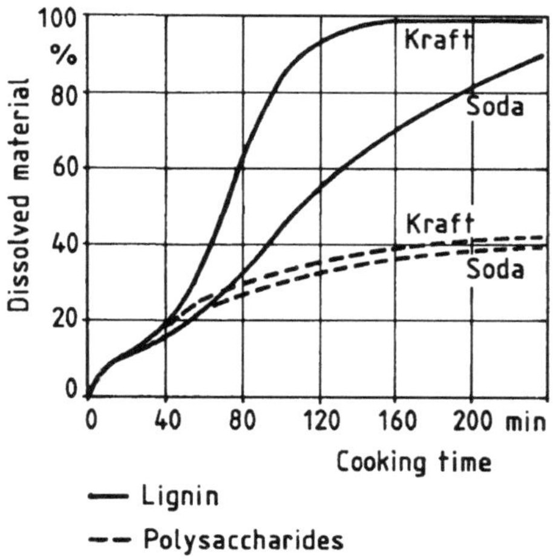

Fig. 16–11. Course of lignin and polysaccharide dissolution in kraft and soda pulping (Stone, Clayton 1960).

as Na_2O. The applied sulfidity varies with changing alkali charge, cooking temperature and several other factors (Sondell 1982). The sulfide charge for hardwoods is generally lower (15–20%) than for softwoods (25–35%). The effect of sulfide in kraft pulping as compared to soda pulping is demonstrated in Fig. 16–11, showing that the delignification rate is much faster in kraft pulping, which reaches 90% delignification within half the time as in soda pulping. The dissolution of polysaccharides is similar for both processes.

These briefly introduced parameters in alkaline pulping are highly dependent on each other and only an optimum co-ordination of all important wood and chemical parameters and technical equipment yields the desired pulp quality (Annergren 1981b).

16.4.3. The Recovery of Pulping Chemicals

The regeneration of the black liquor to fresh white liquor is an integrated and economically necessary part of both the sulfate and the soda process. As already mentioned, a well-established chemical recovery system is one of the advantages of sulfate pulping as compared to conventional sulfite pulping. The importance of the recovery line within the process may be seen from the fact that more than 25% of the investment capital for a new kraft mill is used for the chemical recovery (Harris 1974).

The total recovery process includes numerous chemical and technical facts (Wenzl 1960; Rydholm 1965; Lengyel, Morvay 1973; Bryce 1980a). The special chemical recovery cycle covers four main aspects:

- recovery of pulping chemicals
- reduction of water-pollution by burning the organic matter in the spent liquor

- generation of process heat
- recovery of valuable by-products.

The principal steps of the recovery line are:

- evaporation of the black liquor
- incineration of the concentrated liquor
- causticizing the furnace smelt
- regeneration of the lime.

Fig. 16–12 shows the chemical recovery cycle including the most important chemical reactions.

The weak black liquor coming from the pulp washers has to be evaporated to a solid content of at least 60% in order for it to be combusted in the recovery furnace. This is usually performed in two principal steps. The first is a multiple-effect evaporation by boiling the liquor in 5 to 6 evaporation stages, using the vapour of one stage to heat the next stage. The second step is a direct-contact evaporation by bringing flue gas from the recovery furnace directly into contact with the black liquor to reach a final solid concentration of up to 65%. During the multi-effect evaporation the tall oil soaps are recovered by skimming, and further processed to yield tall oil (→ 18.7.). Prior to evaporation, the black liquor can be oxidized by air or oxygen to reduce the sodium sulfide content (with regard to odourous emissions) and to increase the alkalinity, thus retaining acidic compounds in solution.

The recovery furnace is the heart of the recovery system, and in a modern furnace the combustion of the spent liquors from a pulp production of more than 1 000 t/day is feasible. Makeup sodium sulfate and minor chemicals such as Na_2CO_3 or Na_2SO_3 are added to the concentrated black liquor to compensate for the lost sulfidity. Sometimes elementary sulfur is also added together with the mud of the furnace gas clarification. The recovery furnace has two main functions:

- the combustion of the dissolved organic wood material to generate heat, which can be transformed to process steam
- the production of an inorganic smelt, which is dissolved to yield the so-called green liquor.

The main chemical reactions taking place during the incineration are outlined in Fig. 16–12. The organic material is oxidized to carbon dioxide. Sodium carbonate is formed by reaction of free sodium hydroxide as well as sodium oxide with carbon dioxide. The added sodium sulfate is reduced to sodium sulfide. Thus the green liquor mainly contains sodium carbonate and sodium sulfide. An example of the composition of a green liquor in comparison with a white liquor is given in Table 16–6.

After clarifying the green liquor by removing undissolved material (so-called dregs), the causticizing reaction is carried out to convert the sodium carbonate to sodium hydroxide by treatment with calcium hydroxide (slaked lime). After clar-

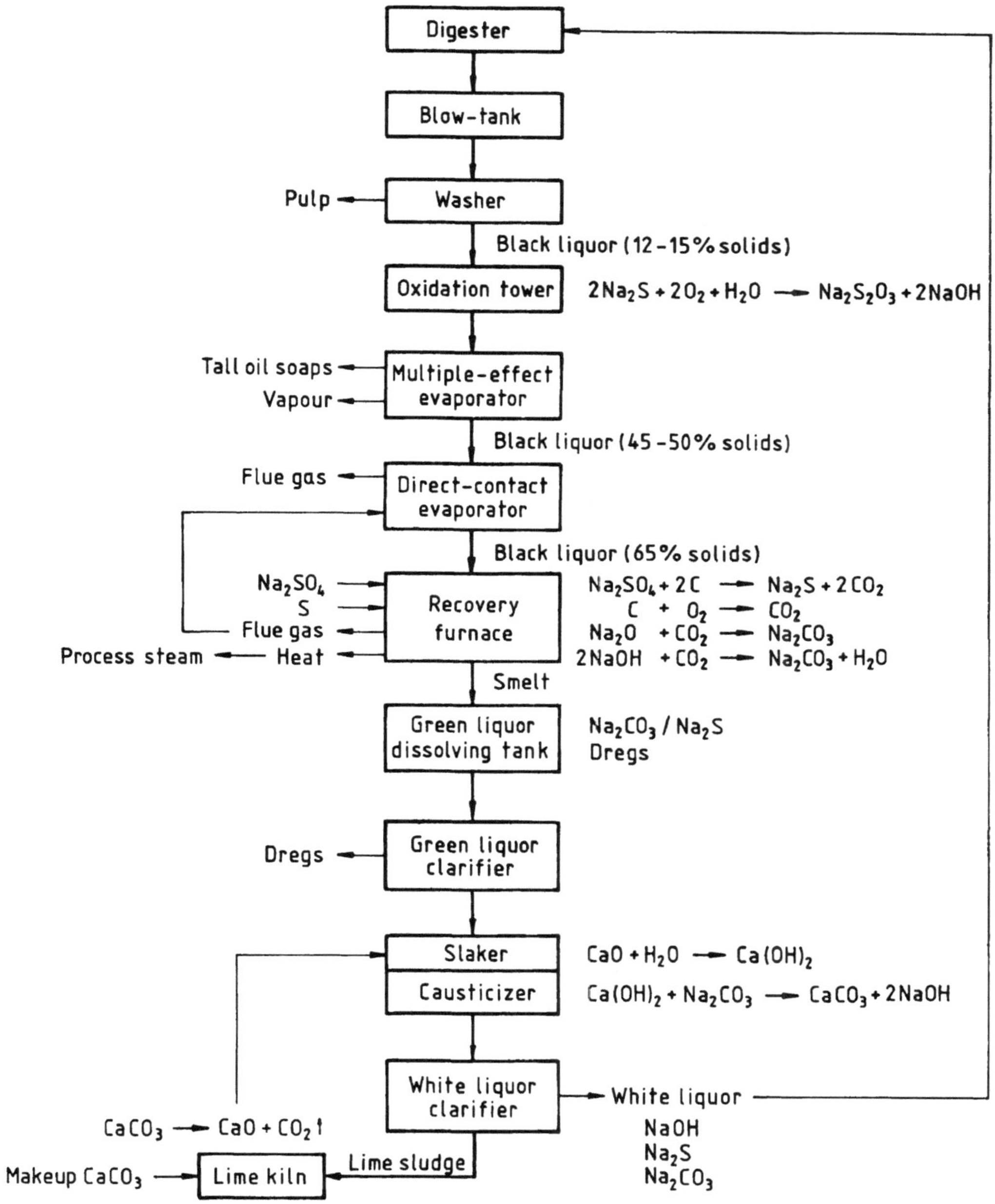

Fig. 16–12. The chemical recovery cycle in kraft pulping.

ifying, the resultant white liquor is ready for use as fresh cooking liquor in the digester.

The lime is regenerated by incinerating the calcium carbonate (lime burning) to calcium oxide and subsequently slaking the oxide with water from the green liquor. Additionally makeup calcium carbonate is added to compensate for the losses incurred.

As a means of considerably simplifying the chemical recovery cycle auto-causticizing has been suggested (Kiiskilä, Virkola 1978; Kiiskilä, Valtonen 1979; Janson 1980; Boehmer 1981). Auto-causticizing refers to the formation of a strong alkali from a weaker alkali without introducing a new agent in the existing chemical system. The conventional lime burning, slaking and causticizing can be omitted by application of certain amphoteric metal oxides (e.g. Al_2O_3, TiO_2, Fe_2O_3, B_2O_3). They react with the weak alkaline sodium carbonate to form a mixed metal oxide with expulsion of carbon dioxide according to the exemplary equation:

$$Na_2CO_3 + Fe_2O_3 \rightarrow 2\ NaFeO_2 + CO_2 \uparrow$$

By dissolving the sodium ferrite in water, sodium hydroxide is formed, and the ferric oxide recycled by precipitation:

$$2\ NaFeO_2 + H_2O \rightarrow Fe_2O_3 + 2\ NaOH.$$

16.4.4. Properties of Alkaline Pulps

The chemical composition and the physical (mechanical) properties of alkaline pulps depend on the wood species and the pulping conditions. The chemical composition of the pulp is a basic factor, affecting the pulp yield, the behaviour during further processing (e.g. bleaching), but also pulp strength properties and colour. Thus, besides morphological fibre properties, the polysaccharide reactions in alkaline medium (→ 11.1.) and the degree of delignification determine the character of a pulp. The amount of residual lignin (usually expressed by the kappa number, →3.2.9.) is the criterion for whether the pulp is to be used as an unbleached grade or bleached for printing paper qualities.

Unbleached alkaline pulps usually have a dark colour (low brightness), which is mainly caused by chromophoric groups in the residual lignin, formed during the alkaline cook (→ 11.2.). The lignin portions redeposited during the last stage of alkaline cooking contribute especially strongly to low brightness values (Surewicz 1962). The residual lignin contents of kraft pulps to be bleached are usually about 2% (based on pulp) in the case of hardwoods, or 3–4% in the case of softwoods. In contrast to the early days of kraft pulping, today effective bleaching procedures are in use for these types of pulps (→16.7). Unbleached grades have higher lignin contents, which may exceed 10% in high-yield kraft pulping. As mentioned earlier, these pulps require mechanical defibration for fibre liberation.

Excellent strength properties are characteristic of kraft pulps. Therefore they are the preferred pulps in strong paper grades, such as the liner in corrugating boards, or bag and wrapping papers. Superior strength values are however only reached by softwood kraft pulps, due to the characteristics of softwood fibres (Table 2–2) (Annergren et al. 1963; Rydholm 1965). The typical short-fibred hardwoods yield pulps with strength properties which cannot fulfil the requirements of linerboard grades. Therefore in these grades only small amounts of hardwood kraft pulp (10–20%) are blended with softwood pulp. With regard to bleached quality, how-

Fig. 16–13. Unbleached pine kraft pulp. SEM micrograph.

ever, hardwood kraft pulp is an excellent fibre material for printing papers with good surface and opacity properties. Required strength values are reached by addition of softwood kraft or sulfite pulps.

Kraft pulps will remain the dominant component in strong paper grades, though neutral sulfite-AQ (NS-AQ) and alkaline sulfite (AS) pulps promise pulp qualities comparable to those of conventional kraft pulps (→ 16.5.4.) (Wong 1981). Fig. 16–13 gives an impression of the strong fibres of an unbleached pine kraft pulp. The difference between the fibre material of chemical and mechanical pulps is evident (→ Figs. 16–5, 16–7).

16.4.5. Additives in Alkaline Pulping

For several years numerous research and application activities have been carried out to overcome the disadvantages of soda and sulfate pulping, mainly with regard to low pulp yields and environmental problems, respectively. Apart from polysulfides in kraft pulping and the action of hydrogen sulfide in pretreatment steps, some hundreds of potential compounds were suggested as pulping additives. Only a few of them, however, proved to meet at least part of the aims of those modified alkaline pulping processes:

- accelerating the delignification rate
- increasing the delignification selectivity
- increasing pulp yields
- improving pulp properties
- eliminating or reducing air pollution problems in kraft pulping.

Polysulfide pulping, which is used industrially here and there, is based on the action of polysulfides, formed e.g. by addition of elemental sulfur to the cooking liquor or by sulfide oxidation with air or oxygen (Teder 1969). The essential benefit of this process is an increase in pulp yield up to 6%, due to polysaccharide stabilization by oxidation of reducing end groups (Alfredsson et al. 1963; Casebier, Hamilton 1965; Lasmarias, Peterson 1980). But the trend to lower sulfidity levels in kraft pulping and the requirements of odour control have reduced the application of this modified alkaline process.

A pretreatment stage using hydrogen sulfide prior to a kraft cook has a comparable effect on the pulp yield combined with less alkali need and no additional sulfur loading as in polysulfide pulping. (Procter, Styan 1974). The main practical problems arise with the safe handling of hydrogen sulfide gas at elevated temperatures and pressures without air pollution (Vinje, Worster 1969). Polysulfide and hydrogen sulfide pulps have strength properties comparable to kraft pulps, except for lower tear strength. At high yield levels (above 50%) the strength values are even higher (Ahlgren et al 1975).

Another reductive polysaccharide stabilization can be performed by addition of sodium borohydride to alkaline cooking liquors. The effect of yield increase depends mainly on the quantities of added borohydride (in the range of 1–2%, based on wood) and the optimum conditions of applying this chemical, e.g. in a pretreatment stage (Aurell, Hartler 1963). In any case an industrial application is limited by the relatively high amounts of this expensive chemical required for noticeable improvements. The same reasons apply to the industrial application of other investigated chemicals like hydrazine or several amines. The latter are fairly unstable at high temperatures and must be used in high charges (10–40%) to yield sufficient effects.

The most spectacular and widely investigated field of additive pulping is alkaline pulping with anthraquinone (AQ) or its derivatives. Since the fundamental discovery that catalytic amounts (0.05–0.25%, based on wood) of anthraquinone-2-monosulfonate (AMS), other derivatives, but especially of anthraquinone itself have unrivaled effects in alkaline pulping (Bach, Fiehn 1972; Holton 1977), a huge number of publications have appeared dealing with the chemical and technological aspects of pulping with AQ and other additives (Holton, Chapman 1977; Fleming et al. 1978; Košiková et al. 1980; Kubes et al. 1980; Fossum et al. 1980a, b; Raubenheimer 1981; Abbot, Bolker 1982).

The benefits of AQ pulping generally include increased delignification rates, selectivity and velocity as well as reduced alkali charges and improved pulp properties and yields. The chemical mechanism of alkaline AQ pulping is fairly well known, with AQ being described as a redox catalyst in the liquor system (Löwendahl, Samuelson 1977; Fleming et al. 1978; Kubes et al. 1980). The lignin reactions are influenced by suppression of secondary condensation reactions and additional support of degradation by cleavage of linkages. The main action on the polysaccharides is the oxidative stabilization of reducing end-groups against the endwise peeling reaction. More details about the reaction mechanism are given in chapter 11.

The main emphasis was first placed on the application of AQ in soda pulping in order to improve this nonsulfur process with regard to pulping rates and pulp properties and yields.

Numerous results demonstrate that softwood soda-AQ pulping of bleachable grades becomes comparable to conventional kraft pulping with regard to yields and cooking times. But the pulp is harder to bleach and the strength properties are generally worse or at least no better than those of kraft pulps (Kubes et al. 1980; Fossum et al. 1980b). The accelerated delignification rate in softwood soda-AQ pulping was shown to be a result of an increased delignification of the secondary walls and middle lamellae of both earlywood and latewood, with a considerably higher delignification rate in the middle lamellae (Bruun et al. 1979b; Gädda, Bruun 1981; Gädda 1982).

The experiences of AQ additions in soda pulping were also transferred to kraft pulping and also proved to be successful in mill trials. By addition of 0.05% AQ, the cooking time was reduced by 25–35% and the alkali charge by 5% while yields increased by 1.5–3% at the same residual lignin level and comparable strength properties were obtained as in conventional kraft pulping (Holton, Chapman 1977; Goel et al. 1980). Fig. 16–14 demonstrates the marked effect of AQ on the yields in kraft pulping.

Pine kraft-AQ pulps were cooked down to very low kappa numbers (13–17) without loss of yield and with strength properties (except tear strength) comparable to normal kraft pulps (McDonough, Van Drunen 1980).

Both soda-AQ and kraft-AQ pulping of hardwoods show promising results. In soda-AQ pulping the delignification rates and the strength properties are raised to the level of kraft pulping, this in addition to yield increases and reduced cooking times.

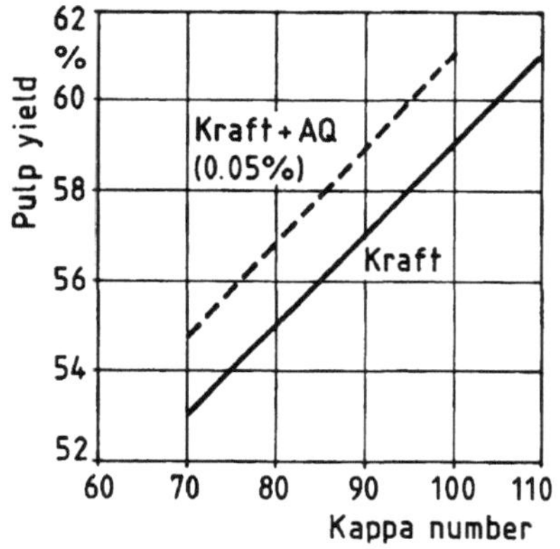

Fig. 16–14. Relationship between yield and kappa number in kraft and kraft-AQ laboratory trials (Goel et al. 1980).

Hardwood kraft-AQ pulps show increased yields (+ 1%) with comparable brightness and overall strength properties (Kubes et al. 1980). The residual amounts of AQ in pulps lie between 1–50 mg/kg pulp, as determined by gas chromatography/mass spectroscopy (GC/MS), high pressure (performance) liquid chromatography (HPLC), thin-layer chromatography (TLC), and activity determination of C^{14}-labelled AQ (Currah 1979; Faix, Pfütze 1980; Mortimer, Fleming 1981).

Based on investigations using primary aliphatic amines and diamines (e.g. monoethanolamine (MEA) as well as ethylenediamine (EDA) as additives in soda pulping, combinations of small amounts of amines with AQ (diaminoanthraquinones (DAAQ), AQ-EDA) were also tested in soda processes (Kubes, Bolker 1978; Mac Leod et al. 1979a, b; Kubes et al. 1980; Abbot, Bolker 1982). The most important effects in soda-AQ-EDA pulping are bleaching advantages and an increase in tear strength (MacLeod et al. 1981).

Very good pulp characteristics were reported for AQ-catalyzed neutral sulfite and semi-alkaline sulfite processes (NS-AQ, SAS-AQ) applied to pine and spruce wood (Virkola et al. 1979, 1981; Raubenheimer, Eggers 1980; Anonymous 1980b; Wong 1981).

Though the field of pulping additives is not yet completely developed, it can be stated on the basis of present knowledge that only anthraquinone will take hold as an important, commercially applied additive in alkaline pulping, due to its superior effects on the pulping procedures and the resulting pulps. It has already been approved as a pulping additive in Canada and the U.S.A. (Mortimer, Fleming 1981), and an advanced industrial application can be expected in the future. Though experiments with many other nonsulfur types of compounds such as phenazine, benzocinnoline or fluorene will not reveal the same catalytic efficiency as AQ, they are useful for a better understanding of the principal mechanisms in alkaline pulping with additives (Kubes et al. 1980; Fleming et al. 1980; Eckert, Amos 1980).

16.4.6. Nonsulfur Alkaline Processes

Environment protection is one of the most urgent reasons for developing sulfur-free chemical pulping processes as alternatives to sulfite and kraft pulping. Nonsulfur pulping covers numerous delignification procedures, some of which are mentioned in the section concerning unconventional pulping procedures (→ 16.6.). A comprehensive literature review was given by Weiner and Pollock (1977).

In this section a short overview is given of four alkaline nonsulfur processes:

- the traditional soda process and its AQ-modification
- the two-stage soda-oxygen process
- single-stage oxygen pulping
- the alkaline-AQ-peroxide process.

Soda pulping has already been introduced as a possible alternative to kraft pulping, especially as a future AQ-catalyzed process. It is a suitable process for hardwood pulping to yield the furnish for paper grades with low strength requirements, and is one of the most important processes for nonwood fibre pulping (Saad et al. 1975; Misra 1980; Jakate et al. 1981). The present chances for industrial soda-AQ pulping are limited to existing soda pulp mills or to kraft mills, which may be forced to switch over to a nonsulfur process because of environmental pressure (Holton, Chapman 1977). It is noteworthy in this connection that even in the soda process, which uses sodium carbonate as a makeup chemical, sulfide may be produced in the recovery furnace by conversion of accumulating sulfur traces derived from wood, water and fuel oil (Palenius 1976).

The typical two-stage soda-oxygen pulping process is mainly suitable for hardwoods but also for softwoods, though with less selectivity. It involves a high-yield soda cook, followed by a mechanical defibration and a concluding oxygen bleaching in alkaline medium. The first stage has the greatest influence on final pulp yields and properties. The yield after the cooking stage should be in the range of 60–65% to reach high final yields after the bleaching procedure, and to achieve optimal refining conditions with regard to energy input and fibre protection. The yields of bleached pulps are comparable to bleached kraft pulp yields, while the strength values are somewhat lower (Worster et al. 1971; Marton, Leopold 1973; Palenius 1976; Nakamura, Matsuura 1975; Kratzl 1978; Abson, Stockman 1979). A practical aspect of this process is the fact that it can be run in established kraft pulp digestion and recovery equipment with an added refining section and an oxygen bleaching plant.

This advantage does not exist for the single-stage oxygen process and its modifications. The delignification is carried out in slightly alkaline solutions (pH 7–9) of sodium hydroxide, sodium carbonate or hydrogen carbonate (also in the presence of hydrogen peroxide) at 140–150 °C and high oxygen pressures in the range of 20–40 bar. The pressure cooking is carried out at low consistency to maintain the temperature level (Abrahamsson, Samuelson 1975; Jamieson et al. 1975; Palenius 1976; Renard et al. 1975; McKelvey et al. 1978; Brodén, Simonson 1981).

To avoid insufficient oxygen penetration into the wood very thin chips (< 1.5 mm) must be used. The pulps are generally obtained in high yields and good brightness, but the strength properties are usually not comparable to those of kraft pulps. Strength improvements can be obtained by sodium hydrogen carbonate precooking, increased carbon dioxide content in the gas phase, and by addition of potassium iodide or other metal compounds (magnesium carbonate, magnesium chloride, calcium chloride) as cellulose protectors (Abrahamsson, Samuelson 1973, 1974; Minor, Sanyer 1974, 1975; Landucci, Sanyer 1975; Fujii, Hannah 1978).

The latest modification of alkaline nonsulfur pulping is the alkali-peroxide (ALCAPER) process (Lachenal et al. 1980). Delignification is carried out in two stages with a refining step in between. The first one is a high-yield soda-AQ cook down to kappa numbers of 50–60. The second delignification step is performed as a medium-to-high consistency (> 10%) hydrogen peroxide bleaching (less than 0.5% H_2O_2 based on wood), reducing the kappa number down to about 30. The obvious advantages of this process are:

- 2–5% higher yields than in kraft pulping
- better mechanical characteristics and higher brightness as compared to kraft and soda-AQ pulps
- low AQ amounts (less than 0.1%)
- elimination of the air pollution problems of kraft pulping
- easy and low-cost substitution within existing kraft mills by installation of a refiner section and a hydrogen peroxide bleaching tower.

16.5. Sulfite Chemical Pulping

16.5.1. The Sulfite Systems

From a chemical standpoint sulfite pulping liquors vary with respect to the possible forms of sulfur dioxide in aqueous solution and the type of base added to this system. The reaction of sulfur dioxide with water principally yields:

- dissolved SO_2 and/or sulfurous acid (H_2SO_3)
- bisulfite (hydrogen sulfite) (HSO_3^-)
- monosulfite (SO_3^{2-}),

according to the equilibrium sequence:

$$SO_2 \rightleftharpoons SO_2\ (\text{dissolved}) \overset{H_2O}{\rightleftharpoons} H_2SO_3 \rightleftharpoons H^+ + HSO_3^- \rightleftharpoons H^+ + SO_3^{2-}.$$

In sulfite pulping the composition of the cooking liquor is characterized by the terms free, combined and total sulfur dioxide, expressed as g SO_2/100 ml of liquor (Tappi Standard T pm–79). The liquor is described more precisely by the actual content of sulfurous acid, bisulfite and monosulfite, derivable from the above-mentioned SO_2-values.

In the presence of alkali (e. g. sodium hydroxide) bisulfite and monosulfite are formed from the dissolved SO_2:

$$H_2O + SO_2 \rightleftharpoons H_2SO_3\ (\text{or } SO_2 \cdot H_2O)$$
$$H_2SO_3 + NaOH \rightleftharpoons NaHSO_3 + H_2O$$
$$NaHSO_3 + NaOH \rightleftharpoons Na_2SO_3 + H_2O$$

The resulting pH level of the liquor depends on the concentrations of sulfurous acid, monosulfite and bisulfite. Fig. 16–15 demonstrates the pH range of sulfite systems (1.5–9), revealing the inflection points of bisulfite at about pH 4.5 and of

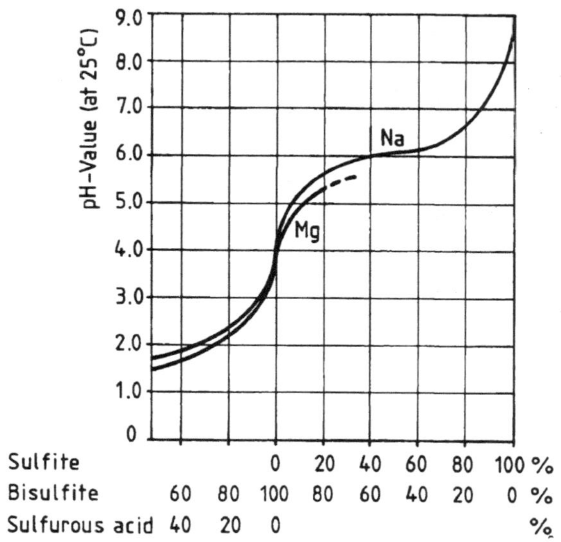

Fig. 16–15. The different composition of sulfite solutions, depending on the pH level (Bryce 1980b).

monosulfite at about pH 9. The real pH levels in sulfite pulping, applying temperatures up to 180 °C, are somewhat different, as the pH values generally increase with increasing temperatures (Schöön, Wannholt 1969).

The pH values are additionally influenced by the type of base used in the cooking liquor. The following bases are normally used in sulfite pulping: calcium, magnesium, sodium and ammonia. Calcium is the classical base in acidic sulfite pulping involving both advantages and severe disadvantages. The advantages are the low price and the good availability of limestone as the dominant raw material. The drawbacks are the limited solubility of calcium sulfite, scaling problems and the lack of a suitable recovery system. The low solubility of calcium sulfite at elevated temperatures requires an excess of free SO_2 to keep the pH value below 3, preventing the formation of calcium sulfite from calcium hydrogen sulfite.

A much better solubility is obtained with magnesium as a base, covering the pH range up to about 5. The outstanding advantage of magnesium-based sulfite pulping is the possibility of combusting the waste liquor to yield magnesium oxide and sulfur dioxide for producing fresh cooking liquor.

As sodium sulfite and sodium bisulfite are soluble under conditions applied in sulfite pulping, sodium is a highly attractive base. Furthermore it is quite easy to prepare the pulping liquors, usually by starting with sodium hydroxide in solution or less frequently with solid sodium carbonate. The sodium recovery was a drawback of this base as compared to magnesium, but there are several commercial recovery systems available today (Wong 1980, 1981).

Ammonia, the other monovalent base, is comparable to sodium with regard to the solubility of the sulfites, but the handling of the aqueous ammonia solutions and the

Table 16–7: Dependence of sulfite pulping data on the choice of the base (according to Bryce 1980b)

	Calcium	Magnesium	Sodium	Ammonium
pH Range for digestion	Below 2	Below 5	0–14	0–14
Pulping rate	Intermediate	Intermediate	Slow	Fast
Level of screenings	Moderate	Moderate	Low	Low
SO_2 absorption	Complex	Relatively simple	Simple	Simple
Scaling tendency	High	Moderate	Low	Low
Liquor incineration	Difficult	Simple	Complex	Simple
Recovery	No recovery of SO_2 and base	Recovery of SO_2 and base	Recovery of SO_2 and base	Recovery of SO_2; No recovery of base

chemical recovery require different equipment and systems in ammonia-based processes. A typical characteristic of ammonia-based sulfite pulping is very rapid delignification. In Table 16–7 the four bases are compared with regard to some important process data.

16.5.2. Sulfite Processes

Though the general trend in chemical pulp capacity expansion shows kraft pulping ahead of the others there are some factors which may improve the future of sulfite pulping, involving the well-known advantages of sulfite pulps over kraft pulps:

- higher yields at a given kappa number, resulting in lower wood consumption
- higher brightness of unbleached pulps
- higher flexibility of bleaching and bleaching without chlorine
- fewer pollution problems
- lower installation capital costs
- higher flexibility in pulp yields and grades.

Whereas formerly sulfite pulping was mainly represented by the acid calcium bisulfite process, today a number of different and modified sulfite processes are known. Sulfite processes are characterized by the composition of the cooking liquor, which influences the cooking pH and the choice of the base. They can be grouped into five principal types:

- acidic sulfite (bisulfite) pulping
- bisulfite pulping
- multi-stage sulfite pulping
- neutral sulfite pulping
- alkaline sulfite pulping.

The most important processes will be introduced after a short description of the general steps in sulfite pulping as outlined in Fig. 16–16 (Bryce 1980b).

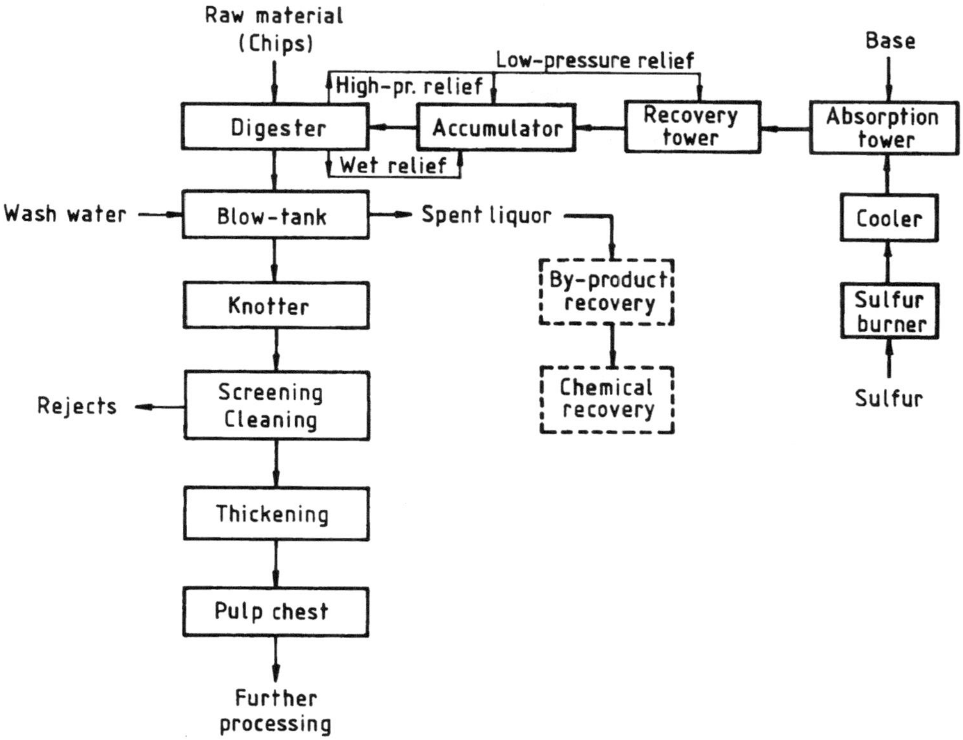

Fig. 16–16. Simplified flowchart of sulfite pulping.

The digester is filled with the maximum quantity of chips, whose quality has a much more pronounced influence on the final pulp quality than in kraft pulping (→ 16.5.3.). To improve the penetration of the cooking liquor into the chips different techniques have been worked out. Commonly the chips are presteamed during the filling procedure to remove the air, but evacuation or pressure impregnation can also be effective (Aurell et al. 1958).

Sulfur dioxide, the essential chemical involved in all sulfite processes, may be prepared by burning pure sulfur or iron sulfide (pyrites) in rotary or spray burners in the presence of air, or obtained at least additionally in liquid form. If the sulfur dioxide is prepared by combustion the hot gas must be cooled down to 30–70 °C. To prepare the sulfite cooking liquor several systems are in use. The typical process in acidic calcium bisulfite pulping is to feed sulfur dioxide gas and water through calcium carbonate rocks (limestone) in the absorption tower, in order to have them react according to the equations:

$$SO_2 + H_2O \rightarrow H_2SO_3$$
$$2\,H_2SO_3 + CaCO_3 \rightarrow Ca(HSO_3)_2 + CO_2 + H_2O$$

Other absorption systems include packed towers, where the cooled sulfur dioxide contacts with water and the base (e. g. $Mg(OH)_2$, NH_4OH, $NaOH$), or most recently so-called moving-bed towers consisting of 4 stages which are partially filled with hollow plastic spheres, causing high turbulence and thus better contact between the gas and the liquid phase (Turbulent Contact Absorber, TCA; Bryce 1980b). In any case the resulting liquor is pumped through the low-pressure recovery tower and finally reaches the high-pressure vessel (accumulator), where the actual cooking liquor is prepared by combining water vapour and relieved sulfur dioxide with the fresh tower liquor ('tower acid'). Before being fed into the digester the cooking liquor is heated to the starting temperature, which depends on the type of process used.

Digestion can be carried out continuously or as a discontinuous batch process of which the most common form is batch cooking. The batch digesters in sulfite pulping are between 70 and 350 m^3 in volume. The liquor is heated by direct steam injection or by indirect heating using a heat exchanger to reach the maximal cooking temperature, which varies between 125 and 180 °C, depending on the special process and the desired pulp type (Table 16–8). The cooking pressure varies between 5 and 7 bar. To ensure a homogeneous distribution of the cooking liquor, it is circulated within the digester by adding steam, but more commonly by means of a liquor circulation pump. Before the final cooking temperature is reached part of the liquor is removed (side relief), followed by a SO_2 gas relief from the top of the digester (top relief), both being returned to the accumulator. The cooking temperature is controlled by steam, and by reducing the pressure when the cook is finished

Table 16–8: Sulfite pulping processes (according to Wong 1980; Sjöström 1981)

Process	pH Range	Bases	Active reagent	Max. temp. °C	Cooking time at max. temp. h	Pulp application
Acidic (bi)sulfite	1–2	Ca^{2+}, Mg^{2+}, Na^+, NH_4^+	HSO_3^-, H^+	125–145	3–7	Dissolving pulp, newsprint, writing paper, xerographic paper, tissue
Bisulfite (Arbiso, Magnefite)	3–5	Na^+, Mg^{2+}, NH_4^+	HSO_3^-, H^+	150–170	1–3	Newsprint, printing paper, writing paper, coated paper, tissue
Multi-stage sulfite						
Stora process		Na^+				
Stage 1	6–8		HSO_3^-, SO_3^{2-}	135–145	2–6	Glassine, greaseproof paper, copying paper, tissue
Stage 2	1–2		HSO_3^-, H^+	125–140	2–4	
Sivola process		Na^+				
Stage 1	3–4		HSO_3^-, H^+	140–150	2–3	Dissolving pulp, printing paper, newsprint
Stage 2	7–10		HSO_3^-, SO_3^{2-}	140–160	1–3	
or:		Na^+				
Stage 1	6–8		HSO_3^-, SO_3^{2-}	120–140	2–3	Dissolving pulp, newsprint, magazine paper
Stage 2	1–2		HSO_3^-, H^+	135–145	3–5	
Stage 3	6–10		HO^-	160–180	2–3	
Neutral sulfite (NSSC)	5–7	Na^+, (NH_4^+)	HSO_3^-, SO_3^{2-}	160–180	0.25–3	Corrugating medium
Alkaline sulfite	9–13	Na^+	SO_3^{2-}, HO^-	160–180	3–5	Kraft-type pulp

the digester content is discharged into the blow-tank at reduced pressure of about 2 bar. While the spent liquor is being removed, wash water is introduced. The ready-washed pulp is screened and cleaned to select the rejects, and finally thickened. The ready unbleached pulp may be further bleached, dried or added to a paper-machine furnish. The spent liquor may be used for the recovery of by-products (→ 18.6.) or combusted after evaporation to yield heat and to recover the base, if possible (Table 16–7).

Some data on important sulfite processes are given in Table 16–8, demonstrating the flexibility of sulfite pulping today. The classical process, the acidic calcium bisulfite process, applies a cooking liquor of $Ca(HSO_3)_2$ with a high excess of free sulfur dioxide, which is not consumed during cooking, but keeps the pH at about 1.2–1.5. As the free, highly acidic SO_2 penetrates much more quickly into the wood chips than the liquor, the temperature increase during the heating period to reach the final temperature must be slow. Otherwise early lignin condensation reactions will take place resulting in incomplete delignification. Fig. 16–17 shows the course of temperature during acid sulfite cooking in comparison with bisulfite pulping. The final cooking temperature is generally kept between 125–135 °C for the production of paper pulps, and up to 145 °C in dissolving pulp production. The total cooking cycle lasts up to 12 h, one of the disadvantages of this process. Other drawbacks are scaling problems, the lack of a chemical recovery system, and environmental problems (Virkola 1975; Göttsching 1980b). With regard to the wood raw material no species with a high resin content (e. g. pines and many hardwoods) or chips with considerable amounts of bark can be used, due to condensation reactions of lignin with extractive components preventing a sufficient sulfonation (→ 10.3.2.).

The use of soluble bases can solve some problems of the calcium bisulfite process, but their application was restricted by high chemical costs as long as no suitable

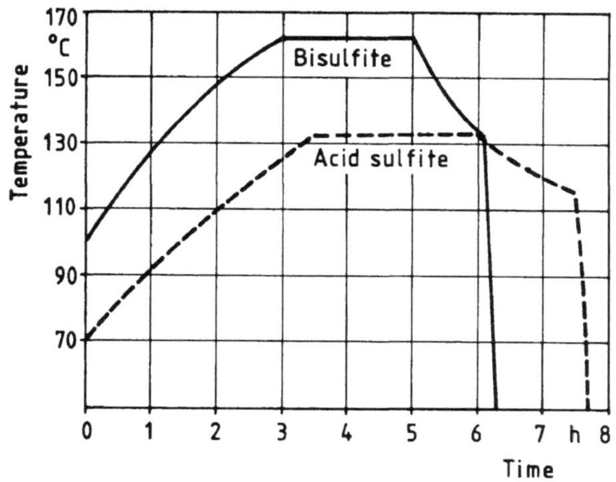

Fig. 16–17. Course of temperature during typical acidic sulfite and bisulfite cooks (according to Bryce 1980b).

recovery systems for sodium and ammonia were available. Though recovery systems are known today, including ion-exchange techniques (Lengyel, Morvay 1973; Rahm 1976; Wong 1980), the advantages of bisulfite and multi-stage sulfite pulping have restrained the acidic processes. But they are still important for the production of dissolving pulps and pulps for newsprint as well as for paper grades where no high demands are made on strength properties (Wong 1980, 1981; Magister 1981).

The bisulfite processes are characterized by cooking pH values of 3–5 and the use of true bisulfite liquors without excess sulfur dioxide. Sodium and magnesium are used as bases, but in principle ammonia is also suitable (Ernest, Hannan 1967). In comparison with acidic sulfite pulping the maximum cooking temperature is somewhat higher (150–170 °C) and the cooking time essentially shorter (1–3 h). The final temperature can, however, be reached even faster as the penetration of chemicals into the chips is better and the starting liquor temperature is higher (90–100 °C).

The main commercial processes are the Arbiso process using sodium as a base and the magnesium-based Magnefite process. The magnesium bisulfite process offers many advantages concerning technical aspects of pulping and recovery (e. g. no escape of sulfur dioxide during blowing, continuous pulping, no scaling during evaporation, easy chemical recovery). Additionally the process is regarded as highly flexible with regard to the wood species used. In laboratory experiments even extractive-rich pine, larch and Douglas fir were pulped successfully, but there are still problems in industrial pulping when the amount of pine wood exceeds about 50% (Patt et al. 1974; Patt, Hoffmann 1977). Mg-bisulfite pulping was modified by pretreatments with $Mg(OH)_2$, $MgSO_4$ or NaOH (Croon 1963; Nelson 1974; Patt, Hoffmann 1977). The yields of unbleached bisulfite pulps are up to 6% higher than those of comparable kraft pulps. Additionally these pulps have higher brightness values and bleaching advantages. Concerning the strength properties, however, kraft pulps remain unrivalled (Patt, Hoffmann 1977).

To combine the advantages of acidic and alkaline pulping, and to obtain a higher flexibility of pulp yields and qualities, numerous multi-stage processes have been suggested (Rydholm 1960, 1965; Virkola et al. 1963). Only two types using sodium as a base have attained commercial importance. The first one is the Stora process starting with a sulfite-bisulfite mixture at pH between 6 and 8, followed by an acidic step at low pH (1–2) by adding sulfur dioxide and/or acidic liquor (Söderquist 1955; Lagergren 1964). With this type of sulfite pulping even extractive-rich pines can be pulped without difficulty because the lignin sulfonation takes place in alkaline medium, avoiding interfering side reactions with extractive components. Due to a glucomannan stabilization in the first stage the yields are higher than in classical sulfite pulping, but the stabilization is only effective in the case of softwoods with high mannan contents (Annergren et al. 1961; Czirnich, Patt 1976). The second type of multi-stage sulfite processes (Sivola (Rauma) processes) includes an initial bisulfite stage (pH 3–4), followed by a cooking stage at pH 7–10 (Bryce, Tomlinson 1962; Laine et al. 1979; Wong 1980). A three-stage modification applying an acidic stage between the bisulfite and alkaline stage is especially suitable for the production of softwood dissolving pulps (Ulmanen et al. 1963).

Neutral sulfite with a small excess of alkali is mainly applied in the production of semichemical pulps (→ 16.3.1.) with sodium as the predominant base, but recently considerable interest has arisen in the AQ-catalyzed neutral sulfite process (NS-AQ), applying cooking pH values between 8 and 10 (Wong 1981). The use of neutral ammonium sulfite for pulping *Pinus massoniana* was reported by Hou et al. (1981).

One of the most promising advances in sulfite pulping is the alkaline sulfite process (AS) using sodium sulfite and sodium hydroxide in combination at pH levels up to 13. The process combines the advantages of kraft pulping (namely strong pulps and no limitation to wood species) with sulfite pulping characteristics such as high yields of bright and well bleachable pulps and fewer or no odour problems (Ingruber, Allard 1973; Ikonopisova et al. 1979).

16.5.3. Influencing Factors in Sulfite Pulping

As shown in the previous section pulping conditions vary widely in sulfite pulping. Therefore the number of variables is much higher and more complex than in kraft pulping. Sulfite pulping is influenced by the following main factors (Rydholm 1965; Lengyel, Morvay 1973; Bryce 1980b):

- wood: wood species
  wood quality
  wood chips properties
- impregnation conditions
- cooking liquor composition: pH value
  SO_2 concentration and amount
- digestion parameters: temperature
  time
  pressure.

Sulfite processes, and the acidic processes in particular, are much more sensitive to wood species than the alkaline kraft process. In the calcium-based acidic sulfite process only small amounts of pine heartwood and bark can be tolerated because otherwise phenolic components condense with lignin under the acidic conditions, preventing the delignification reactions (→ 10.3.2.). If soluble bases are applied in acidic sulfite pulping the raw material basis can be expanded to some extent, but an essential improvement was made possible by the bisulfite processes, and the multi-stage processes starting with a high pH level.

Generally softwoods (spruce, fir, hemlock and pine) are used for the production of paper pulps. For dissolving pulps hardwoods, e. g. beech or birch, are preferred. Dense hardwoods are more easily pulped with soluble bases in acidic processes and especially with bisulfite and multi-stage processes (Rydholm 1960; Utaka et al. 1965; Stark, Eichinger 1979). Final pulp yields and qualities are much more strongly affected by low wood quality (e. g. portions of decayed wood) than in kraft pulping.

The impregnation of wood chips is a combined effect of liquor penetration and diffusion of the dissolved cooking chemicals in the soaked chips. The penetration is mainly affected by the applied pressure and to a lesser extent by the temperature. The diffusion rate is mainly determined by the concentration of chemicals and the accessible total cross-sectional pores area (Sjöström 1981). The penetration is additionally influenced by the type of wood (softwood, hardwood; sapwood, heartwood) and the chip size. To ensure optimum impregnation with chemicals the average chip thickness should be about 5 mm for the classical acidic calcium-based process and about 8 mm in bisulfite pulping. The chip length influences the penetration and the final pulp quality as well. On the one hand short chips are penetrated more quickly by the cooking liquor, on the other hand they generally include a greater proportion of cut and damaged fibres, reducing the pulp strength properties (Hartler 1963). The liquor-to-wood ratio during the impregnation phase is normally about 5:1 and is lowered after completed impregnation to about 3:1 or 4:1 by side relief.

The chemical composition of the cooking liquor is characterized by the pH value and the amounts of free, combined and total sulfur dioxide, factors which are dependent on each other. As worked out by Ingruber and Allard (1967) the pH is the dominant variable influencing the pulping rate and the pulp yield and quality as well, followed by the cooking temperature and the amount of sulfur dioxide. The pH level is especially important in bisulfite pulping, because the hydrogen concentration, which determines the hydrolysis reactions, is low and must be compensated by increased temperature (Table 16–8). Due to the retention of polyoses the yields of bisulfite pulps are higher than in acidic sulfite pulping. Optimum brightness of unbleached pulps and best softwood pulp strength properties are obtained at bisulfite pH level of 4. Hardwood bisulfite pulps show increased strength values at pH values of 5–7 (Hartler et al. 1961a).

The increase of the combined sulfur dioxide charge (SO_2 as monosulfite + $\frac{1}{2}$ SO_2 as bisulfite) in the range from 4–9% based on wood, results in higher yields at a given kappa number, increased brightness, increased breaking length, but decreased tear strength (Hartler et al. 1961b). The control of the total SO_2-concentration is more important in bisulfite processes than in acidic sulfite pulping, where the excess free sulfur dioxide maintains a high hydrogen ion concentration, determining the pulping rate.

The influence of the maximum cooking temperature is different in acidic sulfite and in bisulfite pulping. In acidic sulfite processes a high temperature (above 150 °C) causes a significant loss of polyoses, a fact used in the production of dissolving pulps. At low temperatures between 125 and 135 °C the retention of polyoses is high. At higher pH levels of bisulfite cookings the required maximum temperatures lie between 150 and 170 °C, but the polyoses retention is still high. Also in bisulfite pulping higher yields can be obtained by lowering the cooking temperature.

The cooking time is influenced by the cooking temperature and the liquor composition as well. Cooking times at maximum temperature are thus generally lower in bisulfite pulping. Prolonged times decrease yields and strength properties in acidic sulfite and in bisulfite pulping.

The cooking pressure regulates the amount of free sulfur dioxide and therefore the pressure in the acidic calcium bisulfite process must remain above the steam pressure of the liquor to maintain the level of free sulfur dioxide, preventing calcium sulfite precipitation at higher temperatures.

16.5.4. Properties of Sulfite Pulps

The different sulfite processes offer the production of pulps a wide variety of chemical composition and papermaking properties, e. g. with regard to opacity and strength values. The residual lignin content may vary between 10–15% in unbleached high-yield sulfite pulps, but is generally in the range of 3–5% in the case of softwood and 1–3% in hardwood pulps. The different pulping conditions affect the removal and the retention of polyoses as well (→ 16.5.2., 16.5.3.). In acidic sulfite pulping large portions of polyoses are removed in addition to a DP-reduction of the cellulose. The remaining polyoses in softwood pulps are mannan and xylan whose galactose and arabinose side-groups, respectively, are split off. The residual mannan content is especially high in pulps from two-stage processes with a neutral first step (Stora-type processes) and lowest if the first step or the whole process is performed under acidic conditions (Annergren et al. 1961). Therefore the acidic processes are preferred for the production of dissolving pulps from hardwoods with high alpha-cellulose contents above 90%. Hardwood sulfite paper pulps have very low strength properties and are used mainly for high opacity printing papers.

A high polyoses content at a given kappa number results in high yields and is an essential advantage with regard to the fibre-bonding properties. Stora-type pulps can therefore be combined and even beaten together easily and quickly with short-fibre pulps to reach good opacity characteristics, but are also used as long-fibre paper furnish. Thus the application range is wide, covering copying, wrapping and tissue papers as well as transparent paper grades such as glassine or grease-proof papers (Lagergren 1964).

Magnefite softwood and hardwood pulps were shown to be a suitable furnish for very different paper grades such as printing, writing and base paper for coating as well as newsprint, partly replacing sulfate pulp (Stark, Eichinger 1979).

Table 16–9: Comparison of unbleached Na-sulfite pulps with kraft pulp from *Pinus banksiana* (Bryce 1980b)

	Permanganate No.	Yield %	Beating time min	Tensile m	Burst index kPa · m²/g	Tear index mN · m²/g
Bisulfite – neutral sulfite	22.2	48.5	18	11 400	7.26	10.2
Bisulfite – acidic sulfite	18.0	49.8	26	10 000	6.38	11.8
Neutral sulfite – acidic sulfite	22.1	54.8	7	9 000	5.00	7.35
Bisulfite	18.4	50.8	15	10 400	6.38	9.22
Kraft	17.4	45.5	34	12 900	9.81	13.7

Strength properties at 500 ml CSF

Table 16–10: Comparison of kraft and alkaline sulfite pulp data (according to Ingruber, Allard 1973)

	Yield %	Kappa No.	Brightness GE	C.S.F. ml	Specific volume cm³/g	Breaking length m	Burst index kPa · m²/g	Tear index mN · m²/g
Kraft, 166 °C	47.2	26.8	28.9	300	1.31	14 200	10.9	9.2
Alkaline sulfite, 175 °C, pH 8.0	43.9	43.6	37.9	300	1.30	14 300	10.9	9.6
Alkaline sulfite, 175 °C, pH 9.5	45.5	35.8	33.0	300	1.29	13 000	9.3	11.0

The brightness values of unbleached sulfite pulps are generally higher than in kraft pulping, even at high yields. An exception are alkaline sulfite pulps which are as dark as kraft pulps. All typical sulfite pulps can be beaten much more easily and with lower refining power than kraft pulps to reach the maximum tensile strength (Giertz 1963).

The strength values of sulfite pulps are principally lower than those oft kraft pulps as demonstrated for unbleached softwood pulps in Table 16–9. Only the alkaline sulfite pulps attain the strength properties of kraft pulps (Table 16–10; Ingruber, Allard 1973; Wong 1980, 1981).

Excellent papermaking properties, superior to those of all conventional sulfite pulps and comparable to kraft pulp properties, are also reported for softwood neutral sulfite-AQ pulps. This is especially noticeable because the good strength properties are obtained in combination with high yields of 48–52% for bleached pulps (Virkola et al. 1981; Wong 1981).

Fig. 16–18 shows an unbleached acidic Ca-bisulfite pulp from spruce wood.

16.6. Unconventional Pulping

The term unconventional pulping encompasses well-known pulping principles which are, however, only rarely applied commercially, and numerous delignifica-

Fig. 16–18. Unbleached spruce sulfite pulp. SEM micrograph.

tion procedures worked out on laboratory or pilot-plant scale. As some delignifying agents are also involved in pulp bleaching (→ 16.7.) or in laboratory delignification and lignin determination (→ 3.2.6., 3.2.9.), the term pulping is questionable for certain procedures. Most unconventional delignification processes use very special and sometimes unorthodox chemicals. Therefore their industrial application is often limited by high costs of chemicals and special equipment requirements.

Table 16–11 gives a selective rather than comprehensive survey of unconventional pulping procedures. Earlier literature on this topic is given by Rydholm (1965). Descriptions of nitric acid and hydrotropic pulping and its numerous modifications are given elsewhere (Sandermann 1963; Rydholm 1965; Lengyel, Morvay 1973). These procedures, which have been tested in principle as full-scale processes, are especially suitable for nonwood fibres, but also for hardwoods. Organosolv pulping seems to be a viable future pulping alternative because of the relatively low capital investment required for a new mill, the absence of pollution problems, and the advantage of obtaining polyoses and lignin easily and largely unchanged for further high-value utilization (→18.6.).

16.7. Bleaching of Pulps

16.7.1. Bleaching Principles and Chemicals

Bleaching is not a modern invention but a very old technique used to fade textiles with the aid of sunlight and/or chemicals such as potash and later hypochlorite and chlorine (Naujoks 1979). The early bleaching of paper made from rags and other non-wood fibres was also performed in this way. A new period of bleaching history began at the end of the nineteenth century with the industrial bleaching of wood pulps first with hypochlorite and later with chlorine, partly in combination, and

Table 16–11: Unconventional pulping procedures

Procedure	Process data	References
Nitric acid pulping	HNO_3 (1–60%); 15–150 °C; 0.25–3 h; extraction with aqueous NH_4OH and/or NaOH	Wither, Captein (1960) Brink (1961) Brink et al. (1962) Slavik, Kuniak (1961) Kalisch (1967) Ba et al. (1980)
Hydrotropic pulping	Na-m-xylene sulfonate, Na-p-toluene sulfonate, Na-benzoate, Na-salicylate etc. in H_2O or org. solvents; 150–170 °C; 0.5–6 h	Chene et al. (1961) Robert, Noisillier (1961) Schwenzon (1966) Procter (1971) Bland et al. (1978) Nelson (1978)
Organosolv pulping	Ethanol/H_2O (1:1); 180–215 °C; 20–60 min	Kleinert (1971, 1974) Baumeister, Edel (1980)
	Methanol/H_2O (1:1); 150–210 °C; 30–180 min, 2 steps, alkaline medium	Anonymous 1983
	Butanol/H_2O or Phenol/H_2O (1:1); 180–205 °C; 2–12 h	April et al. (1979, 1982)
Alkali-methanol pulping	4% NaOH in 40% methanol; 140–160 °C; 30–120 min	Nakano et al. (1981)
Phenol pulping	Phenol (+ H_2O, HCl); 160–170 °C; 3–4 h	Schweers, Rechy (1972) Schweers (1974)
Holopulping	Defibration of alkali-impregnated chips; ClO_2-delignification; extraction with alkali; disk refining	Anonymous (1969) Whitney et al. (1969) Cox, Worster (1971) Nicholls et al. (1976)
Acetic acid pulping	CH_3COOH/conc.HCl/acetone (7:1:2); 70 °C; 3–6 h	Koval, Slavik (1963)
DMSO pulping	DMSO/H_2SO_4(99:1, V/V); 130–160 °C	Balhar (1965)
	NO_2/SO_2/H_2S or Cl_2 in DMSO or DMF; → 140 °C; 0.5–3 h	Clermont, Bender (1961)
	DMSO/EDA, DMSO/hydrazine; DMSO/CH_3COOH, H_2SO_4 or HCl; 145–176 °C; 20 min–3 h	Nahum, Pellegrini (1970)
NO_2 delignification	NO_2 (2–3%) in coal oil or CCl_4; room temp.; 10–30 min; extraction with 1% NaOH	Bender et al. (1972)
Vapour-phase SO_2-process	Sat. SO_2 vapour; → 110 °C	Mamers, Grave (1974)
Explosion-CO_2-process	Aqueous CO_2-solution; 160–200 °C; 50 bar; pressure reduction defibration	Mamers et al. (1976)
Autohydrolysis delignification	Autohydrolysis (175–220 °C); 4–120 min; also in the presence of aromatic compounds (e.g. 2-naphtol); extraction with dioxane/H_2O; defibration	Lora, Wayman (1978, 1980) Wayman, Lora (1978, 1979)
Ketone delignification	Acetone, methyl-ethyl ketone, cyclohexanone or ketone-ammonia mixtures; 175–210 °C; 60–150 min	DeHaas, Lang (1974)
Formaldehyde pulping	HCHO (25–50%); 130–200 °C; 5–200 min	Zachariasen (1964)

with an intermediate extraction step with alkali. Though the fundamentals of early bleaching procedures are still valid, the development of pulp bleaching techniques within this century has led to a large number of bleaching chemicals applied in numerous and highly specific processes today (Lorås 1980).

The principal aim of pulp bleaching is to increase brightness. As the light-absorbing chromophoric components in unbleached pulps are predominantly functional groups of degraded and altered residual lignin (→ 13.1.3.), bleaching can be performed either by converting and stabilizing chromophoric groups without loss of substance (lignin-preserving bleaching) or by removing the lignin (lignin-removing bleaching). Along with the lignin other compounds (extractives and ash components, polyoses) and insufficient delignified particles (shives, bark specks) may also be at least partly removed (Axegard, Jonsson 1979; Axegard 1980). Therefore bleaching can additionally be regarded as a purification process which is used especially in the case of dissolving pulp production to obtain a pure pulp with high alpha-cellulose content (Rydholm 1965).

Bleaching results in changes of the optical pulp properties light absorption, light scattering and reflectance, expressed in terms such as brightness, whiteness or opacity. The most important practical value for characterizing the colour of a pulp is its brightness, which can be described and determined in slightly different manners using several standards and testing methods (Tappi Standard T 217 m–48, T 218 os–75; SCAN-C 11:75; ISO 3688-1977 E). The most commonly used brightness value represents the reflectance factor of blue light (357 or 360 nm) of a pulp sheet (in %), based on the reflectance of magnesium oxide (100% brightness) as a standard sample. The increase of brightness between an unbleached and bleached pulp is generally called the brightness gain.

Table 16–12: Bleaching chemicals

	Oxidizing chemicals		Reducing chemicals	
Industrially important	Chlorine	Cl_2	Sodium dithionite	$Na_2S_2O_4$
	Sodium hypochlorite	$NaOCl$	Zinc dithionite	ZnS_2O_4
	Calcium hypochlorite	$Ca(OCl)_2$	Sodium bisulfite	$NaHSO_3$
	Chlorine dioxide	ClO_2		
	Hydrogen peroxide	H_2O_2		
	Sodium peroxide	Na_2O_2		
	Oxygen	O_2		
Less important or not commercially applied	Ozone	O_3	Sulfur dioxide	SO_2
	Sodium chlorite	$NaClO_2$	Sodium borohydride	$NaBH_4$
	Peracetic acid	CH_3CO_3H	Calcium dithionite	CaS_2O_4
	Chlorine monoxide	Cl_2O	Aluminium dithionite	$Al_2(S_2O_4)_3$
	Thioglycolic acid	$CH_2SHCOOH$		
	Hydrogen	H_2		
	Potassium permanganate	$KMnO_4$		

Pulp bleaching chemicals can be classified into oxidizing and reducing agents. Table 16–12 includes the mass bleaching chemicals, but also chemicals used for experimental bleaching such as potassium permanganate or chlorine monoxide (Krause 1971; Bolker, Liebergott 1972). With regard to environmental aspects it is sometimes useful to classify bleaching procedures into those that apply chlorine or chlorine-containing compounds and those that do not.

Industrial lignin-removing processes include multi-stage bleaching sequences adapted to the special pulp type and combining the different oxidizing and reducing abilities of the bleaching chemicals (Tables 16–14, 16–15). Degraded lignin and other reaction products are extracted during intermediate alkaline washing stages. Bleaching with peroxide, oxygen or dithionite requires additional chemicals for buffering (e.g. sodium silicate), sequestering (e.g. ethylenediamine tetracetic acid (EDTA)) or stabilizing (e.g. magnesium salts).

Due to the numerous bleaching chemicals and sequences the influencing factors in pulp bleaching processes are very different, but all processes have the following important conditions in common:

- charge of chemicals
- bleaching consistency
- bleaching time and temperature.

Modern bleaching processes exclusively use continuous systems with bleaching towers, which have replaced the old batch processes (e.g. Holländer bleaching). In traditional bleaching towers the pulp and the bleaching liquid move uniformly. Recent developments include displacement or dynamic bleaching and washing in continuous diffusers (relative movement between pulp and liquid), and gas-phase bleaching (Liebergott, Yorston 1965; Rapson, Anderson 1966; Fiehn 1975; Rapson 1979). Common to all bleaching procedures is the necessity of a thorough mixing of the pulp with the bleaching liquor and an intensive washing of the pulp to remove the solubilized reaction products.

The chemical reactions taking place with the pulp components during bleaching are described in chapters 10 and 11.

16.7.2. Bleaching of Mechanical Pulps

Bleaching of mechanical pulps is a necessary prerequisite for the increasing utilization of unconventional wood species or saw-mill by-products with lower starting brightness than the traditional well-debarked 'white' spruce logs for groundwood production. Only pulps with high brightness can be used as furnish for high-quality fine paper grades.

To preserve the advantage of high yields in mechanical pulping, the bleaching procedures are usually of the lignin-preserving type, using oxidative or reducing chemicals, or a combination of both. While sulfur dioxide, sodium sulfite and bisulfite used to be the traditional chemicals for bleaching of stone groundwood, the dominating bleaching chemicals for all types of mechanical pulps today are peroxide and dithionite (hydrosulfite) (Lindahl, Norberg 1980; Lorås 1981).

Mechanical pulp bleaching is carried out as a single-stage process either with peroxide or dithionite, or as a two-stage process with peroxide followed by the reductive dithionite step. Single-stage bleaching is often preferred because the benefits of the second stage can often be obtained more easily and more economically e.g. by applying higher peroxide concentrations (Schröter 1969; Lorås 1980). In single-stage reductive bleaching sodium dithionite has increasingly replaced zinc dithionite because zinc is known to be toxic to fish. The typical tower bleaching is performed at low consistencies of about 4%, dithionite concentrations of 0.5–1.0% (based on pulp), pH levels of 5–6, temperatures between 20 and 60 °C, and total bleaching times of 1–2 hours (Rapson et al. 1965). As heavy metal impurities catalyze the decomposition of dithionite, sequestering agents (e.g. diethylenetriamine pentacidic acid, DTPA, or ethylenediamine tetracetic acid, EDTA) must be added.

Bleaching of groundwood pulp is mostly carried out as tower bleaching while refiner mechanical and thermomechanical pulps may also be bleached during the fibre separation within the first refiner stage (bleaching at fibre separation) or in the second refiner stage (refiner bleaching) (Lindahl, Norberg 1980; → 16.2.2.).

Peroxide bleaching with sodium peroxide, hydrogen peroxide or mixtures thereof is the only oxidative method which is industrially applied to mechanical pulps. The most frequently used peroxide is hydrogen peroxide in alkaline medium at pH values between 10 and 11. Additional chemicals are sodium silicate, magnesium salts and chelating agents with functions for buffering, stabilizing, and improving brightness gain (Schröter 1968; Berndt 1972; Jensen 1973; Krüger 1979). In peroxide bleaching of mechanical pulps the parameters causing activation of the peroxide (sodium hydroxide concentration, temperature) and stabilization (concentration of sodium silicate and DTPA) must be well adjusted, in addition the choice of consistency and peroxide concentration, in order to reach an optimum brightness gain with no decrease of strength properties (Süss et al. 1980). In the case of groundwood bleaching the NaOH concentrations are 1–1.5% at temperatures of 40–50 °C (Schröter 1976). TMP and chemi-refiner mechanical pulps are much more difficult to bleach and require somewhat altered conditions because of their higher temperatures (60–70 °C) after refining. In contrast to dithionite bleaching the consistency in peroxide bleaching must be high (15–30%) to maintain a high activation level.

Final brightness values in peroxide bleaching of spruce groundwood pulps are generally in the range of 80%, both for the single-stage and the two-stage processes. While final brightness values of TMP in single reductive dithionite bleaching are limited to about 65%, peroxide bleaching can reach at least 70% with brightness gains in the range of 10–15 bleaching points and even more (Krüger et al. 1978; Krüger 1979; Süss et al. 1980; Lorås 1981).

Ozone is a very effective but not specific bleaching agent. Nevertheless it has been used for bleaching of groundwood and thermomechanical pulps with good results, especially in combination with a peroxide step after a low-consistency ozone treatment (Soteland, Kringstad 1968; Liebergott 1971; Berndt 1972; Soteland 1974; Lindholm 1977; Allison 1979). Ozone was shown to influence the bonding potential by activating the fibre surfaces, thus improving the beatability and the tensile index (Kibblewhite et al. 1980; Samuelson et al. 1981).

A combined treatment with peroxide followed by ozone revealed improved post-refining conditions of refiner mechanical pulp rejects (Heitner et al. 1981).

High-yield semichemical sulfite and bisulfite pulps (→ 16.3.3.) can be bleached with peroxide and dithionite with good bleach response if the yields of the unbleached pulps are in the range of 85–90% for softwoods or somewhat lower for hardwoods. Typical NSSC pulps with fairly high brightness values (40–50%) are mainly used unbleached. A considerable brightness gain can be achieved by a lignin-preserving bleaching. NSSC pulps for lignin-removing bleaching are generally cooked down to lower yields of 60–70% (Giertz 1961; Kurdin 1980; Lorås 1980).

16.7.3. Bleaching of Chemical Pulps

The aim of bleaching chemical pulps is to remove the residual lignin after the cooking process to obtain so-called full-bleached pulps with brightness levels above 90% or semi-bleached qualities with brightness values in the range of 60–70%. The residual lignin content is determined indirectly by standardized methods describing the bleachability of a pulp (→ 3.2.9.).

Lignin-removing bleaching is predominantly carried out today in multi-stage procedures with oxidative stages combined with normally at least one alkaline extraction step. Generally sulfite and bisulfite pulps are easier to bleach than kraft pulps, and only extensive multi-stage bleaching has enabled a satisfactory bleaching of dark alkaline-cooked pulps. The multi-stage processes for sulfite pulps are usable for pulps from softwoods and hardwoods. In the case of alkaline pulps hardwood pulps generally require fewer stages than softwood pulps.

To describe the different bleaching sequences briefly special symbols are used (Table 16–13). Tables 16–14 and 16–15 give an impression of the multiplicity of industrial and experimental bleaching today. The application of a certain sequence depends on numerous factors such as type of unbleached pulp, desired final brightness, but also on technical aspects such as existing equipment and prices or availability of chemicals.

Table 16–13: Bleaching nomenclature (Anonymous 1977b; Lorås 1980)

Stages	Chemicals	Symbols
Chlorination	Cl_2	C
Alkaline extraction	NaOH	E
Hypochlorite	NaOCl + NaOH	H
Chlorine dioxide	ClO_2	D
Peroxide	Na_2O_2 + NaOH	P or P/E
	H_2O_2 + NaOH	
Oxygen	O_2 + NaOH	O
Chlorination with small amounts of ClO_2	Cl_2 + (ClO_2)	C_D
Sequential bleaching without intermediate washing	ClO_2/Cl_2	D/C
	Cl_2/NaOCl + NaOH	C/H
	ClO_2/NaOCl + NaOH	D/H
Bleaching with a mixture of Cl_2 and ClO_2	Cl_2 + ClO_2	C + D
Chlorination at low concentration	Cl_2	(C)
Gas-phase bleaching	Cl_2	C_g
	ClO_2	D_g
Ozone	O_3	Z
Acid	e.g. CH_3CO_3H	A

Table 16–14: Common industrial bleaching sequences (Rydholm 1965; Anonymous 1977b; Krüger et al. 1978; Lorås 1980)

Sulfite and bisulfite pulps	
Three stages	C–E–H
Four stages	C–E–H–H
	C–E–H–D
	C–E–D–H
	C–C–E–H
	C–H–E–H
	H–C–E–H
	C–E–D–D/H
	C+D–E–H–D
	E–C–H–D
Five stages	C–E–H–D–H
	C–C–E–H–H
Kraft pulps	
Three stages	C–E–H
(semi-bleached)	D/C–O–D
Four stages	C–E–H–D
(partly semi-bleached)	C–E–H–P
	C–E–H–H
	C–H–E–H
	C–D–E–D
	O–C–E–H
	O–C–E–D
	O–D–E–D
	O–D–O–D

Five stages	C–E–H–P–D
	C–E–H–D–P
	C–E–H–E–H
	C–E–D–E–D
	C–E–D–P–D
	C–E–H–E–D
	C–H–D–E–D
	D–E–D–E–D
	C–C/H–E–H–H
Six stages	C–H–E–D–E–D
	C–E–H–D–E–D
	C–E–H–E–H–D
	C–E–H–D–P–D
	C–E–H–E–D–P
	C+D–E–H–D–E–D
	C–E–H–D–E–D
	O–C–E–D–E–D
	O–C+D–E–D–E–D
	O–D–E–D–E–D
	O–C–D–E–H–D
Seven stages	C–H–H–D–E–D–P

Table 16–15: Non-established bleaching sequences (Anonymous 1976b; Krüger et al. 1978; Lorås 1980)

Reduced chlorine application	(C)–P–H
	(C)–P–D–H
	(C)–P–H–D–H
Peroxide replacing chlorine	P–D–P
	P–D–H
	P–H–H
	P–H–D
	D–P–D
	P–H–D–H
	P–D–P–D
Oxygen bleaching	O–P
	O–D
	O–H
	O–P–D
	O–D–P
	O–C–P
	O–H–P
	O–C–P–D
	O–D–P–D
	O–C_g–E–D_g
	O–D_g–E–D_g
Ozone bleaching	Z–E–P
	Z–E–Z
	Z–E–Z–P
Peracetic acid	P–A–P
	A–E–A–E–A

In most commercial processes chlorination is still the first bleaching step, and is also called the prebleaching stage. Chlorine converts the residual lignin in water- and alkali-soluble degradation products (→ 10.3.3.), and is therefore generally followed by an alkaline extraction step to remove those components.

Chlorination is carried out at low consistency (3–4%) and low temperatures of 20–40 °C for 30–60 min. The chlorine concentration is an important factor since when concentrations are too high oxidation reactions will also take place with the polysaccharides, reducing the strength properties. Higher temperatures up to 60 °C were shown to be acceptable in medium-consistency (about 10%) and high-consistency chlorination (30–35%), in gas-phase chlorination, and whenever chlorine dioxide is additionally used (Hinrichs 1962; Liebergott, Yorston 1965; Gullichsen 1976).

As chlorination causes the largest number of environmental problems with the bleach plant effluents many attempts have been made to replace chlorine or to reduce the amounts of chlorine used or the chlorinated products in the effluents. This can be accomplished e.g. by cooking the pulps down to low kappa numbers, thus reducing the chlorinated organic load in the effluents, or by replacing part of chlorine by chlorine dioxide, leading to more oxidation products than highly toxic chlorinated lignin fragments (Lindgren 1971; Anonymous 1977b). Replacing chlorine altogether by peroxide or oxygen bleaching enables a processing of effluents with few problems with regard to environmental protection.

The sodium hydroxide applied in the alkaline extraction stage is not a bleaching agent as such. Its main effects are the removal of lignin degradation products in combination with a neutralization of acidic components formed during the prebleaching step. In normal paper pulp bleaching 1–2% alkali (based on pulp) is used at 50–60 °C for 30–60 min at medium-consistency level (10–18%). In dissolving pulp production most of the residual polyoses must also be extracted at this stage. Therefore higher concentrations of sodium hydroxide (up to 5%) are applied at considerably higher temperatures up to 100 °C (hot alkali treatment) and higher consistencies (up to 35%) for 3–5 hours.

In principle, hypochlorite can be used in a single-stage bleaching or in the first stage of multi-stage bleaching, but is mostly applied after chlorination and extraction (C-E-H). As hypochlorite has a severe degrading effect on cellulose in the neutral pH range and at very low kappa numbers it must always be applied under alkaline conditions (Rapson 1956; Mori et al. 1974). Usually sodium hydroxide is added to sodium hypochlorite and bleaching is performed at medium consistency, at about 40 °C and bleaching times of 1–3 hours, depending on the chosen temperature, consistency and the preceding stages.

Chlorine dioxide has long been known to be an excellent delignifying and bleaching agent, but it is also a difficult chemical when used in large-size industrial processes due to its high reactivity in the gas phase and its toxidity. Nevertheless chlorine dioxide is gradually displacing chlorine in the first stage of multi-stage bleaching, whereas formerly it was used in the final stages. This development is the result of several advantages of chlorine dioxide, e.g. higher brightness, improved strength properties, lower chemical consumption and a substantial decrease in the BOD of the effluents (Fergus 1973; Rapson 1979; Wintzer 1980; Reeve, Rapson 1981; Bäckström, Germgård 1981; Germgård 1982). Chlorine dioxide bleaching is generally performed at low-to-medium consistency, at pH values of 3–5, and at low temperatures in the first stage or at about 70 °C in intermediate or final stages for 3–5 hours.

Apart from its application in bleaching mechanical pulps, peroxide is also established today in several industrial bleaching sequences for chemical pulps. It is mainly used in the latter stages in combination with chlorine dioxide, yielding increased brightness values and stability. More recently the traditional sodium hydroxide extraction (E) is sometimes replaced by an alkaline peroxide stage (P or P/E) combining bleaching and extraction in one stage, and resulting in a brightness gain without an additional stage. By increasing the application of hydrogen peroxide the amounts of chlorine bleaching chemicals are reduced resulting in decreased chloride load of the effluents. Drawbacks are still the high price of peroxide and the necessary additives for stabilization, the latter being not as important here as they are in mechanical pulp bleaching. Peroxide bleaching is usally performed at medium-to-high consistency at 60–80 °C for 2–4 hours (Delattre 1971, 1974; Krüger, Traser 1976; Hoffmann, Patt 1979; Kindron 1979; Lorås 1980).

Oxygen bleaching (or oxygen delignification) is one of the most thoroughly investigated processes in the field of pulping and bleaching of the last 20 years. As oxidation is the essential reaction in lignin-removing bleaching it is quite reasonable to aim at using oxygen as the cheapest oxidizing agent for bleaching. But as it is not a selective lignin-degrading chemical pulps cannot be bleached to high brightness exclusively with oxygen without considerable attack on the polysaccharides, resulting in rather poor strength properties. Thus the common practice in mill-scale bleaching today is to remove about one-half of the residual lignin in unbleached pulps by oxygen, and to finish with conventional bleaching sequences (e.g. O-C-E-D-E-D or O-D-E-D) (Jamieson, Smedman 1974; Rapson 1979).

Many other bleaching sequences after oxygen treatment are under investigation (Table 16–15). More than 16 plants throughout the world are in operation today, the first having been founded in South Africa in 1970 (Rowlandson 1971; Rapson 1979).

The main practical advantage of oxygen bleaching is the fact that the effluents from the oxygen step can be processed within the normal kraft recovery system (Rapson 1979). The special conditions in oxygen bleaching are not established yet. In principle oxygen bleaching is a gas-phase process at pressures usually between 4 and 8 bar in alkaline medium, performed at high consistencies of 20–30% and temperatures of 90–140 °C, depending on the alkali used. Sodium hydroxide and sodium carbonate are the common alkalis for sulfate pulps, magnesium hydroxide and carbonate for Mg-sulfite and bisulfite pulps. Results from numerous basic research activities and mill trials demonstrate the potential of this process in a future with increasing air and water pollution control (Croon, Andrews 1971; Chang et al. 1974a, b; Eachus 1975; Kalisch 1975; Makkonen, Ranua 1975; Christensen 1975; Rapson 1979; Elton et al. 1980; Augustin 1980; Lorås 1980).

The use of ozone as a bleaching agent for chemical pulps is still in the developing phase, but one process using ozone instead of chlorine in the prebleaching step was recently reported to be ready for commercialization (Rothenberg et al. 1975; Kamishima et al. 1976; Singh 1982).

Apart from optimizing bleached pulp qualities the future developments in bleaching will be directed mainly towards environmental aspects with regard to non-toxic bleaching agents, techniques for the characterization of bleach plant effluents and their purification and processing in closed cycles (Lindström 1977; Hardell, de Sousa 1977; Wiley et al. 1978; Pfister, Sjöström 1979; Lorås 1980; Gregor 1981; Voss et al. 1981; Sjöström et al. 1982).

Literature

Abbot, J. and Bolker, H. I. 1982, Tappi *65,* No. 2, 37–40

Abrahamsson, K. and Samuelson, O. 1973, Svensk Papperstid. *76,* 480–485

Abrahamsson, K. and Samuelson, O. 1974, Svensk Papperstid. *77,* 693–698

Abrahamsson, K. and Samuelson, O. 1975, Svensk Papperstid. *78,* 417–421

Abson, D. and Stockman, V. E. 1979, Tappi *62,* No. 2, 69–72

Ahlgren, P., Olsson, L. A. and Vikström, B. 1975, Svensk Papperstid. *78,* 95–102

Akhtaruzzaman, A. F. M. and Virkola, N. E. 1979, Pap. Puu *61,* 578–586; 629–634

Alfredsson, B., Samuelson, O. and Sandstig, B. 1963, Svensk Papperstid. *66,* 703–706

Allison, R. W. 1979, Appita *32,* 279–284

Amidon, T. E. 1981, Tappi *64,* No. 3, 123–126

Annergren, G. 1981a, Pap. Puu *63,* 195–210

Annergren, G. 1981b, Technical Aspects of Pulping Chemistry. In: The Ekman Days 1981. Int. Symp. Wood Pulp. Chem., Stockholm, Vol. 2, pp. 3–11

Annergren, G., Croon, I., Enström, B. F. and Rydholm, S. A. 1961, Svensk Papperstid. *64,* 386–393

Annergren, G., Rydholm, S. and Vardheim, S. 1963, Svensk Papperstid. *66,* 196–210

Anon. 1969, Chem. Eng. News *47,* No. 21, 30–32

Anon. 1977a, Scanning Electron Microsopy (SEM) Reveals Differences Between Thermomechanical and Stone Groundwood Pulps. STFI-News; Swedish Forest Prod. Res. Lab., Stockholm

Anon. 1977b, Umwelt und Degussa. Umweltfreundliche Bleiche von Sulfitzellstoffen. Degussa, Frankfurt

Anon. 1978, Tappi *61,* No. 11, 27–28

Anon. 1980a, Tappi *63,* No. 11, 27–31

Anon. 1980b, Tappi *63,* No. 11, 21

Anon. 1983, Das MD Organosolv-Zellstoffverfahren. MD Verwaltungsgesellsch., München

April, G. C., Kamal, M. M., Reddy, J. A., Bowers, G. H. and Hansen, S. M. 1979, Tappi *62,* No. 5, 83–85

April, G. C., Bhaoocha, R., Sheng, J. and Hansen, S. 1982, Tappi *65,* No. 1, 41–44
Arjas, A. and Perälä, E. 1980, Entwicklungstendenzen im Energieverbrauch bei der Herstellung verschiedener Holzstoffe. Preprints XIX. Eucepa-Conference, Vol. I., 16:1–16:26
Atack, D. and May, W. D. 1962, Pulp Pap. Mag. Can. *63,* T10–T20
Atack, D., Heitner, C. and Stationwala, M. I. 1978, Svensk Papperstid. *81,* 164–176
Atack, D., Heitner, C. and Karnis, A. 1980, Svensk Papperstid. *83,* 133–141
Auchter, R. J. 1976, Tappi *59,* No. 4, 50–53
Augustin, H. 1980, Papier *34,* 439–445
Augustin, H. and Rahmani, M. 1975, Holz Roh-Werkst. *33,* 26–31
Aurell, R. and Hartler, N. 1963, Tappi *46,* 209–215
Aurell, R., Stockman, L. and Teder, A. 1958, Svensk Papperstid. *61,* 937–944
Axegard, P. 1980, Svensk Papperstid. *83,* 345–351; 398–404
Axegard, P. and Jonsson, U. 1979, Svensk Papperstid. *82,* 310–313
Ba, H. V., Ishizu, A. and Nakano, J. 1980, Mokuzai Gakkaishi *26,* 12–15
Bach, B. and Fiehn, G. 1972, Zellst. Pap. *21,* No. 1, 3–7
Bäckström, T. and Germgård, U. 1981, Svensk Papperstid. *84,* 27–33
Balhar, L. 1965, Sb. Vysk. Prac. Odboru Celulozy Pap. *10,* 9–18
Baumeister, M. and Edel, E. 1980, Papier *34,* No. 10A, V9–V18
Bender, F., Clermont, L. P. and Bowden, A. W. 1972, Pulp Pap. Mag. Can. *73,* No. 7, T 175–T 179
Berndt, W. 1972, Wbl. Papierfabr. *100,* 634–640
Bland, D. E., Skicko, J. and Menshun, M. 1978, Appita *31,* 374–378
Blom, U. A., Nevalainen, P. and Arhippainen, B. 1981, Development of the ALVA Prehydrolysis Process. II. Mill Scale Application. Tappi Proc. Pulp. Conf., Denver, 401–416
Böhmer, V. J. 1981, A New Causticizing Process. Symp. Wood Pulp Manufacture: From Cell Wall Ultrastructure to Latest Pulping Innovations. CSIR, Pretoria, S. A., Paper No. 6
Bolker, H. J. and Liebergott N. 1972, Pulp Pap. Mag. Can. *73,* No. 11, 82–88
Botes, E. 1981, Tappi *64,* No. 6, 81–84
Boyer, R. Q. 1960, Tappi *43,* 688–698
Brecht, W. and Weiss, H. 1957, Papier *11,* 82–89
Brink, D. L. 1961, Tappi *44,* 256–262
Brink, D. L., Merriman, M. M. and Schwerdtfeger, E. J. 1962, Tappi *45,* 315–326
Brodén, Å. and Simonson, R. 1981, Pap. Puu *63,* 787–790
Bruun, H., Gädda, L. and Vartiainen, H. 1979a, Tappi *62,* No. 12, 110–111
Bruun, H., Gädda, L. and Storsjö, M. 1979b, Tappi *62,* No. 4, 65–68
Bryce, J. R. G. 1980a, Alkaline Pulping. In: Pulp and Paper. Chemistry and Chemical Technology (Casey, J. P., Ed.), Vol. I., 3rd Ed. Wiley-Intersci., New York, pp. 377–492
Bryce, J. R. G. 1980b, Sulfite Pulping. In: Pulp and Paper. Chemistry and Chemical Technology (Casey, J. P., Ed.), Vol. I., 3rd Ed. Wiley-Intersci., New York, pp. 291–376
Bryce, J. R. G and Tomlinson, G. H. 1962, Pulp Pap. Mag. Can. *63,* T 355–T 360
Bryne, J. R. and Voelker, M. H. 1960, Tappi *43,* 261A–267A
Cameron, D. W., Jessup, B., Nelson, P. F., Raverty, W. D., Samuel, E. and Vanderhoek, N. 1981, The Response of Pines and Eucalypts to NSSC-AQ Pulping. In: The Ekman Days 1981. Int. Symp. Wood Pulp. Chem. Stockholm, Vol. 2, pp. 64–71
Casebier, R. L. and Hamilton, J. K. 1965, Tappi *48,* 664–669
Casey, J. P. 1980, Pulp and Paper. Chemistry and Chemical Technology (Casey, J. P., Ed.), Vol. I., 3rd Ed. Wiley-Intersci., New York
Chang, H., Gratzl, J. S. and McKean, W. T. 1974a, Tappi *57,* No. 5, 123–126
Chang, H., McKean, W. T. and Gratzl, J. S. 1974b, Papier *28,* No. 10A, V 17–V 24
Chang, H., Sinkey, J. P. and Yan, J. F. 1979, Tappi *62,* No. 9, 103–106
Chase, A. I. 1978, Pulping, Biomass and Nutrient Studies of Woody Shrubs and Shrub Sizes of Tree Species. Bull., Life Sci. and Agriculture Exp. Stat., University of Maine, No. 749

Chene, M., Robert, A. and Noisillier, G. 1961, French patent 1 283 320
Christensen, P. K. 1975, Nor. Skogind. *29,* 84–88
Clermont, L. P. and Bender, F. 1961, Pulp Pap. Mag. Can. *62,* T 28–T 34
Cochrane, J. A. 1956, Pap. Trade J. *140,* No. 17, 30–38
Coletti, J. P. and Buongiorno, J. 1980, Wood Fiber *12,* No. 4, 233–242
Collicutt, S. A., Frazier, W. C., Holmes, G. W., Joice, P., Mackie, D. M. and Torza, S. 1981, Tappi *64,* No. 6, 57–61
Copeland, W. H. 1977, Tappi *60,* No. 8, 34
Corradini, F. 1979, Pulp Pap. Int. *21,* No. 12, 49–51
Cox, L. A. and Worster, H. E. 1971, Tappi *54,* 1890–1892
Crönert, H. 1966, Papier *20,* 691–699
Croon, I. 1963, Svensk Papperstid. *66,* 1–5
Croon, I. and Andrews, D. H. 1971, Tappi *54,* 1893–1898
Currah, J. E. 1979, Tappi *62,* No. 8, 73–76
Czirnich, W. and Patt, R. 1976, Holzforschung *30,* 19–27
Dahm, H. P. 1958, Nor. Skogind. *11,* 405–410
Daniell, W. F. 1980, Stone Groundwood. In: Pulp and Paper. Chemistry and Chemical Technology (Casey, J. P., Ed.), Vol. I., 3rd Ed. Wiley-Intersci., New York, pp. 168–197
DeHaas, G. G. and Lang, C. J. 1974, Tappi *57,* No. 5, 127–130
Delattre, M. 1971, Pap. Puu *53,* 117–127
Delattre, M. 1974, Appita *28,* 89–98
Dillard, B. M., Gilmer, R. J. and Kennedy, J. D. 1976, U.S. patent 3 954 533; cit. in Marteny, W. W. 1980
Eachus, S. W. 1975, Tappi *58,* No. 9, 151–154
Eckert, R. C. and Amos, L. W. 1980, Tappi *63,* No. 11, 89–93
Einspahr, D. W. and Harder, M. L. 1980, Tappi *63,* No. 9, 121–124
Ekström, T. 1976, Pulp Pap. Int. *18,* No. 6, 37–40
Elton, E. F., Magnotta, V. L., Markham, L. D. and Courcheme, C. B. 1980, Tappi *63,* No. 11, 79–82
Erins, P., Cinite, V., Jakobsons, M. and Gravitis, J. 1976, Appl. Polym. Symp. No. *28,* 1117–1138
Ernest, P. M. and Hannan, S. M. 1967, Tappi *50,* No. 12, 110A–116A
Eskilsson, S. 1972, Svensk Papperstid. *75,* 397–402
Eskilsson, S. 1974, Svensk Papperstid. *77,* 165–174
Eskilsson, S. and Hartler, N. 1973, Svensk Papperstid. *76,* 63–70
Ezzat, S. 1979, Cell Chem. Technol. *13,* 229–240
Faix, O. and Pfütze, E. 1980, Holzforschung *34,* 38–39
FAO 1977, Pulp and Paper Capacities. Survey 1976–1981. FAO, Rome
Fergus, B. J. 1973, Tappi *56,* No. 1, 114–117
Fiehn, G. 1975, Zellst. Pap. *24,* 6–14
Fleming, B. I., Kubes, G. J., MacLeod, J. M. and Bolker, H. I. 1978, Tappi *61,* No. 6, 43–46
Fleming, B. I., MacLeod, J. M., Kubes, G. J. and Bolker, H. I. 1980, Tappi *63,* No. 12, 112–113
Flowers, A. G., Pearson, A. J. and Tyler, A. G. 1979, Appita *33,* 195–200
Foelkel, C. E. B. and Zvinakevicius, C. 1980, Tappi *63,* No. 3, 39–42
Fossum, G., Hägglund, S. and Lindqvist, B. 1980a, Svensk Papperstid. *83,* 430–435
Fossum, G., Hägglund, S. and Lindqvist, B. 1980b, Svensk Papperstid. *83,* 455–460
Frederiksson, B. and Hoglund, H. 1977, Appita *31,* 365–370
Fujii, J. S. and Hannah, M. A. 1978, Tappi *61,* No. 8, 37–40
Gädda, L. 1982, Pap. Puu *64,* 5–18
Gädda, L. and Bruun, H. 1981, Pap. Puu *63,* 717–723
Gavelin, G. and Lunden, L. 1980, Svensk Papperstid. *83,* 416–419
Genco, J. M., Krishnagopalan, A. and Shaffer, M. M. 1978, Tappi *61,* No. 5, 107–109

Germgård, U. 1982, Pap. Puu *64,* 76–82
Giertz, H. W. 1961, Tappi *44,* 1–4
Giertz, H. W. 1963, Papier *17,* 191–196
Giertz, H. W. 1974, Chemimechanical Pulping of Hardwoods. Proc. Int. Symp. on New Forestry Resources for the Pulp and Paper Ind., Eucepa, Madrid, pp. 244–276
Giertz, H. W. 1977, Basic Wood Raw Material Properties and Their Significance in Mechanical Pulping. Proc. Int. Mechanical Pulp. Conf., Helsinki 1977, Vol. 1
Ginzel, W., Saad, S. M. and Stegmann, G. 1969, Holzforschung *23,* 152–163
Goel, K., Ayroud, A. M. and Branch, B. 1980, Tappi *63,* No. 8, 83–85
Goring, D. A. I. 1963, Pulp Pap. Mag. Can. *64,* T 517–T 527
Gorsler, H. 1980, Papier *34,* 518–522
Göttsching, L. 1980a, Tappi *63,* No. 3, 50–54
Göttsching, L. 1980b, Papier *34,* 129–133
Gregor, C. H. 1981, Papier *35,* 281–285
Gruber, E. 1980, Wbl. Papierfabr. *108,* 53–60
Gullichsen, J. 1976, Tappi *59,* No. 11, 106–109
Hagemeyer, R. W. 1976, Tappi *59,* No. 4, 46–49
Hardell, H. L. and de Sousa, F. 1977, Svensk Papperstid. *80,* 110–120
Harder, M. L., Einspahr, D. W. and Parham, R. A. 1978, Tappi *61,* No. 11, 121–122
Harris, J. F. 1974, The Role of Total Process Concepts in Evaluating Pulping Research. Tappi Non-Sulfur Pulping Symp., Madison, pp. 161–167
Hartler, N. 1963, Svensk Papperstid. *60,* 696–700
Hartler, N. 1978, Dtsch. Pap. Wirtsch. 1978, No. 1, 109–114
Hartler, N. 1980, Pulp. Pap. Mag. Can. *81,* T 119–T 124
Hartler, N., Lindström, C. and Stade, Y. 1977, Svensk Papperstid. *80,* 121–124
Hartler, N., Stockman, L. and Sundberg, O. 1961a, Svensk Papperstid. *64,* 33–49
Hartler, N., Stockman, L. and Sundberg, O. 1961b, Svensk Papperstid. *64,* 67–85
Hatton, J. V. 1977, Tappi *60,* No. 8, 107–110
Hauan, S., Høydahl, H. E., Lorås, V. and Bøhmer, E. 1975, Papier *29,* 269–274
Heitner, C., Atack, D. and Karnis, A. 1981, Pap. Puu *63,* 53–61
Helberg, B. E., Neuberger, E. D. and Simmons, W. B. 1976, Tappi *59,* No. 5, 52–63
Helberg, B. E., Neuberger, E. D. and Simmons, W. B. 1977, Tappi *60,* No. 2, 70–77
Hergert, H. L. 1976, Tropical Wood as a Raw Material for Dissolving Pulp Manufacture. Proc. Int. Conf. Man-Made Fibres Dev. Countries, Bombay, pp. 224–231
Hinrichs, D. D. 1962, Tappi *45,* 765–770
Hoffmann, P. and Patt, R. 1979, Holzforschung *33,* 167–173
Holton, H. H. 1977, Pulp Pap. Mag. Can. *78,* T 218–T 223
Holton, H. H. and Chapman, F. L. 1977, Tappi *60,* No. 11, 121–125
Holzer, W. F., Henderson, J. T., West, W. B. and Kenneth, F. B. 1962, Tappi *45,* 208–213
Horn, R. A. and Auchter, R. J. 1976, Appl. Polym. Symp. No. *28,* 529–536
Hou, Y., Cao, B., Xie, G., Wang, J., Kong, L., Cao, S., Liu, G. She, G. 1981, Chem. and Ind. of For. Prod. *1,* No. 2, 13–23; No. 4, 21–30
Hunt, K. 1981, Tappi *64,* No. 3, 135–139
Huusari, E. and Syrjaenen, A. 1980, Pulp Pap. Mag. Can. *81,* T 63–T 66
Ikeda, K., Kondo, R. and Kondo, T. 1979, Jap. Tappi *33,* 606–613
Ikonopisova, B., Valtchev, V. and Valtcheva, E. 1979, Cell. Chem. Technol. *13,* 645–649
Ingruber, O. V. and Allard, G. A. 1967, Tappi *50,* 597–614
Ingruber, O. V. and Allard, G. A. 1973, Pulp Pap. Mag. Can. *74,* T 354–T 369
Jakate, D. N., Venkobarao, G., Swamy, V. S. R., Swamy, C. V., Gopichand, K. and Sarma, G. S. R. P. 1981, Tappi *64,* No. 6, 124–125
Jamieson, A. G. and Smedman, L. Å. 1974, Tappi *57,* No. 5, 134–136

Jamieson, A. G., Samuelson, O. and Smedman, L. Å. 1975, Tappi *58,* No. 2, 68–71
Janson, J. 1980, Svensk Papperstid. *63,* 392–395
Jayme, G. and Branscheid, F. 1959, Papier *13,* 284–286
Jayme, G. and Krause, T. 1961, Wbl. Papierfabr. *89,* 518–522
Jensen, R. 1973, Allg. Pap. Rundsch. 1973, 276–284
Jensen, W. 1976, Cell. Chem. Technol. *10,* 627–635
Jensen, W. 1979a, Cell. Chem. Technol. *13,* 501–509
Jensen, W. 1979b, Papier *33,* No. 10A, V42–V56
Jett, J. B. and Zobel, B. J. 1975, Tappi *58,* No. 1, 92–96
Kalisch, J. H. 1967, Tappi *50,* No. 12, 44A–51A
Kalisch, J. 1975, Pulp Pap. Int. *17,* No. 3, 53–69
Kamishima, H., Fujii, T. and Akamatsu, I. 1976, Jap. Tappi *30,* 381–391
Kärnä, A. 1980, Pressure Groundwood (PGW), Status of Development and Use. Preprints XIX. Eucepa-Conf., Vol. I, pp. 19:1–19:44
Keays, J. L. 1974, For. Prod. J. *24,* No. 11, 13–16
Keays, J. L. 1975, Tappi *58,* No. 11, 90–95
Kibblewhite, R. P., Brookes, D. and Allison, R. W. 1980, Tappi *63,* No. 4, 133–136
Kiiskilä, E. and Virkola, N. E. 1978, Pap. Puu *60,* 129–132
Kiiskilä, E. and Valkonen, N. 1979, Pap. Puu *60,* 505–510
Kindron, R. R. 1979, Tappi *62,* No. 9, 67–70
Kleinert, T. N. 1965, Tappi *48,* No. 8, 447–451
Kleinert, T. N. 1971, U.S. patent 3 585 104
Kleinert, T. N. 1974, Tappi *57,* No. 8, 99–102
Klemm, K. H. 1957a, Neuzeitliche Holzschlifferzeugung. Dr. Sändig-Verlag, Wiesbaden
Klemm, K. H. 1957b, Holzschliff. In: Ullmanns Encyklopädie der technischen Chemie, Vol. 8, 3rd Ed. Verlag Chemie, Weinheim, pp. 560–579
Klemm, K. H. 1978, Wbl. Papierfabr. *106,* 405–414
Kleppe, P. J. 1970, Tappi *53,* No. 1, 35–47
Kobo, T., Holokawa, J., Kobayashi, T. and Kobayashi, Y. 1981, Holzforschung *35,* 125–128
Koffinke, R. A. 1980, Tappi *63,* No. 11, 51–54
Košiková, B., Janson, J., Pekkala, O. and Sågfors, P. E. 1980, Pap. Puu *62,* 229–236
Koval, J. and Slavik, I. 1963, Pap. Celul. *18,* 215–216
Kratzl, K. 1978, Holzforsch. Holzverwert. *30,* 31–32
Krause, T. 1971, Cell. Chem. Technol. *5,* 83–92
Krüger, H. 1979, Papier *33,* No. 10A, V1–V9
Krüger, H. and Traser, W. 1976, Papier *30,* 389–395
Krüger, H., Kuzel, P. and Schwab, H. 1978, Einsatz von Wasserstoffperoxid und seinen Derivaten in Bleichprozessen, zur Desinfektion und im Umweltschutz. In: Wasserstoffperoxid und seine Derivate (Weigert, W., Ed.). Hüthig-Verlag, Heidelberg, pp. 111–126
Kubes, G. J. and Bolker, H. I. 1978, Cell. Chem. Technol. *12,* 621–645
Kubes, G. J., Garner, B. C. and Bolker, H. I. 1979, Svensk Papperstid. *82,* 196–201
Kubes, G. J., Fleming, B. I., MacLeod, J. M. and Bolker, H. I. 1980, Wood Sci. Technol. *14,* 207–228
Kurdin, J. A. 1980. Refiner Mechanical and Thermomechanical Pulping. In: Pulp and Paper. Chemistry and Chemical Technology (Casey, J. P., Ed.), Vol. I., 3rd Ed. Wiley-Intersci., New York, pp. 197–252
Lachenal, D., de Choudens, C. and Monzie, P. 1980, Tappi *63,* No. 11, 59–62
Lachenal, D., Bourne, C., de Choudens, C. and Monzie, P. 1979, Tappi *62,* No. 5, 53–57
Lagergren, S. 1964, Svensk Papperstid. *67,* 238–243
Laine, J. E., Turunen, J. and Rasanen, R. H. 1979, Tappi *62,* No. 5, 65–68
Landucci, L. L. and Sanyer, N. 1975, Tappi *58,* No. 2, 60–63

Lasmarias, V. B. and Peterson, R. C. 1980, Cell. Chem. Technol. *14*, 479–496
Law, K. N. and Koran, Z. 1979, Wood Sci. *12*, 106–112
Leach, J. M. and Thakore, A. N. 1976, Tappi *59*, No. 2, 129–132
Leask, R. A. 1968, Tappi *51*, No. 12, 117A–120A
Leask, R. A. 1981, Svensk Papperstid. *84*, No. 14, 28–35
Lee, G. and Gauvin, W. H. 1958, Tappi *41*, 110–116
Lengyel, P. and Morvay, S. 1973, Chemie und Technologie der Zellstoffherstellung. Güntter-Staib Verlag, Biberach/Riss
Liebergott, N. 1971, Pap. Trade J. *100*, No. 31, 28–29
Liebergott, N. and Yorston, F. H. 1965, Tappi *48*, No. 1, 20–24
Lindahl, A. and Norberg, P. H. 1980, Pulp Pap. Mag. Can. *81*, T138–T141
Lindgren, B. O. 1971, Svensk Papperstid. *74*, 57–63
Lindholm, C. A. 1977, Pap. Puu *59*, 17–26
Lindström, K. 1977, Papier *31*, 517–525
Lora, J. H. and Waymann, M. 1978, Tappi *61*, No. 6, 47–50
Lora, J. H. and Waymann, M. 1980, Can. J. Chem. *58*, 669–676
Lorås, V. 1980, Bleaching. In: Pulp and Paper. Chemistry and Chemical Technology (Casey, J. P., Ed.), Vol. I., 3rd Ed. Wiley-Interscі., New York, pp. 633–764
Lorås V. 1981, Svensk Papperstid. *84*, No. 14, 36–45
Louden, L. and Weiner, J. 1976, Odors and Odor Control. Bibliography No. 267, Inst. of Paper Chem., Appleton, Wi.
Löwendahl, L. and Samuelson, O. 1977, Svensk Papperstid. *80*, 549–551
MacLeod, J. M., Kubes, G. I., Fleming, B. I. and Bolker, H. I. 1979a, Alkaline Pulping with Diamineanthraquinones. Proc. Symp. Can. Wood Chem., Harrison Hot Springs, pp. 1–5
MacLeod, J. M., Kubes, G. J., Fleming, B. I. and Bolker, H. I. 1979b, Cell. Chem. Technol. *13*, 793–801
MacLeod, J. M., Kubes, G. J., Fleming, B. I. and Bolker, H. I. 1981, Tappi *64*, No. 6, 77–80
Madan, R. N. 1981, Holzforsch. Holzverwert. *33*, 52–55
Magister, G. 1981, Zellst. Pap. *30*, 201–206
Makkonen, H. and Ranua, H. 1975, Papier *29*, No. 10A, V25–V32
Mamers, H. and Grave, N. C. 1974, Appita *28*, 159–163
Mamers, H., Watson, A. J. and Grave, N. C. 1976, Appita *29*, 356–362
Mannström, B. 1972, Tappi *55*, No. 4, 551–555
Markant, H. P. 1960, Tappi *43*, 699–702
Marteny, W. W. 1980, Semichemical Pulping. The Neutral Sulfite Semichemical or NSSC Process. In: Pulp and Paper. Chemistry and Chemical Technology (Casey, J. P., Ed.), Vol. I., 3rd Ed. Wiley-Intersci., New York, pp. 252–291
Marton, R. and Leopold, B. 1973, Appita *27*, 112–118
Marton, R., Amidon, T. E. and Koeppicus, R. 1976, Tappi *59*, No. 12, 107–112
Marton, R., Goff, S., Brown, A. F. and Granzow, S. 1979, Tappi *62*, No. 1, 49–53
Marton, R., Tsujimoto, N. and Eskelinen, E. 1981, Tappi *64*, No. 8, 71–74
Marttala, T., Hänninen, E., Nikula, M. and Arhippainen, B. 1981, Development of the Ava Prehydrolysis Process. I. Process Development. Tappi Proc. Pulp. Conf., Denver, pp. 401–416
McDonough, T. J. and Van Drunen, T. J. 1980, Tappi *63*, No. 11, 83–87
McGovern, J. N. 1980, Introduction to Pulping. In: Pulp and Paper. Chemistry and Chemical Technology (Casey, J. P., Ed.), Vol. I., 3rd Ed. Wiley-Intersci., New York, pp. 161–167
McGovern, J. N. 1981, Tappi *64*, No. 3, 129–132
McKelvey, R. D., Van Drunen, V. J. and Nicholls, G. A. 1978, Tappi *61*, No. 12, 41–46
Minor, J. L. and Sanyer, N. 1974, Tappi *57*, No. 5, 120–122
Minor, J. L. and Sanyer, N. 1975, Tappi *58*, No. 3, 116–119
Misra, D. K. 1980, Pulping and Bleaching of Nonwood Fibers. In: Pulp and Paper. Chemistry and

Chemical Technology (Casey, J. P., Ed.), Vol. I., 3rd Ed. Wiley-Interesc., New York, pp. 504–568
Mori, K., Suzuki, S., Okuda, K. and Yamamoto, S. 1974, Mokuzai Gakkaishi *20,* 577–579
Morin, M. J. 1974, Tappi *57,* No. 6, 133–135
Mortimer, R. D., Fleming, B. I. 1981, The Analysis of Anthraquinone in Pulping Liquors and Pulp Products. In: The Ekman Days 1981. Int. Symp. Wood Pulp. Chem., Stockholm, Vol. 4, pp. 128–131
Nahum, L. S. and Pellegrini, P. 1970, Svensk Papperstid. *22,* 725–730
Nakamura, M. and Matsuura, H. 1975, Tappi *58,* No. 4, 98–102
Nakano, J., Daima, H., Hosoya, S. and Ishizu, A. 1981, Studies on Alkali-Methanol Cooking. In: The Ekman Days 1981. Int. Symp. Wood Pulp. Chem., Stockholm, Vol. 2, pp. 72–77
Naujoks, E. 1979, Die Kulturgeschichte der natürlichen Textilfasern und ihre Bleichverfahren. Umschau f. d. Textilind., Degussa, Hanau
Nelson, P. J. 1974, Appita *28,* 232–236
Nelson, P. J. 1978, Appita *31,* 437–442
Nicholls, G. A., Jamieson, R. G. and Van Drunen, V. J. 1976, Tappi *59,* 84–87
Nolan, W. J. 1957, Tappi *40,* No. 3, 170–190
Nyholm, P. and Virtanen, U. 1976, Pulp Pap. Int. *18,* No. 6, 41–47, 57
Oniśko, W. 1980, Holztechnologie *21,* 165–171
Palenius, I. 1976, Pap. Puu *58,* 550–556
Palenius, I. 1980, Schlußfolgerungen aus dem Eucepa-Symposium in Helsinki 1980. Preprints XIX. Eucepa Symp., Helsinki, Vol. I., pp. 15:1–15:17
Pamén, A. L. 1960, Svensk Papperstid. *63,* 550–555
Patt, R. 1978. Trends in Kraft Pulp and Paper Consumption – The Interaction of Technological and Economic Aspects. Proc. Symp. Complete Tree Utilization of Southern Pine, New Orleans (McMillin, C. W., Ed.), FPRS
Patt, R. and Hoffmann, P. 1977, Papier *31,* No. 10A, V1–V9
Patt, R., Tröger, F. K. and Czirnich, W. 1974, Papier *28,* 1–6
Peltonen, J. and Rahkila, P. 1979, Papier *33,* No. 10A, V35–V42
Pfister, K. and Sjöström, E. 1979, Pap. Puu *61,* 619–622
Phillips, F. H., Logan, A. F. and Balodis, V. 1979, Tappi *62,* No. 3, 77–81
Procter, A. R. 1971, Pulp Pap. Mag. Can. *72,* No. 8, 67–74
Procter, A. R. and Styan, G. E. 1974, Tappi *57,* No. 10, 123–126
Rahm, H. 1976, Papier *30,* No. 10A, V24–V33
Rahmani, M. 1978, Papier *32,* 185–188
Rapson, W. H. 1956, Tappi *39,* 284–295
Rapson, W. H. 1979, Tappi *62,* No. 6, 14–17
Rapson, W. H. and Anderson, C. B. 1966, Tappi *49,* 329–334
Rapson, W. H., Wayman, M. and Anderson, C. B. 1965, Pulp Pap. Mag. Can. *66,* T 255–T 271
Raubenheimer, S. J. 1981, Anthraquinone as a Pulping Additive. Symp. Wood Pulp Manufacture: From Cell Wall Ultrastructure to Latest Pulping Innovations. CSIR, Pretoria, S. A., Paper No. 5
Raubenheimer, S., Eggers, S. H. 1980, Papier *34,* No. 10A, V19–V23
Reeve, D. W. and Rapson, W. H. 1981, Tappi *64,* No. 9, 141–144
Renard, J. J., Mackie, D. M. and Bolker, H. I. 1975, Pap. Puu *57,* 786–804
Reside, D. A. and Garvin, B. 1978, Tappi *61,* No. 5, 65–67
Riese, W. 1957, Zellst. Pap. *6,* 145–148
Riese, W. 1962, Zellst. Pap. *11,* 216–222
Robert, A. and Noisillier, G. 1961, ATIP Bull. 2, 102–115
Roffael, E. and Parameswaran, N. 1978, Holzforschung *32,* 113–119
Rogers, I., Mahood, H., Servizi, J. and Gordon, R. 1979, Pulp Pap. Mag. Can. *80,* T 286–T 290
Rothenberg, S., Robinson, D. H. and Johnsonbaugh, D. K. 1975, Tappi *58,* No. 8, 182–185

Rothenberg, S., Shaw, J. and Durst, W. B. 1981, Pap. Puu *63,* 111–119
Rowlandson, G. 1971, Tappi *54,* 962–967
Runkel, R. O. H. 1962, Wbl. Papierfabr. *90,* 570–575
Runkel, R. O. H. and Patt, K. F. 1958, Halbzellstoffe. Güntter-Staib Verlag, Biberach/Riss
Rydholm, S. A. 1960, Papier *14,* 535–541
Rydholm, S. A. 1965, Pulping Processes. Intersci. Publ., New York, London, Sydney
Rydquist, J., Lundén, L. and Gavelin, G. 1981, Svensk Papperstid. *84,* No. 10, 18–23
Saad, S. M. and Fahmy, M. S. 1979, Holzforschung *33,* 28–31
Saad, S. M., Zimaity, M. T. and Fahmy, M. 1975, Holzforschung *29,* 177–180
Sakai, K. and Kondo, T. 1976, Svensk Papperstid. *79,* 563–569
Samuelsson, L., Mjöberg, P. J. and Hartler, N. 1981, Svensk Papperstid. *84,* R 110–R 115
Sandermann, W. 1956, Grundlagen der Chemie und chemischen Technologie des Holzes. Akad. Verlagsgesellsch. Geest & Portig KG, Leipzig
Sandermann, W. 1963, Chemische Holzverwertung. BLV Verlagsgesellsch., München
Sarkanen, K. V., Hrutfiord, B. F., Johanson, L. N. and Gardner, H. S. 1970, Tappi *53,* No. 5, 766–783
Schöön, N. H. and Wannholt, L. 1969, Svensk Papperstid. *72,* 489–492
Schröter, H. 1968, Wbl. Papierfabr. *96,* 725–729
Schröter, H. 1969, Wbl. Papierfabr. *97,* 1023–1026
Schröter, H. 1976, Die Holzschliffbleiche. Güntter-Staib Verlag, Biberach/Riss
Schweers, W. 1974, Chem. Tech. *4,* No. 8, 490–493
Schweers, W. and Rechy, M. 1972, Papier *26,* 585–590
Schwenzon, K. 1966, Zellst. Pap. *15,* 202–204
Silander, R. 1981, Pap. Puu *63,* 65–68
Singh, R. P. 1982, Tappi *65,* No. 2, 45–48
Sjöström, E. 1981, Wood Chemistry. Fundamentals and Applications. Academic Press, New York, London
Sjöström, E., Rådeström, R. and Lindström, K. 1982, Svensk Papperstid. *85,* R7–R13
Slavik, I. and Kunjak, L. 1961, Pap. Celul. *16,* 169–170
Söderquist, R. 1955, Svensk Papperstid. *58,* 706–712
Sondell, K. 1982, Tappi *65,* No. 3, 111–112
Soteland, N. 1974, Pulp Pap. Mag. Can. *75,* 91–96
Soteland, N. and Kringstad, K. 1968, Nor. Skogind. *22,* 46–52
Stade, Y. 1976, Zellst. Pap. *25,* 299–305
Stark, H. and Eichinger, R. 1979, Wbl. Papierfabr. *107,* 51–56
Stevens, W. P. and Marton, R. 1966, Tappi *49,* No. 10, 451–456
Stone, J. E. 1957, Tappi *40,* No. 7, 539–541
Stone, J. E. and Clayton, D. W. 1960, Pulp Pap. Mag. Can. *61,* T 307–T 313
Sundholm, J. and Mannström, E. 1982, Pap. Puu *64,* 8–13
Surewicz, W. 1962, Tappi *45,* 570–578
Süss, H. U., Krüger, H. and Schmidt, K. 1980, Papier *34,* 433–438
Teder, A. 1969, Svensk Papperstid. *72,* 294–303
Ulmanen, T., Rasanen, R. and Rantanen, M. 1963, Tappi *46,* 151A–153A
Utaka, G., Oku, K., Matsuura, H. and Sakai, A. 1965, Tappi *48,* 273–281
VDP 1981, Papier 1981, Ein Leistungsbericht der Deutschen Zellstoff- und Papierindustrie. Verband Deutscher Papierfabriken e. V. (Ed.), Bonn
Venter, J. S. M. 1981, Pap. South. Africa *1,* No. 5, 5–10
Vettenranta, I. 1978, Pulp Pap. Mag. Can. *79,* T12–T14
Vikström, B. and Hammar, L. A. 1979, Svensk Papperstid. *82,* 171–177
Vikström, B. and Nelson, P. 1980, Tappi *63,* No. 3, 87–91
Vinje, M. G. and Worster, H. E. 1969, Tappi *52,* 1341–1345

Virkola, N. E. 1975, Papier *29,* No. 10A, V14–V25
Virkola, N. E., Järvelä, O. and Vartiainen, V. 1963, Pap. Puu *45,* 457–481
Virkola, N. E., Kettunen, J. and Yrjälä, I. 1979, Pap. Puu *61*, 685–700
Virkola, N. E., Pusa, R. and Kettunen, J. 1981, Tappi *64,* No. 5, 103–107
Voss, R. H., Wearing, J. T. and Wong, A. 1981, Pulp Pap. Mag. Can. *82,* T65–T71
Wayman, M. and Lora, J. H. 1978, Tappi *61,* No. 6, 55–57
Wayman, M. and Lora, J. H. 1979, Tappi *62,* No. 9, 113–114
Wegener, G. 1981, Pulping Processes – Latest Developments. Symp. Wood Pulp Manufacture: From Cell Wall Ultrastructure to Latest Pulping Innovations. CSIR, Pretoria, S. A., Paper No. 4
Wegener, G. 1983, Naturwiss. Rundsch. *36,* 153–157
Weiner, J. and Pollock, V. 1977, Nonsulfur Pulping. IPC Bibliographic Series No. 275, Appleton, Wi.
Welte, M. and Patt, R. 1978, Papier *32,* 101–107
Wenzl, H. F. J. 1960, Holzforschung *14,* 4–21
Whitney, R. P., Thompson, N. S., Nicholls, G. A. and Han, S. T. 1969, Pulp Pap. *43,* No. 8, 68–70
Wiley, A. J., Dambruch, L. E., Parker, P. E. and Dugal, H. S. 1978, Tappi *61,* No. 12, 77–80
Wintzer, P. 1980, Chem. Ing. Tech. *52,* 392–398
Wither, R. P. and Captein, H. A. 1960, For. Prod. J. *10,* 174–177
Wong, A. 1980, Tappi *63,* No. 4, 53–57
Wong, A. 1981, Pulp Pap. Int. *23,* 55–59
Worster, H. E., Pudek, M. F. and Harrison, R. E. 1971, Pulp Pap. Mag. Can. *72,* No. 12, 69–73
Zachariasen, K. A. 1964, U.S.patent 3 124 503

17. Derivatives of Cellulose

17.1. The Various Kinds of Cellulose Derivatives

Cellulose is mainly utilized in a more or less purified condition in the form of paper and paperboard. But derivatives having properties different from those of cellulose are also used in numerous fields. In some cases the derivatives are only intermediates for transforming the state of cellulose (e.g. for the production of threads and foils).

Cellulose can be regarded as a polyalcoholic compound as each glucose unit within the molecular chain bears three hydroxylic groups (→ 4.2.1.). Cellulose, however, is a solid-state polyalcohol, most of its OH-groups being linked by hydrogen bonds (→ 4.2.4.). Nevertheless the OH-groups may react like those from other alcohols forming various derivatives. Thus the formation of esters with inorganic and organic acids, ethers with other alcohols, alcoholates with bases, oxidation products (acids), halogenides, amines, complexes etc. is possible (Fig. 17–1).

Many of the reactions resulting in the various derivatives have to start as heterogeneous reactions because of the insolubility of cellulose. During several of these reactions the cellulose is dissolved and further reaction steps may occur homogeneously. The dissolution of cellulose as an iron tartratic or cadmium ethylenediamine complex is an example of reactions of that type. Other reactions also cause fundamental changes of the properties of cellulose. Esters are soluble in organic

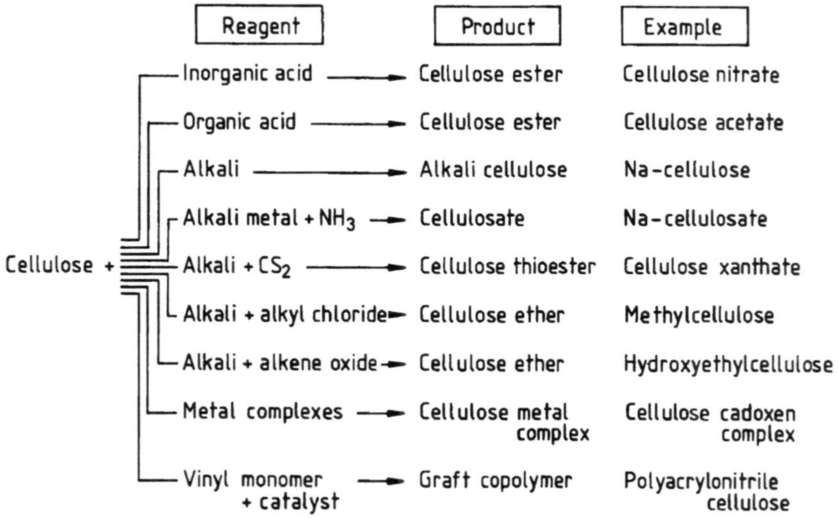

Fig. 17–1. Schematic representation of reagents and products of cellulose derivation.

solvents, ethers in water. The change of the crystalline lattice by the formation of alkali cellulose was described previously (→ 4.3.2.).

From the technical point of view the most important cellulose derivatives are the esters and ethers with a wide range of applications. Alkali-celluloses play an important role as intermediates for the formation of certain cellulose esters and ethers. Oxidative and halogenous derivatives may be undesirable by-products produced during bleaching of pulps (→ 10.2.4.). A natural amine is chitin, the material of the supporting tissue of insects, spiders and crustaceans. Chitin consists of molecular chains made up of β-(1→ 4)-glycosidic linked acetyl glucosamine units.

Earlier summaries, particularly of cellulose esters and ethers, are given by Lieser (1953), Ott et al. (1954), Wurz (1961) and Bikales and Segal (1971). In recent years survey chapters were published in Ullmanns Encyklopädie der technischen Chemie (Eicher, Fischer 1975; Kehren, Reichle 1975; Balser, Iseringhausen 1975) and by Sjöström (1981).

17.2. Cellulosates and Alkali-Celluloses

Cellulose changes its crystalline lattice if it is treated with aqueous alkali (→ 4.3.2.). Depending on the kind of the alkali cation and the alkali concentration different degrees of swelling are obtained. The swelling process is additionally influenced by the condition of cellulose. A hornification caused by high-temperature drying hinders the access of the alkali to the cellulose fibres (Schmiedeknecht, Claus 1966a). As during swelling water is also included in addition to alkali in the cellulose structure, the kind of compound formed between cellulose and alkali has not yet been clarified.

Most of the studies concerned with the formation of alkali cellulose were carried out with sodium hydroxide. The content of alkali in Na-cellulose I increases with increasing alkali concentration. With 12% NaOH the content is 0.68, with 16% NaOH 0.87 and with 19% NaOH a maximum value of 1 mole per mole $C_6H_{10}O_5$ is reached (Geiger, Schneider 1966). The water content is about 3 moles/mole $C_6H_{10}O_5$. The ratio of the components can be expressed in a pseudostoichiometric formula of $(C_6H_{10}O_5)$ (NaOH) $(H_2O)_3$. The treatment of native cellulose with 18% NaOH in methanol resulted in an alkali absorption according to the formula $(C_6H_{10}O_5)_6$(NaOH) (Aboul-Seoud 1968). The addition of water increased the alkali absorption, and with 1 part water to 9 parts methanol the composition corresponded to $(C_6H_{10}O_5)_2$(NaOH).

The chemical structure of Na-cellulose is explained by a hydratation complex in which a water molecule in the NaOH-hydrate is replaced by cellulose (Chedin, Marsaudon 1955) (Fig. 17–2a). Geiger and Schneider propose a mixed structure of

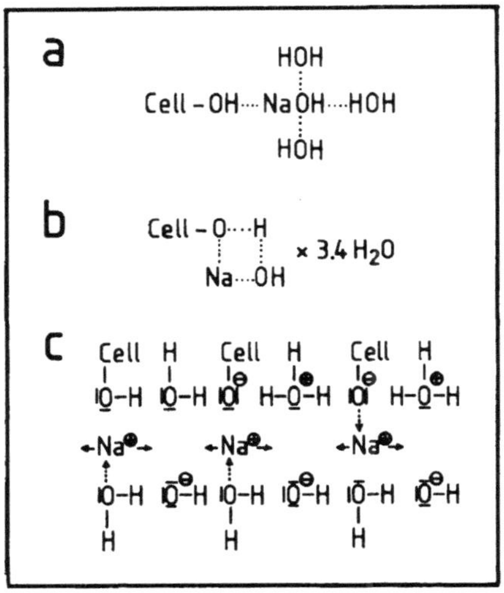

Fig. 17–2. Models of the molecular structure of alkali cellulose.

alcoholate (cellulosate) and addition compound (Fig. 17–2b). From the measurement of the diamagnetic susceptibility Dietl and Schwarzhans (1966) concluded that the Na-cellulose obtained with metallic sodium in liquid ammonia is a pure cellulosate, whereas Na-cellulose obtained with aqueous NaOH is characterized by an oxonium compound containing Na^{+}-ions with limited mobility as shown in Fig. 17–2c.

Cellulosates have also been obtained with other alkali metals (K, Li), NH_3 and Ca, using liquid ammonia as reaction medium (Dietl, Schwarzhans 1966; Bredereck, Vlachopoulos 1980). Ammonia causes an intracrystalline swelling of the cellulose which determines the velocity of reaction. An activation of cellulose with ammonia also reduces the alkali concentration necessary for a lattice conversion (Philipp 1981).

The system K/NH_3 can only react at the surface of the lattice as the solvated K-ions cannot penetrate the lattice planes because of their size. The reaction with Na is slow; that with Li is faster and yields a 2,3,6-tri-O-lithium cellulose.

By treating cellulose with sodium methylate and sodium butylate in water-free methanol or butanol, respectively, Na-cellulosates with degrees of substitution (DS) between 0.5 and 1.7 were obtained (Schwenker et al. 1961; Avny, Rebenfeld 1968). Bredereck (1970) succeeded in converting cellulose (rayon) to K-cellulosate with a DS of 2 by a reaction with potassium methylate in DMSO/methanol. The course of this exchange reaction is favoured by the considerably higher basicity of potassium or caesium alcoholate in DMSO compared with alcohol. This effect is not true of Na- and Li-alcoholates because of the formation of ion pairs. Quarternary ammonium cellulosates (tetramethyl-, benzyltrimethylammonium cellulosates) with substitution degrees of 0.5 and 0.7 were obtained by the reaction of the

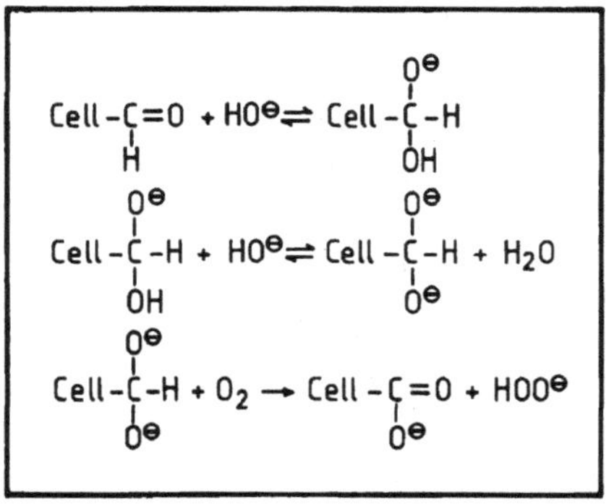

Fig. 17–3. Reactions of the formation of perhydroxyl ions by alkali cellulose in the presence of oxygen.

respective methoxides with cellulose in methanol or DMSO (Bredereck, Dau 1980).

Alkali celluloses are important intermediates in the production of regenerated cellulose and cellulose ethers. The enlargement of the lattice-plane distances and the at least partial replacement of OH-groups by ONa-groups increase the reactivity of cellulose or make certain chemical changes possible, respectively.

Alkali celluloses are sensitive to oxygen, i. e. the molecular chain lengths are reduced during storage in air. According to Mattor (1963) the aldehyde end-groups of cellulose behave as a Lewis acid in alkaline medium and form complexes which react by hydride ion transfer with oxygen to form hydroperoxyde ions (Fig. 17–3). Perhydroxyl ions are able to split cellulose chains in an alkaline medium, and there is one rupture of a glycosidic bond for each perhydroxyl ion (Kleinert 1956). The cleavage of the cellulose chains occurs statistically and the portion of short molecules increases with the time of aging (Brestkin, Čočieva 1968; Jayme, El-Kodsi 1972; Jovanovic et al. 1974).

The reduction of chain lengths during aging (ripening) of alkali cellulose is utilized for improving the workability (e. g. solubility, spinning properties) of the consequent products. The alkali cellulose is stored at constant, slightly elevated temperature (max. 28 °C) until a certain DP is reached. Also a short aging at higher temperatures, or no aging – if the cellulose is treated oxidatively (e.g. with hypochlorite or periodic acid) – can be applied (Kleinert 1972; Mašura 1974).

17.3. Cellulose Esters

17.3.1. Fundamentals of Esterification

As mentioned above hydroxyl groups are important for the chemical conversion of an alcohol, and hence of cellulose. Alcoholic OH-groups are polar groups which can be displaced by nucleophilic groups or compounds in a strong acidic solution. The first step during this nucleophilic substitution is the formation of an oxonium ion (Fig. 17–4a). This mechanism is due to the reaction with inorganic acids, whereas the reaction with an organic acid runs according to a nucleophilic addition (Fig. 17–4b). The addition of the alcohol to the organic acid can be facilitated by means of acid catalysis, by which at first a proton is added to the electronegative oxygen of the carboxylic group. Now the carbon atom of this group is more positively polarized, and is thus more accessible for the nucleophilic alcohol molecule (Fig. 17–4c). All steps within these reactions also run in the reverse direction, i. e. the esterification is an equilibrium reaction. The reverse process, termed saponification, can be suppressed by disturbing the equilibrium in such a way that the ester formation is promoted by binding the reaction water.

The presence of three OH-groups at each glucose unit makes the formation of mono-, di- or triesters possible. The mutual linkages of the OH-groups by hydrogen bonds within the supramolecular structure of cellulose are partially or totally cleaved during the esterification. The introduced ester groups push the cellulose chains asunder, and thus the structure is strongly changed or even destroyed.

The formation of cellulose esters is theoretically possible with all inorganic and organic acids. The most important of them are cellulose nitrate, cellulose xanthate and cellulose acetate because of their technological and commercial utilization. Cellulose sulfate, nitrite, phosphate, tricarbanilate, and some esters with higher carbonic acids have also gained in importance.

17.3.2. Cellulose Nitrate

Cellulose nitrate is one of the most important esters as it is produced commercially for various purposes. Depending on the nitration level cellulose nitrate is used for

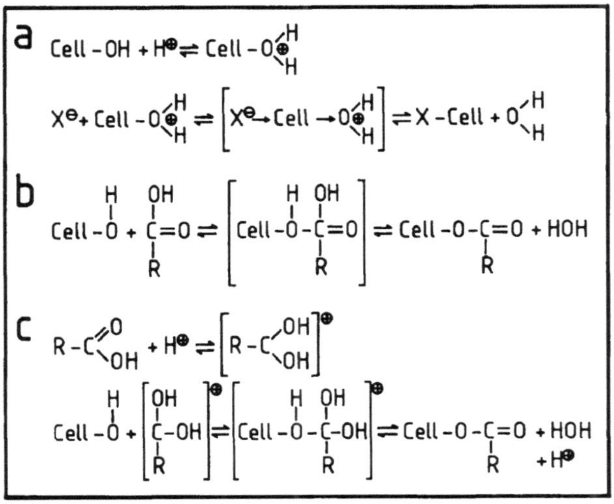

Fig. 17–4. Reaction mechanism of the esterification of cellulose.

plastics (celluloid), lacquers, adhesives, and explosives (smokeless powder, dynamite). In addition to the degree of polymerization (DP) the products differ in the degree of substitution (DS) which can be determined from the content of nitrogen. Additionally, differently substituted cellulose nitrates vary in their solubility in organic solvents (Table 17–1).

The application of aqueous nitric acid with concentrations below 75% results only in a poor esterification of cellulose. With 77.5% HNO_3 about 50% of the OH-groups can be esterified (DS = 1.5). With anhydrous HNO_3 cellulose dinitrate (DS = 2) is obtained. For a higher level of nitration the application of acid mixtures (HNO_3–H_2SO_4, HNO_3–H_3PO_4, HNO_3-acetic anhydride) is necessary (Lieser 1953; Barsha 1954).

Nitration mixtures consisting of aqueous nitric and sulfuric acid are common on a commercial scale. The cellulosic starting materials are chemical grade pulp or cotton linters which are torn up and disaggregated prior to treatment with mixed acids in the nitration vessel (Fig. 17–5). Variations of the nitrating acid, the treating time and the temperature determine different qualities of the resulting cellulose nitrate. The separation of cellulose nitrate from the residual nitrating acid occurs in centrifuges. Washing and cooking with water is apt to remove adhering acid (prestabilization). A subsequent cooking under pressure reduces the chain length of the molecules and equalizes the distribution of the NO_2-groups (stabilization). For technical purposes collodion has to be free of water. Therefore the water is displaced by alcohol (methanol, ethanol, propanol or butanol) in a centrifuge (alcoholizing). Collodion is sold in an alcohol-containing fibrous condition (mostly 35%), or as chips (Anon. 1970).

In the laboratory, where cellulose nitrate plays an important role in the determination of molecular weight, various nitration mixtures are applied. Very often mixtures of nitric acid, phosphoric acid, and phosphorus pentoxide are used, with varying ratios of the components (Marx-Figini 1961; Timell 1963; Philipp, Linow 1965). In order to obtain trinitrate Bennett and Timell (1955) treated cotton linters with various nitrating mixtures containing acetic acid and acetic anhydride, or phosphorus pentoxide and nitrogen pentoxide in addition to nitric acid. A maximum nitration level (14.14% N) was found using a mixture of nitric and acetic acid and acetic anhydride in the proportion of 43:32:25 at 0 °C. Temperature seems to have an important influence on the degradation of the nitrates.

Table 17–1: Various types of cellulose nitrate

Nitrogen content %	Degree of substitution	Common solvents	Application
10.5–11.1	1.8–2.0	Ethanol	Plastics, lacquers
11.2–12.2	2.0–2.3	Methanol, esters, acetone, methylethyl ketone	Lacquers, adhesives
12.0–13.7	2.2–2.8	Acetone	Explosives

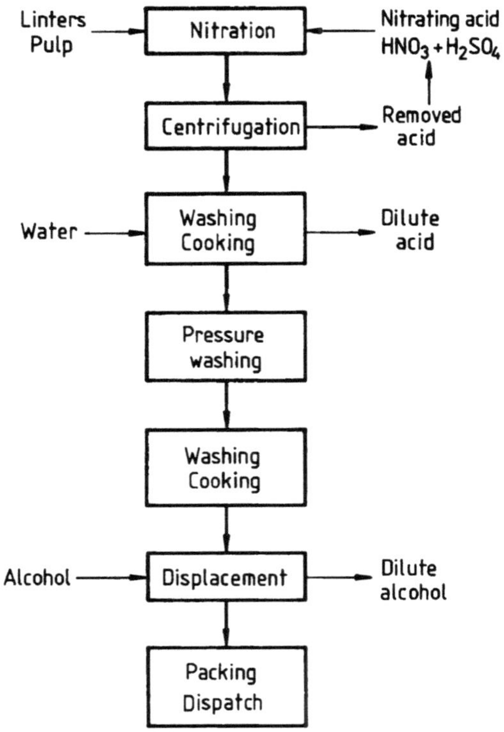

Fig. 17–5. Simplified flowchart of the technical production of cellulose nitrate.

A comparison of different nitrating mixtures containing phosphoric acid and phosphoric anhydride, applied to various celluloses, resulted in an identical nitration level (13.8% N) and no degradation if the temperature was kept at 0 °C (Marx-Figini 1961). Very stable reaction products with nitrogen contents of 13.8–14% were obtained by Thinius and Thümmler (1967) who esterified cellulose with anhydrous nitric acid in chlorinated hydrocarbons, particularly dichloromethane. In this reaction the nitrating components are nitronium ions which are formed during the dissociation of HNO_3 in the organic solvent and which react very rapidly with the hydroxyl groups of the cellulose.

In order to obtain small cellulose nitrate particles with a nitrogen content of 12.1% Belfort and Wortz (1966) treated microcrystalline cellulose with an aqueous mixture of nitric and sulfuric acid. 98% nitric acid was applied for the nitration of methylcellulose (→ 17.4.3.) resulting in a minimum degree of substitution of 1.8 (Bobinski, Carignan 1967).

For determining the degree of polymerization of celluloses from wood the method of direct nitration of wood was applied in several studies. By avoiding degradation effects by the delignification and alkaline extraction during the isolation of cellulose, high DP values are obtained from cellulose nitrate extracted with acetone from nitrated wood. Thus Timell (1956, 1957; Goring, Timell 1962) determined DP's of 5 000–10 000 for celluloses of various woods. For several years wood nitration and the influence of various parameters have been studied by Poller and Patscheke (1969, 1970, 1971, 1979; Poller 1968). Though they found a nitric acid-rich nitrating mixture with low degradation effects on cellulose, sufficient yields of nitrates can be extracted from wood only after intensive treatment, i. e. long periods of time and higher temperatures. By extrapolating DP's determined in cellulose which had been isolated after different nitrating times to nitrating time 0, the DP of native cellulose was evaluated (Patscheke, Poller 1980). Thus for spruce cellulose a DP of about 12 000 was obtained.

Freshly prepared cellulose nitrate has to be stabilized, i. e. it is necessary to remove traces of residual acid and of ester groups others than nitrate deriving from the acids applied additionally (sulfate, phosphate, acetate). In the technical range various patented treatments with aqueous magnesium nitrate, nitric acid or organic acids and amines are known (Hiatt, Rebel 1971).

In the laboratory stabilization is performed with boiling water or methanol (Gagnon et al. 1961; Philipp, Linow 1965). From the thermal degradation of stabilized cellulose nitrate Millett et al. (1961) calculated its activation energy as 157.4 kJ/mole.

Studies of cellulose-nitrate solutions in esters (ethyl acetate) and acetone show that there exist gel particles which are not indicated by viscosimetric measurements (Schurz, Tritthart 1966; Holt et al. 1976). The particles consist of incompletely substituted cellulose structures which are denser and more compact than dissolved molecular clusters. During aging the dissolved molecules are degraded and the particles increase in size (Zipper et al. 1969; Dautzenberg et al. 1972). This phenomenon is explained by an association of the cellulose molecules at the sections where denitration by aging occurs.

Particles representing molecular clusters of cellulose nitrate were made visible in the electron microscope after freeze-drying highly dilute solutions in acetone (Ruscher, Zuchold 1964). The distribution of particle diameters showed maxima at 37 and 97 nm. Bittiger et al. (1969) observed fibrils of varying diameters independent of concentration, solvent and DP of cellulose trinitrate. The high resolution micrograph of Fig. 17–6 confirms this finding.

17.3.3. Esters with Other Inorganic Acids

The treatment of cellulose with aqueous sulfuric acid results only in a very poor yield of cellulose sulfate. Apart from a large bulk of various degradation products, chain fragments with a maximum DS of 1.5 are obtained. This explains why several sulfatation mixtures have been proposed (Hiatt, Rebel 1971). Among them are sulfuric acid/sulfur trioxide, sulfuric acid in liquid sulfur dioxide, sulfuric acid/carboxylic acid, chlorosulfonic acid/sulfur trioxide, and sulfur trioxide in DMF. The reaction mechanism can be described as an addition of the strongly electrophilic

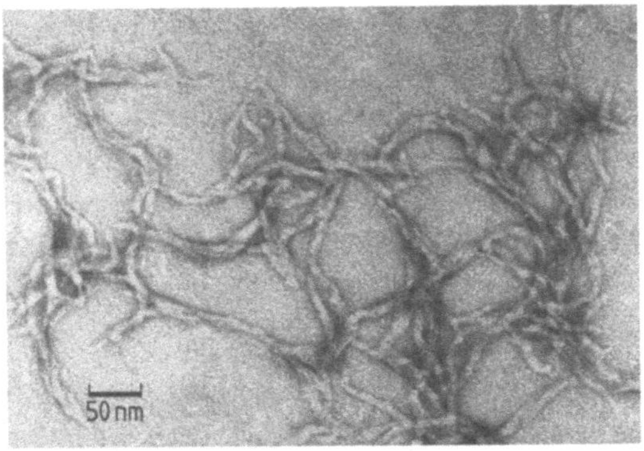

Fig. 17–6. Fibrillar structure of cellulose nitrate from a highly dilute solution. Negative staining, TEM micrograph.

SO_3 to the hydroxyl groups of cellulose followed by a decomposition of the intermediate oxonium ion (Fig. 17–7). Another way of obtaining cellulose sulfate is a trans-esterification reaction between butyl sulfate and cellulose in concentrated H_2SO_4 (Delsemme 1967).

Cellulose sulfates are soluble in water and fairly soluble in DMF, provided their DP's are not too high. They are used as thickeners for lacquers and printing inks. Cellulose sulfates are ionogenic compounds, and thus are able to form salts and have ion-exchange properties. Various studies are concerned with the properties and behaviour of cellulose sulfate and other esters as ion-exchangers (Petropavlovskij 1973; Dautzenberg, Philipp 1974; Lawton, Philipps 1975). A cross-linking with epichlorohydrine prior to sulfatation makes the cellulose sulfate insoluble in water (Pastyr, Kuniak 1972).

There is some interest in cellulose esters containing phosphorus (phosphates, phosphites and derivatives) because of their potential use as textile flame-retardants (Hiatt, Rebell 1971). The ion-exchange properties of cellulose phosphate have also been investigated (Lawton, Phillips 1975). Cellulose phosphate can be obtained by treating cellulose with an alcoholic solution of phosphoric acid and phosphorus pentoxide or with phosphoric acid in molten urea (Usmanov et al. 1966; Katsuura, Fujinami 1968).

In recent years cellulose nitrite has gained some importance. Schweiger (1974) treated cellulose with dinitrogen tetroxide (N_2O_4) or nitrosylchloride (NOCl) in dimethylformamide (DMF) or dimethyl acetamide (DMAC). During the reaction

$$\text{Cell-OH} + SO_3 \rightleftharpoons \left[\text{Cell-}\overset{\text{H}}{\underset{\oplus}{\text{O}}}\text{-}SO_3^{\ominus}\right] \rightleftharpoons \text{Cell-O-}SO_3^{\ominus} + H^{\oplus}$$

Fig. 17–7. Reaction mechanism of cellulose with SO_3.

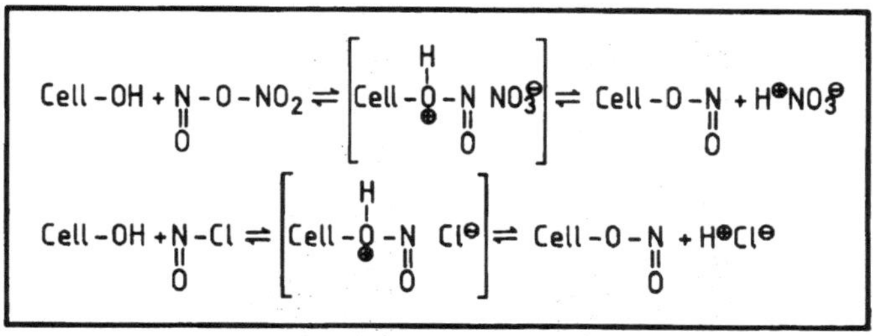

Fig. 17–8. Reaction mechanism of the formation of cellulose nitrite.

cellulose is solvated by the formation of nitrite ester. The reactive isomer of N_2O_4 is nitrosylnitrate. The most probable reaction is an electrophilic displacement with DMF or DMAC as proton acceptor (Fig. 17–8).

Cellulose nitrite esters seem to be very sensitive to water as a chain degradation down to a DP of 200 within 3 hours was observed in the presence of water (Clermont, Bender 1972; Mislovičová, Pašteka 1974). Cellulose nitrite is a very reactive compound which allows easy displacement by other ester groups (transesterification). Thus homogeneous acetylation with acetic anhydride/pyridine yielded acetates with DS ranging from 0 to 2 depending on the conditions involved (Månsson, Westfelt 1980). The easy preparation of cellulose monosulfate, nitrate, and a mixed nitrite-nitrate using cellulose nitrite as an intermediate was shown by Schweiger (1974) and Clermont and Bender (1972). Experiments were also performed to produce films and fibres starting from cellulose nitrite solutions (Schweiger 1974; Hergert et al. 1978). The regeneration medium is alcohol, and by using different alcohols various fibre shapes can be produced.

The synthesis of titanic esters has also been described (Predvoditelev, Bakšeeva 1972). They are derivatives of the theoretical orthotitanic acid ($Ti(OH)_4$). The cellulose titanic esters contain 3–5% Ti, do not burn or glow, and show hydrolytic resistance to mild alkali.

17.3.4. Cellulose Xanthate

Cellulose xanthate, though an important intermediate in the production of regenerated cellulose, is somewhat difficult to classify. It is an ester of dithiocarbonic acid (xanthogenic acid) which is a theoretical inorganic acid existing only in the form of its organic esters and corresponding salts.

The production of cellulose xanthate starts from alkali cellulose which is treated with carbon disulfide. The product resulting from this reaction is sodium cellulose xanthate. The reaction mechanism is complicated as there are various reactions running side-by-side (Fahmy, Fadl 1964; Hovencamp 1966). The first reaction is obviously the formation of dithiocarbonate, which is very instable and reacts spontaneously with cellulose and decomposes to form trithiocarbonate, sulfide and carbonate (Fig. 17–9). Additionally carbonyl sulfide (COS) is formed as an instable intermediate.

At the beginning the xanthation of cellulose proceeds very quickly in the less ordered regions. The xanthation of the ordered regions follows at a low velocity with carbon disulfide, which was absorbed in the structural elements during the first step (Fahmy et al. 1964; Pašteka 1967). An increase in temperature accelerates the reaction, and the amount of absorbed CS_2 decreases. According to Pašteka (1967) xanthation occurs between a solid phase and an intermediate liquid phase of hydrated CS_2. If the content of free NaOH in the alkali cellulose is reduced by neutralization, e. g. with gaseous SO_2, the degree of esterification increases, and the formation of by-products is suppressed. The system NaOH/CS_2 is also a heterogeneous one, and the reaction between them is determined by the alkali concentration (Claus, Schmiedeknecht 1966).

Under technical conditions a DS of 0.5–0.6 is reached which is equivalent to a γ-value (number of xanthate groups per 100 glucose units) of 50–60 (Kehren, Reichle 1975). In laboratory experiments γ-values of 100–300 have been obtained. The hydroxylic groups along the cellulose chains are not uniformly substituted by xanthate groups. The xanthate groups can be linked to all three hydroxyl groups of a glucose unit though their reaction velocity of formation and cleavage is different. Thus the C2–OH ist the most reactive group but it is more unstable than the C6–OH.

During xanthation a reduction of the DP of cellulose has also been observed. The oxidative chain splitting is accelerated by elevated temperature or partial oxygen pressure (Philipp, Dautzenberg 1960). Certain by-products of xanthation (e.g. Na_2S) may accelerate degradation by activating the oxygen; others (e.g. Na_2S_2) may impede degradation.

Primary reactions:

$$CS_2 + Na^{\oplus}OH^{\ominus} \rightarrow HCS_2O^{\ominus}Na^{\oplus}$$
$$HCS_2O^{\ominus}Na^{\oplus} + \text{Cell-OH} \rightarrow \text{Cell-OCS}_2^{\ominus}Na^{\oplus} + H_2O$$
$$CS_2 + \text{Cell-O}^{\ominus}Na^{\oplus} \rightarrow \text{Cell-OCS}_2^{\ominus}Na^{\oplus}$$
$$HCS_2O^{\ominus}Na^{+} + Na^{\oplus}OH^{\ominus} \rightarrow CS_2O^{2-} + 2\,Na^{\oplus}$$

Secondary reactions:

$$CS_2 + CS_2O^{2-} \rightarrow COS + CS_3^{2-}$$
$$COS + 3\,OH^{\ominus} \rightarrow CO_3^{2-} + SH^{\ominus} + H_2O$$
$$CS_2 + SH^{\ominus} \rightarrow CS_3H^{\ominus}$$
$$CS_3H^{\ominus} + OH^{\ominus} \rightarrow CS_3^{2-} + H_2O$$
$$CS_2O^{2-} + 2\,OH^{\ominus} \rightarrow 2\,SH^{\ominus} + CO_3^{2-}$$

Fig. 17–9. Primary and secondary reactions of the formations of cellulose xanthate.

Several studies are concerned with optimizing xanthation and the properties of the resulting xanthate by a slight hydroxyalkylation of the alkali cellulose (Lukanoff, Philipp 1967; Ioleva, Papkov 1968; Jašunskaja, Konovalova 1968).

Though various processes have been developed for a continuous manufacture of xanthate and viscose (Gohlke et al. 1963; Treiber 1966; von Horstig 1969), most processes run discontinuously. After ripening the shredded alkali cellulose is filled into a slowly rotating drum (baratte) which is evacuated after closure. The calculated amount of CS_2 is sucked into the drum through a hollow axle perforated inside the baratte, thus distributing the carbon disulfide uniformly. Part of the CS_2 is evaporated immediately and the rest is dispersed in the alkali cellulose. In most cases the reaction temperature is 26–32 °C, and the reaction time 1–1.5 hours. At the end of the reaction the drum is aerated with nitrogen, and the residual carbon disulfide is exhausted. The orange-yellow sodium cellulose-xanthate is emptied into the dissolving vessel where 40% NaOH is added. The xanthate is dissolved while being stirred and cooled, forming a viscous yellow solution which is called viscose (Wängberg, Treiber 1968; Kehren, Reichle 1975) (Fig. 17–10).

The process as described is the so-called gas-phase xanthation or dry xanthation. An alternative procedure is the emulsion xanthation or wet xanthation which begins in the same manner as dry xanthation. After adding the total amount of CS_2 the

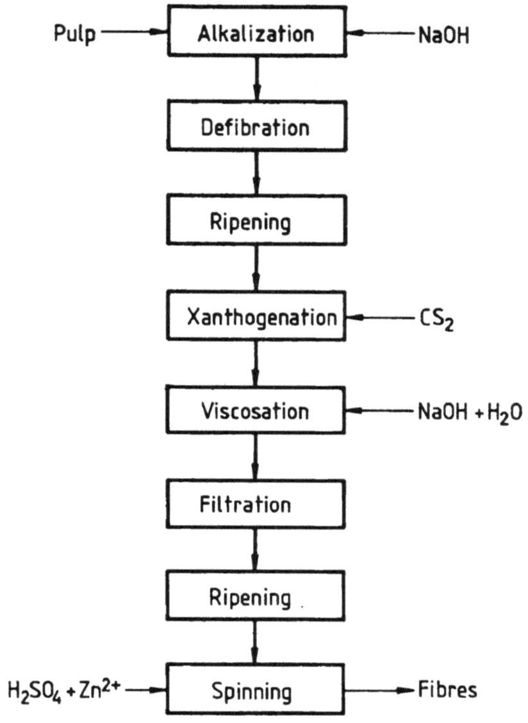

Fig. 17–10. Simplified flowchart of the viscose process.

whole calculated amount of caustic soda solution is added, and the further xanthation is performed in solution (Fahmy, Fadl 1964; Wängberg, Treiber 1968). Xanthation in benzene solution and the addition of organic bases as accelerators have also been proposed (Vasilev, Majboroda 1966, 1967). The advantages of this procedure are a higher reaction velocity, a more uniform distribution of CS_2 within the alkali cellulose, a higher uniformity of the cellulose xanthate, and a better solubility. An acceleration of the xanthation process and uniform products are also obtained when the aged alkali cellulose is submitted to an additional treatment with alkali prior to sulfidation (Sihtola, Nizovsky 1969; Sihtola, Ratanen 1972).

Viscose is subjected to an aging (ripening) process of several days before the cellulose is regenerated in the form of fibres or films. During ripening the γ-value is reduced, but the xanthate groups are distributed more uniformly along the cellulose molecules (Philipp, Chu 1965; Bär et al. 1966). Xanthate groups are split off according to a first order reaction; the decomposition reaction of the C2- and C3-xanthate groups runs much more quickly than that of the C6-xanthate groups (Bandel 1960; Lyselius, Samuelson 1961).

Part of the freed CS_2 is linked by rexanthation predominantly at the C6. The reduced rexanthation capability in C2- und C3-position is explained by an increased protolyzation probability of these free alcoholic groups (Elmgren 1965). The high stability of the C6-xanthate was also indicated by NMR studies of model compounds (Forsén et al. 1966). The other part of CS_2 is spent by the formation of dithiocarbonate. The velocity of dexanthation is influenced positively by temperature increase and negatively by alkali concentration (Fig. 17–11) (Easterwood, Mueller 1960; Dolby, Samuelson 1965; Papkov et al. 1966).

The regeneration of sodium cellulose-xanthate occurs in an acidic precipitating bath containing sulfates. The viscose is forced through a spinning jet (fibres) or a slit

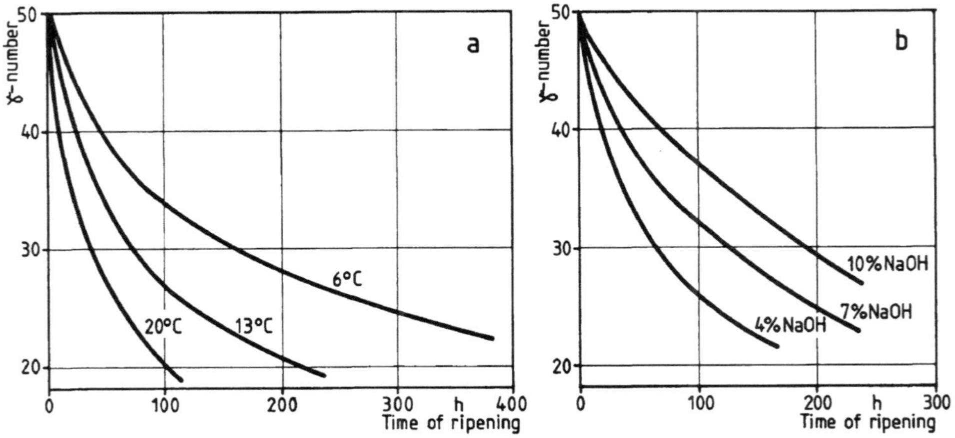

Fig. 17–11. Influence of temperature and alkali concentration on the ripening of cellulose xanthate (according to Lyselius and Samuelson 1961).

(cellophane) into the bath where the viscose is coagulated and decomposed. The salt predominantly causes coagulation while the acid decomposes the xanthate. The latter reaction runs in two steps with different velocity constants:

- formation of cellulose xanthogenic acid
- decomposition of the xanthogenic acid and regeneration of cellulose.

The first step is a bimolecular reaction between dissociated xanthate groups and hydrogen ions (Törnell 1966) (Fig. 17–12). The decomposition of cellulose xanthogenic acid is the slowest, and therefore the velocity-determining reaction, as compared to the rapid ion reactions (equilibrium sodium xanthate – xanthogenic acid) (Jost, Ludwig 1966, 1967). The activation energy of cellulose xanthogenic acid is lower (32 kJ/mole) than that of the sodium salt (58–63 kJ/mole) in acidic solution.

The coagulation determines the properties of the fibres so that for several products the decomposition of xanthate is delayed by variation of the salt concentration in the precipitating bath. Even the addition of sodium sulfate has a delaying effect. A strong delay is caused by zink sulfate (Törnell 1967). The reaction is explained by the formation of a Zn^{2+}-complex and/or a cross-linking of the xanthate molecules (Fig. 17–12). The Zn-xanthate compounds are decomposed more slowly by acid than Na-xanthate.

A number of secondary reactions produce inorganic sulfur compounds H_2S, CS_2, CO_2 and colloidal sulfur in the precipitating bath (Kehren, Reichle 1975).

Various methods have been applied for electron microscopic investigations of cellulose xanthate and the regeneration of cellulose from viscose solutions. As demonstrated with ultrathin sections the cell walls of wood and cotton fibres are bulked during xanthation, and finally, the cell wall structure is largely destroyed (Morehead 1963). Surface replicas from freeze-etched cellulose xanthate show a wide-

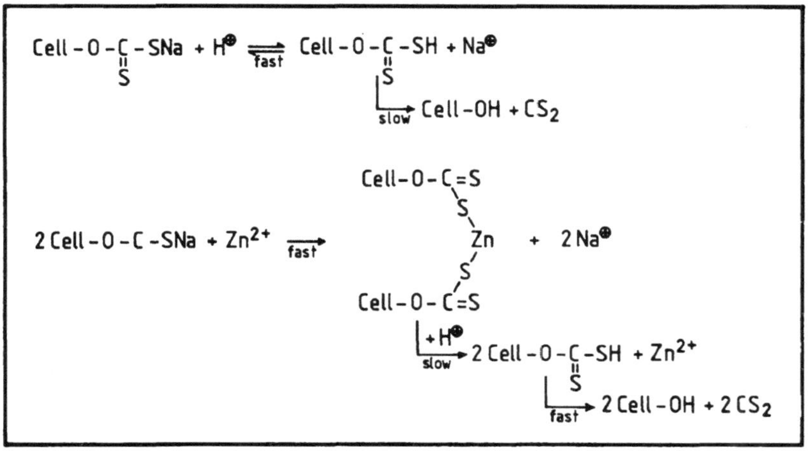

Fig. 17–12. Reactions of the regeneration of cellulose from viscose in the spinning bath.

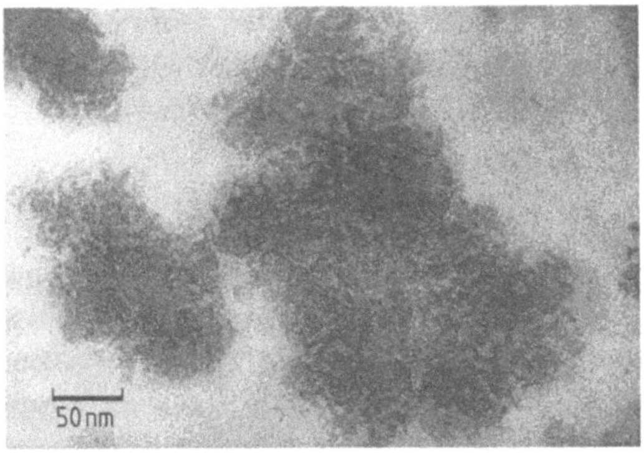

Fig. 17–13. Formation of loosely arranged fibrils in viscose from a highly dilute solution. Negative staining, TEM micrograph.

meshed network with fine fibrillar structures within the meshes (Gröbe et al. 1966; Maron et al. 1967). By applying precipitating media of increasing acidity various steps of coagulation could be obtained (Iovleva et al. 1968; Bandurjan et al. 1971). The coagulation begins with the formation of a homogeneous gel, followed by a curdled structure, and finally a disintegration into fibrils. The transition from a homogeneous gel to loosely arranged fibrils can be recognized in Fig. 17–13.

The structures are influenced by the degree of ripening of the viscose. During the spinning process the fibrils are orientated longitudinally by stress (Jayme, Balser 1967; Gröbe 1968). Depending on the conditions of coagulation and the presence of modifiers, holes or net-like structures may originate from the splitt-off of CS_2 and other decomposition products within the fibres or films (Jayme, Balser 1965; Coalson et al. 1968; Gröbe 1968).

17.3.5. Cellulose Acetate

Cellulose acetate is the most important ester deriving from an organic acid. It is used for manufacturing lacquers, plastics, films and fibres. As compared to cellulose nitrate it is considerably less inflammable. The technical properties of cellulose acetate are determined by the degree of substitution which is responsible for its compatibility with plasticizers and lacquer resins as well as for its solubility in solvents (Table 17–2). A second criterion is the degree of polymerization, ex-

Table 17–2: Various types of cellulose acetate

Acetyl content %	Degree of substitution	Common solvents	Application
13.0–18.6	0.6–0.9	Water	–
22.2–32.2	1.2–1.8	2-Methoxy ethanol	Plastics, lacquers
36.5–42.2	2.2–2.7	Acetone	Fibres, photographic films
43.0–44.8	2.8–3.0	Chloroform	Fabrics, foils, fibres

pressed by the viscosity, which influences the mechanical behaviour of the products and their workability.

The viscosity of the cellulose ester is reduced during the esterification even after triester formation. The viscosity reduction is influenced by various factors; a defined viscosity can be obtained by controlling acid-to-anhydride ratio, catalyst concentration and temperature (Malm et al. 1961a).

The processes applied on a commercial scale can be subdivided as follows:

- acetylation in a homogeneous system (solution acetates),
- – using glacial acetic acid as solvent (acetic acid process),
- – using dichloromethane as solvent (methylene chloride process);
- acetylation in a heterogeneous system (fibre acetates).

In the homogeneous system the cellulose acetate is dissolved during the reaction whereas the formation of fibre acetates occurs in the presence of non-solvents, e.g. carbon tetrachloride, benzene, toluene. In the latter case a reduction of the acetyl content by hydrolysis is not possible (Eicher, Fischer 1975).

For obtaining a more rapid and uniform ester formation it is necessary to pretreat cellulose with water or acetic acid. The velocity of acetylation of pre-swollen cellulose is about three times higher than of non-swollen cellulose (Michie 1966). Pre-swelling obviously opens the way for the acetylating medium to reach the ordered regions more easily. Furthermore, pretreatments with dilute H_2SO_4, ammonia solution and ethylene diamine accelerate acetylation whereas pretreatment with NaOH has no influence (Fahmy, Koura 1967; Manning, Fuller 1969; Koura, Ibrahum 1974; Koura 1977).

Though in principle esterification is also possible with acetyl chloride generally acetic anhydride is used as an acetylating medium. The acetylation reaction needs the presence of a catalyst; sulfuric acid and perchloric acid have proved to be the most effective, and they are the only ones applied in technical processes (Malm et al. 1961b).

The reaction is formally a substitution of one, two or three hydroxyl groups of the glucose unit in the presence of an acidic catalyst according to the mechanism shown in Fig. 17–2c. But there is evidently a more complicated mechanism involved as transesterification occurs as well, or even exclusively. Esterification runs more quickly with sulfuric or perchloric acid than with acetic acid in the reaction medium (Malm et al. 1961b; Delsemme 1967; Garves 1979). In a second, slower reaction the SO_3H- or ClO_3-groups, respectively, are replaced by CH_3CO-groups (Fig. 17–14a). For obtaining a pure cellulose acetate it is important that the foreign ester groups are exchanged completely, a process which can be controlled by the acid ratio and temperature.

An acetylation mechanism via acetyl sulfuric acid or acetyl perchloric acid, respectively, has been discussed (Laamanen, Sihtola 1964; Sihtola, Laamanen 1964). It is

a $$\text{Cell-OH} + \text{HO-SO}_3\text{H} \rightleftharpoons \text{Cell-O-SO}_3\text{H} + \text{H}_2\text{O}$$
$$\text{Cell-O-SO}_3\text{H} + (\text{CH}_3\text{-CO-})_2\text{O} \rightarrow \text{Cell-O-CO-CH}_3 + \text{CH}_3\text{-CO-O-SO}_3\text{H}$$

b $$(\text{CH}_3\text{-CO-})_2\text{O} + \text{HO-SO}_3\text{H} \rightarrow \text{CH}_3\text{-CO-O-SO}_3\text{H} + \text{CH}_3\text{-COOH}$$

c $$\text{Cell-OH} + \text{H}^{\oplus} \rightarrow \text{Cell-OH}_2^{\oplus}$$
$$\text{Cell-OH}_2^{\oplus} + \text{CH}_3\text{-CO}^{\oplus} + {}^{\ominus}\text{O-SO}_3\text{H} \rightleftharpoons \text{Cell-O-SO}_3\text{H} + \text{CH}_3\text{COOH}_2^{\oplus}$$

Fig. 17–14. Various reaction mechanisms of the formation of cellulose acetate in the presence of sulfuric acid.

known that sulfuric acid reacts with acetic anhydride according to equation b in Fig. 17–14. Sihtola and Laamanen (1964) propose a mechanism according to equation c in Fig. 17–14. They also succeeded in identifying the acetylium cation in the reaction mixture by NMR spectroscopy. Other studies, however, showed that the esterification of cellulose runs more quickly with sulfuric acid and acetic anhydride than with acetyl sulfuric acid, and that reaction conditions which restrain the formation of acetyl sulfuric acid are more favourable (Akim 1967; Tanghe, Brewer 1968). The same seems to be true of the system perchloric acid/acetic anhydride.

In a homogeneous system the hydrogen bonds are cleaved very quickly so that a selective esterification at the primary OH-groups occurs as evaluated by IR spectroscopy (Samadjieva, Dimov 1971). Native hydrogen bonds are cleaved until an acetate of a DS of 2.8 is reached. During the reaction new hydrogen bonds are formed between acetate carboxyl groups and cellulose OH-groups. A study of the substituents in a commercial cellulose acetate (DS = 2.4) gave evidence of an approximately equal substitution at C2 and C3, whereas the substitution at C6 was somewhat lower (Björndal et al. 1971).

The starting materials for the technical production of cellulose acetate are cotton linters or chemical grade pulp. After activation with acetic acid during which temperatures up to 50 °C are applied, the cellulose is treated with glacial acetic acid (as a solvent for triacetate), a surplus of acetic anhydride, and sulfuric acid in a cooled kneader. By controlling the temperature the degradation of the cellulose chains is regulated to obtain the desired viscosity. The reaction is finished when the cellulose acetate is completely dissolved in the reaction medium (Fig. 17–15). Subsequently the reaction is interrupted by the addition of dilute acetic acid. If triacetate is not the desired product, the degree of acetylation is standardized (hydrolysis), and equalized by elevating the temperature and regulating the acidity. After cleaning the solution by filtration the cellulose acetate is precipitated by the addition of dilute acetic acid with vigorous stirring. Finally the precipitate is washed with water, centrifuged or pressed for removing the water, and dried (Eicher, Fischer 1975).

The use of dichloromethane saves H_2SO_4 because of its better solution power for triacetate, results in better temperature control because of its lower boiling point,

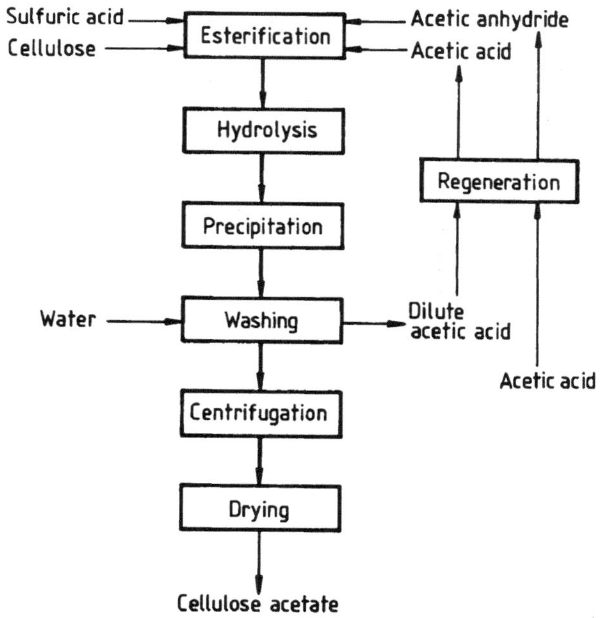

Fig. 17–15. Simplified flowchart of the cellulose acetate process.

and saves dilute acetic acid. The heterogeneous process runs best with perchloric acid as catalyst, and is restricted to the production of triacetate.

For manufacturing fibres cellulose triacetate is dissolved in dichloromethane/methanol (9:1), 2.5-acetate in acetone/ethanol (8:2). The solutions are pressed through jets. The consolidation of the threads occurs by evaporation of the solvent in a preheated air stream (dry spinning process) (Kehren, Reichle 1975). Products of cellulose-2.5-acetate and triacetate differ mainly in moisture absorption and decomposition temperature (Table 17–3).

Various experiments have been made to vary the acetylation conditions. In particular the solvent was varied. Thus Hall and Horne (1973a) succeeded in rapidly forming cellulose triacetate with acetic anhydride in pyridine with acetyl chloride as a catalyst. The acetylation of cellulose occurs with acetyl chloride; the spent acetyl

Table 17–3: Properties of 2.5-acetate and triacetate fibres (Kehren, Reichle 1975)

Properties	2.5-Acetate	Triacetate
Tensile strength (cN/dtex)	1.0–1.5	1.0–1.5
Extension (%)	25–30	25–30
Density (g/cm^3)	1.33	1.30
Moisture absorbance (%)		
(65% rel. humidity, 20 °C)	6–6.5	4–4.5
Water retention (%)	25–28	16–17
Decomposition point (°C)	225–250	310–315
Degree of polymerization	~300	~300

chloride is reformed by transesterification of the acetic anhydride with pyridine hydrochloride, which is formed during the reaction.

A low melting mixture (75 °C) of N-ethyl pyridinium chloride and pyridine dissolves cellulose very easily. The addition of acetic anhydride to a cellulose solution results in a very rapid formation of cellulose triacetate at 85 °C (Husemann, Siefert 1969). The solvent system dimethyl sulfoxide (DMSO)/paraformaldehyde (PF) in combination with pyridine, acetic anhydride and acetyl chloride has also been used for the acetylation of cellulose (Seymour, Johnson 1978; Shiraishi et al. 1978).

Modified cellulose acetates containing chloral groups were obtained with acetic anhydride or acetyl chloride in DMF/chloral/pyridine mixture (Clermont, Manery 1974; Ishizu et al. 1981) (Fig. 17–16).

Several studies are concerned with secondary reactions taking place during acetylation which cause difficulties in the further manufacturing of cellulose acetates. In the first place there is a turbidity (haze) in acetone solution. It originates from gel particles. The particles contain fibre fragments which are poorer in acetyl content and richer in polyoses, particularly in xylan (Borgen 1963; Conca et al. 1963; Golubev et al. 1966). The formation of the gel particles is explained by an aggregation of xylan acetate to less acetylated cellulose or by a higher deacetylation of cellulose in such an aggregate. Gardner and Chang (1974) found that the acetates of native polyoses dissolve in acetone clearly, but acetates of polyoses modified by pulping result in turbide solutions. Particularly sulfate-pulp xylans and sulfite-pulp mannans form crystalline acetates which do not dissolve clearly.

A second problem is the discolouration of cellulose acetate particularly deriving from alkali-treated pulps. It could be shown that this difficulty is caused by oxidative changes at the cellulose molecules. Dialdehyde and keto groups in degradation products result in coloured compounds during acetylation, e.g. 5-acetooxymethyl furfural (Sihtola et al. 1967; Turunen et al. 1969; Krkǒska, Reiser 1971). The acetylation reaction of cellulose is not influenced by the presence of polyoses

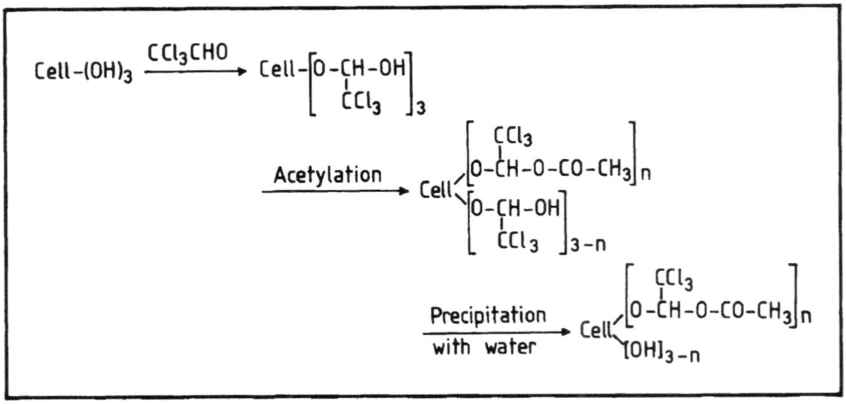

Fig. 17–16. Reaction mechanism of the formation of modified cellulose acetate with chloral.

whereas lignin impedes the reaction velocity and reduces the degree of acetylation (Hernardi, Völgyi 1974).

Solutions of cellulose acetate have also been used for determining the DP of cellulose and the DP distribution. For fractionation by precipitation the choice of the precipitating medium is important: precipitation according to the DP is obtained with non-polar solvents, whereas polar solvents precipitate according to the acetyl content (Bischoff, Philipp 1966; Howard, Parikh 1966).

In the electron microscope solutions of cellulose acetate show aggregates of fibrillar structures if the concentration is not too low (Fig. 17–17). Volynskij et al. (1970) found that the increase of intrinsic viscosity and of the sedimentation constant runs parallel to the formation of fibrillar structures in solution.

After super-cooling dilute solutions of cellulose triacetate in nitromethane / butanol Manley (1963) and Bittiger et al. (1969) obtained square-shaped crystals with screw-like dislocations. The lamellae growing in spirals had a thickness of about 18 nm in which the chain molecules were orientated perpendicularly to the lamellae. The study of the striation and the deviations of the shape, however, gave evidence of a crystal growth in the direction of the 100 planes (Patel, Patel 1969). Later it was shown that cellulose acetate crystals obtained by this procedure contain nitromethane in the crystal lattice (Chanzy et al. 1971).

Crystals of triacetate from cellulose II (CTA II) without solvent inclusion were studied by Chanzy (Chanzy et al. 1973; Chanzy, Roche 1974; Roche 1977). Depending on the solvent needle-like or rhombic crystals were obtained. X-ray and electron diffraction analysis resulted in an orthorhombic unit cell of the space group $P\ 2_1$, and the dimensions a = 2.468 nm, b = 1.054 nm (fibre axis), c = 1.152 nm. A model is proposed which consists of antiparallel pairs of parallel CTA II chains (Fig. 17–18). The crystal structure is kept together by means of hydrogen bonds between acetyl groups and C2-, C3- and C6-OH. Infrared spectroscopic studies

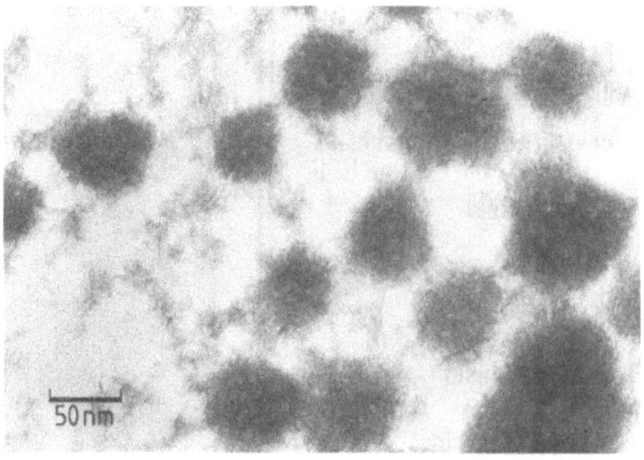

Fig. 17–17. Aggregates of fine fibrillar structures of cellulose acetate prepared from a highly dilute solution. Negative staining, TEM micrograph.

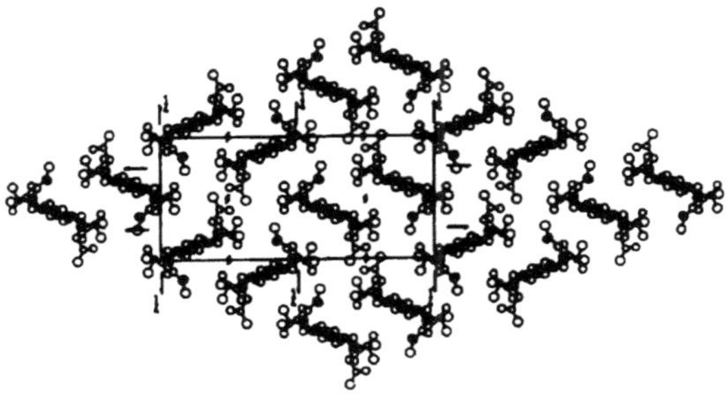

Fig. 17–18. Model of a cellulose triacetate crystal (CTA II) with the arrangement of the molecules (Chanzy, Roche 1974).

have shown, however, that even in crystalline cellulose acetate all OH-groups are accessible for deuterium (Jeffries 1963).

From X-ray diffraction diagrams of heterogeneous acetylated ramie fibres Stipanovic and Sarko (1978) calculated the unit cell of cellulose I triacetate (CTA I). According to these authors the unit cell contains two parallel chains, and has the following dimensions: a = 2.363 nm, b = 1.043 nm (fibre axis), c = 0.27 nm.

Laboratory experiments for improving certain properties (dyeing, strength, plasticity) are concerned with mixed esters, cross-linked acetates, or graft-polymerized acetates (Rijke, Prins 1962; Kiefer, Touey 1965; Rogovin 1967). A partial acetylation of pulp was proposed for improving strength and wet strength properties of paper (Herdle, Griggs 1965; Fahmy, El-Kalyoubi 1970).

17.3.6. Esters with Other Organic Acids

Many studies are concerned with the esterification of cellulose with various organic acids (Fig. 17–19). Generally satisfying yields of esters are obtained by esterification with acid anhydrides or acid chlorides in an indifferent solvent (e.g. pyridine).

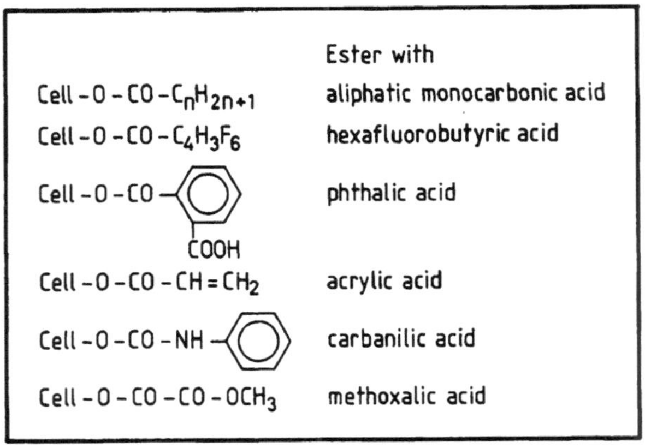

Fig. 17–19. Cellulose esters with various organic acids.

Table 17-4: Properties of cellulose triesters (Malm et al. 1951)

Ester	C Atoms	Behaviour on heating: Shrinking point °C	Melting point °C	Char point °C	Water tolerance value*	% Moisture regain: 25% r.h.	50% r.h.	75% r.h.	95% r.h.	Density g/cm^3	Tensile strength N/mm^2	Specific rotation 25 °C, 589 nm: $CHCl_3$	C_6H_6
Cellulose**	0	–	–	–	–	5.4	10.8	15.5	30.5	1.52	–	–	–
Acetate	2	–	306	315	54.4	0.6	2.0	3.8	7.8	1.28	0.74	−20.9	–
Propionate	3	229	234	315	26.9	0.1	0.5	1.5	2.4	1.23	0.50	−19.0	−39.6
Butyrate	4	178	183	315	16.1	0.1	0.2	0.7	1.0	1.17	0.32	−15.7	−37.4
Valerate	5	119	122	315	10.2	0	0.2	0.3	0.6	1.13	0.19	−12.2	−33.3
Caproate	6	84	94	315	5.88	0	0.1	0.2	0.4	1.10	0.14	− 9.6	−30.0
Heptylate	7	82	88	290	3.39	0	0.1	0.2	0.4	1.07	0.11	− 5.6	−22.5
Caprylate	8	82	86	315	1.14	0	0.1	0.1	0.2	1.05	0.09	− 4.8	−19.6
Caprate	10	87	88	310	–	0	0.1	0.2	0.5	1.02	0.07	− 2.0	−16.6
Laurate	12	89	91	315	–	0	0.1	0.1	0.3	1.00	0.06	− 1.0	−13.0
Myristate	14	87	106	315	–	0	0.1	0.1	0.2	0.99	0.06	0	−13.0
Palmitate	16	90	105	315	–	0	0.1	0.1	0.2	0.99	0.05	0	− 7.8

* ml of water required to start precipitation of ester from 125 ml of an acetone solution of 0.1% concentration

** Starting cellulose, prepared by deacetylation of commercial, medium-viscosity cellulose acetate (40.4% acetyl content)

For obtaining cellulose formiate an esterification with formic acid is possible, but this compound is unstable. Nair and Mehta (1967) applied a heterogeneous formylation for studying the accessibility and the supramolecular structure of cotton.

The synthesis and properties of esters with higher aliphatic acids have been described previously and are summarized by Malm and Hiatt (1954). A summary of various properties of these esters is given in Table 17–4. Though the aliphatic esters of cellulose have desirable properties (e.g. low melting points, high water resistance) only a few of them are produced in large quantities, mostly as mixed esters. Thus cellulose acetopropionate and acetobutyrate are applied for films, lacquers and plastics. Commercial products vary in their content of acetyl and propionyl or butyryl groups, and their degree of hydrolysis depends on the intended purpose. The esterification is performed with mixtures of acetic anhydride and propionic or butyric anhydride, respectively, and sulfuric acid as a catalyst (Eicher, Fischer 1975).

The favourable properties of cellulose valerate and propiovalerate in regard to low melting point, high moisture resistance, heat stability and compatibility with resins and plasticizers are emphasized by Mench et al. (1966). Cellulose caprate also has restricted application as a cement for lenses and optical systems.

For superior dyeing of acetate fibres Rogovin (1967) proposes the introduction of ester groups of higher aliphatic acids or hexafluorobutyric acid.

Starting with cellulose which has been pre-swollen in ethylene diamine, and using an esterificating medium of anhydrides of phthalic acid, Δ^4-cis-tetrahydrophthalic acid, cyclohexane-1,2-cis-dicarboxylic acid, succinic acid or glutaric acid in DMSO or DMF, Garves (1972) succeeded in synthesizing dicarboxylic esters of cellulose. Esters of cellulose with unsaturated acids which are of interest because of their reactive double bonds, were obtained by dissolving cellulose in a melt of ethylpyridiniumchloride and DMF, and by adding acrylic or methacrylic acid (Pohjola et al. 1976).

The treatment of cellulose with phenylisocyanate in pyridine or in DMF with triethylene diamine as a catalyst yields cellulose tricarbanilate, a compound of scientific interest (Burchard, Husemann 1961; Hall, Horne 1973a). The reaction results in a stoichiometric triderivative with a molecular weight more than triple that of the starting cellulose. Cellulose tricarbanilate has good solubility in many organic solvents. It is used mainly for determining the degree of polymerization and the DP-distribution of cellulose (Valtasaari, Saarela 1975; Daňhelka, Kössler 1976; Schroeder, Haigh 1979; Kössler et al. 1981). Bittiger and Husemann (1964) found rod-like structures of cellulose tricarbanilate in the electron microscope. These structures were interpreted as single-molecular crystals. Depending on the DP and concentration fibrils and hexagonal crystals were also observed (Bittiger et al. 1969).

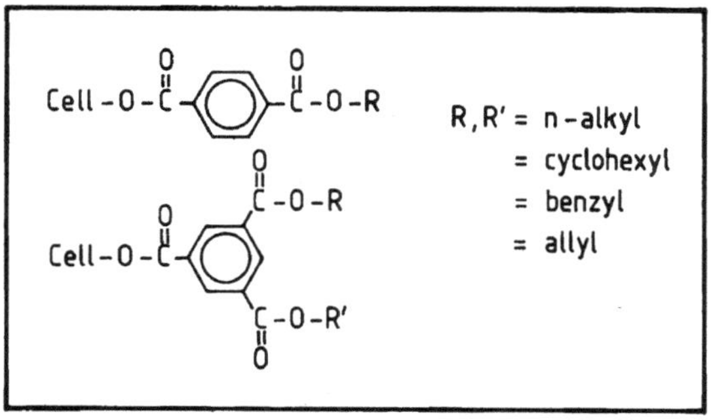

Fig. 17–20. Cellulose esters with terephthalic halfesters and trimesinic diesters.

The formation and behaviour of cellulose methoxalate was studied by Rebek (Rebek et al. 1973; Rebek, Jurkowitsch 1975, 1977). The esterification of cellulose occurs with methoxalic anhydride in benzene / pyridine at room temperature resulting in a trisubstituted product. The ester is soluble in dilute ammonia solution by forming an NH_4-salt of oxalic acid ester with partial saponification of the methoxyl groups. The salt could be precipitated, and after dialysis cellulose oxalate has been obtained.

For making cotton accessible for certain disperse dyes Baumann et al. (1981) esterified the fibres with terephthalic halfesters and trimesinic diesters by means of a reaction with the respective acid chlorides (Fig. 17–20).

17.4. Cellulose Ethers

17.4.1. Fundamentals of Etherification

The general reaction of ether formation runs similar to that of esterification, i. e. by the intermediate formation of an oxonium ion. A surplus of the alcoholic component results in the formation of an ether (Fig. 17–21a). As in cellulose the OH-groups are not very easily accessible a reaction according to this scheme with a soluble alcohol does not give satisfactory products. Thus the starting material has to be converted to alkali cellulose (alkali-consuming process, ether synthesis accord-

Table 17–5: Cellulose ethers obtained by alkali-consuming processes

Formula	Name	Symbol
Cell–O–CH_3	Methylcellulose	MC
Cell–O–C_2H_5	Ethylcellulose	EC
Cell–O–C_3H_7	Propylcellulose	PC
Cell–O–C_6H_5	Benzylcellulose	
Cell–O–CH_2–COOH	Carboxymethylcellulose	CMC

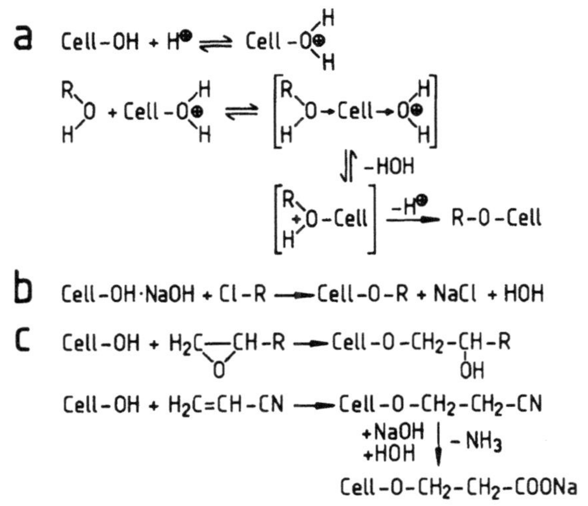

Fig. 17–21. a) Reaction mechanisms of cellulose etherification.
b) Etherification of cellulose with alkali consumption.
c) Etherification of cellulose without alkali consumption.

ing to Williamson), or at least the cellulose has to be pre-swollen in alkali (process without alkali consumption).

The first process runs as a reaction of alkalized cellulose with alkyl halogenide (Fig. 17–21b). This process is applied to ethers like methyl cellulose, ethyl cellulose, carboxymethylcellulose etc. (Table 17–5). Side reactions result in alcohols (methanol, ethanol) and dialkyl ethers.

In the process without alkali consumption only small amounts of NaOH are needed for swelling or widening the lattice, respectively, and for reactivating the etherifying medium. The reaction is an addition reaction, and the etherifying medium may be an epoxide or an α,β-unsaturated compound (Fig. 17–21c). Products obtained by these reactions are hydroxylalkylcelluloses, cyanoethylcellulose and carboxyethylcellulose (Table 17–6).

As the etherifying components of the latter reactions are highly reactive compounds various secondary reactions take place apart from the main reaction. These result in polymeric products or an extension of the ether groups.

Table 17–6: Cellulose ethers obtained by processes without alkali consumption

Formula	Name	Symbol
Cell–O–CH_2–CH_2OH	Hydroxyethylcellulose	HEC
Cell–O–CH_2–CH(OH)–CH_3	Hydroxypropylcellulose	HPC
Cell–O–CH_2–CH(OH)–C_2H_5	Hydroxybutylcellulose	HBC
Cell–O–CH_2–CH_2–CN	Cyanoethylcellulose	
Cell–O–CH_2–CH_2–COOH	Carboxyethylcellulose	

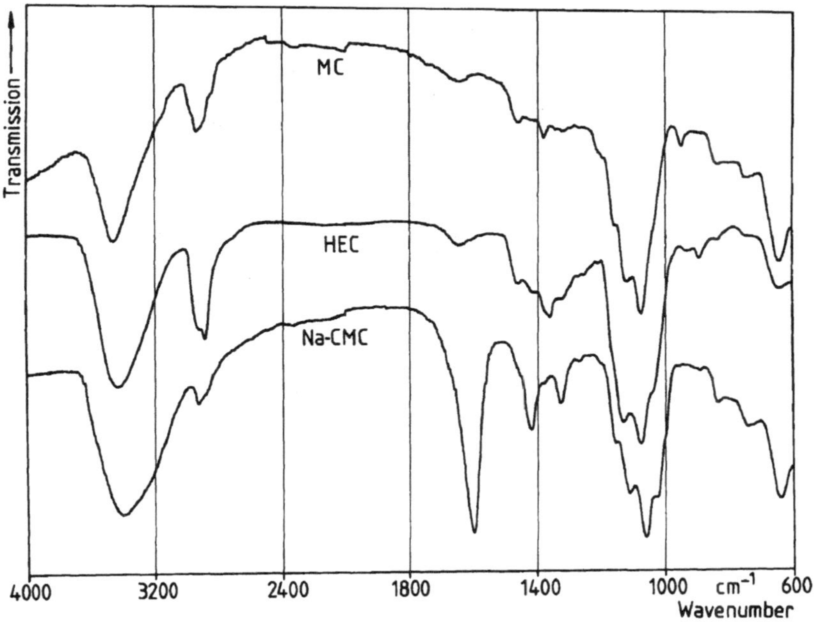

Fig. 17–22. Infrared spectra of three cellulose ethers.

Though having the same cellulose backbone, cellulose ethers can be distinguished by IR spectroscopy as the ether group results in characteristic absorption bands (Friese 1981) (Fig. 17–22).

17.4.2. Solution Properties and Application

The introduction of ether groups into the cellulose molecule results in a swellability or solubility even in cold water. The type of substituents as well as the degree and uniformity of substitution determine these properties of the cellulose ethers. With hydrophilic substituents a solubility in water is reached at a relatively low degree of substitution (DS) which is maintained up to complete substitution (DS = 3). Hydrophobic substituents show alkali and water solubility at low DS, and solubility in organic solvents at high DS (Table 17–7) (Balser, Iseringhausen 1975). The substi-

Table 17–7: Solubility of cellulose ethers in dependency on the substituents and the degree of substitution (according to Balser and Iseringhausen 1975)

Cellulose ether	DS for solubility in		
	4% NaOH	cold water	organic solvent
Methylcellulose	0.4–0.6	1.3–2.6	2.5–3
Ethylcellulose	0.5–0.7	0.8–1.3	2.3–2.6
Hydroxyethylcellulose	0.5	0.5–1	
Na-carboxymethylcellulose	0.5	0.5–1.2	
Cyanoethylcellulose			2.0
Benzylcellulose			1.8–2.0

tution of the OH-groups by ether groups also causes an increase in the molecular weight of the cellulose depending on the degree of substitution and the size of the ether groups (Table 17–8).

Jullander (1965) gives an example of several steps during the methylation of cellulose:

- A low amount of methyl groups results only in chemically modified cellulose with no change in solubility, i. e. the products are soluble only in the usual cellulose solvents.
- A higher degree of methylation (0.4–0.6) increases swellability in water and the products are soluble in NaOH of moderate concentration (4–5%).
- A DS of 1.3–2.6 effects water-soluble products which form solutions of high viscosity. The water solubility of methyl cellulose is explained by cleavage and disturbance of the hydrogen bonds of non-substituted OH-groups, by means of which these groups become accessible to solvatisation by water.
- Degrees of substitution of 2.5–3.0 increase the hydrophobic character. The products are water-insoluble, but soluble in organic solvents.

The solubility of alkylcelluloses decreases with increasing temperature. Methylcellulose flocculates from solution at 45–65 °C and dissolves again during cooling. Ethylcellulose with an ethoxyl content of 17–20% (DS = 0.7–0.8) becomes instable even at 30 °C in solution. The flocculation temperature (gel temperature) of alkylcelluloses can be increased by the introduction of hydroxyalkyl groups. Hydroxyalkyl- and carboxyalkylcelluloses do not show a flocculation at higher temperatures.

Various salts decrease the viscosity of cellulose ether solutions if applied in low concentration. The application of high salt concentrations may effect a flocculation which can be reversibly dissolved by dilution.

An irreversible flocculation of cellulose ethers in neutral or weak acidic solutions occurs with tannins, salts of complex acids and some dyes.

Particularly the water-soluble cellulose ethers yield stable solutions containing molecules of varying dispersities. There are various studies concerned with the rheological properties of cellulose ether solutions (Scherer et al. 1960; Simionescu et al. 1967; Elliott 1969; Klug 1971).

The rheological behaviour as well as other properties, such as film formation and adhesive strength, are very important for the commercial application of cellulose

Table 17–8: Percentage of weight increase of cellulose by substitution with alkoxyl groups (Balser, Iseringhausen 1975)

DS	$-OCH_3$	$-OC_2H_5$	$-OC_3H_7$
1	17.6	23.6	28.9
2	32.6	41.2	47.9
3	45.6	54.8	61.4

ethers. They are used for emulsifying, dispersing and stabilizing agents in the cosmetic, pharmaceutic, food, chemical and plastic industries, as auxiliary material for manufacturing paper and textiles, cement and concrete, as roughcast and thickener in printing inks and lacquers, as a glue particularly for wallpapers and in sized colours, and as protecting films and foils (Balser, Iseringhausen 1975; Balser, Szablikowski 1981).

Other types of cellulose ethers are highly swellable but insoluble products which are applied in hygienic papers and tissues or as additives for improving the humidity of soils. These products are obtained by cross-linking. Cellulose and its derivatives can be cross-linked by treatment with formaldehyde, methylol urea, epichlorohydrine, acrylic amines, triacines, metal chelates etc. (Kiefer, Touey 1965; Tesoro, Willard 1971; Holst 1978).

17.4.3. Alkylcelluloses

Alkylcelluloses are mainly produced by the reaction of alkyl chlorides with alkali cellulose. Methylcellulose may also be obtained from methylation with dimethyl sulfate. This process is, however, restricted to laboratory experiments. On an industrial scale Na-cellulose is converted with methyl chloride (Savage 1971; Balser, Iseringhausen 1975).

The reaction is carried out in an autoclav where the alkali cellulose is treated with gaseous CH_3Cl at 90–110 °C (gas circulating process) or with liquid CH_3Cl at 60–70 °C. In a continuous process the alkali cellulose is pressed with a double-screw press through a horizontal, continuous reactor where the reaction with CH_3Cl takes place. In all processes the by-products methanol and dimethylether are condensed outside the reactor in the circuit of the etherifying medium.

Philipp et al. (1979) studied the reaction of alkali cellulose with methyl chloride. A technical grade alkali cellulose (30% NaOH and 31–32% cellulose according to a

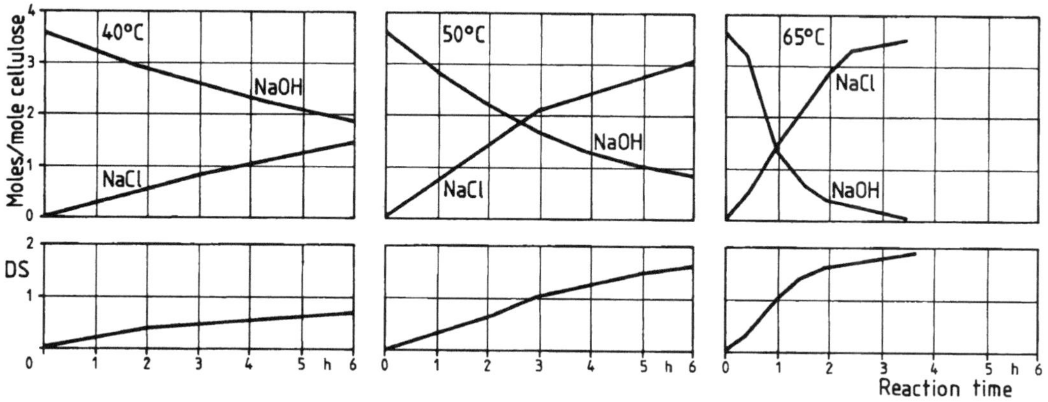

Fig. 17-23. Influence of temperature on the consumption of NaOH, the formation of NaCl, and the DS during the formation of methylcellulose (according to Savage 1970).

Table 17–9: Relative velocity of reaction at the OH-groups of C2, C3 and C6 (Jullander 1965)

Derivative	k_2 : k_3 : k_6
Methylcellulose	5 : 1 : 2
Ethylcellulose	4.5 : 1 : 2
Carboxymethylcellulose	2 : 1 : 2.5
Hydroxyethylcellulose	3 : 1 : 10
Cyanoethylcellulose	3 : 1 : 3

molar ratio of anhydroglucose: NaOH of 3.8) shows an increase of DS proportional to NaOH consumption. On an average 46% of the converted CH_3Cl was consumed for the methylation of cellulose. The highest DS value reached under these conditions was 1.8. The conversion velocity is determined by the temperature, the pressure of gaseous CH_3Cl and the accessibility of the alkali in the cellulose (Fig. 17–23). Two-thirds of the NaOH are converted relatively quickly, whereas the conversion of the residual NaOH occurs slowly.

According to Savage (1970) the methylation rate of cellulose begins slowly, increases to a maximum value after about 1 hour and decreases again. The formation rate of by-products is high at the beginning and decreases later. During ethylation the formation of by-products is low at the beginning and runs later on to a maximum.

The most reactive group during methylation and ethylation is the OH-group at C2 (Table 17–9). This effect is explained by the fact that the C2-OH is the most acidic hydroxyl group in the cellulose and therefore usually reacts with alkoxide ions (Croon 1960). The dissociation of the OH at C3 should be considerably depressed by the inductive effect to the alkoxide ion at C2. Nevertheless the conversion rate at the C3 increases when the C2 is methylated. The C6 is methylated according to its exposed position.

Bittiger et al. (1969) observed relatively thick fibrils of methylcellulose in the electron microscope. The high-resolution micrograph of Fig. 17–24 shows two phases in

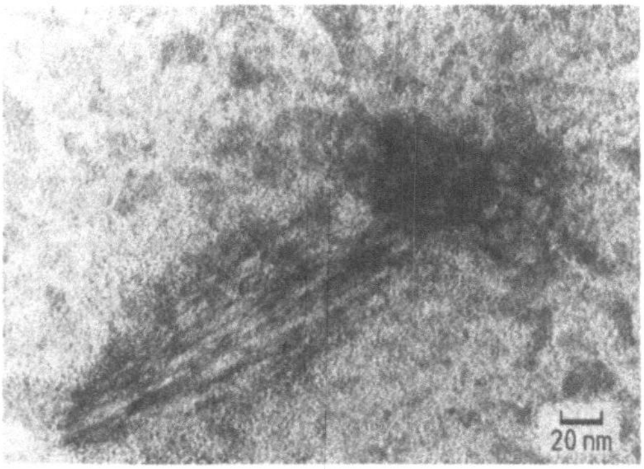

Fig. 17–24. Loose cellulose bundle and molecular cords of methylcellulose from a highly dilute solution. Negative staining, TEM micrograph.

a technical product (DS = 1.5): a loosely arranged fibrillar bundle of obviously low substituted cellulose, and very fine, probably molecular cords of obviously higher substituted cellulose.

17.4.4. Carboxymethylcellulose

According to its multiple applications carboxymethylcellulose (CMC) and its sodium salt (Na-CMC) are commercially produced in much higher amounts than any other cellulose ether. It is produced by the reaction of monochloroacetic acid or sodium monochloroacetate with alkali cellulose.

In the technical process the alkali cellulose is treated with Na-monochloroacetate in a kneader and then in a rotating drum (Fig. 17–25). After dispersion in methanol and neutralization with hydrochloric acid, the product is filtered, washed and dried. A continuous process starts from pulverized bleached pulp which is sprayed with NaOH and subsequently with monochloroacetic acid in a rotating reactor. After drying a technical grade Na-carboxymethylcellulose is obtained with a purity of about 68% (Balser, Iseringhausen 1975).

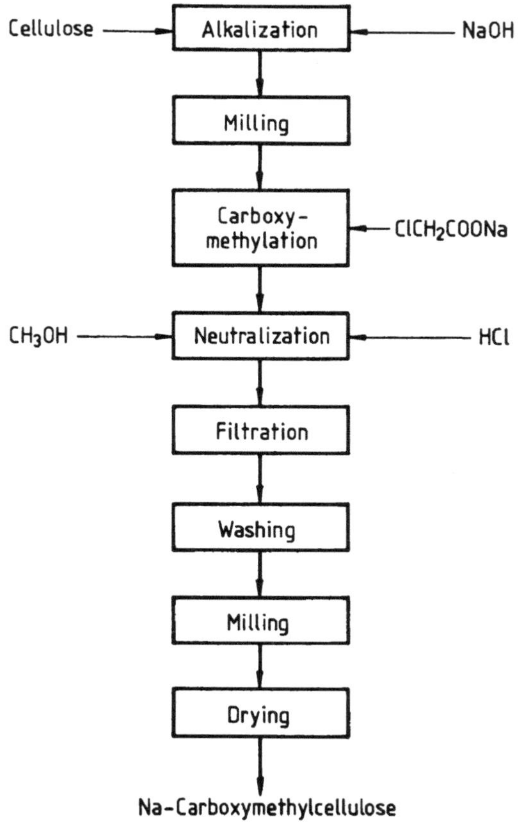

Fig. 17–25. Simplified flowchart of the technical production of Na-carboxymethylcellulose.

There are various studies on optimizing and promoting the carboxymethylating process. An important influence is attributed to the alkalizing step (Fahmy, Mansour 1966; Olaru et al. 1978; Dautzenberg et al. 1980). A molar ratio of NaOH: cellulose of about 1 has proved optimal regarding the attainable degree of substitution, which is in the range of 0.4–1.0. The reaction velocity with monochloroacetic acid is improved by the addition of organic solvents (benzene, toluene, ethanol, isopropanol).

Experiments using alternative starting materials (unbleached pulp, wood, pulp from bagasse, corn cobs or cotton stalks) resulted in comparable CMC-qualities (Fahmy, Mansour 1966; Durso 1976). The degree of substitution and the properties of the CMC-solution (viscosity, thixotropy) are influenced by the type of starting pulp.

During carboxymethylation the OH-groups at C6 show the highest reactivity followed by the C2–OH. The ratio of the relative reactivity was determined to be $k_2:k_3:k_6 = 2:1:2.5$ (Croon, Purves 1959) (Table 17–9).

In aqueous solution Na-CMC tends to form aggregates which are caused by hydrogen bonds between the molecules (Dautzenberg et al. 1978). In cellulose solvents such as cadoxene Na-CMC behaves like a randomly coiled molecule in a good solvent (Brown et al. 1963a). Fig. 17–26 shows coiled Na-CMC molecules within an aggregate prepared from a highly dilute aqueous solution.

A slight carboxymethylation of pulp and rayon improves the strength properties (Nelson, Kalkipsakis 1964; Alince 1976).

17.4.5. Hydroxyalkylcellulose

The most important compounds of this type which are produced on a commercial scale are hydroxyethyl- (HEC) and hydroxypropylcellulose (HPC). They are obtained by the reaction of gaseous ethene oxide or propene oxide onto Na-cellulose

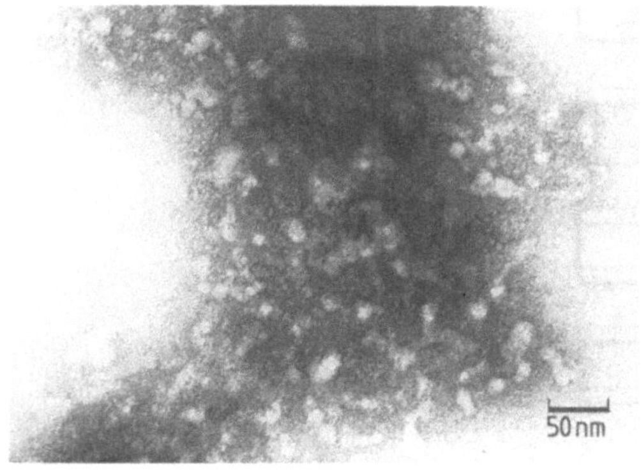

Fig. 17–26. Coiled Na-CMC molecules from a highly dilute solution. Negative staining, TEM micrograph.

in processes similar to those used for methylcellulose (→ 17.4.3.). Though about 2.5 moles of alkene oxide are randomly linked to one anhydroglucose unit the degree of substitution is only 0.5 because of the formation of polyoxyalkene chains (Balser, Iseringhausen 1975). In the laboratory a hydroxyethylation is also possible with liquid ethene chlorohydrine (Ward et al. 1967).

Hydroxyalkylcelluloses are soluble in water, dilute alkali or organic solvents, and have thermoplastic and filmforming properties (Samuel 1969; Gipstein, Wellisch 1973). The reaction of ethene oxide with cellulose in the presence of NaOH is mainly influenced by the alkali concentration (Ramnäs, Samuelson 1968, 1973). The rate of substitution is proportional to the hydroxyde concentration within the molar range of 0.4–1.5 (Fig. 17–27). At higher NaOH concentrations the reaction rates at C2 and C3 are virtually independent of the alkali concentration, which possibly can be explained by a partial blockade of the secondary hydroxyl groups.

The behaviour of hydroxyethylcellulose in solution was studied by Brown (Brown et al. 1963b; Friedberg et al. 1963). In aqueous solution HEC behaves as an inflexible, highly extended chain, while in cadoxene it exhibits properties typical of flexible molecules. Both water and dimethyl sulfoxide are solvents for HEC, but in mixture a strong dipol-dipol interaction diminishes the solvent-HEC interaction and the hydroxyethylcellulose aggregates.

Various studies demonstrate that a slight hydroxyethylation of pulp improves certain strength properties (breaking length, folding endurance, stretch) and the thermal stability, but reduces opacity (Plunguian 1961; Ward et al. 1967, 1968, 1969).

In recent years mixed ethers have become increasingly important. Mixed ethers are obtained by simultaneous treatment of alkali cellulose with methyl chloride and

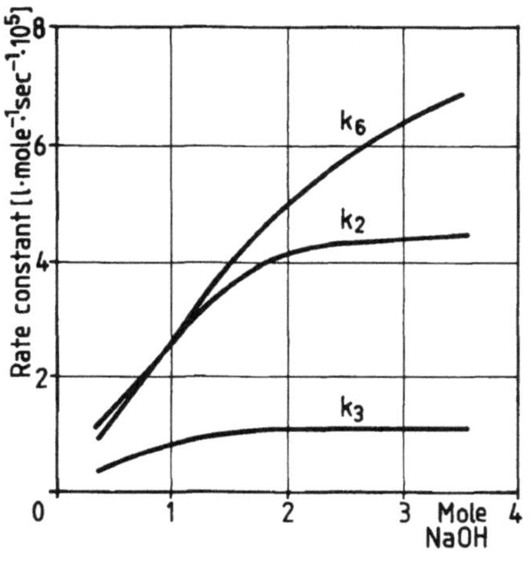

Fig. 17–27. Substitution rate at C2, C3, and C6 during hydroxyethylation of cellulose depending on NaOH concentration (according to Ramnäs and Samuelson 1973).

ethene or propene oxide. By varying the amounts of both reagents the resulting products (HEMC, HPMC) can be directed to specific properties (Table 17–10) (Savage 1971).

17.4.6. Other Cellulose Ethers

Cyanoethylcellulose is produced by the reaction of acrylonitrile with cellulose in the presence of NaOH. The solubility of the resulting products depends on the degree of substitution. A completely substituted cyanoethylcellulose has a nitrogen content of 13%, and is soluble in acetone. For a solubility in alkali a DS of 0.25–0.5 and a uniform distribution of the substituents are necessary (Schleicher et al. 1974; Koura et al. 1977). The importance of a sufficient activation of cellulose for obtaining a uniform substitution at low DS's was emphasized by Lukanoff et al. (1974, 1979). The most important property of cyanoethylcellulose is its high dielectric constance. Thus it is used as insulating material and in capacitors (Bikales 1971).

Cyanoethylcellulose is saponified to carboxyethylcellulose if the reaction mixture contains a surplus of alkali (Fig. 17–21c).

Tritylcellulose (triphenylmethylcellulose) is prepared by heating regenerated cellulose with triphenylmethyl chloride in pyridine (Savage 1971). According to Hall and Horne (1973b) the product is esterified exclusively at the C6, as was proved by tosylation (esterification with p-toluenesulfonic acid) and removement of the tosyl groups by iodine. A homogeneous tritylation was performed by Hagiwara et al. (1981). They dissolved cellulose in DMSO which contained SO_2 and diethyl amine, added trityl chloride and pyridine, and kept the mixture at 50 °C with shaking.

Table 17–10: Properties and property variations of water soluble cellulose ethers (Balser, Szablikowski 1981)

Properties	CMC		MC		HEC	
	existing	influenced by	existing	influenced by	existing	influenced by
Anion activity	+		–	CM	–	CM
Solubility, cold	+	DS, US	+	US	+	US
Solubility, hot	+	DS, US	–	US, HA	+	US
Surface activity	(+)		++		+	
Salt resistance Me^{+}	(+)	DS, US, HA	(–)	US, HA	++	
Salt resistance Me^{++}	–	HA, M	+	HA	++	
Viscosity, rheology	+	DP, DS, US	+	DP, DS, US	+	DP, DS, US
Water retention	+	M	+	HA, M	+	M
Heat stability		M		M		M

CM = carboxymethylation; DP = degree of polymerization
HA = hydroxyethylation; DS = degree of substitution
M = modification; US = uniformity of substitution

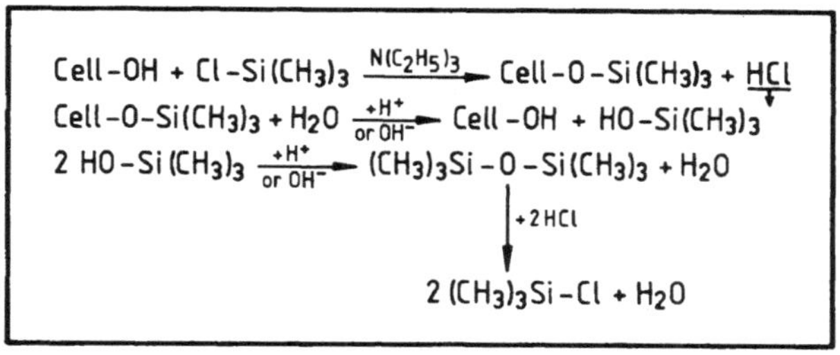

Fig. 17–28. Reactions of trimethylsilylation of cellulose and the regeneration of trimethylsilylcellulose.

Smirnova et al. (1967) synthesized a 2,3-anhydro-6-O-tritylcellulose by saponifying trityltosylcellulose with KOH and Na-methylate in absolute methanol. The compound contained α-oxidic rings.

Other cellulose ethers with aromatic groups – phenylcellulose, benzylcellulose and benzhydrylcellulose (diphenylmethylcellulose) – are mainly of scientific interest. Only benzylcellulose (phenylmethylcellulose) was applied for a time as a basic substance for lacquers.

By treating alkali cellulose with ethyl sulfonate sulfoethylcellulose is prepared. Pastyr and Kuniak (1971) started from cross-linked microcrystalline cellulose which was etherified with Na-chloroethyl sulfonate. Thus they obtained a product with good ion-exchanging properties.

Trimethylsilylcellulose (TMS-cellulose) was produced by Greber and Paschinger (1981) by treating cellulose with trimethylchlorosilane in the presence of tertiary amines (Fig. 17–28). TMS-cellulose is soluble in tetrahydrofurane and can be easily desilicated, thus being a possible intermediate for the production of regenerated cellulose.

17.5. Graft Copolymers of Cellulose

Graft copolymers are compounds consisting of a backbone of a natural polymer bearing side-chains of synthetic polymers which are linked covalently to the backbone. By grafting the cellulose is modified in its behaviour and its properties.

A great many studies are concerned with graft polymerization of cellulose, and results have been summarized in several articles (Rogovin 1960; Krässig, Stannett 1965; Atlas, Mark 1969; Stannett, Hopfenberg 1971; Phillips et al. 1972). In most experiments vinyl polymers (polyvinyl chloride, polysterene, polymethacrylate etc.) were grafted onto cellulosic material such as pulp, cotton, rayon, bagasse and paper. By this derivation dry and wet strengths, surface properties, chemical resistance etc. can be improved (Grigorjan, Rogovin 1968; Rogovin 1974; Bu et al.

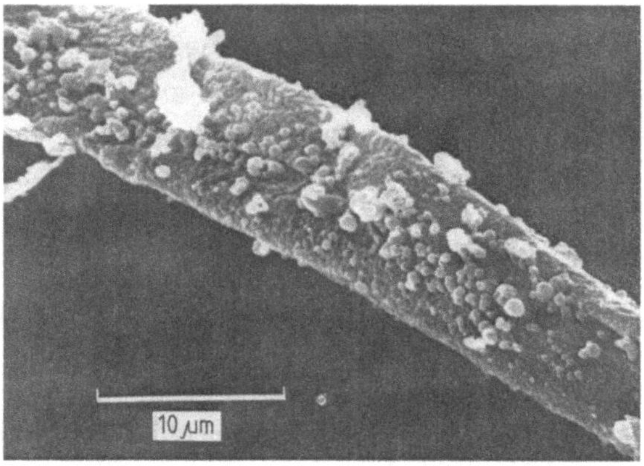

Fig. 17–29. Cotton fibre grafted with methyl methacrylate in a biological (never dried) condition. REM micrograph (by courtesy of F. Mobarak).

1977; Mobarak 1982). Fig. 17–29 shows a cotton fibre grafted with methyl methacrylate in a never-dried condition. By surface grafting cellulose fibres can be enveloped in polyethylene and polypropylene (Chanzy et al. 1968, 1975; Dankovics et al. 1969).

The properties of regenerated cellulose can also be varied by grafting the intermediate cellulose xanthate (Dimov, Pawlov 1969; Kokta, Valade 1972; Rogovin 1974; Morin et al. 1975). Also other cellulose derivatives (e.g. cellulose acetate) were converted to graft copolymers (Wellons et al. 1967; Simionescu, Mihailescu 1970; Yasukawa et al. 1972).

The grafting reactions can be subdivided into three principal groups:

- grafting initiated by radical polymerization methods,
- grafting initiated by ion interaction,
- grafting by condensation or addition reactions.

The radical polymerization route is the technique used most frequently. The radicals necessary for starting and running the reaction can be produced by chemical means, irradiation or mechanical means (Krässig 1971; Stannett, Hopfenberg 1971).

For grafting vinyl compounds to cellulose via radical polymerization the free-radical chain transfer reaction is the one used most commonly. Starting from an easily radicalizable compound (e.g. peroxides) the formation of polymer chains with radical sites is initiated (Fig. 17–30). These radical sites are transferred to the cellulose molecule where the formation of a graft co-polymer chain starts. As can be seen from Fig. 17–30 the graft polymerization is accompanied by reactions forming homopolymers and cross-links between grafted cellulose molecules. In general the homopolymerization exceeds the graft polymerization (Fig. 17–31).

The undesired formation of homopolymers can be suppressed by introducing radical sites into the cellulose molecules prior to grafting. The radicalization of cellulose

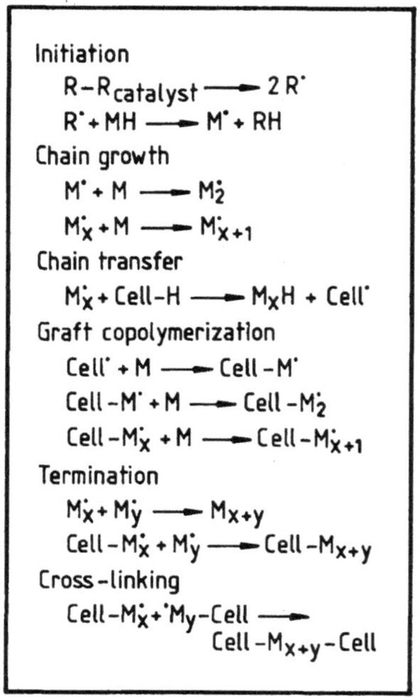

Fig. 17–30. Reactions of the free-radical transfer polymerization.

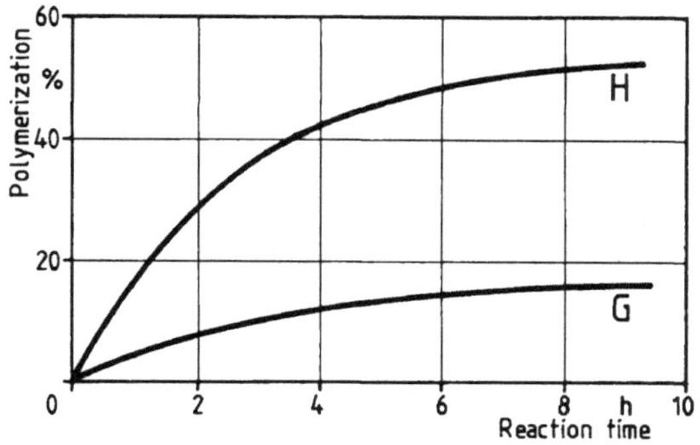

Fig. 17–31. Run of homopolymerization (H) and graft copolymerization (G) during grafting of softwood pulp with vinyl chloride (according to Cornell 1962).

can be performed by introducing halogen-substituted ester groups, mercaptyl groups and diazo groups (Sankalia et al. 1962; Rebek et al. 1967a). Radical sites at the cellulose molecule can also be produced by the reaction with persulfate or the redox systems H_2O_2-Fe^{2+}, and peracetic acid-Fe^{2+} (Sankalia et al. 1962; Morin et al. 1967; Simionescu, Ungureanu 1967; Ogiwara, Kubota 1969; Hatakeyama, Rånby 1975; Rånby 1981). The radicals ($SO_4^{\cdot-}$, HO˙, HOO˙, acetoxy radical) formed in these systems (Fig. 17–32) react with monomers and cellulose by hydrogen abstraction.

An oxidative formation of radical sites can also be obtained by treatment with cerium-IV salts or manganese-IV oxide (Cornell 1962; Dimov, Lazarova 1968;

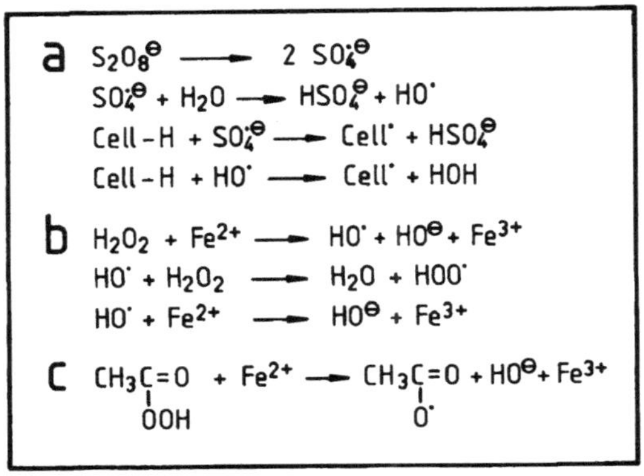

Fig. 17–32. Initiation of radical sites at the cellulose molecule applying persulfate (a), H_2O_2-Fe II (b), and peracetic acid-Fe II.

Hebeish, Mehta 1968; Uhlig, Teichmann 1968, 1971, 1972; Rao, Kapur 1969; Mansur, Schurz 1973). The oxidative splitting of a proton from the cellulose molecule is assumed to produce the radical site (Fig. 17–33a). Using the cerium method Nakamura et al. (1969) grafted large amounts of acrylic esters onto cross-linked cellulose fibres and obtained products with elastomeric properties. Rånby (1978, 1981) applied a manganese-III pyrophosphate and proposed two different ways of radical formation (Fig. 17–33b).

Particularly for grafting polyalkenes onto cellulose fibres reactive centres are produced by the fixation of catalyst systems (VCl_3-Al-triethylate, $TiCl_4$-Al-triethylate) at the cellulose surface (Chanzy et al. 1968; Dankovics et al. 1969).

During the grafting of lignocellulose (bagasse) with methyl methacrylate (MMA) using sodium hydrogensulfate as a catalyst, Mobarak (1982) found a lowering of the apparent activation energy by the lignocellulose and thus an accelleration of the polymerization of the MMA as compared to experiments without lignocellulose.

Another method of initiating a graft polymerization is the irradiation by γ-rays or UV light. Cellulose can be irradiated before the monomer is added. Thus the formation of homopolymers is suppressed. In the presence of air peroxide groups are formed at the cellulose which initiate grafting; in the absence of air the directly

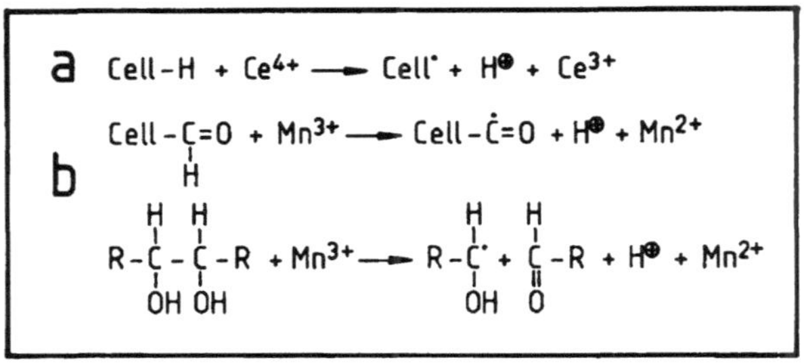

Fig. 17–33. Initiation of radical sites at the cellulose molecule applying Ce-IV (a) and Mn-III salts (b).

formed radical sites can react with the monomers (Fig. 17–34) (Atlas, Mark 1969; Krässig 1971; Lawrence, Verdin 1973). The yield of grafted molecules is influenced by the irradiation energy. Blouin et al. (1968) found 4 600 to 44 000 molecular reactions per 100 eV at a radiation dose of 5 kJ/kg. The grafting reaction can be accellerated by increasing the number of peroxide groups of cellulose by adding H_2O_2 during the irradiation (Kobayashi 1961). The γ-irradiation also effects a cleavage of the cellulose chains, forming radical sites at the C1 (Wellons, Stannet 1965; Sakurada et al. 1972a, b) (→ 13.2.4.) An average of 6 cleavages per 1 000 glucose units was determined at a dose of 100 kJ/kg.

Compared with the radical processes the ionic polymerization is of minor importance. The reactions are carried out in inert solvents (THF, DMF, DMSO), and start from cellulose alkoxides (Schwenker, Pascu 1963; Feit et al. 1964; Avny, Rebenfeld 1968b). The steps of the reaction are shown in Fig. 17–35. Regarding cellulose as a Lewis-base, catalytic sites can be introduced by adding boron trifluoride (Rausing, Sunner 1962). At these sites the polymerization of a monomer is initiated.

Only a few studies have also been made of condensation and addition reactions. There are early studies by Rogovin (1960) during which low-molecular polyethers (e.g. composed of ethylene glycol and adipic acid) were condensed to the cellulose molecule. Rebek et al. (1967b) grafted siloxanes onto cellulose by condensation reactions. An addition reaction is represented by the grafting of polyethylene imine to cellulose (Cooper, Smith 1960).

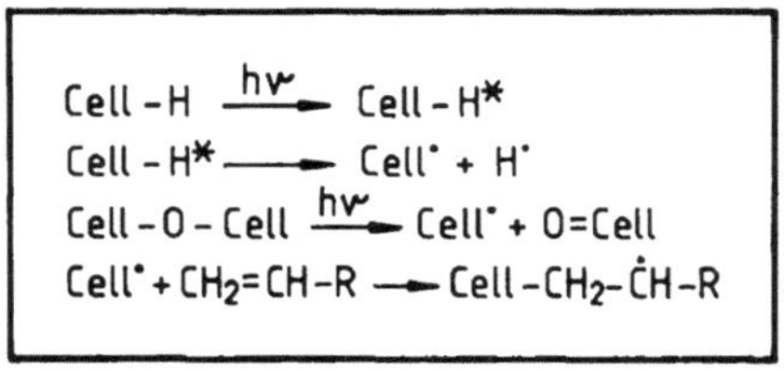

Fig. 17–34. Initiation of grafting by irradiation in the absence of oxygen.

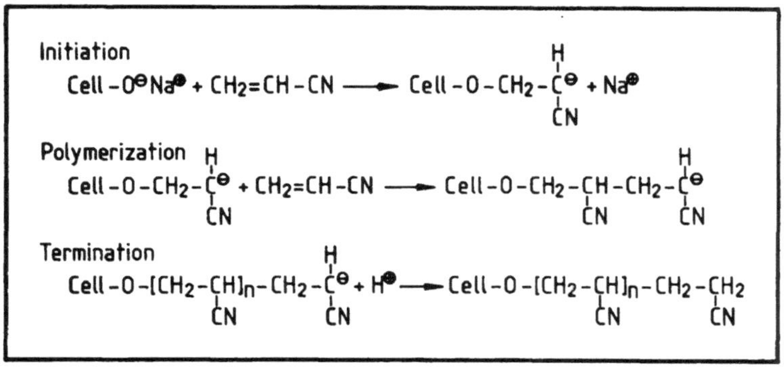

Fig. 17–35. Reaction steps of ionic polymerization.

Literature

Aboul-Seoud, A. 1968, Faserforsch. Textiltechn. *19,* 228–230
Akim, E. 1967, Pure Appl. Chem. *14,* 475–480
Alince, B. 1976, Svensk Papperstid. *79,* 259–262
Anon. 1970, Walsroder Collodiumwolle. Wolff Walsrode
Atlas, S. M. and Mark, H. 1969, Cell. Chem. Technol. *3,* 325–345
Avny, Y. and Rebenfeld, L. 1968a, Text Res. J. *38,* 599–605
Avny, Y. and Rebenfeld, L. 1968b, Text Res. J. *38,* 684–692
Balser, K. and Iseringhausen, M. 1975, Celluloseäther. In: Ullmanns Encyklopädie der technischen Chemie, 4th Ed., Vol. 9. Verlag Chemie, Weinheim, pp. 192–212
Balser, K. and Szablikowski, K. 1981, Papier *35,* 578–585
Bandel, W. 1960, Holzforschung *14,* 111–117
Bandurjan, S., Iovleva, M. and Papkov, S. 1971, Faserforsch. Textiltechn. *22,* 477–481
Bär, H. J., Dautzenberg, H. and Philipp, B. 1966, Faserforsch. Textiltechn. *17,* 551–558
Barsha, J. 1954, Inorganic Esters. In: Cellulose and Cellulose Derivatives (Ott, E., Spurling, H. M. and Grafflin, M. W., Eds.). Intersci. Publ., New York, London, pp. 713–762
Baumann, H., Moradi, A. R. and Korte, E. 1981, Papier *35,* 563–570
Belfort, A. M. and Wortz, R. B. 1966, Ind. Eng. Chem. Prod. Res. Dev. *5,* 115–118
Bennett, C. F. and Timell, T. E. 1955, Svensk Papperstid. *58,* 281–286
Bikales, N. M. 1971, Ethers from α, β-Unsaturated Compounds. In: Cellulose and Cellulose Derivatives (Bikales, N. M. and Segal, L., Eds.). Wiley-Intersci., New York, London, Sydney, Toronto, pp. 811–833
Bikales, N. M. and Segal, L. 1971, Cellulose and Cellulose Derivatives. Wiley-Intersci., New York, London, Sydney, Toronto
Bischoff, K. H. and Philipp, B. 1966, Faserforsch. Textiltechn. *17,* 115–121, 395–400
Bittiger, H. and Husemann, E. 1964, Makromol. Chem. *75,* 222–224, *80,* 239–241
Bittiger, H., Husemann, E. and Kuppel, A. 1969, Cell. Chem. Technol. *3,* 477–488
Björndal, H., Lindberg, B. and Rosell, K. G. 1971, J. Polym. Sci., Pt. C, No. 36, 523–527
Blouin, F. A., Cannizzaro, A. M., Arthur, J. C. and Rollins, M. L. 1968, Text. Res. J. *38,* 811–816
Bobinski, J. and Carignan, P. 1967, J. Appl. Polym. Sci. *11,* 727–734
Borgen, G. 1963, Nor. Skogind. *17,* 194–199
Bredereck, K. 1970, Papier *24,* 921–926
Bredereck, K. and Dau, T. B. P. 1980, Angew. Makromol. Chem. *89,* 167–181
Bredereck, K. and Vlachopoulos, G. 1980, Angew. Makromol. Chem. *84,* 81–96
Brestkin, J. V. and Čočieva, M. M. 1968, Sow. Beitr. Faserforsch. Textiltechn. *5,* 70–75
Brown, W., Henley, D. and Öhman, J. 1963a, Makromol. Chem. *62,* 164–182
Brown, W., Henley, D. and Öhman, J. 1963b, Makromol. Chem. *64,* 49–67
Bu, K. J., Dumitriu, S. and Simionescu, C. I. 1977, Cell. Chem. Technol. *11,* 655–660
Burchard, W. and Husemann, E. 1961, Makromol. Chem. *44–46,* 358–387
Chanzy, H. D. and Roche, E. J. 1974, J. Polym. Sci., Polym. Phys. Ed. *12,* 2583–2586
Chanzy, H., Côté, W. A. and Marchessault, R. H. 1968, Text. Res. J. *83,* 583–591
Chanzy, H., Roche, E. and Vuong, R. 1971, Kolloid-Z. Z. Polym. *248,* 1034–1035
Chanzy, H. D., Roche, E. J. and Vuong, R. K. 1973, J. Polym. Sci., Polym. Phys. Ed. *11,* 1859–1861
Chanzy, H., Revol, J. F., Guiroy, G. and Quere, J. 1975, Svensk Papperstid. *78,* 258–260
Chédin, J. and Marsaudon, A. 1955, Makromol. Chem. *15,* 115–160
Claus, W. and Schmiedeknecht, H. 1966, Faserforsch. Textiltechn. *17,* 404–411
Clermont, L. P. and Bender, F. 1972, J. Polym. Sci., Pt. A, *10,* 1669–1677
Clermont, L. P. and Manery, N. 1974, J. Appl. Polym. Sci. *18,* 2773–2784
Coalson, R. L., Day, A. and Marchessault, R. H. 1968, Holzforschung *22,* 190–198

Conca, R. J., Hamilton, J. K. and Kircher, H. W. 1963, Tappi *46,* 644–648
Cooper, W. and Smith, R. K. 1960, Makromol. Chem. *40,* 148–160
Cornell, R. H. 1962, Tappi *45,* No. 7, 145 A–150 A
Croon, I. 1960, Svensk Papperstid. *63,* 247–257
Croon, I. and Purves, C. B. 1959, Svensk Papperstid. *62,* 876–882
Danhelka, J. and Kössler, I. 1976, J. Polym. Sci., Polym. Chem. Ed. *14,* 287–298
Dankovics, A., Erdélyi, J. and Koltai, M. 1969, J. Appl. Polym. Sci. *13,* 1809–1824
Dautzenberg, H. and Philipp, B. 1974, Faserforsch. Textiltechn. *25,* 521–524
Dautzenberg, H., Linow, K. J. and Philipp, B. 1972, Faserforsch. Textiltechn. *23,* 141–148
Dautzenberg, H., Dautzenberg, H. and Linow, K. J. 1978, Faserforsch. Textiltechn. *29,* 593–598
Dautzenberg, H., Philipp, B. and Zöbisch, B. 1980, Zellst. Pap. *29,* 156–159
Delsemme, A. 1967, Pure Appl. Chem. *14,* 481–487
Dietl, A. and Schwarzhans, K. H. 1966, Papier *20,* 609–615
Dimov, K. and Lazariva, R. 1968, Faserforsch. Textiltechn. *19,* 68–75
Dimov, K. and Pawlov, P. 1969, Papier *23,* 809–814
Dolby, I. and Samuelson, O. 1965, Svensk Papperstid. *68,* 136–140
Durso, D. F. 1976, Svensk Papperstid. *79,* 50–51
Easterwood, M. and Mueller, W. A. 1960, J. Appl. Polym. Sci. *4,* 16–24
Eicher, T. and Fischer, W. 1975, Celluloseester. In: Ullmanns Encyklopädie der technischen Chemie, 4th Ed., Vol. 9. Verlag Chemie, Weinheim, pp. 227–246
Elliott, J. H. 1969, J. Appl. Polym. Sci. *13,* 755–764
Elmgren, H. 1965, Ark. Kemi *24,* 237–282
Fahmy, Y. and El-Kalyoubi, S. 1970, Cell. Chem. Technol. *4,* 613–619
Fahmy, Y. A. and Fadl. M. H. 1964, Svensk Papperstid. *67,* 101–109, 279–285
Fahmy, Y. and Koura, A. 1967, Cell. Chem. Technol. *1,* 301–312
Fahmy, Y. and Mansour, O. 1966, Indian Pulp Pap. *21,* 143–145, 148–151, 627–639
Fahmy, Y. A., Mustafa, A. and Fadl, M. H. 1964, Svensk Papperstid. *67,* 573–578
Feit, B. A., Bar-nun, A., Lahav, M. and Zilkha, A. 1964, J. Appl. Polym. Sci. *8,* 1869–1888
Forsén, S., Garegg, P. J., Lindberg, B. and Petterson, E. 1966, Acta Chem. Scand. *20,* 2763–2770
Frieberg, F., Brown, W., Henley, D. and Öhman, J. 1963, Makromol. Chem. *66,* 168–179
Friese, P. 1981, Fresenius Z. Anal. Chem. *305,* 337–346
Gagnon, P. E., Newbold, B. T. and Thomas, J. 1961, Tappi *44,* 749–752
Gardner, P. E. and Chang, M. Y. 1974, Tappi *57,* No. 8, 71–75
Garves, K. 1972, Tappi *55,* 263
Garves, K. 1979, Cell. Chem. Technol. *13,* 299–305
Geiger, E. and Schneider, W. 1966, Faserforsch. Textiltechn. *17,* 487–492
Gipstein, E. and Wellisch, E. 1973, J. Appl. Polym. Sci. *17,* 2783–2790
Gohlke, B., Schmiedeknecht, H. and Claus. W. 1963, Faserforsch. Textiltechn. *14,* 51–58
Golubev, V. M., Vasilev, B. V., Bilimova, E. S. and Sokolova, L. N. 1966, Sow. Beitr. Faserforsch. Textiltechn. *3,* 577–580
Goring, D. A. I. and Timell, T. E. 1962, Tappi *45,* 454–460
Greber, G. and Paschinger, O. 1981, Papier *35,* 547–554
Grigorjan, R. G. and Rogovin, Z. A. 1968, Sow. Beitr. Faserforsch. Textiltechn. *5,* 27–29
Gröbe, A. 1968, Svensk Papperstid. *71,* 636–645
Gröbe, A., Maron, R. and Purz, H. J. 1966, Faserforsch. Textiltechn. *17,* 457–461
Hagiwara, I., Shiraishi, N., Yokota, T., Norimoto, M. and Hayashi, Y. 1981, J. Wood Chem. Technol. *1,* 93–109
Hall, D. M. and Horne, J. R. 1973a, J. Appl. Polym. Sci. *17,* 3727–3732
Hall, D. M. and Horne, J. R. 1973b, J. Appl. Polym. Sci. *17,* 2891–2896
Hatakeyama, H. and Rånby, B. 1975, Cell. Chem. Technol. *9,* 583–596
Hebeish, A. and Mehta, P. C. 1968, J. Appl. Polym. Sci. *12,* 1625–1647

Herdle, L. E. and Griggs, W. H. 1965, Tappi *48,* No. 7, 103 A–107 A
Hergert, H. L., Hammer, R. B. and Turbak, A. F. 1978, Tappi *61,* No. 2, 63–67
Hernadi, S. and Völgyi, P. 1974, Cell. Chem. Technol. *8,* 89–99
Hiatt, G. D. and Rebel, W. J. 1971, Esters. In: Cellulose and Cellulose Derivatives (Bikales, N. M. and Segal, L., Eds.). Wiley-Intersci., New York, London, Sydney, Toronto, pp. 741–784
Holst. A. 1978, Papier *32,* V 7–V 12
Holt, C., Mackie, W. and Sellen, D. B. 1976, Polymer *17,* 1027–1034
von Horstig, W. 1969, Papier *23,* 397–405
Hovenkamp, S. G. 1966, Faserforsch. Textiltechn. *17,* 100–105, 139–142
Howard, P. and Parikh, R. S. 1966, J. Polym. Sci., Pt. A–1, *4,* 407–418
Husemann, E. and Siefert, E. 1969, Makromol. Chem. *128,* 288–291
Ishizu, A., Isogai, A. and Nakano, J. 1981, Synthesis of Cellulose Derivatives in Homogeneous Systems, Acetylation in a Dimethylformamide-Chloral-Pyridine Mixture. In: The Ekman Days 1981, Int. Symp. Wood Pulp. Chem., Stockholm, Vol. 4, pp. 121–124
Iovleva, M. M. and Papkov, S. P. 1968, Sow. Beitr. Faserforsch. Textiltechn. *5,* 300–301
Iovleva, M. M., Bandurjan, S. I. and Papkov, S. P. 1968, Sow. Beitr. Faserforsch. Textiltechn. *5,* 466–488
Jašunskaja, A. G. and Konovalova, E. M. 1968, Sow. Beitr. Faserforsch. Textiltechn. *5,* 297–299
Jayme, G. and Balser, K. 1965, Papier *19,* 572–581
Jayme, G. and Balser, K. 1967, Pure Appl. Chem. *14,* 395–408
Jayme, G. and El-Kodsi, G. 1972, Papier *26,* 590–599
Jeffries, R. 1963, Polymer *4,* 375–389
Jost, H. and Ludwig, J. 1966, Faserforsch. Textiltechn. *17,* 194–201
Jost, H. and Ludwig, J. 1967, Faserforsch. Textiltechn. *18,* 274–278
Jovanovic, S., Palma, A. and Schulz, G. V. 1974, Makromol. Chem. *175,* 3503–3522
Jullander, I. 1965, Papier *19,* 166–168, 224–229
Katsuura, K. and Fujinami, T. 1968, J. Chem. Soc. Japan Ind. Chem. Sect. *71,* No. 5, 771–772
Kehren, M. L. and Reichle, A. 1975, Cellulosefasern, Chemiefasern. In: Ullmanns Encyklopädie der technischen Chemie, 4th Ed., Vol. 9. Verlag Chemie, Weinheim, pp. 213–226
Kiefer, J. E. and Touey, G. P. 1965, Ind. Eng. Chem. Prod. Res. Dev. *4,* 253–256
Kleinert, T. N. 1956, Pulp Pap. Mag. Can. *57,* 91–94, 102
Kleinert, T. N. 1972, Holzforsch. Holzverw. *24,* 61–66
Klug, E. D. 1971, J. Polym. Sci., Pt. C, No. 36, 491–508
Kobayashi, Y. 1961, J. Polym. Sci. *51,* 359–372
Kokta, B. V. and Valade, J. L. 1972, Tappi *55,* 366–369
Kössler, I., Daňhelka, J., Netopilík, M., Samková, M. and Katuščáková, G. 1981, Svensk Papperstid. *84,* R 137–R 140
Koura, A. 1977, Faserforsch. Textiltechn. *28,* 417–419
Koura, A. and Ibrahum, A. 1974, Faserforsch. Textiltechn. *25,* 57–60
Koura, A., Lukanoff, B., Philipp, B. and Schleicher, H. 1977, Faserforsch. Textiltechn. *28,* 63–65
Krässig, H. 1971, Svensk Papperstid. *74,* 417–428
Krässig, H. A. and Stannett, V. 1965, Adv. Polym. Sci. *4,* 111–156
Krkoška, P. and Reiser, V. 1971, Cell. Chem. Technol. *5,* 437–442
Laamanen, L. and Sihtola, H. 1964, Pap. Puu *46,* 159–166
Lawrence, K. D. N. and Verdin, D. 1973, J. Appl. Polym. Sci. *17,* 2653–2666
Lawton, J. B. and Phillips, G. O. 1975, Text. Res. J. *45,* 4–13
Lieser, T. 1953, Kurzes Lehrbuch der Cellulosechemie. Gebr. Borntraeger, Berlin
Lukanoff, B. and Philipp, B. 1967, Faserforsch. Textiltechn. *18,* 407–413
Lukanoff, B., Schleicher, H. and Philipp, B. 1974, Faserforsch. Textiltechn. *25,* 339–341
Lukanoff, B., Philipp, B. and Schleicher, H. 1979, Cell. Chem. Technol. *13,* 417–427
Lyselius, A. and Samuelson, O. 1961, Svensk, Papperstid. *64,* 735–739, 815–818

Malm, C. J. and Hiatt, G. D. 1954, Organic Esters. In: Cellulose and Cellulose Derivatives (Ott, E., Spurlin, H. M. and Grafflin, M. W., Eds.). Intersci. Publ. New York, London, pp. 763–824
Malm, C. J., Mench, J. W., Kendall, D. L. and Hiatt, G. D. 1951, Ind. Eng. Chem. *43,* 688–691
Malm, C. J., Glegg, R. E., Tanghe, L. J. and Thompson, J. 1961a, Tappi 44, 669–678
Malm, C. J., Tanghe, L. J. and Schmitt, J. T. 1961b, Ind. Eng. Chem. *53,* 363–367
Manley, R. S. J. 1963, J. Polym. Sci., Pt. A, *1,* 1875–1892
Manning, J. H. and Fuller, I. H. 1969, Tappi *52,* 612–616
Mansour, O. Y. and Schurz, J. 1973, Svensk Papperstid. *76,* 415–418
Månsson, P. and Westfelt, L. 1980, Cell. Chem. Technol. *14,* 13–17
Maron, R., Gröbe, A. and Purz, H. J. 1967, Svensk Papperstid. *70,* 715–718
Marx-Figini, M. 1961, Makromol. Chem. *50,* 196–219
Mašura, V. 1974, Cell. Chem. Technol. *8,* 21–25
Mattor, J. A. 1963, Tappi *46,* 586–591
Mench, J. W., Fulkerson, B. and Hiatt, G. D. 1966, Ind. Eng. Chem. Prod. Res. Dev. *5,* 110–115
Michie, R. I. C. 1966, Pulp. Pap. Mag. Can. *67,* T 461–T 468
Millett, M. A., Seborg, R. M., Zoch, L. L. and Masuelli, F. J. 1961, Tappi *44,* 636–647
Mislovičová, D. and Pašteka, M. 1974, Cell. Chem. Technol. *8,* 481–486
Mobarak, F. 1982, personal communication
Morehead, F. F. 1963, Tappi *46,* 524–530
Morin, B. P., Livšic, R. M. and Rogovin, Z. A. 1967, Sow. Beitr. Faserforsch. Textiltechn. *4,* 410–414
Morin, B. P., Stančenko, G. I., Samojlov, V. I. and Rogovin, Z. A. 1975, Faserforsch. Textiltechn. *26,* 382–387
Nair, G. P. and Mehta, P. C. 1967, Cell. Chem. Technol. *1,* 137–152
Nakamura, Y., Negishi, M. and Kakinuma, T. 1969, J. Polym. Sci., Pt. C, No. 23, 629–645
Nelson, P. F., and Kalkipsakis, C. G. 1964, Tappi *47,* 107–110
Ogiwara, Y. and Kubota, H. 1969, J. Appl. Polym. Sci. *13,* 1613–1620
Olaru, N., Steinberg, S., Schneider, E. and Asandai, N. 1978, Cell. Chem. Technol. *12,* 373–379
Ott, E., Spurlin, H. M. and Graffling, M. W. 1954, Cellulose and Cellulose Derivatives. Intersci. Publ., New York, London, Vol. 1, pp. 673–1026
Papkov, S. P., Uchanova, Z. V. and Michajlov, N. V. 1966, Sow. Beitr. Faserforsch. Textiltechn. *3,* 355–360
Pašteka, M. 1967, Cell. Chem. Technol. *1,* 689–698
Pastyr, J. and Kuniak, L. 1971, Cell. Chem. Technol. *5,* 17–22
Pastyr, J. and Kuniak, L. 1972, Cell. Chem. Technol. *6,* 249–254
Patel, G. N. and Patel, R. D. 1969, Polymer *10,* 932–938
Patscheke, G. and Poller, S. 1980, Cell. Chem. Technol. *14,* 3–12
Petropavlovskij, G. 1973, Faserforsch. Textiltechn. *24,* 49–57
Philipp, B. 1981, Holztechnologie *22,* 108–112
Philipp, B. and Chu, C. H. 1965, Faserforsch. Textiltechn. *16,* 244–249, 349–352
Philipp, B. and Dautzenberg, H. 1960, Faserforsch. Textiltechn. *11,* 374–382
Philipp, B. and Linow, K. J. 1965, Zellst. Pap. *14,* 321–326
Philipp, B., Bischoff, K. H. and Loth, F. 1979, Cell. Chem. Technol. *13,* 23–33
Phillips, R. B., Quere, J., Guiroy, G. and Stannett, V. T. 1972, Tappi *55,* 858–867
Plungian, M. 1961, Atip Bull. *1961,* 277–286
Pohjola, L., Riala, R. and Tammela, V. 1976, Pap. Puu *58,* 198–200
Poller, S. 1968, Faserforsch. Textiltechn. *19,* 124–129
Poller, S. and Patscheke, G. 1969, Cell. Chem. Technol. *3,* 489–498
Poller, S. and Patscheke, G. 1970, Faserforsch. Textiltechn. *21,* 8–13
Poller, S. and Patscheke, G. 1971, Faserforsch. Textiltechn. *22,* 310–315
Poller, S. and Patscheke, G. 1979, Holztechnologie *20,* 216–219

Predvoditelev, D. A. and Bakšeeva, M. S. 1972, Sow. Beitr. Faserforsch. Textiltechn. *9,* 361–364
Ramnäs, O. and Samuelson, O. 1968, Svensk Papperstid. *71,* 829–831
Ramnäs, O. and Samuelson, O. 1973, Svensk Papperstid. *76,* 569–571
Rånby, B. 1978, New Methods for Graft Copolymerization onto Cellulose and Starch. In: Modified Cellulosics (Rowell, R. M. and Young, R. A., Eds.). Academic Press, New York, San Francisco, London, pp. 171–195
Rånby, B. 1981, Chemical Mechanism of Graft Copolymerization to Cellulosic Fibers. In: The Ekman Days 1981, Int. Symp. Wood Pulp. Chem., Stockholm, Vol. 4, pp. 111–117
Rao, S. R. and Kapur, S. L. 1969, J. Appl. Polym. Sci. *13,* 2649–2656
Rausing, G. and Sunner, S. 1962, Tappi *45,* No. 1, 203 A–205 A
Rebek, M. and Jurkowitsch, B. 1975, Papier *29,* 89–93
Rebek, M. and Jurkowitsch, B. 1977, Papier *31,* 54–58, 372–374
Rebek, M., Schurz, J. and Popp, W. 1967a, Holzforschung *21,* 58–63
Rebek, M., Schurz, J. and Spörk, H. 1967b, Monatsh. Chem. *98,* 1161–1173
Rebek, M., Wickenhauser, R. and Jurkowitsch, B. 1973, Papier *27,* 255–257
Rijke, A. M. and Prins, W. 1962, J. Polym. Sci. *59,* 171–190
Roche, E. 1977, Aspects du Polymorphism de la Cellulose et du Triacetate de Cellulose Relation entre la Structure moleculaire et la Morphologie cristalline. Doctor Thesis, Université de Grenoble
Rogovin, Z. A. 1960, J. Polym. Sci. *48,* 443–456
Rogovin, Z. A. 1967, Pure Appl. Chem. *14,* 523–533
Rogovin, Z. A. 1974, Tappi *57,* No. 7, 65–68
Ruscher, C. and Zuchold, H. 1964, Faserforsch. Textiltechn. *15,* 626–631
Sakurada, I., Okada, T. and Kaji, K. 1972a, J. Polym. Sci., Pt. C, No. 37, 1–25
Sakurada, I., Okada, T. and Ikada, Y. 1972b, Cell. Chem. Technol. *6,* 35–48
Samadjieva, V. and Dimov, K. 1971, Cell. Chem. Technol. *5,* 443–449
Samuels, R. J. 1969, J. Polym. Sci., Pt. A-2, No. 7, 1197–1258
Sankalia, S. M., Chaudhuri, D. K. R. and Hermans, J. J. 1962, Can. J. Chem. *40,* 2249–2255
Savage, A. B. 1970, Papier *24,* 916–920
Savage, A. B. 1971, Ethers. In: Cellulose and Cellulose Derivatives (Bikales, N. M. and Segal, L., Eds.). Wiley Intersci., New York, London, Sydney, Toronto, pp. 785–809
Scherer, P. C., Tanenbaum, A. and Levi, D. W. 1960, J. Polym. Sci. *43,* 531–535
Schleicher, H., Lukanoff, B. and Philipp, B. 1974, Faserforsch. Textiltechn. *25,* 179–183
Schmiedeknecht, H. and Claus, W. 1966, Faserforsch. Textiltechn. *17,* 276–282
Schroeder, L. R. and Haigh, F. C. 1979, Tappi *62,* No. 10, 103–105
Schurz, J. and Tritthart, H. 1966, Polymer *7,* 475–477
Schweiger, R. G. 1974, Tappi *57,* No. 1, 86–90
Schwenker, R. F. and Pascu, E. 1963, Tappi *46,* 665–672
Schwenker, R. F., Kinoshita, T., Beuling, K. and Pascu, E. 1961, J. Polym. Sci. *51,* 185–198
Seymour, R. B. and Johnson, E. L. 1978, J. Polym. Sci., Polym. Chem. Ed. *16,* 1–11
Shiraishi, N., Katayama, T. and Yokota, T. 1978, Cell. Chem. Technol. *12,* 429–443
Sihtola, H. and Laamanen, L. 1964, Papier *18,* 600–606
Sihtola, H. and Nizovsky, B. 1969, Tappi *52,* 501–504
Sihtola, H. and Rantanen, T. 1972, Cell. Chem. Technol. *6,* 71–83
Sihtola, H., Blomberg, L. and Laamanen, L. 1967, Pure Appl. Chem. *14,* 547–554
Simionescu, C. and Mihailescu, S. 1970, Cell. Chem. Technol. *4,* 23–35
Simionescu, C. and Ungureanu, C. 1967, Cell. Chem. Technol. *1,* 33–44
Simionescu, C., Uglea, C. and Feldman, D. 1967, Cell. Chem. Technol. *1,* 199–209
Sjöström, E. 1981, Wood Chemistry, Fundamentals and Applications. Academic Press, New York, London, Toronto, Sydney, San Francisco
Smirnova, G. N., Galbraich, L. S. and Rogovin, Z. A. 1967, Cell. Chem. Technol. *1,* 11–22
Stannett, V. T. and Hopfenberg, H. B. 1971, Graft Copolymers. In: Cellulose and Cellulose Derivatives

(Bikales, N. M. and Segal, L., Eds.). Wiley Intersci., New York, London, Sydney, Toronto, pp. 907–936
Stipanovic, A. J. and Sarko, A. 1978, Polymer *19,* 3–8
Tanghe, L. J. and Brewer, R. J. 1968, Anal. Chem. *40,* 350–353
Tesoro, G. C. and Willard, J. J. 1971, Crosslinked Cellulose. In: Cellulose and Cellulose Derivatives (Bikales, N. M. and Segal, L., Eds.). Wiley Intersci., New York, London, Sydney, Toronto, pp. 835–875
Thinius, K. and Thümmler, W. 1967, Makromol. Chem. *99,* 117–125
Timell, T. E. 1956, Svensk Papperstid. *59,* 1–11
Timell, T. E. 1957, Svensk Papperstid. *60,* 836–842
Timell, T. E. 1963, J. Polym. Sci., Pt. C., No. 2, 109–116
Törnell, B. 1966, Svensk Papperstid. *69,* 658–663, 695–702
Törnell, B. 1967, Svensk Papperstid. *70,* 449–457, 489–499
Treiber, E. 1966, Chemiefasern *16,* 757–762
Turunen, J., Masalin, E., Sjöholm, R. and Bruun, H. 1969, Pap. Puu *51,* 803–811
Uhlig, E. and Teichmann, R. 1968, Faserforsch. Textiltechn. *19,* 238
Uhlig, E. and Teichmann, R. 1971, Faserforsch. Textiltechn. *22,* 610–617
Uhlig, E. and Teichmann, R. 1972, Faserforsch. Textiltechn. *23,* 148–156
Usmanov, K. U., Akramova, D., Ganiev, B. Z., Khakimov, I. K. and Yunusov, M. V. 1966, Strukt. Modif. Khlopk. Tsellyul., No. 3, 108–117
Valtasaari, L. and Saarela, K. 1975, Pap. Puu *57,* 5–10
Vasilev, A. V. and Majboroda, V. I. 1966, Sow. Beitr. Faserforsch. Textiltechn. *3,* 352–355
Vasilev, A. V. and Majboroda, V. I. 1967, Sow. Beitr. Faserforsch. Textiltechn. *4,* 11–14
Volynskij, A. L., Orlova, T. M., Bakeev, N. F. and Kargin, V. A. 1970, Sow. Beitr. Faserforsch. Textiltechn. *7,* 353–355
Wängberg, L. and Treiber, E. 1968, Svensk Papperstid. *71,* 621–635
Ward, K., Murray, M. L., Morak, A. J. and Voelker, M. H. 1967, Tappi *50,* 462–466
Ward, k., Morak, A. J., Gillesie, R. H. and Voelker, M. H. 1968, Tappi *51,* 218–223
Ward, K., Swenson, H. A., Wink, W. A. and Greisinger, R. J. 1969, Tappi *52,* 2336–2341
Wellons, J. D. and Stannett, V. 1965, J. Polym. Sci., *3 A,* 847–857
Wellons, J. D., Williams, J. L. and Stannett, V. 1967, J. Polym. Sci., Pt k-1, *6,* 1341–1357
Wurz, O. 1961, Celluloseäther – Herstellung und Anwendung. Roether, Darmstadt
Yasukawa, T., Sasaki, Y. und Murakami, K. 1972, Makromol. Chem. *153,* 323–326
Zipper, P., Krigbaum, W. R. and Kratky, O. 1969, Kolloid. Z. *235,* 1281–1287

18. Utilization of Wood and Wood Components for Chemicals and Energy

18.1. Introductory Remarks

In recent years a world-wide increase of interest can be observed in the field of chemical and energetic utilization of wood and biomass (e.g. Schweers 1975; Goldstein 1975a, 1979, 1980a, b; Saeman 1977; Büsing et al. 1978; Sarkanen, Tillman 1979; Bratt 1979; Frühwald, Liese 1980; Weismann 1980; Hajny 1981; Falkehag 1981; Wegener 1982). The general term biomass – in the broadest sense any mass produced by biosynthesis – is restricted within this chapter to wood and forestry biomass, though many processes can also be applied in principle to agricultural plants and residues.

The alternative utilization of renewable plant material will be a necessary future task for industrial and developing countries, though obviously each country will face its own special problems in this regard. Two general aims of utilization are:

- production of chemical and energetic products from renewable sources to replace expensive energy and petrochemicals from exhausted reserves of the fossil raw materials oil, gas and coal
- better utilization of waste material from forestry, the wood and pulp industries, and of discarded wood products. There are economic reasons for using the valuable raw material wood more efficiently and for facing the stricter requirements of environmental control.

Within this chapter a short survey is given of the present status and future possibilities of wood utilization in this field without emphasis on technical aspects. More detailed discussions are found in the cited literature and particularly in a recently published book which reviews this topic comprehensively (Goldstein 1981).

The principal routes of direct wood and plant biomass conversion are shown in Fig. 18–1. The best-known and simplest process is combustion, directly yielding heat. Other thermal procedures are carbonization (pyrolysis), gasification and liquefaction, resulting in fuels and chemicals, respectively. In wood saccharification the main reaction is the hydrolysis of the polysaccharides cellulose and polyoses (hemicelluloses) to monomeric units for further utilization (e.g. fermentation of glucose to ethanol, → 18.4., 18.5.). Lignin remains as an acid-insoluble residue. The most important principle of wood conversion is defibration by chemical and/or mechanical treatments to yield fibrous material and the resulting liquors and wastes. The pulping processes and the production of cellulose fibres and derivatives are discussed in the chapters 16 and 17, respectively.

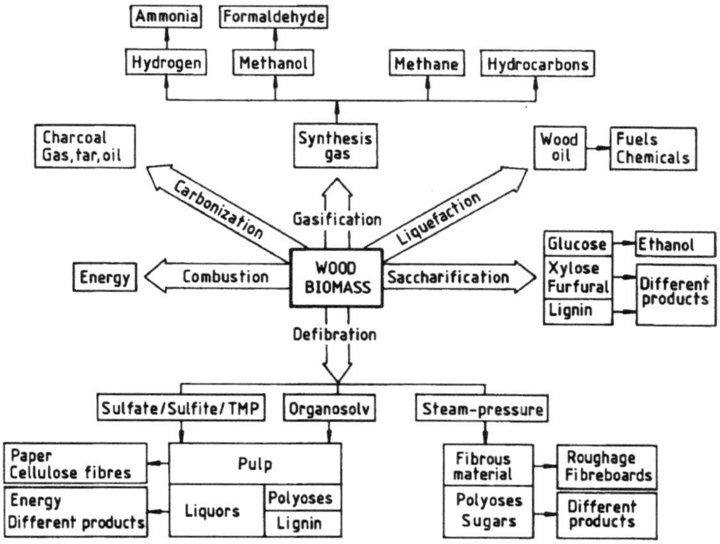

Fig. 18–1. Thermal and chemical techniques for direct wood and biomass conversion

18.2. Thermal Degradation

18.2.1. Combustion

The combustion process produces heat directly in different practical yields depending on the combustion system used. The released energy can be quantified by the calorific value which has an average value of 19 MJ/kg for absolutely dry wood (Voeste 1981). The practical calorific values are predominantly determined by the water content and to a lesser extent by the amount of ash and extractives as well as by the size of the combusted material (e.g. saw dust, chips, logs). Advantages of wood as a combustion material are generally low ash and extremely low sulfur content (Ince 1979; Kollmann 1981).

The wood used for combustion often comprises considerable amounts of bark. The calorific values of barks from different tree species are on an average somewhat higher than the values for wood (18.7–22.7 MJ/kg) (Harder, Einspahr 1978). In practical bark combustion the energy input for the predrying of the generally wet bark is high. Additionally higher ash contents than in wood and dirt reduce the calorific values in practical bark combustion.

18.2.2. Pyrolysis (Carbonization)

The conversion of wood to charcoal is one of the oldest processes used by mankind. Today the technology of charcoal production is important in the industrial and developing countries (Brocksiepe 1976; Paddon, Harker 1980).

The pyrolytic decomposition of wood in the absence of air or oxygen with final temperatures of about 500 °C yields three general groups of substances (→ 12.5.):

- solid components
- volatile, condensable compounds
- volatile, non-condensable gases.

The most important product of carbonization is charcoal. The estimated world-wide industrial production has a magnitude of some 2.4 Mio. t (Büsing et al. 1978). The average practical yields in industrial charcoal production are about 35%, related to wood, and depend on factors such as wood species and wood size, carbonization system, processing time and final temperature. Important criteria for the utilization of charcoal are the properties listed in Table 18–1.

Beech wood *(Fagus sylvatica)* is the dominant raw material for the production of charcoal in Europe. Minor amounts of oak (*Quercus* spec.), ash (*Fraxinus* spec.), alder (*Alnus* spec.) and maple (*Acer* spec.) are also used. In North America, apart from the above-mentioned species, hickory (*Carya* spec.), elm (*Ulmus* spec.), sycamore (*Platanus* spec.) and some softwoods are in use. In South America and South Africa charcoal is produced mainly from eucalyptus (*Eucalyptus* spec.) (Brocksiepe 1976). The behaviour of various tropical wood species during pyrolysis at different conditions was investigated by Petroff and Doat (1978).

Industrial carbonization is carried out predominantly in large-volume retorts with capacities up to 100 m^3 of wood. Different discontinuous and continuous systems are in use (e.g. DEGUSSA system, SIFIC system). Detailed descriptions of the state of industrial charcoal technologies are given elsewhere (Brocksiepe 1976; Büsing et al. 1978; Welling 1979; Marutzky 1981).

Charcoals, with their different properties, offer a wide range of applications. The market is divided into private use (barbecue coal and briquettes) and industrial

Table 18–1: Important properties of charcoal (according to Brocksiepe 1976)

Density	0.45 g/cm^3 (beech)
Net density	1.38–1.46 g/cm^3
Porosity	70%
Inner surface	50 m^2/g
Strength properties	Compressive strength $\parallel$: 26 N/mm^2 $\perp$: 6 N/mm^2, rad. $\perp$: 2 N/mm^2, tang.
Bulk weight	180–220 kg/m^3
Moisture content	5–8%
Carbon content	80–90%
Ash content	1–2%
Calorific value	29–33 MJ/kg
Volatiles	10–18%

applications. In the latter case charcoal is often processed to activated carbon which is used e.g. in metallurgy, water purification, chemical synthesis and various other purposes (Anonymous 1978a). Apart from charcoal, gas, tar and oil, wood vinegar and wood alcohol are typical products of wood carbonization (Wienhaus 1979). The yields of these products depend on the composition of the starting material and especially on the pyrolysis conditions. Due to the high oxygen and hydrogen content of wood and lignocellulosic material the proportions of liquid and gaseous pyrolysis compounds are much higher than in the case of coal pyrolysis.

The crude pyrolytic tar and oil fractions are very complex mixtures composed mainly of light and heavy oil components which can be used for impregnation purposes and in medical applications. After distillation a wood pitch is received as residue. The heavy oil fraction can be processed to creosot. The main component of this product is guaiacol, which is used for pharmaceutical purposes because of its antiseptic properties. The phenolic constituents of the pyrolytic tar may also be applied as adhesives for plywood (Soltes 1980). The wood alcohol (wood spirit) contains about 60% methanol with a couple of impurities (→ 12.5.). It is used as a solvent and for denaturation of ethanol. The wood vinegar fraction (→ 12.5.) can be purified to acetic acid and food vinegar. The decision whether or not to purify and utilize these co-products depends on economical considerations and especially on waste water regulations (Büsing et al. 1978). The non-condensable wood gas (composed of carbon dioxide, carbon monoxide, methane, hydrogen and other hydrocarbons) with a calorific value of about 8.9 MJ/m^3 is used in industrial charcoal production for predrying the wood feed and as a flush gas for retorts (Brocksiepe 1976).

For wood and lignocellulosic residues, also in mixture with municipal solid waste and other organic waste material, pyrolytic processes applying higher temperatures up to 800 °C are in use. In this case mainly low-molecular-weight compounds and higher proportions of gases are produced (Shafizadeh et al. 1976; Maloney 1978; Knight 1979).

18.2.3. Gasification

Wood gasification at temperatures of about 1 000 °C yields a wood gas whose composition depends on the process conditions and the water content of the starting material. The gasification can be carried out in principle as a pyrolytic process or in the presence of air, oxygen, and with additional steam. In the case of wood and lignocellulosic material high amounts of oxygen and hydrogen cause very difficult reactions resulting in a more complex composition of the gas than in the case of coal or municipal solid waste gasification. Though the principles of gasification are well-known from coal gasification the process design of gasifier systems for wood and heterogeneous biomass is still under development in pilot plants, however ready for industrial implementation (Soltes 1980).

Technological data, composition of the gases derived from different systems and the various gas purification systems are reviewed elsewhere (e.g. Brink et al. 1976; Maloney 1978; Knight 1979; Reed, Jantzen 1979; Rowell, Hokanson 1979; Marutzky 1981). General advantages of technical wood and biomass gasification are low need of oxygen and additional steam as well as low sulfur content. Table 18–2 shows the crude wood gas composition from three important gasification processes. Table 18–3 gives an example of the hydrocarbon fraction of wood gas, including important chemicals such as ethene and ethine in small amounts.

By gasification with air a so-called producer gas is received. This gas consists of carbon dioxide, carbon monoxide, methane and hydrogen besides considerable amounts of nitrogen (up to 50%). It can be used as a low-energy gas (calorific value about 5.2 MJ/m^3) for industrial boiler systems and in private burners (Goldstein 1980a). 2.5–3 kg of absolutely dry wood mass can replace about 1 l petroleum or 0.9 l diesel oil (Marutzky 1980).

Gasification in the presence of pure oxygen and additional steam yields so-called water gas, which contains only small amounts of nitrogen resulting in a much higher calorific value of about 11 MJ/m^3 (Marutzky 1980). It can also be used for energetic

Table 18–2: Composition of crude wood gas from different gasification processes

Crude gas component	Process PUROX[1] % (Vol.)	 MOORE-CANADA[2] % (Vol.)	 BATELLE[3] % (Vol.)
H_2	26.0	18.3	20.3
CO	40.0	22.8	23.7
CO_2	23.0	9.2	12.4
CH_4	5.0	2.5	2.7
O_2	0.5	0.5	0.6
N_2	0.5	45.8	40.3
Other hydrocarbons	5.0	0.9	0.6

[1] Gasification with oxygen (Rowell, Hokanson 1979)
[2] Gasification with air (Rowell, Hokanson 1979)
[3] Gasification with air (Hammond et al. 1974)

Table 18–3: Hydrocarbons in wood gas (Maloney 1978)

Hydrocarbon	% (Vol.)
Ethene	2.01
Pentanes	1.09
Ethine	0.76
Propene	0.34
Butene	0.34
Ethane	0.25
Butane	0.17
Propane	0.04

purposes, but interest largely centres on converting it into so-called synthesis gas by cleaning and enriching with hydrogen. The resulting gas is composed of carbon monoxide and hydrogen only, and is suitable for processing fuels and chemicals (Fig. 18–1). Synthesis gas was also produced from kraft and sulfite pulping liquors (Prahačs et al. 1971, → 18.6.2.). Probably the most interesting processing of synthesis gas is catalytic conversion to methanol at high temperatures (450 °C) and high pressure (200 bar) (Katzen 1978). The prerequisite for high methanol yields from wood-derived synthesis gas is high purity and a CO/H_2-ratio of close to 1:2.

Methanol is a very clean fuel which can be used directly for energetic purposes, as a petrol additive or after conversion to gasoline (Seiffert, Held 1981) or diesel oil (Anonymous 1981a). On the other hand methanol is a high-grade solvent and a valuable raw material for further chemical products such as formaldehyde, acrylics, insecticides or fungicides (Rowell, Hokanson 1979). Apart from methanol ammonia can be processed from synthesis gas via hydrogen (Prahačs et al. 1971) as well as methane or higher aliphatic hydrocarbons by catalytic conversion according to the Fischer-Tropsch process (Goldstein 1980a).

18.2.4. Liquefaction

The principal aim of wood liquefaction (hydrogenation) is to obtain high-grade oils with high yields of hydrocarbons and phenols besides gaseous compounds. The oil products are designated as feedstock for further syntheses according to the conversion techniques from petroleum to petrochemicals (PPRIC 1975; Goldstein 1980a) or for energy purposes. Up to now only laboratory results have been worked out applying various experimental conditions. In most of the studies water and/or oil are used as solvents. Hydrogen or carbon monoxide are applied in static or flow reactors at pressures up to 280 bar, temperatures in the range of 250–400 °C and in the presence of different catalysts such as Raney-nickel or sodium carbonate (Garves 1982).

Acetone was used as an organic solvent in the supercritical range (250–340 °C, 250 bar) resulting in a liquefaction of 98% of the wood sample (Köll, Metzger 1978). In the same high-pressure and high-temperature apparatus (HP-HT-flow reactor) birch wood (*Betula* spec.) was degraded in different alcohol/water mixtures at 250 °C. Using 40% ethanol as a reacting agent the process is comparable to ethanol pulping (→ 16.6.) yielding a low-DP cellulose as a residue while the lignin is totally dissolved. Applying water-to-ethanol ratios lower than 30:70 the wood dissolves completely. From the solution glucose could be recovered in yields of 70%, related to cellulose, in addition to 10–15% hydroxymethylfurfural (Köll et al. 1979; Metzger et al. 1981).

A dynamic hydrothermal degradation of poplar wood *(Populus tremuloides)* and spruce wood *(Picea abies)* was achieved in water at about 230 bar and temperatures of 150–360 °C. 94.1% of the poplar wood and 82.5% of the spruce wood were soluble with maximal degradation yields at 180 °C, 270 °C and 340 °C, which could be related to degraded polyoses, cellulose and lignin in this sequence. By this technique a selective production of sugars, furfural and phenolic lignin degradation compounds seems possible (Bobleter, Concin 1979; Bobleter, Binder 1980).

18.3. Wood Saccharification

Hydrolysis of wood polysaccharides (cellulose, polyoses) to sugars is the principal chemical reaction of the saccharification processes. The main product of saccharification is glucose derived from cellulose and partly from mannan. Acid lignin is left as a residue in addition to other by-products such as acetic acid, methanol or furfural.

Though hydrolysis is a well-known chemical reaction in the presence of acids (→ 10.2.), industrial saccharification implies many technical and economical problems e.g. with acid-stable equipments, acid recovery and final glucose yields. These problems are not yet solved in spite of the numerous processes that have been developed since the beginning of this century and applied commercially in Germany, Switzerland and the U.S.A. Industrial wood saccharification is carried out today only in the U.S.S.R. in more than 40 factories (Humphrey 1976) as it is generally considered uneconomical in the industrial countries of the western world.

Most of the processes proposed call for sulfuric or hydrochloric acid in different concentrations. Detailed descriptions of wood saccharification processes are given elsewhere (Eickemeyer, Henneke 1967; Wenzl 1970; Büsing et al. 1978; Goldstein 1980b; Hajny 1981; Stenzenberger 1981). Generally dilute acids are applied at elevated temperatures while saccharification procedures with concentrated acids are carried out at room temperature. Table 18–4 gives a short overview of some data from important processes. The only commercially tested technologies are the Scholler-Tornesch process applying dilute sulfuric acid (0.4% H_2SO_4) and the Bergius-Rheinau process using superconcentrated hydrochloric acid (41% HCl). The principal Scholler process is a discontinuous system of percolating hydrolysis steps with the dilute acid. To minimize the decomposition of the formed sugars they are removed as quickly as possible from the reaction zone. In the U.S.A. the Madison process was developed during World War II as a continuous system on the principles of the Scholler-Tornesch process (Hajny 1981).

The processes using concentrated acids have the advantage of low reaction temperature, but the costs for the corrosion resistant equipment are very high. Hydrochloric acid is the preferred acid because it is volatile and thus recoverable by

Table 18–4: Saccharification processes (according to Büsing et al. 1978, Maloney 1978, Goldstein 1980b)

Process	Prehydrolysis		Hydrolysis		Product yields per 100 kg of wood (oven dry)
	Acid Conc.	Temp. °C	Acid Conc.	Temp. °C	
Bergius-Rheinau 1919			HCl 41–45%	20–25	12 kg glucose 33 kg lignin
Bergius-Rheinau 1948	HCl 1%	130	HCl 41%	20–25	31 kg glucose 30 kg lignin 10.7 l ethanol
Rheinau-Udic 1958	HCl 35%	20–25	HCl 41%	20–25	30 kg glucose 11 kg xylose 30 kg lignin 22 kg molasse (80%)
Noguchi-Institute 1953	HCl 3.5% + steam	100–130	HCl 38% + HCl gas		30 kg glucose 10 kg xylose 24 kg lignin 24 kg molasse 4.5 kg acetic acid
Scholler-Tornesch 1926	H_2SO_4 1.2–1.5%	140–150	H_2SO_4 0.4%	130–180	28 kg glucose 30 kg lignin 19 l ethanol
Madison 1944	H_2SO_4 0.5–0.6%	130–135	H_2SO_4 0.5–0.6%	150–190	24.6 l ethanol
Hokkaido 1948	H_2SO_4 1.5% + steam	180–185	H_2SO_4 80%	20–25	28 kg glucose 7 kg xylose 24 kg lignin 6 kg molasse

distillation. Apart from the commercially viable Bergius process several modifications were developed, which differ mainly in the prehydrolysis step (Büsing et al. 1978). The main problem in applying concentrated sulfuric acid is the acid recovery. In the Hokkaido process the acid recovery problem was solved by dialysis (Maloney 1978).

In most of the saccharification processes a prehydrolysis step is included to remove the more easily hydrolyzable polyoses prior to the main hydrolysis of the cellulose. This is especially important in the case of hardwoods and annual plants with high xylan content. Table 18–5 gives some data on prehydrolysis procedures. They are either directed to furfural or to crystalline xylose for processing e.g. xylitol (→ 18.5.). In the latter case the temperatures must be kept below 130 °C, otherwise furfural is formed. The largest potential for the utilization of glucose is fermentation to ethanol, but other valuable products can also be derived from glucose (→ 18.4.).

The enzymatic or microbial hydrolysis of wood (→ 14.) is still far from a practical solution. The rates of hydrolysis are very low due to the low accessibility caused by

Table 18–5: Prehydrolysis processes (according to Stenzenberger 1981)

Process	Acid/ conc.	Temp. °C	Pressure bar	Time h	Product yields per 100 kg of wood (oven dry)
Quaker Oats 1922	H_2SO_4 5–10%	145–175	4	6–8	9 kg furfural, methanol, acetone
Savo-Rosenlew 1954	Formed CH_3COOH	220–240	15	1–2	6 kg furfural, methanol
Natta 1954	HCl 3%	20–25	atm	3–8	13–16 kg furfural 4–7 kg acetic acid, methanol
Silvichem 1970	CH_3COOH	135		0.4–1	Residue for further processing, furfural, xylitol, sorbitol
Lignin-Chemie-Waldhof 1975	$SO_2 + H_2O$	130		2	8 kg xylose 2.8 kg yeast

the high-crystalline parts of cellulose and especially by the high lignin content. Investigations on suitable lignin degrading enzyme systems and combined mechanical pretreatments will lead to new types of successful and economical wood saccharification procedures (→ 18.4.) (Humphrey 1979; Goldstein 1980b; Tanaka et al. 1980; Fähnrich et al. 1981).

18.4. Chemical Products from Cellulose

The situation of wood cellulose utilization is quite different from the utilization of polyoses and lignin. While the present and future major use of cellulose is the production of fibrous material for processing paper, rayon or cellulose derivatives, portions of the polyoses and generally most of the lignin are obtained as degraded and dissolved compounds in pulping liquors. This organic material in the liquors can serve as an energy source or for further utilization and processing (→ 18.5., 18.6.).

But in principle cellulose can also be used after degradation to its monomeric units. The production of glucose is the first and most important step of cellulose conversion to low-molecular-weight compounds and offers a wide field of different chemicals, including the potential to synthesize new polymers which presently are produced on the basis of natural oil and gas by petrochemistry (Fig. 18–2).

The conversion of cellulose to glucose can be accomplished by acid-catalyzed, enzymatic or microbial hydrolysis. Though the glycosidic bonds in solubilized or amorphous cellulose can be easily hydrolyzed by all types of acid and enzymatic hydrolysis, there are two general obstacles to practical cellulose hydrolysis. As already mentioned (→ 18.3.) the first restricting factor is the presence of lignin in lignocellulosic materials, which can reduce the accessibility of cellulose to enzy-

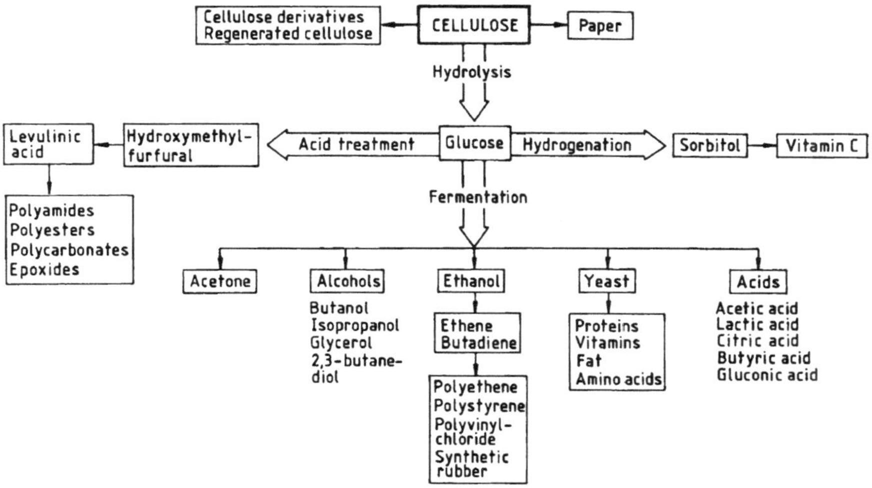

Fig. 18–2. Chemical products derivable from cellulose

matic attack to a minimum. In addition cellulose itself is highly resistant to chemical as well as enzymatic hydrolysis as a result of its crystalline organization within the native cell wall (→ 4.3.).

Apart from the more or less classical wood saccharification processes and their modifications (→ 18.3.) new processes are under investigation in pilot plant studies. They are mainly concerned with the fast and continuous hydrolysis of cellulosic waste materials such as wastepaper, paper-rich municipal garbage, but also sawdust. Maximum yields of 60% glucose were obtained in a single-stage process applying 0.5% sulfuric acid for only 20 sec at high temperatures in a twin-screw extruder device (Anonymous 1979, 1980a; Rugg, Brenner 1980). Another successful system works with a plug-flow reactor (Thompson, Grethlein 1979). Both techniques are based on the optimum combination of high temperatures with short reaction times to reach high glucose yields (Saeman 1979). The major disadvantage of simple one-stage systems is the lack of fractionation of the reaction products (glucose, pentoses, acetic acid), which reduces the economical utilization of the hydrolyzates. In addition the required conditions for cellulose hydrolysis yield high amounts of sugar decomposition products (Saeman 1980). A Japanese process involves small amounts of dry concentrated hydrochloric acid at 30–40 °C for only a few seconds in several impregnation zones (Anonymous 1978b).

Instead of dilute mineral acids organic acids such as maleic acid were also tested with good results. The higher costs of maleic acid are offset by lower costs for equipment and replacement (Anonymous 1978b). Cellulose-containing materials such as pulp and paper wastes, cellulose derivatives as well as coated regenerated cellulose foils can be hydrolyzed with glucose yields up to at least 90% by a multi-step treatment with decreasing concentrations of trifluoroacetic acid without applying pressure (Balser, Fengel 1979; Fengel, Wegener 1979).

The alternative to acid hydrolysis of cellulose is enzymatic hydrolysis carried out either with isolated enzyme systems or with cellulose degrading microorganisms (→ 14.4.). An increasing number of laboratory and pilot-scale investigations are concerned with this topic but no practical and economical technology has been worked out yet (Wilke, Yang 1975; Su, Paulavicius 1975; Dietrichs et al. 1978; Humphrey 1979; Fähnrich et al. 1981; Esterbauer et al. 1981). To overcome the problems of low cellulose accessibility especially in enzymatic hydrolysis various chemical and physical pretreatments have been investigated. A review of this topic was given e.g. by Millett et al. (1976). Apart from the above-mentioned prehydrolysis procedures in connection with wood saccharification processes other chemical pretreatments can be applied to render cellulosic material more susceptible to hydrolysis. These processes include swelling with alkaline solutions or ammonia as well as treatments with steam or gaseous and aqueous sulfur dioxide, respectively (Goldstein 1980b). As the digestibility of agricultural cellulosic materials (e.g. straw) but also of hardwoods (e.g. aspen or beech saw dust) is significantly increased by these treatments the resulting fibre material was found to be also useful as animal fodder (Dietrichs et al. 1978; Kaufmann 1979; Hajny 1981).

Another technique is to dissolve cellulosic materials in cellulose solvents such as cadoxene or iron sodium tartrate solutions and to precipitate amorphous cellulose which is reactive and thus easily hydrolyzable by enzymes (Ladisch et al. 1978).

Mechanical and physical pretreatments such as vibratory milling or grinding (Polčin, Bezuch 1977; Millett et al. 1979; Shimizu 1980), defibration (Thonart et al. 1979) or electron beam radiation (Anonymous 1978b) were shown to decrystallize and to depolymerize cellulose, by which hydrolysis rates and yields are generally increased. But due to the energy required for these procedures on an industrial scale none of them is economically feasible today.

As a practical method of biodegrading lignin has not yet been developed, lignin must be removed in full, or at least in part, from highly lignified biomass such as softwoods by means of delignifying processes. But if the resulting cellulose residue has pulp quality it is generally not designed for glucose production but for commercial high-value pulp purposes. Though enzymatic hydrolysis is capable of converting pulp cellulose completely to glucose in laboratory experiments under ideal conditions, practical yields in enzymatic biomass hydrolysis are still much lower, reaching 20–40%. Apart from low accessibility other restrictions are long reaction times, inhibition and inactivation of enzymes by accumulated hydrolysis products in the hydrolyzates, and high costs for enzymes and their recycling (Goldstein 1980b).

As outlined in Fig. 18–2 the most important pathway to a further utilization of glucose is fermentation. Depending on the type of yeast and the purity of the sugar substrate, fermentation can be directed to yield ethanol, fodder yeast (single cell protein, SCP), organic acids, alcohols and acetone. From solutions containing crude glucose mass products such as ethanol or fodder yeast are generally produced while pure glucose solutions are preferred for conversion to special chemicals.

Ethanol fermentation of glucose and other hexose sugars is a well-known process which is best described by the Embden-Meyerhof-Parnas scheme (Butschek 1965; Hajny 1981). The theoretical yield of this anaerobic process, which is carried out generally with the yeast *Saccharomyces cerevisiae*, is 51% of which 85–95% can be reached by commercially tested techniques (Goldstein 1976). The sugar solutions for ethanol production can be derived from pulping liquors as well as from wood and cellulose hydrolysis processes. Practical ethanol fermentation of pulping liquors is only carried out with softwood sulfite spent liquors with a high content of hexose sugars. Ethanol production was a well established technology in Europe for a long time with e.g. 33 plants in Sweden in 1945, but its importance has decreased in the last two decades (Herrick, Hergert 1977).

High energy costs are the main restricting factor for separating alcohol by steam distillation from the fermentation liquor followed by fractional distillation. Today these costs exceed 40% of the total process costs (Humphrey 1979). Thus the resulting prices of wood-derived ethanol are not competitive with ethanol produced by hydrogenation of ethene or with grain fermented ethanol. On the other hand two new spirit mills were installed in Finland in 1977/78 with alcohol capacities of 8 000 t/year each, combined with a production of 7 000 t/year of carbon dioxide generated in the fermentation process (Jensen 1979).

With regard to the liquors of new sulfite pulping processes for production of high-yield pulps (→ 16.3.3.) the amount of fermentable hexoses is decreased. On the other hand the potential of hexose sugars for ethanol fermentation is increased by about 40% in softwood dissolving pulping as compared to paper pulp production (Forss 1974).

A simultaneous enzymatic hydrolysis of cellulose and fermentation of the resulting glucose to ethanol was recently reported. The process was applied to different cellulosic waste materials including bark and pulping effluent streams (Anonymous 1978b).

Future techniques which are under investigation now also aim at converting pentoses to ethanol. A commercial application would reduce the production costs for ethanol considerably (Edemar et al. 1981). Investigations in this direction were carried out e.g. by Viikari et al. (1981) who succeeded in converting xylose to ethanol by fermentation with a strain of the fungus *Fusarium* spec. Ljungdahl et al. (1981) used two anaerobic thermophilic bacteria (*Clostridium thermocellum* and *Thermoanaerobacter ethanolicus*) in a mixed culture for hydrolysis and fermentation of xylose to ethanol in the presence of cellulosic material.

A comprehensive literature survey, representing the state of techniques for the production of ethanol from biomass and their economy, was recently published (Cheremisinoff, Cheremisinoff 1981).

Ethanol can be used as liquid fuel, but is favoured as a gasoline additive (10–25%) or as a blend for diesel oil (Goldstein 1980a; Seiffert, Held 1981). In the field of chemistry ethanol is an important solvent and represents a potential chemical feedstock for the production of ethene and butadiene. Ethanol can be dehydrogenated to ethene in commercially proved yields of 96% or converted to butadiene in yields of 70% (Goldstein 1976). In spite of high prices for oil the industrial production of these products today is based exclusively on petroleum. As ethene is the most important commercial organic chemical for the production of petrochemicals and plastics (e.g. ethene oxide, ethene chloride, ethene glycol, vinyl chloride, styrene, polyethene) as well as butadiene to produce synthetic rubber the conversion of cellulose to glucose and ethanol probably represents the most promising pathway for the future production of chemicals deriving from cellulose.

Glucose fermentation can also be directed to the production of fodder yeast (single cell protein) to be used as food and flavouring agent but also to provide protein, amino acids, nucleic acids and vitamins. Torula yeast *(Torula utilis)* is the most frequently used organism for this fermentation which can be applied to sugar solutions deriving from saccharification procedures or to sulfite spent liquors (→ 18.5.) (Hajny 1981).

In recent years there has been an increasing interest shown in using celluloses as a substrate for protein production. As in the case of the corresponding ethanol fermentation mixed cultures of a cellulose hydrolyzing organism (e.g. *Sporotrichum pulverulentum)* and a protein producing yeast (e.g. *Torula utilis*) are required (Ek, Eriksson 1975). Thermophilic bacteria (*Thermoactinomyces* spec.) were also applied for protein production. They have a high cellulolytic activity and grow rapidly producing single cell protein by degrading up to 70% of cellulosic substrates (Hägerdal et al. 1979)

Finally the fermentation of glucose can be directed to butanol, isopropanol or polyols such as ethene glycol or glycerol. Moreover acetone and organic acids such as acetic acid, citric acid, butyric acid or lactic acid are obtainable by fermentation. A recent review of these fermentation techniques and their potential applications was given by Hajny (1981).

Glucose can be converted to sorbitol by means of catalytic alkaline hydrogenation (Holz 1965). Sorbitol can be used as a dietetic sweetener and as starting material for the synthesis of ascorbic acid (vitamin C) (Dannhäuser 1974; Funk 1975; Dietrichs 1977). Fructose syrup with high sweetening potential is produced by enzymatic conversion of either glucose or sorbitol (Gams 1976).

Acid catalyzed treatments of glucose at different conditions yield 5-hydroxymethylfurfural which is also a degradation product in wood saccharification (Harris 1975). Hydroxymethylfurfural is not produced commercially today but the potential utilization of this chemical is based on its bifunctional structure which allows numerous reactions to yield intermediates for important plastics such as polyesters, polyamides, polycarbonates, epoxides, furan resins and nylon (McKibbins et al. 1962; Büsing et al. 1978). The commercial production of the degradation product levulinic acid was carried out in the U.S.A. till 1965 on the basis of saw dust, but there is no market for this chemical today (Herrick, Hergert 1977).

Summarizing the role of wood cellulose for the production of non-fibrous products, it can be stated that this most abundant and renewable natural substance offers in principle a large potential for chemicals and foodstuff in a future with decreasing reserves of fossil materials and increasing requirements for human food.

18.5. Utilization of Polyoses

The polyoses represent the second polysaccharidic part of wood with higher proportions in hardwoods (25–35%) than in softwoods (15–25%). In hardwoods pentosans (mainly xylans) are the dominant polyoses while most of the softwood polyoses are hexosans (mainly mannans) (→ 5.1.).

The last-mentioned distinction is important for two reasons in the utilization of polyoses. First, the xylans of hardwoods are more easily extractable and hydrolyzable than softwood mannans, which are partly in close association with cellulose. This fact is used in prehydrolysis processes of wood saccharification (→ 18.3.), in dissolving pulp manufacture (→ 16.), and in steaming processes such as the steam-

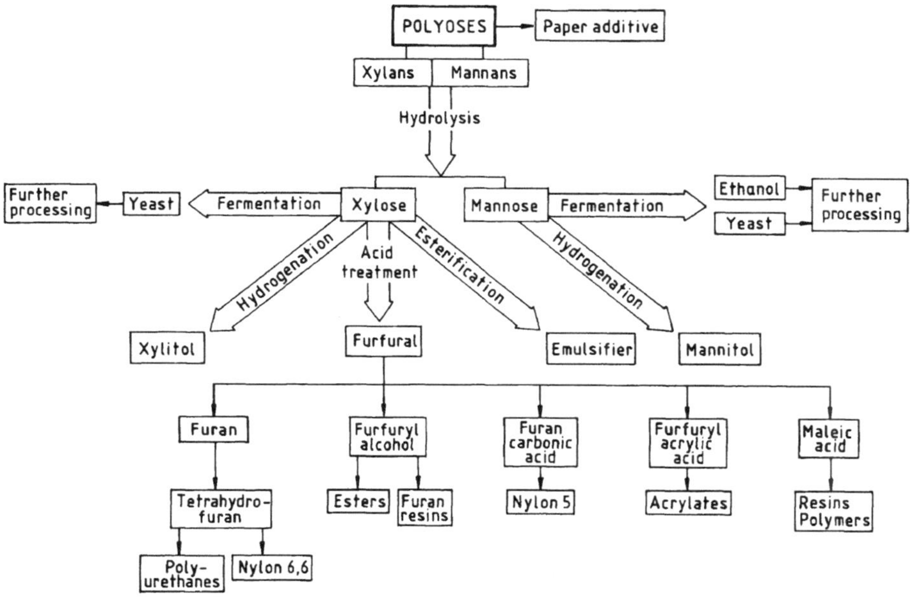

Fig. 18-3. Chemical products derivable from polyoses

ing-extraction process (Dampf-Druck-Extraktionsverfahren, Dietrichs et al. 1978, 1979). Secondly, for further processing of sugars derived from polyoses, it is decisive that today only the hexoses are fermentable industrially to yield ethanol while from both hexoses and pentoses yeast and protein can be produced, respectively. As already pointed out the practical conversion of pentose sugars to ethanol may be possible in the future. Figure 18–3 demonstrates important pathways and products in the present and potential utilization of polyoses.

The application of polyoses preserving their polymeric character and properties is more or less restricted to their function as natural adhesives in pulp and paper, improving the fibre-fibre bonds and strength, respectively. Isolated polyoses were added to different pulps from rice straw, hardwoods and softwoods as well as to paper and boards, resulting in significant improvements of the strength properties (Mobarak et al. 1973; Sihtola, Blomberg 1975). Studies of water-soluble larch arabinogalactan (→ 5.5), which can be extracted from larch species in amounts up to 25%, showed suitable properties as a non-toxic material for the production of stable, low-viscosity emulsions, and can be used as a tablet binder (Nazareth et al. 1961a, b). A literature review of investigations of arabinogalactans including examples of application was given by Adams and Douglas (1963).

Corresponding to cellulose the most important potential of the polyoses for the production of chemicals is based on their monomeric sugar components obtained by hydrolytic procedures. Sugar mixtures from hardwood polyoses contain xylose as the predominant sugar while mannose is the main sugar deriving from softwood polyoses. In both cases minor amounts of other sugars (glucose, galactose, arabinose), uronic and aldonic acids besides acetic acid (derived from the acetyl groups) are present in the hydrolyzates.

In commercial practice the spent liquors from sulfite pulping processes yield suitable substrates for the fermentation of sugars to ethanol and yeast, respectively.

Table 18–6: Monosaccharides and acetic acid in spent sulfite liquors (Forss, Passinen 1976)

	Spruce (*Picea abies*) %	Birch (*Betula verrucosa*) % of dry matter	Aspen (*Populus tremula*) %
Galactose	2.6	0.6	0.0
Glucose	2.6	1.1	0.5
Mannose	11.0	6.4	3.1
Xylose	4.6	21.1	24.3
Arabinose	0.9	0.0	1.5
Glucuronic acid	1.0	1.6	1.2
Total monosaccharides	22.7	30.8	30.6
Acetic acid	3.1	8.0	

Typical monosaccharide compositions and the content of acetic acid in spent sulfite liquors from different wood species are given in Table 18–6.

The considerable amounts of hexose sugars in softwood spent sulfite liquors can be fermented to ethanol, while hardwood liquors or residual sugars in softwood liquors after ethanol fermentation are used for yeast production. In the U.S.A. about 9 000 t/year of fodder yeast (SCP) are produced by fermentation of spent sulfite liquors with *Torula* yeast (McGovern 1980). The crude protein content of the obtained yeast depends on the amount of added nutrients and the growing conditions, but is generally about 50%. The vitamin content of yeast grown on wood sugar solutions is comparable to that of brewer's yeast. The wood-derived protein has a high nutrition quality comparable to the quality of meat and milk protein with regard to the essential amino acids (Hajny 1981).

In Finland the Pekilo process is used for continuous fermentation of spent sulfite liquors by the microorganism *Paecilomyces varioti* in the magnitude of 10 000 t/year (Forss et al. 1971; Jensen 1979). The fibrous yeast obtained has a protein content up to 60% and is officially approved as cattle fodder. The composition of Pekilo protein and its amino acid pattern is shown in Table 18–7.

Table 18–7: Composition of Pekilo protein and its amino acid pattern (Forss, Passinen 1976)

	%
Dry matter	96
Crude protein	55–60
Nucleic acids	10–11
Crude fat	1
Ash	5
	g/16 g N
Aspartic acid	9.3
Threonine	4.1
Serine	4.3
Proline	3.9
Glutamic acid	11.5
Glycine	4.9
Alanine	6.4
Valine	5.1
Methionine	1.9
Isoleucine	4.8
Leucine	7.4
Tyrosine	3.5
Phenylalanine	4.0
Lysine	6.5
Histidine	1.9
Arginine	5.3
Cystine	1.0
Tryptophan	1.1

Though the organism can consume hexoses as well as pentoses it was shown in shake flask experiments that the hexose sugars are fermented more easily and quickly than the pentoses (Forss, Passinen 1976). The fermentation process reduces the biological oxygen demand of the spent liquors (especially by consumption of acetic acid) thus contributing to the solution of pulp mill effluent problems and providing desugared liquors for lignin sulfonate applications (→ 18.6.1.).

A suitable source of xylose-rich sugar solutions is the recently developed steaming-extraction process (Dietrichs et al. 1978, 1979). Using a relatively simple technology, hardwood chips or chopped annual plants and plant residues, respectively, are treated for a few minutes with saturated steam at 180–200 °C. The resulting material is subsequently defibred, and finally washed with water or dilute alkali solutions. The final fibre material, still containing more than two-thirds of the total lignin, has a high chemical and biochemical reactivity. It can be used as roughage, fibreboard material or as raw material for acid and/or enzymatic hydrolysis. In this case glucose is obtained nearly exclusively in addition to a highly accessible lignin residue. The soluble polyoses components are mainly xylose and xylan fragments. In the enzymatically hydrolyzed water washings the xylose content was up to 70%, while in the hydrolyzed alkali extracts even up to 85% xylose (values based on total sugar content in the hydrolyzates) were found (Dietrichs et al. 1978). Such xylose-rich solutions can be purified to crystalline xylose in yields of 8%, based on total raw material. Solutions with lower contents of xylose and xylan fragments may be used as substrates for protein and enzyme production, or can deliver feeding wood sugar molasses (Schmidt et al. 1979). The production of 90 000 t/year of a liquid molasses concentrate as an additive for cattle feed derived from hardboard production in the U.S.A. was reported (Galloway 1975).

Crystalline xylose can be converted to xylitol by catalytic hydrogenation, which is a commercial process in Finland (Herrick, Hergert 1977). Xylitol has proved to be a very good non-cariogenic sweetener for diabetics with comparable sweetness to glucose (Henneke 1970; Dietrichs 1977; Haidegger 1977).

Another interesting wood sugar polyol is mannitol, obtained as a co-product in yields of about 20% in commercial sorbitol production. It is formed either by hydrogenation of fructose or mannose derived from glucose epimerization during mild alkaline hydrogenation of glucose or invert sugar (Herrick, Hergert 1977). Numerous other special chemicals for direct application or further processing can be derived in principle from xylose (Funk 1975).

From a chemical standpoint furfural, the dehydrogenation product of xylose, is the most interesting compound derivable from xylans. In principle, two steps are necessary to yield furfural from xylan: hydrolysis and acid catalyzed dehydrogenation. The world-wide production capacity is estimated to be about 170 000 t/year (excluding the COMECON countries) and is exclusively based on agricultural residues such as corn cobs, sugar cane bagasse or oat hulls. These materials allow high yields of furfural in the range of 15–23% on a dry raw material basis. In comparison, the yields obtained from hardwoods and softwoods are considerably lower, ranging between 6 and 11% (Maloney 1978).

The classical furfural process was established by the Quaker Oats Company and was followed by numerous modified prehydrolysis procedures in connection with wood saccharification (→ 18.3.). Other traditional sources for furfural production from wood are prehydrolysis procedures of kraft and soda dissolving pulp production as well as acidic sulfite pulping of hardwoods (Herrick, Hergert 1977).

The conditions of the steaming-extraction process described above for hardwoods can also be directed to produce considerable amounts of furfural (up to 10%), if higher temperatures and longer reaction times are applied (Dietrichs et al. 1979).

Due to its characteristic properties furfural is used as an industrial solvent, disinfectant or preservative, in solvent refining of petrol oils and as a reactive solvent in phenolic resin production and processing. Furfural has been superseded as a raw material in nylon synthesis by the application of oil-derived butadiene. The chemical structure of furfural, including the furan ring and the aldehyde group, offers a large potential for furan chemistry, which is briefly outlined in Fig. 18–3 and has been extensively described by Büsing et al. (1978).

One of the numerous consecutive compounds of furfural is furfuryl alcohol, obtained by hydrogenation. It can be polymerized by acidic condensation to so-called furan resins. These products are liquids with variable viscosities and are applied e.g. in the production of reinforced plastics (Radcliffe, Lens 1973). Polyurethanes and Nylon 6,6 can be produced via the intermediates furan and tetrahydrofuran (THF). THF is a valuable solvent, especially for polyvinyl chloride (PVC), and a primary product for several furan compounds. The same is true of tetrahydrofurfuryl alcohol and dihydropyran which can serve as starting materials for plastics such as polyesters and polyamides.

As outlined in this chapter there are several promising techniques for using these polysaccharides either for special chemicals or for mass products. Though these products are generally not competitive with oil-derived chemicals today, new isolation and conversion technologies may contribute to an increased utilization in the future.

18.6. Utilization of Technical Lignins

18.6.1. Polymeric Products

The present and future utilization of lignin is an extensive and increasingly important field, as indicated for example by the world-wide increase over the last two decades in the number of basic and technological publications as well as patents (Glasser et al. 1976). One reason for this development can be seen in the increasing appreciation of renewable raw materials. Lignin as a raw material is still far from being utilized intensively. This is true in spite of its large potential for different purposes due to its chemistry, properties and the huge amounts of it deriving world-wide from the pulping processes (some 50 million t per year). The field of lignin utilization can be divided into four general groups:

- lignin as remaining component in mechanical, high-yield semichemical and unbleached chemical pulps
- lignin as fuel
- lignin as polymeric product
- lignin as a source of low-molecular-weight chemicals.

Leaving lignin partly in pulp and paper is a means of utilizing it directly as a fibre-accompanying substance, thus obviating isolation and subsequent utilization problems. The high lignin content in high-yield pulps served e.g. for grafting hydrophilic polymers such as acrylic acid to increase strength properties (Krause et al. 1973). Further progress in lignin preserving bleaching and new techniques for hydrophilizing the residual lignin in pulps will determine the future of pulps with high lignin content (→ 16.3.).

The typical technical lignins are derived from the kraft or soda process (called sulfate or kraft lignin and soda lignin, respectively) or from sulfite pulping (sulfite lignin or lignin sulfonates) (→ 16.).

Technical hydrolysis lignin deriving from wood saccharification is important today only in the U.S.S.R. Organosolv lignin from sulfur-free pulping may become a future raw material with properties far superior to those of the above-mentioned technical lignins.

The main use of lignin today is still as an energy source. Most kraft lignin is used for energy purposes because the recovery of the process chemicals is based on the incineration of the spent black liquors. The calorific value of the organic material in the spent liquor (23.4 MJ/kg) is an important economical factor with regard to high and increasing prizes for gas and oil, though kraft lignin can be used alternatively for high-value non-fuel purposes (Barton 1978).

In the case of sulfite pulping the combustion of the spent liquors is possible if sodium, magnesium or ammonium are used as bases. Stricter water-pollution regulations have generally intensified the utilization of sulfite liquors as an energy source. Liquors from the traditional acidic calcium-based process cannot be incinerated without difficulties associated with evaporator scaling.

The utilization of lignin as a polymeric material or as a starting material for the production of low-molecular-weight chemicals can be divided into technologies which are applied commercially today and those which have not yet been worked out beyond the laboratory or at most the pilot-plant scale.

Though lignin is a macromolecule with suitable properties for many technical purposes the market for lignins or lignin-derived products is still small, related to its potential size. The production of lignin sulfonates amounted to about 350 000 t for 1982 in Western Europe (Kvisgaard 1982). Among the causes of these restrictions in comparison with synthetic products from mineral oil and gas are the following:

- complex chemical structure of lignins and lignin derivatives
- inhomogeneity and polydispersity of lignins
- high amounts of impurities
- considerable sulfur contents in kraft lignins and lignin sulfonates
- high costs for purifying and processing the crude liquors.

Nevertheless there are several established applications of alkali lignins, whole spent sulfite liquors and lignin sulfonates as polymeric material which are introduced briefly within this chapter without emphasis on their technological aspects. More detailed descriptions of commercial utilization of lignins are given by Wiley (1961), Pearl (1967, 1969), Hoyt and Goheen (1971), Maloney (1978) and Bryce (1980a, b).

Alkali lignin can be recovered in high yields from spent liquors by acidifying (to pH values of 8–9) and filtering off the precipitated lignin as acid salt. The weight-average molecular weights of pine and hardwood kraft lignins were determined to be 3 500 and 2 900, respectively with low degrees of polydispersity ($\bar{M}_w/\bar{M}_n$) of 2.2 and 2.8, respectively (Marton, Marton 1964). On the other hand gel filtration chromatography of pine kraft lignin shows a broad distribution with molecular weights between a few hundred up to more than 100 000 (Anonymous 1981b).

After dissolution of the salt tar in water and reprecipitation with hot dilute sulfuric acid the resulting lignin becomes soluble only in alkaline solutions. This fact is a severe restriction for technical applications. To avoid the disadvantage of water insolubility, alkali lignins can be transformed to water-soluble sulfonates by sulfonation. This reaction yields products with widely varying degrees of sulfonation, and special solubilities in different solvents can be produced, depending on the demand of the final utilization (Pearl 1969; Anonymous 1981b). In addition to sulfonation, kraft lignins can also be modified for example by etherification, esterification, nitration, chlorination, oxidation or demethylation (Hoyt, Goheen 1971).

Lignin sulfonates are the main component of sulfite spent liquors (about 60–70%) in addition to polysaccharidic fragments, sugars, uronic and aldonic acids, acetic acid and other degradation products (see also Table 18–6). Lignin sulfonates represent polydisperse systems with molecular weights for softwood lignin sulfonates from a few hundred up to 200 000 and more (Forss, Fremer 1965) (→6.4.1.).

The spent sulfite liquors can be used directly as dilute or concentrated liquids or as dry solids. For many purposes purified lignin derivatives are preferred which are produced either as sulfonic acids or more frequently as salts. The separation of the sulfonates from the accompanying substances in the spent liquor can be accomplished by several means. The carbohydrate content is diminished decisively by fermentation processes (→ 18.5.) leaving behind relatively pure sulfonates. Another well-known technique to obtain sugar-free basic salts of calcium lignin sulfonates in high yields of 90–95% is a two-stage lime precipitation according to the Howard process (Pearl 1969; Hoyt, Goheen 1971). The sugars and other low-molecular-weight material in sulfite liquors can also be removed in principle by ion-exchange chromatography (Felicetta et al. 1959), ion-exclusion chromatography (Perret et al. 1976), gel permeation chromatography (Jensen et al. 1966), ultrafiltration (Bar-Sinai, Wayman 1976) or electrodialysis (Dubey et al. 1965).

Lignin sulfonates may be modified by converting for example calcium sulfonates to sulfonates of other bases. In this way a large number of other types of sulfonates such as iron, zinc or chromium sulfonates can be produced by adding a soluble sulfate of the desired cation followed by precipitation of the insoluble calcium sulfate. Starting with magnesium or sodium sulfonates ion-exchange resins are necessary for the conversion (Hoyt, Goheen 1971). Lignin sulfonates can be desulfonated to some extent by special treatments with sodium hydroxide or ammonia to obtain water-soluble products with increased reactivity.

The technical lignins and modified products have found a wide field of application. Table 18–8 shows the diversity of applications without being complete. Utilization is generally based on the dispersing, adhesive (binding) and surface-active properties of the lignin products. With regard to the intended application several lignin products are offered with very special chemical and physical properties. The products are characterized e.g. by the degree of sulfonation, number of functional groups (carboxyl, carbonyl, alcoholic and phenolic hydroxyl groups), the average molecular weight, molecular weight distribution and surface tension value (Anonymous 1981b, 1982).

Whole sulfite spent liquors or crude lignin sulfonates have found considerable application in gravel-road surface stabilization (causing dust abatement), in general soil stabilization, as mineral binding and as binding agents for animal food pellets. In the last case the content of sugars and inorganic material contributes to the nutritive value of the animal fodder. This application in particular would have a large-scale potential if new markets could be developed (Herrick, Hergert 1977).

Traditional applications of lignin sulfonates involve their utilization as tanning agents or as tanning aids in combination with chrome tanning agents, as foundry core and ore binders and as protective colloids for preventing scale formation in steam boiler and feed line systems.

New fields of utilization were set up by purified or processed lignin sulfonates and alkali lignins with improved properties. The high dispersant ability of purified lignin sulfonates is used in oil well-drilling muds for controlling the fluidity and the rheo-

Table 18–8: Utilization of alkali lignins, lignin sulfonates and modified lignins (Hoyt, Goheen 1971; Herrick, Hergert 1977; Anonymous 1981b)

Dispersants for:	Carbon black
	Insecticides
	Herbicides
	Pesticides
	Clays
	Dyestuffs
	Pigments
	Ceramics
Emulsifiers, stabilizers, extenders for:	Soils
	Road surfaces
	Asphalt
	Waxes
	Oil in water
	Rubbers
	Soap
	Latex
	Fire foam
Metal sequestrants for:	Industrial water
	Agricultural micronutrients
Additives for:	Drilling muds
	Concrete
	Cement grinding
	Industrial cleaners
	Tanning agents
	Rubbers
	Vinylplastics
Binder and adhesives for:	Animal-feed pellets
	Printing inks
	Minerals
	Laminates
	Foundry cores
	Ores
Coreactant for:	Urea-formaldehydes
	Phenolics
	Furans
	Epoxides
	Urethanes
Others:	Protein coagulants
	Protective colloids in steam boilers
	Ion-exchange resins
	Oxygen scavenger
	Components in negative plate expanders for storage batteries

logical properties of the muds, combined with a stabilization and a sequestering effect for contaminating metal ions, thus preventing flocculation.

Lignin sulfonates and sulfonated kraft lignins are both used as additives in the cement and concrete industries. They serve for example as a grinding aid for cements or as a concrete additive to improve mixing properties and to permit the extension of the setting time (Sergeeva et al. 1979). The compressive strength and durability of the cured concrete is increased and in the case of pre-stressed concrete a lignin component is generally added to improve the adhesive strength between steel and concrete. Predominantly, sodium salts of kraft lignins are used as anionic and cationic stabilizers and emulsifiers for asphalt-, wax- or oil-in-water emulsions. Due to the ionic character of the technical lignins they can be processed to ion exchange resins.

The dispersing properties of lignin sulfonates are used in numerous applications as dispersants for ceramics, clays, dyestuffs, pigments, insecticides and carbon black (Goring 1961). For standard pesticides and herbicides complete surfactant systems with dispersing and wetting properties are available (Anonymous 1981b, 1982).

Processed residues after alkaline oxidation of lignin sulfonates to produce vanillin are used as excellent dye and pigment dispersants as well as for thermosetting resin components for decorative paper laminates (Herrick, Hergert 1977). A large number of miscellaneous applications are described in the literature without contributing to the future utilization of technical lignins as mass products (Hoyt, Goheen 1971; Maloney 1978).

A promising future market is expected to be the utilization of kraft lignin in reinforced styrene-butadiene rubber (SBR), replacing carbon black (Falkehag 1975; Glasser 1981).

Perhaps the most important future application will be based on the adhesive properties of lignin products for thermosetting resins and plastics. Intensive studies were made in using lignin as a co-reactant in phenol-formaldehyde adhesives for wood composites such as plywood, particleboards and fibreboards (Shen 1974; Roffael, Rauch 1974; Roffael 1976; Shen et al. 1981). Plywood as well as particleboards and fibreboards produced with lignin portions replacing up to 70% of the phenol component showed identical results with regard to the required strength and weather-resistance properties of conventionally glued products for exterior use (Forss, Fuhrmann 1979). Boil-tested waferboards were produced using exclusively ammonia based spent sulfite liquor as a thermosetting resin binder (Shen, Calvé 1979). Spent sulfite liquor can also be cross-linked to form suitable particleboard resins by oxidative coupling with hydrogen peroxide in the presence of sulfur dioxide (Nimz, Hitze 1980).

A prerequisite for applications in the field of phenolic resins is a purified lignin material with a sufficient number of reactive groups. Generally the reactivity of lignins is hindered by methoxyl groups and the aliphatic side chains. Therefore the resin properties of lignins are improved by demethylation, thus increasing the phenol content as shown in experiments with plywood from several wood species (Gupta, Sehgal 1978, 1979). Low-molecular-weight fragments require large amounts of phenol and formaldehyde to yield suitable resins. Therefore the substitution of more than 20% of phenol gives acceptable results only if high-polymer fractions of lignin sulfonates are used (Forss, Fuhrmann 1976). Deviating results were obtained when lignin sulfonates were used as the only adhesive. The low-molecular-weight fraction of the lignin sulfonates was shown to yield better bonding qualities and shorter curing times as compared to crude lignin sulfonates (Shen, Calvé 1979).

Another future aspect of lignin utilization is its conversion to polyols by oxyalkylation, which can be further processed to form polyurethane foams, adhesives and coatings (Hsu, Glasser 1976; Schweers 1979). In particular sulfur-free, less degraded and less condensed organosolv lignins may be suitable materials for this high-value utilization (Schweers, Meier 1979; Baumeister, Edel 1980). Polyurethanes of good qualities were obtained with lignin contents between 20 and 30%. The direct utilization of organosolv lignin as an extender for polyvinyl plastics was investigated, but without acceptable results (Schweers 1979).

As wood hydrolysis plants are operating only in the U.S.S.R. today, the literature concerned with the utilization of hydrolysis lignin is hardly accessible. This acid lignin was suggested for comparable applications as described in the case of alkali lignins and lignin sulfonates. The dominant utilization is combustion for energy purposes, but its application as a rubber and resin additive, soil stabilizer and co-reactant for phenol-formaldehyde resins was also reported (Hoyt, Goheen 1971).

18.6.2. Low-Molecular-Weight Chemicals

Due to the principally aromatic and aliphatic character of lignin it may serve as a source for a number of chemicals now derived from petroleum and natural gas. A large-scale mass production of low-molecular-weight products from technical lignins in competition with petrochemicals is restricted today by economical and technological considerations. The degradation of lignin macromolecules to simple aromatic and aliphatic compounds requires energy-intensive and therefore expensive processes with generally low yields of pure chemicals. With increasing energy costs the products potentially derivable from waste lignins will become more expensive too.

With regard to the technological problems the degradation processes are complicated by condensed chemical structures of technical lignins, which cause poor solubility.

Additionally the lignins deriving from pulping processes are obtained together with solubilized carbohydrates and considerable amounts of inorganic material, especially sulfur. For that reason the lignins must either be desulfurated for many processes or sulfur-resistant catalysts are required. Fractionation and purification procedures are necessary after degradation because the material obtained is a complex mixture of several components.

The principal routes of breaking down lignin to monomeric or low-molecular-weight compounds as demonstrated in Fig. 18–4 are:

- alkaline oxidation or hydrolysis
- alkali fusion
- nucleophilic alkaline demethylation
- pyrolysis
- hydrogenolysis.

The chemicals obtained can be roughly classified as follows:

- unspecific products such as coal, oil, tar and pitch
- gases such as carbon monoxide, carbon dioxide, hydrogen
- phenol and substituted phenols
- benzene and substituted benzenes
- other saturated and unsaturated hydrocarbons
- organic sulfur chemicals
- organic acids.

There are several comprehensive reviews concerned with the principles, technologies and processing of lignin conversion to low-molecular-weight products (e.g. Goheen 1971; Goldstein 1975b; Herrick, Hergert 1977). Therefore only a short

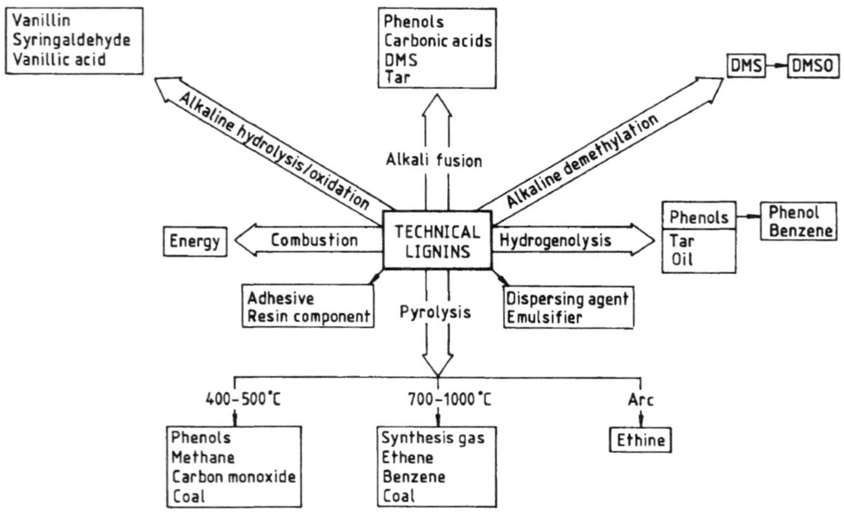

Fig. 18–4. Chemical products derivable from technical lignins

summarizing discussion of the most important chemicals and their utilization will follow in this chapter.

The alkaline degradation of lignin can be performed either as hydrolysis with sodium hydroxide or as alkaline oxidation in the presence of oxygen, metal oxides or organic oxygen compounds. Depending on the process conditions various compounds are obtained. The most important substituted phenol with respect to commercial utilization is vanillin. This flavouring agent was traditionally derived from the tropical vanilla bean *(Vanilla planifolia)* and there are several pathways described for the synthesis of vanillin starting with guaiacol or eugenol (Goheen 1971). The commercial production of lignin-derived vanillin started in about 1937 (Pearl 1969). Several processes were developed employing softwood lignin sulfonates after concentration of the liquor and removal of most of the carbohydrates by fermentation. The alkaline hydrolysis according to the Howard-Smith process is based for example on heating the liquors for 2 to 12 hours at 100–165 °C in the presence of sodium hydroxide and recovering the formed vanillin by extraction with benzene after treatment of the reaction solution with carbon dioxide. The purification can be carried out by vacuum distillation and recrystallization. Other processes are based on oxidation with metal oxides or oxygen pressure in lime-treated solutions, instead of expensive sodium hydroxide. The reaction is carried out in dilute medium at temperatures up to 225 °C and short reaction times of less than 2 hours (Töppel 1961; Goheen 1971). The commercial production of vanillin is only economical from softwood lignin sulfonates since in the case of hardwoods syringaldehyde is also formed which cannot be separated economically from vanillin (Töppel 1961; Pearl 1969). The practical industrial yields of vanillin from softwood sulfite pulping liquors are in the range of 5–10%, related to lignin. Higher yields with extreme values of about 30% were reported from laboratory experiments (Herrick, Hergert 1977). The yields generally depend on the starting material, type of catalysts and the applied alkali. Numerous modified procedures are reported in the scientific and patent literature (Goheen 1971).

Vanillin can also be produced from alkaline lignins but the yields are substantially lower. Therefore industrial production is limited exclusively to softwood lignin sulfonates. Apart from the main product vanillin several interesting by-products can be recovered such as calcium oxalate, vanillic acid, acetovanillone, syringaldehyde or acetosyringon.

Vanillin is predominantly used as a flavouring agent but the market for this purpose is very limited (world production 1982: 7 400 t; Kvisgaard 1982). A future large-scale utilization of vanillin or vanillin derivatives can only be expected in the field of organic synthesis if the price can compete with oil-derived chemicals.

Other applications of vanillin related chemicals are for example the use of ethyl vanillate as a UV absorber in sunburn preparations and plastics, as a food preservative, and as a medicine component. The diethylamide of vanillic acid is used as an analeptic agent for respiration and blood pressure control. A very special application is the use of vanillin for the synthesis of L-dihydroxyphenylalanine ('L-dopa') applied in the case of Parkinson's disease (Pearl 1969). A large number of other applications of vanillin and related compounds have been suggested especially in the pharmaceutical field (Goheen 1971).

Polyesters can be produced from vanillin with good fibre-forming properties. The polymers are formed by condensation of the diester derivative or the hydroxy ethyl ether of vanillic acid. However the products are not economically competitive with polymers from conventional polyester production (Erä, Hanulla 1974; Lange, Kordsachia 1981).

At present apart from vanillin only dimethylsulfide (DMS) and dimethylsulfoxide (DMSO) are produced on an industrial scale. The principal reaction involved in producing DMS from lignin is a nucleophilic demethylation. Sulfide ions are required for the formation of methyl mercaptan (MM) as an intermediate compound. In a second step the mercaptide ion attacks another methoxyl group, forming DMS. The production of DMS is generally accomplished from kraft black liquors, mainly because alkali and sulfide are already involved as process chemicals. In principle DMS and MM can also be generated from soda lignin or purified spent sulfite liquors.

In the commercial process founded in the U.S.A. in 1960 elemental sulfur is added to concentrated black liquor and the mixture is heated to 200–250 °C. The DMS and the volatile by-products (mainly H_2O, MM, H_2S) thus formed are flashed and condensed. The crude DMS is purified by passing through sodium hydroxide and by final distillation. The practical yields of purified DMS are about 3% (based on lignin) deriving from methoxyl contents of about 13–20% in softwood and hardwood kraft lignins.

DMS and MM are used as odorants for natural gas but also as solvents and reaction media in oganic syntheses.

The most important derivative of DMS is the oxidation product DMSO, a liquid with excellent dissolving properties. Therefore it serves for example as a solvent for biological and pharmaceutical materials, in the production of synthetic fibres, as a selective extraction medium in petrochemistry and as a carrier substance for herbicides, insecticides, drugs and various medicinal chemicals. Due to its chemical and physical properties DMSO can also be used as a reactant and catalytic solvent in numerous chemical reactions (e.g. Hearon et al. 1962).

As vanillin and DMS are produced only in low yields from pulping spent liquors the manufacture of both products does not interfere with the utilization of waste lignins as a fuel or for other purposes.

By pressure heating in the presence of sodium hydroxide and sodium sulfide at temperatures of 250–290 °C kraft black liquors were degraded in laboratory experiments to a number of interesting products including ether-soluble phenols such as pyrocatechol, protocatechuic acid, acetic, formic and oxalic acid, DMS, MM, unidentified degradation compounds and a demethylated lignin residue (Enkvist et al. 1962, Enkvist, Linfors 1966). Comparable results were obtained by treating a high-molecular-weight fraction of kraft lignin with 4% NaOH for 30 min at 300 °C. Phenol, catechol, guaiacol and their alkyl derivatives were obtained in a yield of 11%, related to lignin. The addition of sodium sulfide causes increased demethylation reactions of guaiacyl compounds and lower yields of identifiable phenols (Clark, Green 1968). Though several techniques of alkali fusion have been worked out, no industrial application has yet been realized for them (Allan, Mattila 1971).

The same is true of the techniques of pyrolysis and hydrogenolysis (pyrolysis in the presence of hydrogen). As in the case of wood carbonization, lignins can also be pyrolyzed. Due to the higher carbon content the amounts of coal and tar are generally higher than in wood pyrolysis. The specific composition of the pyrolyzate depends on the lignin material and the pyrolysis conditions with the final temperature having a dominant influence. The general products derivable from lignin pyrolytic degradation can be classified into 4 groups:

- pyrolytic coal
- pyrolytic tar (containing numerous phenolic compounds)
- aqueous distillate (water, methanol, acetic acid, acetone etc.)
- gaseous compounds (mainly carbon monoxide, carbon dioxide, methane, ethane).

Low-temperature lignin pyrolysis at temperatures between 400 and 500 °C generally yields a mixture of phenols and substituted phenols besides methane, carbon monoxide and lignin coal (Goheen et a. 1976).

Industrial pyrolysis processes are only reported from the U.S.S.R. where hydrolysis lignins from wood saccharification are considered to be valuable material for the production of phenolic compounds and activated lignin carbon. Yields of monophenols up to 10% and of lignin coal up to 60% (both based on lignin) were obtained by pyrolysis of hydrolysis lignin in anthracene oil under reduced pressure and temperatures of 440–460 °C. The composition of the phenolic fraction was shown to be dependent on the starting material. In the case of hydrolysis lignins from agricultural residues (e.g. sun-flower shells, corn dobs) cresols were the dominant phenolic components, while using softwood lignin guaiacol amounted to about 50% of the phenolic mixture (Epshtein et al. 1977; Wienhaus et al. 1980). The yields of phenols can be increased by a secondary pyrolysis of the condensable tar compounds or by addition of metals (Allan, Matilla 1971).

By high-temperature pyrolysis of different pulping spent liquors (kraft, sulfite, NSSC) at 700–1 000 °C a crude synthesis gas and activated carbon can be produced simultaneously. In addition high-value unsaturated hydrocarbons such as ethene and benzene were obtained from kraft black liquor in yields up to 6% each (based on dry liquor substance) (Prahačs et al. 1971). Organosolv lignins deriving from an ethanol-water pulping procedure were recently pyrolyzed in a fluidized-sand-bed laboratory test plant yielding 21.5% phenolic compounds of which about two-thirds (14.1%) were monomeric phenols (Kaminsky et al. 1980).

The electric arc treatment of technical lignins with helium as a sweep gas yielded up to 14% of ethine as a major reaction product besides low amounts of methane and ethene (Hearon et al. 1964).

Hydrogenolysis is a well-known technique for structural studies of lignins (→ 6.3.1.) and a suitable process to produce phenols from lignins in relatively high yields (e.g. Goheen 1971; Hrutfiord 1971; Goldstein 1975b; Wienhaus et al. 1976; 1980).

Laboratory experiments showed that hydrogenation of hydrochloric acid lignin at 250 °C in the presence of complex compounds of transition metals (iron, cobalt, nickel) yields up to 36% of monomeric phenolic degradation compounds (Schweers 1969).

The technical realization of phenol production from waste lignins by hydrogenolysis is generally restricted by the necessity of expensive high-pressure equipment, energy-intensive reactions, expensive catalysts, and finally by the problems of separating and purifying single compounds from the complex phenolic mixtures. The only hydrogenolysis procedure tested on a semi-technical scale is the Noguchi process (Goheen 1966; Oshima et al. 1966). In this process inexpensive sulfur-stable metal sulfides are used as catalysts and pressures up to 100 bar are applied at 370–430 °C for 0.5–4.0 hours. From desulfonated, deashed spent sulfite liquor up to 23% monophenols, based on lignin, can be obtained. The main components are cresols in addition to phenol, ethyl phenols, propyl phenols, 2,4-xylenol and others.

In connection with the Noguchi process a catalytic dehydro-dealkylation procedure was described for alkylated phenols resulting in 40–50% phenol and 30% benzene, related to total alkyl phenol content (PPRIC 1975, Anonymous 1980b). Interesting results were recently reported by Wienhaus et al. (1980) who hydrogenated sulfite and sulfate waste lignins in brown-coal tar-oil and obtained up to 13.3% phenolic compounds from sulfite lignins and 10.5% from sulfate lignin. Based on these results from small-scale experiments a technical process scheme for hydrogenolysis of sulfate lignins was suggested.

To avoid technical difficulties and economical disadvantages the utilization of mixed lignin-degradation phenols was investigated analogously to coal tar (Goldstein 1975b). Lignin tars containing higher amounts of phenolic components than coal tar may be a suitable material for example as an antioxidant, and as a preservative with fungicidal, herbicidal and insecticidal properties (Goldstein 1975b; Hewgill, Legge 1976)

The key to a future wood-derived chemical industry is expected to be a multi-product plant which produces valuable chemicals by several processes including e.g. ethanol, furfural, phenols, yeast and some other by-products (Katzen 1978). But the restricting factors for these industries today are the price of the raw material – either waste wood or wood from special energy or chemistry plantations – and the problems involved in obtaining sufficient quantities of raw material required for chemical processes in competition with wood as a raw material for pulp, paper and boards (e.g. Noack, Frühwald 1981; Wegener 1982).

18.7. Utilization of Extractives

As the extractive components of wood (→ 3.2.4., 7.) and bark (→ 9.2.7.) comprise a large number of chemical compounds there was an early interest in utilizing these substances for different purposes, e.g. as tanning agents, dyes, perfuming agents or naval stores. Today some extractives of wood and bark are still or again valuable sources for special products obtainable from trees in addition to products derived from the macromolecular wood components (Zinkel 1975).

From the standpoint of practical utilization the extractives may be divided into the following groups:

- naval stores
- wood extractives obtained by means of solvent extraction
- bark extractives obtained by means of solvent extraction
- foliage chemicals.

Fig. 18–5 gives a short survey of these low-molecular-weight silvichemicals derived from living trees, wood, bark and foliage. The utilization of the so-called technical foliage (a mixture of leaf and twig material) is becoming a topic of increasing interest in connection with whole-tree or full-tree utilization (Eskilsson, Hartler 1973; Keays 1974). Among others valuable substances such as essential oils, leaf protein, chlorophylls and carotenoids as well as animal fodder can be obtained from foliage by different techniques. These aspects shall not be discussed here but are described elsewhere (Hannus, Pensar 1973; Barton 1978, 1981; Barton, MacDonald 1978; Law et al. 1978). The utilization of extractives was also discussed by Gardner and Hillis (1962), Hillis and Swan (1962), Herrick and Hergert (1977) and Maloney (1978), covering the patent literature and the role of extractives in pulp and paper production as well.

The use of wood resins in connection with the construction of early sailing ships led to the general term naval stores for rosin, turpentine and tall oil which are the most important group of extractives today with regard to commerical production. These naval stores are obtained either by tapping living pine trees (gum naval stores), by

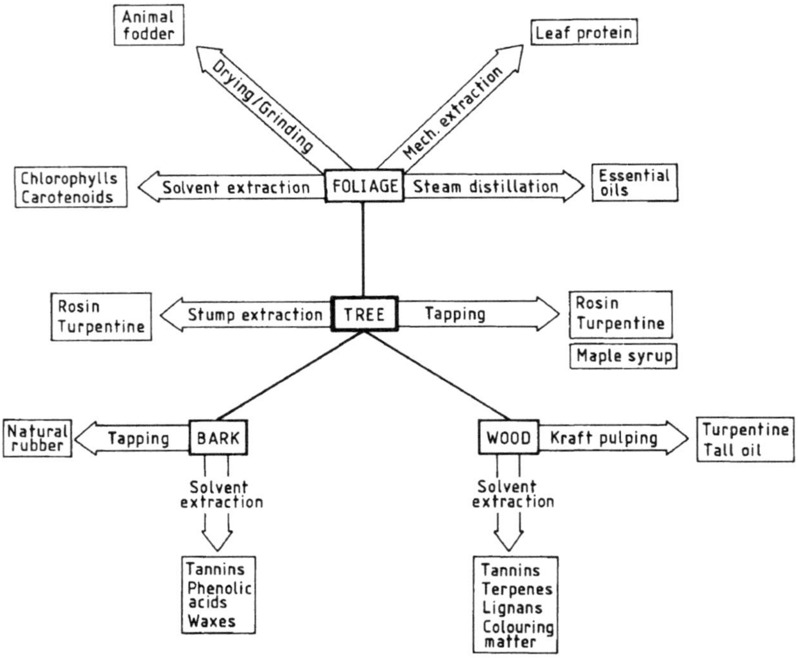

Fig. 18–5. Chemical products derivable from extractives

solvent extraction of stumps of coniferous trees or as a by-product of pine and softwood kraft pulping (wood naval stores).

The most traditional way of producing naval stores is by tapping living pines (*Pinus* spec.), yielding so-called gum or oleoresin with the two main components rosin and turpentine. In the U.S.A., where the largest naval stores industry is established, slash pine *(Pinus elliottii)* and longleaf pine *(Pinus palustris)* are used nearly exclusively for oleoresin harvesting. Elsewhere in the world several other pine species are used such as Austrian pine *(Pinus nigra)* in Austria, pinaster pine *(Pinus pinaster)* in France, Asiatic longleafed pine *(Pinus longifolia)* in India or mercus pine *(Pinus merkusii)* in Indonesia and the Philippines (Sandermann 1960).

The yields of crude oleoresin (crude gum) amount to between 2.5 and 4 kg per year and tap. Increased resin production can be induced by growing new genetic pine types (Franklin et al. 1970) but special success has been reported for treatments with herbicides like paraquat, diquat, ethrel or others. In pine trees thus treated, forming so-called pine lightwood, the amounts of turpentine and rosin are 7 to 8 times higher than in untreated wood. These treatments are not yet applied in commercial oleoresin production because of environmental regulations (Drew, Roberts 1978; Zinkel, McKibben 1978; Zavarin et al. 1978).

The turpentine yield of tapped oleoresin is generally between 18 and 25%. The oil is composed of α-pinene (60–70%), β-pinene (20–35%) and other terpenes such as camphene or 3-carene (5–12%). The wood turpentine obtained by extraction of stumps and roots differs from tapped turpentine with regard to higher contents of α-pinene (75–80%) (Herrick, Hergert 1977).

While the production of oleoresin by tapping and wood extraction is generally stagnant the production of tall oil and sulfate turpentine in connection with kraft pulping of pines and other coniferous woods remains the dominant source for naval stores with increasing production rates. In the U.S.A., the world's largest producer of kraft pulp, some 430 000 t of refined tall oil, tall oil fatty acids and rosin, as well as 90 million l of sulfate turpentine were produced in 1976 (Campbell 1977). In New Zealand about 8 000 t of tall oil are produced by kraft pulping of radiata pine *(Pinus radiata)* (Uprichard 1978).

The fatty acids (mainly occurring in wood as esters) and resin acids are saponified during alkaline pulping and are recovered by skimming the soap from the black liquor as so-called tall oil soap. The crude tall oil is obtained by acidifying the liquor, thereby converting the acid salts into free acids. Most of the crude oil is refined by vacuum distillation. The yields as well as the composition of tall oils depend on several factors such as wood species, heartwood proportion, wood age, wood felling time, geographical site, chip-storage time and pulping conditions. The yields are normally between 30 and 50 kg/t of pulp for many pine species in North America and Scandinavia, but much higher yields of more than 100 kg/t of pulp were reported from slash pine pulping in the southern part of the U.S.A. (Sandermann 1960; Ellerbe 1973).

The main constituents of crude tall oil are resin acids (rosin), fatty acids and neutral substances (unsaponifiable material). Average proportions are 30–50% of resin acids and fatty acids, respectively, besides about 10% of unsaponifiable substances (Sandermann 1960; Bryce 1980b). Important resin acids are abietic acid, neoabietic acid and palustric acid. The main fatty acids are oleic and linoleic acid. Hydrocarbons, higher alcohols and sterols make up the neutral part of tall oil (Table 18–9).

The volatile terpenes of wood are condensed after sulfate pulping as sulfate turpentine in yields of 3–6 l/t of pulp (Uprichard 1978; Bryce 1980b). The magnitude of sulfate turpentine yields is mostly dependent upon the chip-storage time besides the factors mentioned in connection with tall oil yields. Turpentine losses of about 80% after 30 weeks of chip storage in piles were reported (Somsen 1962). In general sulfate turpentine has a composition comparable to the turpentine in tapped oleoresin. While formerly turpentine was produced by wood extraction and tapping alone, today more than 80% of the total turpentine production in the U.S.A. is derived from kraft pulping (McGovern 1980).

The utilization of naval stores covers a wide range of applications (Zinkel 1975; Herrick, Hergert 1977; Barton 1978; Uprichard 1978; McGovern 1980). The crude tall oil is used e.g. as core oil, flotation agent, and for the production of surface-active agents. The resin acid fraction of tall oil is mainly applied in sizing paper to

Table 18–9: Composition of crude tall oil fractions (Agnello, Barnes 1960)

	% of the fraction
Resin acids	
Abietic acid	20–25
Palustric acid	8–10
Neoabietic acid	22–27
Dihydroxyabietic acid	4–5
Isodextropimaric acid	3–4
Unidentified	29–43
Fatty acids	
Oleic acid	46–48
Linoleic acid	43–45
Linolenic acid	1–2
Saturated acids	6–8
Unsaponifiables	
Phytosterols	25–35
Higher alcohols	5–15
Hydrocarbons	35–60

control the water absorptivity, but is also used as a component in synthetic adhesives and surface coatings. Finally, resin acids are applied in the manufacture of synthetic rubber, paints and varnishes, and in the synthesis of chemicals and pharmaceuticals.

Fatty acids are used in alkyd resin production, as a detergent and soap component, and as intermediate chemicals.

Turpentine was originally used as a valuable solvent for oil- and resin-based paints. Today the major use for turpentine components such α- and β-pinene is the manufacture of perfumary chemicals mainly used as additives in soaps and detergents. Many flavour chemicals such as menthol, myrcene or camphor can be synthesized from turpentine components (Derfer 1963, 1966; Ansari 1970). Minor amounts of turpentine are used in insecticides, disinfectants and adhesives.

Another valuable product from living trees is the sap of the sugar maple *(Acer saccharum)*. It is commercially harvested in Canada in amounts of some 13 000 t/year, representing about 75% of the world's supply of maple syrup products (Barton 1978). The main components of the maple syrup are vanillin, syringaldehyde, dihydroconiferyl alcohol and isomaltol (Underwood et al. 1969).

Latex ist another natural product which is tapped commercially from the bark of the gum tree *(Hevea brasiliensis)*. This tree species is cultivated in many tropical countries and provides about 99% of the world's production of natural rubber which amounted to 2.9 million t in 1970. In comparison with some corresponding 5 million t of synthetic rubber in 1970 it becomes evident that natural rubber is still an important product. Latex is converted to rubber by vulcanization and the rubber is mainly used after addition of pigments, fillers, softening agents and other additives (Neumüller 1973).

Though numerous extractives can be isolated by solvent extraction from wood and bark as well as from leaves, fruits and roots, only some groups of these substances are in commercial use. The most important group of compounds, which is used for several purposes, is the polyphenolics. From a chemical standpoint this group can be divided into hydrolyzable tannins and condensed tannins (phlobaphenes) as well as phenolic acids. These substances are generally extractable with water at temperatures between 80–120 °C from the heartwood and/or bark of many trees. The most important sources of tannin extracts are the wood of quebracho *(Schinopsis lorentzii, Schinopsis balansae)* and chestnut *(Castanea sativa)*, the bark of several acacia species (e.g. *Acacia decurrens, Acacia mearnsii)*, and the cupules of some Eastern Mediterranean oak species (e.g. *Quercus aegilops*) (Faber 1978; Pizzi et al. 1981).

The chestnut tannins belong to the group of hydrolyzable tannins and contribute only 5–10% of the total industrial world production. The acacia bark extracts (called wattle or mimosa) and the wood extracts from quebracho are condensed tannins and are produced in a magnitude of more than 250 000 t/year world-wide (Pizzi et al. 1981). Condensed tannins can also be obtained from the bark of several pine species (e.g. *Pinus radiata, Pinus patula, Pinus elliottii, Pinus taeda*), hemlock species *(Tsuga canadensis, Tsuga heterophylla)* and Douglas fir *(Pseudotsuga menziesii)*.

The traditional application field of tannins is the manufacture of leather. The above-mentioned production figures demonstrate the present world-wide role of natural tanning agents in spite of the dominant application of synthetic tanning agents (Faber 1978). Minor amounts of tannins are used in suspensions of clays, minerals, pigments, dyes and pesticides (Barton 1978).

The most promising utilization at present and in the future is the substitution of phenol in phenol-formaldehyde resins for the production of plywood, particleboards and laminated beams. There is much experience in this field in Australia, New Zealand and the U.S.A. but the industrial application of wattle tannin-based adhesives is well established only in South Africa. Today, in this country, the total amount of exterior-grade particleboard and plywood production is based on wattle tannins replacing phenol and partly urea in formaldehyde resins (Pizzi 1980; Pizzi et al. 1981). Modified tannins from the bark of *Pinus brutia* were recently applied successfully as components in formaldehyde resins (Ayla, Weissmann 1982).

Crude phenolic acid extracts from softwood barks were investigated with regard to their suitability as dispersing agents, drilling mud components, and as a starting material for alkaline fusion and hydrogenolysis to yield e.g. phloroglucinol and catechin (Goldstein 1975b).

Plicatic acid, belonging to the group of lignans (→ 7.2.3.) is extractable from western red cedar heartwood *(Thuja plicata)* in yields up to 5%. This extractive has proved useful for applications such as complexing and electro-refining metals. The flavonoids dihydroquercetin and quercetin which are extractable from the wood and bark of Douglas fir *(Pseudotsuga menziesii)* and western larch *(Larix occidentalis)* (→ 7.2.3.) are potentially suitable as antioxidants, dyes, fungicides and pharmaceuticals, but markets for these products have yet to be established (Herrick, Hergert 1977).

Waxes can be extracted from the bark of conifers and deciduous trees. As far as their chemistry is concerned they are esters of fatty acids with higher alcohols (→ 7.2.2., 7.3.2.). From a technical standpoint crude waxes may also include some other components such as alcohols, steroids and dicarboxylic acids. Some commercial importance attaches to the wax from Douglas fir bark *(Pseudotsuga menziesii)*, which is extractable in yields up to 7%. It is used in polishes, ski waxes, lubricants and soaps (Herrick, Hergert 1977). Douglas fir bark waxes with improved qualities were suggested for coating fruit (Barton 1978).

Some tropical woods have a high content of coloured extractives which are extractable with water, alcohol or ether (→ 7.3.4.). Independently of the wood species involved, these woods are classified into groups such as blue wood, red wood or yellow wood. Well-known wood species are:

blue wood	– Campèche *(Haematoxylon campechianum)*
red wood	– Pernambuc *(Caesalpinia echinata)*
	– Brazilette *(Haematoxylon brasiletto)*
	– Sibukau *(Caesalpinia sappan)*
yellow wood	– Fustik *(Chlorophora tinctoria)*.

Together with numerous plant dyes the wood-derived colours were the most important dyestuffs for cotton, wool, leather or skin till the advance of synthetic colours. Except for some special applications these substances have no industrial importance today (Puth 1962).

Literature

Adams, M. F. and Douglas, C. 1963, Tappi *46,* 544–548

Agnello, L. A. and Barnes, E. O. 1960, Ind. Eng. Chem. *52,* No. 9, 726–732

Allan, G. G. and Mattila, T. 1971, High Energy Degradation. In: Lignins. Occurrence, Formation, Structure and Reactions (Sarkanen, K. V. and Ludwig, C. H., Eds.). Wiley-Intersci., New York, pp. 575–596

Anon. 1978a, Holz Roh-Werkst. *36,* 454

Anon. 1978b, Chem. Week, Apr. 5, 41–42

Anon. 1979, Chem. Eng. News, Oct. 8, 19–20

Anon. 1980a, For. Prod. J. *30,* No. 8, 48

Anon. 1980 b, Chem. Eng. News, Nov. 3, 35–36

Anon. 1981a, S. Afr. Dig., Aug. 21, 11

Anon. 1981b, Westvaco Lignochemicals and Their Applications. Westvaco Chem. Div., Polychem. Dep., Charleston Heights, S.C.

Anon. 1982, Lignin Chemicals. Borregaard A. S., Spec. Chem. Sect., Sarpsborg, Norway

Ansari, H. R. 1970, Flavour Ind. *1,* No. 4, 252–262

Ayla, C. and Weissmann, G. 1982, Holz Roh-Werkst. *40,* 13–18

Balser, K. and Fengel, D. 1979, German patent DE 27 27 552

Bar-Sinai, Y. L. and Wayman, M. 1976, Tappi *59,* No. 3, 112–114

Barton, G. M. 1978, Chemicals from Trees – Outlook for the Future. Special Paper, 8th World Forestry Congress, Jakarta, Indonesia

Barton, G. M. 1981, Foliage. In: Organic Chemicals from Biomass (Goldstein, I. S., Ed.). CRC Press, Inc., Boca Raton, Flo., pp. 249–275

Barton, G. M. and MacDonald, B. F. 1978, Tappi *61,* No. 1, 45–48

Baumeister, M. and Edel. E. 1980, Papier *34,* No. 10A, V9–V18

Bobleter, O. and Binder, H. 1980, Holzforschung *34,* 48–51

Bobleter, O. and Concin, R. 1979, Cell. Chem. Technol. *13,* 583–593

Bratt, L. C. 1979, Pulp Pap. *53,* No. 6, 102–108

Brink, D. L., Charley, J. A., Faltico, G. W. and Thomas, J. F. 1976, The Pyrolysis-Gasification-Combustion Process: Energy Considerations and Overall Processing. In: Thermal Uses and Properties of Carbohydrates and Lignins (Shafizadeh, F., Sarkanen, K. V. and Tillman D. A., Eds.). Academic Press, New York, San Francisco, London, pp. 97–125

Brocksiepe, H. G. 1976, Holzverkohlung. In: Ullmanns Encyklopädie der technischen Chemie, Vol. 12, 4th Ed. Verlag Chemie, Weinheim, pp. 703–708

Bryce, J. R. G. 1980a, Sulfite Pulping. In: Pulp and Paper. Chemistry and Chemical Technology (Casey, J. P., Ed.), Vol. I, 3rd Ed. Wiley-Intersci., New York, pp. 291–376

Bryce, J. R. G. 1980b, Alkaline Pulping. In: Pulp and Paper. Chemistry and Chemical Technology (Casey, J. P., Ed.), Vol. I, 3rd Ed. Wiley-Intersci., New York pp. 377–492

Butschek, G. 1965, Sprit. In: Ullmanns Encyklopädie der technischen Chemie, Vol. 16, 3rd Ed. Verlag Chemie, Weinheim, pp. 110–161

Büsing, J., Dietrichs, H. H., Schweers, W., Weissmann, G., Scheiblich, R., Stenzenberger, H. and Roffael, E. 1978, Chemisch-technologische Grundlagen zur Nutzung von Holz und Holzabfallstoffen als Chemierohstoff. BMFT-Forschungsbericht 01VQ 266

Campbell, D. E. 1977, Pulp Pap. *51,* No. 14, 138–139

Cheremisinoff, N. P. and Cheremisinoff, P. N. (Eds.) 1981, Gasohol Sourcebook. Literature Survey and Abstracts. Ann Arbor Science Publ. Inc., Ann Arbor, Mi.

Clark, I. T. and Green, J. 1968, Tappi *51,* No. 1, 44–48

Dannhäuser, K. 1974, Chem. Rundsch. *27,* No. 40, 19

Derfer, J. M. 1963, Tappi *46,* 513–517

Derfer, J. M. 1966, Tappi *49,* No. 10, 117A–120A

Dietrichs, H. H. 1977, Wege zur biochemischen Nutzung von Holz. In: Mitt. der BFH, No. 118, Hamburg, Reinbek, pp. 129–135

Dietrichs, H. H., Sinner, M. and Puls, J. 1978, Holzforschung *32,* 193–199

Dietrichs, H. H., Sinner, M. and Puls, J. 1979, Gewinnung von Chemikalien, Zuckern und Faserstoffen aus Lignocellulosen mit dem Dampf-Druck-Extraktionsverfahren und dem Organosolv-Aufschluß. In: Verwertung von nachwachsenden Rohstoffen. Symp. Verbindungsstelle Landwirtsch.-Industrie e.V., Essen

Drew, J. and Roberts, D. R. 1978, For. Prod. J. *28,* No. 9, 21–26

Dubey, G. A., McElhinney, T. R. and Wiley, A. J. 1965, Tappi *48,* No. 2, 95–98

Edemar, L. G., Eriksson, K. E. and Johnsrud, C. 1981, Technical and Economical Aspects on Ethanol Production Based on Lignocellulosic Materials. In: The Ekman Days 1981, Int. Symp. Wood Pulp. Chem., Stockholm, Vol. 5, pp. 121

Eickemeyer, R. and Hennecke, H. 1967, Holz Zentralbl. *93,* No. 86, 1374–1375

Ek, M. and Eriksson, K. E. 1975, Appl. Polym. Symp. No. 28, 197–203
Ellerbe, R. W. 1973, Pap. Trade J. *157,* No. 26, 40–43
Enkvist, T. and Lindfors, T. 1966, Pap. Puu *48,* No. 11, 639–648
Enkvist, T., Turunen, J. and Ashorn, T. 1962, Tappi *45,* 128–135
Epshtein, Y. V., Akhimina, E. I. and Raskin, M. N. 1977, Khim. Drev. No. 6, 24–44
Erä, V. and Hannula, J. 1974, Pap. Puu *56,* 489–496
Eskilsson, S. and Hartler, N. 1973, Svensk Papperstid. *76,* No. 2, 63–70
Esterbauer, H., Jungschaffer, G. and Schurz, J. 1981, Holzforschung *35,* 129–135
Faber, K. 1978, Gerbstoffe. In: Ullmanns Encyklopädie der technischen Chemie, Vol. 16, 4th Ed. Verlag Chemie, Weinheim, pp. 128–137
Fähnrich, P., Irrgang, K., Rütten, B. and Schimz, K. L. 1981, Papier *35,* No. 10A, V1–V9
Falkehag, S. I. 1975, Appl. Polym. Symp. No. 28, 247–257
Falkehag, I. 1981, Direction in Research on the Alternative Uses of Wood Components. In: The Ekman Days 1981. Int. Symp. Wood Pulp. Chem., Stockholm, Vol. 4, pp. 3–8
Felicetta, V. F., Lung, M. and McCarthy, J. L. 1959, Tappi *42,* 496–502
Fengel, D. and Wegener, G. 1979, Hydrolysis of Polysaccharides with Trifluoroacetic Acid and its Application to Rapid Wood and Pulp Analysis. In: Hydrolysis of Cellulose: Mechanisms of Enzymatic and Acid Catalysis (Brown, Jr. R. D. and Jurasek, L., Eds.). Adv. Chem. Ser. 181, ACS, Washington, D. C., pp. 145–158
Forss, K. 1974, Pap. Puu *56,* 174–178
Forss, K. and Fremer, K. E. 1965, Pap. Puu *47,* 443–454
Forss, K. and Fuhrmann, A. 1976, Pap. Puu *58,* 817–824
Forss, K. and Fuhrmann, A. 1979, For. Prod. J. *29,* No. 7, 39–43
Forss, K. and Passinen, K. 1976, Pap. Puu *58,* 608–618
Forss, K. G., Gadd, G. O., Lundell, R. O. and Williamson, H. W. 1971, Finnish patent 44366
Franklin, E. C., Taras, M. A. and Volkman, D. A. 1970, Tappi *53,* 2302–2304
Frühwald, A. and Liese, W. 1980, Naturwiss. Rundsch. *33,* 497–505
Funk, H. F. 1975, Appl. Polym. Symp. No. 28, 145–152
Galloway, G. 1975, Pulp Pap. *49,* No. 9, 104–105
Gams, T. C. 1976, Stärke *28,* 344–349
Garves, K. 1982, Holz Roh-Werkst. *40,* 41–44
Gardner, J. A. F. and Hillis, W. E. 1962, The Influence of Extractives on the Pulping of Wood. In: Wood Extractives and Their Significance to the Pulp and Paper Industries (Hillis, W. E., Ed.). Academic Press, New York, London, pp. 367–403
Glasser, W. G. 1981, For. Prod. J. *31,* No. 3, 24–29
Glasser, W. G., Hsu, O. H. H. and Nakano, J. 1976, Svensk Papperstid. *79,* No. 16, 527–530
Goheen, D. W. 1966, U.S. patent 3 253 044
Goheen, D. W. 1971, Low Molecular Weight Chemicals. In: Lignins. Occurrence, Formation, Structure and Reactions (Sarkanen, K. V. and Ludwig, C. H., Eds.). Wiley-Intersci., New York, pp. 797–831
Goheen, D. W., Orle, J. V. and Wither, R. P. 1976, Indirect Pyrolysis of Kraft Black Liquors. In: Thermal Uses and Properties of Carbohydrates and Lignins (Shafizadeh, F., Sarkanen, K. V. and Tillman, D. A., Eds.). Academic Press, New York, San Francisco, London, pp. 227–244
Goldstein, I. S. 1975a, Science *189,* 847–852
Goldstein, I. S. 1975b, Appl. Polym. Symp. No. 28, 259–267
Goldstein, I. S. 1976, Biotechnol. Bioeng. Symp. *6,* 293–301
Goldstein, I. S. 1979, Unasylva *31,* No. 125, 2–9
Goldstein, I. S. 1980a, Tappi *63,* No. 2, 105–108
Goldstein, I. S. 1980b, Tappi *63,* No. 9, 141–143
Goldstein, I. S. 1981, Organic Chemicals from Biomass. CRC Press, Inc., Boca Raton, Flo.

Goring, D. A. I. 1961, The Physical Chemistry of Lignin. In: Proc. of the Wood Chemistry Symp., Montreal. Butterworths, London, pp. 233–254

Gupta, R. C. and Sehgal, V. K. 1978, Holzforsch. Holzverwert. *30,* 85–87

Gupta, R. C. and Sehgal, V. K. 1979, Holzforsch. Holzverwert. *31,* 7–9

Härgerdal, B., Ferchak, J., Pye, E. K. and Forro, J. R. 1979, The Cellulolytic Enzyme System of *Thermoactinomyces.* In: Hydrolysis of Cellulose: Mechanisms of Enzymatic and Acid Catalysis (Brown, Jr. R. D. and Jurasek, L., Eds.). Avd. Chem. Ser. 181, ACS, Washington, D. C., pp. 331–345

Haidegger, E. 1977, Stärke *29,* 430–435

Hajny, G. J. 1981, Biological Utilization of Wood for Production of Chemicals and Foodstuffs. For. Prod. Lab. Res. Paper FPL 385, U.S.D.A.

Hammond, V. L., Murdge, L. K., Allen, C. H. and Schiefelbein, G. F. 1974, Pulp Pap. *48,* No. 2, 4–57

Hannus, K. and Pensar, G. 1973, Pap. Puu *55,* 509–517

Harder, M. L. and Einspahr, D. W. 1978, Tappi *61,* No. 12, 87–88

Harris, J. F. 1975, Appl. Polym. Symp. No. 28, 131–144

Hearon, W. M., MacGregor, W. S. and Goheen, D. W. 1962, Tappi *45,* 28A–36A

Hennecke, H. 1970, Papier *24,* 9–13

Herrick, F. W. and Hergert, H. L. 1977, Utilization of Chemicals from Wood. In: The Structure, Biosynthesis, and Degradation of Wood. Recent Advances in Phytochemistry (Loewus, F. A. and Runeckles, V. C., Eds.), Vol. 11. Plenum Press, New York, London, pp. 443–515

Hewgill, F. R. and Legge, F. 1976, Wood Sci. Technol. *10,* 125–129

Hillis, W. E. and Swain, T. 1962, The Influence of Extractives on the Color of Groundwood and Newsprint. In: Wood Extractives and Their Significance to the Pulp and Paper Industries (Hillis, W. E., Ed.). Academic Press, New York, London, pp. 405–419

Holz, H. 1965, Kohlenhydrate. In: Ullmanns Encyklopädie der technischen Chemie. Vol. 9, 3rd Ed. Verlag Chemie, Weinheim, pp. 637–678

Hoyt, C. H. and Goheen, D. W. 1971, Polymeric Products. In: Lignins. Occurrence, Formation, Structure and Reactions (Sarkanen, K. V. and Ludwig, C. H., Eds.). Wiley-Intersci., New York, pp. 833–865

Hrutfiord, B. F. 1971, Reduction and Hydrogenolysis. In: Lignins. Occurrence, Formation, Structure and Reactions (Sarkanen, K. V. and Ludwig, C. H. Eds.). Wiley-Intersci., New York, pp. 487–509

Hsu, O. H. H. and Glasser, W. G. 1976, Wood Sci. *9,* No. 2, 97–103

Humphrey, A. E. 1976, Single Cell Proteins: A Case Study in the Utilization of Technology in the Soviet System. In: Soviet Science and Technology (Thomas, J. R. and Kruse-Vaucienne, U. M., Eds.). George Washington Univ., Washington, D. C., pp. 358–373

Humphrey, A. E. 1979, The Hydrolysis of Cellulosic Materials to Useful Products. In: Hydrolysis of Cellulose: Mechanisms of Enzymatic and Acid Catalysis (Brown, Jr. R. D. and Jurasek, L., Eds.). Adv. Chem. Ser. 181, ACS, Washington, D. C., pp. 25–53

Ince, P. J. 1979, How to Estimate Recoverable Heat Energy in Wood or Bark Fuels. For. Prod. Lab. Gen. Tech. Report FPL 29, U.S.D.A.

Jensen, W. 1979, Cell. Chem. Technol. *13,* 501–509

Jensen W., Fogelberg, B. C., Forss, K., Fremer, K. E. and Johanson, M. 1966, Holzforschung *20,* 48–50

Kaminsky, W., Schweers, W. and Schwesinger, H. 1980, Holzforschung *34,* 73–76

Katzen, R. 1978, Chemicals from Trees – Outlook for the Future. Special Paper, 8th World Forestry Congress, Jakarta, Indonesia

Kaufmann, W. 1979, Gewinnung von Futtermitteln durch das Dampf-Druck-Extraktionsverfahren. In: Verwertung von nachwachsenden Rohstoffen. Symp. Verbindungsstelle Landwirtsch. – Industrie e. V., Essen

Keays, Y. L. 1974, For. Prod. J. *24,* No. 11, 13–16

Knight, J. A. 1979, Pyrolysis of Wood Residues with a Vertical Bed Reactor. In: Progress in Biomass Conversion (Sarkanen, K. V. and Tillman, D. A., Eds.), Vol. I. Academic Press, New York, London, pp. 87–115

Kollmann, F. 1981, Holz Zentralbl. 107, No. 19, 319–320

Köll, P., Metzger, J. 1978, Angew. Chem. *90,* 802–803

Köll, P., Brönstrup, B. and Metzger, J. 1979, Holzforschung *33,* 112–116

Krause, T., Gahin, M. and Schurz, J. 1973, Papier *27,* 639–642

Kvisgaard, H. J. 1982, Wood, a Versatile and Renewable Chemical Feedstock. ECMRA Conference 1982, Oslo

Ladisch, M. R., Ladisch, C. M. and Tsao, G. T. 1978, Science *201,* 743–745

Lange, W., Kordsachia, O. 1981, Holz Roh-Werkst. *39,* 107–112

Law, K. N., Lo, S. N. and Koran, Z. 1978, Wood Sci. *11,* No. 2, 91–96

Ljungdahl, L. G., Carreira, L. and Wiegel, J. 1981, Production of Ethanol from Carbohydrates Using Anaerobic Thermophilic Bacteria. In: The Ekman Days 1981. Int. Symp. Wood Pulp. Chem., Stockholm, Vol. 4, pp. 23–28

Maloney, G. T. 1978, Chemicals from Pulp and Wood Waste. Production and Applications. Chem. Technol. Rev. No. 101. Noyes Data Corp., Park Ridge, N. J., pp. 141–159

Marton, J. and Marton, T. 1964, Tappi *47,* 471–476

Marutzky, R. 1980, Verkohlung, Pyrolyse und Vergasung von Holz und pflanzlichen Reststoffen. In: Heizen mit Holz (Bossel, U., Ed.) 2nd. Ed. Solentec GmbH, Adelebsen, pp. 177–204

Marutzky, R. 1981, Holz Zentralbl. *107,* No. 19, 315–317

McGovern, J. N. 1980, Silvichemicals . In: Pulp and Paper. Chemistry and Chemical Technology (Casey, J. P., Ed.). Vol. I, 3rd Ed. Wiley-Intersci., New York, pp. 492–504

McKibbins, S. W., Harris, J. F., Saeman, J. F. and Neill, W. K. 1962, For. Prod. J. *12,* 17–23

Metzger, J., Köll, P. and Brönstrup, B. 1981, Nachr. Chem. Tech. Lab. *29,* No. 11, 762–764

Millett, M. A., Baker, A. J. and Satter, L. D. 1976, Biotechnol. Bioeng. Symp. *6,* 125–153

Millett, M. A., Effland, M. J. and Caulfield, D. F. 1979, Influence of Fine Grinding on the Hydrolysis of Cellulosic Materials – Acid vs. Enzymatic. In: Hydrolysis of Cellulose: Mechanisms of Enzymatic and Acid Catalysis (Brown, Jr. R. D. and Jurasek, L., Eds.). Adv. Chem. Ser. 181, ACS, Washington, D. C., pp. 71–89

Mobarak, F., El-Ashmawy, A. E. and Fahmy, Y. 1973, Cell. Chem. Technol. *7,* 325–335

Nazareth, M. R., Kennedy, C. E., Narayanan, V. L. and Bhatia, V. N. 1961a, J. Pharm. Sci. *50,* 560–563

Nazareth, M. R., Narayanan, V. L. and Bhatia, V. N. 1961b, J. Pharm. Sci. *50,* 564–567

Neumüller, O. A. 1973, Römpps Chemie-Lexikon Vol. 3, 7th Ed. Franckh'sche Verlagshandlung, Stuttgart, pp. 1737–1741

Nimz, H. H. and Hitze, G. 1980, Cell. Chem. Technol. *14,* 371–382

Noack, D. and Frühwald, A. 1981, Allg. Forstz. No. 36, 913–920

Oshima, M., Kashima, K., Watanabe, H., Kubo, T. and Tabata, H. 1966, Bull. Chem. Soc. Japan *39,* 2750–2767

Paddon, A. R. and Harker, A. P. 1980, Charcoal Production Using a Transportable Metal Kiln. Rural Techn. Guide 12. Trop. Prod. Inst., London

Pearl, I. A. 1967, The Chemistry of Lignin. Marcel Dekker Inc., New York

Pearl, I. A. 1969, Tappi *52,* 1253–1260

Perret, J. M., Garceau, J. J., Pineault, G. and Lo, S. N. 1976, Pulp Pap. Mag. Can. *77,* No. 11, T 228–T 230

Petroff, G. and Doat, J. 1978, Bois For. Trop. No. 177, 51–64

Pizzi, A. 1980, J. Macromol. Sci. – Rev. Macromol. Chem. C. 18, No. 2, 247–315

Pizzi, A., Scharfetter, H. and Kes, E. W. 1981, Holz Roh-Werkst. *39,* 85–89

Polčin, J. and Bézuch, B. 1977, Wood Sci. Technol. *11,* 275–290

PPRIC 1975, Feasibility Study of Production of Chemical Feedstock from Wood Waste. Pulp Paper Res. Inst. Can. Report OSY4-0093, Pointe Claire, Que.
Prahačs, S., Barclay, H. G. and Bhatia, S. P. 1971, Pulp Pap. Mag. Can. *72,* No. 6, T 199–T 213
Puth, M. 1962, Holz Zentralbl. *88,* 743–744; 787–788
Radcliffe, A. T., Lens, T. J. 1973, Kunststoffe *63,* 854–857
Reed, T. B. and Jantzen, D. 1979, Introduction. In: A Survey of Biomass Gasification. Vol. II: Principles of Gasification. Solar Energy Res. Inst. Golden, Co., pp. 1–10
Roffael, E. 1976, Beiträge zur Verwendung von alkalischen Phenolformaldehydharzen und Ligninsulfonaten bei der Verleimung von Holzspänen. WKI-Bericht No. 8, Braunschweig
Roffael, E. and Rauch, W. 1974, Holz Zentralbl. *100,* No. 96, 1461–1462
Rowell, R. M. and Hokanson, A. E. 1979, Methanol from Wood: A Critical Assessment. In: Progress in Biomass Conversion (Sarkanen, K. V. and Tillman, D. A., (Eds.), Vol. I. Academic Press, New York, London, pp. 117–144
Rugg, B. and Brenner, W. 1980, Continuous Acid Hydrolysis of Waste Cellulose for Ethanol Production. Proc. of the 7th Energy Technol. Conf. and Exposition, Washington, D. C.
Saeman, J. F. 1977, Energy and Materials from the Forest Biomass. In: Clean Fuels from Biomass and Wastes. Proc. Symp. Orlando, F., pp. 153–168
Saeman, J. F. 1979, Key Factors in the Hydrolysis of Cellulose. In: Biomass as a Non-Fossil Fuel Source. Preprints Symp. Vol. 24, No. 2. ACS, Div. of Petrol. Chem. Inc.
Saeman, J. F. 1980, Assessment of Dilute Acid Hydrolysis of Cellulose. In: Proc. Bio-Energy '80 World Congress and Exposition, Atlanta, Ga. pp. 162–164
Sandermann, W. 1960, Naturharze, Terpentinöl, Tallöl. Springer-Verlag, Berlin, Göttingen, Heidelberg
Sarkanen, K. V. and Tillman, D. A. (Eds.) 1979, Progress in Biomass Conversion, Vol. I. Academic Press, New York, London
Schmidt, O., Puls, J., Sinner, M. and Dietrichs, H. H. 1979, Holzforschung *33,* 192–196
Schweers, W. 1969, Holzforschung *23,* 5–9
Schweers, W. 1975, Papier *29,* 1–8
Schweers, W. 1979, Verwertung von Ethanol-Lignin. In: Verwertung von nachwachsenden Rohstoffen. Symp. Verbindungsstelle Landwirtsch. – Industrie e. V., Essen
Schweers, W. and Meier, D. 1979, Holzforschung *33,* 15–18
Seiffert, U. and Held, W. 1981, Chem. Ing. Tech. *53,* No. 2, 82–89
Sergeeva, V. N., Tarnarutskii, G. M., Gribanova, N. V. and Telysheva, G. M. 1979, Khim. Drev. No. 3, 3–12
Shafizadeh, F., Sarkanen, K. V. and Tillman, D. A. (Eds.) 1976, Thermal Uses and Properties of Carbohydrates and Lignins. Academic Press, New York, London
Shen, K. C. 1974, For. Prod. J. *24,* No. 2, 38–44
Shen, K. C. and Calvé, L. 1979, Ammonium-based Spent Sulfite Liquor for Waferboard Binder. Forintek Can. Corp., East. For. Prod. Lab., Ottawa, Ont.
Shen, K. C., Fung, D. P. C. and Calvé, L. 1981, Ammonium-based Spent Sulfite Liquor Binder System for Waferboard Manufacture. Forintek Can. Corp., East. Lab., Techn. Rep. 509 ER
Shimizu, K. 1980, Mokuzai Gakkaishi *26,* 488–493
Sihtola, H. and Blomberg, L. 1975, Cell. Chem. Technol. *9,* 555–560
Soltes, E. J. 1980, Tappi *63,* No. 7, 75–77
Somsen, R. A. 1962, Tappi *45,* 623–628
Stenzenberger, H. D. 1981, Papier *35,* No. 10 A, V9–V18
Su, T. M. and Paulavicius, I. 1975, Appl. Polym. Symp. No. 28, 221–236
Tanaka, R., Yaku, F., Muraki, E. and Koshijima, T. 1980, Cell. Chem. Technol. *14,* 859–868
Thompson, D. R. and Grethlein, H. E. 1979, Ind. Eng. Chem. Prod. Res. Dev. *18,* No. 3, 166–169
Thonart, P., Paquot, M. and Mottet, A. 1979, Holzforschung *33,* 197–202
Töppel, O. 1961, Papier *15,* 81–93

Underwood, J. C., Filipic, V. C. and Bell, R. A. 1969, J. Assoc. Offic. Agr. Chem. *52,* No. 4, 717–719

Uprichard, J. M. 1978, Turpentine and Tall Oil from Radiata Pine: Valuable By-Products from Kraft Pulping. Special Paper, 8th World Forestry Congress, Jakarta, Indonesia

Viikari, L., Linko, M. and Enari, T. M. 1981, Ethanol from Cellulosic Materials. In: The Ekman Days 1981. Int. Symp. Wood Pulp. Chem., Stockholm, Vol. 4, pp. 18–22

Voeste, T. 1981, Holz Zentralbl. *107,* No. 25, 391–394

Wegener, G. 1982, Holz Roh-Werkst. *40, 181–185; 209–214; 241–245*

Weismann, A. 1980, Nutzung und Nutzungsmöglichkeiten nachwachsender Rohstoffe zur Gewinnung von Energie und chemischen Grundstoffen. Bericht Projektgr. 63, BML, Bonn

Welling, J. 1979, Stand der Technologie der Holzverkohlung. DiplomaThesis, Universität Hamburg

Wenzl, H. F. J. 1970, The Acid Hydrolysis of Wood. In: The Chemical Technology of Wood. Academic Press, New York, London, pp. 158–252

Wienhaus, O. 1979, Holztechnologie *20,* 144–148

Wienhaus, O., Fischer, F. and Schiene, R. 1976, Zellst. Pap. *25,* 109–116

Wienhaus, O., Schiene, R. and Fischer, F. 1980, Zellst. Pap. *29,* 125–128

Wiley, P. R. 1961, Tappi *44,* 22 A–26 A

Wilke, C. R. and Yang, R. D. 1975, Process Development and Design Studies for the Enzymatic Hydrolysis of Cellulose. In: Enzymatic Hydrolysis of Cellulose. Proc. Symp. SITRA, Helsinki, pp. 485–506

Zavarin, E., Wong, Y. and Zinkel, D. F. 1978, Lightwood as a Source of Unusual Naval Stores Chemicals. In: Proc. of Lightwood Res. Coordinating Council Annual Meeting, pp. 19–30

Zinkel, D. F. 1975, Appl. Polym. Symp. No. 28, 309–327

Zinkel, D. F. and McKibben, C. R. 1978, Chemistry of Naval Stores from Pine Lightwood – A Critical Review. In: Proc. of Lightwood Res. Coordinating Council Annual Meeting, pp. 133–156

Subject Index

Bold-faced numbers: see also following page(s)
F means figure, T means table

Species Index (Wood, Plants)

Species Index (Bacteria, Fungi)

Fengel/Wegener
Wood
Walter de Gruyter, Berlin · New York 1989

Errata

Page	Paragraph, Table, Figure	Line	Correction
2	2	1	replace *Sanderman* by *Sandermann*
6	1	5	replace *Sakai* by *Saiki*
12	Table 2-2	10	replace *mm* by *μm* (in the line vessels, length)
20	3	4	replace *pits have* by *pits may have*
25		12	replace *Sakai* by *Saiki*
75	Table 4-3		add *and light scattering* at the end of the table heading
77	3	2	replace *coordinative* by *covalent*
98	Fig. 4-30b		replace *101* by $10\bar{1}$
136	5	2	replace *Glaser* by *Glasser*
136	Fig. 6-4		replace $-e$ by $-e^-$
144	Fig. 6-7		replace *C–O–* by *HC–O–* (between unit 3 and unit 13)
155	1	3	replace $\overline{M}_w/\overline{M}_w$ by $\overline{M}_w/\overline{M}_n$
180		15	replace *L 47-L56* by *247-256*
184	Fig. 7-1		replace 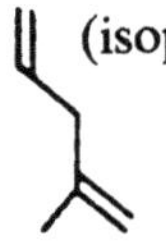 (isoprene) by 
205	2	7	replace *Quercus alba* by *Quercus rubra*
211	Fig. 7-29		replace *red* by *blue*
223		2	replace *Hansen* by *Hausen*
242	3	3	delete *sieve cells and*
248	1	5	delete *arabino-*
263	Fig. 9-12		replace 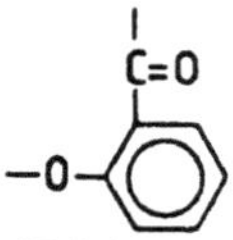 (Trichocarpin) by 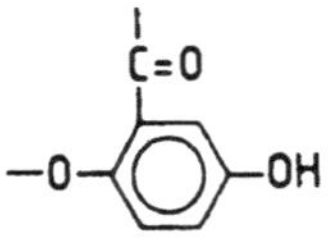
285	Fig. 10-19		replace *lignin aqueous* by *lignin by aqueous*
335	Fig. 12-16		replace *REM* by *SEM*
336	Fig. 12-17		replace *Kollman* by *Kollmann*
358	3	12	replace *Lignin-OH* $\xrightarrow{hv}$ by *Lignin-OH* $\xrightarrow{h\nu}$

Page	Paragraph, Table, Figure	Line	Correction
375	1	2	replace *Acomycetes* by *Ascomycetes*
385	Fig. 14-7		replace *REM* by *SEM*
418	Table 16-2		replace *Forts. Table 16-2* by *Table 16-2 (continued)*
430	4	2	replace *preceded* by *succeeded*
516	Fig. 17-19		replace *REM* by *SEM*
555	4	10	replace *Swan* by *Swain*
606		24	delete *phenols 205* (3rd column)
607		16	under tannins 208 add: *phenols, lignans 205*
613		1	replace *(Wood, Plants)* by *(Bacteria, Fungi)* (column title)

Bei Fragen zur Produktsicherheit wenden Sie sich bitte an:
If you have any questions regarding product safety,
please contact:

Walter de Gruyter GmbH
Genthiner Straße 13
10785 Berlin
productsafety@degruyterbrill.com